STRUCTURE AND FUNCTION OF THE ASPARTIC PROTEINASES

Genetics, Structures, and Mechanisms

ADVANCES IN EXPERIMENTAL MEDICINE AND BIOLOGY

STRUCTURE AND FUNCTION OF THE ASPARTIC PROTEINASES

Genetics, Structures, and Mechanisms

Edited by

Ben M. Dunn

University of Florida–College of Medicine
Gainesville, Florida

PLENUM PRESS • NEW YORK AND LONDON

Library of Congress Cataloging-in-Publication Data

Aspartic Proteinase Conference on Structure and Function of the
  Aspartic Proteinases: Genetics-Structures-Mechanisms (1990 : Sonoma
  County, Calif.)
   Structure and function of the aspartic proteinases : genetics,
  structures, and mechanisms / edited by Ben M. Dunn.
       p.   cm. -- (Advances in experimental medicine and biology : v.
  306)
   "Proceedings of the Aspartic Proteinase Conference on Structure
  and Function of the Aspartic Proteinases: Genetics-Structure
  -Mechanisms, held September 22-27, 1990, in Sonoma County,
  California."--T.p. verso.
   Includes bibliographical references and index.

  ISBN-13: 978-1-4684-6014-8   ISBN-13: 978-1-4684-6012-4
  DOI: 10.1007/978-1-4684-6012-4

   1. Aspartic proteinases--Congresses.   I. Dunn, Ben M.   II. Title.
  III. Series.
   [DNLM: 1. Aspartic Proteinases--genetics--congresses.   2. Aspartic
  Proteinases--physiology--congresses.   QU 136 A838s 1990]
  QP609.A86A86   1990
  574.19'256--dc20
  DNLM/DLC
  for Library of Congress   91-39487
                                                               CIP

Proceedings of the Aspartic Proteinase Conference on Structure and
Function of the Aspartic Proteinases: Genetics-Structure-Mechanisms,
held September 22-27, 1990, in Sonoma County, California

© 1991 Plenum Press, New York
Softcover reprint of the hardcover 1st edition 1991
A Division of Plenum Publishing Corporation
233 Spring Street, New York, N.Y. 10013

This work is dedicated to the loving memory of
Ralph Lee Dunn and Arthur Edward Kugler

Financial Support for the Meeting was Provided by the Following Sponsors

ABBOTT LABORATORIES

AGOURON PHARMACEUTICALS, INC.

APLIED BIOSYSTEMS

BIOMEGA, INC.

BRISTOL-MYERS SQUIBB COMPANY

DEVELOPMENTAL THERAPEUTICS BRANCH, DIVISION OF AIDS, NIAID

DUPONT CENTRAL RESEARCH & DEVELOPMENT: VIRAL DISEASES GROUP

GENENCOR

GENENTECH

GLAXO RESEARCH LABORATORIES

HOFMANN-LAROCHE, INC.

LILLY RESEARCH LABORATORIES

MERCK SHARP & DOHM RESEARCH LABORATORIES

MONSANTO

PFIZER CENTRAL RESEARCH

SMITHKLINE BEECHAM PHARMACEUTICALS

SQUIBB INSTITUTE FOR MEDICAL RESEARCH

THE UPJOHN COMPANY

WARNER LAMBERT/PARKE-DAVIS PHARMACEUTICAL RESEARCH DIVISION

PREFACE

In September, 1990, a group of 160 scientists from 19 countries and 21 of the United States met at the Red Lion Inn in Rohnert Park, Sonoma County, California. The purpose of this meeting was to share new information from recent research on the Aspartic Proteinases. This book is a compilation of the information transferred in that forum.

The Aspartic Proteinases include all those enzymes from the "fourth" class of proteolytic enzymes, the first three being the Serine, Cysteine and Metalloproteinases. Of course, all the scientists in attendance at the Sonoma Aspartic Proteinase Conference would agree that our current level of understanding of the structure and function of the Aspartic Proteinase class of enzymes is clearly first class. The reasons for this require a bit of historical perspective.

The group of scientists who are engaged in study of this family of enzymes first met as a separate entity in 1976, in Norman, Oklahoma, at a meeting organized by Jordan Tang of the Oklahoma Medical Research Foundation. This was an exciting time, as the first crystal structures of some of these enzymes were described by Blundell, James and Davies. During that conference, the relationship between the two halves of the mammalian and fungal enzymes was recognized and this has provided a structural foundation for analysis of the retroviral enzymes, which came later. A book was published by Plenum Press documenting this conference,[1] and the current book is an update to that important work.

The second open meeting of this group took place in 1984, in Prague, Czechoslovakia, and was organized by Vladimir Kostka, as a FEBS workshop. At this time, much new information was described and the importance of the development of enzyme inhibitors was emphasized, mostly in relation to the human enzyme renin. The broad spectrum of occurrence of enzymes of this family was also discussed at this meeting and the first efforts at detailed active site investigations were reported. Again, a book detailing the advances in this field was published,[2] this time by Walter de Gruyter, and, with the earlier work of Tang, has stood as the reference work for this field for the past five years.

A third conference was organized in 1988 by Bent Foltmann and held in Elsinore, Denmark. The rapidly accelerating pace of scientific advance was clearly evident at this important meeting. Most significantly, this meeting saw the first reports of the intimate relationship between the retroviral processing proteinases and the Aspartic Proteinases. Although this meeting provided an excellent forum for scientific interchange, it did not result in a publication. Hence, it was decided that a follow-up conference should be held in 1991 and the American members of this international group took on the responsibility for hosting the meeting. Due to the extremely rapid pace of discovery in the retroviral area, our planning was pushed up by one full year, leading to the September, 1990 date for the conference detailed in this book.

Two additional meetings are part of the history of this "family." A workshop was organized in 1982 by Tom Blundell and John Kay, and held at Birkbeck College in London. This meeting was attended by a smaller group of international scientists, around 30, but was important for the intensity of the discussions and for the emphasis on crystal structures provided by the host Department of Crystallography at Birkbeck. A second workshop was organized by Mike Samloff in 1985 under the sponsorship of the Chugai Pharmaceutical Company of Japan and held in Tokyo. Again, approximately 25 international scientists attended this event and exchanged information with each other and with the host Japanese scientists while enjoying the legendary hospitality of Tokyo.

In arranging the 1990 conference, some thought was given to the order in which topics would be discussed. While earlier meetings had been organized to present crystallographic information, kinetic studies, inhibitor design or proenzyme activation events and so on of all enzymes together, the 1990 conference was arranged to present all the information dealing with one particular subgroup of this family together as an intensive, day-long concentration. Thus, the meeting was divided into four major sub-topics. The first of these is the Gastric Mammalian Proteinases and includes pepsins, gastricsins and chymosins. The second category is the Microbial Aspartic Proteinases, including fungal, yeast and bacterial enzymes. These two sub-topics were discussed on the first two days of the conference, reflecting the historical position of these two groups of enzymes. However, as the reader can judge from the following, these two "traditional" areas within the Aspartic Proteinase field continue to provide exciting new discoveries and insights of importance to the whole family. Third, the Non-Gastric Mammalian Proteinases were described. This important group includes cathepsin D, the lysosomal enzyme, cathepsin E, the *non*-lysosomal, but intracellular enzyme and renin. The design of renin inhibitors has employed many chemists and biologists over the past 15 years and has yielded many critical lessons in inhibitor design that are being utilized today in the rush to develop anti-viral compounds. The biological functions of cathepsin D and cathepsin E are topics of intense interest today and the interaction of potential drugs in the human body with both those enzymes as well as the gastric enzymes has major consequences for successful therapy. Finally, the fourth section of this conference was devoted to the retroviral enzymes, especially that from the Human Immunodeficiency Virus, HIV PR. The exciting advances in our understanding of that enzyme were highlighted by a brilliant series of lectures, culminating in the detailed description of a variety of enzyme-inhibitor complexes in multi-color slides.

This volume includes manuscripts arising from the lectures delivered at the conference arranged in the same daily sequence as their presentation. In a few instances, the order of the manuscripts in the book is different than the order of presentation at the meeting. In addition, shorter manuscripts summarizing information presented in poster format are also included. The success of this meeting was in large part due to the very high percentage of attendees who presented information in one format or the other. The afternoon poster sessions provided an excellent opportunity for in-depth discussion of both the morning lectures and of the information presented in poster format. As the conference organizer, it was especially gratifying to see the intensity with which all conferees devoured the information presented in poster format and engaged each other in heated, but collegial discussions. As seen in the Table of Contents, the book includes the poster manuscripts for each session immediately following the manuscripts describing the plenary lectures. This organization expands and enriches the information presented.

It is a pleasure to acknowledge the assistance of many individuals with the organization and operation of this conference. Jordan Tang and David Davies provided essential advice and guidance in selecting speakers. It was especially important that all the individuals mentioned earlier in this introduction as being involved in organization of the preceding conferences were able to participate and each one chaired one of the sessions. Kirk Hayenga was indispensable as the local organizer, and was assisted by Mick Ward and Katherine Kodama. A special debt of gratitude is due to my research group from the University of Florida, who not only gave excellent poster presentations, but helped with all aspects of the organization and operation of the meeting. This group, Bill Farmerie, Paula

Scarborough, Todd Lowther, Chetana Rao and Wieslaw Swietnicki, also had to endure my multiple personalities during the editing process. Bill Farmerie is responsible for indoctrinating me in the intricacies of Microsoft Word. A special thanks goes to Paula Scarborough for proof-reading every word of the book and providing essential advice on style and substance. My scientific colleague and good friend, John Kay, of the University of Wales, was extremely important in recommending new topics and speakers and for "reacting" to some of my more outlandish ideas.

Finally, the following page is a list of the groups that have provided financial support that was necessary to bring this meeting together. Without their foresight and generosity, none of this would have come about.

It is hoped that this volume will provide a convenient source for information on individual proteins as well as on the relationships between the members of this unique class of enzymes.

Ben M. Dunn
January, 1991
Gainesville, Florida

REFERENCES

1. "Acid Proteases: Structure, Function and Biology," Jordan Tang, ed., Advances in Experimental Medicine and Biology, Vol. 95, Plenum Press, New York (1977).
2. "Aspartic Proteinases and Their Inhibitors: Proceedings of the FEBS Advanced Course No. 84/07," Vladimir Kostka, ed., Walter de Gruyter, Berlin (1985).

CONTENTS

MICROBIAL ASPARTIC PROTEINASES: PLENARY LECTURES

NON-GASTRIC MAMMALIAN PROTEINASES: PLENARY LECTURES

NON-GASTRIC MAMMALIAN PROTEINASES: POSTER REPORTS

RETROVIRAL ASPARTIC PROTEINASES: POSTER REPORTS

STUDIES ON PEPSIN MUTAGENESIS AND RECOMBINANT

RHIZOPUSPEPSINOGEN

X. L. Lin, M. Fusek, Z. Chen, G. Koelsch, H. P. Han, J. A. Hartsuck, and J. Tang

Protein Studies Program, Oklahoma Medical Research Foundation
University of Oklahoma Health Science Center
Oklahoma City, Oklahoma 73104

INTRODUCTION

During the past fifteen years, many chemical and three-dimensional structures of aspartic proteases have been determined. It is now clear that, in spite of some diverse substrate specificities, these enzymes are alike in many ways, including their folding patterns, active center structures, and mechanisms of zymogen activation. Because the catalytic apparatuses of these enzymes are nearly identical, it is obvious that this group of enzymes shares a common catalytic mechanism. Although a number of catalytic mechanisms have been proposed for the aspartic proteases after consideration of kinetic and structural results (Fruton, 1976; James et al., 1977; Davies, 1990), a consensus on the mechanism has not emerged. One of the reasons for the uncertainty is that static structural information suggests, but does not provide, direct evidence for a mechanism of catalysis. On the other hand, kinetic experiments suggest reaction schemes but do not pinpoint structural components which are operational. It seems clear that additional experimental information is necessary to link structural information to reaction schemes.

We have been trying to fill this gap with a combination of structural information, site-directed mutagenesis, and kinetics. We have chosen porcine pepsin as the subject of our mechanistic studies because the structural information (Andreeva et al., 1984; Cooper et al., 1990; Abad-Zapatero et al., 1990) as well as extensive kinetic data (Fruton, 1976) are already available. For the mechanistic studies, our approach to pepsin mutagenesis is to make conservative changes of amino acids at or near the active site which produce altered enzymic activities but which are not destructive enough to alter the gross protein conformation. The analyses of enzymic activities of pepsin mutants may then provide insights to the catalytic mechanism. Some of the results of this study are summarized below.

Described in the following are also results of a study on rhizopuspepsinogen. Like other fungal aspartic proteases, rhizopuspepsin is produced as an active enzyme by the fungus *Rhizopus chinensis*. So far, a naturally occurring precursor of an aspartic protease has not been found in fungi. Because rhizopuspepsin is an important structure-function model of aspartic proteases, we have tried to obtain an active zymogen of this protease by expressing its cDNA. The successful expression of recombinant rhizopuspepsinogen has permitted us to study the properties of its activation. These results are described in this article.

Structure and Function of the Aspartic Proteinases
Edited by B.M. Dunn, Plenum Press, New York, 1991

Table 1. Expression of pepsinogen in *E. coli*

Intracellular Locations	Vectors	Promoter	Form	Approximate Level
Cytosol	pKK223-3	*tac*	Inclusion	5 mg/L
	pGBT-T19	*tac*	bodies	25 mg/L
Periplasmic space	pIN-III-Omp	*lpp-lac*	Soluble	5 mg/L
	+Nt extension	*lpp-lac*	Soluble	30 mg/L
Cytosol	pET-3b	*T7*	Inclusion bodies	500 mg/L

RECOMBINANT PEPSINOGEN AND PEPSIN

In order to efficiently carry out mutagenesis studies of pepsin, an effective expression system for the recombinant enzyme is essential. Several *E. coli* expression systems which have been developed in our laboratory are summarized in Table 1. In early mutagenesis studies, porcine pepsinogen cDNA was expressed under the control of *tac* promoter and the recombinant zymogen and enzymes were refolded and purified to homogeneity (Lin *et al.*, 1989). Improved systems were developed later which include the expression and export of pepsinogen as active protein to the periplasmic space of *E. coli* and the synthesis of pepsinogen as 'inclusion bodies' under the control of the bacteriophage *T7* promoter (Table 1). The latter system gives the best yield, allows simple purification procedures (unpublished results), and is presently the system of choice in our laboratory. The purified recombinant pepsinogen and pepsin have kinetic parameters essentially the same as that of the native zymogen and enzymes (Lin *et al.*, 1989).

MUTAGENESIS STUDIES OF PEPSINOGEN AND PEPSIN

As discussed above, the purpose of the mutagenesis studies of pepsin was to gain insights into the mechanisms of catalysis and zymogen activation. Results from two different types of mutations are described below.

Hydrogen Bond Mutants

The first type of mutation was designed to remove individual hydrogen bonds near the active site. The effects of these mutations were analyzed with kinetic measurements to assess the consequences. The most interesting mutants of this group are T218A and S35A, where Thr and Ser at positions 218 and 35 are changed to Ala. From the crystal structure of pepsin (Andreeva *et al.*, 1984; Cooper *et al.*, 1990; Abad-Zapatero *et al.*, 1990), it is known that the hydroxyl group of Thr_{218} in pepsin is H-bonded to the active site Asp_{215}, and likewise Ser_{35} is H-bonded to the active site Asp_{32} (see diagram in Figure 1). Thus the removal of the OH groups, and the corresponding H-bonds, may affect the ionization of the active site carboxyl groups. The effect on the active-site pK_a's in the mutants may be experimentally determined. It should be mentioned that the same hydrogen-bonding

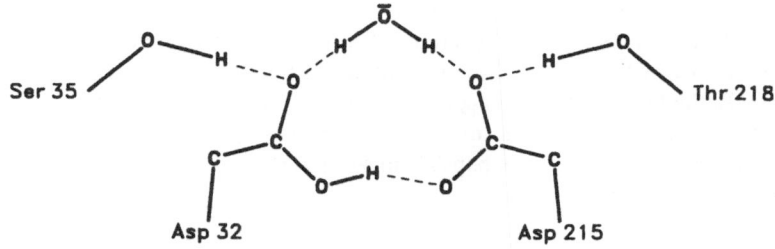

Figure 1. Schematic representation of the hydrogen bond network in the active site of pepsin. A water molecule is situated medially between the Asp_{32} and Asp_{215}, the two catalytic aspartic acid residues, and hydrogen bonds to a carboxylate oxygen of each. The sidechain hydroxyl group of Ser_{35} donates a proton to the "outer" carboxyl oxygen of Asp_{32}. A symmetrically related hydrogen bond is accomplished by the sidechain hydroxyl of Thr_{218} donating a proton to the "outer" carboxyl oxygen of Asp_{215}. Asp_{32}, in its unionized state, donates a proton to form a hydrogen bond with Asp_{215}.

arrangements are present in the structures of most of the aspartic proteases, including rhizopuspepsin, penicillopepsin, endothiapepsin, and chymosin. Therefore, the pepsin mutants are representative of many other aspartic proteases.

We first obtained the k_{cat}/K_m values for pepsin and the mutants in the pH range of 1.5 to 4.5 using as substrate Lys-Pro-Ala-Glu-Phe-Phe(NO_2)-Ala-Leu. From these data, the two active site pK_a values of native and mutant enzymes were calculated from a nonlinear regression method (Knowles, 1976). Our data yielded for native pepsin pK_1 of 1.50 and pK_2 of 3.95. In the mutant T218A, pK_1 increased by 0.76 to 2.26, but pK_2 was not significantly changed at 3.79. Mutant S35A on the other hand had increases in both pK_a's in comparison to the native pepsin: pK_1 2.04 (+0.54) and pK_2 4.79 (+0.84). Because the mutation of Thr_{218} to Ala increased pK_1 but had little effect on pK_2, it is suggested that pK_1 of pepsin active center is intimately related to the ionization of Asp_{215}. It further suggested that within the catalytic pH range, Asp_{215} is ionized (Fig. 2). Since Asp_{215} is H-bonded to a water molecule which most likely serves as the nucleophile in the pepsin catalysis, the negative charge on Asp_{215} can migrate to the water oxygen by resonance and thus facilitate the nucleophilic attack on the substrate's carbonyl carbon (Figure 2). The loss of H-bonding from Ser_{35} to Asp_{32} in mutant S35A increased both the pK_1 and pK_2. This is not entirely surprising since the distance measurement of the active site atoms in rhizopuspepsin (Suguna *et al.*, 1987) and endothiapepsin (Pearl & Blundell, 1984) suggests that one of the carbonyl oxygens of Asp_{32} is H-bonded to the nucleophilic water (Figure 2). The fact that pK_1 is increased in both mutations also suggests that this lower pK_a represents mainly the protonation and deprotonation of the nucleophile water. pK_2 must then represent the ionization of Asp_{32}, so that the role of Asp_{32} would be mainly in the protonation of the carbonyl oxygen of the substrate (Figure 2).

The question of whether the observed pK_a shifts are the consequence of H-bond removal and are not due to local conformational changes is, of course, impossible to answer at the present. However, the catalytic mechanism proposed from the distance measurements of active site atoms in the high resolution structures of rhizopuspepsin (Suguna *et al.*, 1987) and endothiapepsin (Pearl & Blundell, 1984) are in essential agreement with our conclusions. Nevertheless, there is a need for substantiation of pK_a values in the enzyme from kinetic studies with other substrates and by direct measurements with (for example) NMR. Some of these studies are underway in our laboratory.

Another mutant of interest is S219A. The removal of the hydroxyl group from Ser$_{219}$ caused an increase of K$_m$ by about 10-fold in the pH range of 2 to 4.5 without affecting the values of k$_{cat}$. In the crystal structures of aspartic proteases and their inhibitors (Davies, 1990), the OH group of Ser$_{219}$ is hydrogen bonded to the amide nitrogen of residue S$_3$. Our results suggest that this interaction is an early event in the formation of the Michaelis-Menten complex between pepsin and substrate.

Active Site Mutant

A second type of mutation is the change of the active site aspartic acids. Mutation of Asp$_{32}$ to Ala was made in order to answer two questions. First, the intramolecular cleavage of pepsinogen to form pepsin (Al-Janabi *et al.*, 1972; Marciniszyn *et al.*, 1976) has been assumed to be carried out in pepsinogen by the pepsin active site. No direct evidence exists, however, to support this assumption. If the assumption is correct, the active site mutant D32A would be expected to negate the pepsin activity and pepsinogen activation to the same extent. Second, the transfer of energy gained from substrate binding to tight transition-state binding is thought to provide the residual activities of the active-site mutants of subtilisin (Carter & Wells, 1988). If this is also true for pepsin, one might see a higher residual activity for active-site mutants because of a higher transition-state binding energy acquired from binding of a larger substrate of 8 side chains in enzyme specificity pockets.

Mutant D32A pepsinogen did not convert to pepsin in acid solutions (Lin *et al.*, 1989). So the mutant zymogen is not capable of intramolecular activation. This result suggests that pepsin active site is involved in the intramolecular cleavage of the native pepsinogen to form pepsin.

The proteolytic activities of mutant D32A pepsinogen were assayed using a highly sensitive assay (Lin *et al.*, 1989) in order to measure residual activity. Initially, we did find activity which was 10^5-fold less in specific activity than that of native pepsin. Although this was the level of activity expected for residual activity of the active site mutants (Carter & Wells, 1988), we decided to determine the number of active sites in mutant D32A pepsinogen using pepstatin titration. We observed that the proteolytic activity of D32A mutant was completely inhibited with 10,000:1 molar ratio of mutant pepsinogen:pepstatin. This result

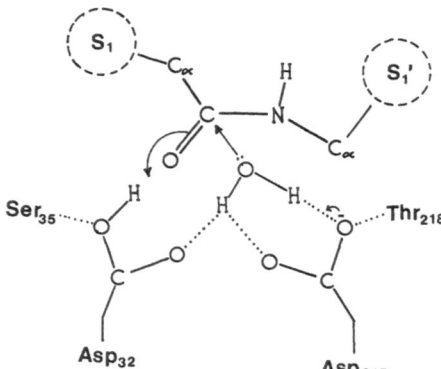

Figure 2. Schematic representation of a proposed catalytic mechanism for peptide bond hydrolysis in pepsin. S$_1$ and S$_1$' denote the sidechains of the substrate. The unionized Asp$_{32}$ donates a proton to the carbonyl oxygen of the scissile bond, resulting in a withdrawl of electrons away from the carbon center. A water molecule, activated by donation of protons to the catalytic aspartic acid residues, is then favorable for nucleophilic attack.

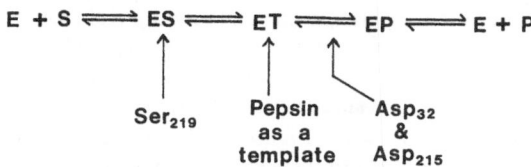

Figure 3. The roles of various mutated pepsin side chains in peptide bond hydrolysis. The hydroxyl group of Ser_{219} is proposed to be involved in the formation of the enzyme-substrate complex ES. Asp_{32} is not necessary for binding of substrate, as its absence forms a 'template' pepsin. However, the absence of the side chain of Asp_{32} does not allow formation of product.

suggested that only 1 out of 100,000 molecules of D32A mutant was fully active while all the rest were not capable of catalyzing proteolysis. We confirmed this finding with a partial denaturation experiment. D32A pepsinogen was incubated at pH 8.0 for 1 hour at 24°C. Then, the pH was adjusted to 6.0, where the mutant zymogen is stable. The alkaline incubation caused a denaturation of 40% of the protein as demonstrated by monoQ FPLC analysis (Lin *et al.*, 1989). However, 100% of the proteolytic activity was recovered after the alkaline incubation. This result indicated that the residual activity was in the form of native pepsinogen which is known to be stable at pH 8.0. Finally, when D32A pepsinogen was incubated at pH 2.0 for 20 minutes at 24°C and then repurified for the mutant pepsinogen on a monoQ column in FPLC, the D32A pepsinogen recovered (about 50%) possessed no proteolytic activity. The residual proteolytic activity was observed in monoQ chromatography at a position corresponding to native pepsin but where very little protein was present. Since the condition of acid incubation was sufficient to convert all native pepsinogen to pepsin and since the majority of D32A pepsin was unchanged, these results support the explanation that the residual proteolytic activity was due to the presence of native pepsinogen in D32A mutant protein. All three lines of evidence outlined above established the fact that mutant D32A had no measurable proteolytic activity. The residual activity is probably due to the error of protein synthesis in *E. coli* which placed an Asp at position 32 instead of the coded Ala in a small fraction of molecules. The hazards of residual activity of enzyme mutant due to synthetic errors have been previously mentioned by Schimmel (1989). It is known that the probability for the error of protein synthesis at individual amino acids can be as high as 1/1,000 (Schimmel, 1989). Since only the replacement of Ala with Asp at position 32 of pepsin would be productive, the probability of the productive synthetic error would be near that for the observed residual activity.

The correct folding of mutant D32A was established from the tight binding of this mutant to pepstatin. We measured K_d between the mutant D32A and pepstatin in an experiment in which D32A protein of different amounts was added to a mixture of constant ratio of pepsin to pepstatin (Lin *et al.*, 1989). The binding of D32A protein to pepstatin caused increases of pepsin activity, and, from these increases, the K_d was estimated to be 5 x 10^{-10} M. This value indicated that the binding strength of D32A to pepstatin is only slightly less than that of pepstatin to native pepsin, 5 x 10^{-11} M (Workman & Burkitt, 1979). Since pepstatin is a transition-state analogue inhibitor, these results suggest that pepsin can serve as a transition state binding template without the involvement of the carboxyl group of the active site Asp_{32}.

The roles of various mutated pepsin side chains in the overall catalytic scheme is illustrated in Figure 3. The hydroxyl group of Ser_{219} is proposed to be involved in the formation of the enzyme-substrate complex ES. The main argument to support this is that K_m approximates K_s for pepsin substrates (Fruton, 1976). The pepsin active site apparently can provide a template for the tight transition state binding (ET) without the involvement of the active site Asp_{32}. This observation, in general, conforms with the transition-state theory of enzyme catalysis (Kraut, 1988). However, in contrast to the case of subtilisin (Carter & Wells, 1988), without the active site side chain of Asp_{32}, pepsin is unable to proceed further in the catalytic reaction to EP as shown in Figure 3. For this reason, the role of the active site aspartates is suggested to be involved in the acceleration of the product formation (EP) by a mechanism depicted in Figure 1.

RECOMBINANT RHIZOPUSPEPSINOGEN

A rhizopuspepsinogen cDNA was constructed from two partial clones. One of the clones contained the sequence of rhizopuspepsin and part of the pro region (Delaney et al., 1987). The other clone contains the pre, pro, and part of the enzyme regions (Wong, R. N. S., Delaney, R. and Tang, J., unpublished results). The sequences of pre, pro, and the N-terminal regions of the enzyme are shown in Figure 4. Since the boundary between the pre and pro regions is not known, we constructed three different vectors to express rhizopuspepsinogen (Chen et al., 1990). A vector using a tac promoter expressed rhizopuspepsinogen in E. coli cytosol as inclusion bodies at a level of about 1.5 mg/L culture. After refolding from an urea solution (Lin et al., 1989) and purification to homogeneity, this zymogen (cRpg) was shown to contain 40 residues in its pro region (Figure 4). The second expression construction utilized a vector pINompA3 (Ghrayeb et al., 1984) in which an omp leader directed the translocation of folded rhizopuspepsinogen into the periplasmic space of E. coli. The expression of this recombinant zymogen (pRpg) was about 40 mg/L culture. pRpg has a pro region with 51 amino acid residues. The third expression system utilizes the T7 bacteriophage promoter in a vector pET-3a developed by Studier (Studier et al., 1990). In this construction, the zymogen, tRpg, contains 51 residues of the pro region and is fused to 16 residues derived from the vector. The level of the tRpg zymogen produced from this system was about 500 mg/L of culture. After refolding and purification by a one-step chromatography on a Sephacryl S100 column, the yield was about 50 mg/L of original culture.

All three forms of recombinant rhizopuspepsinogen are activated in acid to form effective enzymes. Removal of the activation peptide has been demonstrated by gel electrophoresis. Kinetic study of the activation of pRpg and tRpg has been performed. The conclusions concerning both of these species are identical. The rhizopuspepsin is not specifically denatured at a pH range of 7-8, as is the case for porcine pepsin. Consequently, a milk clotting assay (McPhie, 1976) was used to quantify the rhizopuspepsin formed as a function of time. This was possible because at pH 5.3, where the milk clotting assay is performed, rhizopuspepsin activity can be measured even though activation is negligibly slow. At low pH (below 2.5) and dilute zymogen concentration (less than 1 mg/ml), the rate of appearance of hydrolytic activity conforms to a first order kinetic scheme. As additional confirmation of the unimolecular activation reaction mechanism, the first order rate constant was measured at different Rpg concentrations. Identical first order constants were observed for 1 mg/ml Rpg and for 0.5 mg/ml Rpg. The maximal rate of activation occurs near pH 2. Furthermore, above pH 3, bimolecular, rhizopuspepsin-catalyzed activation, can be demonstrated. Our proof of the bimolecular reaction mechanism was two-fold. First, addition of preformed rhizopuspepsin accelerated the activation process. Second, the time course of the appearance of peptic activity had a sigmoid shape and could be fit to a mixed kinetic scheme (Al-Janabi et al., 1972). From this analysis, first and second order rate constants could be extracted.

For those familiar with the activation kinetics of porcine pepsinogen, the activation kinetics of Rpg bring about déjà vu. Both enzymes are activated at a maximal rate near pH 2.

```
-TG ACA TTT ACT CTC AAC TCT TCT TGT ATC GCA ATT GCT GCA CTG GCT GTC GCA GTT    56
(Met)Thr Phe Thr Leu Asn Ser Ser Cys Ile Ala Ile Ala Ala Leu Ala Val Ala Val
    -67                         -60                                        -50

AAC GCT GCC CCT GGA GAA AAG AAA ATC AGC ATT CCT TTA GCA AAG AAT CCC AAC TAC   103
Asn Ala Ala Pro Gly Glu Lys Lys Ile Ser Ile Pro Leu Ala Lys Asn Pro Asn Tyr
                                -40

AAG CCT AGT GCT AAG AAC GCC ATT CAA AAG GCT ATT GCA AAG TAC AAC AAG CAC AAG   163
Lys Pro Ser Ala Lys Asn Ala Ile Gln Lys Ala Ile Ala Lys Tyr Asn Lys His Lys
    -30                             -20

ATT AAT ACT TCT ACT GGT GGT ATT GTC CCT GAT GCT GGT GTT GGT ACT GTC CCA ATG   217
Ile Asn Thr Ser Thr Gly Gly Ile Val Pro Asp Ala Gly Val Gly Thr Val Pro Met
    -10                             -1  1
                                        |
                                N-terminus of
                                rhizopuspepsin

ACT GAT TAC .......
Thr Asp Tyr ......
    10
```

Figure 4. Amino acid and cDNA sequences of the 'prepro' region of rhizopuspepsinogen. The 'prepro' region consists of 68 amino acids, which are marked by negative numbers in reverse direction from the N-terminus position of rhizopuspepsin. The first nucleotide is known to be an A for the initiation (Met) at -68 in the genomic sequence of *Rhizopus chinensis* (Delaney, R., personal communication). The amino acid and cDNA sequence of rhizopuspepsin has been published previously (Delaney *et al.*, 1987). The N-termini positions of pRpg and cRpg are shown by single and double arrowheads respectively. tRpg contains in its N-terminus the sequence of Met-Ala-Ser-Met-Thr-Gly-Gly-Gln-Gln-Met-Gly-Arg-Gly-Ser-Ile-His- followed by pRgn sequence. This extra sequence is derived from the expression vector and the recombinant process.

Moreover, when one compares the actual values of the activation rate constants, the similarity is almost uncanny. At pH 2, k_1 is 1.0 ± 0.2 min^{-1} for pRpg, 2.4 min^{-1} for tRpg, and 1.2 ± 0.1 min^{-1} for porcine pepsinogen under identical conditions. At pH 3.5, k_1 and k_2 are 0.015 min^{-1} and 0.035 (mg/ml)$^{-1}$ min^{-1} for tRpg; 0.07 min^{-1} and 0.87 (mg/ml)$^{-1}$ min^{-1} for porcine pepsinogen. Surprisingly, this similarity is valid for zymogens from the various expression systems even though their activation peptides are of different lengths. This work is the first characterization of a fungal aspartic protease zymogen. To us it was unexpected that the activation characteristics of the Rpg are so very similar to those of porcine pepsinogen.

ACKNOWLEDGEMENTS

The authors wish to dedicate this paper for the 75th birthday of Professor Raul E. Trucco. We also wish to acknowledge the contributions of Dr. Ricky N. S. Wong and X. J. Wang in the early phase of this work, the information on unpublished rhizopuspepsinogen gene sequence from Dr. Robert Delaney, and technical assistance from Ms. Azar Dashti and Jeff Loy. This work was supported by NIH Grant AM-01107.

REFERENCES

Abad-Zapatero, C., Rydel, T. J. & Erickson, J. W., 1990, *Proteins* 8:62-81.

Al-Janabi, J., Hartsuck, J. A. & Tang, J., 1972, *J. Biol. Chem.* 247:4628-4632.

Andreeva, N. S., Zdanov, A. S., Gustchina, A. E. & Fedorov, A. A., 1984, *J. Biol. Chem.* 259:11353-11365.

Carter, P. & Wells, J. A., 1988, *Nature* **332:** 564-568.

Chen, Z., Han, H. P., Wang, X. J., Koelsch, G., Lin, X. L., Hartsuck, J. A. & Tang, J., 1990, *J. Biol. Chem.* (submitted for publication).

Cooper, J. B., Khan, G., Taylor, G., Tickle, I. J. & Blundell, T.L., 1990, *J. Mol. Biol.* **214:**199-222.

Davies, D. R., 1990 *Annu. Rev. Biophys. Chem.* **19:**89-215.

Delaney, R., Wong, R. N. S., Meng, G. -Z., Wu, N. -H. & Tang, J., 1987, *J. Biol. Chem.* **262:**1461-1467.

Fruton, J.S., 1976, *Adv. Enzymol.* **44:**1-36.

Ghrayeb, J., Kimura, H., Takahara, M., Hsiung, H., Masui, Y. & Inouye, M., 1984, *EMBO J.* **3:**2437-2442.

James, M. N. G., Hsu, I. -N. & Delbaere, T. J., 1977, *Nature* **267:**808-813.

Knowles, J. R., 1976, *CRC Crit. Rev. Biochem.* **4:**165-173.

Kraut, J., 1988, *Science* **242:**533-540.

Lin, X. L., Wong, R. N. & Tang, J., 1989, *J. Biol. Chem.* **264:**4482-4489.

Marciniszyn, J. Jr., Huang, J. S., Hartsuck, J. A. & Tang, J., 1976, *J. Biol. Chem.* **251:**7095-7102.

McPhie, P., 1976, *Anal. Biochem.* **73:**258-261.

Pearl, L. & Blundell, T., 1984, *FEBS Lett.* **174:**96-101.

Schimmel, P., 1989, *Acc. Chem. Res.* **22:**232-233.

Studier, W. F., Rosenberg, A. H., Dunn, J. J. & Dubendorff, J. W., 1990, *Methods Enzymol.* **185:**60-89.

Suguna, K., Bott, R. R., Padlan, E. A., Subramanian, E., Sheriff, S., Cohen, G. H. & Davies, D. R., 1987, *J. Mol. Biol.* **196:**877-900.

Workman, R. J. & Burkitt, D. W., 1979, *Arch. Biochem. Biophys.* **194:**157-164

INHIBITOR BINDING INDUCES STRUCTURAL CHANGES IN PORCINE PEPSIN

Cele Abad-Zapatero,[+] T. J. Rydel,[+] D. J. Neidhart,[*] J. Luly,[†]
and J. W. Erickson[+]

[+]Laboratory of Protein Crystallography, D-47E, AP-9A
[†]Cardiovascular Chemistry, D-47L
  Abbott Laboratories
  Abbott Park, Illinois   60064

[*]Department of Chemistry
  Michigan State University
  East Lansing, Michigan   48824

INTRODUCTION

Crystal structures of several fungal aspartic proteinases have been refined at high resolution: penicillopepsin[1], *Endothia parasitica* pepsin[2] and *Rhizopus chinensis* protease.[3] The structure of a recombinant form of human renin has been solved[4] at 2.5 Å and the refined structures of two other mammalian aspartic proteinases have been reported recently: chymosin at 2.3 Å resolution[5], and the monoclinic form of porcine pepsin at 2.3 and 1.8 Å resolution[6,7], as well as the original hexagonal crystal form[8].

Structural studies on enzyme/inhibitor complexes with the fungal enzymes provided the first insights into substrate binding and on the ligand-induced structural alterations in the aspartic proteinase family. Initial X-ray diffraction studies of the binding of pepstatin to rhizopuspepsin[9] and penicillopepsin[10] indicated that a ß-hairpin, or "flap" (residues 72-81), undergoes a conformational change and rigidifies, owing to the formation of hydrogen bonds between the inhibitor and the distal portion of the flap[9-11]. Small conformational changes in the flap region were also found in the crystal structures of endothiapepsin complexed with several renin inhibitors[12,13].

Recently Sali *et al.*[14] observed a rigid body movement of a subdomain (residues 190-303) within the carboxy domain of endothiapepsin, described as a 4.1° rotation combined with 0.3 Å translation. The existence of a flexible subdomain within the carboxy lobe of pepsin, as well as all fungal aspartic proteinases, has also been proposed on the basis of the structural superposition of porcine pepsin with the three fungal enzymes[6]. Similar conclusions were independently reached by other investigators[7,8]. The significance and role of this relative screw motion in substrate binding, catalysis and activation of the zymogen are still to be ascertained.

The knowledge acquired in the study of the inhibited fungal enzymes has not been supplemented with similar studies based on mammalian systems since there have been no crystallographic structures available of such complexes. This paper describes the refined

Structure and Function of the Aspartic Proteinases
Edited by B.M. Dunn, Plenum Press, New York, 1991

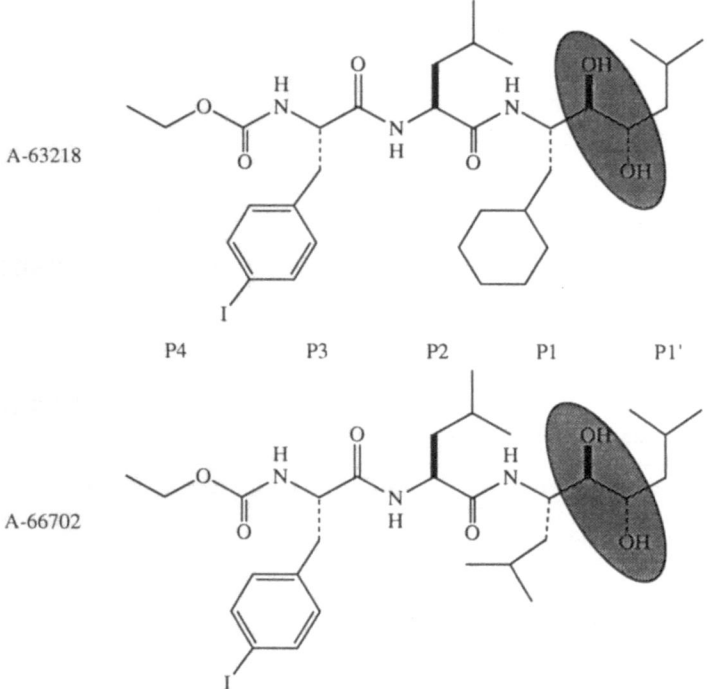

Figure 1. Schematic representation of the peptidomimetic inhibitors of porcine pepsin used in this study. Both are based on the "glycol" motif as a mimic of the tetrahedral transition state and were initially designed as human renin inhibitors[15].

Table 1. Summary of Stereochemical Properties for the Refined Structures of the Pepsin/Inhibitor Complexes

Final Value	Target Value	A63218	A66702
R-factor		0.173	0.173
Resolution range (Å)		5.0-2.2	5.0-1.8
Average $|F_{obs}| - |F_{calc}|$	3.95	2.66	4.51
Distances (Å)+			
i - (i + 1)	0.020	0.009	0.017
i - (i + 2)	0.040	0.031	0.037
i - (i + 3)	0.050	0.032	0.048
Planarity (Å)	0.020	0.009	0.014
Chiral volume (Å³)	0.150	0.120	0.171
No. of reflections		10,328	20,353
Protein atoms		2,429	2,429
Water molecules		273	268

+i - (i + 1): bond distance.
 i - (i + 2): next nearest neighbor distance of 3 bonded atoms defining a bond angle.
 i - (i + 3): the planar 1 - 4 distance.

structure of porcine pepsin complexed with two small, potent, peptidomimetic renin inhibitors, and analyzes the conformational changes induced in the structure of the mammalian enzyme upon inhibitor binding.

MATERIALS AND METHODS

Peptidomimetic inhibitors of human renin containing the glycol hook at the C-terminal were synthesized by methods described previously[15]. The para-iodo substituent was introduced to facilitate the crystal structure solution by Patterson Methods. The structures of the two inhibitors used in the present study are shown in Figure 1. Crystals of the enzyme/inhibitor complexes were obtained by a small variation of the protocol used to grow crystals of the native enzyme[6]. The enzyme was dissolved in distilled water at approximately 20 mg/ml and a concentrated ethanol solution of the corresponding inhibitor added to achieve a stoichiometric enzyme/inhibitor (1:1) ratio. Crystals of both complexes were orthorhombic (Space Group $P2_12_12_1$) and essentially isomorphous, with average cell constants a = 124.0, b = 65.0 and c = 36.2 Å. Diffraction data for the pepsin/A63218 complex were collected by diffractometer to a nominal resolution of 2.2 Å. A higher resolution (1.8 Å) data set was collected for the pepsin/A66702 using a Nicolet-Xentronics area detector. The structure of the pepsin/A63218 complex was solved first by molecular replacement methods using an early, partially refined (R = 0.34) pepsin model in combination with packing analysis facilitated by the use of TABLES[16]. This solution was adopted for the other complex and both structures were refined by a combination of the constrained-restrained (CORELS), restrained (PROLSQ) and simulated annealing (XPLOR) methods. The number of reflections of each data set as well as the refinement results are presented in Table 1. Full details of the data collection, structure solution and refinement will be given elsewhere.

Structural alignments were performed using the superposition method of Rossmann and Argos[17] using the following input parameters: E1 = E2 = 5.00; PCUT = 0.1; ESIG = 1.0; and TSIG = 0.5. The algorithm always converged within a few cycles. The angles of rotation were calculated from the trace of the final rotation matrices and the translation vectors correspond to the relative movement of the centers of mass of the two different atom clusters being compared. In addition to the structures refined in this work, three other sets of atomic coordinates were available for porcine pepsin from the Protein Data Bank[18], two for the monoclinic crystal form (3PEP) and (4PEP), refined at 2.3 and 1.8 Å respectively, and one (2PEP) for the hexagonal form refined at 2.1 Å. For convenience, the coordinate data sets corresponding to the pepsin inhibitor complexes studied here will be referred to as 5PEP (Pepsin/A63218) and 6PEP (Pepsin/A66702).

RESULTS AND DISCUSSION

The current refined structure of porcine pepsin complexed with A63218 consists of 2429 protein atoms, the complete structure of the inhibitor, and 273 solvent molecules assumed to be water molecules. The crystal structure of the isomorphous pepsin/A66702 complex gave an independent structure of the polypeptide chain, the entire inhibitor, and 268 putative water molecules. Table 1 presents the summary of the stereochemical properties for the refined structures of both complexes. The quality of the final electron density map can be judged by the electron density corresponding to the two different inhibitors (Figures 2a and b).

At this point we do not consider the solvent structure around the enzyme completely established. In the present model all the solvent molecules have been considered to be water.

However, a preliminary analysis of the binding energies of a methyl probe around the enzyme, using the program package GRID[19], indicates that there are approximately twenty potential methyl binding sites in the entire enzyme (interaction energy <-4.0 kcal/mol). Several of these putative sites were examined in conjunction with F_o-F_c and $2F_o$-F_c electron density maps and we concluded that some of them could be intepreted as ethanol molecules bound to the enzyme. Further refinement of the revised solvent structure is underway.

Inhibitor Binding

Since the inhibitors differ only at the P_1 sidechain, isobutyl in A66702 vs. cyclohexylmethyl in A63218 (Figure 1), their mode of binding was expected to be very

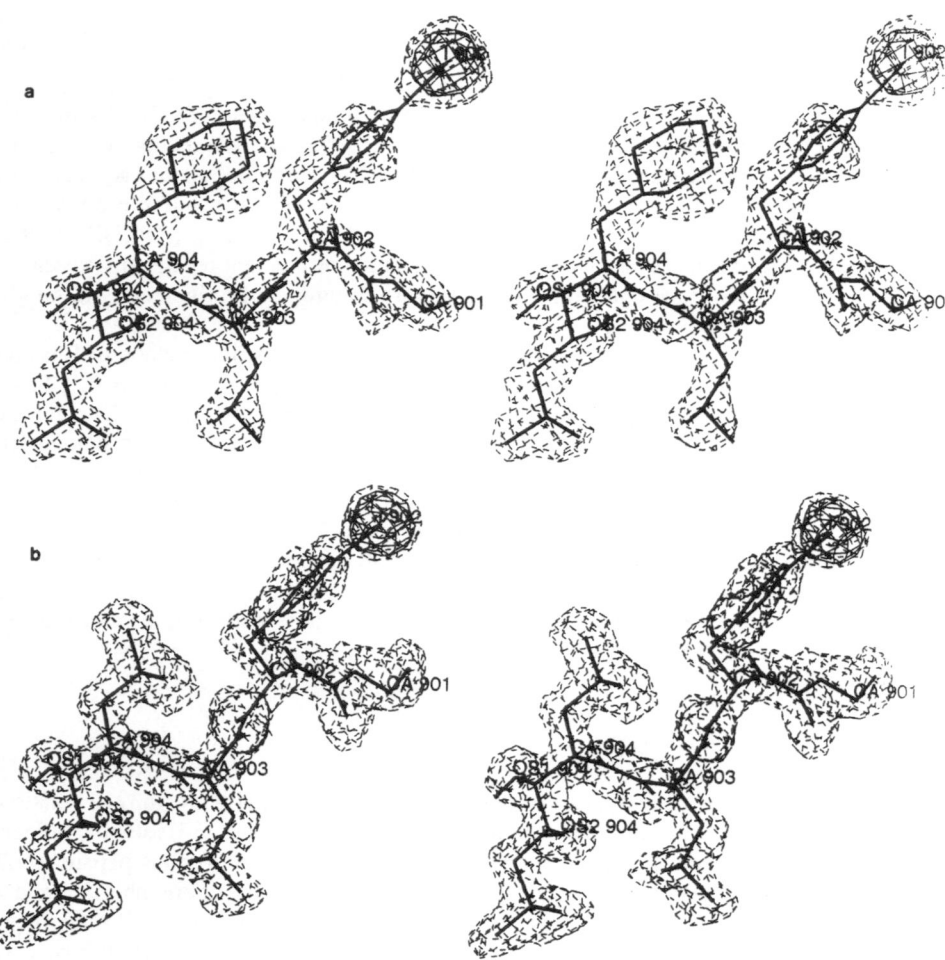

Figure 2. Stereo diagram of the electron density corresponding to the inhibitors in the final $2F_o$-F_c maps. Contour level is 1σ (dashed line). The 6σ level (continuous) is also shown to illustrate the position of the iodine atom within the Iodo-Phe substituent. a) pepsin/A63218, data from 5 to 2.2 Å; b) Pepsin/A66702, data from 5 to 1.8 Å.

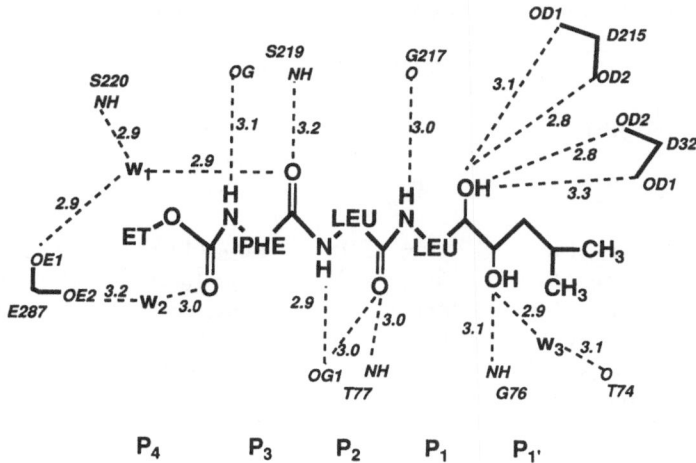

Figure 3. Schematic representation of the hydrogen bonds formed between pepsin and the inhibitor A66702. The mode of binding of the other inhibitor (A63218, not shown) is analogous. The distances between acceptor and donor atoms are given in Å. The active site subsites are labeled. The approximate positions of water molecules. are denoted by the labels w_1, w_2, w_3

similar. Both peptidomimetics bind in an extended conformation (Figures 2a and b) similar to the mode of binding found in many other aspartic proteinase inhibitors, including pepstatin. The ensuing discussion will refer to the pepsin/A66702 complex for which data are available at higher resolution. Figure 3 shows a schematic representation of the hydrogen bonding interactions between A66702 and pepsin. The side chain carboxylate of Glu_{287} is involved in two, water-bridged, hydrogen bonds to the CO groups of the P_3 and Etoc (ethoxycarbonyl, pseudo P_4) subtituents. The former is mediated by a tightly bound (B<10 Å^2) water bridging Glu_{287} (OE1), Ser_{220} (NH) and the carbonyl oxygen from the Iodo-Phe residue of the inhibitor. The hydrogen bonds between the P_3 residue of the inhibitor and Ser_{219} are conserved among the fungal aspartic proteinase/inhibitor complexes. Cooper and coworkers had predicted this interaction to be equivalent to the one observed among the fungal enzymes, where residue 219 is a Threonine[8]. The carbonyl oxygen of conserved Gly_{217} hydrogen bonds to the NH group of the P_1. This interaction is present in all known aspartic proteinase/inhibitor complexes. However, the inhibitors used in this study are too short to provide the dyad-related hydrogen bond involving the carbonyl of conserved residue Gly_{34} and the NH group of a P_2' residue[12].

The hydrogen bonds on the flap side utilize both the main chain and side chain atoms of Thr_{77} which together grip both the NH and CO groups of the P_2 residue (Figure 3). The NH of Gly_{76} interacts with the second hydroxyl group of the glycol moiety. This hydroxyl also forms a water-mediated hydrogen bond to the carbonyl oxygen of Thr_{74}. The other hydroxyl of the glycol, corresponding to the reduced carbonyl of the P_1 substituent, displaces the conserved water molecule shared by the active site aspartates in the native pepsin structure, and is directly hydrogen bonded to both aspartates[6-8].

In addition to the hydrogen bonding interactions, there are 27 non-bonded, direct contacts (< 4.0 Å) between the inhibitor and the protein and seven additional contacts mediated through solvent molecules (Table 2). There were no interactions between the inhibitor and the putative C-domain flap in pepsin that encompasses residues 290-299. This region includes a four residue insertion (292-295) relative to the fungal enzymes present in

Table 2. Non-Bonded Contacts in the Pepsin/Inhibitor Complexes

Inhibitor Subsites

ETO	IPH	LEU	GL1*	GL2*
P_4	P_3	P_2	P_1	P_1'
Glu_{13} (3.1)	Tyr_{75} (4.0)	Thr_{77} (2.0)	Ile_{30} (4.0)	Tyr_{189}(3.4)
Ser_{219} (3.5)	Gly_{76} (3.0)	Thr_{218}(3.2)	Asp_{32}(2.8)	Ile_{213} (3.8)
Leu_{220}(4.0)	Thr_{77} (3.5)	Met_{289}(3.9)	Gly_{34}(3.0)	Asp_{215}(2.4)
w_{403}	Phe_{111}(3.7)	Ile_{300} (3.5)	Tyr_{75}(3.3)	Gly_{217}(2.5)
w_{409}	Ala_{115}(3.8)		Gly_{76}(3.0)	Thr_{218} (2.9)
w_{418}	Gly_{217}(3.8)		Thr_{77}(3.7)	Ile_{300} (3.9)
w_{420}	Thr_{218}(3.4)			
	Ser_{219}(2.1)			w_{704}
	w_{403}			
	w_{418}			
S_4	S_3	S_2	S_1	S_1'

Pepsin Subsites

* GL1 and GL2 denote (respectively) the P_1 and P_1' subsites of the alkyl-diol inhibitor (Figure 1). Numbers in parentheses are distances from the amino acid residue to the corresponding inhibitor subsite in Ångstroms.

pepsin, renin, cathepsin D and yeast proteinase A among the known aspartic proteinases. It had been predicted to be involved in inhibitor binding[7,8] (see below).

Structural Changes of the Enzyme

The structure of the liganded enzyme differs in one major respect from the structure of the native pepsin. The outermost structural elements of the amino- and carboxy-domain lobes move inward, resulting in a more compact structure. The end result of the overall movement of both domains is a tight grip on the inhibitor.

The conformational changes induced by ligand binding were compared to the background of differences observed among the three-dimensional models of porcine pepsin available from the Protein Data Bank (2PEP, 3PEP and 4PEP). The results of the least squares superposition of the three pepsin structures are summarized in Table 3A. The hexagonal form of pepsin (2PEP) deviates more from either of the monoclinic forms than they vary between themselves. The differences between 3PEP and 4PEP are localized to the regions of weakest electron density where the polypeptide chain is probably disordered (278-280; 292-297). Excluding these segments, the rms deviation between them is 0.36 Å for the 318 C_α pairs. Residues 241-245 could not be seen in the electron density map of the hexagonal form of pepsin so that coordinates for only 321 amino acids are available for data set 2PEP. Excluding the highly variable loops indicated above and the missing residues, the rms deviation between either 3PEP, or 4PEP, and 2PEP is approximately 0.50 Å (313 pairs). This difference is 40% higher than the variation observed between the two monoclinic forms of pepsin (rms of 0.36 Å for 318 pairs). The portion of the polypeptide chain referred to as the "flap" (residues 72-81) is closed further down in the hexagonal structure than in the monoclinic forms of pepsin, and accounts for most of the variation observed. These changes are very localized and are probably due to the weak crystal packing forces acting on an intrinsically flexible part of the molecule. Overall, these results indicate that 0.5 Å may be

Table 3A. Overall Superposition Among Available Pepsin Structures

The first row of each entry of the table is the root mean square deviation (in Å) between the structures denoted by the corresponding row and column of the matrix. The integer number in the second row of each entry, specifies the number of C_α pairs in the comparison as specified in a), b) and c). Entries above the diagonal are comparisons between all available coordinates for the corresponding structures.

	2PEP[†]	3PEP[§]	4PEP[¶]
2PEP		0.92 321[a]	0.55 321[a]
3PEP	0.56 313[c]		0.78 326
4PEP	0.50 313[c]	0.36 318[b]	

Table 3B. Overall Superposition Between Pepsin Structures and Pepsin/Inhibitor Complexes

	3PEP[§]	4PEP[¶]	5PEP	6PEP
3PEP		0.78 326	0.92 326	1.0 326
4PEP	0.36 318[b]		0.93 326	1.0 326
5PEP	0.83 318[b]	0.78 318[b]		0.21 326
6PEP	0.91 318[b]	0.87 318[b]	0.20 318[b]	

[†] 2PEP Cooper et al. [8], [§]3PEP Abad-Zapatero et al. [6], [¶] 4PEP Sielecki et al. [7]
[a] residues 1-240; 246-327; 2PEP vs. 1-239; 245-326 of 3PEP or 4PEP
[b] residues 1-277; 281-291; and 297-326 for both
[c] residues 1-240; 246-278; 282-292; 298-327; 2PEP vs. 1-239; 245-277, 281-291; 297-326; of 4PEP

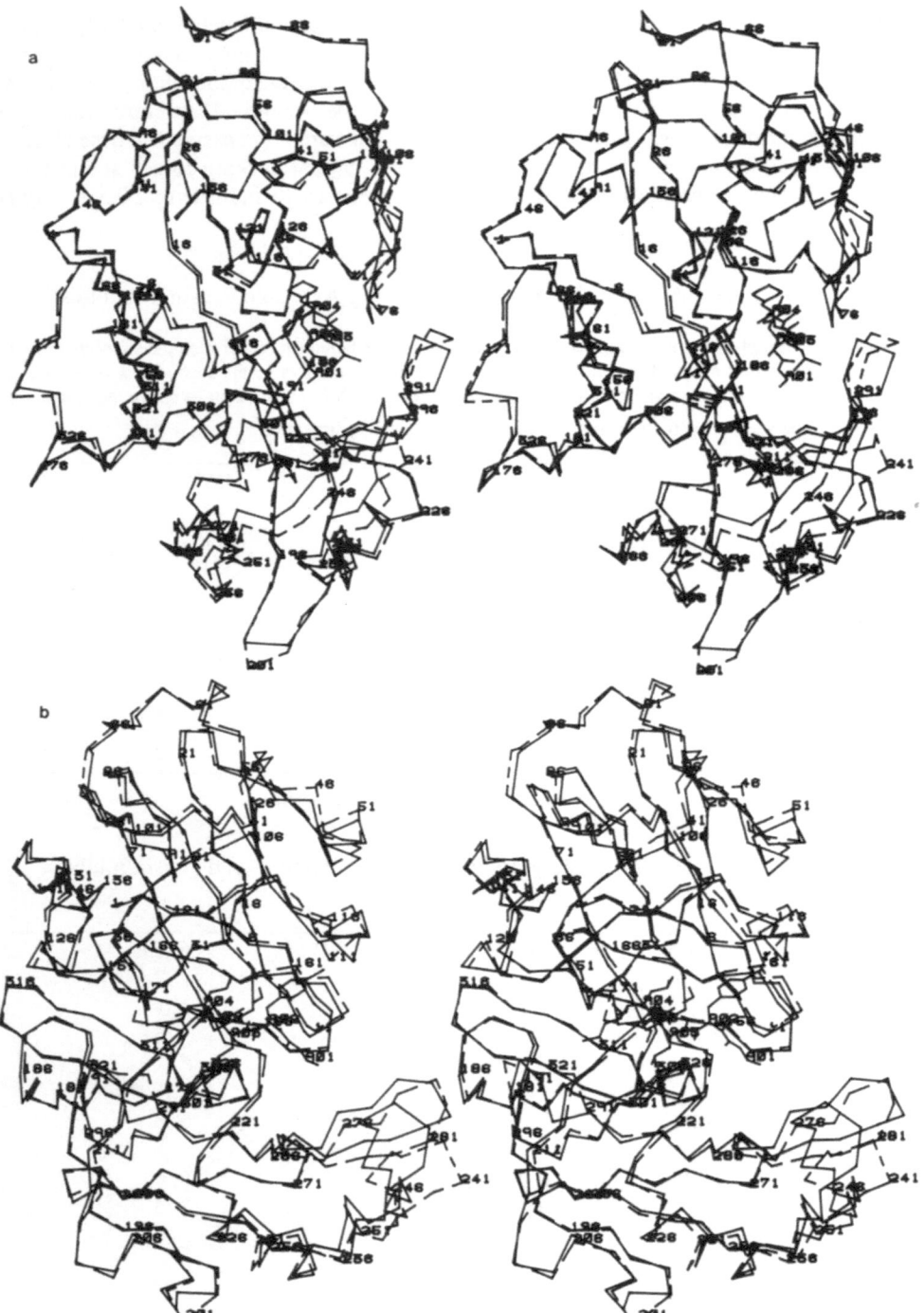

Figure 4. Stereo diagram of the polypeptide backbone fold for the liganded (line) and unliganded (dash) pepsin structure, for the two oligopeptide inhibitors, after the overall least squares superposition of the core residues (Table 4). a) pepsin/A63218, b) Pepsin/A66702. Two different views are presented: a) emphasizes the inhibitor cavity and the displacement of the flexible subdomains at the amino and carboxy end; the view in b) is approximately along the interdomain dyad. It illustrates the extended conformation of the inhibitor and the position of the "glycol" moiety in relation to the active site aspartates (Asp$_{32}$, Asp$_{215}$).

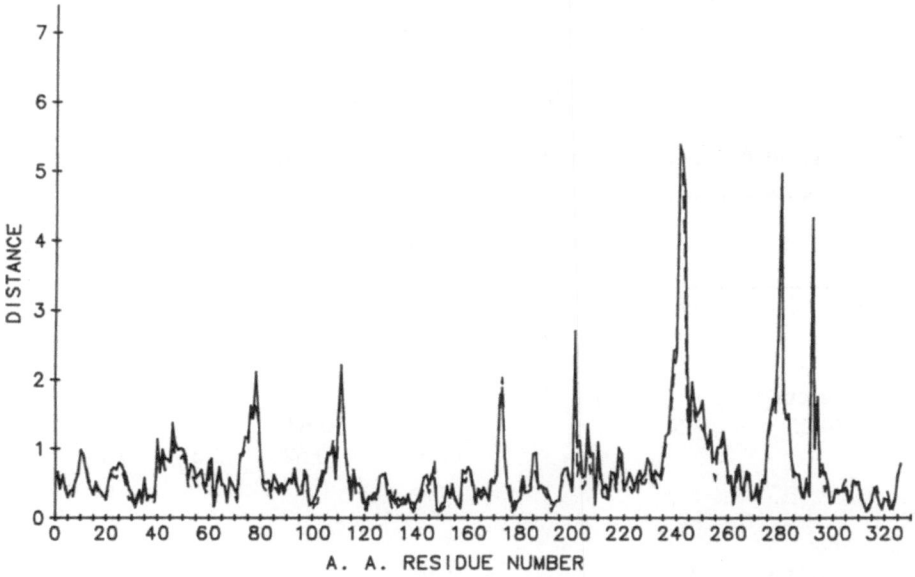

Figure 5. Plot of the C_α-C_α distance vs. residue number for liganded and unliganded pepsin after the overall least squares superposition. Continuous line 3PEP vs. 6PEP (Pepsin/A66702); dashed 3PEP vs. 5PEP (pepsin/A63218).

considered an upper bound for the background rms deviation between pepsin structures refined in different laboratories and crystallized in different space groups.

The results from similar overall comparisons between either of the monoclinic pepsins and the two liganded forms are presented in Table 3B and illustrated in Figures 4 and 5. The overall rms deviation of the monoclinic native pepsins against the orthorhombic inhibited forms ranges from 0.78 to 0.91 Å for 318 C_α pair and is approximately twice as large as could be expected from the refinement uncertainty. In contrast, the relative difference between the two isomorphous liganded structures (5PEP and 6PEP) is only 0.20 Å. The results of the overall least squares superposition of either 5PEP or 6PEP on the 4PEP pepsin structure revealed a core of approximately 190 residues which superposed to less than 0.30 Å and two smaller clusters of residues which departed significantly in each domain (Table 4 and Figures 4 and 5).

The most structurally altered residues in the amino domain include the ß-hairpin known as the "flap" (ßE3,ßF1; 69-82) and the adjacent ßG3 strand (103-106), which form sheet IV[6,7], and the residues in the two helices αA(48-52), and αB(110-114). These regions, which we will refer to as the "flap subdomain", move in a concerted fashion relative to the native structure by an approximate rotation of 3° and a translation of 1 Å (Table 4). This simple rigid body motion accounts for at least 25% of the total rms deviation. Local main chain conformational changes account for the remaining differences. For instance, the backbone residues 109-112 move by about 2 Å to accommodate the iodine atom of the Iodo-Phe residue resulting in a distorted helical conformation (Figure 4b). The side chain of Phe_{111} in this region also swings away to avoid close contact with the inhibitor. These and other localized changes necessary to accomodate the bulky P_3 substituent may partly account for the weak affinity ($IC_{50} \approx 10\ \mu M$) of these inhibitors for pepsin. The residues at the tip of the flap (73-79) also alter their backbone conformations in order to optimize their hydrogen

Table 4. Comparison of the Structures of Native and Inhibited Pepsin

A) Pepsin vs. pepsin/A63218

Description	Number of Residues	rms (Å) before	after	displacement rot (°)	trans. (Å)[c]
flap subdomain 41-57; 69-82; 103-116	45	0.85[a] 0.78[b]	0.67 0.62	2.6 2.6	1.0 1.1
core 4-40; 58-68; 83-102;117-195; 210-220; 299-326	186	0.48 0.38	0.43 0.30	0.0 0.0	0.6[d] 0.6
flexible subdomain 196-209; 221-238; 245-277; 282-291	75	1.10 1.07	0.62 0.49	3.1 2.6	1.7 1.6

B) Pepsin vs. Pepsin/A66702

Description	Number of Residues	rms (Å) before	after	displacement rot (°)	trans. (Å)[c]
flap subdomain 41-57; 69-82; 103-116;	45	0.89 0.84	0.71 0.67	2.6 2.6	1.1 1.2
core 4-40; 58-68; 83-102;117-195; 210-220; 299-326	186	0.51 0.42	0.43 0.30	0.0 0.0	0.7[d] 0.7
flexible subdomain 196-209; 221-238; 245-277; 282-291	75	1.30 1.27	0.67 0.53	3.6 3.6	1.9 1.8

[a] rms deviation against pepsin data set 3PEP, [b] rms deviation against pepsin data set 4PEP, [c] relative displacement of the centers of mass of the two clusters of atoms, [d] this rms deviation reflects the small translational shifts resulting from optimizing the superposition of the 186 core residues after superposing the complete polypetide chains.

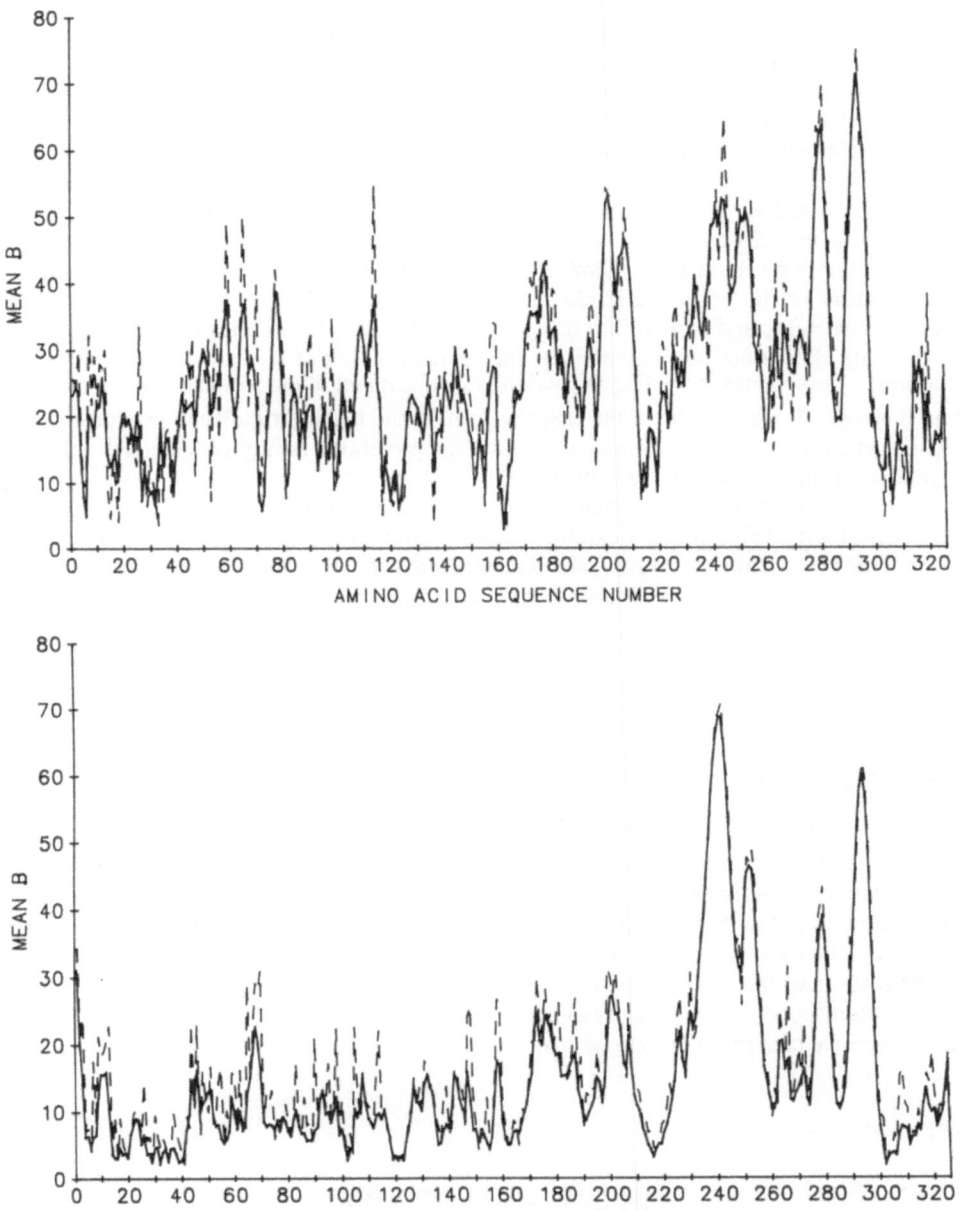

Figure 6. Plot of the average temperature factors for the main chain (line) and side chain (dashed) vs. the residue number. Top) Unliganded pepsin (refinement described in ref. 6). Bottom) Pepsin/A66702.

bonding potential with the inhibitor (Figure 4a). These changes observed upon inhibitor binding in the flap "subdomain" indicate that there has been a concerted motion of this region toward the inhibitor to ensure favorable enzyme-ligand interactions.

The 75 residues (196-209; 221-238; 245-277; 281-291) within the flexible subdomain in the C-terminal lobe[6-8] moved by a rotation of 3.6° and a translation of 2.0 Å in relation to the core of the native enzyme. This rigid body motion accounts for at least 50% of the rms deviation observed between the two forms of the enzyme in the flexible subdomain region. The amount of rotation is similar to that found by Sali *et al.*[14] in their analysis of the endothiapepsin-CP69,799 complex. The remaining residues of the flexible subdomain (278-280; 292-297; and 239-244) appear to exhibit large local structural differences but their precise conformations are uncertain.

The temperature factor of the residues participating in the ligand binding in the flap subdomain decreased by about 50% in the complexed pepsin structure (Figure 6a and b). However, the mobility of the residues pertaining to the flexible subdomain at the carboxy end diminished only by about 10% . Similar to the case for the native pepsin structure (6-8), the conformation of the polypeptide chain between 292 and 297 is poorly defined in both complex structures and is therefore probably disordered. This implies that this part of the polypeptide chain does not play a major role in the binding of these short peptidomimetic inhibitors. Compounds extending beyond the P_1' substituent might form specific interactions with this most flexible region of the carboxy domain and thereby rigidify its conformation in pepsin , as well as in other members of the aspartic proteinase family which contain the four residue insertion (e.g., renin and cathepsin D).

The localization of conformational changes in these two subdomains of the molecule is consistent with their intrinsic flexibility as observed in the different members of this family of enzymes[1-6]. Thus, inhibitor-induced conformational changes, similar to the ones described here in the amino and carboxy domains of pepsin, might also be expected to occur in other mammalian aspartic proteinases of biomedical interest. The structures of the pepsin-inhibitor complexes presented here can provide an initial framework for future analysis of enzyme/inhibitor interactions among the mammalian aspartic proteinases.

SUMMARY AND CONCLUSIONS

The refined structures of two isomorphous pepsin/inhibitor complexes demonstrate that significant conformational changes take place upon ligand binding for a mammalian representative of the aspartic proteinase family. These differences can be attributed mostly to the concerted rigid body movements of two separate clusters of residues relative to a central core. One cluster in the amino domain comprises the flap, the adjacent ß strand (sheet IV) and helices, as well as the interconnecting loops. The other, larger cluster is in the carboxy end and corresponds approximately to the flexible subdomain described previously. Similar conformational changes are proposed to occur in renin and cathepsin D.

REFERENCES

1. M. N. G. James and A. R. Sielecki, *J. Mol. Biol.* **163**:299 (1983).
2. T. Blundell, J. Jenkins, L. Pearl, T. J. Sewell, L. Pearl, J. B. Cooper, I. J. Tickle, B. Veerapandian and S. P. Wood, *J. Mol Biol.* **211**:919 (1990).
3. K. Suguna, R. R. Bott, E. A. Padlan, E. Subramanian, S. Sheriff, C. H. Cohen and D. R. Davies, *J. Mol. Biol.* **196**:877 (1987).
4. A. R. Sielecki, K. Hayakawa, M. Fujinaga, M. E. P. Murphy, M. Fraser, A. K. Muir, C. T. Carilli, J. A. Lewicki, J. D. Baxler and M. N. G. James, *Science* **243**:1346 (1989).

5. G. L. Gilliland, E. L. Winborne, J. Nachman and A. Wlodawer, *Proteins: Structure, Function and Genetics.* **8**:82 (1990).

6. C. Abad-Zapatero, T. Rydel and J. W. Erickson, *Proteins: Structure, Function and Genetics* **8**:62 (1990).

7. A. R. Sielecki, A. A. Fedorov, A. Boodhoo, N. S. Andreeva and M. N. G. James. *J. Mol. Biol.* **214**:43 (1990).

8. J. B. Cooper, G. Kahn, G. Taylor, I. J. Tickle and T. L. Blundell. *J. Mol. Biol.* **214**:199 (1990).

9. R. Bott, E. Subramanian and D. R. Davies *Biochemistry* **21**:6956 (1982).

10. M. N. G. James, A. Sielecki, F. Salituro, D. H. Rich and T. Hofmann, *Proc. Natl. Acad. Sci.* **79**:6137 (1982).

11. K. Suguna, E. A. Padland, C. W. Smith, W. D. Carlson and D. R. Davies, *Proc. Natl Acad. Sci.* **84**:7009 (1982).

12. S. I. Foundling, J. Cooper, F. E. Watson, A. Cleasby, L. H. Pearl, B. L. Sibanda, A. Hemmings, S. P. Wood, T. L. Blundell, M. J. Valler, C. G. Norey, J. Kay, J. Boger, B. M. Dunn, B. J. Leckie, D. M. Jones, B. Atrash, A. Hallett and M. Szelke. *Nature* **327**:349 (1987).

13. T. L. Blundell, J. Cooper, S. I. Foundling, D. M. Jones, B. Atrash and M. Szelke, *Biochemistry* **26**:5585 (1987).

14. A. Sali, B. Veerapandian, J. B. Cooper, S. I. Foundling, D. J. Hoover and T. L. Blundell, *EMBO J.* **8**:2179 (1989).

15. J. R. Luly, N. BaMaung, J. Soderquist, A. K. L. Fung, H. Stein, H. D. Kleinert, P. A. Marcotte, D. Egan, B. Bopp, I. Merits, G. Bolis, J. Greer, T. Perun and J. J. Plattner, *J. Med. Chem.* **31**:2264 (1988).

16. C. Abad-Zapatero and T. J. O'Donnell, *J. Appl. Cryst.* **20**:532 (1987)

17. M. G. Rossmann and P. Argos, *J. Biol. Chem.* **250**:7525 (1975).

18. F. C. Bernstein, T. F. Koetzle, G. F. Williams, E. F. Meyer, Jr., M. D. Brice, J. R. Rodgers,O. Kennard, T. Shimanouchi and M. Tasumi, *J. Mol. Biol.* **112**:535 (1977).

19 P. J. Goodford, *J. Med. Chemistry* **28**:849 (1985).

FUNCTIONAL IMPLICATIONS OF THE THREE-DIMENSIONAL STRUCTURE OF

BOVINE CHYMOSIN

Gary L. Gilliland, Maureen Toner Oliva, and Jonathan Dill

Center for Advanced Research in Biotechnology
The Maryland Biotechnology Institute
University of Maryland
and
The National Institute of Standards and Technology
9600 Gudelsky Drive
Rockville, Maryland 20850

INTRODUCTION

Chymosin (EC 3.4.23.4, formerly rennin) is one of the primary enzymes used to initiate milk clotting for cheese production (MacKinlay & Wake, 1971). This process begins with the specific cleavage of the $Phe_{105}*Met_{106}$ peptide bond of κ-casein by this enzyme (Jolles *et al.*, 1968). The sequence of this cleavage site is

-His-Pro-His-Pro-His-Leu-Ser-Phe*Met-Ala-Ile-Pro-Pro-Lys-Lys-.

98   105 106   112

A number of studies with synthetic peptides, which were designed based on the 103-108 κ-casein sequence, have been undertaken to determine the kinetic parameters of chymosin (e.g., Visser & Rollema, 1986), and recently studies by Visser and coworkers (1987) have been performed which provide information concerning the substrate specificity of this enzyme. Based upon the results of these kinetic and model building studies, it was proposed that residues 103-108 fit snugly into the active site cleft and that the addition of the 98-102 sequence, -His-Pro-His-Pro-His-, assisted in the positioning of the 103-108 peptide segment into the active site by its favorable electrostatic interactions with residues on the protein surface.

Recently, the crystal structure of recombinant bovine chymosin has been determined to 2.3 Å resolution and compared with the structures of three fungal aspartic proteinases (Gilliland *et al.*, 1990). The enzyme has an irregular shape with overall dimensions of 40 x 50 x 65 Å. Chymosin has the eukaryotic aspartic proteinase fold which consists primarily of ß-sheets composed of both parallel and anti-parallel ß-strands. There are only a few short α-helices interspersed among the ß-strands. The enzyme structure consists of N- and C-terminal domains with similar arrangements of secondary structural elements. A deep cleft containing the active site aspartates separates the two domains. The orientation of atoms of the carboxylate groups of the active site aspartates and the position of Wat-411 located between these two groups is nearly identical to positions of corresponding atoms found in other aspartic proteinases.

In this paper the results of a preliminary comparison of chymosin with the recently determined structures of pepsin (Abad-Zapatero et al., 1990; Sielecki et al., 1990; Cooper et al., 1990) are described and discussed. These results are also compared with those from the previous comparisons of chymosin with the three fungal aspartic proteinases. Substrate binding is examined by model building substrates and substrate analogs into the active site cleft of the structure determined from X-ray studies and comparing these results with the structures of inhibitor-aspartic proteinase complexes which have been previously reported. Results of electrostatic calculations are presented which indicate that chymosin is much more polarized than other aspartic proteinases, and the role that this property may have in substrate cleavage is considered. A complete description of the environment around Glu_{109} is also given, and the possible role of this residue in pH dependent activation of the enzyme is discussed.

EXPERIMENTAL PROCEDURES

Chymosin coordinates

The coordinates of recombinant chymosin B [1CMS, the Brookhaven Protein Data Bank identifier (Bernstein et al., 1977)] were obtained from the recent crystal structure determination reported by Gilliland and coworkers (1990). The crystals were grown by vapor diffusion from macroseeded droplets containing 10 mg/ml enzyme, 22.5-27.5% saturated sodium chloride, 0.05 M MES at pH 6.0 equilibrated against 1 ml of reservoir solution containing 45-55% saturated sodium chloride with 0.05 M MES at pH 6.0. The crystals are of space group I222 with a = 72.7 Å, b = 80.3 Å and c = 114.8 Å. These crystals were grown under conditions reported for the natural enzyme (Berridge, 1945), and they are isomorphous with those described in earlier diffraction studies (Bunn et al., 1970). X-ray data collection for the structure determination was performed with a Siemens electronic area detector mounted on an Elliot GX-21 rotating anode. The determination of crystal orientation and the integration of reflection intensities were performed with the XENGEN program system (Howard et al., 1987). The structure was determined by molecular replacement to 2.3 Å. The molecular replacement solution utilized the structure of bovine pepsinogen (Remington, unpublished data) which had been modified to eliminate non-homologous regions of the structure for the search model. Once the rotation and translation function solutions were obtained, the structure was refined by restrained least-squares using the program PROFFT (Finzel, 1987) interspersed with model rebuilding with the program FRODO (Jones, 1978) implemented on an Evans & Sutherland PS390 graphics system. The model was adjusted and solvent added by the interpretation of $2F_o - F_c$ and $F_o - F_c$ difference maps and complete omit maps (Bhat, 1988). The final crystallographic R-factor for the chymosin structure was 0.165 at 2.3 Å resolution with a root-mean-square deviation of bonded distances from ideality of 0.020 Å.

Superposition of Structures

The superpositioning of chymosin with other aspartic proteinase molecules was performed with the ALIGN program (Satow et al., 1986). Only the positions of the C_α atoms of the polypeptide backbone were considered during the superpositioning procedure. In all comparisons involving chymosin, chymosin was the reference structure. The structure of recombinant chymosin has been compared with four other high resolution aspartic proteinase structures which have been deposited in the Brookhaven Protein Data Bank. They include rhizopuspepsin (2APR) at 1.8 Å resolution (Suguna *et al.*, 1987b), penicillopepsin (2APP) at 1.8 Å resolution (James & Sielecki, 1983), endothiapepsin (4APE) at 2.1 Å resolution (Pearl & Blundell, 1984), pepsin (3PEP) at 2.3 Å resolution (Abad-Zapatero *et al.*, 1990), pepsin (4PEP) at 1.8 Å resolution (Sielecki *et al.*, 1990), and pepsin (5PEP) at 2.3 Å resolution (Cooper *et al.*, 1990).

The results of the comparisons were examined using the programs FRODO (Jones, 1978) running on an Evans & Sutherland PS390 graphics system and PV (Tung, Bacon & Gilliland, unpublished program) implemented on Silicon Graphics 4D workstations.

Substrate and Substrate Analog Model Building

The initial position of the substrate and substrate analogs was based upon the position of the reduced inhibitor in the active site of rhizopuspepsin (Suguna *et al.*, 1987a). The structure of the rhizopuspepsin-reduced inhibitor complex (3APR) was superimposed on the structure of chymosin (1CMS) as previously described. The substrate or substrate analog backbone and sidechain atoms were positioned as closely as possible to the positions of corresponding atoms of the rhizopuspepsin reduced inhibitor.

Model building of the substrate and substrate analogs into the active site of chymosin was assisted by the use of the QUANTA program of Polygen Corporation. This program was also used to perform energy minimization calculations. The CONTACT routine of the program FRODO (Jones, 1978) was used to define the enzyme-substrate contact regions. Only interactions between pairs of atoms which were not covalently bonded to one another and which were within 4.0 Å of each other were considered.

Macromolecular Dipole Calculations

The magnitude of the dipole moment, p, of a macromolecule is calculated as follows:

$$p = \left(\left(\sum_n (x_n - x_{cc})c_n\right)^2 + \left(\sum_n (y_n - y_{cc})c_n\right)^2 + \left(\sum_n (z_n - z_{cc})c_n\right)^2\right)^{1/2}$$

The charges on the atoms are assumed to be point charges. The magnitude of point charge n is represented as c_n at the position within the molecule at coordinates x_n, y_n, and z_n. The coordinates of the center of charge of the molecule are represented by x_{cc}, y_{cc}, and z_{cc}. They are calculated as follows:

$$x_{cc} = \frac{\sum_n |c_n| x_n}{\sum_n |c_n|} \; ; \; y_{cc} = \frac{\sum_n |c_n| y_n}{\sum_n |c_n|} \; ; \; z_{cc} = \frac{\sum_n |c_n| z_n}{\sum_n |c_n|} \; .$$

The directions given for the dipole moments are relative to the position of the charge center of the protein and are defined by the following:

$$\cos \alpha = \frac{\sum_n (x_n - x_{cc})c_n}{p} \; ; \; \cos \beta = \frac{\sum_n (y_n - y_{cc})c_n}{p} \; ; \; \cos \gamma = \frac{\sum_n (z_n - z_{cc})c_n}{p} \; .$$

Solvent screening effects are not included in the calculations (Moult *et al.*, 1985).

Table 1. Root-mean square differences (Å) between the aligned structures of acid proteinases

	2APR	2APP	4APE	3PEP	4PEP	5PEP	
1CMS	1.21	1.39	1.71	0.93	0.85	0.89	rms(Å)
	295	292	295	307	300	304	pairs
	104	92	84	188	187	189	identities
2APR		1.21	1.22	1.36	1.32	1.39	rms(Å)
		294	294	297	296	297	pairs
		125	121	118	118	116	identities
2APP			1.10	1.52	1.51	1.47	rms(Å)
			300	292	291	290	pairs
			167	94	94	96	identities
4APE				1.58	1.60	1.66	rms(Å)
				281	282	285	pairs
				83	82	85	identities
3PEP					0.34	0.47	rms(Å)
					315	306	pairs
					326	324	identities
4PEP						0.40	rms(Å)
						312	pairs
						324	identities

RESULTS AND DISCUSSION

Comparison with other enzymes

Although chymosin has been previously compared with the fungal aspartic proteinases rhizopuspepsin, penicillopepsin, and endothiapepsin (Gilliland *et al.*, 1990), no comparisons were made with other gastric aspartic proteinases such as pepsin. Recently, however, three high resolution structures of porcine pepsin have become available (Abad-Zapatero *et al.*, 1990; Sielecki *et al.*, 1990; Cooper *et al.*, 1990). We report here preliminary results of a comparison of these three structures with that of chymosin. Summary statistics for these comparisons are presented in Table 1 along with results of the comparison of chymosin and pepsin with the fungal aspartic proteinases. That the differences between chymosin and all of the pepsins are much less than the differences between chymosin and the fungal proteinases is a reflection of the almost two-fold increase in sequence homology among the mammalian proteinases. The largest differences in all of the structures are the surface loop regions which connect the secondary structural elements. In particular, the regions (in chymosin) include residues 10-14, 75-80, 160-162, 241-245, 249-252, and 291-295; several of these regions have poorly defined electron density and high temperature factors in the chymosin structure.

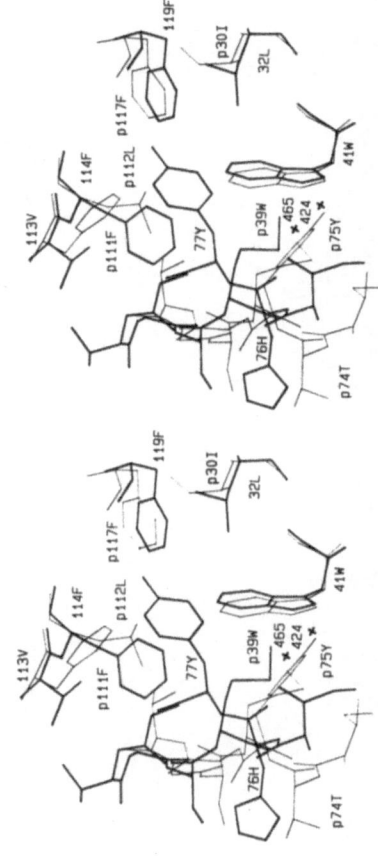

Figure 1. A stereoscopic view of the flap region of recombinant chymosin (heavy line) with the homologous structure of pepsin (thin line) superimposed. Residues of both recombinant chymosin and pepsin are labeled, with the pepsin labels preceded with a 'p'.

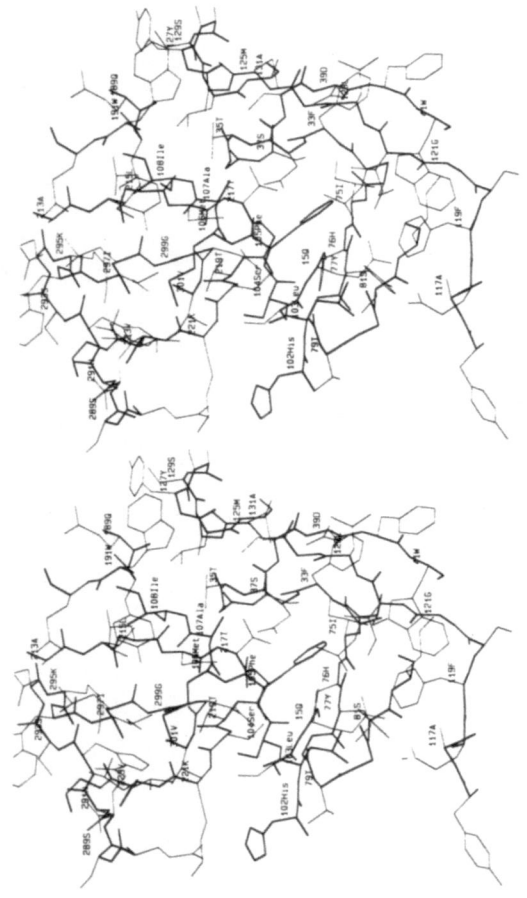

Figure 2. A stereoscopic view of the synthetic peptide substrate bound to the active site of chymosin. The substrate is shown in heavy lines as is the backbone of chymosin while the chymosin side chains are shown in thin lines.

The most notable difference between chymosin and all of the aspartic proteinases involved in the comparisons was the position of the flap (residues 73-85 in chymosin), a loop on the surface of the protein that forms part of the the substrate binding site. The difference in conformation may be a result of the interaction of Tyr_{77} with the hydrophobic residues Phe_{119} and Leu_{32}. The residue corresponding to Leu_{32} is aspartate in rhizopuspepsin and endothiapepsin and asparagine in penicillopepsin. Phe_{119} corresponds to asparagine in rhizopuspepsin and penicillopepsin and to isoleucine in endothiapepsin. In chymosin the position of Tyr_{77} is stabilized by hydrophobic interactions with Leu_{32} and by aromatic-aromatic interaction with Phe_{119}. The aromatic side chain interactions of Tyr_{77} with Phe_{119} is similar to that found for other pairs of aromatic side chains (Burley & Petsko, 1985).

The position of the flap in the pepsin structure is similar to that seen in the fungal proteinases. In the pepsin structure the OH of Tyr_{75} hydrogen bonds to NE1 of Trp_{39}. This is quite different from what is seen in chymosin despite the fact that there are only two differences in the flap sequences, His_{76} and Gln_{83} of chymosin, both corresponding to threonines in pepsin. These two residues are located on the surface and are not directly involved in interactions which would assist in positioning of the flap. Therefore, the orientation and conformation of the flap may be the result of differences in contacts which result from other sequence differences of the two proteins. In addition to the differences described above for the flap tyrosine, there are three sequence changes which may affect the flap position: Leu_{32}, Val_{113}, and Phe_{114} of chymosin corresponding to Ile_{30}, Thr_{77}, and Leu_{112} of pepsin, respectively. Another factor which may contribute to the variation in flap orientation and conformation is crystal packing interactions. Pepsin and chymosin crystals have different space group and unit cell parameters resulting in different arrangements of molecules within the crystal lattice. The flaps of both molecules are involved in symmetry contacts (Gilliland et al., 1990; Sielecki et al., 1990). The differences in conformation of the flap regions of pepsin and chymosin is shown in Figure 1.

Enzyme specificity

The eukaryotic aspartic proteinases active site clefts can accomodate seven to eight amino acid residues. A number of crystallographic reports of eukaryotic aspartic proteinase-inhibitor complexes have characterized the interactions at many of the subsite binding pockets (Bott et al., 1982; Andreeva et al., 1984; James et al., 1985; Blundell et al., 1987; Cooper et al., 1987; Foundling et al., 1987; James & Sielecki, 1987; Suguna et al., 1987a). The environment of the S_1 and S_1' substrate binding subsites of chymosin were examined by comparison of the structure with those of the fungal proteinases complexes which have been reported (Gilliland et al., 1990). The structurally equivalent residues of the S_1 and S_1' chymosin subsites are in general similar to those found in the fungal proteinases. However, there is one major difference in the S_1 subsites of chymosin and the fungal proteinases, a substitution of a hydrophobic amino acid residue Leu_{32} for polar residues Asp_{33} and Asn_{31} in the rhizopuspepsin and penicillopepsin structures, respectively. In pepsin the corresponding residue is Ile_{30}, again a hydrophic residue.

In order to assist in understanding the principles of substrate specificity for this enzyme, model building studies of chymosin complexed with a fragment of the natural substrate have been undertaken. The sequence of the heptapeptide is His-Leu-Ser-Phe∗Met-Ala-Ile corresponding to residues 102 to 108 of κ-casein. The residues of chymosin which interact (i.e., are less than 4.0 Å from substrate atoms) with P_4 through P_3' are listed in Table 2. This peptide and its hypothetical positioning within the active site are shown in Figure 2.

The two subsites, S_1 and S_1', are shallow pockets within the active site cleft. The S_1 subsite, in which Phe_{105} binds, is blocked by the position of Tyr_{77}. This residue must

Table 2. Chymosin residues in contact with residues of an ideal heptapeptide substrate with a sequence identical to that of the κ-casein cleavage site

Substrate	Chymosin residues		
P_4	His_{102}	S_4	Ser_{220}, Gln_{288}
P_3	Leu_{103}	S_3	Ser_{14}, Gln_{15}, Tyr_{77}, Gly_{218}, Thr_{219}, Ser_{220}
P_2	Ser_{104}	S_2	Gly_{78}, Thr_{219}
P_1	Phe_{105}	S_1	Leu_{32}, Asp_{34}, Gly_{36}, Tyr_{77}, Gly_{78}, Phe_{119}, Ile_{122}, Asp_{216}, Gly_{218}, Thr_{219}
P_1'	Met_{106}	S_1'	Gly_{36}, Tyr_{190}, Asp_{216}, Thr_{219}, Glu_{290}, Ile_{297}
P_2'	Ala_{107}	S_2'	Gly_{36}, Ser_{37}, Tyr_{190}
P_3'	Ile_{108}	S_3'	Tyr_{190}

move, and thus, a significant movement of the flap must occur, to accomodate the phenylalanine side chain of the substrate if binding is similar to that found in the aspartic protease-inhibitor complexes mentioned above. The shallow pocket forms as a result of the extension of the flap over two ß-strands. The sidechain bound to the S_1 subsite lies over the ß-strands and is partially covered by the flap. This pocket is quite hydrophobic with the plane of the phenylalanine aromatic ring parallel to the plane of the carboxylate oxygen atoms of the active site aspartate, Asp_{34}. The S_1' subsite has a similar architecture to the S_1 subsite. The pocket is not quite as hydrophobic as that for S_1; there is an additional charged residue, Glu_{290}, in close proximity to the κ-casein Met_{106} side chain. The loop corresponding to the flap at this subsite is composed of residues 289 to 294. This loop is smaller than the corresponding loop found in pepsin which is composed of residues 291 to 298 (Sielecki *et al.*, 1990). Subsite S_2 and subsite S_2' accomodate Ser_{104} and Ala_{107}, amino acid residues with relatively small side chains, and therefore have relatively few contacts (Table 2). The S_3 and S_3' accomodate Leu_{103} and Ile_{108}, respectively. The modeling indicates that the S_3 subsite is quite hydrophobic, consistent with the type of amino acid residue it accommodates. The S_3' subsite on the other hand does not appear to make extensive contacts with the protein even though the peptide residue Ile_{108} is quite hydrophobic.

Model building of idealized oligopeptides with various sizes and composition into the active site of chymosin indicates that electrostatic interactions play a role in substrate binding and specificity. Substrate with the sequence Leu-Ser-Phe*Met-Ala-IleOMe, corresponding to κ-casein sequence 103 to 108, fits quite snugly into the active site. The contacts involve primarily van der Waals interactions; there are no polar or electrostatic interactions stabilizing the complex except for hydrogen bonds formed between atoms of the substrate main chain and chymosin residues. When the sequence is extended to Pro-His-Leu-Ser-Phe∗Met-Ala-IleOMe, corresponding to κ-casein sequence 101 to 108, the same interactions occur with chymosin as seen with the shorter peptide, but now the peptide histidine (κ-casein His_{102}) is located less than 4.0 Å from OE1 and OE2 of Glu_{288} in chymosin. This electrostatic interaction would favor binding which is supported by the results of studies by Visser and coworkers (1987) who found a greater than 15-fold increase in the k_{cat}/K_m value for the longer peptide. A substrate of equivalent size but with proline substituted for histidine binds less tightly. Even with the proline substitution, however, it binds more tightly than shorter peptides. There is also a suggestion that if a longer peptide were constructed which incorporated the His-Pro-His-Pro-His sequence of κ-casein (residues 98-102), the additional histidines may interact electrostatically with Asp_{279} and Glu_{280} of chymosin, providing additional binding energy.

Chymosin A and B Isozymes

There are two isozymes of chymosin which differ by a single amino acid at position 244. Chymosin A, which has a significantly higher specific activity (Foltmann, 1960), has an aspartate at this position; Chymosin B has a glycine. This residue is positioned near the midpoint of a loop, composed of residues 241-249, on the surface of the molecule . This region was difficult to fit during the structure determination (Gilliland et al., 1990), and was found in the crystal structure to lie close to a symmetry related molecule. It has been postulated by Safro and Andreeva (The 18th Linderstrom-Lang Conference on Aspartic Proteinases, Abstracts p. 80) based on their model building studies and the biochemical studies of Visser and coworkers (1987) that the increased activity of the A isozyme may be the result of the enhanced binding affinity of κ-casein, owing to favorable electrostatic interactions between His_{102} of κ-casein and Asp_{244}. (As mentioned above, His_{102} of κ-casein is also close to Asp_{288}.) This hypothesis was investigated by examination of the model of chymosin complexed with the peptide Pro-His-Leu-Ser-Phe∗Met-Ala-IleOMe corresponding to residues 101 to 108 of the κ-casein molecule. The histidine residue corresponding to position 102 in κ-casein is found in the model to be less than 9.0 Å from C_α of Gly_{244}. When aspartate of chymosin is substituted for glycine, the Asp_{244} OD1 and OD2 atoms are less than 4.0 Å from the histidine of the substrate, indicating the potential for stronger electrostatic interactions between the substrate and chymosin A.

Charge Distribution and Dipole Moments

A comparison of the sequences of the eukaryotic aspartic proteinases quickly reveals that chymosin has a larger ratio of positively to negatively charged residues. In addition to the N- and C-termini, there are 54 charged residues on the surface of chymosin at neutral pH. The chymosin molecule has a net charge of -12 with the N- and C-terminal domains contributing -5 and -7, respectively. In the N-terminal domain, which has the lower net negative charge, six of the positive charges are located in the region from residues 48 to 62 forming a positive patch near one end of the molecule (Figure 3). This group of positive residues is not found in other aspartic proteinase molecules.

To determine to what extent this patch of positive residues polarizes chymosin with respect to other aspartic proteinase molecules, electrostatic calculations were performed to determine the orientations and magnitudes of the dipole moments of these enzymes. The results are shown in Table 3. These data indicate that for chymosin the dipole is oriented nearly parallel to the long axis of the molecule with the center of charge very near the center of the active site. The positive end of the dipole moment is of course centered on the positive

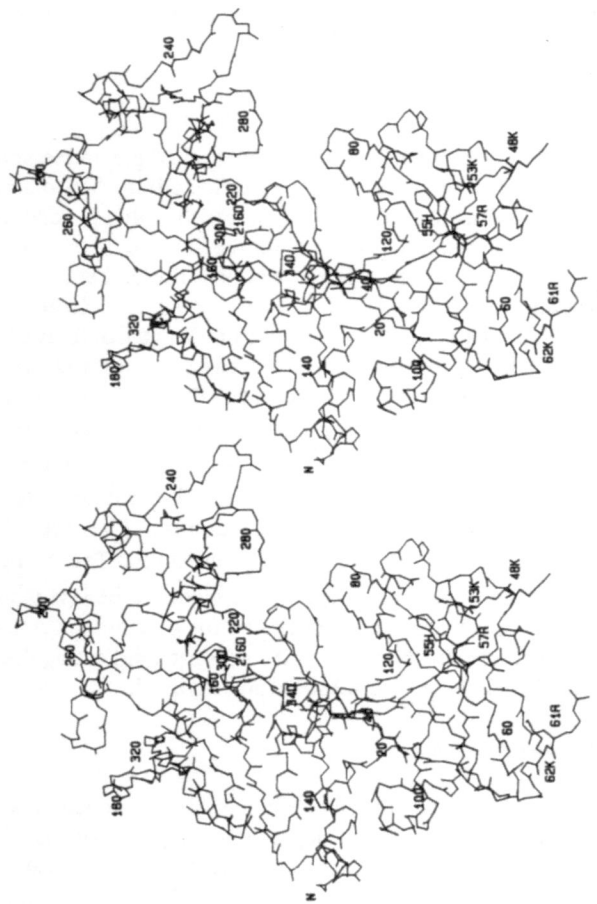

Figure 3. A stereoscopic view of the backbone atoms of recombinant bovine chymosin with every twentieth Cα position labeled. The atoms of side chains of the positively charged residues in the region from 48 to 62 in the amino acid sequence are shown and labeled.

Table 3. Acid Proteinase Dipole Moments

Enzyme	Dipole Moment		In the Direction of			Charge Center at		
	e-Å	Debye	cos a	cos ß	cos γ	x	y	z
Chymosin (1CMS)	188	901	-0.240	0.255	0.937	18.2	27.0	25.3
Penicillopepsin (2APP)	125	601	0.727	0.150	-0.671	22.0	13.7	16.0
Endothiapepsin (4APE)	129	620	-0.433	0.176	0.548	31.7	-4.3	5.8
Rhizopuspepsin (2APR)	108	519	0.320	-0.7091	0.522	38.3	59.2	102.2
Pepsin (4PEP)	151	727	0.651	0.594	0.473	9.0	18.2	17.1

N-terminal patch. The orientation of all other dipoles for the aspartic proteinases in Table 3 are quite different. In addition, the magnitude of the dipole for chymosin is nearly one and one half times larger than that calculated for the other enzymes. This indicates that chymosin would be polarized in a different direction and to a greater extent than would other enzymes. The electric dipole of chymosin may play an important role in determining the efficiency of cleavage of the $Phe_{105}*Met_{106}$ peptide bond of κ-casein. The electrostatic field associated with the molecular dipole may cause the orientation of a higher proportion of chymosin molecules in a manner which facilitates binding to substrate; this may be particularly important if the phospholipids on the micelle surface, to which κ-casein is associated, are negatively charged. Alternatively, the patch of positive residues on one end of chymosin may have favorable electrostatic interactions with negatively charged residues of other protein components of the casein micelles. This would insure that chymosin binds the substrate in the correct orientation for cleavage of the $Phe_{105}*Met_{106}$ peptide bond of κ-casein.

Buried Glu$_{109}$

In the chymosin structure there is a completely buried glutamate residue, Glu_{109}, near the N-terminal positive patch of surface residues. This negatively charged residue is conserved in all mammalian gastric enzymes and several of the fungal enzymes including rhizopuspepsin. The carboxylate oxygen atoms of this residue hydrogen-bond to a water molecule, Wat-404. The distances between Wat-404 and OE1 and OE2 of Glu_{109} are 3.0 and 3.5 Å, respectively. This water has a relatively low thermal or B factor of 7.3 Å^2 and hydrogen bonds to the oxygen atoms of Asp_{120} and Wat-417 with bond distances of 2.6 and 2.9 Å, respectively.

The oxygen ligands of Wat-404 have an average B value of 11.3 Å^2 and are disposed in an almost trigonal planar arrangement. The carboxylate oxygen atoms of Glu_{109} are considered as a single bidentate ligand of the water molecule. Since Wat-404 interacts directly with OE1 and OE2 of Glu_{109}, it may be a positively charged ion such as sodium or hydronium. If it is a positive ion, it is more likely to be hydronium rather than sodium since the distances to the oxygen atoms are longer than those usually associated with sodium ions. Above and below Wat-404 are His_{55} and Trp_{41}, both at distances greater than 4.0 Å and

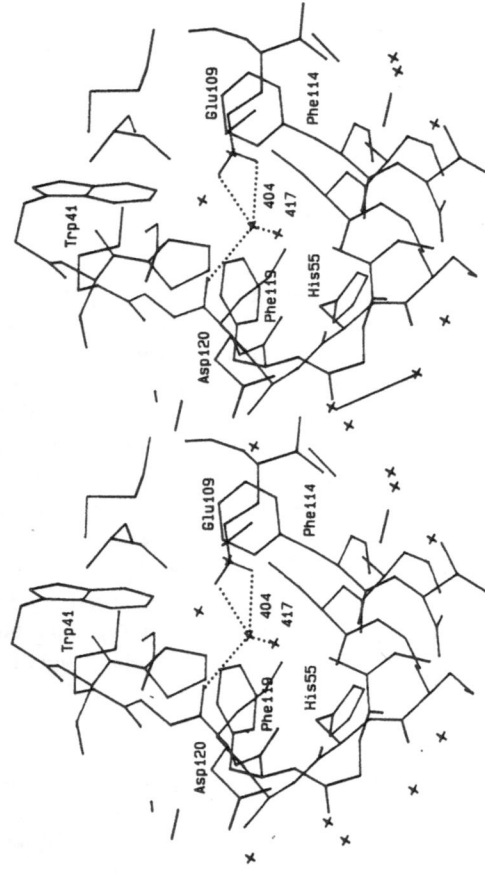

Figure 4. A stereoscopic view of the environment of the buried Glu$_{109}$ of recombinant bovine chymosin. Possible hydrogen bonds are shown with dashed lines.

neither of which interact with the other ligands associated with Wat-404. The environment of Glu_{109} is shown in Figure 4.

We would like to suggest that Glu_{109} plays a role in the pH dependent activation of the chymosin enzyme. Acid proteinases are presumably synthesized at neutral pH. At this pH the Glu_{109} side chain would normally be negatively charged and highly solvated and, thus, be unlikely to bury itself in the hydrophobic interior of the protein. When the protein enters the acidic environment of the gastric lumen, the carboxylate group of Glu_{109} would become protonated allowing the side chain to bury itself inside the protein. If this is the case, the difference in folding induced by the protonation state of Glu_{109} may alter the activity of chymosin, providing a mechanism to protect the organism prior to the secretion of the enzyme into the stomach.

SUMMARY

Many aspects of the structure of chymosin are quite unique even though structure comparisons indicate a high degree of structural homology with other eukaryotic aspartic proteinases. The structural homology is shown to be directly related to the sequence homology which varies from 30 to 60%. The recent structures of pepsin (Abad-Zapatero *et al.*, 1990; Sielecki *et al.*, 1990; Cooper *et al.*, 1990) have allowed the first preliminary comparisons of two different gastric enzymes. These structures are quite similar, even more so than the structures of the fungal proteinases. However, unlike chymosin, the position of Tyr_{77} in the flap of pepsin is similar to that found in the fungal aspartic proteinases despite the fact that pepsin is more similar in the flap sequence and the S_1 binding site to chymosin than to the fungal proteinases.

Attempts at obtaining crystals complexed with substrate analogs which are suitable for diffraction studies have been unsuccessful. Therefore, substrate binding has been examined by model building substrates and substrate analogs into the active site cleft of the structure determined from X-ray studies. The model complexes have been compared with the structures of inhibitor-aspartic proteinase complexes which have been previously reported. The results reported here indicate that there are valid reasons why the natural substrate, κ-casein, binds and is cleaved between positions 105-106. The positively charged histidine residues (98, 100, and 102) of κ-casein, which are located prior to the cleavage site, appear to be able to interact with negatively charged residues of chymosin which are quite distant from the active site. These residues include Glu_{288}, Asp_{279}, and Glu_{280} of chymosin. The latter two residues are approximatly 20 and 25 Å from the center of the active site. These studies also suggest that the difference in activities of the A and B isozymes of chymosin may be due to the increased binding affinity of the substrate as a result of strong electrostatic interactions with Asp_{244} of chymosin and positively charged His_{102} of the substrate.

An examination of the charged amino acid residues of the chymosin structure has produced two interesting observations. First, there is an asymmetric distribution of charged residues; the N-terminal domain has a smaller net negative charge than the C-terminal domain. This is due to a patch of positive charges on the surface located in the region from residues 48 to 62. Electrostatic calculations in which overall dipole moments were estimated for each of the eukaryotic aspartic proteinases have been performed. The results indicate that chymosin is much more electrically polarized than other aspartic proteinases, and that the orientation of the overall macromolecular dipole is quite different than that of other eukaryotic aspartic proteinases. Second, a completely buried glutamate residue, Glu_{109}, is located within the N-terminal domain. It is proposed that this residue plays a role in pH dependent activation of the enzyme.

REFERENCES

Abad-Zapatero, C., Rydel, T. J., & Erickson, J., 1990, Revised 2.3 Å structure of procine pepsin: evidence for a flexible subdomain, *Proteins: Struc. Func. Gen.*, **8**:62.

Andreeva, N. S., Zdanov, A. S., Gustchina, A. E.,& Fedorov, A. A., 1984, Structure of ethanol-inhibited porcine pepsin at 2-Å resolution and binding of the methyl ester of phenylalanyl-diiodotyrosine to the enzyme, *J. Biol. Chem.*, **259**:11353.

Bernstein, F. C., Koetzle, T. F., Williams, G. J. B., Meyer, E. F., Jr., Brice, M. D., Rogers, J. R., Kennard, O., Shimanouchi, T., & Tasumi, M., 1977, The protein data bank: a computer-based archival file for macromolecular structures, *J. Mol. Biol.*,**112**:535.

Berridge, N. J., 1945, The purification and crystallization of rennin, *Biochem. J.*, **39**:179.

Bhat, T. N., 1988, Calculation of an OMIT map, *J. Appl. Crystallogr.*, **21**:279.

Blundell, T. L., Cooper, J., Foundling, S. I., Jones, D. M., Atrash, B., & Szelke, M., 1987, On the rational design of renin inhibitors: X-ray studies of aspartic proteinases complexed with transition state analogues, *Biochemistry*, **26**:5585.

Bott, R., Subramanian, E., & Davies, D. R., 1982, Three-dimensional structure of the complex of the *Rhizopus chinensis* carboxyl proteinase and pepstatin at 2.5-Å resolution, *Biochemistry*, **21**:6956.

Bunn, C. W., Camerman, N., T'sai, L. T., Moews, P. C., & Baumber, M. E., 1970, X-ray diffracton studies of rennin crystals, *Phil. Trans. Roy. Soc.*, **B257**:253.

Burley, S. K., & Petsko, G. A., 1985, Aromatic-aromatic interaction: a mechanism of protein structure stabilization, *Science*, **229**:23.

Cooper, J. B., Khan, G., Taylor, G., Tickle, I. J., & Blundell, T.L., 1990, Three-dimensional structure of the hexagonal crystal form of porcine pepsin at 2.3 Å resolution, *J. Mol. Biol.*, **214**:199.

Cooper, J., Foundling, S., Hemmings, A., Blundell, T., Jones, D. M., Hallett, A.,& Szelke, M., 1987, The structure of a synthetic pepsin inhibitor complexed with endothiapepsin, *Eur. J. Biochem.*, **169**:215.

Finzel, B. C., 1987, Incorporation of fast Fourier transforms to speed restrained least-squares refinement of protein structures, *J. Appl. Crystallogr.*, **20**:53.

Foltmann, B., 1960, Chromatographic purification of prorennin, *Acta Chem. Scand.*, **14**:2247.

Foundling, S. I., Cooper, J., Watson, F. E., Cleasby, A., Pearl, L. H., Sibanda, B. L., Hemmings, A., Wood, S.P., Blundell, T.L., Valler, M.J., Norey, C. G., Kay, J., Boger, J., Dunn, B. M., Leckie, B. J., Jones, D. M., Atrash, B., Hallett, A., & Szelke, M., 1987, High resolution X-ray analyses of renin inhibitor-aspartic proteinase complexes, *Nature*, **327**:349.

Gilliland, G. L., Winborne, E. L., Nachman, J.,& Wlodawer, A., 1990, The Three-Dimensional Structure of Recombinant Bovine Chymosin at 2.3 Å Resolution, *Proteins: Struc.Func. Gen.*, **8**:82.

Howard, A. J., Gilliland, G. L., Finzel, B. C., Poulos, T. L., Ohlendorf, D. H., & Salemme, F. R., 1987, The use of an imaging proportional counter in macromolecular crystallography, *J. Appl. Crystallogr.*, **20**:383.

James, M. N. G., & Sielecki, A. R., 1983, Structure and refinement of penicillopepsin at 1.8 Å resolution, *J. Mol. Biol.*, **163**:299.

Jolles, J., Alias, C., &Jolles, P., 1968, The tryptic peptide with rennin-sensitive linkage of cow's κ-casein, *Biochim. Biophys. Acta*, **168**:591.

James, M. N. G., Sielecki, A. R.,& Hofmann, T., 1985, X-ray diffraction studies on penicillopepsin and its complexes: the hydrolytic mechanism, in "Aspartic Proteinases and Their Inhibitors," Kostka, V., ed., New York, Walter de Gruyter.

Jones, T. A., 1978, A graphics model bulding and refinement system for macromolecules. *J. Appl. Crystallogr.*, **11**:268.

MacKinlay, A. G., & Wake, R. G., 1971, κ-casein and its attack by rennin(chymosin), in: "Milk Proteins," Vol 2. H. A. McKenzie, ed., New York, Academic Press.

Moult, J., Sussman, F., & James, M. N. G., 1985, Electron density calculations as an extension of protein structure refinement-Streptomyces griseus protease A at 1.5 Å resolution, *J. Mol. Biol.*, **182**:555.

Pearl, L., & Blundell, T., 1984, The active site of acid proteinases, *FEBS Lett.*, **174**:96.

Sali, A., Veerapandian, B., Cooper, J. B., Foundling, S. I., Hoover, D. J., & Blundell, T. L., 1987, High-resolution X-ray diffraction study of the complex between endothiapepsin and an oligopeptide inhibitor: the analysis of the inhibitor binding and description of the rigid body shift in the enzyme, *EMBO J.*, **8**:2179.

Satow, Y., Cohen, G. H., Padlan, E. A., & Davies, D. R., 1986, Phosphocholine binding immunoglobulin Fab McPC603. An X-ray diffraction study at 2.7 Å, *J. Mol. Biol.*,**190**:593.

Sielecki, A. R., Fedorov, A. A., Boodhoo, A., Andreeva, N., & James, M. N. G., 1990, Molecular and crystal structure of monoclinic porcine pepsin refined at 1.8 Å resolution, *J. Mol. Biol.*, **214**:143.

Suguna, K., Padlan, E. A., Smith, C. W., Carlson, W. D.,& Davies, D. R., 1987a, Binding of a reduced peptide inhibitor to the aspartic proteinase from *Rhizopus chinensis*: Implications for a mechanism of action, *Proc. Natl. Acad. Sci. U.S.A.*, **84**:7009.

Suguna, K., Bott, R. R., Padlan, E. A., Subramanian, E., Sheriff, S., Cohen, G. H., &Davies, D. R., 1987b, Structure and refinement at 1.8 Å resolution of the aspartic proteinase from *Rhizopus chinensis*, *J. Mol. Biol.*, **196**:877.

Visser, S., & Rollema, H.S., 1986, Quantification of chymosin action on nonlabeled κ-casein-related peptide substrates by ultraviolet spectrophotometry: description of kinetics by the analysis of progress curves, *Anal. Biochem.*, **153**:235.

Visser, S., Slangen, C. J., & van Rooijen, P. J., 1987, Peptide substrates for chymosin (rennin). Interaction sequences located outside the (103-108)-hexapeptide region that fits into the enzyme's active-site cleft, *Biochem. J.*, **244**:553.

WHY DOES PEPSIN HAVE A NEGATIVE CHARGE AT VERY LOW pH ? AN ANALYSIS OF CONSERVED CHARGED RESIDUES IN ASPARTIC PROTEINASES

Natalia S. Andreeva[1] and Michael N. G. James[2]

[1]V.A. Engelhardt Institute of Molecular Biology
Academy of Sciences of the USSR
117984 Moscow, USSR

[2]Medical Research Council of Canada Group in
Protein Structure and Function
Department of Biochemistry, University of Alberta
Edmonton, Alberta, Canada T6G 2H7

INTRODUCTION

Pepsin has several properties which are markedly different from those common for other proteins. It has a very low pH optimum for the hydrolysis of different substrates and a high activity at pH 2. This implies a very stable tertiary structure under conditions in which many proteins are fully denatured. These properties of pepsin are critical for its physiological function which takes place in the extreme acid conditions of the gastric lumen. There the unfolded hydrophobic cores of the proteins to be cleaved are exposed to the first hydrolytic attack by pepsin. Another very specific property of pepsin is its extremely low isoelectric point. It has a net negative charge in the range of low pH values including the pH optimum for catalytic activity. The proteins to be cleaved have a net positive charge in this range of pH. These molecular features of pepsin are closely interdependent as all of them are the consequence of the specific arrangements and interaction of the charged groups in the three-dimensional fold of the enzyme. Data on the refined porcine pepsin A structure at 1.8 Å resolution (Sielecki *et al.*, 1990) have pointed towards their explanation. Porcine pepsin has also been refined at 2.3 Å resolution in the monoclinic crystal form by Abad-Zapatero *et al.* (1990) and in the hexagonal crystal form by Cooper *et al.* (1990).

We discuss at first the structural grounds of the electrophoretic behaviour of the enzyme. "Anodic migration" of porcine pepsin even in the most acidic solutions has been described in very early observations (Michaelis & Davidsohn, 1910; Ringer, 1915). Many years later Tiselius *et al.* (1938) confirmed the results of Ringer (1915) in a very careful and detailed analysis of the electrophoretic behaviour of solutions of crystalline porcine pepsin preparations. They showed that pepsin migrates as a homogeneous protein with a net negative charge in the pH range 1.08 to 4.57. Thus, the isoelectric point of pepsin must be less than 1. Electrofocusing experiments have revealed that the isoelectric point of pepsin might be somewhat higher but the presence of a net negative charge at the pH optimum for the catalytic activity is unquestioned.

A simple accounting of the positively and the negatively charged groups of porcine pepsin A cannot explain the presence of the net negative charge on the enzyme molecule at low pH. Although pepsin has only a few positively charged groups (His_{53}, Arg_{307}, Arg_{315}, Lys_{319} and the NH_3^+-terminus) but a large number of negatively charged groups, only two of the potentially negatively charged groups are supposedly ionized at pH < 2.0 (one of the active site carboxyl groups (Asp_{32} and Asp_{215}) and the serine-phosphate on Ser_{68}). Otherwise, the majority of the carboxyl groups would be expected to be protonated in such conditions. At the same time several more carboxyl groups must be charged even in the extreme acid media to balance the overall charge of + 5 coming from the basic residues, and in addition to them, some other negatively charged groups must provide for the movement of the molecule towards the positive pole.

In an attempt to explain this property a detailed analysis of the environment of pepsin carboxyl groups has been performed. During this analysis, it was found that there are several regions conserved both in sequence and in structure in the three-dimensional fold of the aspartic proteinases. Several aspartic acid residues are in special micro-environments which must influence their pK_a values. There are no data available for a quantitative analysis at present. Carboxyl groups with reduced proton affinity or, in contrast, with increased proton affinity due to special hydrogen bonding environment, can be detected. We think that a detailed description of these regions is important, as they represent common features of many non-viral aspartic proteinases. At the same time, the disposition of these carboxyl groups relative to the arrangement of the basic groups makes it possible to explain some of the unusual properties of pepsin.

ANALYSIS AND DISCUSSION

Three aspartic acid residues of porcine pepsin A are in close proximity to three charged basic groups and could retain their net negative charges even at very low pH values. The monoclinic pepsin crystals were obtained from solutions at pH 2. Two positively charged groups of pepsin, His_{53} and Arg_{307}, as well as their negatively charged counterparts, Asp_{118} and Asp_{11}, are highly conserved among non-viral aspartic proteinases. The disposition of these residues and their microenvironments are only slightly different in the different enzymes. However, the general tendency to reduce the proton affinity for the carboxyl groups should be common for all of them and can be detected by the inspection of the individual hydrogen bonding patterns. The conserved internal ion pair Asp_{11}-Arg_{307} has been described in homologous structures of penicillopepsin (Asp_{14}-Lys_{304}, James & Sielecki, 1983), in rhizopuspepsin (Asp_{14}-Lys_{305}, Suguna et al. 1987], endothiapepsin (Asp_{14}-Lys_{305}, Blundell et al., 1990), chymosin (Asp_{13}-Arg_{303}, Gilliland et al., 1990) and recombinant human renin (Asp_{11}-Arg_{308}, Sielecki et al., 1989). In pepsin, as in renin, Asp_{11} does not form a direct hydrogen bond with the guanidinium group of Arg_{307}. However the carboxylate of Asp_{11} is in close proximity to it (Figure 1). Thus the carboxyl group of Asp_{11} should have a reduced proton affinity in that it is the direct recipient of two hydrogen bonds from the amide side chain of Asn_8 and the main chain nitrogen atom of Asp_{159} and could receive a third hydrogen bond from protonated Asp_{159}. His_{53} is sandwiched between Asp_{52} and Asp_{118}. The carboxylate group of Asp_{118} is only partially exposed to solvent and is the recipient of three hydrogen bonds from surrounding protein groups. It is likely that this residue remains negatively charged at low pH.

The ion pair Arg_{315}-Asp_{138} was observed for the first time in pepsin (Andreeva et al., 1984). Both residues are conserved in the vertebrate aspartic proteinases of known sequence, therefore this ion pair should be present in enzymes of this class. As the refined structure has shown, the guanidinium group of Arg_{315} interacts intimately through direct hydrogen-binding with the carboxylate of Asp_{138} (Sielecki et al., 1990). It is clear that in the environment shown in Figure 2 the two charges would be neutralized even below pH 2.

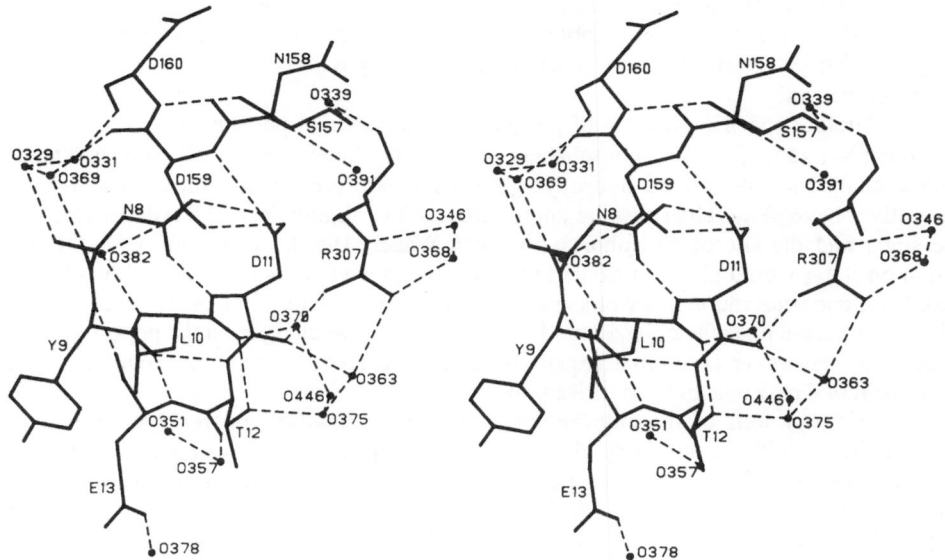

Figure 1 A stereoscopic view of the surroundings of Asp$_{11}$ in monoclinic pepsin crystals studied at pH 2. The immediate hydrogen-bonding environment of Asp$_{11}$ and the proximity of the guanidinum group of Arg$_{307}$ make it possible for the carboxyl group to be negatively charged in spite of the low pH. Hydrogen bonding interactions are denoted by dashed lines.

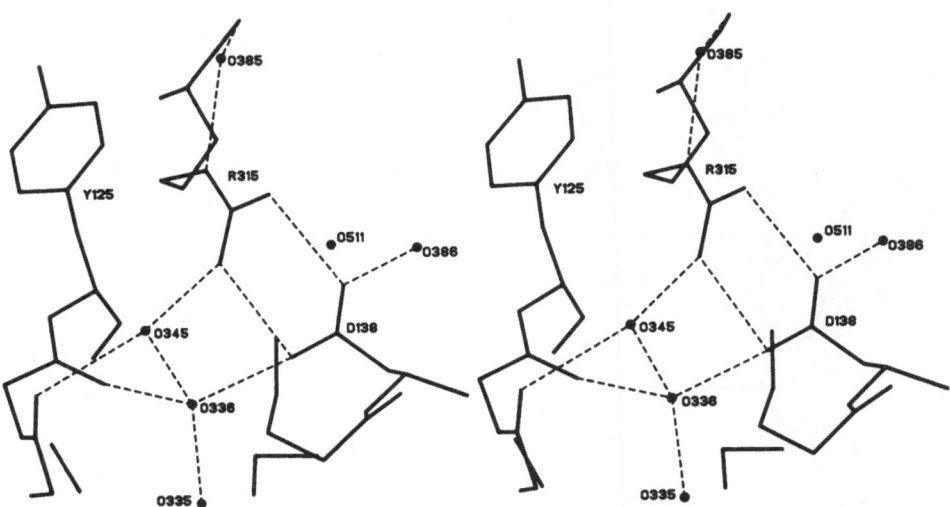

Figure 2 A stereo view of the region containing the ion pair between Arg$_{315}$ and Asp$_{138}$. It is clear that in such environment the two charges are neutralized even at pH 2.

The previously suggested ion pair between the charged NH_3-terminus of pepsin and the carboxyl group of Glu_4 (Andreeva *et al.*, 1984) was not confirmed with the refined data. The NH_3-group of Ile_1 is exposed to solvent, while the carboxyl group of Glu_4 is a recipient of two hydrogen bonds but probably acts as a donor in the interaction with the carbonyl oxygen of Thr_{17}. Thus this group would contribute only weakly, if at all, to a net negative charge on pepsin.

Another specific property of pepsin is the presence of phosphoserine$_{68}$. Although accurate pK_{a1} value for serine phosphate or pepsin is not known, it is likely to be less than 1. Analogously phosphorylated hydroxyl groups that are present in glucose and fructose derivatives have pK_{a1} values that are approximately 1 or slightly less (0.84) depending upon the sugar and the site of phosphorylation (van Wazer, 1958). Thus, the $-CH_2-OPO_3H^-$ group on Ser_{68} would also contribute to the net negative charge on pepsin at low pH. The carboxylate groups and the pair of aspartate residues at the active site, Asp_{32} and Asp_{215}, in addition to serine phosphate Ser_{68} could balance the positive charges on the pepsin molecule at pH < 2. However additional negatively charged carboxyl groups are required for the movement of a molecule to the positive pole.

Investigations of the environments of several aspartic acid residues that do not interact directly with the positively-charged groups in pepsin have resulted in some interesting observations. As often observed, aspartic acid residues are located at the ends of ß-strands and at the junctions of orthogonal ß-sheet layers. Among them Asp_{87}, Asp_{96}, Asp_{303} and Asp_{314} deserve special attention. Asp_{87} is located at the second junction of the B- and the A-layers of the orthogonally packed ß-sheets in the N-terminal domain (Sielecki *et al.*, 1990). It is one of the highly conserved residues among the aspartic proteinases which have been sequenced. This residue is buried and has a conserved surrounding. The environment of Asp_{87} in pepsin is shown in Figure 3. It forms three strong (short) hydrogen bonds, and the

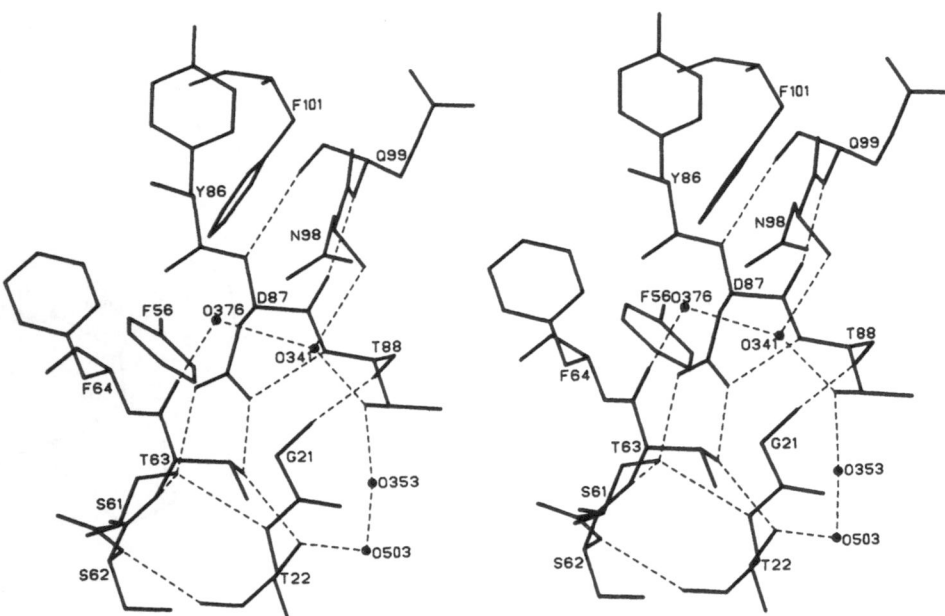

Figure 3 A stereo view of the environment of the buried aspartate side chain of Asp_{87}. As the hydrogen bonding parameters show, it is the recipient of 3 strong (short) hydrogen bonds from $Ser_{61}O^\gamma$, $Thr_{63}O^\gamma$ and $Thr_{88}N$. This aspartate group contributes a full negative charge to the net charge of pepsin at pH 2.

Table 1. Hydrogen bonding for some conserved aspartic acid residues with abnormal pK_a in different aspartic proteinases

Aspartic proteinases and pH of their crystallization

	Pepsin	Pepsinogen	Rhizopus pepsin	Chymosin	Endothia pepsin
	pH 2[a]	pH 6[b]	pH 6[c]	pH 5.6[d]	pH 3.5[e]
Asp_{87}	*Hydrogen bonding distances*				
$O\delta1$ ...... $Thr_{88}N$	2.7	2.7	2.7	2.7	2.6
$O\delta1$ ...... $Thr_{63}O\gamma$	2.6	2.5	2.6	2.7	2.7
$O\delta2$ ...... $Ser_{61}O\gamma$	2.6	2.6	2.6	2.6	2.6
Asp_{303}*	*Hydrogen bonding distances*				
Thr_{216} O ...... $O\delta1$	2.7	---	2.7	2.9	3.1

Parameters of hydrogen bonds of Asp_{87} carboxyl in pepsin

Acceptor Donor	d(D-A)	d(A-H)	α(C-D..A)	α(D-H..A)
$O\delta1$.... $Thr_{88}N$	2.68	1.75	---	153
$O\delta1$ .... $Thr_{63}O\gamma$	2.63	1.63	109	178
$O\delta2$ .... $Ser_{61}O\gamma$	2.64	1.64	106	173

* Side chain of this residue is subjected to conformational changes during conversion from pepsinogen to pepsin; [a] Sielecki *et al.*, 1990; [b] James & Sielecki, 1986; [c] Suguna *et al.*, 1987; [d] Gilliland *et al.*, 1990; [e] Blundell *et al.*, 1990

parameters of these bonds are similar. Asp_{87} acts only as a recipient of hydrogen bonds. This property is common for pepsin, chymosin, rhizopuspepsin, endothiapepsin, penicillopepsin and pepsinogen (Table 1). However, in penicillopepsin the surroundings of Asp_{87} are slightly different from those of others. Nevertheless the presence of the short hydrogen bonds in which Asp_{87} acts as a recipient is common for all of them. The hydrogen-bonded environment of Asp_{87} is relatively unambiguous and it is clear that this residue could contribute a full negative charge to the net charge of pepsin. The comparison of pepsin and pepsinogen structures also support this result. Data presented in Table 1 show the carboxylate group of Asp_{87} does not move during the activation process, which means that the charged state of this group is unlikely to change during conversion from alkaline to acid media.

These considerations show that all aspartic proteinases have buried negatively-charged carboxyl groups in the region of Asp_{87} which preserve their charged states even at low pH. As the comparison of the primary structures shows the substitution of neutral amino acid residues to the basic groups is often observed at the vicinity of Asp_{87} in the three-dimensional structure, however there is no direct interaction between them. Pepsin A has no basic residue in this region.

On the other hand, the hydrogen-bonded environment of Asp_{96}, a neighboring buried residue, is ambiguous. The doubt about its ionization state arises mainly from the orientation of the side-chain amide group of Gln_{99}. If the nitrogen atom of this amide is the one pointing towards the carboxyl of Asp_{96}, then this carboxyl is an acceptor of the hydrogen bonds. In case the positions of the $N_{\varepsilon2}$ and $O_{\varepsilon2}$ are interchanged, this would imply a protonated Asp_{96} is acting as a donor in the interaction with $O_{\varepsilon2}$ of Gln_{99}. The present X-ray structure cannot distinguish between these two possibilities.

Asp_{314} is also a rather conserved residue, especially among the aspartic proteinases of vertebrates. The side chain of Asp_{314} is packed against a non-polar surface on one side (Leu_{150}, Ile_{170}, Trp_{181} and Val_{312}) and a polar region on the other. The turn at Arg_{315} to Asn_{318} causes the polypeptide chain to wrap around the side chain of Asp_{314} and these residues provide three hydrogen bonds to the carboxylate of Asp_{314}. The fact that it is partially buried and the recipient of three hydrogen bonds suggests that it remains ionized even at pH 2 and would have at least a partial negative charge at low pH.

Contrary to the above carboxyl groups the carboxylate of Asp_{303} should have an increased proton affinity. It cannot be detected by the same analysis of the environment of this group in pepsin because pepsin has been studied at pH 2. However the comparison of this environment in pepsin and *Rhizopus chinensis* enzyme studied at pH 6 reveals a high conservation of the three-dimensional structure in the region including this residue. In all cases the Asp_{303} carboxyl group acts as a proton donor to provide a hydrogen bond to the carbonyl oxygen of Thr_{216}.

From this discussion it is not possible to evaluate the total balance of charges in pepsin exactly. What seems clear, however, is that there are sufficient carboxyl groups in pepsin with depressed pK_a values due to their special microenvironment to account for the net negative charge and hence the anodic migration of pepsin even at very low pH values.

How can the stability of pepsin molecules in an extremely acid medium be explained? For most proteins the ratio of positively charged groups to negatively charged groups at neutral pH values is close to 1.0. Thus, at the extremes of the pH scale, as the net charge on the protein is increased, the increasing charge repulsion will destabilize the folded state. The greater charge density on the folded protein can be relieved so the unfolded state should have a lower electrostatic free energy component (Kauzmann, 1954; Tanford, 1961).

Because pepsin has such an extraordinarily high ratio of potentially negatively charged groups to positively charged groups, it has an unusually stable native structure at low pH values. As discussed above, due to the unusual microenvironments of several carboxylate groups, pepsin remains with a slight net negative charge even at very low pH values . The average charge density on pepsin at this low pH is relatively small, and the disposition of charges excludes any significant local deviations from this average value, so the native state is stabilized relative to the unfolded state. However, upon raising the pH to values that are greater than the average pK_a values of most carboxylate groups, the net negative charge on pepsin would become very large and the unfolded state would be stabilized. Indeed, at pH values greater than 6.5 the tertiary structure of pepsin becomes extremely unstable and it denatures irreversibly (Bovey & Yanari, 1960).

In summary one can say that all non-viral aspartic proteinases including pepsin have a number of rather conserved buried or partially buried charged aspartic acid residues with a very low pK_a value. The specific property of pepsin A is a very small number of positively charged groups which cannot balance the net negative charges of these aspartate residues and the serine-phosphate Ser_{68}. The disposition of these groups makes pepsin stable and active in the extremely acid media corresponding to the real physiological conditions of the enzyme action.

ACKNOWLEDGEMENTS

This analysis could be performed only on the basis of a high resolution well refined structure. We are indebted to Dr. Anita Sielecki for her work and to Dr. Alexander Fedorov who participated in the refinement of the monoclinic pepsin structure. The refinement of porcine pepsin was aided by a grant from Cangene Corporation, Toronto. The research was also supported by the MRC of Canada and the Academy of Sciences of the USSR.

REFERENCES

Abad-Zapatero, C., Rydel, T. J. & Erickson, J., 1990, *Proteins: Structure, Function, and Genetics*, **8**:62-81.

Andreeva, N. S., Zdanov, A. S., Gustchina, A. E. & Fedorov, A. A., 1984, *J.Biol. Chem.*, **259**:11353-11365.

Blundell, T. L., Jenkins, J. A., Sewell, B. T., Pearl, L. H., Cooper, J. B., Tickle, I. J., Veerapandian, B. & Wood, S. P., 1990, *J. Mol. Biol.*, **211**:919-941.

Bovey, F. A. & Yanari, S. S., 1960, *in: The Enzymes* (Boyer, P. D., Lardy, H. & Myrbäck, K., eds.), 2nd edn., Vol. 4, pp. 63-92, Academic Press, New York, London.

Cooper, J. B., Khan, G., Taylor, G., Tickle, I. J. & Blundell, T. L., 1990, *J. Mol. Biol.*, **214**:199-222.

Gilliland, G. L., Winborne, E. L., Nachman, J. & Wlodawer, A., 1990, *Proteins: Structure, Function, and Genetics*, **8**:82-101.

James, M. N. G. & Sielecki, A. R., 1983, *J. Mol. Biol.*, **163**:299-361.

James, M. N. G. & Sielecki, A. R., 1986, *Nature*, **319**:33-38.

Kauzmann, W., 1954, *in: The Mechanism of Enzyme Action* (McElroy, W. D. & Glass, B., eds.), p. 70. Johns Hopkins Press, Baltimore, MD.

Michaelis, L. & Davidsohn, H., 1910, *Biochem. Z.*, **28**:1-6.

Ringer, W. E., 1915, *Zeit. Physiol. Chem.*, **95**:195-258.

Sielecki, A. R., A. A. Fedorov, A. Boodhoo, N. S. Andreeva & M. N. G. James., 1990, *J. Mol. Biol.*, **214**:143-170.

Sielecki, A. R., Hayakawa, K., Fujinaga, M., Murphy, M. E. P., Fraser, M., Muir, A. K., Carilli, C. T., Lewicki, J. A., Baxter, J. D. & James, M. N. G., 1989, *Science*, **243**:1346-1351.

Suguna, K., Bott, R. R., Padlan, E. A., Subramanian, E., Sheriff, S., Cohen, G. H. & Davies, D. R., 1987, *J. Mol. Biol.*, **196**:877-900.

Tanford, C., 1961, *In: Physical Chemistry of Macromolecules* (Chapter 7), Wiley, New York.

Tiselius, A., Henschen, G. E. & Svensson, H., 1938, *Biochem. J.*, **32**:1814-1818.

van Wazer, J. R., 1958, *in: Phosphorus and its Compounds*, Inter Science, New York.

X-RAY STRUCTURAL STUDIES OF MAMMALIAN ASPARTIC PROTEINASES

J. B. Cooper and M. P. Newman

Laboratory of Molecular Biology
and ICRF Unit of Structural Molecular Biology
Department of Crystallography
Birkbeck College
Malet Street
London, WC1E 7HX, U.K.

INTRODUCTION

Pepsins are the major digestive enzymes of the alimentary canal and operate optimally at the low pH of the stomach. Pepsin A (EC.3.4.23.1), the major component of gastric juice in pigs, is produced by the mucosal cells of the fundus region of the stomach as an inactive precursor, pepsinogen. On secretion into the acidic lumen of the stomach, cleavage of the propart peptide occurs yielding active pepsin as a single chain molecule.

The neonatal proteinase chymosin (EC.3.4.23.3) is abundant in the fourth stomach of suckling calves (Foltmann, 1981). Chymosin preferentially cleaves the milk protein κ-casein at a single Phe*Met bond with little non-specific proteolysis, inducing instability in milk micelles leading to clotting. This makes chymosin a commercially valuable enzyme in the production of cheese.

Single crystals of porcine pepsin were first obtained by Northrop (1930). Those grown by Dr. J. Philpot in Tiselius's laboratory in Uppsala were the first protein crystals to be studied by X-ray diffraction using the now standard method of capillary mounting to prevent water loss (Bernal and Crowfoot, 1934; Hodgkin, 1970). The crystals were hexagonal; a = b = 67.4 Å, c = 290.1 Å (1 molecule per asymmetric unit), although the c-dimension was not determined correctly until the advent of monochromatic X-ray sources (Perutz, 1949). Although the diffraction extended to high resolution, the crystal structure of this form remained unsolved until recently (Cooper et al., 1990). Previously, the structure of a monoclinic crystal form was determined at high resolution (Andreeva et al., 1984; Sielecki et al., 1990; Abad-Zapatero et al., 1990). The structure of porcine pepsinogen has also been solved at high resolution (James and Sielecki, 1986; Hartsuck and Remington, 1988) and recently the 2.5 Å resolution structure of human renin has been determined (Sielecki et al., 1989).

Crystals of bovine chymosin were first obtained by Bunn et al., (1971) but subsequent attempts to solve the structure by isomorphous replacement were unsuccessful

(Jenkins *et al.*, 1977). More recently, the three dimensional structures of native and recombinant chymosin B have been determined by molecular replacement (Safro *et al.*, 1987; Newman *et al.*, 1991; Gilliland *et al.*, 1990).

The specificity pockets for peptide substrates of pepsin and chymosin may be identified by analogy with other aspartic proteinases such as endothiapepsin (Foundling *et al.*, 1987; Cooper *et al.*, 1989), penicillopepsin (James and Sielecki, 1983) and rhizopuspepsin (Suguna *et al.*, 1987) for which there are high resolution X-ray analyses of inhibitor complexes. These studies have demonstrated the existence of well defined specificity pockets S_6 -$S_{3'}$ (nomenclature of Berger and Schecter, 1970). In general, the primary specificity pockets, S_1 and $S_{1'}$, are specific for hydrophobic residues.

METHODS

X-ray analysis of hexagonal pepsin

The hexagonal crystals used in this analysis were grown from a 280 mg/ml pepsin solution the pH of which was lowered to 3.6 by addition of 0.5M sulphuric acid (Jenkins, 1979). Crystals grew by slow cooling from 35°C in a thermos flask. X-ray data were collected with 1.4 Å radiation at LURE using vee-shaped cassettes and processed using the methods described by Arndt and Wonacott (1977). The final dataset consists of 15,613 unique reflections to 2.34 Å with an R_{merge} of 9.6%.

Coordinates for the three high resolution aspartic proteinase structures (endothiapepsin, penicillopepsin and rhizopuspepsin) were placed in a common orientation and Crowther's cross rotation function (Crowther, 1972) was calculated using 3.0 Å data and radii limits of 4 and 15 Å. From Figure 1 it can be seen that the sharpest feature was obtained with penicillopepsin. The improvement in height/RMS ratio of this peak as the resolution and radius of integration were increased, and its insensitivity to removal of the loops from the model suggested that this was a promising solution. Unfortunately, the other rotation functions (Figure 1) were found to be notably inconsistent. The high symmetry of the Patterson function and the low sequence identity of these models with porcine pepsin (about 30% on average) may be jointly responsible. The rotation function was repeated several times using the rotated penicillopepsin model to refine its orientation.

The TFSGEN program (Tickle, 1985) was used to calculate the translation function for both possible space groups (P6$_1$22 and P6$_5$22). The highest peak was obtained for P6$_5$22 where the strongest features in the map lie along one streak parallel to c (Figure 2). All these peaks, except the highest, caused the symmetry related molecules to crash. Hence, there appeared to be a plausible molecular replacment solution which defined the space group as P6$_5$22.

The structure was refined using RESTRAIN (Haneef *et al.*, 1985) with gradual inclusion of the high resolution data. Sim weighted $2F_o$-F_c and F_o-F_c maps were very encouraging in certain regions and were used to guide the necessary rebuilding. Several refinement rounds were performed.

Coordinates for the proenzyme, porcine pepsinogen (R=17% at 1.65 Å resolution), were kindly provided by Dr. J. Remington (University of Oregon) and were used to guide rebuilding in the regions with poor electron density. The final structure had an R-value of 0.19 at 2.34 Å, and defined the positions of 2425 enzyme atoms with a reasonable range of thermal parameters (Cooper *et al.*, 1990).

X-ray analyses of native and mutant bovine chymosin B

Orthorhombic (I222 or I2$_1$2$_1$2$_1$) crystals of chymosin (one molecule per asymmetric unit) were obtained by Bunn *et al.*, (1971) using microdialysis of a 10 mg/ml protein solution

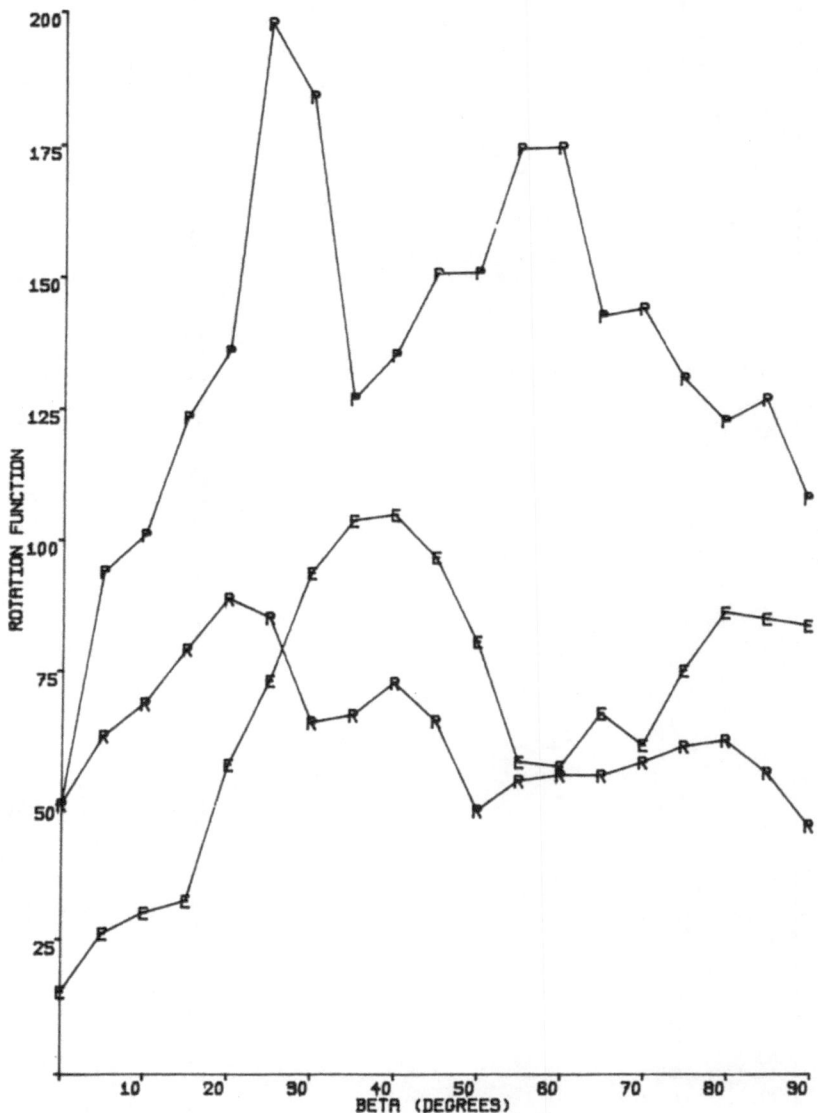

Figure 1. Rotation functions for hexagonal pepsin calculated using endothiapepsin (E), penicillopepsin (P) and rhizopuspepsin (R). Data to 3.0 Å were used with Patterson radii limits of 4 and 15 Å.

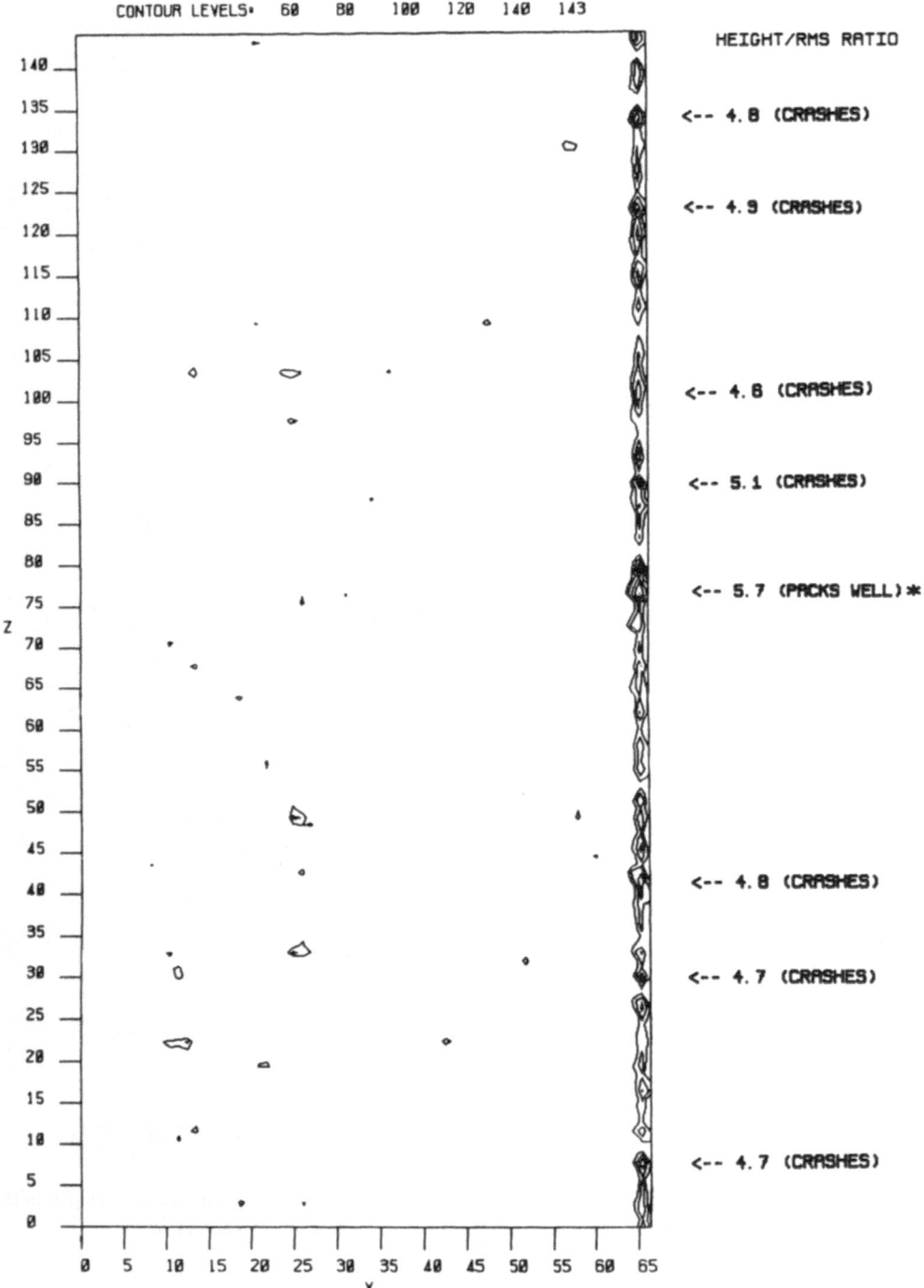

Figure 2. The translation function for hexagonal pepsin calculated using resolution limits of 20 Å and 4 Å assuming P6₅22. The strongest features in the map lie in one streak but only the highest (*) allows reasonable packing of the molecules.

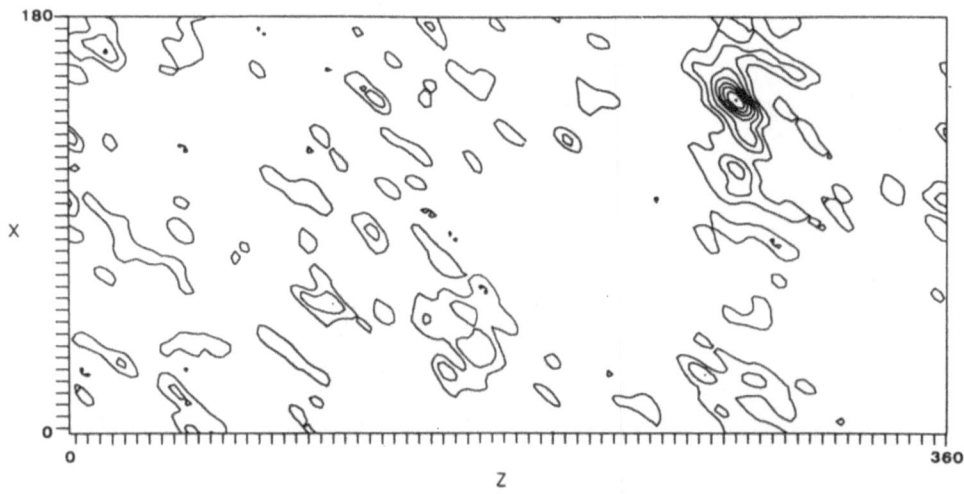

Figure 3. Rotation functions for chymosin using rhizopuspepsin as the search model (Patterson radii: 6-25 Å; resolution limits: 3-20 Å).

in 50 mM sodium phosphate buffer (pH 5.6) with 1 M NaCl. Equilibration against a salt concentration of 2 M NaCl caused slow growth of crystals at 6°C. Diffractometer data for native chymosin were collected on several crystals to a resolution of 2.2 Å (R_{merge} = 8.9%). Data for the isomorphous Val_{111}-Phe mutant crystal were collected by FAST area detector (Arndt, 1985) to 2.0 Å resolution (R_{merge} = 7.2%).

The orientational and translational parameters of chymosin were determined by Safro *et al.*, (1985). However, molecular replacement was repeated using all the available fungal structures in an effort to confirm previous studies (Newman *et al.*, 1991). Placing the search models in a common orientation for calculating the rotation function led to a consistent peak appearing with all models under all conditions of Patterson radius and resolution (Figure 3). The best peak was obtained for the most homologous enzyme, rhizopuspepsin, and even better results were obtained for its structurally conserved core.

The translation function of Tickle (1985) was calculated for the correctly oriented search models in both possible space groups. No significant solution was obtained for $I2_12_12_1$ but a consistent peak appeared for all search models with I222 (Figure 4). This corresponded to a reasonable packing arrangement of the molecules.

Rigid body followed by restrained refinement of a hybrid chymosin model (built from the fungal enzymes) using RESTRAIN (Haneef *et al.*, 1985) led to a promising electron density map for the native data. Subsequent rebuilding based on the pepsinogen structure (Hartsuck & Remington, 1988) and use of the superior mutant X-ray data led to further improvements. The position of the Phe_{111} side chain in the mutant enzyme was determined from a difference Fourier map. The X-PLOR refinement program (Brünger *et al.*, 1989) was used successfully to define the positions of several regions including the active site 'flap' in both native and mutant enzymes. The R-factors for native and mutant enzymes are 16.5% and 19.5%, respectively (Newman *et al.*, 1991; Strop *et al.*, 1990).

RESULTS AND DISCUSSION

Comparison of chymosin and pepsin structures

The peptide backbones of chymosin and pepsin form two topologically similar lobes characteristic of the aspartic proteinase fold observed in the fungal structures (Blundell *et al*,

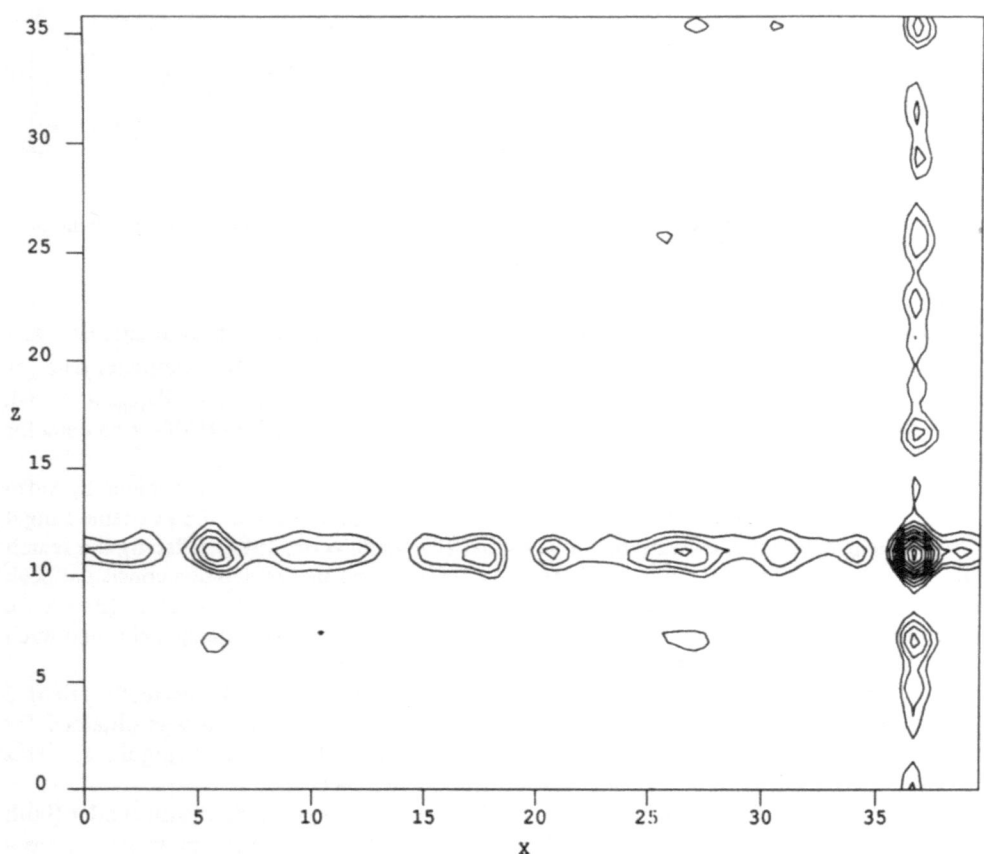

Figure 4. Translation function of chymosin using rhizopuspepsin as the search model (resolution limits: 3.5-24 Å).

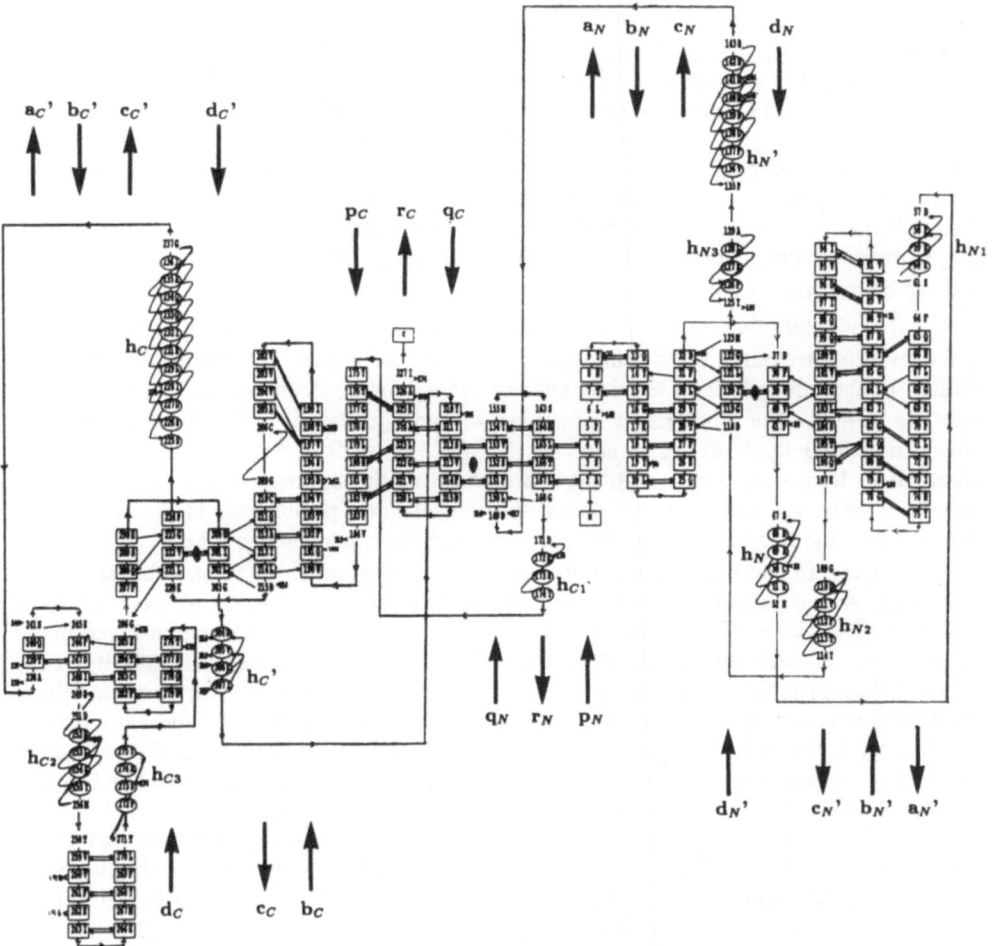

Figure 5. A schematic diagram of the chymosin structure showing the hydrogen bonds stabilizing the molecule. The inter- and intra-lobe two-fold axes are shown as large and small diad markers, respectively. Strands labelled 'p' are equivalent to those called 'a' in text.

this volume). The molecules can be divided into two lobes of approximately 170 residues each, formed by the N-terminal and C-terminal halves of the polypeptide chain. Pepsin and chymosin have almost identical secondary structural elements, consisting largely of ß-sheet with a few small right handed α-helical segments, the longest being helix h_C which spans residues 225-236 in both enzymes. The pattern of secondary structural elements forming the two domains is shown schematically in Figure 5 for chymosin. The sheets and helices are named by analogy to the scheme adopted for endothiapepsin (Blundell *et al.*, 1985; 1990) which emphasizes the intra- and inter-lobe two-fold symmetry. In each lobe strands labelled a, b, c, d are related to a', b', c', d' by the intra-lobe diad and these strands are related to their equivalents in the opposite lobe by the inter-lobe diad e.g. a_N and a_C. The positions of helices h_N, h_N', h_C and h_C' further emphasise the topological intra- and inter-domain two fold symmetry in that they all occur after d strands.

The central underlying sheet of the aspartic proteinase fold is formed by six all anti-parallel ß-strands: a_N, r_N, q_N, q_C, r_C and a_C. These form sheet 3 which resides beneath the strands forming the base of the active site cleft. Upon sheet 3, which consists of strands from both the N- and C-terminal lobes, lie two ß-sheets, one per lobe, each formed by 7 or 8 strands. The pattern of strands forming these sheets, 1_N and 1_C, is similar in both lobes and is related by a topological two-fold axis. Both of these sheets are highly twisted and adopt different three dimensional structures although the topologies are identical. The b, c, b' and c' strands are the most twisted since they fold to form extra sheets, 2_N and 2_C, beneath 1_N and 1_C respectively. The strands named b_C' and c_C' are the most distorted, each being interrupted by turns, short helices and in one case a ß-hairpin.

Strands c_N, d_N', d_N; and d_N', d_N, c_N' in sheet 1_N and c_C, d_C', d_C and d_C', d_C, c_C', in sheet 1_C form structures resembling the Greek letter ψ (Figure 5). These structures are parallel to sheet 3 and the C-terminal ends of the c_C and c_N strands associate to form the catalytic center. The stabilizing arrangement of hydrogen bonds called the fireman's grip (Pearl and Blundell, 1984) occurs here. In this, the hydroxyls of threonines 33 and 216 are involved in hydrogen bonds with the main chain of the opposite lobe. The sequence homology between both domains and with other aspartic proteinases is greatest at this interface as is the two-fold structural pseudo-symmetry between the two lobes.

Chymosin and pepsin have 58% sequence identity and show remarkable structural homology. A least squares superposition of the two enzymes (Figure 6) gave an RMS deviation of 0.9 Å for 301 C_α atoms within 3.5 Å . All three disulfides are conserved in the two enzymes, at positions 45-50, 206-210 and 250-283. Despite the close similarity of the active sites, pepsin displays higher proteolytic activity than chymosin (Martin *et al.*, 1980). The largest structural differences occur in surface loops which have high temperature factors. The largest deletion in chymosin relative to pepsin occurs in the 290-298 loop where chymosin is four residues shorter. In pepsin this loop projects over the active site forming part of the S_1' pocket, which may account, in part, for the differing specificities of the two enzymes. The 290-298 loop is topologically related to a helix (h_{N2}) which occurs in the $c_N'd_N'$ loop and forms part of the S_1 and S_3 pockets. It might be expected from Figure 5 that the loop 240-245, which is related by the inter-lobe diad to the active site flap in the N-terminal lobe, would cover the cleft to some extent. However, in pepsin and chymosin the 240-245 loop adopts a more exposed orientation with high temperature factors. In chymosin this loop has a number of negatively charged side chains: Glu_{245}, Asp_{247}, Asp_{249} and Asp_{251}. It has been suggested that this region of the enzyme may interact with the natural substrate molecule, κ-casein, which has a positively charged region between residues 98 and 102 (Safro *et al.*, 1987).

A study of rigid body movements in endothiapepsin (Šali *et al.*, 1989) showed that there was a tendency of residues 190-302 to move relative to the rest of the molecule. In a

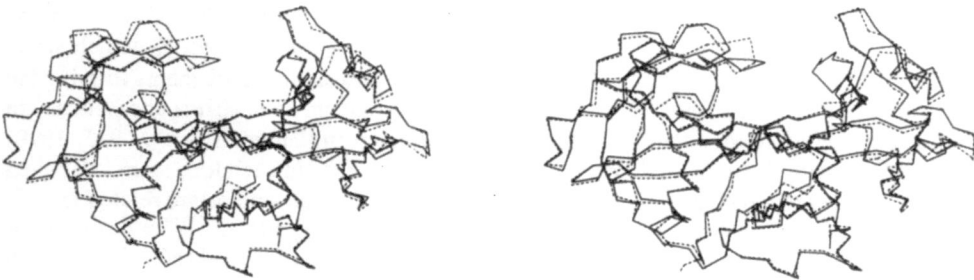

Figure 6. Least squares superposition of pepsin and chymosin in stereo showing the C_α-backbones only (chymosin as dashed lines).

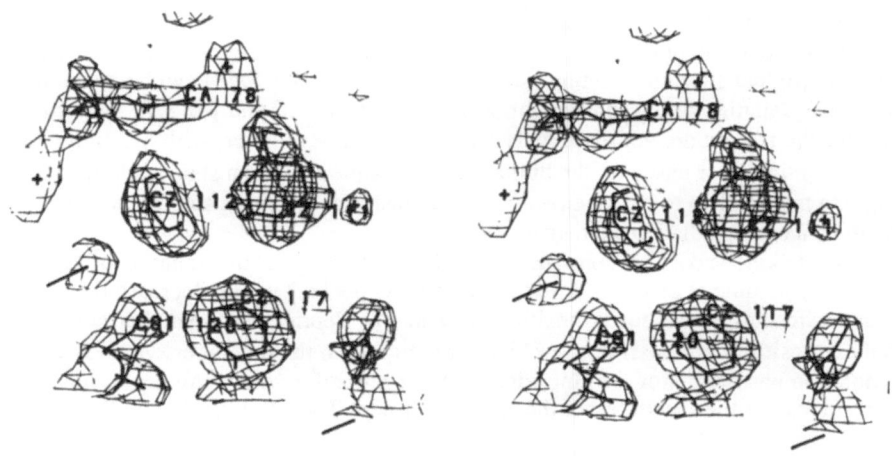

Figure 7. Electron density for the introduced Phe$_{111}$ residue of mutant chymosin.

separate rigid body comparison of chymosin and pepsin, only a slight movement was found. This may not be surprising in view of the high sequence homology and implies that crystal packing has little effect on the relative orientation of the two rigid group in aspartic proteinases.

Comparison of the native and Val $_{111}$-Phe mutant chymosin

The Val$_{111}$-Phe chymosin mutant was produced to examine the effects of a substitution in the active site helix (h$_{N2}$) on the specificity of the enzyme (Strop *et al.*, 1990). In the mammalian enzymes, residue 111 (pepsin numbering) lies at the junction between S$_1$ and S$_3$. This position is occupied by phenylalanine in pepsin but substituted by a valine in chymosin. Hence, this residue may play an important role in substrate recognition. The introduced phenylalanine side chain is on the edge of S$_1$ with a large solvent accessible area and fairly high thermal factors (mean B$_{iso}$ = 42 Å^2) although the electron density defines the ring clearly (Figure 7). No conformational change in the main chain at the site of the mutation is indicated. The RMS deviation for C$_\alpha$-atoms between the native and the mutant structures is 0.27 Å, indicating no global alterations in the structure.

In the mutant enzyme, the active site flap (residues 72-79) appeared to occupy two different conformations in which the side chain of Tyr$_{75}$ adopted orientations separated by 120° about χ_1. In one orientation the side chain partially blocks the entrance to the S$_1$ pocket. In the other, more highly occupied conformation, the tyrosine side chain hydrogen-bonds to the side chain of Trp$_{39}$, as it does in all the other known structures. In contrast, only the orientation in which the S$_1$ pocket is partially blocked occurs with the native enzyme. This is probably because unfavorable crystal contacts would result, were the flap to adopt the more usual conformation. Removal of the γ-methyl of Val$_{111}$ and substitution of a larger aromatic group in the mutant enzyme alters steric interactions with the tip of the flap and disfavours the Tyr$_{75}$ side chain packing in the S$_1$ pocket. Thus, the flap in the mutant enzyme adopts a pepsin-like conformation in a larger proportion of the molecules. Hence, the tyrosine ring adopts two orientations with a significant preference for the pepsin-like conformation. In solution, the absence of crystal contacts probably allows the flaps of both native and mutant enzyme to adopt the more usual conformation in which the S$_1$ pocket is left open for substrates.

Kinetic analyses of the native and mutant enzymes with various substrates and inhibitors have been performed (Strop *et al.*, 1990). Wild-type chymosin displayed consistently tighter substrate binding (lower K_m values) for all peptides with phenylalanine or *p*-nitrophenylalanine at P_1 and leucine at P_3 both at pH 5.6 and pH 3.7. In general, K_m values for the mutant are approximately two or three times higher, with k_{cat} little changed. Similarly, an inhibitor based on the bulkier phenylstatine transition state analogue binds less strongly to mutant than to native enzyme. In contrast, isovaleryl pepstatin binds more tightly to the mutant enzyme than to the native.

These kinetic data are consistent with the proposition that the mutation of valine to the larger phenylalanine at position 111 in the S_1 and S_3 pockets would decrease the binding of peptides with bulky residues at positions P_1 and P_3. Peptides with smaller side chains in both these positions are unaffected or bind more tightly to the mutant enzyme. In summary, this mutation has been shown to introduce little perturbation in the chymosin molecule as a whole and to alter the specificity of the enzyme for peptides with large P_1 and P_3 residues.

Comparison of pepsin with its proenzyme

The enzyme and proenzyme structures are very similar as can be seen in Figure 8 where pepsin and pepsinogen (Hartsuck & Remington, 1988) are compared. Most of the differences occur in the vicinity of the cleft which, in pepsinogen, is covered and filled by the propart (1P-44P) and the first 13 residues of pepsin. This stretch of 13 residues adopts completely different conformations in the active and zymogen forms (James & Sielecki, 1986).

The propart forms several salt bridges to the enzyme including one involving the aspartate diad and Lys_{36P} which is wedged between two tyrosine rings (Tyr_9 and Tyr_{37P}), the hydroxyls of which hydrogen-bond to the outer carboxyl oxygens. Of these residues Lys_{36P} is almost invariant (18/19) and Tyr_9 is highly conserved (17/24). The activation peptide blocks the active site in a different manner to peptide inhibitors. However, some of the residues of the cleft which hydrogen-bond to inhibitors, as defined by several inhibitor complexes of endothiapepsin (e.g. Šali *et al.*, 1989), are involved in interactions with amino acids of the activation peptide which form the first loop of fully processed pepsin, for example both NH and OH of Ser_{219} are within hydrogen bonding distance of O_{11}, and O_{217} forms a hydrogen bond to the OH of Thr_{12}. Some of the differences between binding of an inhibitor and the activation peptide, which forms a distorted helix (3_{10}) in the active site, are shown schematically in Figure 9.

Interestingly, the propart residues 2P to 9P structurally align with residues 2 to 9 of pepsin since the first ten amino acids of the propart form ß-strand a_N of pepsinogen. During

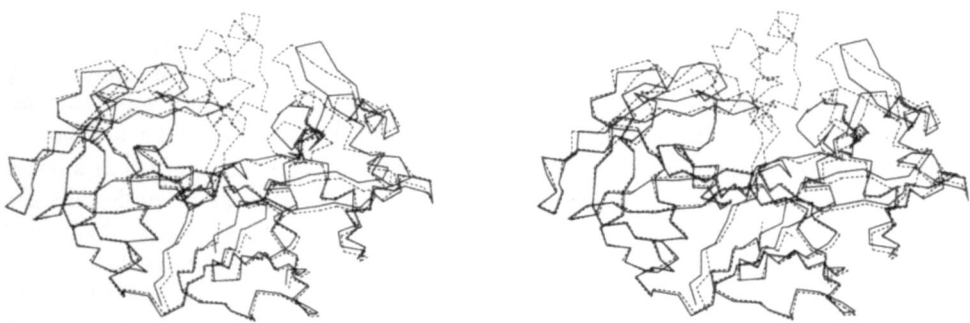

Figure 8. Least squares superposition of pepsin (solid lines) and pepsinogen (dashed lines) in stereo.

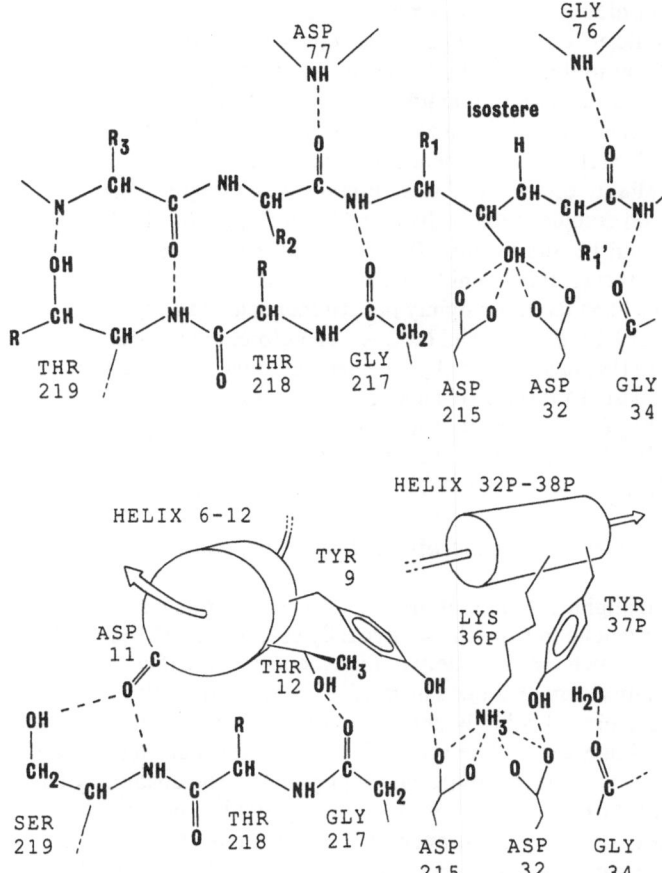

Figure 9. A schematic comparison of inhibitor (top) and propart (bottom) interactions (hydrogen bonds shown dashed).

activation the propeptide is cleaved monomolecularly at positions 16P-17P unleashing active pseudopepsin prior to cleavage of the 44P-1 peptide bond, which may be a bimolecular reaction. Whether residues 3-7 replace the first strand of pepsinogen before or after these cleavage events is unknown. The fact that residues 7, 12 and 13 form part of the S_3 pocket would suggest that the rearrangement may be needed in order to generate active pseudopepsin.

Several stretches of polypeptide near the active site cleft appear to adopt very different conformations in the proenzyme due to interactions with the propart. These include the active site flap (75-80) and loop residues 291-298 which are forced away from the cleft in pepsinogen and residues 240-244 and 277-282 which are closer to the cleft region when the propart is present. Helix h_{N2} (110-114), which contributes to the S_1 and S_3 pockets, is also somewhat perturbed by the propart. In native pepsin the Phe_{111} and Leu_{112} side chains are positioned more deeply in the active site cleft. Both pepsin and pepsinogen have a similar variation in the thermal disorder along the polypeptide chain, for example the regions 240-245, 278-282 and 294-298 have high thermal factors in both structures. It is clear that the regions where pepsin and pepsinogen differ significantly correspond to the most flexible regions of both proteins. The inherent flexibility of these loops and the active site helix (h_{N2})

may facilitate unfolding of the propart during activation. Some of these flexible regions may also undergo conformational changes upon substrate binding.

In addition to these local differences due to the propart, there appears to be a rigid body shift of part of the C-terminal lobe away from the cleft region. This mobile rigid body of aspartic proteinases was defined as residues 190-302 using difference distance maps (Sali et al., 1989). The relative movement is best described by a 5.8 degrees screw rotation and a 0.6 Å slide parallel to the rotation axis. Figures of 4.1 degrees and 0.3 Å were obtained for endothiapepsin on complexation with one inhibitor (Sali et al., 1989). It has been argued that these effects could be due to the different crystal environments of the molecules in each comparison. However, this is unlikely since the same analysis for the two crystal forms of pepsin (hexagonal and the more tightly packed monoclinic form) gave a smaller relative rigid body movement of 1.9 degrees and 0.02 Å. Therefore, the aspartic proteinase fold probably has a tendency to flex in a concerted way when the active site cleft is occupied by a peptide be it inhibitor, propart or presumably substrate. Recent kinetic experiments on the temperature dependence of k_{cat}/K_m (Hofmann et al., 1988) and older data by Fruton (1976) on the change of k_{cat}/K_m as the S_3 pocket is filled also suggest a conformational change affecting the activation barrier.

Conservation and invariance in the aspartic proteinase family

From an alignment of 24 of the known aspartic proteinase sequences (Blundell et al., 1990) it is clear that apart from the catalytically essential DTG sequences there are a number of other residues which are absolutely or highly conserved within the family. The pepsin and chymosin structures were examined in an effort to account for this conservation.

There are nine absolutely conserved glycine residues most of which occur at turns (e.g. glycines 34 and 217). The presence of side chains at these glycine positions would disrupt the planarity of the catalytic carboxyl groups. The absolutely conserved glycines at 119 and 122 lie at each end of the ß-strand $d_{N'}$ which is beneath the active site cleft. Examination reveals that the presence of a side chain at these positions would disrupt the active site in the vicinity of Trp_{39} and Ser_{35}, both of which are absolutely conserved due to their hydrogen bonding role. Similar arguments apply to the absolutely conserved glycine at position 303, which is at the end of strand $d_{C'}$. A side chain at this position would disrupt the active site in vicinity of Asp_{215}. Two of these glycines, 122 and 303, lie within highly conserved and inter-lobe diad related -hydrophobic- hydrophobic- Gly- sequences which are present in the mammalian, fungal and dimeric retroviral aspartic proteinases. These residues lie within the central strands of the ß-structures ($d_{N'}$ and $d_{C'}$). Their conservation appears necessary to allow a twist in the chain and to avoid major steric disruption of the catalytic site residues.

Many of the absolutely or highly conserved aromatic residues have a side chain hydrogen bonding interaction with other conserved side chains or the main chain of the enzyme, thereby stabilizing the fold of the protein. For example, the side chains of Trp_{190} and Tyr_{275}, both highly conserved, hydrogen-bond with the main chain carbonyls at positions 123 and 257, respectively. In pepsin, the hydroxyl of invariant Tyr_{14} forms hydrogen bonds with the main chain carbonyl of Ser_{156} and the guanidinium group of Arg_{308}, which is a highly conserved residue. Tyr_{14} lies at the end of the activation peptide and may act as an anchor during the refolding which occurs on activation (see above). A similar role for several aromatic residues of trypsinogen has been proposed (Huber & Bode, 1978). However, in *mucor pusillus* pepsin, the tyrosine ring adopts a completely different conformation, although the OH group is still involved in a hydrogen bond to the main chain (M. Newman, unpublished results).

In general, charged residues occur at the surface of proteins and are therefore not expected to be highly conserved. However, there is a number of highly or absolutely

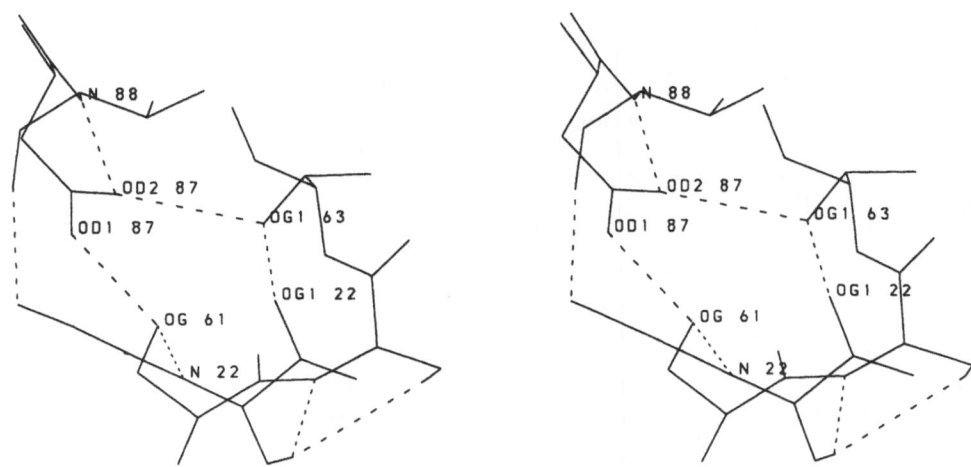

Figure 10. An example of a highly conserved, buried polar interaction. The interaction of Asp_{87} with the side chains of residues 61 and 63 which are invariably Ser or Thr. Hydrogen bonds are shown dashed.

conserved polar residues in addition to the catalytic aspartates. Inspection reveals that most of these residues are buried and are involved in hydrogen bonds to the main chain or other conserved residues. For example, the carboxyl of Asp_{87} (which is conserved in 23/24 sequences) forms hydrogen bonds with the side chains at 61 and 63 which are invariably Ser or Thr (Figure 10). The side chain NH_2 of Gln_{99} (23/24) is hydrogen bonded to the main chain carbonyls of residues 100 and 134. Two more conserved buried side chains forming hydrogen bonds to main chain carbonyl oxygens are those of Asn_{37} (invariant as Asn or Asp) and Ser_{42} (20/24) which interact with the main chain at positions 130 and 104, respectively. Similarly, the Asp_{118} side chain is hydrogen bonded to the main chain NH of residue 54, and the carboxyl of Asp_{304} interacts with the carbonyl of residue 216. Asp_{304} has been implicated by site-directed mutagenesis as one factor in lowering the pH optimum of catalysis (Yamauchi *et al.*, 1988; Mantafounis & Pitts, 1990) via its network interaction with Asp_{215}. Changing Ala_{304} of renin, which is maximally active at neutral pH, to an Asp lowered the optimum pH by approximately 0.5 pH units. The highly conserved residues Arg_{308} and Asp_{11} appear close enough to interact via a salt link, and residue 315, which is absolutely conserved as an Asp or Asn, appears to form a side chain hydrogen bond to the main chain carbonyl of Asp_{149}.

Most of the above residues are buried and appear to be conserved for their hydrogen-bonding role in the core of the protein. Similar arguments could not be applied to Asp_{171} which is very highly conserved (22/24) and lies on the surface, on the opposite side of the enzyme to the active site cleft. The aspartate at position 11 shows a similar degree of exposure and high degree of conservation. Hence, it is difficult to assign structural roles to these residues from the mature enzymes alone. However, in pepsinogen the side chains of Asp_{171} and Asp_{11} were found to form salt bridges with the propart of pepsinogen involving Lys_{3P} and Arg_{13P}, respectively (James and Sielecki, 1986). These residues may therefore be involved in stabilizing the proenzyme at neutral pH and in the acid denaturation of the propart which preceeds the autocatalytic processing of the proenzyme. Tyr_9 is also highly conserved and yet has no side chain hydrogen bonds except to solvent in mature pepsin. It is, however, involved in an interaction with the aspartate diad in pepsinogen (see above). Therefore, a fuller understanding of the conservation and invariance in this family is gained by examining the structures, not only of the mature enzymes, but also the precursor forms.

59

ACKNOWLEDGEMENTS

We would like to thank Professor T. L. Blundell for his comments on this paper. We thank Drs. N. Andreeva and M. Safro for their collaboration in the native chymosin structure determination. We extend thanks to our colleagues Dr's. C. Frazao, J. Jenkins, G. Khan, G. Taylor, I. J. Tickle and S. P. Wood, who assisted greatly in these analyses. We also thank Drs. P. Strop, J. Sedlacek and colleagues for construction, expression and kinetic analysis of the chymosin mutant.

REFERENCES

Abad-Zapatero, C., Rydel, T. J. & Erickson, J., 1990, *Proteins*, 8:62-68

Andreeva, N. S., Zdanov, A. S., Gustchina, A. E. & Fedorov, A. A., 1984, *J. Biol. Chem.*, **259**:11353-11365.

Arndt, U. W., 1985, *Methods in Enzymol.*, 114:472-485.

Arndt, U. W. & Wonacott, A. J., 1977, "The Rotation Method in Crystallography," North-Holland, Oxford.

Berger, A. & Schechter, I., 1970, *Phil. Trans. Roy. Soc.*, B257:249-264.

Bernal, J. D. & Crowfoot, D., 1934, *Nature*, 133:794-795.

Blundell, T. L., Jenkins, J., Pearl, L., Sewell, T. & Pederson, V., 1985, *in:* "Aspartic Proteinases and their Inhibitors", Kostka, V., ed., Walter de Gruyter, Berlin, pp. 151-161.

Blundell, T. L., Jenkins, J. A., Sewell, B. T., Pearl, L. H., Cooper, J. B., Wood, S. P. & Veerapandian, B., 1990, *J. Mol. Biol.*, **211**:919-941.

Bunn, C. W., Moews, P. C. & Baumber, M. E., 1971, *Proc. Roy. Soc. Lond.*, B178:245-258.

Brünger, A. T., Karplus, M. & Petsko, G. A., 1989, *Acta Cryst.*, A45:50-61.

Cooper, J. B., Foundling, S. I., Blundell, T. L., Boger, J., Jupp, R. A. & Kay, J., 1989, *Biochemistry*, **28**:8596-8603.

Cooper, J. B., Khan, G., Taylor, G., Tickle, I. J. & Blundell, T. L., 1990, *J. Mol. Biol.*, **214**:199-222.

Crowther, R. A., 1972, *in:* "The Molecular Replacement Method", Rossmann, M. G., ed., Gordon & Breach, New York, pp. 173-178.

Crowther, R. A. & Blow, D. M., 1967, *Acta Cryst.*, 23:544-548.

Foltman, B., 1981, *Essays in Biochemistry*, 17:52-84.

Foundling, S. I., Cooper, J., Watson, F. E., Cleasby, A., Pearl, L. H., Sibanda, B. L., Hemmings, A., Wood, S. P., Blundell, T. L., Valler, M. J., Norey, C. G., Kay, J., Boger, J., Dunn, B. M., Leckie, B. J., Jones, D. M., Atrash, B., Hallett, A., & Szelke, M., 1987, *Nature (London,* 327:349-352.

Fruton, J. S., 1976, *Adv. Enzymol. Relat. Areas Mol. Biol.*, **44**:1-36.

Gilliland, G. L., Winborne, E. L., Nachman, J. & Wlodawer, A., 1990, *Proteins*, 8:81-101.

Haneef, I., Moss, D. S., Stanford, M. J. & Borkakoti, N., 1985, *Acta Cryst.*, A41:426-433.

Hartsuck, J. & Remington, S., 1988, Brookhaven Protein Databank (entry 1PSG).

Hofmann, T., Allen, B., Bendiner, M., Blum, M., & Cunningham, A., 1988, *Biochemistry* 27:1140-1146.

Hodgkin, D. M. C., 1970, *Phil. Trans. Roy. Soc. Lond.*, B257:65.

Huber, R. & Bode, W., 1978, *Acc. Chem. Res.*, 11:114-121.

James, M. N. G. & Sielecki, A. R., 1983, *J. Mol. Biol.*, 163:299-361.

James, M. N. G. & Sielecki, A. R., 1986, *Nature*, 319:33-38.

Jenkins, J. A., 1979, Ph. D. Thesis. University of Sussex.

Jenkins, J., Tickle, I., Sewell, T., Ungaretti, L., Wollmer, A. & Blundell, T., 1977.. *in:* "Acid Proteases, Structure, Function and Biology", Tang, J., ed., Plenum Press, New York, pp. 43-60.

Mantafounis, D. & Pitts, J., 1990, *Protein Eng.*, 3:605-609.

Martin, P., Raymond, M. -N., Bricas, E. & Ribadeau-Dumas, B., 1980, *Biochem. Biophys. Acta*, 612:410-420.

Newman, M., Safro, M., Frazao, C., Khan, G., Zdanov, A., Tickle, I. S., Blundell, T. L., & Andreeva, N., 1991, *J. Mol. Biol.*, **221**, in press.

Newman, M., Frazao, C., Khan, G., Tickle, I. J., Blundell, T. L., Safro, M., Andreeva, N. & Zdanov, A., 1990, *J. Mol. Biol.* in press.

Northrop, J. H., 1930, *J. Gen. Physiol.*, **13**:739-766.

Pearl, L. H. & Blundell, T. L., 1984, *FEBS Lett.*, **174**:96-101.

Perutz, M., 1949 *Research*, **2**:52-61.

Powers, J. C., Harley, A. D., & Myers, D. V., 1977, *Adv. Exp. Med. Biol.*, **95**:141-157.

Safro, M., Andreeva, N. & Zdanov, A., 1985, *in:* "Aspartic Proteinases and their Inhibitors," Kostka,V., ed., Walter de Gruyter, Berlin, pp. 183-187.

Safro, M. G., Andreeva, N. S. & Blundell, T. L., 1987, *Mol. Biol. (Mosc.)*, **21**:1582-1589.

Šali, A., Veerapandian, B., Cooper, J. B., Foundling, S. I., Hoover, D. J. & Blundell, T. L., 1989, *EMBO J.*, **8**:2179-2188.

Sielecki, A. R., Fedorov, A. A., Boodhoo, A., Andreeva, N. S. & James, M. N. G., 1990, *J. Mol. Biol.*, **214**:143-170.

Sielecki, A. R., Hayakawa, K., Fujinaka, M., Murphy, M. E. P., Faser, M., Muir, A. K., Carilli, C. T., Lewicki, J. A., Baxter, J. D. & James, M. N. G., 1989, *Science,* **243**:1346-1351.

Strop, P., Sedlacek, J., Stys, J., Zaderabkova, Z., Blaha, I., Pavlickova, L., Pohl, J., Fabry, M., Kosta, V., Newman, M., Frazao, C., Shearer, A., Tickle, I. J. & Blundell, T. L., 1990, *Biochemistry*, **29**: 9863-9871.

Suguna, K., Padlan, E. A., Smith, C. W., Carlson, W. D. & Davies, D. R., 1987, *Proc. Natl. Acad. Sci.(USA)*, **84**:7009-7013.

Tickle, I. J., 1985, in "Molecular Replacement, Proceedings of the Daresbury study weekend". Machin, P.,ed., Daresbury Laboratory (DL/SCI/R23), pp. 22-26.

Yamauchi, T., Nagahama, M., Hori, H. & Murakami, K., 1988, *FEBS Lett.*, **230**:205-208.

ASPARTIC PROTEASE INHIBITORS FROM

THE PARASITIC NEMATODE *ASCARIS*

Mark R. Martzen,* Brad A. McMullen,^ Kazuo Fujikawa,^
and Robert J. Peanasky+

*Department of Neurobiology and Anatomy
University of Rochester
Rochester, New York 14642

^Department of Biochemistry
University of Washington
Seattle, Washington 98195

+Department of Biochemistry and Molecular Biology
University of South Dakota School of Medicine
Vermillion, South Dakota 57069

INTRODUCTION

Ascaris suum and *Ascaris hominis* are intestinal endoparasitic nematodes. They live in hostile proteolytic environments and infect pigs and humans, respectively. Over one-quarter of the world's human population (Muller, 1979) and masses of pigs are infected by these parasites. The resulting medical, agricultural and economic effects are significant (Levine, 1980; Mahmoud, 1989).

Multiple isoforms of unique protein protease inhibitors are present in both *A. suum* and *A. hominis*. They inactivate the host's pancreatic digestive enzymes (Peanasky *et al.*, 1987). In addition, *A. suum* inhibitors of the gastric aspartic protease, pepsin, have also been reported (Abu-Erreish & Peanasky, 1974a, b; Martzen *et al.*, 1990).

Protein inhibitors of classes of proteases have been isolated from numerous sources, yet specific physiological functions of many of them are unclear. Aspartic proteases have been isolated from a variety of sources as well (Barrett & McDonald, 1980; Bedi *et al.*, 1983; Kay, 1985; Tang & Wong, 1987) and still the catalytic mechanism of this class of proteases is not completely defined.

In order to address some of these problems we asked the *A. suum* and *A. hominis* systems the following questions. What is the primary structure of the *A. suum* major aspartic protease inhibitor designated PI-3? What secondary/tertiary structural features does this inhibitor possess? What is the relationship between the *A. suum* and *A. hominis* aspartic protease inhibitors? Finally, can the PI-3 molecule form an enzyme-inhibitor complex with pepsin suitable for X-ray crystallographic studies of the aspartic protease catalytic mechanism of action? We report here the answers that we found to these questions.

METHODS

Purification of Inhibitors

A. suum and A. hominis were collected and then fractionated as described in Martzen et al. (1990). Ascaris hominis, the human parasite, was obtained with the cooperation of Dr. Jin Soon Ju at the Korea University Medical College. Inhibitors of pepsin in aqueous extracts were removed by pepsin-AH-Sepharose 4B affinity chromatography and were released above pH 10. The A. suum inhibitors were then resolved into individual isoforms either by chromatofocusing or DEAE-Sephadex chromatography. The small amount of A. hominis starting material precluded additional fractionation of that parasite's inhibitors beyond the affinity chromatography step. Each of these inhibitors was analyzed on PAGE gels and by Western blotting.

Inhibitor Assay

To determine inhibitor activity, pepsin and limiting amounts of pepsin inhibitor were allowed to complex. Free pepsin present at equilibrium was then determined by a standard hemoglobin digestion assay (Abu-Erreish & Peanasky, 1974a). One unit of inhibitor activity was defined as that amount of protein which inactivates 1 µg of pepsin under the prescribed assay conditions. Specific activity is units per mg of inhibitor protein.

Polyacrylamide Gel Electrophoresis

Non-dissociating, discontinuous PAGE was performed by the methods of Ornstein (1964) and Davis (1964). Proteins were separated on 12.5% vertical slab gels at 25°C. The PAGE system stacked at pH 8.3 and separated at pH 9.5. Electrophoresis was at 25 mA for 30 min and then 40 mA for 2 h. Proteins were stained with 1% Naphthol Blue Black (NBB) in 7% acetic acid and destained in 7% acetic acid.

Cytochemical Staining

The method of Furihata et al. (1975) for cytochemical staining of pepsins was adapted for pepsin inhibitors. The average protein load was 40 ng of inhibitor per gel lane. Inhibitors separated on PAGE were incubated in pepsin (50 µg/ml) at pH 1.6 and 37°C for 45 min. The water washed gel was then incubated in 0.65% hemoglobin at pH 1.6 and 25°C for 20 min. Finally, the water washed moist gel was incubated in a closed container at 37°C for 1 h and the gel was then stained with 1% NBB in 7% acetic acid for 5 min. Destaining was overnight in 7% acetic acid. In the presence of pepsin inhibitor, pepsin was inactivated and undigested hemoglobin amplified the stain. In areas where pepsin was active, hemoglobin was digested and the gel was clear.

Western Blotting

Polyclonal antibody to homogeneous PI-3 was developed in New Zealand white rabbits (Peanasky et al., 1984), was purified on protein A-Sepharose CL-4B and used as the primary antibody for Western blots. PAGE gels with inhibitors separated by electrophoresis were equilibrated in 25 mM TRIS/192 mM glycine/pH 8.3 transfer buffer. The proteins were

Table I. Pepsin Inhibitors From *A. suum*

Step	Specific Activity	Percent Recovered
Ultracentrifuged Extract	1	(100)
0-65% $(NH_4)_2SO_4$	4	95
Affinity Chromatography	3070	77
Chromatofocusing Column		
Peak 1	2900	1
Peak 2	4200	14
Peak 3	4240	46
Peak 4	2800	9

Adapted with permission from Martzen, *et al.*, (1990), *Biochemistry*, **29**:7366. Copyright (1990) American Chemical Society

electrophoretically transferred from the gel to nitrocellulose at normal polarity and 100 mA for 5 h. Non-specifically adsorbing sites were saturated with non-fat dry milk. The blocked transfer sheet was washed 6X with 50 mM TRIS/150 mM NaCl/pH 7.4 buffer and the primary antibody applied. Goat anti-rabbit IgG conjugated to horseradish peroxidase was the secondary antibody and was visualized with 4-chloro-l-naphthol substrate.

Primary Structure of PI-3

The primary amino acid sequence of PI-3 was determined by automated Edman degradation of the intact molecule after pyroglutamate aminopeptidase digestion and by automated Edman degradation of chemically and enzymatically generated overlap peptides (Martzen *et al.*, 1990).

Secondary/Tertiary Structure of PI-3

Preliminary estimates of the secondary/tertiary structure of the PI-3 molecule were made from sequence analysis of cystine-containing peptides to determine the disulfide bond pattern of the molecule (Martzen *et al.*, 1990) and from circular dichroism spectra of the inhibitor. Computer generated hydropathy (Kyte & Doolittle, 1982) and net charge distribution analysis (Hopp & Woods, 1981) of the PI-3 molecule by the GENEPRO program (Version 4.1, Riverside Scientific Enterprises, Seattle, WA) and energy minimization (Brooks *et al.*, 1983) by the Polygen/CHARMM program and QUANTA system for graphical modeling were also used to analyze the inhibitor. Finally, pepsin and PI-3 were complexed and crystals of the enzyme-inhibitor complex formed. The unit cell size and space group of the crystals was determined and X-ray crystallographic measurements of the complex are currently in progress in the laboratory of Dr. M. N. G. James.

RESULTS AND DISCUSSION

An outline of a typical pepsin inhibitor preparation from *A. suum* is shown in Table I. The values for the isolation of pepsin inhibitors from *A. hominis,* up to the chromatofocusing step, were virtually identical. Inhibitor desorbed from the pepsin-AH-Sepharose affinity gel had an average specific activity of 3070 units/mg protein. This is a >3000 fold increase in specific activity over the ultracentrifuged extract. Overall recovery of inhibitor by this protocol was 70%.

The major homogeneous pepsin inhibitor, PI-3, was separated from the other pepsin inhibitors by chromatofocusing. This procedure, however, did not completely resolve all of the pepsin inhibitors. These mixtures of pepsin inhibitors could not be separated from each other even by repeated chromatofocusing using different pH ranges. This suggests that the structures of the *A. suum* pepsin inhibitors are similar and amino acid compositions of the individual inhibitors support this view (Abu-Erreish & Peanasky, 1974a). To obtain these inhibitors as homogeneous isoforms, DEAE-Sephadex chromatography employing a complex pH gradient (not shown) was required (Abu-Erreish & Peanasky, 1974a). The distribution of pepsin inhibitors was PI-1 (1%), PI-2 (20%), PI-3 (66%) and PI-4 (13%).

To demonstrate that the NBB stained bands on the PAGE gels were pepsin inhibitors, a new cytochemical staining technique was developed. Pepsin inhibitor proteins from *A. suum* and *A. hominis* were analyzed on both NBB stained PAGE gels (Figure 1a) and cytochemically stained PAGE gels (Figure 1b). Bovine serum albumin and ovalbumin are shown as controls. Other proteins used as controls (but not shown) were alcohol dehydrogenase, hexokinase, rennin, soybean trypsin inhibitor and urease. All of the control proteins used were arbitrarily chosen on the basis of their mobility.

As shown in Figure 1b, the cytochemical technique was very sensitive and none of the control proteins were visualized. PAGE and cytochemical analysis indicates four anodic pepsin inhibitors from *A. suum* and two anodic pepsin inhibitors from *A. hominis*. *A. hominis* pepsin inhibitors released from the affinity gel migrated very near to the PI-2 and PI-3 inhibitors of *A. suum*. Together these two inhibitors represent most of the inhibitor profile of *A. suum*. Neither parasite possessed any cathodic pepsin inhibitors.

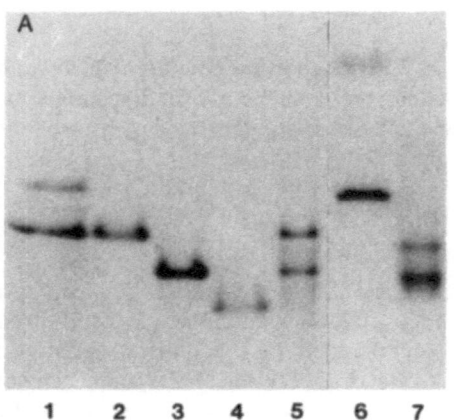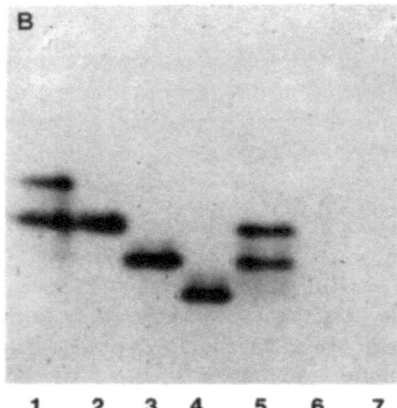

Figure 1. Polyacrylamide gel electrophoresis of pepsin inhibitors and selected control proteins; examination of the gels by naphthol blue-black and specific cytochemical staining. Lanes 1, 2, 3 and 4 are inhibitor fractions obtained from *A. suum*, lane 5 is *A. hominis* pepsin inhibitors after affinity chromatography, lane 6 is serum albumin and lane 7, ovalbumin. (The last two proteins are controls.) In A, the gels were stained with naphthol blue-black and all of the bands are visible. In B, the cytochemical staining procedure was used and only the pepsin inhibitor bands were visible. The procedures are described in Methods.

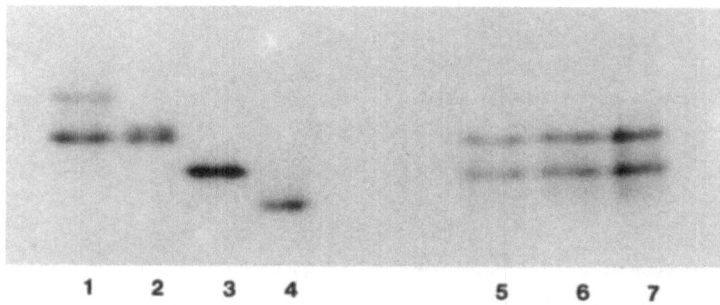

Figure 2. Western blot of *A. suum* and *A. hominis* isoinhibitors of pepsin. The primary antibody was raised in rabbits to pepsin inhibitor 3 from *A. suum*. The secondary antibody was goat anti-rabbit IgG (long and short chains) conjugated to horseradish peroxidase. Lanes 1 to 4 are *A. suum* pepsin inhibitors 1 to 4. Lanes 5 to 7 are increasing concentrations of *A. hominis* inhibitors released from the affinity gel.

Western blotting (Figure 2) shows the antigenic similarity among the *A. suum* pepsin inhibitors and between the *A. suum* and *A. hominis* pepsin inhibitors. Rabbit anti-*A. suum* PI-3 was the primary antibody in this experiment. It is clear that even though the two parasites have somewhat different inhibitor profiles (Figure 1a), all of the inhibitors from both species are recognized by the same primary antibody developed against PI-3. When IgG from non-immunized rabbits was the primary antibody, no inhibitor bands were visualized. This indicates that there is both intra- and inter-species structural similarity between these inhibitors. These inhibitors, which have been conserved through evolution, must serve some physiological function in both parasites.

In the next experiments the primary structure of PI-3 was determined. PI-3 was chosen for this analysis because it was present in the largest amounts and it is similar in structure to the other inhibitors as shown by the cross-reaction between antibody to PI-3 and the other pepsin inhibitors. Table II shows the PTH-amino acid yield at each Edman cycle for each peptide used to establish the sequence of the inhibitor. Figure 3 shows the overlap

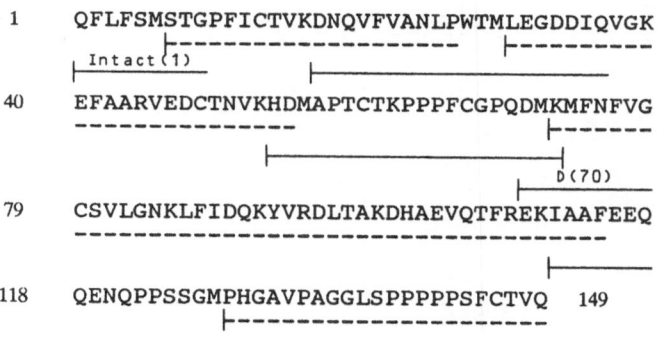

Figure 3. Amino acid sequence of reduced and pyridyl-ethylated *A. suum* pepsin inhibitor 3. The dashed lines represent four of the peptides obtained by treatment with cyanogen bromide (CNBR). The solid lines (unlabeled) represent three of the fragments obtained by treatment with endopeptidase Lys-C (K). Peptide D (70) was obtained from the digest with endoproteinase Asp-N. The peptide sequence, Intact (1), was obtained by sequencing an intact inhibitor. Adapted with permission from Martzen, *et al.*, (1990), *Biochemistry*, **29**:7366. Copyright (1990) American Chemical Society.

Table II. Fragments used to determine the amino acid sequence of *Ascaris* pepsin inhibitor 3. Peptides were obtained after reduction, pyridylethylation and digestion with cyanogen bromide (CNBR), endopeptidase Lys-C (K) and endoproteinase Asp-N (D). The number in parenthesis identifies the position of the first residue of each fragment in the sequence. The yield in pmol of PTH-amino acid at each step in the Edman cycle follows the amino acid. Adapted with permission from Martzen *et al.*, 1990, Biochemistry, **29**:7366. Copyright (1990) American Chemical Society.

Cycle	Intact(1)	CNBR(7)	K(17)	CNBR(30)	K(53)	D(70)	CNBR(72)	K(111)	CNBR(128)
1	Gln	Ser 145	Asp 37	Leu 112	His 96	Asp 78	Lys 50	Ile 780	Pro 134
2	Phe 148	Thr 106	Asn 62	Glu 73	Asp 297	Met 202	Met 60	Ala 969	His 53
3	Leu 159	Gly 154	Gln 61	Gly 122	Met 367	Lys 102	Phe 260	Ala 1055	Gly 129
4	Phe 106	Pro 128	Val 37	Asp 51	Ala 360	Met 163	Asn 115	Phe 621	Ala 136
5	Ser 456	Phe 98	Phe 63	Asp 73	Pro 180	Phe 159	Phe 235	Glu 565	Val 83
6	Met 142	Ile 124	Val 37	Ile 76	Thr 163	Asn 113	Val 208	Glu 434	Pro 121
7	Ser 205	Cys	Ala 27	Gln 71	Cys	Phe 114	Gly 225	Gln 438	Ala 127
8	Thr 148	Thr 40	Asn 19	Val 56	Thr 156	Val 63	Cys	Gln 561	Gly 110
9	Gly 104	Val 46	Leu 34	Gly 92	Lys 92	Gly 81	Ser 66	Glu 354	Glu 113
10	Pro 80	Lys 32	Pro 61	Lys 57	Pro 97		Val 184	Asn 358	Leu 81
11		Asp 29	Trp 9	Glu 34	Pro 143		Leu 127	Gln 320	Ser 61
12		Asn 23	Thr 20	Phe 61	Pro 154		Gly 116	Pro 344	Pro 55
13		Gln 17	Met 18	Ala 61	Phe 70		Asn 91	Pro 326	Pro 47
14		Val 16	Leu 16	Ala 82	Cys		Lys 75	Ser 246	Pro 34
15		Phe 14	Glu 13	Arg 80	Gly 89		Leu 104	Ser 185	Pro 39
16		Val 14	Gly 18	Val 28	Pro 37		Phe 93	Gly 129	Pro 31
17		Ala 18	Asp 9	Glu 24	Gln 38		Ile 118	Met 97	Ser 24
18		Asn 8	Asp 7	Asp 27	Asp 29		Asp 47	Pro 64	Phe 18
19		Leu 8	Ile 10	Cys	Met 21		Gln 94	His 28	Cys
20		Pro 6	Gln 7	Thr 32	Lys 12		Lys 63	Gly 41	Thr 11
21				Asn 18			Tyr 55	Ala 31	Val 8
22				Val 19			Val 60	Val 13	Gln 4
23				Lys 19			Arg 50	Pro 18	
24				His 5			Asp 26	Ala 17	
25				Asp 12			Leu 62	Gly 11	
26							Thr 23	Gly 10	
27							Ala 41	Leu 4	
28							Lys 40		
29							Asp 21		
30							His 7		
31							Ala 34		
32							Glu 10		
33							Val 20		
34							Gln 12		
35							Thr 11		
36							Phe 14		
37							Arg 6		
38							Glu 13		
39							Lys 6		
40							Ile 12		
41							Ala 12		
42							Ala 16		
43							Phe 12		

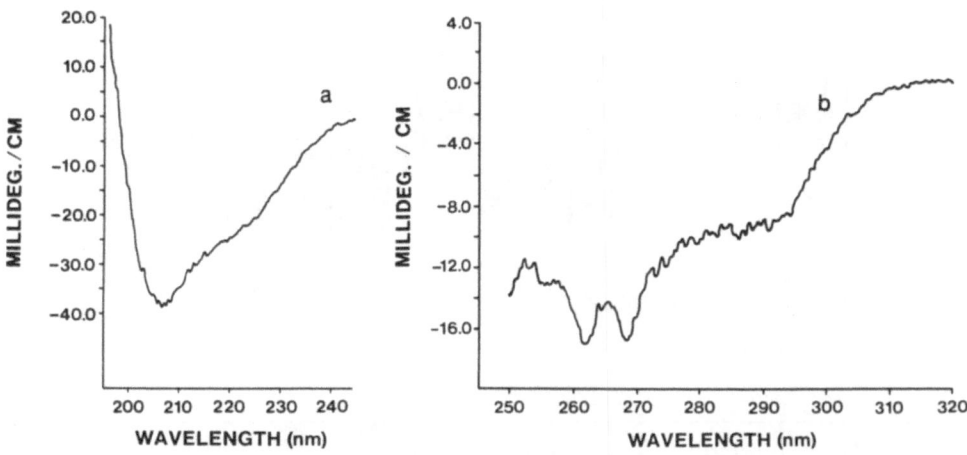

Figure 4. Circular dichroism spectra of *A. suum* pepsin inhibitor-3 in the far-UV (a) and near-UV (b) regions. A Jasco J-500A spectropolarimeter with a Jasco DP-501N data processor and TRIO Model CO-1530A oscilloscope was used. Protein (1.5 mg/ml) was dissolved in 50 mM TRIS/150 mM NaCl (pH 7.4). Measurements were at ambient temperature. Spectra were the average of 8 accumulations smoothed by computer.

peptides and proof of sequence of *A. suum* PI-3 (Martzen *et al.*, 1990). The amino terminus of PI-3 was blocked. Following treatment with pyroglutamate aminopeptidase, the inhibitor could be sequenced. (Glutamine was found at the amino-terminus of another pepsin isoinhibitor which had the same sequence for the first ten residues.) The PI-3 molecule is composed of 149 amino acids and has a molecular weight of 16,396. Several amino acid doublets appear throughout the molecule, and it has a run of five prolines near its C-terminal end. The cysteines of PI-3 are all in disulfide linkage. They follow the pattern: Cys_{13}-Cys_{59}, Cys_{48}-Cys_{66} and Cys_{79}-Cys_{146}. This is the first aspartic protease inhibitor of animal origin that has been sequenced. In addition, it is apparently unique, as it shares no significant sequence homology with any protein currently in the Protein Identification Resource (PIR) data base.

Preliminary attempts at structural estimates of the PI-3 molecule were made by both physical analysis and computer generated models. Circular dichroism spectra of the molecule were taken in both the far-UV (190-240 nm) and near-UV (250-320 nm) regions (Figure 4a, b). The far-UV scan (Figure 4a) suggested the presence of alpha-helix in the molecule's secondary structure with high negative ellipticity at 208 nm and high positive ellipticity at 193 nm. This was a surprising observation as all other *Ascaris* protease inhibitors currently sequenced appear to be composed of beta-sheets and random coils with beta-turns when analyzed by the Chou and Fasman (1978) algorithms.

The CD spectrum from 250-270 nm (Figure 4b) displayed a series of complex ellipticity minima. Minima were observed at 261.8 and 268.5 nm indicating the presence of phenylalanine in more hydrophobic regions of the molecule (Horwitz *et al.*, 1969). Contributions to the CD spectrum from 270-320 nm were difficult to interpret, but corroborated the presence of tyrosine, tryptophan and disulfide bonds in the molecule. Both of these scans will provide a good baseline for further characterization of the inhibitor at different pH's and in complex with pepsin. In addition, analysis at 77°K may allow for more detailed interpretation of regions of ellipticity.

Hydropathy (Figure 5a) and net charge distribution (Figure 5b) of the PI-3 molecule were computer-generated for analysis. The hydropathy analysis indicates hydrophilic regions in both N-and C-terminal halves of the molecule, with the greatest hydrophilicity in the C-

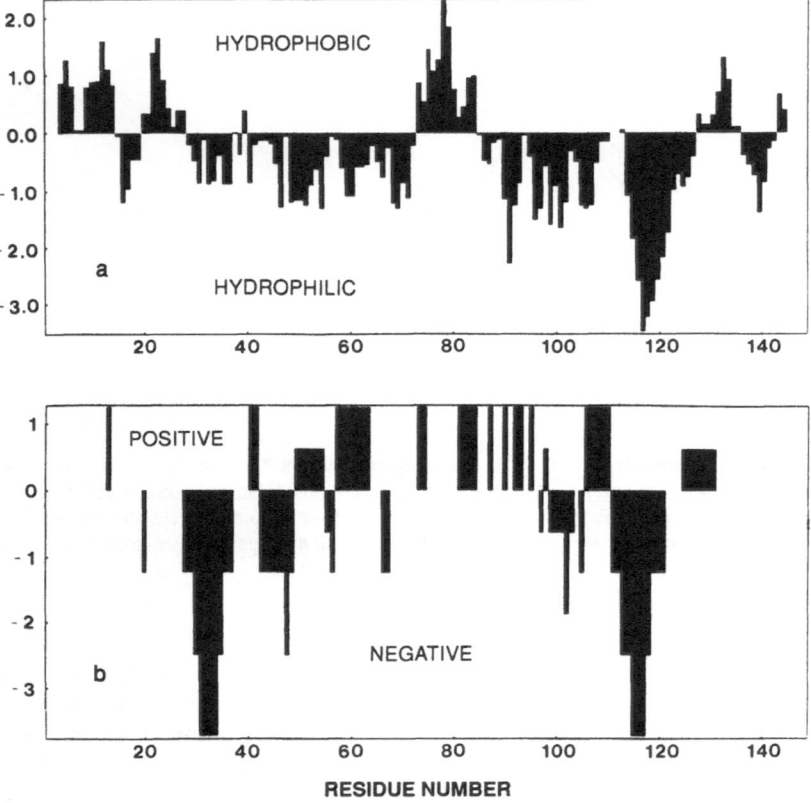

Figure 5. a) Hydropathy plot of the amino acid sequence of *A. suum* pepsin inhibitor 3 generated by the method of Kyte and Doolittle (1982) using a window of 7 residues. b) Net charge distribution of *A. suum* pepsin inhibitor 3 using a window of 7 residues by the method of Hopp and Woods (1981).

terminal end. The two hydrophilic regions are connected by a short stretch of hydrophobic amino acids and there is an additional region of hydrophobic amino acids at the N-terminal end of the molecule. Finally, there are two regions of high negative charge distribution, each corresponding to one of the hydrophilic domains.

The disulfide bridge pattern of PI-3 and the analyses above suggest that the inhibitor is composed of two domains and that these domains are separated in space on the basis of charge. This was confirmed by a preliminary computer representation of the molecule after energy minimization. A two-dimensional representation of the primary amino acid sequence of PI-3 and its disulfide assignments is shown in Figure 6.

Experiments to resolve the three-dimensional structure of the molecule and to identify its inhibitory reactive site are also ongoing. Orthorhombic crystals of the pepsin-PI-3 complex have been made. The unit cell size is a = 98 Å, b = 140 Å and c = 68 Å and the space group is $C222_1$.

In summary, pepsin inhibitors from the intestinal endoparasitic nematodes *A. suum* and *A. hominis* were isolated and analyzed, and the PI-3 molecule of *A. suum* was studied in detail. This is the first reported isolation of pepsin inhibitors from *A. hominis*. Analyses of the two species demonstrated different inhibitor profiles, while Western blotting indicated that there was both intra- and inter-species structural similarity between the inhibitors. Conservation of these inhibitors and their similarity in two intestinal helminths suggests that these protein protease inhibitors play a physiological role in these parasites.

Figure 6. Complete amino acid sequence for *A. suum* pepsin inhibitor 3 with disulfide bridge pairing. Reprinted with permission from Martzen, *et. al.*, (1990), *Biochemistry*, **29**:7366. Copyright (1990) American Chemical Society.

The structure of the PI-3 molecule was solved and its disulfide bridge pattern determined. It was possible to generate crystals of the pepsin-PI-3 complex and X-ray diffraction studies are now in progress. These measurements will define the inhibitory reactive site of the PI-3 molecule and the inhibitor will serve as a useful tool in the study of the catalytic mechanism of action of the aspartic protease class of enzymes.

Finally, what is the function of the pepsin inhibitors from *A. suum* and *A. hominis*? Is pepsin the true physiological target enzyme of the inhibitor, or is the true target another (perhaps neutral) aspartic protease? If pepsin is the true target, then the inhibitors may play their most important role during the migratory phase of the parasite's life cycle. At this time the parasites pass through the stomach twice; first after ingestion on their way to the intestine and second after migration through the liver, heart and lungs of the host before returning to the intestine to become adults. Alternately, the inhibitor may defend against a neutral aspartic protease and act as an immunosuppressant. Lauritzen *et al.* (1984) have demonstrated the inhibition of leukocyte migration *in vitro* by aspartic protease inhibitors. As adults, the parasites reside in the small intestine, where any pepsin present would already be inactivated. If these inhibitors are inducible, they should now be turned off. These pepsin inhibitors were isolated from adults. This suggests they are constitutive proteins.

One additional possibility is that the parasites require *host proteases,* either 1) because they cannot synthesize analogous proteases of their own, or 2) to perform controlled proteolytic events at specific times during growth and development (e.g. molting and larval development), or 3) to present an altered antigenic surface to their hosts by incorporation of host specific markers (i.e. hosts recognize parasites as "self"). The inhibitor system would then be part of the physiological control mechanism for the acquisition, storage and/or use of these proteases (Martzen *et al.*, 1985, 1986; Peanasky *et al.*, 1974, 1987). Clearly, additional questions concerning these interesting and useful proteins need to be answered, and the physiological function(s) of these inhibitors remains to be solved.

ACKNOWLEDGMENTS

Specimens of *Ascaris suum* for this study were collected with the kind permission of John Morrell and Co. (Sioux Falls, SD) at their abattoir. We thank Drs. John Erickson and Gene Homandberg for collecting the circular dichroism data. The encouragement of Dr. Earl Davie and the valuable discussions and advice of Drs. Ralph Bottenus, Steve Leytus and Joost Meijers are gratefully acknowledged. We thank Dianna Olson for her expert editorial assistance in the preparation of this manuscript. Research was supported by NIH grant AI-10992 and HL-16919.

REFERENCES

Abu-Erreish, G. M., & Peanasky, R. J., 1974a, Pepsin inhibitors from *Ascaris lumbridoides*: Isolation, purification and some properties, *J. Biol. Chem.*, **249**:1558.

Abu-Erreish, G. M., & Peanasky, R. J., 1974b, Pepsin inhibitors from *Ascaris lumbricoides*: Pepsin-inhibitor complex, stoichiometry of formation, dissociation, and stability of the complex, *J. Biol. Chem.*, **249**:1566.

Barrett, A. J., & McDonald, J. K., 1980, *"Mammalian Proteases: A Glossary and Bibliography,"* Vol. 1, pp. 303-356, Academic Press, New York.

Bedi, G. S., Balwierczok, J., & Back, N., 1983, Rodent kinin-forming enzyme systems -purification and characterization of an acid protease from Murphy-Sturm lymphosarcoma, *Biochem. Pharmacol.*, **32**:2071.

Brooks, B. R., Bruccoleri, R. E., Olafson, B. D., States, D. J., Swaminathan, S., & Karplus, M., 1983, CHARMM: A program for macromolecular energy, minimization and dynamics calculations, *J. Comput. Chem.*, **4**:187.

Chou, P. Y., & Fasman, G. D., 1978, Empirical predictions of protein conformation, *Annu. Rev. Biochem.*, **47**:251.

Davis, B. J., 1964, Disc electrophoresis - II. Method and application to human serum proteins, *Ann. N.Y. Acad. Sci.*, **121**:404.

Furihata, C., Sasajima, K., Kazama, S., Kogure, K., Kawachi, T., Sugimura, T., Tatematsa, M., & Takahashi, M., 1975, Changes in pepsinogen isozymes in stomach carcinogenesis induced in rats by N-Methyl-N^1-nitro-N-nitrosoguanidine, *J. Natl. Cancer Inst.*, **55**:925.

Hopp, T. P., & Woods, K. R., 1981, Prediction of protein antigenic determinants from amino acid sequences, *Proc. Natl. Acad. Sci. U.S.A.*, **78**:3824.

Horwitz, J., Strickland, E. H. & Billups, C., 1969, Analysis of vibrational structure in the near-ultraviolet circular dichroism and absorption spectra of phenylalanine and its derivatives, *J. Amer. Chem. Soc.*, **91**:184.

Kay, J., 1985, Aspartic proteinases and their inhibitors, *in:* "Aspartic Proteinases and Their Inhibitors," Kostka, V., ed., pp. 1-18, de Gruyter, Berlin.

Kyte, J., & Doolittle, R. F., 1982, A simple method for displaying the hydropathic character of a protein, *J. Mol. Biol.*, **157**:105.

Lauritzen, E., Moller, S., and Leerhoy, J., 1984, Leucocyte migration inhibition *in vitro* with inhibitors of aspartic and sulfhydryl proteinases, *Acta. Path. Microbiol. Immunol. Scand. Sect. C*, **92**:107.

Levine, N. D., 1980, "Nematode Parasites of Domestic Animals and of Man," pp. 256-295, Burgess Publishing Co., Minneapolis.

Mahmoud, A. A. F., 1989, Parasitic protozoa and helminths: Biological and immunological challenges, *Science*, **246**:1015.

Martzen, M. R., Geise, G. L., Hogan, B. J., & Peanasky, R. J., 1985, *Ascaris suum*: Localization by immunochemical and fluorescent probes of host proteases and parasite proteinase inhibitors in cross sections, *Exp. Parasitol.*, **60**:139.

Martzen, M. R., Geise, G. L., & Peanasky, R. J., 1986, *Ascaris suum*: Immunoperoxidase and fluorescent probe analysis of host proteases and parasite proteinase inhibitors in developing eggs and second stage larvae, *Exp. Parasitol.*, **61**:138.

Martzen, M. R., McMullen, B. A., Smith, N. E., Fujikawa, K., & Peanasky, R. J., 1990, Primary structure of the major pepsin inhibitor from the intestinal parasitic nematode *Ascaris suum*, *Biochemistry*, **29**:7366.

Muller, R., 1979, Nematodes, *in:* "Parasites and Western Man," Donaldson, R. J., ed., p. 90, University Park Press, Baltimore.

Ornstein, L., 1964, Disc electrophoresis - I. Background and theory, *Ann. N.Y. Acad. Sci.*, **121**:321.

Peanasky, R. J., Abu-Erreish, G. M., Gaush, C. R., Homandberg, G. A., O'Heeron, D., Linkenheil, R. K., Kucich, U., & Babin, D. R., 1974, Proteinase inhibitors from *Ascaris lumbricoides*: Properties and their physiological role, *in:* "Bayer Symposium V: Proteinase Inhibitors," Fritz, H., Tschesche, H., Greene, L. J., & Truscheit, E., eds., pp. 649-666, Springer-Verlag, Berlin.

Peanasky, R. J., Bentz, Y., Paulson, B., Graham, D. L.,& Babin, D. R., 1984, The isoinhibitors of chymotrypsin/elastase from *Ascaris lumbricoides*: Isolation by affinity chromatography and association with the enzymes, *Arch. Biochem. Biophys.*, **232**:127.

Peanasky, R. J., Martzen, M. R., Homandberg, G. A., Cash, J. M., Babin, D. R., & Litweiler, B., 1987, Proteinase inhibitors from intestinal parasitic helminths: Structure and indication of some possible functions, *in:* "Molecular Paradigms for Eradicating Helminthic Parasites," MacInnis, A. J., ed., pp. 349-366, Alan R. Liss, New York.

Tang, J., & Wong, R. N. S., 1987, Evolution in the structure and function of aspartic proteases, *J. Cell. Biochem.*, **33**:53.

NONSPECIFIC ELECTROSTATIC BINDING OF SUBSTRATES AND INHIBITORS

TO PORCINE PEPSIN

Petr Kuzmič, Chong-Qing Sun, Zhi-Cheng Zhao, and Daniel H. Rich

School of Pharmacy and Department of Chemistry
University of Wisconsin-Madison
425 N. Charter St.
Madison, Wisconsin   53706

INTRODUCTION

Porcine pepsin has long been recognized to have primary and secondary specificity for hydrophobic amino acids, and the binding of low molecular weight inhibitors (e.g., aliphatic alcohols) is also dominated by hydrophobic interactions[1]. However, the long standing 'hydrophobic dogma' in pepsin catalysis was seriously challenged when Pohl and Dunn[2] described a series of polycationic oligopeptides that are among the most reactive synthetic substrates known. These highly hydrophilic molecules should not be well accommodated in a hydrophobic active site. Moreover, the steady state kinetic parameters were remarkably sensitive to the acidity of the medium. As the pH was increased from pH 3 to 6, the specificity number k_{cat}/K_m increased by three to four orders of magnitude and approached values close to the diffusion limit. The catalytic turnover number k_{cat} was much less affected by the increase in pH, and the Michaelis constant K_m decreased accordingly, by three to four orders of magnitude. A typical pH profile of steady state kinetic parameters, for the substrate Lys-Lys-Ala-Lys-Phe-Phe(NO_2)-Arg-Leu, is shown in Figure 1.

Two hypotheses were suggested[2] to explain the high reactivity of the polycationic substrates and the pronounced effects of pH. First, it was proposed that the positively charged residues could form tight ion pairs with carboxylate residues in the enzyme. These *specific* electrostatic interactions - in the extended catalytic cleft - would stabilize the Michaelis complex and thereby decrease K_m. Alternatively, it was suggested that the substrate could initially interact with a patch of negatively charged residues on the enzyme surface, outside the active site (*nonspecific* electrostatic binding). The nonspecific ionic association would then be followed by diffusion of the ligand into the active site. *A priori,* both hypotheses are plausible. Nonspecific electrostatic binding is consistent with the fact that porcine pepsin contains 43 acidic and only 4 basic residues, so that between pH 3 and pH 6 there will be a large increase in negative surface charge. Specific electrostatic binding could be rationalized if some of the negatively charged residues were part of the active site, and formed ion pairs with the bound substrate.

Structure and Function of the Aspartic Proteinases
Edited by B.M. Dunn, Plenum Press, New York, 1991

The aim of the present study was to distinguish between the two hypotheses by using positively charged pepsin inhibitors. We reasoned that inhibition constants are less complex quantities than substrate kinetic parameters and provide a more direct measure of the binding thermodynamics. We found that the binding of polycationic ligands to pepsin is dominated by nonspecific electrostatic interactions and that the minimal kinetic mechanism includes seven steps.

RESULTS AND DISCUSSION

Monocationic Analogs of Pepstatin

Analogs of the the subnanomolar inhibitor Iva-Val-Val-Sta-Ala-Iaa[3] (Table 1) were prepared as electrostatic active site probes that span binding subsites S_4 through S_3'. In the determination of the inhibition constants, the enzyme was preincubated with the inhibitor before the chromogenic substrate (Lys-Pro-Ala-Glu-Phe-Phe(NO_2)-Arg-Leu)[2] was added, so that the time dependent processes observed in preliminary experiments were eliminated. If there were electrostatic interactions between the lysine sidechains and carboxylate residues in the active site, then the binding of the charged pepstatin analogs should be tighter when compared to Iva-Val-Val-Sta-Ala-Iaa. The data in Table 1 indicate the opposite trend; the introduction of positive charge into positions P_4 through P_3' causes a decrease in the binding affinity in comparison with the neutral parent compound. The most pronounced losses of ΔG_b (3.2; 5.7 kcal/mol) were observed for the P_3 and P_1 substitutions. Positions P_4, P_2, and P_2' were moderately affected, and the introduction of lysine into the position P_3' had no effect.

The effects of positive charge in subsites P_2 and P_3 were examined in more detail (Table 2) by replacing the valine residues in the parent peptide with three pairs of unbranched aliphatic aminoacids. The length of the alkyl spacer $(CH_2)_n$ was identical in each pair (Abu - Dab; Nva - Orn; Nle - Lys), but one residue was neutral, while the other carried a positive charge. The results listed in Table 2 indicate that in subsites P_2 and P_3 the hydrophobic

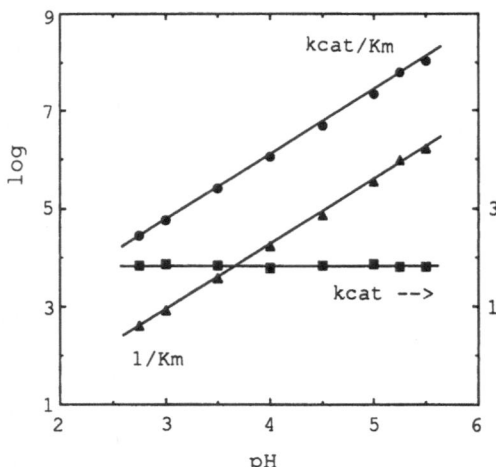

Figure 1. Dependence of the steady-state kinetic parameters for Lys-Lys-Ala-Lys-Phe-Phe(NO_2)-Arg-Leu on pH (originally reported in reference 2).

Table 1 Effect of Monocationic Side chain in Pepstatin Analogs on Inhibition Constants and Relative Binding Energies

Inhibitor[a]	K_i, nM	$\Delta\Delta G_b$, kcal/mol[b]
Iva-Val-Val-Sta-Ala-Iaa	0.10	0.0
Boc-*Lys*-Val-Val-Sta-Ala-Iaa	0.22	+ 0.5
Iva-*Lys*-Val-Sta-Ala-Iaa	19.2	+ 3.2
Iva-Val-*Lys*-Sta-Ala-Iaa	6.3	+ 2.6
Iva-Val-Val-*Ly*Sta-OEt[c]	> 1000	> 5.7
Iva-Val-Val-Sta-*Lys*-Iaa	0.72	+ 1.2
Iva-Val-Val-Sta-Ala-*Lys*-OMe	0.10	0.0

[a] Abbreviations: Iaa = isoamyl amide; Iva = isovaleric acid; LySta = (2S,3S)-2,8-diamino-3-hydroxy octanoic acid; Sta = statine; [b] T 37°C, pH 4.00 (I = 100 mM); [c] data taken from reference 4.

interactions are more important than electrostatic effects. First, all charged inhibitors bind less strongly than the corresponding neutral analogs. Second, as the sidechain of the charged aminoacids is shortened and the positive charge is 'pulled' inside the active site, the binding of inhibitors decreases even more. Third, similar losses in binding energy occur as the side chain of the neutral analogs is shortened. The removal of each methylene group contributes approximately equally to the decrease in binding energy (Figure 2). The linearity of the plots in Figure 2 quantitatively shows that the binding energy in S_2 and S_3 is derived mainly from hydrophobic interactions, and that subsite S_3 is more sensitive to hydrophobic effects[5] (slope +0.95) than S_2 (slope +0.55).

Table 2. Effect of Side-chain Length and Charge on Inhibition Constants and Relative Binding Energies of Iva-Xxx-Yyy-Sta-Ala-Iaa [a]

	Xxx		Yyy	
	K_i, nM	$\Delta\Delta G_b$, kcal/mol	K_i, nM	$\Delta\Delta G_b$, kcal/mol
Nle	0.005	0.0	0.030	0.0
Nva	0.017	+ 0.8	0.089	+ 0.7
Abu	0.031	+ 1.1	0.20	+ 1.2
Lys	19.2	+ 5.1	6.3	+ 3.3
Orn	72.0	+ 5.9	7.2	+ 3.4
Dab	13.8	+ 4.9	21.2	+ 4.0

[a] Abbreviations: Abu = 2-aminobutyric acid; Dab = 2,4-diamino butyric acid; Nva = norvaline.

The effect of pH on the inhibition constants of all lysine analogs was examined (Figure 3a, b). If there were specific electrostatic interactions in the active site, the binding of charged inhibitors should increase with an increase in pH, due to increased ionization of enzyme carboxylates. However, the analogs with lysine in positions P_2 and P_3' were not sensitive to variations in pH. The remaining charged inhibitors actually showed slightly weaker binding as the alkalinity of the medium increased. For the P_3-lysine derivative, the sudden loss of binding above pH 5 could be due to a conformational change in the enzyme molecule because a similar loss of binding energy above pH 5 occurred with the P_3-norleucine inhibitor (Figure 3b), which does not contain any pH sensitive functional groups. In summary, the results obtained with the pepstatin analogs listed in Tables 1 and 2 indicate that the binding of these inhibitors to pepsin is dominated by hydrophobic, rather than electrostatic interactions within the extended catalytic cleft.

Polycationic Statine-Containing Inhibitor

The high reactivity and pronounced pH sensitivity of the polycationic pepsin substrates[2] could be caused by nonspecific electrostatic binding to a part of the negatively charged protein surface outside the active site, and inhibitor 1, Lys-Lys-Ala-Lys-Sta-Arg-Leu, was prepared to test this hypothesis. As the theoretical model for the analysis of data, we used the Debye-Hückel equation (1), which has been previously used to analyze nonspecific electrostatic interactions in acetylcholine esterase[6]. The use of the Debye-Hückel theory, as well as other, more rigorous, models of electrostatic interactions in biomolecules is thoroughly documented in numerous reviews.[7]

$$\log k_1 \;=\; \log k_1^{\circ} \;+\; \frac{1.18\, Z_E Z_L \sqrt{I}}{1 + 0.329\, d \sqrt{I}} \tag{1}$$

Equation (1) predicts that the bimolecular rate constant for association of charged molecules (k_1, $M^{-1}sec^{-1}$) depends on the charges on the enzyme (Z_E) and on the ligand (Z_L), as well as on the ionic strength of the medium. The parameter d is the *average* interionic distance measured in Ångstroms, and k_1' is the limiting value of k_1 at zero ionic strength. The bimolecular association rate of a negatively charged enzyme and a positively charged ligand will decrease with increasing ionic strength.

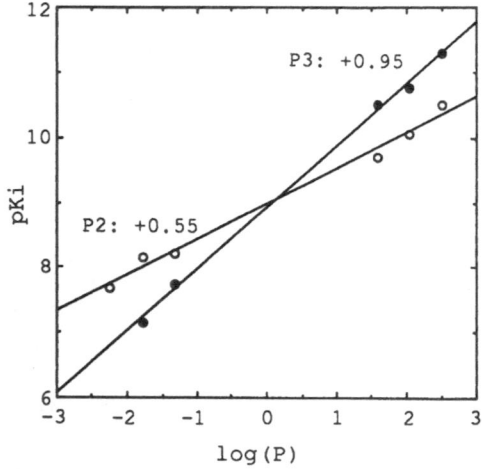

Figure 2. Correlation of the inhibitory activity for Iva-Xxx-Val-Sta-Ala-Iaa (P_3) and Iva-Val-Yyy-Sta-Ala-Iaa (P_2) with the hydrophobicity constant log (P).

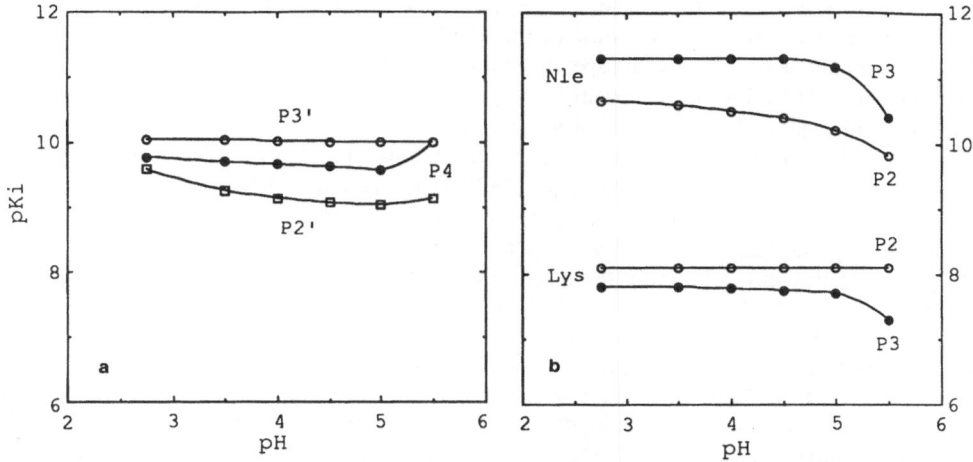

Figure 3. a) pH dependence of the inhibitory activity for Iva-Val-Val-Sta-Ala-Lys (P3'), Iva-Val-Val-Sta-Lys-Iaa (P2'), and Lys-Val-Val-Sta-Ala-Iaa (P4). b) pH dependence of the inhibitory activity for Iva-Val-Nle-Sta-Ala-Iaa (P2-Nle), Iva-Val-Lys-Sta-Ala-Iaa (P2-Lys), Iva-Nle-Val-Sta-Ala-Iaa (P3-Nle), and Iva-Lys-Val-Sta-Ala-Iaa (P3-Lys).

At pH 4.0 and ionic strength 100 mM, peptide **1** showed slow onset of inhibition and the inhibition constant changed from 1100 ± 100 nM for the initial enzyme-inhibitor complex, to 130 ± 15 nM for the tightened complex. Both values indicate significantly weaker binding compared to the monocationic inhibitors. The apparent first order rate constant was dependent on the ionic strength of the buffer. At low ionic strength the formation of the tight complex was faster (5 mM, halftime 0.1 min) than at high salt concentration (500 mM, halftime 0.6 min). In preliminary experiments, the slow phase in binding of inhibitor **1** was also found to be dependent on pH. The formation of the tightened enzyme-inhibitor complex was faster in more alkaline buffers where the surface charge on pepsin is expected to be more negative.

Stepwise binding of the polycationic inhibitor **1** is consistent with the reported time-dependent binding of synthetic fragments of the positively charged propart peptide[8], L-V-K-V-P-L-V-R-K-K-S-L-R-G-N-L (single-letter aminoacid code, pepsinogen residues 1-16) to pepsin. The initial event in the stepwise binding was identified as "a simple collisional process giving a complex with the peptide loosely bound to the surface of the protein." The "rapid surface binding that is mainly electrostatic in nature is followed by a slower second phase in which the peptide finds the active site. The interaction at this stage is dominated by hydrophobic binding."[8] It was observed that the initial binding was enhanced as pH was increased. However, the proteolytic activity of pepsin was not diminished upon formation of the initial complex, which indicates that the binding occurred outside the active site.

The apparent inhibition constant K_i for the tightly bound complex between inhibitor **1** and pepsin is strongly dependent on the ionic strength of the buffer (Figure 4). As the ionic strength increased from 2 mM to 200 mM, the binding at pH 4.55 decreased by a factor of 30 (from K_i 0.9 nM to K_i 30 nM). Nonlinear least squares fit of the binding data to the Debye-Hückel equation (1) provided estimates of parameters d and $Z_E Z_L$. The average interionic distance d was 26 ± 8 Å, and the estimated Coulombic product $Z_E Z_L$ was -19 ± 6. If we assume that all four positively charged residues in the inhibitor are involved in the electrostatic binding, the kinetically significant charge on the enzyme molecule is approximately -5. The net charge on the protein surface is probably considerably higher, because the total number of electrically uncompensated carboxylate residues in the primary sequence is 39, and at the indicated pH at least half of these residues will be in the ionized state.

The relevance of the kinetic properties of inhibitor **1** to the corresponding polycationic substrate, Lys-Lys-Ala-Lys-Phe-Phe(NO$_2$)-Arg-Leu, was tested by examining (a) the dependence of the substrate specifity number k_{cat}/K_m on the ionic strength, and (b) the dependence of the inhibition constant K_i on pH. When the ionic strength of the buffer was varied between 50 and 450 mM (pH 4.55), the substrate turnover number remained constant (68 ± 4 sec^{-1}). On the other hand, the specificity number k_{cat}/K_m decreased by an order of magnitude, from 21×10^6 M^{-1}sec^{-1} to 1.8×10^6 M^{-1}sec^{-1}, and the Michaelis constant increased proportionally. Thus in Figure 5, the effect of increasing sodium chloride concentration on steady state parameters resembles a competitive inhibition pattern. The proper meaning of these results lies in establishing the sensitivity of the apparent bimolecular rate constant k_{cat}/K_m to ionic strength, according to the Debye-Hückel theory.

The inhibition constant for compound **1** is strongly dependent on pH (Figure 6), in contrast with the kinetics of the monocationic inhibitors which were pH insensitive. An increase in pH from 2.7 to pH 5.0 (at constant ionic strength 100 mM) caused a pronounced increase in the inhibitory activity; K_i decreased from 4.9 μM to 7.1 nM. On the logarithmic scale, the changes in K_i with pH were linear, and the proportionality constant was identical with the dependence of $\log(k_{cat}/K_m)$ on pH for the corresponding substrate. These results establish identical sensitivity of both quantities (Figure 7) to variations in molecular surface charge on pepsin, induced by changes in pH. Thus the physical model for stepwise, nonspecific electrostatic association appears to be valid for both the inhibitor **1** and the corresponding substrate. In both cases, the elementary rate constant most sensitive to charge effects is the bimolecular association rate constant k_1, according to eq. (1).

Minimal Kinetic Mechanism For Aspartic Proteinases

The involvement of the stepwise binding of substrates has important implications for the minimal kinetic mechanism, which has to be considered for proteolytic enzymes. If the substrate and the enzyme initially form a loosely bound ionic encounter complex E·S, which later transforms itself into the reactive Michaelis complex, then it is necessary to postulate the existence of multiple complexes also on the part of the products; the essential ionic character of ligands S, P, and Q remains unchanged in the hydrolytic reaction. The minimal kinetic mechanism then contains seven elementary steps, and seven distinct enzyme species (Scheme 1). Corresponding analytical formulas for the turnover number k_{cat} and the specificity

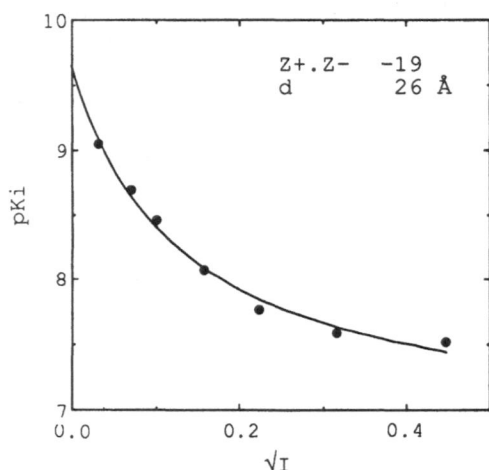

Figure 4. Dependence of the inhibition constant for Lys-Lys-Ala-Lys-Sta-Arg-Leu on the ionic strength of the buffer (5 mM sodium acetate, pH 4.55).

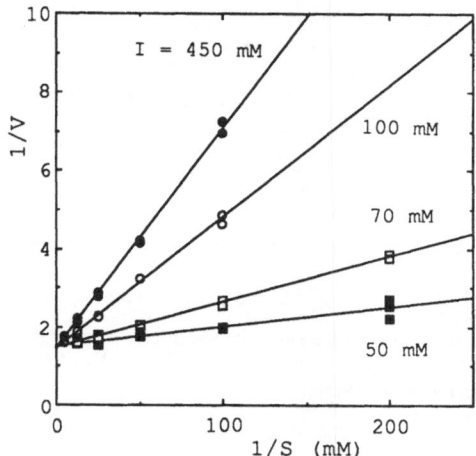

Figure 5. Lineweaver-Burk plots for the pepsin-catalyzed hydrolysis of Lys-Lys-Ala-Phe-Phe(NO₂)-Arg-Leu at varied ionic strength (5 mM sodium acetate, pH 4.55).

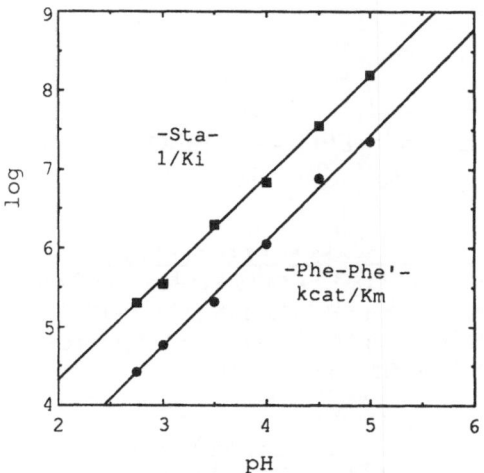

Figure 6. pH dependence of the inhibition constant for Lys-Lys-Ala-Lys-Sta-Arg-Leu (log $1/K_i$) and the specificity number for Lys-Lys-Ala-Lys-Phe-Phe(NO₂)-Arg-Leu (log k_{cat}/K_m).

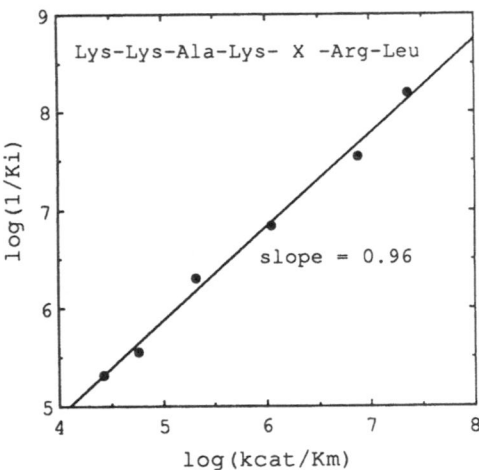

Figure 7. Linear correlation between the substrate specificity number (log k_{cat}/K_m) for Lys-Lys-Ala-Lys-Phe-Phe(NO$_2$)-Arg-Leu and the inhibition constant (log $1/K_i$) for Lys-Lys-Ala-Lys-Sta-Arg-Leu, measured at different pH values. Data taken from Figure 6.

number k_{cat}/K_m [equations (2) and (3)] were derived by using Cleland's partition analysis.[9] The expression for the Michaelis constant is obtained simply by dividing equation (2) by equation (3).

Stepwise association of ligands with enzymes is considered important to most microscopic kinetic theories. In a comprehensive review on dynamical simulation of rate constant in protein-ligand interactions, Case[10] pointed out that the first (bimolecular) step is likely to have the greatest influence on the overall association rate, especially in the case of

Scheme 1

$$E \underset{k_2}{\overset{k_1}{\rightleftharpoons}} E \cdot S \underset{k_4}{\overset{k_3}{\rightleftharpoons}} ES \underset{k_6}{\overset{k_5}{\rightleftharpoons}} EPQ \underset{k_8}{\overset{k_7}{\rightleftharpoons}} EP \cdot Q \xrightarrow{k_9} EP \underset{k_{12}}{\overset{k_{11}}{\rightleftharpoons}} E \cdot P \xrightarrow{k_{13}} E$$

$$k_{cat} = \frac{k_3\,k_5\,k_7\,k_9\,k_{11}\,k_{13}}{\begin{aligned}&k_4\,k_6\,k_8\,k_{11}\,k_{13} + k_4\,k_6\,k_9\,k_{11}\,k_{13} + k_4\,k_7\,k_9\,k_{11}\,k_{13}\\&+ k_5\,k_7\,k_9\,k_{11}\,k_{13} + k_3\,k_6\,k_8\,k_{11}\,k_{13} + k_3\,k_6\,k_9 k_{11}\,k_{13} + k_3\,k_7\,k_9\,k_{11}\,k_{13}\\&+ k_3\,k_5\,k_8\,k_{11}\,k_{13} + k_3\,k_5\,k_9\,k_{11}\,k_{13} + k_3\,k_5\,k_7\,k_{11}\,k_{13} + k_3\,k_5\,k_7\,k_9\,k_{11}\\&+ k_3\,k_5\,k_7\,k_9\,k_{12} + k_3\,k_5\,k_7\,k_9\,k_{13}\end{aligned}} \quad (2)$$

$$\frac{k_{cat}}{K_m} = \frac{k_1\,k_3\,k_5\,k_7\,k_9}{k_2\,k_4\,k_6\,k_8 + k_2\,k_4\,k_6\,k_9 + k_2\,k_4\,k_7\,k_9 + k_2\,k_5\,k_7\,k_9 + k_3\,k_5\,k_7\,k_9} \quad (3)$$

small ligands. The stepwise, nonspecific electrostatic association of polycationic substrates with porcine pepsin at nonphysiological pH then represents a striking illustration of this phenomenon. The uniqueness of pepsin and presumably also other gastric proteases - in their total surface charge being continuously adjustable by changes in pH - could make these enzymes one of the convenient test systems for studies of electrostatic association between biomolecules.

Rich and Northrop[11] considered a six step mechanism for proteases from the point of view of conformational changes induced in the enzyme upon ligand binding. A flexible part of the supersecondary structure in all aspartic proteinases, the 'flap', shows pronounced spatial displacement in crystal structures of free enzymes compared to complexes with inhibitors.[12] The flexible domain effectively entraps the inhibitor in the active site, and it is very likely that the binding of substrates occurs in the similar fashion. The binding of the substrate and the conformational change in the enzyme-substrate complex are often considered as kinetically distinct, sequential processes (see for example a recent kinetic study of the arabinose binding protein, ABP, by Vermesch et al.[13]). In such case the closed and open conformers of the enzyme-substrate complex would exist as distinct enzyme forms. If these were accounted for in the kinetic mechanism for proteases, the scheme indicated above would contain 10 enzyme species and 20 primary rate constants.

Polycationic Substrates

We have experimentally established the stepwise binding of the polycationic inhibitor 1, and the parallelism between $\log(K_i)$ and $\log(k_{cat}/K_m)$ for the corresponding substrate. Based on the sensitivity of the inhibition constant to the ionic strength, we propose that the first step in the bimolecular association is determined by nonspecific electrostatic interactions, which affect primarily the rate constant k_1 [equation (1)]. It is interesting to simulate the changes in k_1 with pH and compare the effects produced on substrate kinetic parameters with the experiment. It is assumed that ionization constants of the 43 carboxylate residues on the enzyme surface overlap within a wide range of values, in dependence on the microscopic environment. Consequently the changes in the molecular surface charge [Z_E in eq. (1)] will be linear within a wide range of pH, and the logarithm of k_1 will increase linearly.

Figure 8 depicts a pseudo 'free energy profile', in which the bimolecular association barrier (reflected in k_1) and the free energy of the loose ionic complex E·S (reflected in equilibrium dissociation constant k_2/k_1) was lowered due to nonspecific electrostatic attraction. The x-axis is labeled according to percentile distribution of individual enzyme forms, and the displacement on the y-axis is proportional to the logarithm of individual rate constants. The simulated effects on the kinetic parameters of a polycationic substrate are shown in Figure 9. The change in the bimolecular association segment had no effect on k_{cat}. The specificity number k_{cat}/K_m showed a dramatic increase, and the Michaelis constant decreased accordingly. The simulated pattern is identical with the pH profile shown in Figure 1, and the predictions of the Debye-Hückel theory based on the nonspecific electrostatic association are satisfied.

CONCLUSION

Binding of polycationic ligands to pepsin at nonphysiological pH (pH > 1.0) occurs in a stepwise fashion. The first step is a fast, nonspecific electrostatic association, followed by slower surface diffusion into the active site. Secondary enzyme-ligand interactions within the extended catalytic cleft are dominated by hydrophobic rather than electrostatic forces. The minimal kinetic mechanism for proteases, which takes into account the effects of nonspecific binding, contains seven elementary steps and fourteen primary rate constants. The nonspecific interactions determine kinetic properties of both substrates and inhibitors *via* changes in the bimolecular association rate constant k_1 (Scheme 1). In the case of the polycationic inhibitors, the changes in k_1 are reflected in the apparent inhibition constant $K_i = k_2k_4/k_1k_3$; an increase in the bimolecular association rate produces an increase in the inhibitory effect. An alternative description would be that the nonspecific binding causes an increase in the 'effective molarity'[14] of the ligand in close vicinity of each enzyme molecule,

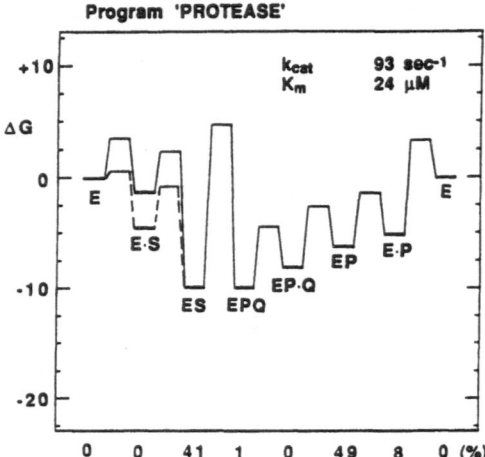

Figure 8. Schematic representation of the pseudo-'free energy profile' for the nonspecific association of an enzymatic substrate. The displacement on the 'ΔG' axis is proportional to the logarithm of rate constants for individual steps. Numerical values at the horizontal axis indicate the percentage of enzyme forms present at the steady state.[9]

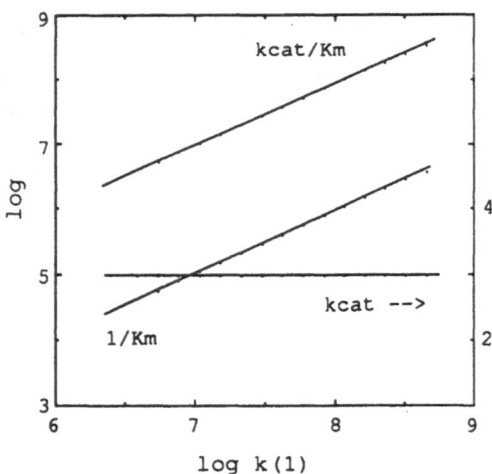

Figure 9. Changes in the steady-state kinetic parameters that accompany the lowering of the energy barrier for nonspecific enzyme-substrate association, due to a change in experimental conditions such as pH (simulated by the broken line in Figure 8). The parameters were calculated from rate constants by the method of partition analysis.[9]

and that the inhibitory effect is enhanced due to the increased local concentration of the inhibitor.

In the case of the polycationic substrates, the changes in k_1 are reflected in the apparent bimolecular rate constant k_{cat}/K_m (and in the Michaelis constant K_m), while the turnover number k_{cat} is unaffected. It is instructive to consider the relationship between the Michaelis constant K_m and the dissociation constant of the enzyme-substrate active complex ES. Even for the kinetically simplest case of an enzymatic reaction,

$$E + S \xrightleftharpoons[k_2]{k_1} ES \xrightarrow{k_3} E + P$$

the Michaelis constant $(K_m = K_s + k_3/k_1)$ is more complex then the reciprocal substrate binding constant $(K_s = k_2/k_1)$. The two quantities become numerically equal only under the conditions of rapid equilibrium $(k_1, k_2 \gg k_3)$. For some pepsin substrates[1] K_m appears to be equal K_s, but any extrapolation is dangerous, especially with 'fast' substrates. For example, a substrate reported in the literature[2] showed K_m 220 x 10^{-9} M and k_{cat} 50 sec^{-1}. Assuming that k_1 approaches the diffusion limit 10^9 M^{-1}sec^{-1}, the estimate for the 'off' rate constant k_2 obtained under the rapid equilibrium approximation $(K_m = K_s, k_{cat} = k_3)$ is 220 sec^{-1}. This value is comparable with k_{cat} within an order of magnitude, so that the rapid equilibrium approximation is clearly self-contradictory.

The involvement of nonspecific enzyme-substrate association only adds complexity to the issue, because when the binding of the substrate occurs in a stepwise fashion, then the apparent binding constant itself is a kinetically complex variable. Kinetic significance of the nonspecific binding to enzymes is probably underestimated at the present time, although it was previously considered in theory.[15] Detailed studies of nonspecific interactions in enzymatic systems, which can in principle arise from a variety of intermolecular forces other than electrostatic, will add to our understanding of substrate specificity and inhibitor binding.

REFERENCES

1. J. Fruton, *Adv. Enzymol. Related Areas Mol. Biol.* **44**:1-36 (1976).

2. J. Pohl and B. M. Dunn, *Biochemistry* **27**:4827-4834 (1988).

3. D. H. Rich and M. S. Bernatowicz, *J. Med. Chem.* **25**:791-795 (1982).

4. F. C. Salituro, N. Agarwal, T. Hofmann and D. H. Rich, *J. Med. Chem.* **30**:286-295 (1987).

5. Logarithmic partition coefficients logP for the sidechains (beginning with C_β) were calculated from atomic contributions reported in (a) A. K. Ghose and G. M. Crippen, *J. Comput. Chem.* **7**:565-577 (1986); (b) P. Furet, A. Sele and N. C. Cohen, *J. Mol. Graphics* **6**:182-189 (1988); (c) D. Eisenberg and A. D. McLachlan, *Nature (London)* **319**:199-203 (1986).

6. H. J. Nolte, T. L. Rosenberry and E. Neumann, *Biochemistry* **19**:3705-3711 (1980).

7. (a) E. Neumann, in "Topics of Bioelectrochemistry and Bioenergetics" (G. Milazzo, Ed.), John Wiley, New York, pp. 113-160 (1981). (b) E. Neumann, in "Structural and Functional Aspects of Enzyme Catalysis" (H. Eggerer, R. Huber, Eds.), Springer-Verlag, Berlin (1981). (c) E. Neumann, in "Modem Bioelectrochemistry" (F. Gutmann, K. Keyzer, Eds.), Plenum Press, New York, pp. 97-175 (1986). (d) E. Neumann, *Prog. Biophys. Molec. Biol.* **47**:197-231 (1986). (e) N. K. Rogers, *Progr. Biophys. Molec. Biol.* **48**:37-66 (1986).

8. B. M. Dunn, B. Parten, M. Jimenez, C. E. Rolph, M. Valler and J. Kay, in "Aspartic Proteinases and their Inhibitors" (V. Kostka, Ed.), Walter de Gruyter, Berlin, pp. 221-243 (1985).

9. W. W. Cleland, *Biochemistry* **14**:3220-3224 (1975).

10. D. A. Case, *Prog. Biophys. Molec. Biol.* **52**:39-70 (1988).

11. D. H. Rich and D. B. Northrop, in "Computer Aided Drug Design" (T. J. Perun and C. L. Probst, Eds.), Marcel Dekker, New York, p. 185-250 (1989).

12. M. Miller, J. Schneider, B. K. Sathyanarayna, M. V. Toth, G. R. Marshall, L. Clawson, L. Selk, S. B. H. Kent, and A. Wlodawer, *Science (Washington, D.C.)* 246:1149-1152 (1989).

13. P. S. Vermesch, J. J. G. Tesmer, D. D. Lemon and F. A. Quiocho, *J. Biol. Chem.* 265:16592-16603 (1990).

14. (a) P. Haberfield and J. J. Cincotta, *J.Org.Chem.* 55:1334-1338 (1990). (b) A. J. Kirby, *Adv.Phys.Org.Chem.* 17:183 (1980).

15. (a) K. C. Chou and S. P. Jiang, *Sci. Sin.* 17:664-680 (1974). (b) K. C. Chou, *Sci. Sin.* 19:505-528 (1976). (c) K. C. Chou and S. Forsén, *Biophys. Chem.* 12:255-263 (1980). (d) K. C. Chou and G. P. Zhou, *J. Amer. Chem. Soc.* 104:1409-1413 (1982).

ANALYSIS OF THE PROMOTER OF A HUMAN PEPSINOGEN A GENE

P. H. S. Meijerink, J. P. Bebelman, G. Pals, F. Arwert, R. J. Planta,[*]
A. W. Eriksson and W. H. Mager[*]

Institute of Human Genetics and [*]Biochemical Department
Vrije Universiteit
P.O. Box 7161, 1007 MC
Amsterdam, The Netherlands

INTRODUCTION

Pepsinogen A (PGA), inactive precursor of the aspartic proteinase pepsin A (E.C. 3.4.23.1), is encoded on the human genome (chromosome 11q12-13) by a multigene family[1]. Electrophoretic separation of the PGA isozymogens from urine or gastric mucosa reveals three main fractions: Pg3, Pg4 and Pg5. The electrophoretic patterns of PGA show a large, genetically-determined, inter-individual heterogeneity. These differences can be explained by the presence of a number of haplotypes, containing different numbers and types of PGA genes[2]. The study of PGA gene clusters by RFLP analysis previously enabled us to define the haplotypes corresponding to certain phenotypes[3]. In this paper we describe part of our studies aimed at identifying *cis*-acting elements and *trans*-acting factors involved in the tissue-specific transcriptional regulation of PGA gene expression.

EXPERIMENTAL AND DISCUSSION

The nucleotide sequence of about 1.7 kb of the 5'-flanking region of the PGA gene encoding Pg5 was analyzed. This gene, characterized by an *Eco*RI-generated DNA fragment of 15.0 kb[3], was purified from cosmid cgHGP23[1]. The results are shown in Figure 1. A computer search revealed several sites of homology with consensus recognition sequences for general transcription factors.

To elucidate putative *cis*-acting elements and corresponding *trans*-acting factors involved in the transcription activation of PGA gene expression, we applied band shift assays and DNase I footprinting. Band shift assays using a 171 bp *Pvu*II+*Xho*I generated proximal promoter fragment (Figure 1, position -145 to +26) and nuclear extracts of HeLa cells, porcine gastric mucosa or liver, revealed several protein-DNA complexes. In general, all three extracts yielded similar complexes, although differences in intensity occurred. Notably, retardation analysis using gastric nuclear extract showed two additional fast migrating and

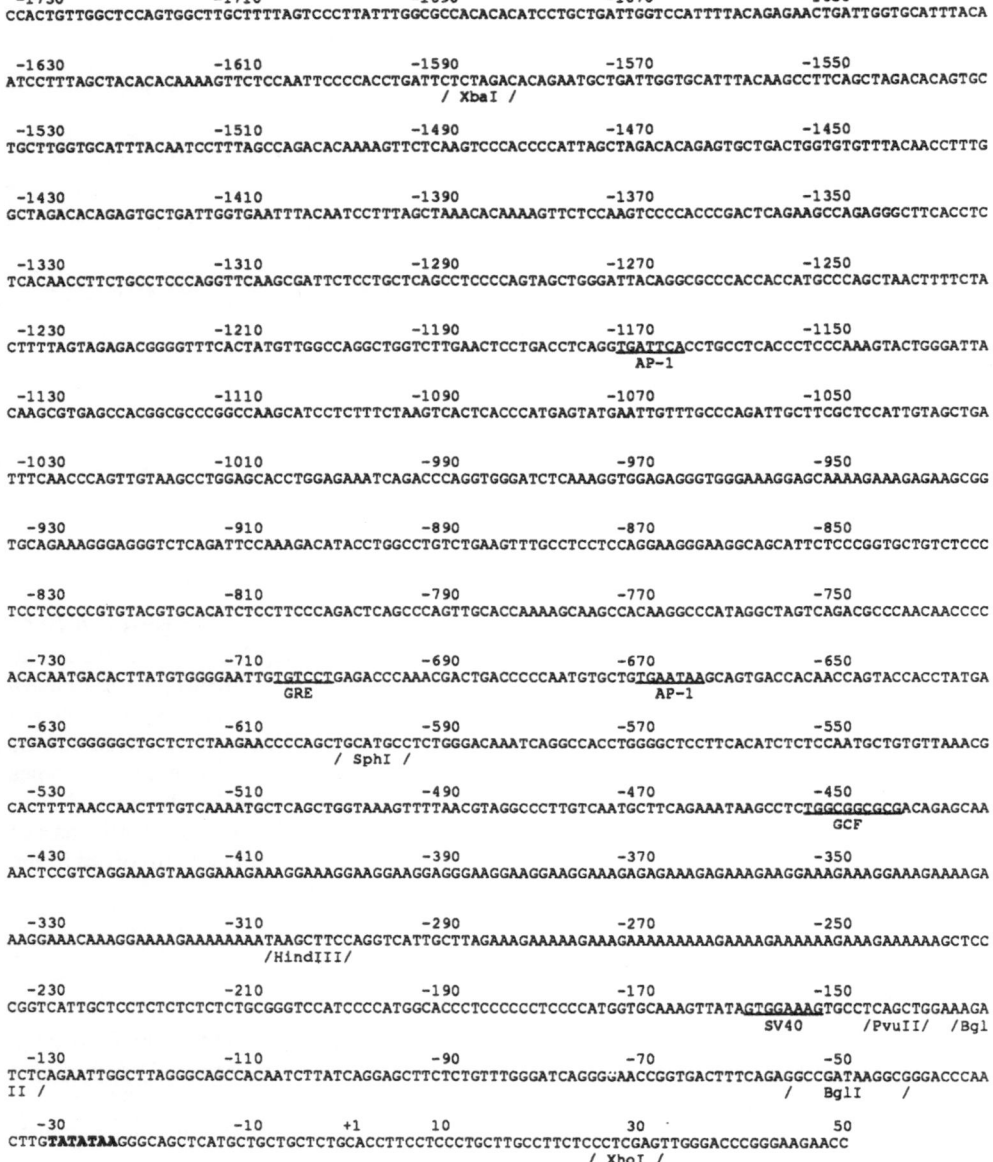

```
          -1730                 -1710                 -1690                 -1670                 -1650
CCACTGTTGGCTCCAGTGGCTTGCTTTTAGTCCCTTATTTGGCGCCACACACATCCTGCTGATTGGTCCATTTTACAGAGAACTGATTGGTGCATTTACA

          -1630                 -1610                 -1590                 -1570                 -1550
ATCCTTTAGCTACACACAAAAGTTCTCCAATTCCCCACCTGATTCTCTAGACACAGAATGCTGATTGGTGCATTTACAAGCCTTCAGCTAGACACAGTGC
                                              / XbaI /

          -1530                 -1510                 -1490                 -1470                 -1450
TGCTTGGTGCATTTACAATCCTTTAGCCAGACACAAAAGTTCTCAAGTCCCACCCCATTAGCTAGACACAGAGTGCTGACTGGTGTGTTTACAACCTTTG

          -1430                 -1410                 -1390                 -1370                 -1350
GCTAGACACAGAGTGCTGATTGGTGAATTTACAATCCTTTAGCTAAACACAAAAGTTCTCCAAGTCCCCACCCGACTCAGAAGCCAGAGGGCTTCACCTC

          -1330                 -1310                 -1290                 -1270                 -1250
TCACAACCTTCTGCCTCCCAGGTTCAAGCGATTCTCCTGCTCAGCCTCCCCAGTAGCTGGGATTACAGGCGCCCACCACCATGCCCAGCTAACTTTTCTA

          -1230                 -1210                 -1190                 -1170                 -1150
CTTTTAGTAGAGACGGGGTTTCACTATGTTGGCCAGGCTGGTCTTGAACTCCTGACCTCAGGTGATTCACCTGCCTCACCCTCCCAAAGTACTGGGATTA
                                                         AP-1

          -1130                 -1110                 -1090                 -1070                 -1050
CAAGCGTGAGCCACGGCGCCCGGCCAAGCATCCTCTTTCTAAGTCACTCACCCATGAGTATGAATTGTTTGCCCAGATTGCTTCGCTCCATTGTAGCTGA

          -1030                 -1010                  -990                  -970                  -950
TTTCAACCCAGTTGTAAGCCTGGAGCACCTGGAGAAATCAGACCCAGGTGGGATCTCAAAGGTGGAGAGGGTGGGAAAGGAGCAAAAGAAAGAGAAGCGG

           -930                  -910                  -890                  -870                  -850
TGCAGAAAGGGAGGGTCTCAGATTCCAAAGACATACCTGGCCTGTCTGAAGTTTGCCTCCTCCAGGAAGGGAAGGCAGCATTCTCCCGGTGCTGTCTCCC

           -830                  -810                  -790                  -770                  -750
TCCTCCCCCGTGTACGTGCACATCTCCTTCCCAGACTCAGCCCAGTTGCACCAAAAGCAAGCCACAAGGCCCATAGGCTAGTCAGACGCCCAACAACCCC

           -730                  -710                  -690                  -670                  -650
ACACAATGACACTTATGTGGGGAATTGTGTCCTGAGACCCAAACGACTGACCCCCAATGTGCTGTGAATAAGCAGTGACCACAACCAGTACCACCTATGA
                                 GRE                               AP-1

           -630                  -610                  -590                  -570                  -550
CTGAGTCGGGGGCTGCTCTCTAAGAACCCCAGCTGCATGCCTCTGGGACAAATCAGGCCACCTGGGGCTCCTTCACATCTCTCCAATGCTGTGTTAAACG
                                        / SphI /

           -530                  -510                  -490                  -470                  -450
CACTTTTAACCAACTTTGTCAAAATGCTCAGCTGGTAAAGTTTTAACGTAGGCCCTTGTCAATGCTTCAGAAATAAGCCTCTGGCGGCGCGACAGAGCAA
                                                                            GCF

           -430                  -410                  -390                  -370                  -350
AACTCCGTCAGGAAAGTAAGGAAAGAAAGGAAAGGAAGGAAGGAGGGAAGGAAGGAAGGAAAGAGAGAAAGAGAAAGAAGGAAAGAAAGGAAAGAAAAGA

           -330                  -310                  -290                  -270                  -250
AAGGAAACAAAGGAAAGAAAAAAAATAAGCTTCCAGGTCATTGCTTAGAAAGAAAAAGAAAGAAAAAAAAAGAAAAGAAAAAAGAAAGAAAAAAGCTCC
                                   /HindIII/

           -230                  -210                  -190                  -170                  -150
CGGTCATTGCTCCTCTCTCTCTCTGCGGGTCCATCCCCATGGCACCCTCCCCCCTCCCCATGGTGCAAAGTTATAGTGGAAAGTGCCTCAGCTGGAAAGA
                                                                          SV40      /PvuII/  /Bgl

           -130                  -110                   -90                   -70                   -50
II /   TCTCAGAATTGGCTTAGGGCAGCCACAATCTTATCAGGAGCTTCTCTGTTTGGGATCAGGGGAACCGGTGACTTTCAGAGGCCGATAAGGCGGGACCCAA
                                                                                     /  BglI  /

            -30                  -10                   +1          10                   30                   50
CTTGTATATAAGGGCAGCTCATGCTGCTGCTCTGCACCTTCCTCCCTGCTTGCCTTCTCCCTCGAGTTGGGACCCGGGAAGAACC
                                                            / XhoI /
```

Figure 1. Nucleotide sequence of the 5'-flanking and leader region of the human pepsinogen A 15.0 gene. +1 = transcription start site[4]. Some putative protein-DNA binding sites and relevant restriction sites are indicated.

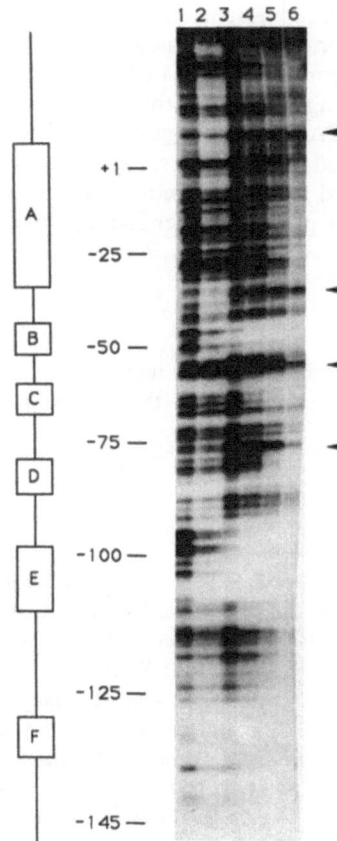

Figure 2. Footprint analysis of the proximal part of the promoter of the pepsinogen A 15.0 gene. A 171 bp *Pvu*II+*Xho*I-generated fragment (non-coding strand) was end-labelled using Klenow polymerase and then incubated with increasing amounts (lanes 1-6: 0, 1, 2.5, 5, 10 and 15 μg respectively) of protein from porcine gastric mucosa nuclei. Nuclear extracts were prepared from freshly obtained porcine gastric fundic mucosa, essentially as described by Gorski et al.[5] Regions protected from DNaseI cleavage are boxed. Arrows indicate hypersensitive sites.

possibly gastric mucosa-specific protein-DNA complexes. One complex band disappeared after digestion of the 171 bp probe fragment with *Bgl*I at position -49 prior to using it as a probe.

As a first step toward the identification of the protein binding sites a DNase I footprint analysis was performed. To that end, the 171 bp *Pvu*II+*Xho*I fragment was incubated in the presence of increasing amounts of porcine gastric mucosa nuclear extract. The B-region (Figure 2) contains the *Bgl*I-site used to divide the proximal promoter fragment into two probes for gel retardation. This data may correspond to the disappearance of one DNA-protein complex after digestion of the 171 bp probe fragment with *Bgl*I. We propose that the elements detected in this way form part of the basal promoter of the human PGA 15.0 gene.

To establish the functional role of these promoter regions in the transcription activation of the PGA 15.0 gene, we attempted to apply a transient expression assay using recombinant plasmids containing various fragments of the 5'-flanking region fused to the CAT reporter gene. However, these constructs yielded only very low CAT activity after transfection of HeLa cells or primary cultures of gastric chief cells from either monkey or pig. The control plasmid, containing an RSV promoter, did show high transcriptional activity in all three cell types. The low CAT activity in gastric chief cells may be due to the fact that the 1.6 kb of upstream DNA, used in our studies, region lacks a tissue-specific transcriptional enhancer. On the other hand it might reflect the need for species-specific transcription factors. These possibilities are presently under investigation.

ACKNOWLEDGMENTS

We thank Dr. H. van Kruining for technical assistance during part of the work, which was partly financially supported by the Ane and Signe Gyllenbergs Foundation, Helsingfors, Finland.

REFERENCES

1. B. Zelle, M. P. J. Evers, P. C. Groot, J. P. Bebelman, W. H. Mager, R. J. Planta, J. C. Pronk, S. G. M. Meuwissen, M. H. Hofker, A. W. Eriksson, and R. R. Frants, Genomic structure and evolution of the human pepsinogen A multigene family, *Hum. Genet.* **78**:79 (1988).
2. R. R. Frants, J. C. Pronk, G. Pals, J. Defize, B. D. Westerveld, S. G. M. Meuwissen, J. Kreuning, and A. W. Eriksson, Genetics of urinary pepsinogen: A new hypothesis, *Hum. Genet.* **65**:385 (1984).
3. J. P. Bebelman, M. P. J. Evers, B. Zelle, R. Bank, J. C. Pronk, S. G. M. Meuwissen, W. H. Mager, R. J. Planta, A. W. Eriksson, and R. R. Frants, Family and population studies on the human pepsinogen A multigene family, *Human. Genet.* **82**:142 (1989).
4. K. Sogawa, Y. Fujii-Kuriyama, Y. Mizukami, Y. Ichihara, and K, Takahashi. Primary structure of human pepsinogen gene, *J. Biol. Chem.* **258**:5306 (1983).
5. K. Gorski, M. Carneiro, and U. Schibler, Tissue-specific in vitro transcription from the mouse albumin promoter, *Cell* **47**:767 (1986).

SEPARATION AND CHARACTERIZATION OF HUMAN PEPSINOGENS AND

PEPSINS BY HIGH-RESOLUTION DISCONTINUOUS ELECTROPHORESIS

W. Wnuk and E. Loizeau

Division of Gastroenterology
University Hospital of Geneva
Geneva, Switzerland

INTRODUCTION

The mucosal lining of the human stomach contains 2 immunologically distinct types of secretory zymogens, pepsinogen A (PgA) and pepsinogen C (PgC).[1] The difference between the primary structure of PgA and PgC is about 50%.[2,3] Both PgA and PgC consist of molecular variants that differ in net ionic charge. This heterogeneity is due to post-translational modifications (isoforms) and to differences in the primary structure (isozymogens); the latter may be about 1% within the group of PgA.[4] The polymorphism in human pepsinogens and their pepsin counterparts raises the question as to whether some isoenzymes or isoforms are more mucolytic than others. Pepsin A1 (called also pepsin 1), a group of the most acidic isoforms of pepsin, displays some features[5,6] suggesting that this class of pepsins is more "ulcerogenic" than others, but data are currently limited. Studies on the mucolytic properties of the individual isoenzymes and isoforms as well as a detailed examination of the secretion of each form of pepsin(ogen) within the stomach are required to give some insight into this question. To date, such studies have not been easy to undertake, since there was no suitable method for the separation and quantification of different forms of pepsin. Here, we report the multiphasic buffer systems that create a low pH during discontinuous electrophoresis in polyacrylamide gels. These electrophoretic systems "stack" and separate pepsinogens and pepsins at pH close to their pI values, exhibiting a high resolving power for these proteins.

METHODS

The buffer systems were calculated from theory by Jovin[7]. The recipes given below were computed by means of the program described. The system used for separations of pepsinogens was the following: catholyte, 20 mM cacodylic acid, 10 mM KOH, pH 6.2; stacking gel (3%T, 20%C) buffer, resolving gel (11%T, 3%C) buffer and anolyte, 140 mM ß-alanine, 10 mM HCl, pH 5.0 (operative pH 5.5). The system employed for fractionations of pepsins was the following: catholyte, 20 mM propionic acid, 10 mM KOH, pH 4.8; stacking gel (3%T, 20%C) buffer, resolving gel (12%T, 3%C) buffer and anolyte, 60 mM glycine, 42 mM HCl, pH 2.4 (running pH 3.7). For measurements of molecular size, the Rf values at various gel concentrations and Ferguson plots were used as described by Chrambach[8]. The pepsin isoenzymes and isoforms were identified by means of bidimensional electrophoresis. In the first dimension, pepsinogens were separated by electrophoresis at pH 5.5 and then activated in the gel by acidification (pH 2.1, 30 min). The pepsins obtained from pepsinogens were resolved by electrophoresis at pH 3.7 in the second dimension. In all experiments, the same set of samples was applied to two identical gels. After electrophoresis, one gel was stained with silver. The second gel was incubated in acid-hemoglobin and then stained with Coomassie to visualize the bands of proteolytic activity.

RESULTS

When the extract of fundic gland mucosa with a common pepsinogen A phenotype (AA) was fractionated by electrophoresis at pH 5.5, pepsinogen was resolved into at least 9 components, 4 major (PgC, PgA5, PgA4, PgA3) and at least 5 minor (PgC', PgA6, PgA5', PgA4', PgA3' or PgA2) (Figure 1, a). To date, the minor pepsinogen isoforms PgA5' and PgA4', that represent presumably post-translational modifications of PgA5 and PgA4, could not be identified routinely because they co-migrated with the major isozymogens[9]. Another minor component, designated here as PgA6, seems to be a new form of pepsinogen A. In contrast to the rare molecular variant, Pg5s,[10] which also migrates behind the major isozymogen PgA5, PgA6 was found in most human gastric mucosal extracts examined so far. Moreover, a distinctive group of 10 to 12 minor pepsinogen isoforms, migrating much slower than the major isozymogens, was also observed. As determined by electrophoresis using Ferguson plots, the relative molecular mass of each minor isoform is about twice as high as those of each major isozymogen (74,000 vs. 40,000). The slow moving pepsinogens constitute presumably a group of dimeric isoforms (PgAd).

When human gastric juices were fractionated by discontinuous electrophoresis at pH 3.7, pepsin was resolved into as many as 19 proteolytic bands (Figure.1, b). The pepsin components were identified by means of bidimensional electrophoresis. The isozymogens separated in the first dimension were activated and then resolved in the second dimension (Figure.1, a and c). Most of the activated pepsinogens emerged as doublets. One component in each doublet most likely represents a complex of pepsin with its activation peptide. The pepsinogen components PgC, PgC', PgA6, PgA4 and PgA4', each of them being a precursor of one distinct pepsin isoenzyme or isoform, were converted to pepsins PC, PC', PA6, PA4 and PA4', respectively. The two major isozymogens PgA5 and PgA3, both gave rise to pepsins which were indistinguishable by electrophoresis at pH 3.7 (PA3). Similarly, the two minor components PgA5' and PgA3' (PgA2) were converted to one single isoform of pepsin (PA3' or PA2). As for the group of putative pepsinogen dimers, PgAd, their pepsin counterparts appeared as a cluster of at least 3 distinct isoforms. The analysis of Ferguson plots indicated that their molecular mass is higher than that of other pepsins (65,000 vs. 35,000). These results show that the dimeric structure may be maintained in the pepsin isoforms (PAd) originating from pepsinogen dimers. Besides the pepsin components described above, a group of 8 to 10 minor isoforms (PA1) was also found. These pepsins appeared as a series of the most anodal proteolytic bands after electrophoresis at pH 3.7. The pepsins PA1 moved in polyacrylamide gels as proteins with a molecular mass of about 35,000. Their inactive precursors (PgA1) migrated with and were obscured by PgA2 - PgA5 on electrophoresis at pH 5.5.

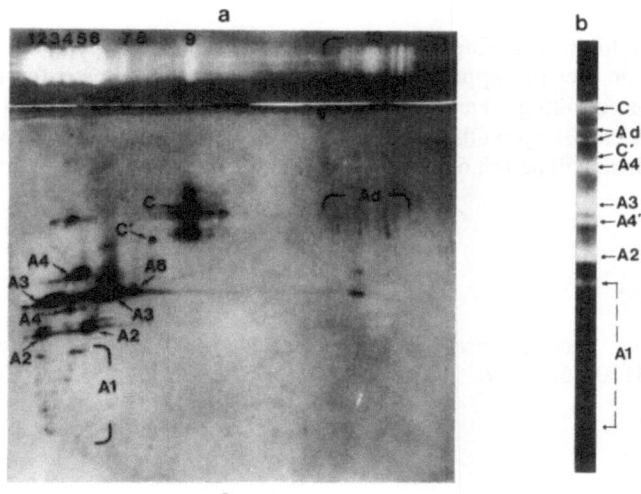

Figure 1. Separation of human pepsinogens and pepsins by discontinuous electrophoresis at low pH. (a) Electrophoresis at pH 5.5 of mucosal extract from body of stomach, followed by activity staining; (b) electrophoresis at pH 3.7 of gastric juice, followed by activity staining; (a, c) bidimensional electrophoresis of extract from fundic gland mucosa; (c) electrophoresis of the activated pepsinogens at 3.7 in the second dimension, followed by silver staining.

Table 1. Conversion of human pepsinogens to pepsins

	Designation	
No[a]	Pepsinogen	Pepsin
	PgA1[b]	PA1[c]
1	PgA2 = PgA3'	PA2 = PA3'
2	PgA3	PA3
3	PgA4'	PA4'
4	PgA4	PA4
5	PgA5'	PA2 = PA3'
6	PgA5	PA3
7	PgA6	PA6
8	PgC'	PC'
9	PgC	PC
10	PgAd[d]	PAd[e]

[a]as numbered in Figure.1a; [b]heterogenous; [c]8-10 components; [d]dimeric, 10-12 components; [e]dimeric, at least 3 components.

CONCLUSION

With its high resolving power for pepsinogens and pepsins, discontinuous electrophoresis at low pH appears to be the technique of choice for studies on the polymorphism of pepsin(ogen) and on the role of pepsin in the pathogenesis of peptic ulcer and related diseases. Based on this method, for instance, a quantitative assay technique can be elaborated, which will permit detailed clinical studies on the secretion of different forms of pepsin.

Acknowledgement

This work was supported by the Swiss National Science Foundation Grants 3.639.087 and 3100-26597.89.

REFERENCES

1. I. M. Samloff, Peptic Ulcer: The Many Proteinases of Aggresion, *Gastroenterology*, **96**:586 (1989).
2. K. Sogawa, Y. Fujii-Kuriyama, Y. Mizukami, Y. Ichihara, and K. Takahashi, Primary Structure of Human Pepsinogen Gene, *J. Biol. Chem.*, **258**:5306 (1983).
3. R. T. Taggart, L. G. Cass, T. K. Mohandas, P. Derby, P. J. Barr, G. Pals, and G. I. Bell, Human Pepsinogen C (Progastricsin). Isolation of cDNA Clones, Localization to Chromosome 6, and Sequence Homology with Pepsinogen A, *J. Biol.Chem.*, **264**:375 (1989).
4. B. Foltmann, Purification, Structure and Activation of Pepsinogens, *in*: "Pepsinogen in Man. Clinical and and Genetic Advances," J. Kreuning, M. Samloff, J. I. Rotter, and A. W. Eriksson, eds.,Alan R. Liss, Inc., New York (1985).
5. V. Walker and W. H. Taylor, Pepsin 1 secretion in chronic peptic ulceration, *Gut* **21**:766 (1980).
6. J. P. Pearson, R. Ward, A. Allen, N. B. Roberts, and W. H. Taylor, Mucus Degradation by Pepsin. Comparison of Mucolytic Activity of Human Pepsin 1 and 3: Implication in Peptic Ulceration, *Gut*, **27**:243 (1986).
7. T. M. Jovin, Multiphasic Zone Electrophoresis. IV. Design and Analysis of Discontinuous Buffer Systems with a Digital Computer, *Ann. N.Y. Acad. Sci.*, **209**:477 (1973).
8. A. Chrambach, "The Practice of Quantitative Gel Electrophoresis," VCH Publishers, Deerfield Beach (1985).
9. J. Defize, G. Pals, J. C. Pronk, R. R. Frants, G. Rimmelzwann, B. D. Westerveld, and A. W. Eriksson, Purification of the Pepsinogen A Isozymogens by Means of High Resolution Ion- Exchange Chromatography. Evidence for Post-translational Modifications, *Scand. J. Clin. Lab. Invest.*, **45**:649 (1985).
10. K. Kishi and T. Yasuda, Newly Characterized Genetic Polymorphism of Uropepsinogen group A (PGA) Using Both Isoelectric Focusing and Immunoblotting, *Hum. Genet.*, **75**:209 (1987).

A HIGHLY INFORMATIVE POLYMORPHISM OF THE PEPSINOGEN C GENE

DETECTED BY POLYMERASE CHAIN REACTION

R. Thomas Taggart,[1] Takeshi Azuma,[1,2] S. Wu,[3] Graeme I. Bell[3]
and Anne M. Bowcock[4]

[1]Department of Molecular Biology and Genetics
Wayne State University School of Medicine
Detroit, Michigan   48201

[2]Department of Preventive Medicine
Kyoto Prefectural University
Kyoto, Japan

[3]Howard Hughes Medical Institute
The University of Chicago
Chicago, Illinois

[4]Department of Pediatrics
The University of Texas
Southwestern Medical Center
Dallas, Texas

INTRODUCTION

Human pepsinogen, the inactive precursor of pepsin, comprises two biochemically and immunologically distinct groups of isozymogens; namely PGA and PGC. PGA has been localized to 11q13 and PGC to 6p21.1→pter.[1-4] Previous studies demonstrated that a region of the human pepsinogen C gene contained a restriction fragment length polymorphism by Southern blot analysis of genomic DNA with a PGC cDNA clone. This RFLP involved a 100 bp insertion or deletion of intron sequence located between exons 7 and 8. Analysis of families segregating for this polymorphism indicated that there is a single human PGC gene located on the short arm of chromosome 6.

A more informative polymorphism detected by polymerase chain amplification was developed by using subclones of a cosmid genomic clone to map the region and to select PCR primers. In this report, we describe the use of sequence data derived from PGC genomic subclones containing the polymorphism for the PCR amplication of the region. We were able to increase the number of alleles detected DNA from 2 to 4 and thereby increase the heterozygosity from 0.26 to 0.49. The strategy we utilized for conversion of the PGC

Structure and Function of the Aspartic Proteinases
Edited by B.M. Dunn, Plenum Press, New York, 1991

insertion - deletion polymorphism detected by Southern blot analysis to a more informative PCR polymorphism may also prove useful for other similar gene markers.

MATERIALS AND METHODS

Southern Analysis of Genomic DNA

The PGC clones used in these studies included PGC 401 (cosmid clone), PGC 3.6 *Eco*RI (genomic subclone), PGC 315 *Msp*I (M13 mp18 genomic subclone), PGC 301 (cDNA clone), PGC 306 (cDNA clone) and have been described previously.[3-5] The PGC cDNA probes employed, PGC 301 and PGC 306, detected exons 2-9 and exons 1-5 respectively.[3,4] The PGC 301 cDNA clone and PGC 315 *Msp*I clone which detect the polymorphic region have been contributed to the American Type Culture Collection. Genomic restriction fragments were subcloned into M13mp18 vectors for sequence analysis. Dideoxy chain termination sequencing procedures were performed using the M13 universal sequencing primer and oligonucleotides specific for the PGC coding sequence.[3,4]

Genomic DNA was isolated, digested with restriction enzymes, separated on agarose gels and subjected to Southern analysis following the hybridization and washing conditions described previously.[3] Probes were prepared from gel slice inserts of genomic and cDNA clones by the random primer method described by Feinberg and Vogelstein.[6] Genomic fragments subcloned into M13mp18 were used to probe genomic Southern blots by specifically labeling the M13 vector using the M13 probe primer method.[7] DNA samples from lymphoblastoid lines of the 40 large reference families obtained from the Human Polymorphism Study Centre (CEPH, Paris) were typed for the SacI RFLP revealed by the PGC 301 probe.

Enzymatic Amplification of Genomic DNA

Genomic DNA (1 µg) was subjected to amplification with 2.5 units of the DNA polymerase from *Thermus aquaticus*. The oligonucleotide primers chosen were upstream, 5'-AGCCCTAAGCCTGTTTTTGG-3', and downstream, 5'-GGCCAGATCT-GCGTGTTTTA-3'. Buffers for the PCR amplification were described previously.[8] The reaction mixture, including 25 pmol of each primer, was subjected to 25 cycles of 30 sec at 94°C, 30 sec at 55°C and 1 minute at 72°C; a final extension at 72°C for 7 min was included. The amplification reactions were performed in a PerkinElmer Cetus Thermocycler (model PCR1000). Alternatively, the reaction mixture was subjected to 25 cycles of denaturation at 94°C for one minute and annealing and extension at 58°C for 6 minutes. One tenth of the reaction mixture (100 µl total volume) was electrophoresed in 2 % regular agarose (Fisher) for 16 - 18 hr at 1 V/cm. Under these conditions, the four alleles were resolved.

DNA Sequencing

Homozygotes for the four different PGC301 alleles detected with PCR were identified among the CEPH individuals. These were: 135002 (allele 1), 1329101 (allele 2), 88401 (allele 3) and 137702 (allele 4). DNA from these individuals was amplified with the PGC oligonucleotides, the amplification products were fractionated in ultra-pure low melting agarose (BRL), excised from the gel, and a second round of amplification was performed with only one oligonucleotide primer. Conditions described for the first round of amplification were performed, but the number of cycles was increased to 40. Single-stranded

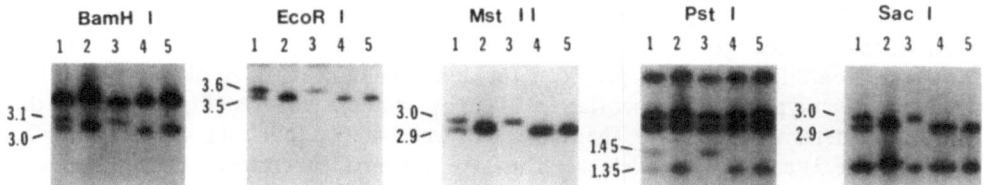

Figure 1. Pepsinogen C insertion - deletion restriction fragment polymorphism. Genomic DNA from five individuals (lanes 1-5) was digested with the restriction enzymes indicated, separated on agarose gels, transferred to nitrocellulose and hybridized to a PGC cDNA probe. Three phenotypes were observed among the individuals; homozygotes for the larger allele (lane 3), homozygotes for the smaller allele (lanes 2,4,5) and a heterozygote containing both the larger and smaller allele (lane 1).

amplification products were purified on a Centricon-30 microconcentrator (Amicon). Single stranded DNA was dried, dissolved in 7 µl of H_2O and sequencing buffer and one pmole of the oligonucleotide primer not used in the single stranded PCR reaction was added. The mixture was placed at 95°C for 10 minutes, 65°C for two minutes and cooled to 50°C for ten minutes. DNA sequencing was then performed as described by Gyllensten and Erlich, with USB sequenase and reagents.[12]

RESULTS AND DISCUSSION

Comparison of PGC alleles detected with Southern blotting and PCR

The PGC RFLP was detected by digestion of genomic DNA with *Eco*RI, *Bam*H I, *Pst* I, *Sac* I and *Mst* II (Oxan I) using cDNA clones as shown in Figure 1. A 100 bp difference in size was found between the allelic fragments with all of these enzymes. The large or small fragments (alleles), revealed by digestion with these enzymes were inherited in a codominant fashion in the families analyzed. These observations indicate that the RFLP involves a rearrangement similar to the deletion - insertion polymorphisms described by Schumm *et al.*[9] in 58 out of 515 polymorphic loci examined. The frequencies of the *Sac*I alleles in 74 unrelated CEPH parents were 0.86 and 0.14. The heterozygosity of this system was 0.26.

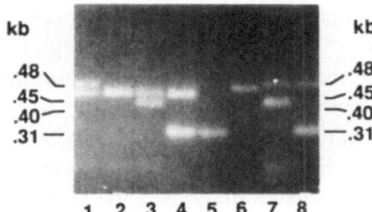

Figure 2. Pepsinogen C polymorphism detected by polymerase chain reaction amplification of genomic DNA. Genomic DNA from eight individuals was subjected to PCR conditions using oligonucleotide primers flanking the polymorphic region of the pepsinogen C gene (lanes 1-8). The sizes of the four alleles detected among a survey of the U.S. individuals is indicated. The three larger alleles (.48, .45, and .40 kb) correspond to the larger allele detected by Southern analysis of genomic DNA (Figure 1). Several additional heterozygous phenotypes were resolved by PCR that appear as homozygous phenotypes on Southern blots (lanes 1, 3 and 7).

When the PGC polymorphic region was amplified with PCR and products were fractionated on agarose gels, they could easily be detected with ethidium bromide staining and UV irradiation. PCR analysis provided for discrimination of three alleles which has previously been typed as the upper (larger) allele by Southern analysis. Figure 2 shows the PGC alleles detected with PCR. The four alleles were: 480 bp (allele 1), 445 bp (allele 2), 400 bp (allele 3) and 315 bp (allele 4). Thus, the 315 bp allele corresponded to the lower SacII fragment, and the 480 bp, 445 bp, and 400 bp alleles corresponded to the upper SacII fragment. Sequence analysis indicated that the size variation involved duplication of repetitive sequences contained within the original MspI 315 genomic subclone. Frequencies for these alleles in 70 unrelated CEPH parents and 16 other unrelated Caucasoids were: .09, .66, .09 and .16 respectively. PCR system increased the heterozygosity of this locus to 0.49. Individuals who appear to be homozygous for these alleles are shown in Figure 2 (alleles 1, 2 and 4 respectively; samples 6, 2 and 5). The segregation of the lower fragment detected with Southern blotting corresponded to the 315 bp PCR allele and the larger RFLP allele corresponded to one or more of the larger three alleles detected with PCR.

Several of the different genotypes that are detected with PCR are shown in Figure 2. This system is specific and reproducible; each allele is equally amplified as long as the number of cycles is not greater than 25 and no other amplification products are detected. Mendelian inheritance of the PGC alleles after PCR amplification of DNA was obtained in CEPH reference families. PGC was localized by linkage analysis to 7 % recombination from HLA DQA and HLA DPB (P >.95), and at least 9 % and 5 % (P >.95) away from D6S8 and D6S28 respectively. These two loci are estimated to be 11 cM and 5 cM respectively from the HLA cluster, and to lie distal to it. PGC lies on the proximal side of HLA and is tightly linked to D6S5 which is estimated to be 15 cM away from the HLA cluster.

Previous studies of the PGC 401 genomic clone suggested that the region containing the polymorphism could be specifically detected within a 315 bp Msp I fragment located the intron between exon 7 and exon 8.[4] We probed genomic DNA with PGC cDNA clones and a series of MspI and TaqI genomic subclones in order to map and identify the potential sequences to be utilized for selection of PCR primers. Only the bands involved in the polymorphism were detected in genomic digests when this 315 bp Msp I genomic fragment was used as a probe whereas the same fragments were detected in genomic DNA digests with a partially overlapping 226 bp Taq I fragment.[4] These mapping studies conducted with M13mp18 subclones utilized the probe primer method to rapidly identify those clones containing the polymorphic region and subsequent sequence analysis provided the information necessary for selection of PCR primers. PCR typing of the polymorphism within the PCG gene is rapid and the results are unambiguous. PCR typing of the CEPH reference families allowed us to demonstrate that this gene is tightly linked to the D6S5 locus which lies proximal to the HLA cluster at a distance of 15 cM. It is interesting to note that the homologous locus in the mouse, urinary pepsinogen locus (Upg-1), is closely linked to H2 on mouse chromosome 17.[10] This suggests that although PGC is contained in a conserved linkage group involving the histocompatibility complex of both species, several rearrangements must have occurred.[10,11]

PCR has been shown to be highly efficient at detecting DNA polymorphisms due to sequence size variation. Alleles detected with PCR have been between 0.4 and 2.5kb, and can differ in size by 50 - 100 base pairs. Several recent studies have described highly informative PCR polymorphisms identified by various dinucleotide repeats that typically involve size variation less than 50 bp. Alleles which cannot be resolved easily by typical Southern analysis can be resolved after specific amplification of regions containing the polymorphism by PCR and subsequent analysis on conventional agarose or on polyacrylamide gels. The method we describe here for conversion of previously identified insertion - deletion RFLP's to PCR polymorphisms may prove useful for the generation of more informative markers.

ACKNOWLEDGEMENTS

This work was supported in part by a grant from the Center for Molecular Biology at Wayne State University, Detroit, Michigan and by NIH Grant GM 28428.

REFERENCES

1. R. T. Taggart, T. K. Mohandas, T. B. Shows and G. I. Bell, A gene complex containing variable numbers of pepsinogen genes is located in the centromeric region of human chromosome 11 and determines the high frequency electrophoretic polymorphism, *Proc. Natl. Acad. Sci. U. S. A.* **82**:6240-6244 (1985).

2. H. Nakai, M. G. Byers, T. B. Shows and R. T. Taggart, Assignment of the pepsinogen gene complex (PGA) to human chromosome 11q13 by *in situ* hybridization, *Cytogenet Cell Genet* **43**:215-217 (1986).

3. R. T. Taggart, L. G. Cass, T. K. Mohandas, P. Derby, P. J. Barr, G. Pals and G. I. Bell, Human pepsinogen C (progastricsin): Isolation of cDNA clones, localization to chromosome 6, and sequence homology with pepsinogen A, *J Biol Chem* **264**:375-379 (1989).

4. G. Pals, T. Azuma, T. K. Mohandas, G. I. Bell, J. A. Bacon, I. M. Samloff, D. A. Walz, P. J. Barr and R. T. Taggart, Human pepsinogen C (progastricsin) polymorphism: Evidence for a single locus located at 6p21.1-pter, *Genomics* **4**:137-145 (1989).

5. T. Azuma, G. Pals and R. T. Taggart, RFLP for the human pepsinogen C gene (PGC), *Nucleic Acids Res* **16**:9372 (1988).

6. A. P. Feinberg and B. Vogelstein, A technique for radiolabelling DNA restriction endonuclease fragments to high specific activity, *Analyt Biochem* **132**:6-13 (1983).

7. J. E. Arrand, Preparation of nucleic acid probes, *in*: "Nucleic Acid Hybridization," B. D. Hames and S. J. Higgins, eds pp. 17-44, IRL Press, Oxford, (1985).

8. A. M. Bowcock, A. Ray, H. Erlich and P. B. Sehgal, Rapid detection and sequencing of alleles in the 3' flanking region of the interleukin-6 gene, *Nucleic Acids Res.* **17**:6855-6864 (1989).

9. J. W. Schumm, R. G. Knowlton, J. G. Braman, D. F. Barker, D. Botstein, G. Akots, V. A. Brown, T. A. Gravius, C. Helms, K. Hsiao, K. Rediker, J. G. Thurston, and H. Donnis-Keller, Identification of more than 500 RFLPs by screening random genomic clones, *Am. J. Hum. Genet.* **42**:143-159 (1988).

10. J. M. Szymura and J. Klein, Linkage of a gene controlling urinary pepsinogen with the major histocompatibility complex of the mouse, *Immunogenet.* **13**:267-271 (1981).

11. A. G. Searle, J. Peters, M. F. Lyon, E. P. Evans, J. H. Edwards and V. J. Buckle, Chromosome Maps of Man and Mouse, III, *Genomics* **1**:3-18 (1987).

12. U. B. Gyllensten and H. D. Erlich, Generation of single-stranded DNA by the polymerase chain reaction and its application to direct sequencing of the HLA-DQA locus, *Proc. Natl. Acad. Sci. U.S.A.* **85**:7652-7656 (1988).

CONSEQUENCES OF INTRAMOLECULAR IONIC INTERACTIONS FOR THE ACTIVATION RATE OF HUMAN PEPSINOGENS A AND C AS REVEALED BY MOLECULAR MODELLING

Ruud A. Bank,[1,2] Robert B. Russell,[2] Gerard Pals,[1]
and Michael N.G. James[2]

[1]Institute of Human Genetics
Free University
P.O. Box 7161, 1007 MC
Amsterdam, The Netherlands

[2]Medical Research Council of Canada
Group in Protein Structure and Function
Department of Biochemistry
University of Alberta
Edmonton, Alberta, Canada T6G 2H7

INTRODUCTION

The multigene family human pepsinogen A (PGA) consists of several isozymogens.[1,2] DNA sequences for PGA-3, 4, and 5 have previously been reported.[3] The substitutions between the various allozymogens (PGA residue numbering according to Evers *et al.*) are

(1) PGA-3: Glu_{43}, Val_{77}, Gln_{207}, Ala_{250}, Leu_{338};
(2) PGA-4: Glu_{43}, Leu_{77}, Lys_{207}, Thr_{250}, Val_{338} and
(3) PGA-5: Lys_{43}, Leu_{77}, Gln_{207}, $Ala_{250,}$ Leu_{338}.

N-terminal amino acid sequences of isozymogens with determined electrophoretic mobilities were reported independently by three different groups.[4-7] The DNA sequence of PGA-3 and 5 could be confirmed, whereas for PGA-4 a different sequence was found.[7] In contrast to PGA, only one gene of human pepsinogen C (PGC) has been found and no genetic variation has been described so far.[8]

Pepsinogens are zymogens as they are the precursors of pepsins. Although zymogens are usually activated upon cleavage by other proteolytic enzymes, pepsinogens only require H^+ ions to initiate the activation process. Under acidic conditions, a conformational change of the N-terminal portion of the zymogen takes place, resulting in an intramolecularly catalyzed hydrolysis that removes an amino acid stretch called the activation segment. The activation segment for PGA is 47 amino acids long whereas that of PGC includes 43 residues. The removal can be either direct (i.e. the cleavage site is the bond between the pepsin moiety and the activation segment, here abbreviated as p-bond) or involves a two-step process (i.e. first a bond in the central region is cleaved - the i-bond - followed by hydrolysis of the p-bond). The activation processes of PGA and PGC were recently studied by Japanese and Danish workers.[5-7,9] Their main results can be summarized as follows.

PGC

After the conformational change has taken place, the next step is cleavage between Phe_{27}-Leu_{28} (residue numbering according to reference 8), leaving an intermediate form of PGC (= i-PGC). The final step is a proteolysis which removes the stretch p28-45. Direct conversion of PGC to pepsin does not take place, neither intra- nor intermolecularly.

PGA

After completing the pH-dependent shift the intramolecular conversion to pepsin can occur in two ways: (1) direct release of the intact activation segment or (2) the formation of an intermediate form (i-PGA) followed by removal of the remaining part of the activation segment. Event (2) is by far the most prominent one. Once a small amount of i-PGA or pepsin is formed, strong intermolecular attack could occur against the remaining pepsinogen and i-PGA.

Interestingly, the rate at which the intermediate form is produced differs not only between PGA and PGC, but also between PGA-3 and PGA-5. The activation rate was found in the order PGC > PGA-3 > PGA-5.

As can be concluded from the primary structure, charge distributions of pepsinogens are highly asymmetric.[3,8] The activation segment is highly positively charged, whereas the pepsinogen moiety has a very low pI. Crystallographic studies on porcine pepsinogen have revealed, that the activation segment is bound to the pepsin portion of the molecule via hydrophobic and ionic attractions. Under acidic conditions, carboxylate groups become protonated resulting in a destabilization of the ionic attractions between the activation segment and pepsin, inducing a conformational change of the former one.[10] In this paper we will discuss the relation between intramolecular ionic interactions and activation rates of the PGA-allozymogens and PGC as revealed by molecular modelling.

MATERIALS AND METHODS

The molecular structures of PGA-3 and PGC were determined with homology modelling, using the crystal structure of porcine pepsinogen A (refined at 1.8 Å resolution, see reference 10) as the template. We have previously described the procedure elsewhere[11] and have discussed the observed molecular properties of the models in relation to their remarkable renal handling. Here we will focus our attention on the intramolecular ionic interactions in an attempt to explain the observed different activation rates of PGA-3, PGA-5 and PGC.

RESULTS AND DISCUSSION

A list of the observed ion-pair interactions, derived from the molecular models of the pepsinogens is given in Table I. It should be stressed that the prediction of these ion-pairs stems from energy minimization in the absence of solvent, hence they are subject to some

Table I. Possible ion-pair interactions in PGA and PGC#

PGA		PGC

Arg_9 . . . Asp_{205}*/Glu_{60}*
Arg_{14} . . . Asp_{58}*
Arg_{15} . . . Glu_{60} Lys_{43} . . . Asp_{49}
Arg_{20} . . . Glu_{19} Lys_{100} . . . Asp_{165}*
Lys_{24} . . . Asp_{289}* Lys_{207} . . . Asp_{206}
Lys_{28} . . . Asp_{25} Arg_{354} . . . Glu_{60}*
Lys_{29} . . . Asp_{25} Arg_{362} . . . Asp_{185}*
His_{30} . . . Glu_{50}*
Arg_{36} . . . Glu_{54}
Lys_{37} . . . Asp_{79}*/Asp_{262}*

Lys_{11} . . . Glu_{15}
Arg_{14} . . . Asp_{54}*/Glu_{332}
Lys_{20} . . . Glu_{19}
Arg_{28} . . . Glu_{25}
Lys_{37} . . . Asp_{75}*/Asp_{260}*
His_{96} . . . Asp_{161}*
Arg_{353} . . . Asp_{349}

#Favorable electrostatic interactions exist in the present models, with the distances between the charged groups ranging from 2.5 to 4.0 Å
*Conserved ion-pair relative to porcine pepsinogen.

degree of uncertainty. The credibility of the models however, is strengthened by the conservation of several ion-pair interactions relative to the template; these residues are marked in Table I with an arterisk. For a discussion of other indicators of the basic correctness of the models in general and the observed potential charge-charge interactions in particular see Bank et al.[11]

Several of the positively charged amino acids are involved in electrostatic interactions with carboxylate groups of either the activation segment or the pepsin portion. At neutral pH, the ionic attractions between the basic residues in the activation segment and the carboxylate groups in the pepsin part bind the activation segment to the pepsin part of the molecule. Interestingly, there are in PGA more possible ion-pairs between the activation segment and pepsin than in PGC. Our simple count of salt bridges between the activation segment and pepsin in the predicted structure of PGA yields an increase of 5 in the number of salt bridges relative to PGC. Although the quantitative estimation of interaction energy between charged residues remains difficult (especially in calculations that include the effects of polarizability and solvent screening) it is very likely that this results in a stronger binding. This might be the reason why PGA is activated at a much slower rate as compared with PGC under similar conditions.

The postulation that an increase in the number of ionic attractions between the activation segment and the pepsin portion is responsible for the delay in the pH-dependent conformational change of the activation segment and hence in the reduction of the activation rate is also confirmed by the model of PGA-3 and PGA-5. At pH 2 PGA-3 activates at a much faster rate than PGA-5.[5-7] According to Evers et al.[3] there are only 2 amino acid substitutions between these allozymogens: Glu_{43} and Val_{77} for PGA-3 whereas for PGA-5 the residues Lys_{43} and Leu_{77} are found. The Val → Leu substitution in the pepsin portion (near the active site) is of a conservative nature. However, the Glu → Lys substitution in the activation segment introduces an extra basic residue in favour of PGA-5. This basic residue gives rise to an extra ion-pair interaction with the pepsin portion: it can interact with Asp_{49}. The Lys_{43}/Asp_{49} interaction in PGA-5 is therefore most likely responsible for the significantly lower activation rate of PGA-5.

Besides differences in the activation rate between PGA-3 and PGA-5, differences are reported for the specificity in the intramolecular attack for the i-bond. For PGA-3 the major cleavage site has been reported between Leu_{23}-Lys_{24} whereas the minor cleavage point is the Asp_{25}-Phe_{26} bond. For PGA-5 the reverse order is seen. It has been postulated that the substitution at residue 43 causes a minor conformational difference between the segments of PGA-3 and 5 resulting in differences in specificity for the cleavage-point of the i-bond.[6] This explanation is not very convincing, as the C-terminal end of the activation segment and the N-terminus of pepsin seem to have few secondary structural features. The weak electron density pattern of the topologically equivalent part of porcine pepsinogen[10] indicates a certain flexibility in this region. A great flexibility in the association of the activation segment of prochymosin, another aspartic proteinase, with its catalytic domain is also observed by McCaman & Cummings[12]: a deletion of seven amino acids at the C-terminus of the propart did not shift the cleavage site for the i-bond. Another hypothesis is that there may be some structural differences in the active site regions of the zymogens that would affect the peptide bond specificity.[7] The Val → Leu substitution (residue 77) near the active site is however of a conservative nature. Although both speculations are theoretically still possible, a more straigthforward explanation is feasible using known events. Initial activation of PGA proceeds mainly via the intramolecular cleavage of the i-bond.[6] The activation process is greatly accelerated by intermolecular attack once a small amount of i-PGA or pepsin is formed. This intermolecular attack occurs mainly at the Asp_{25}-Phe_{26} bond.[6] As the initial activation rate of PGA-5 is greatly decreased as compared with PGA-3, much more time is available for the intermolecular attack resulting in an increase of Phe_{26} i-PGA. In this explanation no differences are seen in the intramolecular cleavage of the i-bond between PGA-3 and 5: both pepsinogens solely hydrolyze the Leu_{23}-Lys_{24} bond.

REFERENCES

1. R. R. Frants, J. C. Pronk, G. Pals, J. Defize, B. D. Westerveld, S. G. M. Meuwissen, J. Kreuning, A. W. Eriksson, Genetics of urinary pepsinogen: a new hypothesis, *Hum. Genet.*, 65:385-390 (1984).
2. R. T. Taggart and I. M. Samloff, Immunochemical, electrophoretical and genetic heterogeneity of pepsinogen I. Characterization with monoclonal antibodies, *Gastroenterology*, 92: 143-150 (1987).
3. M. P. J. Evers B. Zelle, J. P. Bebelman, V. van Beusechem, L. Kraakman, M. J. V. Hoffer, J. C. Pronk, W. H. Mager, R. J. Planta, A. W. Eriksson and R. R. Frants, Nucleotide sequence comparison of five human pepsinogen A (PGA) genes: Evolution of the PGA multigene family, *Genomics*, 4: 232-239 (1989).
4. R. A. Bank, B. C. Crusius, T. Zwiers, S. G. M. Meuwissen, F. Arwert and J. C. Pronk, Identification of a Glu → Lys substitution in the activation segment of human pepsinogen A-3 and 5 isozymogens by peptide mapping using endoproteinase Lys-C, *FEBS Lett.*, 238: 105-108 (1988).
5. B. Foltmann, Activation of human pepsinogens, *FEBS Lett.*, 241: 69-72 (1988).
6. T. Kageyama, M. Ichinose K. Miki, S. B. Athauda, M. Tanji and K. Takahashi, Difference of activation processes and structure of activation peptides in human pepsinogens A and progastricsin, *J. Biochem.*, 105: 15-22 (1989).
7. S. B. P. Athauda , M. Tanji, T. Kageyama and K. Takahashi, A comparative study on the NH_2-terminal amino acid sequences and some other properties of six isozymic forms of human pepsinogens and pepsins, *J. Biochem.*, 106: 920-927 (1989).
8. G. Pals, T. Azuma, T. K. Mohandas, G. I. Bell, J. Bacon, I. M. Samloff, D. A. Walz, P. J. Barr and R. T. Taggart, Human pepsinogen C (progastricsin) polymorphism: evidence for a single locus located at 6p21.1-pter, *Genomics*, 4: 137-145 (1989).
9. B. Foltmann and A. L. Jensen, Human progastricsin. Analysis of intermediates during activation into gastricsin and determination of the amino acid sequence of the propart, *Eur. J. Biochem.* 128: 63-70 (1982).

10. M. N. G. James and A. R. Sielecki, Molecular structure of an aspartic proteinase zymogen, porcine pepsinogen at 1.8 Å resolution, *Nature*, **319**: 33-38 (1986).

11. R. A. Bank, R. B. Russell, G. Pals and M. N. G. James, Model building of human pepsinogens A and C as an aid for explaining their renal handling, Submitted for publication.

12. M. T. McCaman and D. B. Cummings, Unusual zymogen-processing properties of a mutated form of prochymosin, *Proteins Struct. Funct. Genet.*, **3**: 256-261 (1988).

CHARACTERISTICS AND COMPOSITION OF PEPSINS FROM ATLANTIC COD

Asbjorn Gildberg,[1] Ragnar L. Olsen,[1] and Jon B. Bjarnason[2]

[1]Norwegian Institute of Fisheries and Aquculture
9000 Tromso
Norway

[2]Science Institute
University of Iceland
Reykjavik
Iceland

INTRODUCTION

Three pepsins have been purified from the gastric mucosa of Atlantic cod (*Gadus morhua*). The enzymes were denominated Pepsin I and Pepsin IIa and IIb. Pepsin I is a 35.5 kDa neutral protein (pI 6.9), whereas Pepsin IIa and IIb are 34 kDa acidic proteins (pI 4.0 and 4.1, respectively). Apparently, all three pepsins are glycosylated. The cod pepsins resemble bovine cathepsin D in being unable to hydrolyze N-acetyl-L-Phenylalanyl-3,5-diiodo-L-Tyrosine, a dipeptide used as substrate for mammalian pepsins. With hemoglobin as a substrate, they expressed lower catalytic efficiency than porcine pepsin. The apparent substrate affinity of the cod pepsins was substantially higher at pH 3.5 than at pH 2. N-terminal sequence analyses of cod pepsins indicate a significant evolutionary gap between fish and mammalian pepsins. A comparative evaluation of the results obtained indicate that fish pepsins may constitute an intermediate stage in a genetic evolution from cathepsin D to mammalian pepsin.

Pepsins from various fish species have previously been purified (Gildberg, 1988). Most teleost fish species contain two or three major pepsins with an optimum for hemoglobin digestion at pH between 2 and 4.

The Atlantic cod is a teleost fish adapted to temperatures below 10 °C. The present report describes catalytic properties and chemical composition of three pepsins from the gastric mucosa of Atlantic cod. Purification procedure and more experimental details have recently been published elsewhere (Gildberg *et al.,* 1990).

Two of the pepsins have been subjected to N-terminal amino acid sequence analyses, and the structural relationship between these enzymes and mammalian aspartic proteinases is discussed.

Structure and Function of the Aspartic Proteinases
Edited by B.M. Dunn, Plenum Press, New York, 1991

RESULTS AND DISCUSSION

The stomach mucus contained about 5 mg pepsin per g wet weight, and Pepsin I, Pepsin IIa and IIb were obtained in similar amounts. All three enzymes were efficiently inhibited by pepstatin A and in the presence of 10% NaCl, whereas a moderate activity improvement was achieved when 10 mM Ca^{2+} or Cu^{2+} was present. This was consistent with results obtained with porcine pepsin. Substrate specificity studies indicated a closer resemblance of the cod pepsins with bovine cathepsin D than with porcine pepsin. Similar to cathepsin D, the cod pepsins did not hydrolyze the dipeptide pepsin substrate N-acetyl-L-phenylalanyl-3,5-diiodo-L-tyrosine, and opposite to porcine pepsin they digested casein faster than bovine albumin at pH 3.

The kinetic properties for hemoglobin digestion by cod Pepsin I and IIa, porcine pepsin and bovine cathepsin D were compared at different temperatures at pH 2 and 3.5. The cod pepsins expressed higher apparent K_m values than the mammalian enzymes at all temperatures investigated. This was most pronounced with Pepsin I. The K_m values obtained for pepsin I at pH 2 were about 10 times higher than the K_m values determined at pH 3.5. This is the reason why the pH optimum of this enzyme is shifted to a higher value when the substrate concentration is decreased far below K_m (Figure 1). For all the enzymes the lowest K_m was obtained at the lowest temperature (10°C); however, the K_m of the two cod pepsins did not reveal particularly strong temperature dependence such as has been demonstrated with other fish enzymes (Hochachka & Somero, 1968).

The catalytic efficiency of the cod pepsins were not significantly reduced by reduction of the incubation temperature from 37 to 25°C, and only moderately reduced by further reduction of the temperature to 10°C. This certainly reveals some cold adaptation by the cod pepsins.

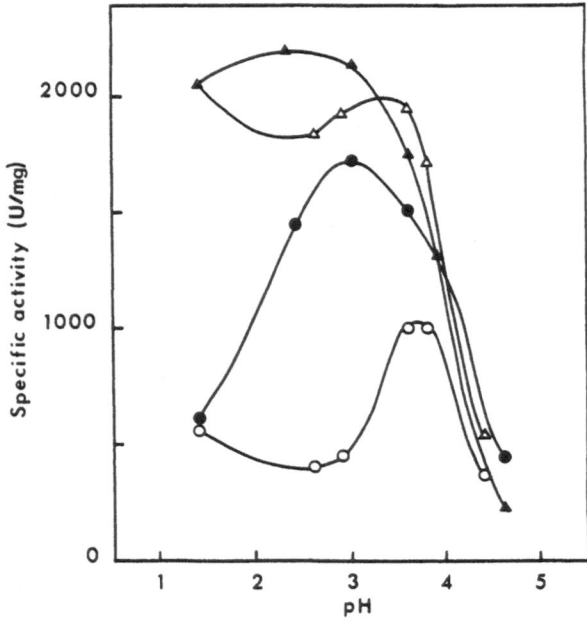

Figure 1. pH dependence of the activity of purified pepsins from Atlantic cod at two substrate concentrations. Open symbols, 1% (150 µM) hemoglobin, filled symbols, 4% (600 µM) hemoglobin. The enzyme incubations were performed for 1 hr at 25°C. Pepsin I, circles; Pepsin IIa, triangles. One unit equals one µmol tyrosine equivalent per hr.

Table 1. N-terminal amino acid sequence of Atlantic cod pepsins compared with porcine pepsin and cathepsin D

Enzyme	Residue numbers
	-2----1---1----2----3----4----5----6----7-----8----9---10---11---12---13---14---15
Cod Pepsin IIa	X---X---X---X--Thr-Glu--X--Met-Thr-Asn-Asp-Ala-Asp-Leu
Cod Pepsin I	Met-Val-Thr-Gly-Glu-Gln-Met-Thr-Asn-Asp-Ala-Asp-Leu-Ala-Tyr-Phe-
Porcine Pepsin[*]	Ala-Leu- Ile-Gly-Asp-Glu-Pro-Leu-Glu-Asn-Tyr-Leu-Asp-Thr-Glu-Tyr-Phe-
Porcine cathepsin D[†]	Gly-Pro- Ile-Pro-Glu-Val-Leu-Lys-Asn-Tyr-Met-Asp-Ala-Gln-Tyr-Tyr-

Enzyme	Residue numbers
	16---17---18--19--20---21---22---23--24---25---26--27---28---29--30--31---32
Cod Pepsin I	Gly-Val- Ile-Ser- Ile- Gly- Thr-Pro-Pro-Glu- Ser-Phe-Lys- Val-Ile-Phe-Asp-
Porcine Pepsin	Gly-Thr- Ile-Gly- Ile-Gly- Thr-Pro-Ala-Gln-Asp-Phe-Thr- Val- Ile- Phe-Asp-
Porcine cathepsin D	Gly-Glu- Ile-Gly-Ile-Gly- Thr-Pro-Pro-Gln-Ser -Phe-Thr- Val-Val-Phe-Asp-

[*]Tang et al., 1973; [†]Huang *et al.*, 1979

Carbohydrate analysis showed that the cod pepsins are glycoproteins. The approximate carbohydrate contents were 1, 2, and 5 % for Pepsin I, Pepsin IIa and IIb, respectively. The latter is similar to the carbohydrate content of cathepsin D from porcine spleen (Huang *et al.*, 1979). The amino acid composition of the cod pepsins, in general, showed similarities with porcine pepsin. However, a relatively high content of basic amino acids in the cod pepsins gave an overall closer resemblance with the amino acid composition of porcine cathepsin D. Pepsin I contains higher amounts of lysine, histidine, and arginine, and lower amounts of aspartic and glutamic acid than Pepsin II. This is obviously the main reason why pepsin I has much higher pI than Pepsin II. The cystine content indicated three possible disulfide linkages per molecule of Pepsin I and four per molecule of Pepsin IIa. The cod pepsins contained about twice as much methionine as porcine pepsin. This resembles the methionine content of porcine cathepsin D.

The N-terminal amino acid sequence analysis revealed about 50 % identity between cod Pepsin I and both porcine pepsin and cathepsin D (Table 1). The short sequence of cod Pepsin IIa, which has been determined, is almost identical to Pepsin I. The sequence comparison indicates a significant evolutionary gap both between the fish pepsins and cathepsin D and between the fish pepsins and mammalian pepsins.

It is well known that certain invertebrates apply cathepsins as digestive enzymes (Vonk & Western, 1984; Gildberg, 1987). Altogether the results presented in this report strengthen the hypothesis presented earlier (Gildberg, 1988) that fish pepsins may constitute an intermediate stage in genetic evolution from cathepsin D to mammalian pepsins.

ACKNOWLEDGEMENTS

We wish to thank Dr. Jay W. Fox for performing the amino acid sequence analysis.

REFERENCES

Gildberg, A., 1987, Purification and characterization of cathepsin D from the digestive gland of the pelagic squid *Iodarodes sagittatus*, *J. Sci. Food Agric.*, **39**:85.

Gildberg, A., 1988, Aspartic proteinases in fishes and aquatic invertebrates, *Comp. Biochem. Physiol.*, **91B**:425.

Gildberg, A., Olsen, R. L., and Bjarnason, J. B., 1990, Catalytic properties and chemical compositon of pepsins from Atlantic cod (*Gadus morhua*), *Comp. Biochem. Physiol.*, **96B**:323.

Hochachka, P. W., and Somero, G. N., 1968, The adaptation of enzymes to temperature, *Comp. Biochem. Physiol.*, **27**:659.

Huang, J. S., Huang, S. S., and Tang, J., 1979, Cathepsin D isozymes from porcine spleens, *J. Biol. Chem.*, **254**:11405.

Tang, J., Sepulveda, P., Marciniszyn, J. Jr., Chen, K. C. S., Huang, W.-Y., Tao, N., Liu, D. & Lanier, J. P., 1973, Amino-acid sequence of porcine pepsin, *Proc. Natl. Acad. Sci. USA*, **70**:3437.

Vonk, H. J. & Western, J. R. H., 1984, Invertebrate proteinases, *in:* "Comparative Biochemistry and Physiology of Enzymatic Digestion," Vonk, H. J., and Western, J. R. H., ed., Academic Press, London.

REDUCTION OF NON-STEROIDAL ANTI-INFLAMMATORY DRUG INDUCED

GASTRIC DAMAGE IN THE RAT BY SOLUBLE PEPSTATIN DERIVATIVES

C. J. Grinham , C. J. Campbell , C. E. Barker and A. Baxter

Department of Biochemistry
Glaxo Group Research Ltd.
Greenford Road , Greenford
Middlesex , United Kingdom

INTRODUCTION

Chronic peptic ulceration is believed to result from an imbalance between aggressive (acid, pepsin) and defensive (bicarbonate, mucus) factors. Acute gastric ulceration often arises as a complication of trauma or surgery. The drugs used to treat peptic ulcers either inhibit acid secretion (e.g. H2-antagonists) or stimulate gastric defences (e.g. prostaglandins, bismuth salts). Neither therapy is ideal because of side effects and problems with recurrence. As there is now mounting evidence that pepsin might be important in the pathophysiology of peptic ulcer disease it was deemed important to study the beneficial effects of pepsin inhibitors.

NSAIDs are the most widely prescribed group of drugs worldwide and although they have significant benefit in alleviating the symptoms of arthritis , relieving pain and reducing inflammation, they can induce serious gastrointestinal ulceration when taken chronically. For example 35% of patients taking these drugs develop gastric lesions within 12 months and it has been calculated that one episode of gastrointestinal hemorrhage occurs for every 6000 NSAID prescriptions.

A rat model of acute NSAID-induced gastric ulceration was developed to study the effects of pepsin inhibitors. Our aim was to correlate the inhibitors' *in vitro* potency against rat pepsins with any *in vivo* protective effects.

METHODS

Rat Model of NSAID-Induced Ulceration

Female random hooded rats (100g), fasted overnight but with access to water *ad libitum*, were dosed (po) with indomethacin (0-56 mM) followed at 4 h by HCl (0.15 M). The rats were sacrificed after a further 2 h, the stomachs excised, opened along the greater curvature, washed in saline (0.9%) and pinned out. Damage was quantified by awarding lesions of 0.5 mm² one point. The erosion index represents the cumulative score for each stomach. A suitable indomethacin concentration was determined to be 56 mM. Subsequent experiments, designed to study the effects of pepsin inhibitors, followed the regimen outlined above except that at 4 h inhibitor or vehicle was dosed. All drugs were administered in 1 ml volumes.

Preparation of Pepsin Inhibitor Solutions (5 mM)

For the *in vivo* studies soluble pepstatin derivatives were dissolved, and pepstatin A suspended, in HCl (0.15 M). Val-D-Leu-Pro-Phe-Val-D-Leu was prepared in HCl (0.15 M) containing DMSO (3%). For the *in vitro* studies the soluble pepstatins were dissolved in water , and pepstatin A and Val-D-Leu-Pro-Phe-Val-D-Leu (1 mM) in aqueous DMSO (10% and 3% respectively).

Inhibition of Rat Pepsins in vitro

Rat gastric mucosal extracts were used as an enzyme source. Mucosal scrapings from excised, washed stomachs were homogenized in phosphate buffered saline, the extract centrifuged (65,000 x g, 0.5 h, 4°C) and the supernatant adjusted to pH 1.7 to activate pepsinogens. Active pepsin was measured at pH 1.7 using [³H]-acetyl hemoglobin substrate as described by Baxter *et al*.

RESULTS AND DISCUSSION

Indomethacin at 56 mM produced an erosion index large enough to enable reliable measurements of lesion reduction to be made. The lesions were dark red and manifested themselves as point or band-like lesions, the latter being prominent on the crests of the mucosal folds. The lesions were located mainly in the corpus but with some in the antrum. The majority were found along the greater curvature. (Diclofenac-induced lesions compared well macroscopically with those induced by indomethacin). The damaging effects of indomethacin were optimized at 4 h and it was found that 2 h was sufficient time for the inhibitors to have a significant effect in reducing the erosion index.

The soluble pepstatin derivative pepstatinyl-Gly-Lys-Lys at 5 mM reduced the erosion index by approximately 60%. Pepstatin A at the same concentration was ineffective suggesting the importance of solubility and perhaps mucosal absorption for *in vivo* efficacy. Figure 1 compares the efficacies of soluble pepstatins in reducing lesions. Both C-terminally extended and N-terminally truncated soluble pepstatins were tested and their *in vivo* protective effects compared with *in vitro* inhibition of rat pepsins. Pepstatinyl-Gly-Lys-Lys, pepstatinyl-Gly-Cysteic acid-Cysteic acid-NH₂ and Lac-Val-Sta-Ala-Sta-NH₂ were all effective in reducing the erosion index ($p \leq 0.002$) and were all potent pepsin inhibitors *in vitro* (IC_{50} = 0.075, 0.032 and 0.13 µM respectively). Lac-Ala-Sta-Ala-Sta-NH₂ was less potent *in vitro* (IC_{50} = 35.5 µM) and not significantly effective *in vivo* ($p = 0.087$). Pepstatin A , although potent *in vitro* (IC_{50} = 0.03 µM) , is poorly soluble and ineffective *in vivo*. Val-D-Leu-Pro-Phe-Val-D-Leu did have a significant effect on the erosion index ($p = 0.033$) although to a lesser extent than the soluble pepstatins. This was reflected in the poor *in vitro* activity (IC_{50} > 100 µM). It is concluded that soluble pepstatin derivatives which inhibit pepsin activity *in vitro* also significantly reduce the erosion index.

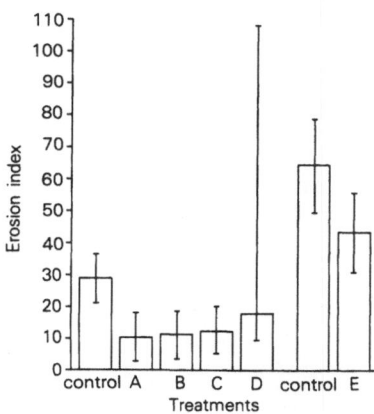

Figure 1. Relative ability of soluble pepstatins to reduce the erosion index. Analysis of variance, after applying the square root transformation to the data, was used to compare treated with control groups of rats (n = 5-7). For each analogue adjusted means were calculated from three (compounds A, B, C and E) or two (compound D) separate experiments. Bars = 95% CI. A) pepstatinyl-Gly-Lys-Lys, B) pepstatinyl-Gly-Cysteic acid-Cysteic acid-NH$_2$, C) Lac-Val-Sta-Ala-Sta-NH$_2$, D) Lac-Ala-Sta-Ala-Sta-NH$_2$, E) Val-D-Leu-Pro-Phe-Val-D-Leu.

A three-fold potentiation of indomethacin-induced lesions by porcine pepsin (3 mg; 7500 U) has also been observed. In this series of experiments, in order to allow potentiation, damage was induced by a sub-maximal dose of indomethacin (28 mM), followed at 4 h by pepsin.

CONCLUSIONS

The rat model of NSAID-induced gastric ulceration has indicated that 1) soluble pepstatin derivatives, which inhibit rat pepsins *in vitro*, are capable of reducing erosions *in vivo* by inhibiting endogenous rat pepsins and that 2) exogenously administered pepsin can potentiate the NSAID-induced damage. Pepsin activity may, therefore, be implicated in the production of NSAID-induced gastric damage.

REFERENCE

Baxter, A., Campbell, C. J., Grinham, C. J., Keane, R. M., Lawton, B. C. & Pendlebury, J. E., 1990, *Biochem. J.*, **267**:665-669.

PRODUCTION OF PROCHYMOSIN IN LACTOCOCCI

Guus Simons, Ger Rutten, Miranda Hornes, Monique Nijhuis
and Martien van Asseldonk

Molecular Genetics Group, Department of Biophysical Chemistry
Netherlands Institute for Dairy Research (NIZO)
P.O. Box 20, 6710 BA EDE
The Netherlands

INTRODUCTION

Lactococci (*Lactococcus lactis* ssp. *lactis* and *L. lactis* ssp. *cremoris*) are used on a large scale as starter cultures for the production of fermented milk products such as cheese, butter and buttermilk. Lactoccocci are cultured in milk based media and, in addition to the milk clotting agent chymosin, are added directly to the cheese vat during cheese manufacturing. It is interesting to use lactococci as a production organism for chymosin since no foreign components or proteins originating from media or the host are added to the cheese vat when chymosin isolated from lactococci is used for cheese manufacturing.

Lactococci are Gram-positive asporogenous prokaryotes with low proteolytic activity and they are able to secrete proteins directly into the medium. This paper summarizes (i) the properties of one of these secreted proteins (i.e., the proteinase *PrtP* of *L. lactis* strain SK11), (ii) the characteristics of the *PrtP* secretion signal and (iii) the various expression and secretion cassettes with bovine prochymosin.

PROTEINASE *PrtP* OF *L. lactis* SK11

Lactococci contain a complex proteolytic system which is required to break down milk proteins during growth in milk. A key enzyme in this process is a cell envelope-located proteinase.[1] The complete nucleotide sequence of proteinase genes of different *L. lactis* strains has been determined[2,3,4] and although they have very different caseinolytic specificities, they are 98% homologous. Their genes encode very large proteins with a calculated molecular weight of more than 200 kDa. These subtilisin-type serine proteinases are synthesized as prepro-proteins and are attached to the cell envelope. The active proteinase does not contain the first N-terminal 187 amino acids as deduced from their nucleotide sequence. At NIZO we determined the complete nucleotide sequence of the proteinase gene (*prtP*) of *L. lactis* ssp. *cremoris* strain SK11, encoding a non-bitter proteinase.[3]

Upstream of the *prtP* gene another gene has been located encoding a *trans*-acting lipoprotein (PrtM) that is required for the activation of the proteinase.[5,6] It is assumed that this protein is involved in the autocatalytic processing of the proteinase. For the construction of secretion vectors we are using the regulatory sequences and coding sequences of the *prtP* gene of *L. lactis* ssp. *cremoris* strain SK11.

THE *prtP* SIGNAL SEQUENCE

Inspection of the N-terminal amino acid sequence of PrtP indicates three potential signal peptidase I cleavage sites at position 19/20, 22/23 and 33/34 respectively.[7] Since it has been generally assumed that signal sequences of Gram-positive organisms like *B. subtilis* and *S. aureus* have signal sequences with a length of 30-35 amino acids,[8] we used a leader sequence of 33 amino acids, consisting of several basic amino acids at the N-terminus followed by a hydrophobic core and a shorter uncharged region ending with an amino acid carrying a small side chain, to test its potential for secretion of a heterologous protein. We fused the α-amylase gene of *Bacillus stearothermophilus* to the *prtP* coding sequence for the first 33 N-terminal amino acids. The construct was transformed to the plasmid-free *L. lactis* strain MG1363[9] by means of electroporation. The *L. lactis* strain MG1363 harboring this gene fusion directed the efficient secretion into the culture medium of active α-amylase[10] indicating that *L. lactis* is able to secrete a heterologous protein into the medium making use of the 33 amino acids leader sequence of the PrtP.

PROCHYMOSIN EXPRESSION PLASMIDS

The various cassettes for expression/secretion of prochymosin and proteinase-prochymosin fusion proteins are shown in Figure 1. As cloning vectors we are using the broad host range pNZ vectors based on the *L. lactis* pSH71 replicon.[11] All constructs were transformed to the plasmid-free *L. lactis* strain MG1363[9] by means of electroporation. No proteinase activity (against casein) could be detected with this strain. The copy number of the plasmids was estimated to be 20 to 30.

The following constructs with prochymosin B cDNA were made:

pNZ410:

For intracellular expression of prochymosin, the prochymosin coding sequence from amino acid 7 (*Bam*HI site) was fused to the coding sequence for the first 9 amino acids of PrtP via a five linker peptide.

pNZ890:

In this construct the prochymosin gene was fused in frame to the putative signal peptide coding sequence of *prtP*. Since prochymosin initiates with an Ala residue, the Ala/Ala signal peptidase I cleavage site at amino acid position 33/34 of the *prtP* coding sequence is retained.

pNZ888 and *pNZ1582*:

These gene fusion constructs were used to determine whether the region downstream of the signal peptide coding region influences the amount of secreted protein. At amino acid position 62/63 another potential signal peptidase I cleavage site was postulated by computer

116

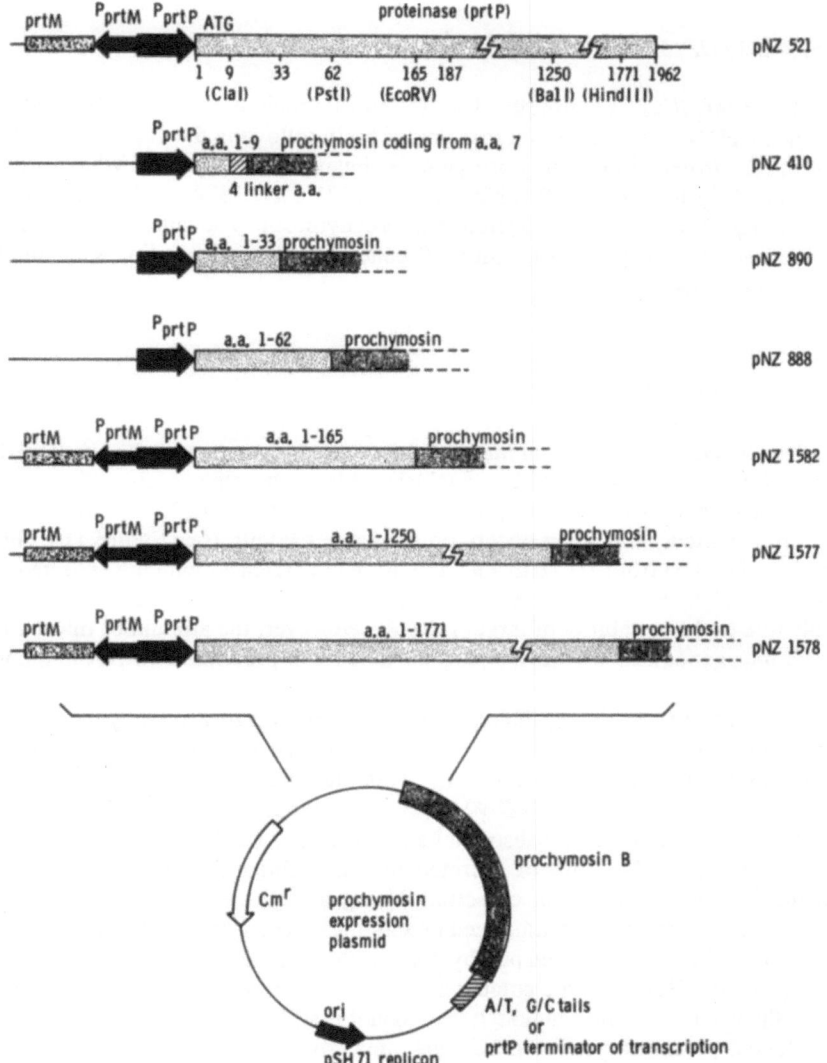

Figure 1. Schematic representation of the construction of the various fusion cassettes for expression/secretion of prochymosin and proteinase-prochymosin fusion proteins in lactococci. P_{prtP} and P_{prtM} represent the promoter sequences of the proteinase gene (*prtP*) and the maturation gene (*prtM*), respectively.

analysis. Fusion of the prochymosin coding sequence with the coding sequence for the first 62 amino acids resulted into plasmid pNZ888. In pNZ1582 the prochymosin is linked to almost the complete pre-pro region (165 versus 187 amino acids) of the *prtP* coding sequence. It has been demonstrated that amino acid extensions of up to 90 aminoacids at the N-terminus of prochymosin do not interfere with the activation process[12].

pNZ1577 and *pNZ1578*:

In *Aspergillus* ssp. improved secretion of prochymosin was obtained when the prochymosin cDNA was fused in frame immediately following the codon for the last amino acid of the *A. awamori* glucoamylase gene.[13] Fusions were made in which prochymosin was preceded by the first 1250 (pNZ1577) or 1771 (pNZ1578) amino acids of the PrtP amino acid sequence to facilitate secretion of prochymosin. The latter construct contained a fusion between almost the full length *prtP* gene (promoter and coding sequence) and the prochymosin cDNA.

PROCHYMOSIN EXPRESSION

The levels of prochymosin mRNA of the various constructs and of *prtP* mRNA were compared by means of Northern blot analysis. All constructs contained the same 5' non-translated sequence. However, in pNZ1582, pNZ1577 and pNZ1578 transcription of the oppositely oriented *prtM* gene occurs on the same cassette (see Figure 1). The mRNA species found ranged from 1.3 kb to 7 kb and were in agreement with the expected size. The amount of mRNA decreased with increasing length of the fusion cassette which is most probably due to the instability of larger mRNA. However, the amount of mRNA produced by the various gene fusions at least equals the amount of proteinase mRNA in the wild-type strain.

Intra- and extracellular prochymosin production was measured by means of Western blot analysis. In *L. lactis* MG1363 harboring pNZ410 only intracellular prochymosin synthesis could be detected, as expected. This material could not be activated to active chymosin at low pH. Plasmid pNZ890 directed the synthesis of extracellular prochymosin although intracellularly still a substantial amount of (unprocessed) prochymosin material could be detected. The ratio of secreted to intracellular prochymosin increased with increasing length of the fusion cassette. Using a leader sequence of 62 amino acids (pNZ888) the majority of the synthesized prochymosin was secreted. No increased secretion efficiency was observed when the prochymosin cDNA was linked to longer leader sequences of the *prtP* gene. However, the amount of secreted material still increased with increasing length of the leader sequence, although the amount is still less when compared to the amount of homologous proteinase produced from the same promoter in the same strain.

ACKNOWLEDGEMENTS

We thank Willem M. de Vos and Roland J. Siezen for critical reading of the manuscript. This work was financed by the Cooperative Rennet and Dye Factory (CSKF) and by the J. Mesdag Foundation, both at Leeuwarden, The Netherlands.

REFERENCES

1. T. D. Thomas and G. G. Pritchard, *FEMS Microbiol. Rev.* **46**:245-268 (1987).
2. J. Kok, K. Leenhouts, A. J. Haandrikman, A. M. Ledeboer and G. Venema, *Appl. Environ. Microbiol.* **54**:231-238 (1985).
3. P. Vos, G. Simons, R. J. Siezen and W. M. de Vos, *J. Biol. Chem.* **264**:13579-13585 (1989).
4. M. Kiwaki, H. Ikemura, M Shimizu-Kadota and A. Hirashima, *Mol. Microbiol.* **3**:359-369 (1989).
5. A. J. Haandrikman, J. Kok, H. Laan, A. Soemitro, A. M. Ledeboer, W. Konings and G. Venema, *J. Bacteriol.* **171**:2789-2794 (1989).
6. P. Vos, M. van Asseldonk, F. van Jeveren, R. J. Siezen, G. Simons and W. M. de Vos, *J. Bacteriol.* **171**:2795-2802 (1989).
7. G. Simons, M. van Asseldonk, G. Rutten, M. Nijhuis, M. Hornes and W. M. de Vos, 5th European Congress on Biotechnology, Copenhagen, Denmark. Munksgaard International Publisher, Copenhagen, in press (1990).
8. G. von Heijne and L. Abrahamsen, *FEBS Lett.* **244**:439-446 (1989).
9. M. J. Gasson, *J. Bacteriol.* **154**:1-9 (1983).
10. G. Simons, M. Nijhuis, A. Braks, G. Rutten and M. van Asseldonk, in preparation (1990).
11. W. M. De Vos, *FEMS Microbiol. Rev.* **46**:281-295 (1987).
12. A. E. Franke, F. S. Kaczmarek, M. E. Eisenhard, K. F. Geoghegan, D. E. Danley, J. R. De Zeeuw, M. M. O'Donnell, M. G. Gollaher, Jr. and L. S. Davidow, *Dev. in Industrial Microbiology*, **29**:43-57 (1988).
13. M. Ward, L. J. Wilson, K. H. Kodama, M. W. Rey and R. M. Berka, *Bio/Technology* **8**:435-440 (1990).

STRUCTURE AND CHROMOSOMAL LOCALIZATION OF THE HUMAN

PROCHYMOSIN PSEUDOGENE

Tönis Örd,[1] Meelis Kolmer,[2] Juhani Jänne,[2] Richard Villems[3]
and Mart Saarma[4]

[1]Institute of Chemical Physics and Biophysics
Estonian Academy of Sciences
Jakobi 2, 202400 Tartu
Estonia

[2]Department of Biochemistry & Biotechnology
University of Kuopio, P.O.B. 6
SF-70211 Kuopio
Finland

[3]Estonian Biocenter
Estonian Academy of Sciences
Jakobi 2, 202400 Tartu
Estonia

[4]Institute of Biotechnology
University of Helsinki
Karvaamokuja 3, SF-00380 Helsinki
Finland

INTRODUCTION

Chymosin (EC 3.4.23.4) is an aspartyl proteinase synthesized in the chief and mucous neck cells of the gastric glands in the fourth stomach of newborn calves (Andren *et al.*, 1982). Chymosin is also found in other mammalian species [for example, in newborn pig (Foltmann *et al.*, 1977; Foltmann *et al.*, 1981), cat (Jensen *et al.*, 1982), seal (Shamsuzzaman & Haard, 1984) and lamb (Baudys *et al.*, 1988; Pungerčar *et al.*, 1990)].

The presence of chymosin in human is a matter of controversy. Malpress (1967) showed that there was no detectable chymosin activity in gastric juice of human infants. Foltmann and Axelsen (1980) found no chymosin-specific immunoreactive protein in stomach of human infants. However, Warner *et al.*, (1986) were able to detect immunologically reactive cells, stained with antichymosin antibodies, in gastric adenocarcinomas of adult patients. Also Henschel *et al.* (1987) claimed that in the gastric

```
hPCψ      MRGLVVFLAVFALSEVNAITRVPLHKGKSLRRALRERRLLEDFLRNHHYAV
bPC       ..C...L.......QGTE...I..Y......K..K.HG......QKQQ.GI
lPC       ..C...L.......QGTE...I..Y...P..K..KE.G......QKQQ.G.
hPGA      .KW.LLLGL.-....CI-MYK...IRK.....T.S..G..K...KK.NLNP


SRKHSSS----GVVASESLTNYLDCQYFGKIYIGTPPQKFTLVFDTGSPDIWVPSVYCN
.S.Y.GF----.E...VP......S.......L.....E..VL.....S.F....I..K
.SEY.GF----.E...VP......S.......L.....E..VL.....S.F....I..K
A..YFPQWEAPTL.DEQP.E....ME...T.G....A.D..V......SNL.......S


SDACQNHQRFDPSKSST*QNMGKSLSIQYGTGSMRGLLGYDTVTVSNIVDPHQTVGLST
.N..K.......R...F..L..P...H......Q.I.............IQ.......
.N..K.......R...F..L..P...R......Q.I.............IQ.......
.L..T..N..N.ED...Y.STSETV..T......T.I......Q.GG.S.TN.IF...E


QEPGDIFTYSEFDGILGLAYPSLASE•SVPVFDNTMQRHLVAQDLFSVYMSR-NDQGSM
.....V...A.......M........Y.I.....M.N..........D.-.G.E..
.....V...A.......M........Y.....M.D.R..........D.-SG....
T...SFLY.AP..........IS.SGAT.....IWNQG..S.......L.ADDQS..V


LTLRAIDLSYYTGSLHWIPMT*QEYWQFTVDSVIIDGVVVACDGGCQAILDTGTSLLVG
...G...P.........V.V.V.Q........T.S.....E...........K...
...G...P.........V.V.L.K........T.S.A....E...........K...
VIFGG..S.......N.V.V.VEG...I....ITMN.EAI..AE.....V.......T.


PGGHILNIQQAIGATAGQYNEFDIDCGRLSSIPTAVFEIHGKKYPLPPSAYTSQDQGFC
.SSD...........QN..G......DN..YM..V...N..M...T............
.SSD...........QN..G......S...M..V...N..M...T.Y.....EEG..
.TSP.A...SD...SENSDGDMVVS.SAI..L.DI..T.N.VQ..V.....IL.SE.S.


TSGFQG----DYSSQQWILGNVFIWEYYSVFDRTNNRVGLAKAV
.....S----ENH..K....D...R........A..L......I
......----ENH.H.....D...R........A..L......I
I.....MNLPTE.GEL....D...RQ.FT....A..Q....PVA
```

Figure 1. Comparison of amino acid sequence deduced for hPCψ with that of bPC (Hidaka et al., 1986), lamb prochymosin (lPC; Pungerčar et al., 1990) and human pepsinogen A (hPGA; Sogawa et al., 1983). Dots signify identity with the amino acid residues found in hPCψ protein. Dashes indicate gaps that have been introduced for alignment. Asterics indicate codons containing frameshift deletion mutations in hPCψ and the solid circle indicates a nonsense mutation in hPCψ at codon 192. The DTG residues corresponding to active site amino acid residues of aspartic proteases are underlined in the hPCψ.

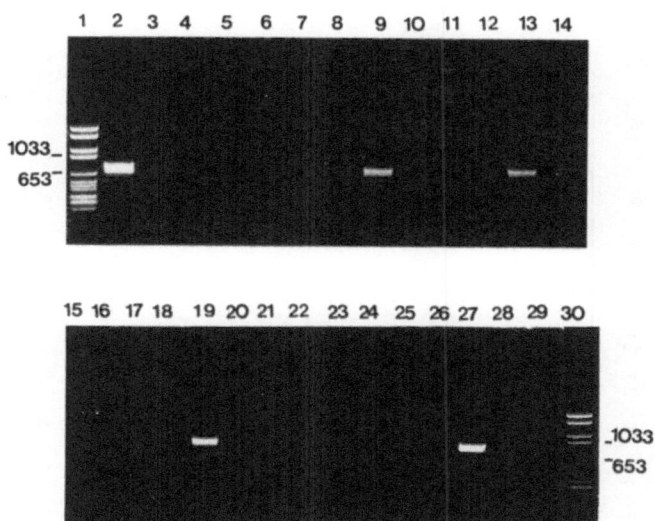

Figure 2. 1.5% agarose gel electrophoresis of PCR products. Lanes 1 and 30, DNA molecular-weight marker (fragment sizes: 2176, 1766, 1230, 1033, 653, 517, 453, 394, 298, 234, 220 and 154 bp); lanes 2-28, BIOS PCRable™ DNAs in the order of appearance in Table I. Single *hPCψ*-specific band is visible on lanes 2, 9,13, 19 and 27. Lane 29, reagent blank (no DNA added to PCR mixture).

juice of five human infants, out of seventeen studied, there was a protein giving a specific immunoreaction with antichymosin antibodies.

We have recently succeeded in the cloning and chromosomal localization of the human genomic region homologous to the bovine prochymosin gene (Örd *et al.*, 1990; Kolmer *et al.*, 1990).

STRUCTURE OF THE HUMAN PROCHYMOSIN PSEUDOGENE

Two different human genomic libraries were screened with bovine chymosin (*bPC*) cDNA probe for the presence of chymosin specific sequences. Overall 5 different recombinant bacteriophage clones containing chymosin specific sequences were isolated. Human sequences homologous to exons of *bPC* gene are distributed in a DNA fragment of 10 kb. Sequencing of the human genomic region revealed a one nucleotide (nt) deletion and a two nt deletion in the human sequence corresponding the bovine prochymosin exons 4 and 6, respectively. There also was a termination codon in the open reading frame corresponding the bovine prochymosin exon 5 (data not shown). Presence of these mutations in human genomic region corresponding to *bPC* gene have rendered it unable to produce a full-length protein homologous to *bPC*. Thus this genomic sequence apparently represents a human prochymosin pseudogene (*hPCψ*). No functional gene has been found so far. The aligned human sequence and the coding part of *bPC* gene share 82 % nucleotide homology. Comparison of the deduced amino acid sequence of *hPCψ* with that of *bPC* gene reveals 76 % homology (identical amino acid at 292 positions out of 381; Figure 1).

Southern blot analysis of human genomic DNA indicates that *hPCψ* is a single copy gene in human genome (data not shown).

However, the possibility remains that an alternative splicing of the human prochymosin pseudogene would lead to the synthesis of a functional prochymosin mRNA.

Table 1. Segregation of Human Chromosomes and *hPCψ* in Human-Rodent Somatic Cell Hybrids

Cell line	Chromosome																								hPCψ
	1	2	3	4	5	6	7	8	9	10	11	12	13	14	15	16	17	18	19	20	21	22	X	Y	hPCψ
(human) UP004	+	+	+	+	+	+	+	+	+	+	+	+	+	+	+	+	+	+	+	+	+	+	+	+	+
324	·	·	·	·	·	·	·	·	·	·	·	·	·	·	·	·	·	+	·	·	·	·	+	·	·
423	·	·	+	·	·	·	·	·	·	·	·	·	·	·	·	·	·	·	·	·	·	·	·	·	·
734	·	·	·	·	·	·	·	·	+	·	·	·	·	·	·	·	·	·	·	·	·	·	·	·	·
750	·	·	·	·	D	·	·	·	·	·	·	·	·	+	+	·	·	+	·	·	·	·	·	·	·
803	·	·	·	+	+	·	·	+	·	·	15	·	·+	+	·	·	·	·	+	·	·	·	+	·	·
860	·	·	15	·	+	·	·	·	·	15	·	·	·	·	·	·	·	·	45	·	+	+	·	·	+
867	60	·	·	·	+	+	·	·	·	·	·	·	+	+	·	·	·	+	+	·	·	·	·	·	·
940	·	·	·	·	+	·	·	·	·	·	·	·	·	·	·	·	·	·	+	+	·	·	·	+	·
212	·	·	·	·	Dg	·	·	·	·	·	·	·	·	·	·	·	·	·	·	·	·	·	·	·	·
507	·	·	+	·	+	·	·	·	·	·	·	+	·	+	·	·	·	·	+	40	+	+	·	+	+
683	·	·	·	·	·	·	·	·	·	·	+	45	+	+	·	·	·	+	+	·	+	25	·	+	+
756	·	·	·	·	D	+	+	·	·	·	·	+	+	45	·	·	·	·	+	+	+	·	·	+	·
811	·	·	·	·	·	·	+	+	·	·	·	·	·	·	·	·	+	+	·	·	·	·	·	·	·
983	·	·	·	·	+	·	·	+	+	+	·	·	·	·	·	·	·	·	+	·	·	·	·	·	·
862	·	·	·	·	+	+	·	+	·	·	·	·	·	·	·	·	+	·	·	·	·	·	·	·	·
909	·	·	·	·	D	·	·	·	·	·	·	·	·	+	+	·	·	·	·	·	+	·	+	·	+
937	+	·	·	·	+	+	·	·	·	·	·	·	+	+	·	·	·	·	·	·	·	·	·	·	·
854	·	+	·	·	D	·	·	+	·	·	·	·	·	·	·	·	·	·	·	·	·	·	·	·	·
904	·	·	·	·	D	+	·	·	·	·	·	+	·	·	·	5	·	·	·	·	+	·	·	+	·
967	·	·	·	·	+	·	·	+	·	·	·	·	·	·	·	+	·	·	·	·	·	·	+	·	·
968	·	·	·	·	+	·	·	+	·	·	·	·	+	·	·	·	·	·	+	·	·	·	·	·	·
1006	·	·	·	55	+	·	+	·	·	·	·	·	+	·	+	·	·	+	+	·	+	·	·	·	·
1049	·	·	·	·	+	·	·	·	·	·	+	·	·	·	·	·	·	·	·	·	·	·	·	·	·
1079	·	·	+	·	D	·	·	·	·	45	·	·	·	·	10	+	·	·	+	·	·	·	·	·	+
1099	+	·	·	·	·	·	·	·	·	·	·	·	+	·	·	·	·	·	+	·	+	+	·	·	·
(hamster)CHO104	·	·	·	·	·	·	·	·	·	·	·	·	·	·	·	·	·	·	·	·	·	·	·	·	·
Percentage discord	4	20	32	24	72	32	24	36	28	28	24	24	24	20	24	28	16	24	20	28	20	16	28	32	

Note. D - Deleted at 5p15.1-5p15.2; Dg - multiple deletions in 5q; Percentage numbers are the percent of the cell population containing the noted chromosome; "+" indicates > 75%

Alternatively, the 5' part of the gene down to the middle of the exon 5 may, in fact, encode a homodimeric aspartyl protease like that described for retroviral proteases (Pearl & Taylor, 1987; Navia *et al.*, 1989).

ASSIGNMENT OF HUMAN PROCHYMOSIN PSEUDOGENE TO CHROMOSOME 1

DNA from 25 different human x hamster somatic cell hybrid and human and hamster genomic DNA (BIOS Corporation, New Haven, CT) were used for amplification of a *hPCψ*-specific sequence. *hPCψ*-specific oligodeoxyribonucleotide primers (5'-ACTGCGGGCCAGTACAATGAG-3' and 5'-GTGGATCTCGAAGACAGCCGT-3') were designed to recognize sequences of the 7th and 8th exon of *hPCψ* on the basis of the lowest homology with the gene of human pepsinogen A (Sogawa *et al.*, 1983), the sequence with putatively highest homology with *hPCψ*. Predicted size of the PCR product was about 650 base pairs.

Reaction conditions were optimized for human genomic DNA and PCR was carried out with a human chromosomal panel. Results are presented on Figure 2 and summarized in Table 1. Single reaction product of the predicted size, was obtained with human genomic DNA control and with DNA from hybrid cell lines 683, 867, 937 and 1099. The lowest percentage of discordancy (4%) was obtained with chromosome 1. All other chromosomes showed clearly higher discordancy, varying from 16% to 72% (Table 1). No PCR product was obtained with hamster genomic DNA or with reagent blank (no DNA added). Incorrect result given by DNA from the cell line 683 (supposedly lacking chromosome 1, Table 1) may not be, after all, false: cell line 683 probably contains human chromosome 1, or fragments of it, not distinguishable by karyotyping.

These results clearly show that *hPCψ* is located on chromosome number 1.

REFERENCES

Andren, A., Björck, L. & Claesson, O., 1982, Immunohistochemical studies on the development of prochymosin- and pepsinogen-containing cells in bovine abomasal mucosa, *J. Physiol.*, **327**:247-254.

Baudyš, M., Erdene,T. G., Kostka, V., Pavlik, M. & Foltmann, B., 1988, Comparison between prochymosin and pepsinogen from lamb and calf, *Comp. Biochem. Physiol.*, **89B**:385-391.

Foltmann, B., Lonblad, P. & Axelsen, N. H., 1977, Demonstration of chymosin (EC 3.4.23.4) in the stomach of newborn pig, *Biochem. J.*, **169**:425-427.

Foltmann, B., Pedersen, V. B., Jacobsen, H., Kauffman, D. & Wybrandt, G., 1977, The complete amino acid sequence of prochymosin, *Proc. Natl. Acad. Sci. USA*, **74**:2321-2324.

Foltmann, B., Jensen, A. L., Lonblad, P., Axelsen, E. & Axelsen, N. H., 1981, A developmental analysis of the production of chymosin and pepsin in pigs, *Comp. Biochem. Physiol.*, **68B**:9-13.

Henschel, M. J., Newport, M. J. & Parmar, V., 1987, Gastric proteases in the human infant. *Biol. Neonate*, **52**:268-272.

Hidaka, M., Sasaki, K., Uozumi, T. & Beppu, T., 1986, Cloning and structural analysis of the calf prochymosin gene, *Gene*, **43**:197-203.

Jensen, T., Axelsen, N. H. & Foltmann, B., 1982, Isolation and partial characterization of prochymosin and chymosin from cat, *Biochim. Biophys. Acta*, **705**:249-256.

Kolmer, M., Örd, T., Alhonen, L., Hyttinen, J.-M., Saarma, M., Villems, R. & Jänne. J., 1990, Assignment of Human Prochymosin Pseudogene to Chromosome 1, *Genomics*, submitted.

Malpress, F. H., 1967, Rennin and the gastric secretion of normal infants, *Nature*, **215**:855-857.

Navia, M. A., Fitzgerald, P. M. D., McKeever, B. M., Leu, C.-T., Heimbach, J. C., Herbar, W. K., Sigal, I. S., Darke, P. L. & Springer, J. P., 1989, Three-dimensional structure of aspartyl protease from human immunodeficiency virus HIV-1, *Nature*, 337:615-620.

Pedersen, V. B., Christensen, K, A. & Foltmann, B., 1979, Investigations on the activation of bovine prochymosin, *Eur. J. Biochem.*, **94**:573-580.

Pearl, L. H. & Taylor, W. R., 1987, A structural model for the retroviral proteases, *Nature*, 329:351-354.

Pungerčar, J., Strukelj, B., Gubensek, H., Turk, V. & Kregar, I., 1990, Complete primary structure of lamb preprochymosin deduced from cDNA, *Nucl. Acids Res.*, 18:4602.

Shamsuzzaman, K. & Haard, N. F., 1984, Purification and characterization of a chymosin-like protease from the gastric mucosa of harp seal (*Pagophilus groenlandicus*), *Can. J. Biochem. Cell Biol.*, 62:699-708.

Sogawa, K., Fujii-Kuriyama, Y., Mizukami, Y., Ichihara, Y. & Takahashi, K., 1983, Primary structure of human pepsinogen gene, *J. Biol. Chem.*, 258:5306-5311.

Warner, T. F., Donnelly, J., Reza-Hafez, G., Renwick, B., Engstrand, D. & Barsness, L., 1986, Immunochemical evidence for gastric proteases in adenocarcinoma of the stomach, *Cancer*, 58:1328-1332.

Örd, T., Kolmer, M., Villems, R. & Saarma, M., 1990, Structure of the human genomic region homologous to the bovine prochymosin-encoding gene, *Gene*, 91:241-246.

AMINO ACID SEQUENCE OF LAMB PREPROCHYMOSIN AND ITS COMPARISON TO OTHER CHYMOSINS

Jože Pungerčar, Borut Štrukelj, Franc Gubenšek, Vito Turk and Igor Kregar

Department of Biochemistry
J. Stefan Institute
Jamova 39, 61000 Ljubljana
Yugoslavia

INTRODUCTION

Chymosin (EC 3.4.23.4) is one of the most important aspartic proteinases used as a milk-clotting enzyme in cheese production. The primary structures of two calf chymosin forms (A and B) are known.[1,2] The only difference in their structures is at position 302 (preprochymosin numbering). More recently, other forms of calf chymosin were also found and sequenced.[3-8] Calf chymosin has been cloned in many laboratories using different expression vector/host systems. Recombinant chymosin is comparable to the wild type enzyme in cheesemaking properties.[9,10] Like traditionally used calf rennet, lamb rennet has also been utilized in cheese making because it has similar high milk-clotting and low proteolytic activities.[11] Recently, we briefly reported on the deduced amino acid sequence of lamb preprochymosin.[12]

In the present paper, we describe cloning and sequencing of a cDNA encoding lamb chymosin and compare its nucleotide and deduced amino acid sequences to sequences of other chymosins.

MATERIALS AND METHODS

Total RNA was isolated from lamb abomasum of a 9 day old animal (*Ovis aries*) by the guanidinium thiocyanate-cesium trifluoroacetate method.[13] Poly(A)$^+$ RNA was purified by affinity chromatography on oligo(dT)-cellulose.[14] Complementary DNA was synthesized by the Amersham cDNA synthesis system and ligated into pUC19 (Pharmacia) using *Eco*RI linkers. A cDNA library was prepared using competent cells *E. coli* DH5α from BRL.[15] The cDNA library was successively screened with two oligonucleotide probes, 5'-d(ATGGCTGAGATCACCAGGATCCCTC)-3' corresponding to the 5'-end of the coding region of calf prochymosin cDNA preceeded by the ATG codon and 5'-d(TCAGATGGCTTTGGCCA)-3' corresponding to the 3'-end of calf chymosin cDNA. The

Structure and Function of the Aspartic Proteinases
Edited by B.M. Dunn, Plenum Press, New York, 1991

longest cDNA insert of about 1300 bp was sequenced using T7 Sequencing kit (Pharmacia) and double-stranded plasmid DNA as a template. DNA manipulations were performed according to standard procedures.[16] Restriction and DNA modifying enzymes were purchased from Boehringer-Mannheim (FRG). Radiolabeled nucleotides were from Amersham (UK). Oligonucleotides were synthesized by the solid-phase phosphoramidite method using an Applied Biosystems 381A synthesizer and purified by electrophoresis on a 20% polyacrylamide gel.

RESULTS AND DISCUSSION

Sequencing data showed that the cDNA insert of 1317 bp contains a full-length copy of lamb preprochymosin mRNA.[12] Restriction sites used for subcloning and sequencing were the same as in calf preprochymosin cDNA[4] with two exceptions: the *Eco*RI site at position 526 was absent but an additional *Pst*I site at position 733 was found. The 5'-noncoding region of lamb preprochymosin cDNA contains 12 nucleotides, whereas the 3'-noncoding region is complete, including a signal for polyadenylation and a 25 residue long poly(A) tail. The open reading frame codes for all 381 amino acids of lamb preprochymosin.

Figure 1 shows an alignment of the deduced amino acid sequence of lamb preprochymosin with other chymosins. From this comparison, lamb chymosin described here can be denoted as the B form because of the presence of glycine at position 302. The cDNA sequence of calf chymosin most similar to our sequence was reported by Maat *et al.*[6] Their deduced amino acid sequence (CC1, Figure 1) was confirmed in our laboratory (J. Pungercar, unpublished). The coding regions for lamb and calf preprochymosin show 95.6% and 94.5% similarity in nucleotide and amino acid sequences, respectively. Lamb prochymosin differs from calf prochymosin[6] in 21 amino acid residues, whereas the 323 amino acid long mature proteins differ in only 17 residues. Both aspartic acid residues in the active site at positions 92 and 274, corresponding to the residues Asp_{32} and Asp_{215} in porcine pepsinogen numbering,[17] are conserved.

However, looking at the differences between lamb and calf proenzymes, the amino acid substitution at position 54 (i.e. p36 residue in porcine pepsinogen numbering) was considered to be of great importance. The lysine residue at position p36 is believed to be essential for the correct folding and autocatalytic cleavage of the proparts at low pH in all zymogens of aspartic proteinases known thus far.[18] The amino acid sequence of lamb prochymosin presented here has at position p36 a glutamic acid residue. Following the proposed activation model based on the conserved Lys_{p36} residue its substitution for an acidic residue would lead to the incorrect folding and inability of the subsequent self-activation of the proenzyme. This rather surprising substitution needs further explanation.

The deduced amino acid sequence of lamb prochymosin is based on the nucleotide sequence of a single full-length cDNA clone. The particular amino acid substitution can be caused by a single nucleotide transition; if the AAG or AAA codon (code for lysine) mutates into the GAG or GAA codon (code for glutamic acid), respectively. There are several possibilities to explain this interesting mutation. First, in lamb two allelic forms of prochymosin could be expressed, the Glu_{p36} and the Lys_{p36} forms. Only the Lys_{p36} one could be self-activated and the Glu_{p36} would be inactive. Second, the lamb could be defective in the synthesis of active chymosin. Third, an incorporation of the wrong nucleotide at the particular site by reverse transcriptase or *E. coli* DNA polymerase I during construction of the cDNA library could be possible.

Until now, the complete amino acid sequences of prochymosins from three species (calf, lamb, man) have been reported. From protein,[2,19] cDNA,[3-6,8] and gene sequencing[7] a few allelic forms of calf prochymosin were found differing at positions 17, 218, 230, 302 and 335 (Figure 1). In man only a single chymosin pseudogene was found with one

```
LC    MRCLVVLLAV FALSQGᐁAEIT RIPLYKGKPL RKALKERGLL EDFLQKQQYG VSSEYSGFᐁGE VASVPLTNYL   70
MLC              ....      .......... .....PX... .....xxxxx xxxxxxxx..  ..........
CC1                        ........S. ......H... .......... I..K......  ..........
CP    .......... ......
CC2              .......S. ......H... .......... I..K......  ..........
CC3              .......S. ......H... .......... I..K......  ..........
CC4              ......T... .......S. ......H... .......... I..K......  ..........
CC5              .......S. ......H... .......... I..K......  ..........
CC6         .... ....*..... .......S. ......H... .......... I..K......  ..........
FC             S... .V.H...S. ......H... .B.xxxxxxx xxxxxxxxDK .SNE..ADF.
HC    ..G...F... ....EVNA. .V.H...S. .R..R..R.. ...RNHH.A  .RKH.SS.V ...ES.....
NM1              ......NT. ......H... G..K.H.VE L.RKR.HS.V ..NEA.....

                                   +
LC    DSQYFGKIYL GTPPQEFTVL FDTGSSDFWV PSIYCKSNAC KNHQRFDPRK SSTFQNLGKP LSIRYGTGSM  140
MLC   ..E..Q....
CC1   .......... .......... .......... .......... .......... .......... ...H......
CC2   .......... .......... .......... .......... .......... .......... ...H......
CC3   .......... .......... .......... .......... .......... .......... ...H......
CC4   .......... .......... .......... .......... .......... .......... ...H......
CC5   .......... .......... .......... .......... .......... .......... ...H......
CC6   .......... .......... .......... .......... .......... .......... ...H......
FC    ..E......I ....Z
HC    .C.......I .....K..LV .....P.I.. ..V..N.D.. Q.......S. ...&..M..S ...Q.....
NM1

LC    QGILGYDTVT VSNIVDIQQT VGLSTQEPGD VFTYAEFDGI LGMAYPSLAS EYSVPVFDNM MDRRLVAQDL  210
CC1   .......... .......... .......... .......... .......... ...I...... .N.H......
CC2   .......... .......... .......... .......... .......... ...I...... .N.H......
CC3   .......... .......... .......... .......... .......... ...I...... .N.H......
CC4   .......... .......... .......... .......... .......... ...I...... .N.H......
CC5   .......... .......... .......... .......... .......... ...I...... .N.H......
CC6   .......... .......... .......... .......... .......... ...I...... .N.H......
HC    R.L....... ......PH.. .......... I...S..... ..L....... .*.......T .Q.H......

                                                                       +
LC    FSVYMDRSGQ GSMLTLGAID PSYYTGSLHW VPVTLQKYWQ FTVDSVTISG AVVACEGGCQ AILDTGTSKL  280
CC1   .......N.. E......... .......... ....V.Q... .......... V......... ..........
CC2   .......D.. E......... .......... ....V.Q... .......... V......... ..........
CC3   .......N.. E......N... .......... ....V.Q... .......... V......... ..........
CC4   .......N.. E......... .......... ....V.Q... .......... V......... ..........
CC5   .......N.. E......... .......... ....V.Q... .......... V......... ..........
CC6   .......N.. E......... .......... ....V.Q... .......... V......... ..........
HC    .....S.ND. ......R... L......... I.M.&.E... .....I.D. V....D.... ........L.

LC    VGPSSDILNI QQAIGATQNQ YGEFDIDCDS LSSMPTVVFE INGKMYPLTP YAYTSQEEGF CTSGFQGENH  350
CC1   .......... .......... .........N ..Y....... .......... S.....DQ.. ......S...
CC2   .......... .......... .........N ..Y....... .......... S.....DQ.. ......S...
CC3   .......... .......... .........N ..Y....... .......... S.....DQ.. ......S...
CC4   .......... .......... .........N ..Y....... .......... S.....DQ.. ......S...
CC5   .......... .........D .........N ..Y....... .......... S.....DQ.. ......S...
CC6   .......... .........D .........N ..Y....... .......... S...G.DQ.. ......S...
HC    ...GGH.... .......AG. .N......GR ...I..A... .H..K...P. S.....DQ.. .......DYS

LC    SHQWILGDVF IREYYSVFDR ANNLVGLAKA I                                          381
CC1   .QK....... .......... .......... .
CC2   .QK....... .......... .......... .
CC3   .QK....... .......... .......... .
CC4   .QK....... .......... .......... .
CC5   .QK....... .......... .......... .
CC6   .QK....... .......... .......... .
HC    .Q.....N.. .W........ T..R...... V
```

Figure 1. Comparison of the deduced amino acid sequence of lamb preprochymosin LC[12] with sequences of other chymosins. The deduced amino acid sequences of different calf chymosin cDNAs CC1, CC3, CC5 and CC6 are from references 6, 3, 4 and 5, respectively. The deduced amino acid sequence of the calf chymosin gene CC4 is from reference 7, confirmed by cDNA cloning.[8] The signal sequence of calf preprochymosin CP is from reference 19 and the sequence of calf prochymosin B CC2 from reference 2, both determined by protein sequencing. The deduced amino acid sequence of the human chymosin pseudogene HC was taken from references 20 and 21. The N-terminal amino acid sequences of prochymosin and chymosin from Mongolian lamb MLC are from reference 24 and those from cat are taken from reference 22. The deduced amino acid sequence of the bovine chymosin homologue NM1 found by cloning of multiple bovine aspartic proteinase genes and sequenced in the region of exon 2 is from reference 28. A dot (·) indicates the same amino acid residue as in lamb preprochymosin, the asterisk (*) denotes a stop codon and "&" indicates deletion of a nucleotide (frameshift mutation). The unknown residue is indicated by "X" and an undetermined residue by "x". Aspartic acid residues in the active site are denoted by a cross (+). Arrowheads indicate the beginning of prochymosin and the mature protein.

nonsense and two frameshift mutations showing 82% nucleotide identity to calf chymosin gene in the coding regions.[20,21] Additionally, the N-terminal amino acid sequences of prochymosins from at least three other animal species were determined: cat,[22] pig[23] and Mongolian lamb.[24] Comparison of the deduced amino acid sequence of prochymosin from lamb (*Ovis aries*) with the N-terminal sequence of prochymosin from Mongolian lamb (*Ovis platyurea*) shows the difference in at least three amino acid residues (Figure 1). However, the suggested disulfide bridge in the propart of Mongolian lamb prochymosin[24] was not found in our deduced amino acid sequence of lamb prochymosin.

A high degree of genetic polymorphism observed in calf chymosin was explained by the presence of a single chymosin gene with many alleles.[7,25,26] Recent data on hybridization of calf chymosin cDNA to bovine genome restriction enzyme fragments showed that chymosin is also a member of a multigene family.[27] A chymosin homologue was found by screening of bovine genomic DNA with chymosin cDNA and sequenced in the region of exon 2.[28] Its deduced protein sequence shows 68% and 64% similarity to calf and lamb chymosin, respectively. Similar data for lamb chymosin have not been reported.

ACKNOWLEDGEMENT

This work was supported by the grants of the Research Council of Slovenia, Federal Secretariat for Development and Krka Pharmaceutical Works. The nucleotide sequence of lamb preprochymosin has been submitted to the EMBL/GenBank database under the accession number X53037.

REFERENCES

1. B. Foltmann, V. B. Pedersen, H. Jacobsen, D. Kauffman and G. Wybrandt, *Proc. Natl. Acad. Sci. USA* **74**:2321-2324 (1977).
2. B. Foltmann, V. B. Pedersen, D. Kauffman and G. Wybrandt, *J. Biol. Chem.* **254**:8447-8456 (1979).
3. T. J. R. Harris, P. A. Lowe, A. Lyons, P. G. Thomas, M. A. W. Eaton, T. A. Millican, T. P. Patel, C. C. Bose, N. H. Carey and M. T. Doel, *Nucl. Acids Res.* **10**:2177-2187 (1982).
4. D. Moir, J. -I. Mao, J. W. Schumm, G. F. Vovis, B. L. Alford and A. Taunton-Rigby, *Gene* **19**:127-138 (1982).
5. K. Nishimori, Y. Kawaguchi, M. Hidaka, T. Uozumi and T. Beppu, *J. Biochem.* **91**:1085-1088 (1982).
6. J. Maat, L. Edens, I. Bom, A. M. Ledeboer, M. Y. Toonen, C. Visser and C. T. Verrips, *in:* "Proc. Third Eur. Congr. Biotechnol.," Vol. 3, pp. 193-199. Verlag Chemie, Weinheim, Germany (1984).
7. M. Hidaka, K. Sasaki, T. Uozumi and T. Beppu, *Gene* **43**:197-203 (1986).
8. T. A.Örd, A. A. Torp and M. I. Kolmer, *Biotekhnologiya* **3**:307-311 (1987).
9. M. L. Green, S. Angal, P. A. Lowe and F. A. O.Marston, *J. Dairy Res.* **52**:281-286 (1985).
10. V. E. Bines, P. Young and B. A. Law, *J. Dairy Res.* **56**:657-664 (1989).
11. M. K. Harboe, *in:* "Aspartic Proteinases and Their Inhibitors," V. Kostka, ed., pp. 537-550. Walter de Gruyter, Berlin, Germany (1989).
12. J. Pungerčar, B. Štrukelj, F. Gubenšek, V. Turk and I. Kregar, *Nucl. Acids Res.* **18**:4602 (1990).
13. P. E. Mirkes, *Anal. Biochem.* **148**:376-383 (1985).
14. M. Aviv and P. Leder, *Proc. Natl. Acad. Sci. USA* **69**:1408-1412 (1972).
15. D. Hanahan, *J. Mol. Biol.* **166**:557-580 (1983).
16. T. Maniatis, E. F. Fritsch and J. Sambrook, "Molecular Cloning. A Laboratory Manual," Cold Spring Harbor Lab., Cold Spring Harbor, NY (1982).
17. M. N. G. James and A. R. Sielecki, *Nature* **319**:33-38 (1986).
18. B. Foltmann, *Biol. Chem. Hoppe-Seyler* **369**:311-314 (1988).
19. Y. Ichihara, K. Sogawa and K. Takahashi, *J. Biochem.* **98**:483-492 (1985).

20. T. A. Örd and J. KH. Piiper, *Dokl. Akad. Nauk SSSR* **301**:761-764 (1988).

21. T. Örd, M. Kolmer, R. Villems and M. Saarma, *Gene* **91**:241-246 (1990).

22. T. Jensen, N. H. Axelsen and B. Foltmann, *Biochim. Biophys. Acta* **705**:249-256 (1982).

23. B. Foltmann, P. Cranwell and A. Turvey, Proc. 18th Linderstroem-Lang Conf. Aspartic Proteinases, p. 82. Elsinore, Denmark. (abstract) (1988).

24. M. Baudys, T. G. Erdene, V. Kostka, M. Pavlik and B. Foltmann, *Comp. Biohem. Physiol.* **89B**:385-391 (1988).

25. W. J. Donnelly, D. P. Carroll, D. M. O'Callaghan and D. Walls, *J. Dairy Res.* **53**:657-664 (1986).

26. E. M. Hallerman, A. Nave, M. Soller and J. S. Bechmann, *J. Dairy Sci.* **71**:3378-3389 (1988).

27. D. J. McConnell, Q. Lu, Y. F. Chen, D. Hughes and D. P. Carroll, *Heredity* **60**:315. (abstract) (1988).

28. Q. Lu, K. H. Wolfe and D. J. McConnell, *Gene* **71**:135-146.(1988).

QUANTUM-CHEMICAL STUDY OF THE CATALYTIC MECHANISM

OF ASPARTIC PROTEINASES

V. K. Antonov and S. L. Alexandrov

M. M. Shemyakin Institute of Bioorganic Chemistry
Academy of Sciences of the U.S.S.R.
Moscow, U.S.S.R

INTRODUCTION

Aspartic proteinases are a group of enzymes functioning by a general base mechanism, the water molecule serving as nucleophile.[1] The X-ray data (for example references 2-4) clearly show that two aspartic acid residues Asp_{32} and Asp_{215} (porcine pepsin sequence) are involved in catalysis. These residues are bound to each other by a hydrogen bond and possess different acid dissociation constants. (pK_a) (for porcine pepsin, 1.5 and 4.7, respectively.[5] The hypothesis on the action mechanism of aspartic proteinases has been put forward[6] which is compatible with most of the available data. The aim of this report is to check the validity of this mechanism by quantum-chemical analysis.

RESULTS AND DISCUSSION

The CNDO/2 method[7] (the "GEOMO" program[8]) has been used. The model system (Figure 1) comprises acetate anion and acetic acid, modelling the Asp_{32} and Asp_{215} residues, respectively. Both carboxyl groups formed the planar conjugate structure with the Asp_{215} $H^{\delta 1}$-proton bound to the Asp_{32} $O^{\delta 2}$-atom by a linear H-bond. Moreover, the C^{γ}-atoms of these acids were fixed rigidly relative to the C'-atom of the N-methylacetamide, modelling a substrate. The substrate amide bond was normal to the plane of the carboxyl conjugate O-atoms; in addition, the amide N-atom was localized in this plane. The carbonyl group of the N-methylacetamide was directed to the carboxyl groups according to the X-ray data.[4] The reacting water O-atom (OW) was localized along the p_z-orbital of the amide carbonyl at the distance equal to 3 Å from the substrate C'-atom (the ground state I). Minimization of the potential energy of this system shows that one of the water protons (H2W) forms a H-bond (near to linear) with the Asp_{32} $O^{\delta 1}$-atom (Table 1). This H-bond has an abnormally short length (1.309 Å) and increases the negative charge on the OW-atom up to -0.432 (this charge in nonbounded H_2O is equal to -0.26[9]). The distance between the $H^{\delta 1}$-proton of Asp_{215} and the carbonyl O-atom was too large (2.99 Å). Therefore this residue can not act as an electrophilic activator of the amide bond. The amide N-atom becomes strongly

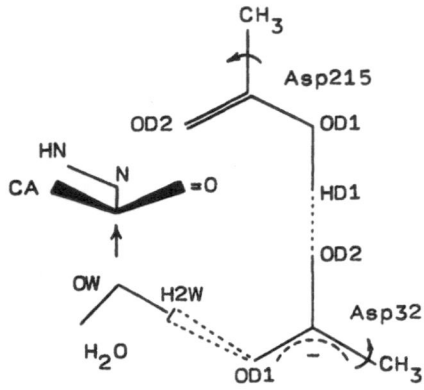

Figure 1. Schematic representation of the model system.

pyramidalized ($\Delta_N = 0.224$ Å; Table 2), and its pK_a value increases up to 3.[9] The direction of the nitrogen lone pair did not require the inversion of the N-atom, as it was supposed in other publications (for example, see reference 4).

Further shortening of the distance between OW- and C'-atoms proceeded without activation barrier with the decrease of the system energy up to 188 kcal per mole. The covalent bond is formed between these atoms (the bond order was equal to 0.79) and the true tetrahedral intermediate (state II) arises, the parameters of which are given in Tables 1 and 2. The effective acceptance of the H2W-proton by the Asp$_{32}$ O$^{\delta 1}$-atom proceeded synchronously with the nucleophilic attack and the stable *syn*-conformation of the protonated Asp$_{32}$ was formed.

Table 1. The parameters of the hydrogen bonds

State	Proton	H-bond donor		H-bond acceptor		Length, Å	Bond order
I	H2W	OW	H_2O	$O^{\delta 1}$	Asp$_{32}$	1.309	0.178
	H$^{\delta 1}$	$O^{\delta 1}$	Asp$_{215}$	$O^{\delta 2}$	Asp$_{32}$	1.567	0.084
II	H2W	$O^{\delta 1}$	Asp$_{32}$	OW	H_2O	1.359	0.165
	H$^{\delta 1}$	$O^{\delta 1}$	Asp$_{215}$	$O^{\delta 2}$	Asp$_{32}$	1.781	--
III	H2W	$O^{\delta 1}$	Asp$_{32}$	OW	H_2O	1.417	0.091
	H$^{\delta 1}$	$O^{\delta 1}$	Asp$_{215}$	=O	N-Methyl-	1.710	0.035
		$O^{\delta 1}$	Asp$_{215}$	N	acetamide	1.425	0.097
IV	H2W	$O^{\delta 1}$	Asp$_{32}$	$O^{\delta 2}$	Asp$_{215}$	1.375	0.165
	H$^{\delta 1}$	$O^{\delta 1}$	Asp$_{215}$	N	N-Methyl-	1.309	0.217
V	H$^{\delta 1}$	N	N-Methyl acetamide	$O^{\delta 1}$	Asp$_{215}$	1.297	0.191

Table 2. The parameters of the N-Methylacetamide amide group

State	Bond length, Å and (Order)			Δ_C, Å	Δ_N, Å	amide N pK$_a$	Total energy, kcal/mole
	C'=O	C'-N	C'-OW				
I	1.272 (1.74)	1.381 (1.13)	3.000 (-)	0.030	0.223	3.4	0.0
II	1.350 (1.12	1.443 (0.91)	1.422 (0.79)	0.436	0.450	6.5	-118.5
III	1.349 (1.11)	1.447 (0.87)	1.417 (0.85)	0.456	0.375	9.6	-120.9
IV	1.354 (1.09)	1.452 (0.84)	1.393 (0.91)	0.477	0.399	7.7	-109.0
V	1.352 (1.12)	1.464 (0.77)	1.389 (0.93)	0.479	0.405	--	-82.7

However, the effective protonation of the amide N-atom in state II by the Asp_{215} $H^{\delta 1}$-proton was sterically impossible, as a result of the very large distance (3.456 Å) between these atoms (although the amide N-atom pK_a value was increased up to 7[9]). The rotation of the Asp_{215} carboxyl group around the C^{β}-C^{γ} bond on 180° allows the interaction of the $H^{\delta 1}$-proton and the N-atom (state III). In this "active" tetrahedral intermediate the $H^{\delta 1}$-proton forms the H-bond both with the carbonyl O-atom and with the amide N-atom. This H-bond is abnormally short (1.425 Å), and reflects the partial orbital overlapping between these atoms (the bond order increased up to 0.097; Table 1). Moreover, this state was energetically favorable by 2.5 kcal per mole relative to the state II, apparently due to the H-bonds with the N-methylacetamide.

The competition between O- and N-atoms for the H-bond formation disappeared in state IV, in which both reactive carboxyl groups were rotated by 180°. In this tetrahedral intermediate the affinity of the $H^{\delta 1}$-proton to the amide N-atom strongly increased, as evidenced by the bond order between these atoms equal to 0.217 (Table 1). Besides, in this state the H-bond between the H2W-proton and the OW-atom is broken, and the one between the carboxyl groups is restored. However, in contrast to state II, in state IV the Asp_{32} residue was the proton donor. The very short length (1.375 Å) of this H-bond increases the affinity of the H2W-proton to the Asp_{215} $O^{\delta 2}$-atom, as evidenced by the bond order between these atoms equal to 0.165 (Table 1). This is important for the further regeneration of the Asp_{32} anionic form. The state IV was unfavorable by 10 kcal per mole. Therefore, the transformation of the state II to the state IV is most likely accompanied by synchronous rotation of both carboxyl groups and decreases the activation barrier up to 6 kcal per mole.

The Asp_{215} carboxyl group in the state IV had the stable *syn*-conformation[10] (in relation to proton affinity). Therefore, the formation of the relatively stable protonated tetrahedral intermediate (state V) was only possible with the synchronous protonation of Asp_{215} $O^{\delta 2}$-atom by the H2W-proton followed by the breakdown of the H-bond between the carboxyl groups. This state was characterized by the unique, abnormally short (1.297 Å) H-bond between the $H^{\delta 1}$-proton and the Asp_{215} $O^{\delta 1}$-atom (Table 1), and this state was energetically unfavorable by 25 kcal per mole due to the absence of the H-bond between acid residues. It seems that the real activation barrier is lower. The *ab initio* calculations of the proton transfer for the simplest system CH_3COOH-CH_3NH_2[11] showed the true value, which did not exceed 5-7 kcal per mole. The C'-N bond order in state V was decreased up to 0.768 (Table 2); this value was typical for the protonated N-atom amides.[9] In this situation no hindrances could prevent the breakdown of state V on the hydrolysis products.

One of the most important peculiarities of a general base catalysis by aspartic proteinases is an absence (on all hydrolysis stages) of the intermediates, in which the substrate is covalently bound with the reactive groups of the active site. In this situation, the pyramidalization of an amide group during the formation of the productive enzyme-substrate complex is determined not only by the secondary enzyme-substrate interactions, but also to a great extent by the abnormal negative charge on the water O-atom.

REFERENCES

1. V. K. Antonov, L. M. Ginodman, L. D. Rumsh, Y. V. Kapitannikov, T. N. Barshevskaya, L. P. Yavashev, A. G. Gurova and L. I. Volkova, Studies of the mechanism of action of proteolytic enzymes using heavy oxygen exchange, *Eur. J. Biochem.*, **117**:195 (1981).
2. M. N. G. James and A. R. Sielecki, Structure and refinement of penicillopepsin at 1.8 Å resolution, *J. Mol. Biol.*, **163**:299 (1983).
3. T. Hoffman and R. S. Hodges, Effect of pH on the activities of penicillopepsin and *Rhizopus* pepsin and a proposal for the productive substrate binding mode in penicillopepsin, *Biochemistry*, **23**:635 (1984).

4. M. N. G. James and A. R. Sielecki, Stereochemical analysis of peptide bond hydrolysis catalyzed by the aspartic proteinase penicillopepsin, *Biochemistry*, **24**:3701 (1985).

5. J. R. Knowles, On the mechanism of action of pepsin, *Phil. Trans. Roy. Soc. London*, **B257**:135 (1970).

6. V. K. Antonov, Specificity and mechanisms of action of proteolytic enzymes, *Bioorgan. Khimiya*, **6**:805 (1980).

7. J. A. Pople and D. L. Beveridge, "Approximate molecular orbital theory," McGraw-Hill, New York (1970).

8. D. Rinaldi, "QCPE Program N290 (GEOMO)," Indiana University, Bloomington (1974).

9. S. L. Alexandrov, V. K. Antonov and P. N. Mel'nikov, Quantum-chemical analysis of the proteolytic enzymes I. The state of cleavable bond of substrates and tetrahedral intermediates, *Mol. Biol.*, **18**:1569 (1984).

10. R. D. Gandour, On the importance of orientation in general base catalysis by carboxylate, *Bioorgan. Chem*, **10**:169 (1981).

11. S. Schneiner, Theoretical studies of proton transfers, *Acc. Chem. Res.*, **18**:174 (1985).

EFFECTS OF VISCOSITY AND SOLVENT DEUTERIUM IDENTIFY MULTIPLE
PARTIALLY RATE-LIMITING STEPS IN THE KINETICS OF PORCINE PEPSIN

Karen L. Rebholz and Dexter B. Northrop

Division of Pharmaceutical Biochemistry
School of Pharmacy
University of Wisconsin
Madison, Wisconsin 53706

INTRODUCTION

The mechanism of porcine pepsin and other aspartic proteinases has been studied for many years; however, the importance of diffusion of reactants to and from the enzyme has not been defined. We have now undertaken viscosity studies to examine the role of diffusion on the rate of catalysis by porcine pepsin. Also, isotope effects have been reported that are difficult to interpret within accepted kinetic mechanisms. Solvent deuterium isotope effect studies were undertaken to show the extent to which proton transfer affects the observed reaction rate and to determine at which step in the reaction proton transfer(s) occur.

EXPERIMENTAL

Porcine pepsin was purchased from Sigma and used without further purification. Leu-Ser-*p*-nitro-Phe-Nle-Ala-Leu-OMe was purchased from Bachem and used without further purification. Lys-Pro-Ala-Glu-Phe-*p*-nitro-Phe-Arg-Leu and Phe-Gly-*p*-nitro-Phe-Phe-OMe were provided by Professor Daniel H. Rich. Buffers were 0.04 and 0.02 M formate with ionic strength adjusted to 0.10 M with KCl. Sucrose and glycerol were added to the desired buffers, and relative viscosities were determined using a viscometer. Percent deuterium in the reaction mixtures was determined by the method of Northrop, *et al.* (in press). Kinetic data were collected on an OLIS modified CARY-14, 15, or 16 interfaced to a computer. Temperature was kept at 25 ± 0.1 °C or 37 ± 0.1 °C with a specially crafted thermojacketed cell holder. Kinetic data were analyzed using RAGASSEK (Regression and Graphical Analysis for Steady-State Enzyme Kinetics) written by DBN and employing the non-linear regression routine of Duggleby (1984).

Structure and Function of the Aspartic Proteinases
Edited by B.M. Dunn, Plenum Press, New York, 1991

RESULTS

Viscosity Effects

The relative viscosity of the assay media was varied from 1.00 to 4.044 by adding up to 1.14 M sucrose. Both V_{max} (40.8 ± 1.4 sec^{-1}) and V/K (4.66 x 10^6 ± 5.1 x 10^5 M^{-1}sec^{-1}) for Lys-Pro-Ala-Glu-Phe-p-nitro-Phe-Arg-Leu decreased as a result of added viscosogen at pH 4.25. No changes in reaction rates were observed with the slow substrate, Phe-Gly-p-nitro-Phe-Phe-OMe, used as a control to verify that the viscosogen did not affect the chemical steps of catalysis. In another control experiment, we found that glycerol (another commonly used viscosogen) inhibited the enzymatic reaction with the slow substrate and produced complex kinetic data which we could not interpret.

pH and Solvent Deuterium Effects

V_{max} and V/K were determined for the substrate Leu-Ser-p-nitro-Phe-Nle-Ala-Leu-OMe in H_2O and D_2O at varying values of pL (i.e., pH or pD). Plots of pL versus log(V_{max}) and log(V/K) are flat in H_2O from pH 3.0 to 5.5. The average values of V_{max} and V/K are 225 ± 6 sec^{-1} and 4.00 x 10^7 ±3.53 x 10^6 M^{-1}sec^{-1}, respectively. These plots are also flat when D_2O is used as the solvent. There is a solvent deuterium isotope effect of 1.502 ± 0.0456 on V_{max}, but no significant effect (1.044 ± 0.320) could be detected on V/K at 25°.

A proton inventory was conducted at pL 4.0 by determining kinetic constants in percentages of D_2O from 0 to 98.99. Non-linear regression analysis of the proton inventory data solving for the number of protons in the Gross-Butler equation (Schowen, 1978) resulted in a value of 1.00 ± 0.38 protons.

DISCUSSION

The results of the viscosity experiments illustrate that the diffusion of reactants to and from the enzyme contributes to the observed reaction rates at low and high concentrations of substrates. The data do not distinguish whether the viscosity effect on V/K is due to decreased rate of association of substrate or decreased rate of dissociation of the first product, but because the value for V/K is much less than expected for a diffusion-controlled reaction (e.g., 10^8 - 10^9 M^{-1}), it seems highly unlikely that the association of substrate is being detected in these experiments.

It is unusual to observe an isotope effect on V_{max} without an accompanying effect on V/K as is reported here for porcine pepsin and as was also reported for recombinant human renal renin (Green et al. 1990).[1] Most solvent isotope effects are seen on both V_{max} and V/K and usually arise from proton transfer during the bond-breaking segment of the reaction. As evidence for this phenomenon we cite proton inventory experiments (Schowen, 1978). While many steps of enzymatic catalysis can potentially generate solvent isotope effects, proton inventories usually detect only one and sometimes two protons as generating significant effects. The inventory experiment similarily supports the involvement of one proton.

V_{max} contains the rate constants for all the segments in an enzymatic reaction after substrate addition while V/K contains only the rate constants through the first irreversible step in the mechanism. As a result, the larger isotope effect is more commonly expressed on V/K due to slow steps downstream that suppress the effect on V_{max}. The relationship between

---

[1] An obvious exception is an enzyme with a ping-pong kinetic mechanism. For proteases, those undergoing the formation and breakdown of an acyl enzyme in which the latter is rate limiting might show kinetic isotope effects on V_{max} but not on V/K.

Scheme I

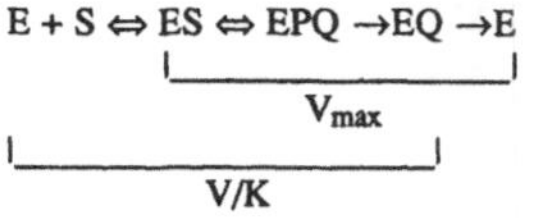

$$E + S \Leftrightarrow ES \Leftrightarrow EPQ \rightarrow EQ \rightarrow E$$

$$V_{max}$$

$$V/K$$

V_{max} and V/K during steady-state is illustrated by the multi-step reaction mechanism shown in Scheme I.

As shown in Scheme II the first irreversible step in pepsin catalysis is likely to be the release of the first product because transpeptidation experiments have shown that catalysis is readily reversible (Clement, 1973).

Scheme II

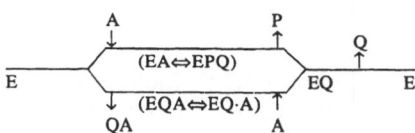

Therefore, both V_{max} and V/K include the bond-breaking steps in the reaction, and if an isotope effect on bond-breaking were being expressed, the effect would be expected to be seen in both V_{max} and V/K. The presence of an effect only on Vmax suggests that the isotope effect does not arise from the bond-breaking steps but rather originates from a later step in the reaction that contributes to V_{max} but not to V/K.

We propose that such a step could be a deprotonation of the enzyme. After both products are released, the enzyme may be in a different form, F, that goes through a proton dependent step to return to form E as shown in Scheme III.

Scheme III

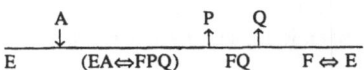

In this model, we propose that after peptide hydrolysis, the catalytically active aspartate carboxyl groups in the active site may be left in a wrong state of protonation that is unable to support a second catalytic turnover. A second explanation is that the release of the second product cannot occur until after a similar proton dependent step has occured as shown in Scheme IV.

Scheme IV

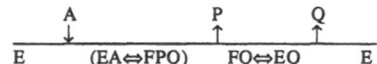

In either case, the partially rate-limiting step that causes the solvent isotope effect comes after the conversion of substrate to products. This finding supports a steady-state kinetic model for pepsin and argues against the rapid equilibrium model commonly invoked with reference to Fruton (1970) who found that K_m equals K_s. It can be shown that K_m and K_s may be equivalent in value but different in origin.

The steady-state model has important implications for the design of aspartic proteinase inhibitors. If the proton dependent step occurs *before* the release of the second product, then EQ is accumulating during steady-state and inhibitors should be designed as analogs to P. Similarly, if the proton dependent step occurs *after* the release of the second product, then form F is accumulating and inhibitors should be designed as analogs to Q.

ACKNOWLEDGEMENT

The authors thank Professor Daniel H. Rich for the gift of two substrates and for partial financial support from grant AR20100 from the National Institutes of Health. KLR was supported by a training grant in cellular and molecular biology (T32 GMO7215).

REFERENCES

Clement, G. E., 1973, Catalytic activity of pepsin, *in*: "Progress in Bioorganic Chemistry," Kaiser, E. T. and Kezdy, F. J., eds, **2**:177-238.

Duggleby, R. G., 1984, Regression analysis of nonlinear Arrhenius plots: An empirical model and a computer program. *Comput. Biol. Med.* **14**:447-455.

Fruton, J. S., 1970, The specificity and mechanism of pepsin action, *Adv. Enzymol. Relat. Areas Mol. Biol.* **33**:401-443.

Green, D. W., Aykent, S., Gierse, J. K. and Zupec, M. E., 1990, Substrate specificity of recombinant human renal renin: Effect of histidine in the P_2 subsite on pH dependence, *Biochemistry* **29**:3126-3133.

Northrop, D. B., Rebholz, K. L. and Benke Marti, K. M., Measuring the gram-atom percent of deuterium in heavy water by absorption spectroscopy, *in:* "Isotope effects in enzyme mechanisms", Cook, P., ed., CRC Press.,In press.

Schowen, K. B. J., 1978, Solvent hydrogen isotope effects, *in:* "Transition states of biochemical processes," Gandour, R. D. and Schowen, R. L., eds., Plenum Press, New York, pp. 225-283.

STRUCTURE - FUNCTION DATABASE FOR ACTIVE SITE BINDING TO THE

ASPARTIC PROTEINASES

Chetana Rao,[1] Paula E. Scarborough,[1] W. Todd Lowther,[1]
John Kay,[2] Brian Batley,[3] Stephen Rapundalo,[3] Sylvester Klutchko,[3]
Michael D. Taylor[3] and Ben M. Dunn[1]

[1]Department of Biochemistry & Molecular Biology
University of Florida
Gainesville Florida   32610

[2]University of Wales
College of Cardiff
Wales
United Kingdom

[3]Parke-Davis Pharmaceutical Research Division
Warner-Lambert Company
2800 Plymouth Road
Ann Arbor, Michigan   48105

INTRODUCTION

Aspartic Proteinases are produced in the human body by a variety of cells. Some of these proteins, for example, pepsin, gastricsin and renin are secreted and exert their effects in the extracellular spaces. Cathepsin D and cathepsin E, on the other hand, are intracellular enzymes.

The objective of our study was to understand the differences in the binding specificities among the family of these enzymes. We examined the dissociation constants of a series of renin inhibitors by quantitating competitive inhibition of hydrolyses of synthetic chromogenic substrates in the pH 3 - 5 range. Renin was studied by RIA at pH 6.0. Inhibitors examined were based on the general structure 4-morpholinylsulfonyl-L-Phe-P_2-(Cyclohexyl)-Ala-[isostere]-P_1-P_2'. The isosteric replacements for the scissile peptide bond included hydroxyethylene, 1,2-diols, 1,3-diols and difluoroketones. P_2 derivatives included His, Asp methyl ester, Lys-ε-(methylamino)thioxomethyl, Gly, allylGly, 2-allylthioGly and various heterocyclic structures.

The synthetic substrates used in the inhibition assays were based on the parent peptide Lys-Pro-Ala-Lys-Phe*Nph-Arg-Leu (Dunn *et al.*, 1986; Pohl & Dunn, 1988). Substitutions at the P_5 and P_4 positions were used to measure the substrate specificity for

Structure and Function of the Aspartic Proteinases
Edited by B.M. Dunn, Plenum Press, New York, 1991

cathepsin E and porcine pepsin. This also will help us understand the specific interactions of amino acid side chains in the active site cleft of these enzymes.

MATERIALS AND METHODS

Series of chromogenic substrates were synthesized by the solid phase method using an Applied Biosystems Model 430A Synthesizer. To study the substrate specificity, synthetic peptides were constructed from the parent peptide Lys-Pro-Ala-Lys-Phe-Nph-Arg-Leu with substitutions at P_4 and P_5 positions.

Stock solutions were made in distilled water and quantified by amino acid analysis. Porcine pepsin and bovine cathepsin D were purchased from Sigma, St. Louis, Missouri. Human pepsin was supplied by Dr. A. P. Ryle of the University of Edinburgh Medical School, Edinburgh, Scotland, U.K. Human cathepsin E was generously given by Dr. I. M. Samloff of the V. A. Medical Center, Sepulvada, CA. Human cathepsin D was prepared from human spleen as described by Afting and Becker (1981). Prof. J. Tang of the Oklahoma Medical Research Foundation provided the human gastricsin.

The active concentrations of cathepsin E and porcine pepsin were determined by titration against the most potent inhibitor of the enzyme. The hydrolyses of the chromogenic substrates were monitored at 37°C, pH range 3-5, in a 0.1 M Na-formate buffer by averaging the decrease in absorbance over a range of 234-324 nm using a Hewlett Packard diode array spectrophotometer. The K_m and V_{max} values were measured using the initial rate at six different substrate concentrations. The K_i values were determined by quantitating competitive inhibition of the chromogenic substrates. Renin was studied by measuring generated angiotensin I with an RIA at pH 6.0.

RESULTS AND CONCLUSIONS FROM INHIBITION STUDIES

The compounds chosen for this study, with the exception of compound 10, were all good renin inhibitors, with K_i values in the nanomolar region (Table 1). A variety of substitutions was chosen to explore selectivity versus the other human aspartic proteinases. Bovine cathepsin D and porcine pepsin are included in Table 1 for comparative purposes. Most of these compounds are potent inhibitors of all enzymes studied (with the exception of compounds 7 - 9 in Table 1), with some exhibiting subnanomolar K_i values.

The 1,3-diol (2) was more selective for human cathepsin D than cathepsin E in contrast to the 1,2-diol (1), which was less potent against both cathepsin D and cathepsin E.

The 3-hydroxyl group yielded a variable effect on the aspartic proteinases. The R stereoisomer (3) was slightly better for renin whereas the S isomer (4) was more effective against porcine pepsin and human cathepsin E.

Substitution of ATM or HIS in the P_2 position (compounds 7 and 9, respectively) produced compounds with little affinity for the aspartic proteinases except for renin. Structures 7 and 9 were the most selective inhibitors for renin found in this study.

An extra methyl group at the P_1' position on the 1,3-diol (compare 2 to 3) produced better inhibition constants for all the aspartic proteinases with the exception of renin. Compound 13 is an excellent across the board inhibitor for all the enzymes tested.

The effect of the P_1' substitution (AEM vs MBA) depends on whether an alcohol or a ketone is present at the P_1' position. Comparing compounds 12 and 13, MBA substitution produced a better inhibitor for porcine pepsin, cathepsin D and cathepsin E, while AEM substitution was superior for human pepsin and gastricsin. In compounds 14 and 15, MBA substitution was preferred for all the enzymes with the exception of cathepsin D.

Table 1. K_i or IC_{50} values (all in nMolar) from inhibition assays with various aspartic proteinases. Renin inhibition values are at pH 7.5. Bovine cathepsin D assays were done with hemoglobin digestion. All other values were derived from studies with chromogenic substrates, as described in the text.

No.	P_1-P_1' Stereo	P_4	P_3	P_2	P_1-P_1'	MR^a	PP^b	HG^c	HP^d	BCD^e	HCD^f	HCE^g
1	SRS	SMO	Phe	ALG	CAD	0.2	16	8	13	9	2.5	17
2	SSR	SMO	Phe	ALG	CAF	0.3	11	19	58	29	0.1	2
3	SSR	SMO	Phe	ALG	CDH	0.2	804	--	77	47.4	5.4	10.7
4	SSS	SMO	Phe	ALG	CDH	0.9	14	--	69	38.9	6.1	1.7
5	SSS	SMO	Phe	ASG(R/S)	CDH	13	50	--	516	142	14.9	2.9
6	SSR	SMO	Phe	ASP(Me)	CDH	0.6	324	--	185	352	12.6	44
7	SSR	SMO	Phe	ATM	CDH	0.3	1223	--	NI	7%@100	208	303
8	SSR	SMO	Phe	GLY	CDH	44	NI	--	NI	1%@100	NI	NI
9	SSR	SMO	Phe	HIS	CDH	1.1	NI	--	NI	2%@1000	NI	851
10	SSS	SMO	Phe	LYS(TMU)	CDH	10%@1nM	54	109	NI	540.5	44.8	14.5
11	SSR	SMO	Phe	ALG	dithiane	0.2	2	--	16	22	0.6	2.2
12	S	SMO	Phe	ALG	FCO-AEM	0.4	2.8	0.9	11	69	18	0.5
13	S	SMO	Phe	ALG	FCO-MBA	0.8	0.3	9	24	19	0.4	0.2
14	SR	SMO	Phe	ALG	FCH-AEM	0.7	NI	--	1239	2250	0.8	21
15	SR	SMO	Phe	ALG	FCH-MBA	0.3	12	13	27	37.0	3.2	0.1

[a]MR = Monkey renin (IC_{50}); [b]PP = porcine pepsin; [c]HG = Human gastricsin; [d]HP = Human pepsin; [e]BCD = bovine cathepsin D; [f]HCD = Human cathepsin D; [g]HCE = Human cathepsin E (determination at pH 4.5); NI = < 50 % inhibition at 4000 nM; Abbreviations: SMO = N-(4-morpholinylsulfonyl); ALG = allylglycine; ASG = 2-allylthioglycine; ATM = 2-amino-4-thiazolylalanine; TMU = (methylamino)thioxomethyl; CAD = 1-(cyclohexylmethyl)-2,3-dihydroxy-5-methylhexyl); CAF = 1-(cyclohexylmethyl)-2,4-dihydroxy-5-methylhexyl); CDH = 1-(cyclohexylmethyl)-2,4-dihydroxyhexyl); FCO = 1-(cyclohexylmethyl)-1-amino-3,3-difluoro-2,4-dioxobutyl; FCH = 1-(cyclohexylmethyl)-1-amino-3,3-difluoro-2-hydroxy-4-oxobutyl; AEM = 2-(4-morpholinyl)ethylamine; MBA = 2-methylbutylamino.

145

Table 2. Substrate specificity for porcine pepsin and human cathepsin E. The structure of the peptide substrates used in this study is X-Y-A-K-F-Nph-R-L, where Nph = *p*-nitrophenylalanine.

X	Y	PORCINE PEPSIN			CATHEPSIN E		
		k_{cat} (sec^{-1})	K_m (μM)	k_{cat}/K_m (μM^{-1}sec^{-1})	k_{cat} (sec^{-1})	K_m (μM)	k_{cat}/K_m (μM^{-1}sec^{-1})
A	P	5.08 ± 0.66	15.0 ± 3.0	0.33 ± 0.08	0.50 ± 0.07	25 ± 3	0.020 ± 0.004
D	P	7.13 ± 0.73	15.0 ± 2.0	0.48 ± 0.08	0.40 ± 0.10	46 ± 16	0.010 ± 0.004
R	P	8.40 ± 0.84	33.0 ± 7.0	0.25 ± 0.06	0.50 ± 0.09	39 ± 11	0.010 ± 0.004
S	P	1.82 ± 0.33	9.0 ± 3.0	0.20 ± 0.08	0.30 ± 0.08	50 ± 12	0.007 ± 0.002
L	P	6.28 ± 0.59	9.0 ± 0.8	0.70 ± 0.09	0.60 ± 0.07	29 ± 2	0.020 ± 0.003
K	P	64.71 ± 10.24	326 ± 58	0.20 ± 0.05	10.51 ± 1.61	150 ± 27	0.070 ± 0.020
K	A	50.14 ± 11.92	168 ± 30	0.30 ± 0.09	1.60 ± 0.20	164 ± 10	0.010 ± 0.001
K	D	7.57 ± 0.93	208 ± 29	0.04 ± 0.01	3.0 ± 0.60	247 ± 68	0.010 ± 0.004
K	R	8.97 ± 1.09	238 ± 20	0.04 ± 0.01	0.50 ± 0.06	116 ± 10	0.040 ± 0.006
K	S	10.61 ± 1.09	124 ± 17	0.09 ± 0.01	2.10 ± 0.40	252 ± 65	0.008 ± 0.003
K	L	10.0 ± 1.11	159 ± 68	0.06 ± 0.03	0.50 ± 0.20	58 ± 38	0.009 ± 0.006

The best inhibitor for cathepsin D was compound 2, which binds 20 times more tightly to cathepsin D than to cathepsin E. For cathepsin E, compounds 13 and 15 both displayed approximately a 30-fold preference for cathepsin E versus cathepsin D.

RESULTS AND CONCLUSIONS FROM SUBSTRATE SPECIFICITY STUDIES

All substrates tested were cleaved at readily measurable rates by both porcine pepsin and human cathepsin E. We have made this comparison of porcine pepsin and human cathepsin E since we are constructing a model of cathepsin E based on the crystal structure of porcine pepsin (Abad-Zapatero et al., this volume).

Lysine in the P_5 position is favored with the highest k_{cat} values of 65 sec^{-1} and 11 sec^{-1} for porcine pepsin and cathepsin E respectively. Increasing the side chain by two atoms as in arginine at position P_5 decreases the k_{cat} 8 fold for porcine pepsin and 20 fold for cathepsin E. The lowest k_{cat} was seen with serine at position P_5 for both the enzymes. Thus, it may be possible that specific side chain interactions at a remote position such as P_5 help optimize catalysis.

The highest k_{cat} values observed were for proline at position P_4 for both the enzymes. The k_{cat} value for alanine was comparable with proline for porcine pepsin. Absence of the ring structure of proline by substitution with leucine decreased the k_{cat} 20 fold for cathepsin E and 6 fold for porcine pepsin. The lowest k_{cat} value was observed for arginine and leucine at position P_4 for cathepsin E in contrast to aspartate at position P_4 for porcine pepsin. No side chain interactions seem to be involved at this position for porcine pepsin, implying that the conformational effect of the Pro residue in the parent peptide is the dominating factor contributing to the rapid hydrolysis observed.

ACKNOWLEDGEMENTS

We gratefully acknowledge the many contributions to our program of peptide synthesis and protein chemistry by Mr. Benny Parten and Mrs. Alicia Alvarez. We are also grateful to our colleagues around the world who have provided samples of enzymes for analysis.

REFERENCES

Abad-Zapatero, C., Rydel, T. J., Neidhart, D. J., Luly, J. and Erickson, J. W., Inhibitor binding induces structural changes in porcine pepsin, in: "Structure and Function of the Aspartic Proteinases: Genetics, Structures and Mechanisms," Ben M. Dunn, ed., Plenum Press, New York.

Afting, E. G. and Becker, M. L., 1981, Two-step affinity-chromatographic purification of cathepsin D from pig myometrium with high yield, Biochem. J. 197:519-522.

Dunn, B. M., Jiminez, M., Parten, B. F., Valler, M. J., Rolph, C. E. and Kay, J., 1986, A systemic series of synthetic chromophoric substrates for aspartic proteinases, Biochem. J. 237:899-906.

Pohl, J. and Dunn, B. M., 1988, Secondary enzyme-substrate interactions: Kinetic evidence for ionic interactions between substrate side chains and the pepsin active site, Biochemistry 27:4827-4834.

INTRODUCTION TO FUNGAL PROTEINASES AND EXPRESSION IN FUNGAL SYSTEMS

Michael Ward and Katherine H. Kodama

Genencor International
180 Kimball Way
South San Francisco, California   94080

INTRODUCTION TO FUNGAL ASPARTIC PROTEINASES

Aspartic proteinase production is widespread among the fungi and examples of both secretion and localization in subcellular organelles are known. For example, the yeast *Saccharomyces cerevisiae* produces aspartic proteinases which reside in the vacuole (proteinase A, analagous to mammalian cathepsin D, Ammerer *et al.*, 1986; Woolford *et al.*, 1986) and within the secretory apparatus (product of the YAP3 gene, Egel-Mitani *et al.*, 1990). It is likely that other fungi produce equivalent proteinases. In addition, yeasts secrete aspartic proteinases such as the barrier proteinase of *S. cerevisiae* (MacKay *et al.*, 1988; this volume) and the PEP1 proteinase of *Saccharomycopsis fibuligera* (Hirata *et al.*, 1988). The genes for all these proteinases have been cloned and DNA sequences determined from which the amino acid sequences have been derived. Many filamentous fungi are also known to secrete aspartic proteinases and these, as well as secretion of heterologous aspartic proteinase by these fungi, will be the major focus of this chapter.

The secreted filamentous fungal proteinases that are active at acid pH can be divided into two classes; those which are pepstatin-sensitive (i.e., true aspartic proteinases or fungal pepsins) and those which are insensitive to this aspartic proteinase inhibitor (carboxyl proteinases). The best studied members of the pepstatin-insensitive class are those from *Scytalidium lignicolum* (Murao & Oda, 1985). Three proteinases (A1, A2, and B), active at acid pH and insensitive to pepstatin, are secreted by this fungus. Proteinases A1 and A2 are also insensitive to the other aspartic proteinase inhibitors diazoacetyl-DL-norleucine methylester (DAN) and 1,2-epoxy-3-(*p*-nitrophenoxy)propane (EPNP). However, proteinase B is sensitive to EPNP but not to DAN (Oda & Murao, 1974; Oda *et al.*,1975). The complete amino acid sequence has been determined for proteinase B (Maita *et al.*, 1984) and modification with various inhibitors has indicated that Glu_{53} and Asp_{98} are active site residues (Tsuru *et al.*, 1986; 1989a; 1989b). The amino acid sequences around these two residues is similar to that around the two active site aspartate residues of other aspartic proteinases. Takahashi *et al.* (this volume) have characterized a pepstatin-insensitive carboxyl proteinase from *Aspergillus niger* var. *macrosporus*. The protein consists of two non-covalently bound polypeptide chains of 39 and 173 residues. The DNA sequence of the cloned gene shows that it has homology with proteinase B from *S. lignicolum* and is initially

Table 1. Source of sequences of secreted filamentous fungal aspartic proteinases

SPECIES	SOURCE OF SEQUENCE	REFERENCE
Aspergillus awamori	Protein	Ostoslavskaya *et al.*, 1986
	Genomic DNA	Berka *et al.*, 1990
Aspergillus oryzae	Genomic DNA	C. L. Carmona, pers. commun.
Rhizopus chinensis	Protein	Takahashi, 1987
" "	Partial cDNA	Delaney *et al.*, 1987
		Chen *et al.*, 1991
Rhizopus niveus API	Genomic DNA	Horiuchi *et al.*, 1988
" " APII	Genomic DNA	Unpublished
		(see Horiuchi *et al.*, 1990)
Rhizomucor miehei	Protein	Bech & Foltmann, 1981
" "	Genomic DNA	Gray *et al.*, 1986
Rhizomucor pusillus	Protein	Baudys *et al.*, 1988
" "	Genomic DNA	Tonouchi *et al.*, 1986
Endothia parasitica	Protein	Barkholt, 1987
Penicillium janthinellum	Protein	Cunningham *et al.*, 1976
Irpex lacteus	Partial cDNA	Kobayashi *et al.*, 1989

synthesized as a single polypeptide with an 11 residue spacer region between the two chains of the mature enzyme.

Some of the secreted, pepstatin-sensitive aspartic proteinases of filamentous fungi are of commercial interest. Most notable are the enzymes (fungal rennets) produced by *Rhizomucor miehei* , *R. pusillus* and *Endothia parasitica* which are used as substitutes for bovine chymosin to clot milk during cheese manufacture. The amino acid sequences of several secreted, pepstatin-sensitive aspartic proteinases from filamentous fungi are known, either as a result of protein sequencing or by deduction from the nucleotide sequence encoding the enzyme (Table 1). The sequences of the mature proteinases are aligned for comparison alongside bovine chymosin and porcine pepsin in Figure 1. There are obvious similarities between all these sequences particularly in the regions of the active site aspartates. The three-dimensional structures of the aspartic proteinases from *Penicillium janthanellum* (penicillopepsin; James & Sielecki, 1983), *Endothia parasitica* (endothiapepsin; Jenkins *et al.*, 1977; Blundell *et al.*, 1985), and *Rhizopus chinensis* (rhizopuspepsin; Bott *et al.*, 1982; Suguna *et al.*, 1987) have been deduced by X-ray crystallography and share obvious homology with those of porcine pepsin (Andreeva *et al.*, 1984) and bovine chymosin (Gilliland *et al.*, 1990). A common ancestry for these enzymes seems, therefore, well established. For recent reviews of this subject area see Tang and Wong (1987) and Thompson (1990).

The *Aspergillus awamori* (*A. niger* var. *awamori*) aspartic proteinase (aspergillopepsin A) gene was cloned using oligonucleotide probes (Berka *et al.*, 1990) designed according to the known protein sequence (Ostoslavskaya *et al.*, 1986). Using this cloned gene it was possible to clone the equivalent gene from *A. oryzae* by cross-hybridization (C. L. Carmona, personal communication). In this latter species two forms of aspartic proteinase have been recognized. These are indistinguishable immunologically and by amino acid composition but can be differentiated electrophoretically by mobility (Tsujita & Endo, 1976). It has been suggested that the two forms are isoenzymes which differ only by their non-covalent association with carbohydrate (Tsujita & Endo, 1977). The published amino acid composition is very similar to the amino acid composition deduced from the nucleotide sequence shown in Figure 1. The cloned *A. oryzae* gene sequence was used as a probe for Southern hybridization analysis of total DNA from this species. Even under non-

```
AAAP:  --SKGSAVTTP-QNND-EEYLTPVTVGKS--TLHLDFDTGSADLWVFSDELPSSEQTG---HDL-YTP--SSSATKLSGYTWD----ISYGDGSSASGDVYRDTVTVGG
AOAP:  --GHGTVVTSP-EPNDI-EYLTPVNIGGT--TLNLDFDTGSADLWVFSEELPKSEQTG---HDV-YKP--SGNASKIAGASWD----ISYGDGSSASGDVYQDTVTVGG
RCAP:  -AGVGTVPMTDY-GNDV-EYYGQVTIGTPGKKFNLDFDTGSSDLWIAST-LCTN----CGSRQTK-YDPKQSSTYQAD-GRTWS---ISYGDGSSASGILAKDNVNLGG
RNAP:  -ASGSVPMVDY-ENDV-EYYGEVTVGTPGIKLKLDFDTGSSDLWVFSTE-LCSS----CSNSHTK-YDPKKSSTYAAD-GRTWS---ISYGDGSSASGILATDNVNLGG
PJAP:  -AASGVATNTPT-AND-EEYITPVTIGGT--TLNLNFDTGSADLWVFSTELPASQQSG---HSV-YNP--SATGKELSGYTWS---ISYGDGSSASGNVFTDSVTVGG
EPAP:  --STGSATTTPIDSLDD-AYITPVQIGTPAQTLNLDFDTGSSDLWVFSSETTASEVDG---QTI-YTPSKTTAKLLSGATWS---ISYGDGSSSSGDVYTDTVSVGG
RMAP:  AAADGSVDTPGYYDFDLEEYAIPVSIGTPGQDFLLLFDTGSSDTWVPH-KGCTKSEGCV--GSRFFDPSASSTFKA----TNYNLNITYGTGGAN-GLYFEDSIAIGD
RPAP:  AEGDGSVDTPGLYDFDLEEYAIPVSIGTPGQDFYLLFDTGSSDTWVPH-KGCDNSEGCV--GKRFFDPSSSSTFKE----TDYNLNITYGTGGAN-GIYFRTSITVGG
ILAP:  --AAGSVPATNQL-VD---YVVNVGVGSPATTYSLLVDTGSSNTWLGAD-----------------KSYVKTSTSSATSDKVSVTYGSG-SFSGTEYTDTVTLGS
                *          *      *      *         *       *             *  ****  **                           ** *    *
CHYM:  ---GEVASVPLTNYLD-SQYFGKIYLGTPPQEFTVLFDTGSSDFWVP-SIYCK-SNACK--NHQRFDPRKSSTFQN----LGKPLSIHYGTGSMQ-GILGYDTVTVSN
PPEP:  -----IGDEPLENYLD-TEYFGTIGIGTPAQDFTVIFDTGSSNLWVP-SVYCS-SLACS--DHNQFNPDDSSTFEA----TSQELSITYGTGSMT-GILGYDTVQVGG

AAAP:  VTTNKQAVEAASKISSEFVQNTAN-----DGLLGLAFSSINTVQPKAQTTFFDTVKSQLDSPL------FAVQLKHDAP-------GVYD--FGYIDDSKYTGSITYT
AOAP:  VTAGGQAVEAASKISDQFVQDKNN-----DGLLGLAFSSINTVKPKPQTTFFDTVKDQLDAPL------FAVTLKYHAP-------GSYD--FGFIDKSKFTGEL---
RCAP:  LLIKGQTIELAKREAASFANGPN-----DGLLGLGFDTITTVRGVKTPMDNLISQGLISRPI------FGVYLGKASNGGG----GEYI--FGGYDSTKFKGSLTTV
RNAP:  LLIKKQTIELAKRESSAFATDVI-----DGLLGLGFNTITTVRGVKTPVDNLISQGLISRPI------FGVYLGKQSNGGG----GEYI--FGGYDSSKFKGSLTTV
PJAP:  VTAHGQAVQAAQQISAQFQQDTNN-----DGLLGLAFSSINTVQPQSQTTFFDTVKSSLAQPL------FAVALKHQQP-------GVYD--FGFIDSSKYTGSLTYT
EPAP:  LTVTGQAVESAKKVSSSFTEDSTI-----DGLLGLAFSTLNTVSPTQQKTFFDNAKASLDSPV------FTADLGYHAP-------GTYN--FGFIDTTAYTGSITYT
RMAP:  ITVTKQILAYVDNVRGPTAEQSPNADIFLDGLFGAAYPD-NTAMEAEYGSTYNTVHVNLYKQGLISSPLFSVYMNTNSGTGEVVFGGVNNTLLGG-DIAYTD------
RPAP:  ATVKQQTLAYVDNVSGPTAEQSPDSELFLDGIFGAAYPD-NTAMEAEYGDTYNTVHVNLYKQGLISSPVFSVYMNTNDGGGQVVFGGVNNTLLGG-DIQYTD------
ILAP:  LTIPKQSIGVASRDSGFDGVDGILGVGPVDLTVGTLSPHTSTSIPTVTDNLFSQGTIPTNLLA----VSFEPTTSESSTN------GEL--TFGATDSSKYTGSITYT
                 *            *           *                              *                          *     *  *
CHYM:  IVDIQQTVGLSTQEPGDVFTYAEF-----DGILGMAYPSLASEYSIP-------VFDNMMNRHLVAQDLFSVYMDR-DGQESML-------TLGAIDPSYYTGSLHWV
PPEP:  ISDTNQIFGLSETEPGSFLYYAPF-----DGILGLAYPSISASGATP-------VFDNLWDQGLVSQDLFSVYLSSNDDSGSVV-------LLGGIDSSYYTGSLNWV

AAAP:  ---DADSSQGYWGFSTDGYSIGDGSSSSSGF--SAIADTGTTLILLDDEIVSAYYEQVSGASGETEAGGYVFSC--STNPPD---------FTVVIGDYKAVVPGKYI
AOAP:  AYADVDDSQGFWQFTADGYSVGKGDAQKAPI--TGIADTGTTLLMLDDEIVDAYYKQVQGAKNDASAGGYVFPC--ETELPE--------FTVVIGSYNAVIPGKHI
RCAP:  PI---DNSRGLWGITVDRATVGTSTVASSFD---GILDTGTTLLILPNN-VAASVARAYGASDNGD-GTYTISCDTSRFKPLV-------FSIN-GASFQVSPDSLV
RNAP:  PI---DNSEGFWGVTVKSTKIGGTTVSASFD---AILDTGTTLLLLPDD-VAAKVARSYGASDNGD-GTYSITCDTSKLQPLV-------FTL--GSSTFEVPSDSL
PJAP:  ---GVDNSQGFWSFNVDSYTA--GSQSGDGF--SGIADTGTTLLLLDDSVVSQYYSQVSGAQQDSNAGGYVFDC--SSSID--------FSVSISGYTATVPGSLI
EPAP:  ---AVSTKGGFWEWTSTGYAVGSGTFKSTSI--DGIADTGTTLLYLPATVVSAYWAQVSGAKSSSSVGGYVFPC--SATLPS-------FTFGVGSARIVIPGDYI
RMAP:  -VMSRYGGYYFWDAPVTGITVDGSAAVRFSRPQAFTIDTGTNFFIMPSSAASKIVKAALPDATETQQG-WVVPCASYQNSKSTISIVMQKSGSSSD-TIEISVPVSKM
RPAP:  -VLKSRGGYYFWDAPVTGVKIDGSDAVSFDGAQAFTIDTGTNFFIAPSSFAEKVVKAALPDATESQQG-YTVPCSKYQDSKTTFSLDLQKSGSSSD-TIDVSVPISKM
ILAP:  PITSTSPASAYWGINQTIRYGSSTSILSST---AGIVDTGTTLT-LIASDAFAKYKKATGAVADNNTGLLRLTTAQYANLQSLFFTIGGQTFELTANAQIWPRNLNTA
                                      ****                              *                                       *
CHYM:  PVTV----QQYWQFTVDSVTISGVVACEGGCQAIL-DTGTSKLVGPSSDIL-NIQQAIGATQN-QYGEFDIDCDNLSYMPTVVFEINGKMYPLTPSAYTSQ------
PPEP:  PVSV----EGYWLQITLDSITMDGETIACSGGCQAIV-DTGTSLLTGPTSAIAINIQSDIGASEN-SDGEMVISCSSIDSLPDIVFTIDGVQYPLSPSAYILQ------

AAAP:  NYAPISTGSSTCFGGIQSNSGLGLS----ILGDVFLKSQYVVFNSEG-PKLGFAAQA------
AOAP:  NYAPLQEGSSTCVGGIQSNSGLGLS----ILGDVFLKSQYVVFDSQG-PRLGFAAQA------
RCAP:  FEEYQGQ----CIAGF--GYGNFDFA---ILGDTFLKNNYVVFNQGV-PEVQIAPVAQ-----
RNAP:  IFEKDGNK----CIAGFAAG-GDLA----ILGDVFLKNNYVVFNQEV-PEVQIAPVAN-----
PJAP:  NYGPSGDG-STCLGGIQSNSGIGFS----IFGDIFLKSQYVVVFDSDGPQLGFAPQA------
EPAP:  DFGPISTGSSSCFGGIQSSAGIGIN----IFGDVALKAAFVVFNGATTPTLGFASK-------
RMAP:  LLPVDQSNE-TCMFIILPDGGNQY-----IVGNLFLRFFVNVYDFGNN-RIGFAPLASAYENE-
RPAP:  LLPVDKSGET-CMFIVLPDGGNQF-----IVGNLFLRFFVNVYDFGKN-RIGFAPLASGYENN-
ILAP:  IGGSASSVYLIVGD-LGSDSGEGLDF---INGLTFLERFYSVYDTTNK-RLGLATTSFTTATSN
               *       *  * *       *  *    *   *
CHYM:  ------DQGFCTSGFQSENHS----QKWILGDVFIREYYSVFDRANNL-VGLAKAI------
PPEP:  ------DDDSCTSGFEGMDVPTSSGELWILGDVFIRQYYTVFDRANNK-VGLAPVA------
```

Figure 1. Alignment of the sequences of the mature forms of filamentous fungal secreted aspartic proteinases. AAAP: *Aspergiluus awamori*; AOAP: *Aspergillus oryzae*; RCAP: *Rhizopus chinensis*; RNAP: *Rhizopus niveus*; PJAP: *Penicillium janthinellum*; EPAP: *Endothia parasitica*; RMAP: *Rhizomucor miehei*; RPAP: *Rhizomucor pusillus*; ILAP: *Irpex lacteus*; CHYM: bovine chymosin; PPEP: porcine pepsin. *, indicates residues common to all fungal sequences. Cysteine residues and arginine residues at potential N-linked glycosylation sites are underlined.

stringent conditions no evidence for additional related genes was found supporting the hypothesis that the two forms of aspartic proteinases are isozymes encoded by a single gene. However, no directly determined amino acid sequence is available for this enzyme so the relationship between the cloned gene and the studied enzyme has yet to be confirmed. In contrast, two separate genes encoding secreted aspartic proteinases have been identified in *Rhizopus niveus* (Horiuchi *et al.*, 1990). There is strong homology between the *Aspergillus awamori* and *A. oryzae* genes as would be expected. Each has three introns at the same positions within the coding sequence, those in *A. oryzae* being surmised by comparison with the *A. awamori* sequence.

Certain features of the primary sequences of the mature filamentous fungal aspartic proteinases are noteworthy. There are variable numbers of cysteine residues in these sequences but their positions, when present, are conserved. Thus, there is the potential for two disulphide bridges in the *Rhizomucor* and *Rhizopus* enzymes, a single disulphide bridge

```
SECRETION SIGNAL PEPTIDES
AAAP: MVVFSKTAALVLGL--S-SAVSA......
AOAP: MVILSKVAAVAVGL--S-TVASA......
RMAP: ML-FSQITSAILLTAASLSLTTA......
RPAP: ML-FSKISSAILLTAASFALTSA......
RNAP: MK-FTLISSCVALAAMTLAVEAA......
RCAP: MT-FTLNSSCIAIAALAVAVNAA......


PROPEPTIDES                                               *
AAAP:       APAPRTRKGFTINQIARPANKTRTINLPGMYA-RS-LLA-KF----GGTVPQSVKEA-A.....
AOAP: LPTGPSHSPHARRGFTINQITRQTARVGPKTASFPAIYSR-ALA-KY----GGTVPAHLKSAVA.....
RMAP:       RPVSKQSESKDKLLALPLTSVSRKFSQTKFGQQQLAEKL----AGLKPFSE-----.....
RPAP:       RPVSKQSDADDKLLALPLTSVNRKYSQTKHG-QQAAEKL----GGIKAF-------.....
RNAP:       PNGKKINIPLAKNNSYKPSAKNALNKALAKYNRR---KVG--SGGITTE-------.....
RCAP:       PGEKKISIPLAKNPNYKPSAKNAIQKAIAKYNKH---KINTSTGGIVPD-------.....
```

Figure 2. Alignment of the secretion signal peptides and propeptides of secreted filamentous fungal aspartic proteinases. AAAP: *Aspergillus awamori;* AOAP: *Aspergillus oryzae;* RMAP: *Rhizomucor miehei;* RPAP: *Rhizomucor pusillus*; RNAP: *Rhizopus niveus*; RCAP: *Rhizopus chinensis*; * conserved lysine residue.

in the *Endothia, Penicillium* and *Aspergillus* enzymes, and no disulphide bridges in the *Irpex* aspartic proteinase. The presence and position of potential asparagine-linked glycosylation sites is extremely variable (see Figure 1). None are present in the *A. awamori* sequence whereas there is a single site in the *A. oryzae* sequence. However, one of the forms of the *A. oryzae* aspartic proteinase studied by Tsujita and Endo (1976) contained no attached carbohydrate, so the N-linked site observed in the sequence may not be used *in vivo*. Two conserved and one non-conserved potential N-linked glycosylation sites are shared by the two *Mucor* enzymes. However, it is probable that only the two conserved sites are actually glycosylated *in vivo* (Aikawa *et al.*, 1990). All the sequences are rich in serine and threonine residues possibly allowing oportunity for O-linked glycosylation.

The mammalian gastric aspartic proteinases are secreted as inactive zymogens which are processed at low pH to yield active, mature enzymes. The secreted fungal proteinases are also presumably synthesized initially as preproenzymes. The secretion signal peptide or pre-region would be removed as the protein enters the endoplasmic reticulum. Consequently, the sequence of this is only known as a result of gene cloning and DNA sequence analysis. Secreted aspartic proteinase zymogens have not been observed in their native fungal hosts. Again, the presence and sequence of the pro-regions could only be deduced from the DNA sequence of cloned genes. The known prepro-regions of filamentous fungal aspartic proteinases are shown in Figure 2. It should be emphasized that, except in the case of *Rhizomucor miehei* aspartic proteinase (RmAP, Hiramatsu *et al.*, 1989), the cleavage site between the secretion signal peptide and the start of the propeptide has merely been inferred by comparison with known signal peptidase recognition sites (Perlman & Halvorson, 1983).

Like their mammalian counterparts, the prosequences of filamentous fungal aspartic proteinases tend to be rich in basic amino acids. A lysine residue identified in Figure 2 has been postulated to interact with the active site aspartate residues in the zymogen molecule (Foltmann, 1988). They are variable in length, the longest being the *A. oryzae* sequence at 58 residues although, since the amino terminal of the mature enzyme is not known, this must be regarded as presumptive.

The cloned *A. awamori* aspartic proteinase gene (*pepA*) has been used to construct a vector to allow deletion of this gene from the fungal genome (Berka *et al.*, 1990). As a result the secreted proteinase activity of deleted strains was greatly reduced demonstrating that this gene encodes the major secreted proteinase under the culture conditions used. However, some residual proteinase activity was secreted by the deleted strains as judged by zones of clotting or clearing on agar plates containing skim milk. The residual activity was only observed at acid pH and was not inhibited by pepstatin, phenylmethyl-sulfonylfluoride (PMSF) or EDTA (inhibitiors of aspartic, serine and metalloproteinases respectively). It was, however, inhibited to some extent by DAN suggesting that some residual proteinase activity may be due to a pepstatin-insensitive carboxyl proteinase. Strains deleted for the *pepA* gene have subsequently been used as a hosts for the expression of various heterologous aspartic proteinases (see below).

HETEROLOGOUS EXPRESSION OF FUNGAL ASPARTIC PROTEINASES

There have been several examples of the expression of filamentous fungal aspartic proteinase genes in heterologous fungal hosts (Gray *et al.*, 1986; Yamashita *et al.*, 1987; Dickinson *et al.*, 1987; Christensen *et al.*, 1988; Horiuchi *et al.*, 1990). Since transformation systems are often not available for the original host organisms such as *Rhizomucor pusillus* and *R. miehei* heterologous systems are necessary in order to study the effects of *in vitro* mutagenesis on enzyme function.

The gene for *Rhizomucor pusillus* aspartic proteinase (RpAP) has been expressed in *S. cerevisiae* under the control of the *GAL7* promoter (Yamashita *et al.*, 1987). The gene encoded the entire *Rhizopus* preproenzyme and the mature enzyme accumulated in the medium at a concentration of more than 150 mg/l. Using culture medium of a relatively high pH it was possible to identify and purify the inactive zymogen form, confirm the presence and amino terminal sequence of the propeptide, and to study activation of the zymogen (Hiramatsu *et al.*, 1989). It was concluded that activation of the RpAP zymogen resembled that of mammalian pepsinogen and prochymosin (Foltmann, 1986; Pedersen *et al.*, 1979). That is, at low pH (below pH 5.0), processing was dependent on RpAP activity. The propeptide was removed predominantly by an intermolecular reaction as a result of cleavage by an already active RpAP molecule although the possibilty of an intramolecular mechanism, particularly during the initial stages of activation of a sample, was not ruled out. However, unlike the mammalian enzymes which form an intermediate processing form at pH 2.0, only direct processing to the mature form was observed for RpAP even at pH 2. An inactive mutant form of RpAP was expressed and it was noted that a native yeast proteinase was also able to remove the propeptide. However, in a *pep4-3* mutant strain which does not produce vacuolar proteinase A the inactive proRpAP was not processed. Thus, the native yeast proteinase able to remove the proregion from proRpAP could be proteinase A or another proteinase which requires proteinase A for activation.

A problem often encountered with the secretion of heterologous proteins by yeast or filamentous fungi is that of hyperglycosylation. The RpAP produced in yeast was seen to be more extensively glycosylated than the natural *Rhizomucor* enzyme (Yamashita *et al.*, 1987). This hyperglycosylation was observed to cause a decrease in milk clotting activity but increased proteolytic activity using acid-denatured hemoglobin as substrate (Aikawa *et al.*, 1990). Treatment with endo-α-N-acetylglucosaminidase H (endo H) to remove the N-linked carbohydrate restored these parameters to levels similar to those of the natural *Rhizomucor* enzyme. Site-directed mutagenesis was employed to alter each of the three potential N-linked glycosylation sites and these mutants were expressed in yeast. Only two of the potential sites (Asn_{79} and Asn_{188}) were found to be glycosylated. Removal of both of these sites decreased the amount of enzyme secreted but the non-glycosylated form obtained had similar properties to the natural *Rhizomucor* enzyme.

Mature and proenzyme forms of *Rhizopus niveus* aspartic proteinase I have also been produced in yeast (Horiuchi *et al.*, 1990). Unlike the RpAP gene, the gene encoding this enzyme contains an intron. It was necessary to remove this intron in order to obtain expression since it has been shown that filamentous fungal introns are not spliced by yeast (Innis *et al.*, 1985). Subsequently, Ashikari *et al.* (1990) have shown that splicing of the

Rhizopus intron in yeast occurred with low efficiency if the sequences within the intron were altered to conform with the consensus sequences of yeast introns.

The gene encoding the prepro-aspartic proteinase from *Rhizomucor miehei* (RmAP) has been expressed in a number of heterologous fungal systems. The gene was cloned by Gray *et al.* (1986) and expressed from its native promoter in *Aspergillus nidulans*. Mature, active proteinase was secreted to the culture medium but the enzyme was hyperglycosylated. An attempt was also made by these authors to express the preproenzyme in *E. coli* using the *trp* promoter. It was concluded that the majority of the enzyme produced accumulated as intracellular, inactive polypeptides in this system. PreproRmAP has also been expressed from its native promoter in *Mucor circinelloides* and secreted, active proteinase obtained (Dickinson *et al.*, 1987). To date the best published yield of secreted RmAP (approximately 3 g/l in a small fermentor) from a heterologous system was reported for expression of the preproRmAP gene in *Aspergillus oryzae* using the *A. oryzae* α-amylase gene promoter (Christensen *et al.*, 1988). Under certain conditions the proform of the enzyme was apparently observed in the culture supernatant. The enzyme was somewhat overglycosylated but no difference in activity was attributed to this.

We have constructed an expression vector which places the coding sequence of preproRmAP under control of the *A. awamori* glucoamylase gene (*glaA*) promoter. This vector has been transformed into either *A. nidulans* or a *pepA*-deleted *A. awamori* host and secretion of active proteinase obtained. Transformants of *A. awamori* apparently secreted up to 0.45 g/l of active RmAP in shake flask culture as measured by a milk clotting assay using authentic RmAP as standard. Again the heterologous RmAP was hyperglycosylated and endo H treatment to remove the N-linked glycosylation caused approximately a two-fold increase in milk clotting activity (i.e., approximately 0.8 g/l total RmAP was actually produced by the transformants). The reduction in milk clotting activity caused by overglycosylation is very similar to that noted for RpAP by Aikawa *et al.* (1990). One of the N-linked glycosylation sites (Asn_{79}) is probably located on the "flap" region overhanging the active site cleft. Aikawa *et al.* suggested that it seemed entirely possible that extensive glycosylation at this site could affect activity of the enzyme. However, they also noted that if

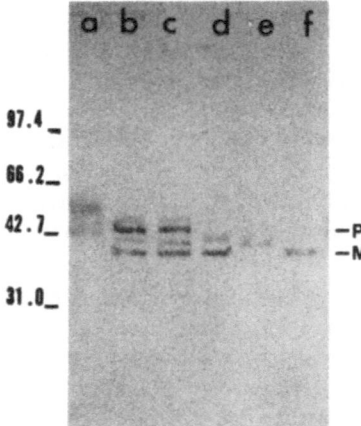

Figure 3. Western immunoblot analysis of RmAP in culture supernatant from an *Aspergillus awamori* transformant. Cell free samples were taken from a 50 ml shake flask culture grown for three days in medium maintained at pH 6.0. Lane a, untreated; lane b, treated with endo H; lane c, incubated at pH 2.0, 30°C for 16 hours and treated with endo H; lane d, incubated at pH 4.8, 30°C for 16 hours and treated with endo H; lane e, untreated authentic *Rhizomucor* RmAP; lane f, authentic RmAP treated with endo H. The sizes of molecular weight standards, in kDa are given on the left. P, proRmAP; M, mature RmAP.

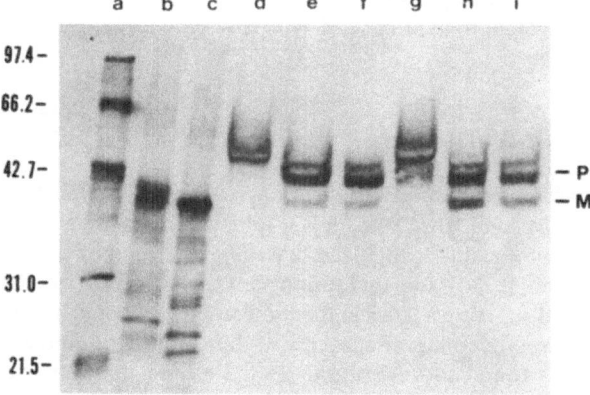

Figure 4. Western immunoblot analysis of inactive mutant RmAP in culture supernatant from an *Aspergillus awamori* transformant. Culture conditions were the same as for Figure 3. Lane a, molecular weight standards (sizes in kDa are given to the left); lane b, untreated authentic *Rhizomucor* RmAP; lane c, authentic RmAP treated with endo H; lane d, untreated *Aspergillus* RmAP after three days culture; lane e, as for lane d but treated with endo H; lane f, as for lane e but incubated at pH 4.8, 30°C for 16 hours; lane g, untreated *Aspergillus* RmAP after six days culture; lane h, as for lane g but treated with endo H; lane i, as for lane h but incubated at pH 4.8, 30°C for 16 hours. P, proRmAP; M, mature RmAP.

this site was destroyed by site directed mutagenesis, overglycosylation of the remaining site (Asn$_{188}$), which is well removed from the active site, could also affect activity. Mutant strains of *A. awamori* have recently been identified which will secrete up to 3 g/l RmAP in shake flask cultures.

As was the case for RpAP produced in yeast, it was possible to recognize proRmAP, in addition to the mature enzyme, in the *A. awamori* culture supernatants after 3 days if the pH was maintained at a comparatively high value (pH 6.0). However, after 7 days under these same conditions or at any time point at pH 4.5 only mature RmAP was observed. The two forms of the enzyme were only clearly distinguishable by Western analysis if the samples were first treated with endo H (Figure 3). The proform of RmAP in cell free samples of culture supernatants disappeared if the pH was dropped to 4.8 and the samples incubated for 16 hours at 30°C, presumably as a result of conversion to the mature form (Figure 3). Interestingly, this processing was not observed at pH 2.0 (Figure 3) whereas activation of RpAP produced in yeast did occur under similar conditions (Hiramatsu *et al.*, 1989).

In vitro mutagenesis was performed in order to change the active site Asp$_{38}$ to a valine residue and thus inactivate the enzyme. This form of RmAP was expressed in either *Aspergillus* species as above and processing of the mutant enzyme examined. As for wild-type RmAP, we observed bands of the expected mobility of both pro- and mature forms of the inactive mutant in shake flask culture supernatants of *A. awamori* transformants at pH 6.0. However, the mutant proRmAP in cell free samples from these cultures was not affected by incubation at pH 4.8, 30°C for 16 hours demonstrating that processing of the inactive mutant did not occur in the absence of *Aspergillus* cells (Figure 4). Again, only apparently mature RmAP was seen in culture supernatants at pH 4.5. We suggest that proRmAP processing can occur autocatalytically but that, as for yeast, a native *Aspergillus* enzyme is produced which is capable of cleaving the prosequence. Processing by the *Aspergillus* enzyme, which may be intracellular or secreted, is minimized at pH 6.0. It is noteworthy that several non-aspartic proteinases, including *A. oryzae* metallo- and serine proteases, are able to cleave the propeptides of prochymosin or pepsinogen (Petrova *et al.*, 1987; Stepanov *et al.*, 1990).

The mature RmAP was purified from pH 4.5 cultures of *A. nidulans* transformants producing either the inactive mutant or wild-type RmAP and the following amino terminal sequences were determined:

WILD-TYPE: DGSVDTPGYYD
INACTIVE: GSVDTPGYYD

The expected amino terminus for authentic RmAP is AAADGSVDTPGYYD and it is possible that in *Aspergillus* the mature RmAP is subject to degradation by an aminopeptidase after initial processing. However, it is recognized that ragged termini of aspartic proteinases can be generated (Foltmann, 1986) and the amino termini observed for RmAP produced in *Aspergillus* have also been reported for authentic RmAP (Piquet *et al.*, 1981). Despite the possibility of imprecise processing or subsequent digestion by aminopeptidase, the RpAP produced in yeast was found to have the expected amino terminus of AEGDGSVDTP (Yamashita *et al.*, 1987).

EXPRESSION OF MAMMALIAN ASPARTIC PROTEINASES IN FUNGI

Bovine chymosin is the preferred enzyme for clotting milk during cheese manufacture but is in short supply. Consequently, there have been numerous attempts to produce this enzyme in microorganisms. Production and secretion of prochymosin has also become something of a model system for heterologous gene expression in a number of organisms. Benefits to the use of prochymosin include simple assay procedures and the fact that peptide fusions to the amino terminus of the propeptide will be removed by the normal autocatalytic activation to yield authentic chymosin at low pH. In many of the examples given below short peptide sequences were added to prochymosin as a result of fusion between the prochymosin cDNA and DNA encoding the promoter and secretion signal peptide of genes native to the organism in question. Efforts to express prochymosin cDNA in *E. coli* have generally led to intracellular accumulation of inactive chymosin in the form of inclusion bodies (Emtage *et al.*, 1983; Nishimori *et al.*, 1984; McCaman *et al.*, 1985; Kawaguchi *et al.*, 1987). Active chymosin can be recovered from the inclusion bodies but this requires denaturation and refolding (Kawaguchi *et al.*, 1984; Marston *et al.*, 1984), an economically unattractive process. However, chymosin is produced commercially in *E. coli* by Pfizer, Inc. and this system has also been used to express mutated forms of prochymosin cDNA (McCaman & Cummings, 1986; 1988; Suzuki *et al.*, 1988; Mantafounis & Pitts, 1990).

Several different yeast species have been employed for the production of bovine chymosin. In *Saccharomyces cerevisiae*, as in *E. coli,* intracellular prochymosin tends to be insoluble and difficult to activate (Mellor *et al.*, 1983; Goff *et al.*, 1984; Moir *et al.*, 1985). Expression of preprochymosin cDNA did not lead to secretion of chymosin, whereas substituting the signal peptide from yeast invertase for the natural chymosin secretion signal allowed secretion of approximately 10% of the total prochymosin made (Moir *et al.*, 1985). Subsequently, "supersecretor" mutants of yeast were identified which secreted up to 50% of the prochymosin synthesized (Smith *et al.*, 1985). The yeasts *Yarrowia lipolytica* (Franke *et al.*, 1988) and *Kluveromyces lactis* (van den Berg *et al.*, 1990) have also been used for expression of prochymosin cDNA. Secretion may be more efficient in these species and commercially viable yields have been obtained from the latter species by Gist-Brocades nv. Apparently, all the prochymosin secreted in these systems can be readily activated to mature chymosin.

Filamentous fungi have also been used as hosts for the production of bovine chymosin. These fungi were considered particularly suitable for this work because they are capable of secreting very large amounts of native enzymes, including native aspartic proteinases in some cases. *Aspergillus awamori* strains have been transformed with vectors

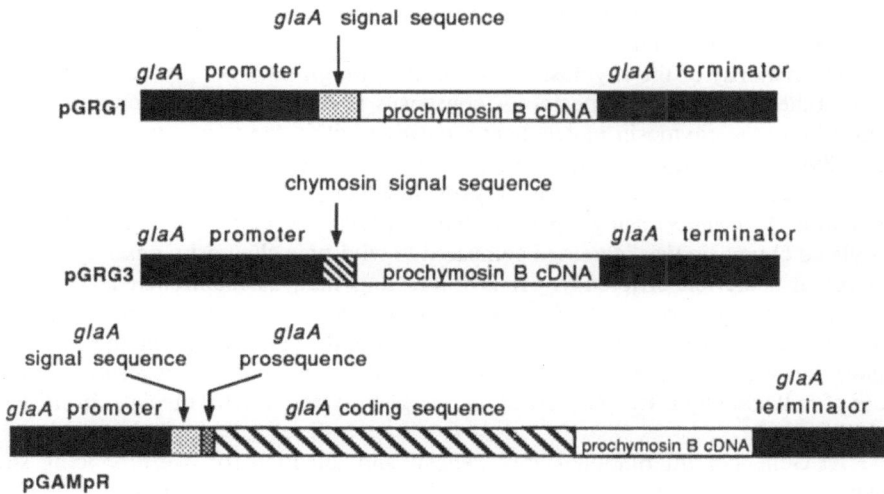

Figure 5. Diagram of the bovine chymosin expression cassettes in vectors used for transformation of *Aspergillus*.

containing the prochymosin B cDNA expression cassettes shown in Figure 5 (Ward, 1989a; 1989b; Ward *et al.*, 1990). The glucoamylase gene (*glaA*) promoter was employed in each construction and the *A. awamori* strain used was a production strain capable of secreting approximately 3 g/l native glucoamylase in shake flask cultures.

Direct expression of prochymosin with either the glucoamylase or the natural chymosin secretion signal sequence (pGRG1 or pGRG3 in Figure 5) gave low levels (up to approximately 15 mg/l) of secreted chymosin. Similar levels of secreted chymosin have been obtained from *A. oryzae* (Boel *et al.*, 1987) and *Trichoderma reesei* (Harrki *et al.*, 1989). Under certain culture conditions, both prochymosin and mature chymosin could be identified in supernatants from the *A. awamori* transformants using Western analysis (Ward *et al.*, 1990). Adjustment of samples from these cultures to pH 2 and incubation for 30 minutes led to loss of prochymosin and the accumulation of a chymosin-specific band of the expected mobility for pseudochymosin. This is the expected autocatalytic activation product for authentic chymosin under these conditions (Pedersen *et al.*, 1979). Production was greater in *pepA* deleted strains (Berka *et al.*, 1990), possibly as a result of reduced chymosin degradation by the native *Aspergillus* aspartic proteinase. Chymosin specific mRNA levels were high in these transformants but pro- and mature chymosin accumulated intracellularly leading to the conclusion that secretion was limiting production. Subsequently, an expression vector (pGAMpR, Figure 5) was constructed in which the entire coding sequence for glucoamylase was fused in frame to the prochymosin cDNA sequence. Transformants with pGAMpR produced much higher levels of extracellular, active chymosin (approximately 150 mg/l in strains not deleted for *pepA*) than previous transformants, probably as a result of improved secretion efficiency for the glucoamylase-chymosin fusion protein compared to chymosin alone (Ward *et al.*, 1990). It was possible to identify both mature chymosin and the fusion protein by Western analysis and show that incubation at pH 2 for 30 minutes caused the release of pseudochymosin from the fusion protein. The formation of pseudochymosin under these conditions gave a concomitant increase in milk clotting activity of the sample. Pseudochymosin formation was inhibited by pepstatin and also occurred with a purified sample of the glucoamylase-chymosin fusion protein (K. Hayenga and M. Murphy, personal communication). These results strongly suggest that processing of the fusion protein can be catalyzed by chymosin itself. However, it cannot be ruled out that a native *Aspergillus* proteinase is also capable of processing the fusion protein as discussed above for *Rhizomucor* aspartic proteinases produced in yeast or *Aspergillus*.

The majority of the chymosin secreted by *Aspergillus* is not glycosylated, as is the case for authentic bovine chymosin. A consensus N-linked glycosylation site has been introduced into the chymosin molecule at a position within the flap region (at residue 74; Ward, 1989b). A glycosylation site exists naturally at the equivalent position in the *Rhizopus* aspartic proteinase molecules. The glycosylated chymosin was produced by direct expression in *A. awamori* using a pGRG3-type expression vector. Production of secreted, glycosylated chymosin was increased compared to wild-type chymosin, possibly as a result of improved secretion efficiency. It is interesting that, as mentioned above, a non-glycosylated mutant form of RmAP was not produced at as high a level as the natural glycosylated form in yeast. The milk clotting activity of the glycosylated chymosin produced in *Aspergillus* was greatly reduced compared to authentic chymosin. This result is similar to the loss of milk clotting activity observed for overglycosylated RmAP obtained from yeast or *Aspergillus*.

At Genencor International, mutagenesis and selection for overproducing strains following transformation has allowed commercially viable yields of chymosin to be obtained from *Aspergillus*.

In addition to bovine chymosin and RmAP, human procathepsin D has also been produced in *Aspergillus awamori* using the *glaA* promoter to drive expression of the cDNA (G. E. Conner, B. M. Dunn, M. W., K. H. K. and R. M. Berka, personnal communication). Production in this host should facilitate investigations on the activation of this aspartic proteinase as well as structure/function studies.

REFERENCES

Aikawa, J., Yamashita, T., Nishiyama, M., Horinouchi, S. & Beppu, T., 1990, *J. Biol. Chem.*, **265**:13955-13959.

Ammerer, G., Hunter, C. P., Rothman, J. H., Saari, G. C., Vallis, L. A. & Stevens, T. H., 1986, *Mol. Cell. Biol.*, **6**:2490-2499.

Andreeva, N. S., Zdanov, A. S., Gustchina, A. E. & Federov, A. A., 1984, *J. Biol. Chem.*, **259**:11353-11365.

Ashikari, T., Amachi, T., Yoshizumi, H., Horiuchi, H., Takagi, M. & Yano, K., 1990, *Mol. Gen. Genet.*, **223**:11-16.

Barkholt, V., 1987, *Eur. J. Biochem.*, **167**:327-338.

Baudys, M., Foundling, S., Pavlík, M., Blundell, T. & Kostka, V., 1988, *FEBS Lett.*, **235**:271-274.

Bech, A.-M. & Foltmann, B., 1981, *Neth. Milk Dairy J.*, **35**:275-280.

Berka, R. M., Ward, M., Wilson, L. J., Hayenga, K. J., Kodama, K. H., Carlomagno, L. P. & Thompson, S. A., 1990, *Gene*, **86**:153-162.

Blundell, T., Jenkins, J., Pearl, L., Sewell, T. & Pedersen, V., 1985, *in*: Aspartic Proteinases and their Inhibitors (Kostka, V., ed.). pp. 151-161. Walter de Gruyter, Berlin.

Boel, E., Christensen, T. & Wöldike, H., 1987, European Patent Application, Publication number 0 238 023

Bott, R., Subramanian, E. and Davies, D. R., 1982, *Biochemistry*, **21**:6956-6962.

Chen, Z., Han, H. P., Wang, X. J., Koelsch, G., Lin, X. Y., Hartsuck, J. A. & Tang, J., 1991, Manuscript in preparation.

Christensen, T., Wöldike, H., Boel, E., Mortensen, S. B., Hjortshoej, K., Thim, L. & Hansen, M. T., 1988, *Bio/technology*, **6**:1419-1422.

Cunningham, A., Wang, H.-M., Jones, S. R., Kurosky, A., Rao, L., Harris, C. I., Rhee, S. H. & Hoffmann, T., 1976, *Can. J. Biochem.*, **54**:902-914.

Delaney, R., Wong, R. N. S., Meng, G., Wu, N. & Tang, J., 1987, *J. Biol. Chem.* **262**:1461-1467

Dickinson, L., Harboe, M., van Heeswijk, R., Strøman, P. & Jepsen, L. P., 1987, *Carlsberg Res. Commun.*, **52**:243-252.

Egel-Mitani, M., Flygenring, H. P. & Hansen, M. T., 1990, *Yeast*, **6**:127-137.

Emtage, J. S., Angal, S., Doel, M. T., Harris, T. J. R., Jemkins, B., Lilley, G. & Lowe, P. A., 1983, *Proc. Natl. Acad. Sci. USA*, **80**:3671-3675.

Foltmann, B., 1986, *in*: Molecular and Cellular Basis of Digestion (Desnuelle, P., Sjostrom, H. & Noren, O., eds.). pp. 491-505.

Foltmann, B., 1988, *Biol. Chem. Hoppe-Seyler*, **369**:311-314

Franke, A. E., Kaczmarek, F. S., Eisenhard, M. E., Geoghegan, K. F., Danley, D. E., De Zeeuw, J. R., O'Donnell, M. M., Gollaher, M. G. Jr. & Davidow, L. S., 1988, *in*: Developments in Industrial Microbiology, Vol. 29 (Pierce, G., ed.). pp. 43-57. Society for Industrial Microbiology, Arlington.

Gilliland, G. L., Winborne, E. L., Nachman, J. & Wlodower, A.., 1990, *Proteins*, **8**:82-101.

Goff, C. G., Moir, D. T., Kohno, T., Gravius, T. C., Smith, R. A., Yamasaki, E. & Taunton-Rigby, A., 1984, *Gene*, **27**:35-46.

Gray, G. L., Hayenga, K., Cullen, D., Wilson, L. J. & Norton, S., 1986, *Gene*, **48**:41-53.

Harrki, A., Uusitalo, J., Bailey, M., Penttil , M. & Knowles, J. K. C., 1989, *Biotechnology*, **7**:596-603.

Hiramatsu, R., Aikawa, J., Horinouchi, S. & Beppu, T., 1989, *J. Biol. Chem.*, **264**:16862-16866.

Hirata, D., Fukui, S. & Yamashita, I., 1988, *Agric. Biol. Chem.*, **52**:2647-2649.

Horiuchi, H., Yanai, K., Okazaki, T., Takagi, M. & Yano, K., 1988, *J. Bacteriol.*, **170**:272-278.

Horiuchi, H., Ashikari, T., Amachi, T., Yoshizumi, H., Takagi, M. & Yano, K., 1990, *Agric. Biol. Chem.*, **54**:1771-1779

Innis, M. A., Holland, M. J., McCabe, P. C., Cole, G. E., Wittman, V. P., Tal, R., Watt, K. W. K., Gelfand, D. H., Holland, J. P. & Meade, J. H., 1985, *Science*, **228**:21-26.

James, M. N. G. & Sielecki, A. R., 1983, *J. Mol. Biol.*, **163**:299-361.

Jenkins, J., Tickle, I., Sewell, T., Ungaretti, L., Wollmer, A. & Blundell, T., 1977, *in*: Acid Proteinases, Structure, Function and Biology (Tang, J., ed.). pp. 43-60. Plenum, New York.

Kawaguchi, Y., Shimizu, N., Nishimori, K., Uozumi, T. & Beppu, T., 1984, *J. Biotechnol.*, **1**:307-315.

Kawaguchi, Y., Kosugi, S., Sasaki, K., Uozumi, T. & Beppu, T., 1987, *Agric. Biol. Chem.*, **51**:1871-1877.

Kobayashi, H., Sekibata, S., Shibuya, H., Yoshida, S., Kusakabe, I. & Murakami, K., 1989, *Agric. Biol. Chem.*, **53**:1927-1933.

MacKay, V. L., Welch, S. K., Insley, M. Y., Manney, T. R., Holly, J., Saari, G. C. & Parker, M. L., 1988, *Proc. Natl. Acad. Sci. USA*, **85**:55-59.

McCaman, M. T. & Cummings, D. B., 1986, *J. Biol. Chem.*, **261**:15345-15348.

McCaman, M. T. & Cummings, D. B., 1988, *Proteins*, **3**:256-261.

McCaman, M. T., Andrews, W. H. & Files, J. G., 1985, *J. Biotechnol.*, **2**:177-190.

Maita, T., Nagata, S., Matsuda, G., Maruta, S., Oda, K., Murao, S. & Tsuru, D., 1984, *J. Biochem.*, **95**:465-475.

Mantafounis, D. & Pitts, J., 1990, *Prot. Eng.*, **3**:605-609.

Marston, F. A. O., Lowe, P. A., Doel, M. T., Schoemaker, J. M., White, S. & Angal, S., 1984, *Biotechnology*, **2**:800-804.

Mellor, J., Dobson, M. J., Roberts, N. A., Tuite, M. F., Emtage, J. S., White, S., Lowe, P. A., Patel, T., Kingsman, A. J. & Kingsman, S. M., 1983, *Gene*, **24**:1-14.

Moir, DS. T., Mao, J., Duncan, M. J., Smith, R. A. & Kohno, T., 1985, *in*: Developments in Industrial Microbiology, Vol. 26 (Underkoffler, L., ed.). pp. 75-85. Society for Industrial Microbiology, Arlington.

Murao, S.& Oda, K., 1985, *in*: Aspartic Proteinases and their Inhibitors (Kostka, V., ed.). pp. 379-399. Walter de Gruyter, Berlin.

Nishimori, K., Shimizu, N., Kawaguchi, Y., Hidaka, M., Uozumi, T. & Beppu, T., 1984, *Gene*, **29**:41-49.

Oda, K. & Murao, S., 1974, *Agric. Biol. Chem.*, **38**:2435-2444.

Ostoslavskaya, V. I., Revina, L. P., Kotlova, E. K., Surova, I. A., Levin, E. D., Timokhina, E. A. & Stepanov, V. M., 1986, *Bioorg. Khim.*, **8**:1030-1047.

Pedersen, V. B., Christensen, K. A. & Foltmann, B., 1979, *Eur. J. Biochem.*, **94**:573-580.

Perlman, D. & Halvorson, H. O., 1983, *J. Mol. Biol.*, **167**:391-409.

Petrova, E. N., Revina, L. P., Timokhina, E. A., Lavrenova, G. I., Vaganova, T. I. & Stepanov, V. M., 1987, *Biokhimiya*, **53**:1389-1396.

Smith, R. A., Duncan, M. J. & Moir, D. T., 1985, *Science*, **229**:1219-1224.

Stepanov, V. M., Lavrenova, G. I., Terent'eva, E., Yu. & Khodova, O. M., 1990, *FEBS Lett.*, **260**:173-175.

Suguna, K., Bott, R. R., Padlan, E. A., Subramanian, E., Sheriff, S., Cohen, G. H. & Davies, D. R., 1987, *J. Mol. Biol.*, **196**:877-900.

Suzuki, J., Sasaki, K., Sasao, Y., Hamu, A., Kawasaki, H., Nishiyama, M., Horinouchi, S. & Beppu, T., 1989, *Prot. Eng.*, **2**:563-569.

Takahashi, K., 1987, *J. Biol. Chem.*, **262**:1468-1478.

Tang, J. & Wong, R. N. S., 1987, *J. Cell. Biochem.*, **33**:53-63.

Thompson, S. A., 1990, *in*: Molecular Industrial Mycology: Systems and Applications (Leong, S. A. & Berka, R. M., eds.). in press. Marcel Dekker, New York.

Tonouchi, N., Shoun, H., Uozumi, T. & Beppu, T., 1986, *Nucl. Acids Res.*, **14**:7557-7568.

Tsujita, Y. & Endo, A., 1976, *Biochim. Biophys. Acta*, **445**:194-204.

Tsujita, Y. & Endo, A., 1977, *J. Biochem.*, **81**:1063-1070.

Tsuru, D., Shimada, S., Maruta, S., Yoshimoto, T., Oda, K., Murao, S., Miyata, T. & Iwanaga, S., 1986, *J. Biochem.*, **99**:1537-1539.

Tsuru, D., Kobayashi, R., Nakagawa, N. & Yoshimoto, T., 1989a, *Agric. Biol. Chem.*, **53**:1305-1312.

Tsuru, D., Naotsuka, A., Kobayashi, R., Yoshimoto, T., Oda, K. & Murao, S., 1989b, *Agric. Biol. Chem.*, **53**:2751-2756.

van den Berg, J. A., van der Laken, K. J., van Ooyen, A. J. J., Renniers, T. C. H. M., Rietveld, K., Schaap, A., Brake, A. J., Bishop, R. J., Schultz, K., Moyer, D., Richman, M. & Shuster, J. R., 1990, *Bio/technology*, **8**:135-139.

Ward, M., 1989a, *in*; Genetics and Molecular Biology of Industrial Microorganisms (Hershberger, C. L., Queener, S. W. & Hegeman, G., eds.). pp. 288-294. American Society for Microbiology, Washington.

Ward, M., 1989b, *in*: Proceedings of the EMBO - Alko Workshop on Molecular Biology of Filamentous Fungi (Nevalainen, H. & Penttil , M., eds.). pp. 119-128. Foundation for Biotechnical and Industrial Fermentation Research, Helsinki.

Ward, M., Wilson, L. J., Kodama, K. H., Rey, M. W. & Berka, R. M., 1990, *Bio/technology* **8**:435-440.

Woolford C. A., Daniels, L. B., Park, F. J., Jones, E. W., van Arsdell, J. N. & Innis M. A., 1986, *Mol. Cell. Biol.*, **6**:2500-2510.

Yamashita, T., Tonouchi, N., Uozumi, T. & Beppu, T., 1987, *Mol. Gen. Genet.* **210**:462-467.

CHARACTERIZATION OF THE BAR PROTEINASE, AN EXTRACELLULAR

ENZYME FROM THE YEAST *SACCHAROMYCES CEREVISIAE*

Vivian L. MacKay, Jacque Armstrong, Carli Yip, Susan Welch,
Kathy Walker, Sherri Osborn, Paul Sheppard, and John Forstrom

ZymoGenetics, Inc.
4225 Roosevelt Way NE
Seattle, Washington   98105

INTRODUCTION

Haploid *S. cerevisiae* cells of the **a** mating type constitutively secrete an extracellular proteinase that cleaves the peptide mating pheromone (α-factor) secreted by mating-type α cells. DNA sequence analysis of the *BAR1* gene that encodes Bar proteinase demonstrated that the primary translation product of 587 amino acids has strong homology to two-domain aspartic proteinases such as pepsin, chymosin, and others, but contains a unique third domain that is not homologous to these enzymes. When produced by wild-type yeast cells, the Bar enzyme exists as a heterogeneous, heavily glycosylated protein with apparent molecular weight >200,000 Da. By producing the proteinase in mutant yeast strains that are defective in glycosylation, we have been able to purify and characterize a homogeneous species. In this paper, we will describe some of the enzyme's physical properties and substrate requirements, as well as present data indicating that the third domain is required for secretion of the proteinase to the culture medium.

BIOLOGY OF BAR PROTEINASE

In heterothallic strains of the yeast *S. cerevisiae*, haploid cells exist as **a** or α mating types. These can grow and divide stably as haploids or can fuse to form stable **a**/α diploids. The conjugation or mating process, which has been studied extensively in the last twenty years, has been shown to require the secretion of and response to mating type-specific peptide pheromones. (For reviews, see Cross *et al.*, 1988; Herskowitz, 1989). Thus, **a** cells constitutively secrete **a**-factor which acts on α cells, and α cells produce α-factor which affects **a** cells. The pheromones act through cell type-specific surface receptors that are coupled to heterotrimeric G proteins. Response to the pheromone of the opposite cell type prepares the cell for fusion *via* several changes: synthesis of mating type-specific surface agglutination factors, alteration of cell wall mannan synthesis, induction or enhancement of

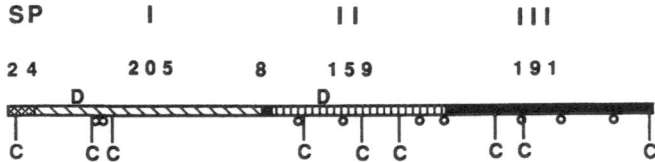

Figure 1. Schematic representation of the primary translation product of the *S. cerevisiae BAR1* gene. SP = putative signal peptide; I, II, and III = three domains of the polypeptide as predicted by sequence homology with aspartyl proteinases; D = predicted active site aspartic acid; C = cysteine; o = potential asparagine-linked glycosylation sites. Amino acids in each domain are indicated above the diagram.

transcription of several mating-specific genes, and arrest of the responding cell in the G1 phase of the cell cycle.

If cells do not mate, they can recover from G1 arrest and resume growth and cell division by one of two mechanisms: an adaptation response to the pheromone or inhibition/inactivation of the pheromone. In the latter case, **Mata** cells were shown to produce an extracellular activity, called Barrier, that acts as an antagonist of α-factor (Hicks & Herskowitz, 1976). As shown in a later section, Barrier activity is a proteinase that cleaves α-factor. Recently, Matα cells have also been shown to have such an activity that inactivates **a**-factor (Steden *et al.*, 1989). Mutant cells which lack these enzymes are super-sensitive to the pheromone of the opposite cell type (Sprague & Herskowitz, 1981; Chan & Otte, 1982; Steden *et al.*, 1989).

CHARACTERISTICS OF BAR PROTEINASE PREDICTED FROM DNA SEQUENCE ANALYSIS OF THE CLONED *BAR1* GENE

Using a *bar1* mutant that lacks Barrier activity, we cloned the gene encoding Bar proteinase by *in vivo* complementation of the mutation (MacKay *et al.*, 1988). The *BAR1* gene was shown to be the structural gene for the enzyme by the ability of *Schizosaccharomyces pombe* (an unrelated yeast) transformed with the *BAR1* gene to secrete the activity. DNA sequence analysis demonstrated that the primary translation product (587 amino acids) has significant homology to a variety of aspartyl proteinases and appears to be organized into several domains (Figure 1). Thus, there is a predicted amino-terminal signal peptide of 24 amino acids (von Heinje, 1986), two domains (205 and 159 amino acids separated by an 8 amino acid linker) with homology to the domains of aspartyl proteinases, particularly around the active site residues (Figure 2), and a third domain of 191 amino acids with no significant homology to these enzymes or other proteins in databases. (Domain boundaries were predicted solely from sequence homology.) Sequence homology between Bar's first two domains and bovine chymosin, for which crystal structure coordinates are available, allows the generation of a hypothetical structure for Bar proteinase (Figure 3). As expected for an aspartyl proteinase, mutation of the active site Asp in the second domain to either Glu or Ala abolished detectable activity, although the protein was still secreted into the culture medium (MacKay *et al.*, 1988). Unlike other aspartyl proteinases, however, the putative active site triad in the second domain is Asp-Ser-Gly, rather than Asp-Thr-Gly.

There are nine cysteine residues in the primary translation product: one in the proposed signal peptide, two in the first domain, three in the second domain, and three in the third domain. At least some of these appear to be disulfide bonded, as boiling partially purified fractions with reducing agents destroys the activity (although boiling in the absence of reducing agents does not). The polypeptide contains nine potential asparagine-linked glycosylation sites, all (or nearly all) of which are utilized. Since wild-type yeast cells generally hyperglycosylate such sites, the protein is secreted to the culture medium as a

```
PDB4AP  -> Endothiapepsin
PDB2AP  -> Penicillopepsin
PDB1CMS -> Chymosin


BAR1   : ltndgTGHLEFlLQHEEEMYYATTLDIGTPSQSLTVLFDTGSADFWVMds (50)
PDB4AP:     STGSATTTPIDSLDDAYITPVQIGTPAQTLNLDFDTGSSDLWVF-- (41)
PDB2AP:     AASGVATNTPTA-NDEEYITPVTIGG--TTLNLNFDTGSADLWVF-- (42)
1CMS   :     GEVASVPLTNYLDSQYFGKIYLGTPPQEFTVLFDTGSSDFWVP-- (43)
                                                       *

BAR1   : snpfclpnsntssysnatyngeevkpsidcrsMSTYNEHRSSTY-QYLEN (99)
PDB4AP: -SSETT-------ASEVDG-------------QTIYTPSKSTTAKLLSGA (69)
PDB2AP: -STELP-------ASQQSG-------------HSVYNP--SATGKELSGY (69)
1CMS   : -SIYCK-------SNA-----------CKNHQRFDPRKSSTF-QNLGK (71)


BAR1   : gRFYITYADGtfADGSWGTETVSINGIDIPNIQFGVAKYAT------tpv (143)
PDB4AP: -TWSISYGDGSSSSGDVYTDTVSVGGLTVTGQAVESAKKVSSSFTEDSTI (117)
PDB2AP: -TWSISYGDGSSASGNVFTDSVTVGGVTAHGQAVQAAQQQISAQFQQDTNN (118)
1CMS   : -PLSIHYGT-GSMQGILGYDTVTVSNIVDIQQTVGLSTQEPGDVFTYAEF (119)


BAR1   : SGVLGIGFPRRESVKGYegapneyypnfpQILKSEKiidvvaySLFLNSP (193)
PDB4AP: DGLLGLAFSTLNTVSPT---------QQKTFFDNAKAS--LDSPVFTADL (155)
PDB2AP: DGLLGLAFSSINTVQPQ---------SQTTFFDTVKSS--LAQPLFAVAL (157)
1CMS   : DGILGMAYPSLASEYSI-----------PVFDNMMNRHLVAQDLFSVYM (157)


BAR1   : D-SGTGSIVF-GAIDESKFSGDLFTFPMVNE--YPT-IVDAPATLAmtiq (238)
PDB4AP: GYHAPGTYNF-GFIDTTAYTGSITYTAVSTKQGFWEWTSTGYAVGS---- (201)
PDB2AP: KHQQPGVYDF-GFIDSSKYTGSLTYTGVDNSQGFWSFNVDSYTAGS---- (202)
1CMS   : DRNGQESMLTLGAIDPSYYTGSLHWVPVTV-QQYWQFTVDSVTISGVVV- (202)


BAR1   : glgaqnksscehETFTTTKYPVLLDSGTSLLNAPKVIadkmasf-vNASY (287)
PDB4AP: -----------GTFKSTSIDGIADTGTTLLYLPATVVSAYWAQVSGAKS (239)
PDB2AP: -------------QSGDGFSGIADTGTTLLLLDDSVVSQYYSQVSGAQQ (238)
1CMS   : --------ACE------GGCQAILDTGTSKLVGPSSDILNIQQ-AIGA-T (239)
                           *

BAR1   : SEEEGIYILDCPV--SVGDVEYNFDFGdlqISVPLSSLILSPETEGSY-C (334)
PDB4AP: SSSVGGYVFPCSA--TLPSFTFGVGSAR--IVIPGDYIDFGPISTGSSSC (283)
PDB2AP: DSNAGGYVFDCST--NLPDFSVSISGYT--ATVPGSLINYGPSGDGST-C (283)
1CMS   : QNQYGEFDIDCDNLSYMPTVVFEINGKM-YPLTPSAYT-----SQDQGFC (283)


BAR1   : GFAVQPTN--DSMVLGDVFLSSAYVVFDLDnykislaqan    (372)
PDB4AP: FGGIQSSAGIGINIFGDVALKAAFVVFNGATTPTLGFASK    (326)
PDB2AP: LGGIQSNSGIGFSIFGDIFLKSQYVVFDSDG-PQLGFAPQA   (323)
1CMS   : TSGFQSENHSQKWILGDVFIREYYSVFDRANN--LVGLAKAI  (323)
```

Figure 2. Sequence conservation between the *BAR1* primary translation product and members of the aspartyl proteinase family. Capitalized letters in the *BAR1* sequence indicate conservation with one or more of the other three enzymes. Active site aspartic acids are marked with asterisks.

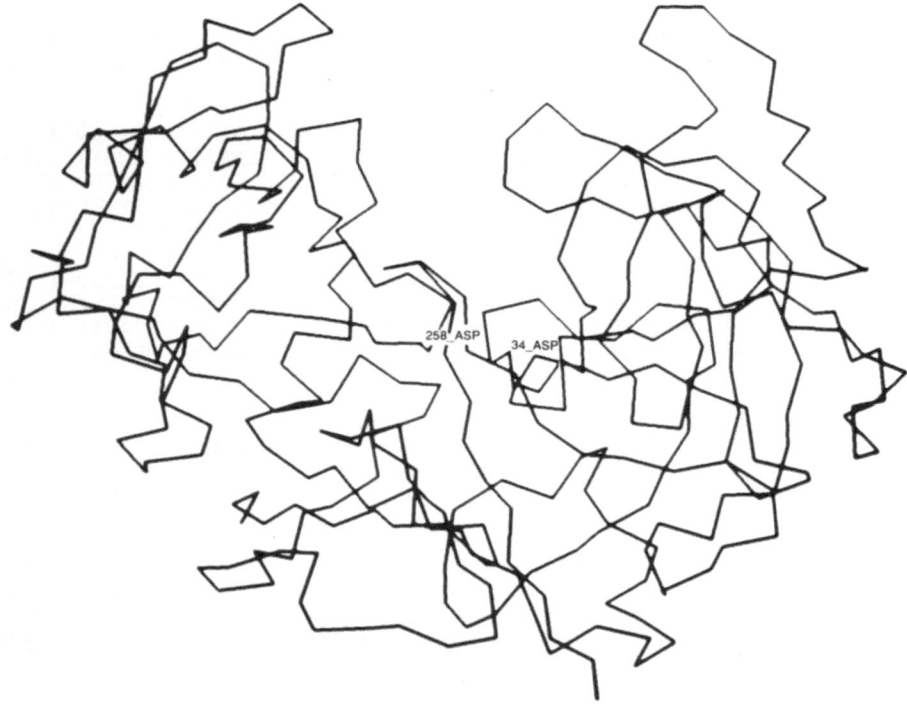

Figure 3. Three-dimensional structure assignment of the Bar proteinase as predicted from sequence homology with bovine chymosin. See Figure 2. First domain is on the right of the active site cleft and second on the left.

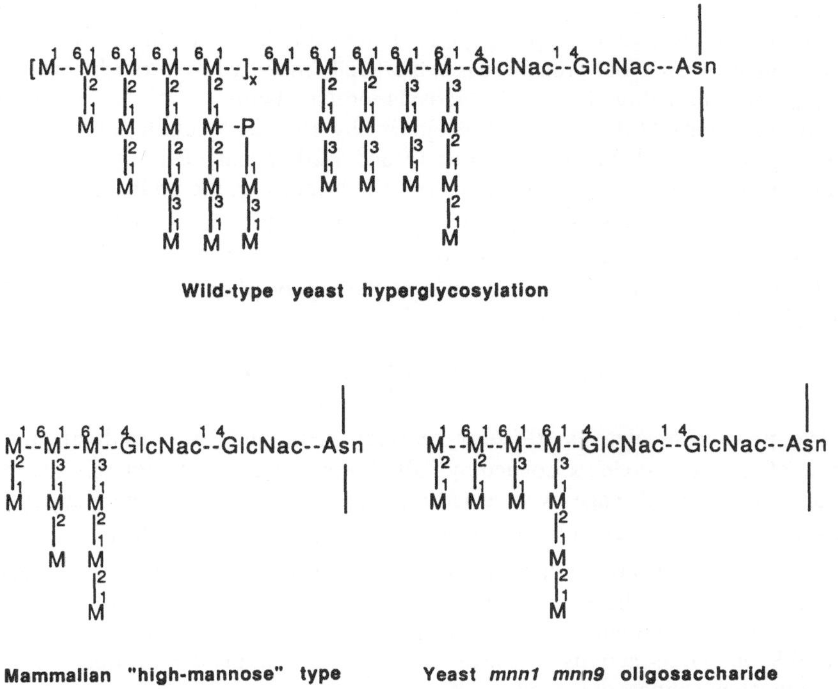

[M--M--M--M--M--]x--M--M--M--M--M--GlcNac--GlcNac--Asn (with branching mannose structures)

Wild-type yeast hyperglycosylation

M--M--M--GlcNac--GlcNac--Asn (with branching mannose structures)

Mammalian "high-mannose" type

M--M--M--M--GlcNac--GlcNac--Asn (with branching mannose structures)

Yeast *mnn1 mnn9* oligosaccharide

Figure 4. Yeast and mammalian asparagine-linked glycosylation. M = mannose; GlcNAc = N-acetylglucosamine; Asn = asparagine.

heterogeneous, heavily glycosylated moiety >200 kDa. In addition, the third domain is quite serine/threonine-rich and may contain substantial O-linked glycosylation (see below).

Sequence analysis of the 5' noncoding region suggested several sequences that might be involved in regulation of the gene, that is, its expression only in Mata cells, not in Matα or a/α cells. Kronstad *et al.* (1987) demonstrated that *BAR1* is transcribed only in a cells because of a 5' binding site for a repressor protein that is present in α and a/α cells. Moreover, *BAR1* transcription and expression of the proteinase are enhanced in a cells incubated with α-factor; two sequences which flank the repressor binding site are responsible for this increase. These sequences are found in the 5' noncoding region of several genes whose transcription is enhanced by α-factor treatment (Kronstad *et al.*, 1987).

CHARACTERIZATION OF THE BARRIER ACTIVITY AS A PROTEINASE

Purification and characterization of the enzyme has been hampered because of the heterogeneous hyperglycosylation of the protein. However, based on the extensive work of C. E. Ballou and co-workers (1990), we have recently developed stable mutant *mnn1 mnn9* yeast strains which add only a homogeneous $Man_{10}GlcNAc_2$ oligosaccharide (similar to mammalian high mannose type) to asparagine-linked glycosylation sites (MacKay *et al.*, 1990; see Figure 4). Expression of Bar proteinase was enhanced approximately 20-fold by replacing the *BAR1* promoter with the strong constitutive promoter from the *TPI1* gene encoding triose phosphate isomerase (Alber & Kawasaki, 1982) and ligating the expression unit into a multicopy number plasmid. Bar proteinase secreted by the *mnn1 mnn9* transformants migrated as a 92-95 kDa homogeneous protein (Figure 5). The predicted molecular weight of the polypeptide (assuming cleavage of the putative signal peptide) is 61.6 kDa, approximately the size of the polypeptide expressed in *E. coli* (data not shown). Thus, even in the mutant strains, Bar proteinase is substantially modified post-translationally by N-glycosylation and probably also by O-glycosylation (see below). We have purified the protein approximately 2000-fold to >90% purity by ethanol precipitation, followed by chromatography on DEAE-Sepharose, mono-Q, and S-200 resins. During purification, we employed a semi-quantitative biological assay which relies on the ability of isolated Bar proteinase to overcome the supersensitivity of *bar1* mutants to α-factor (Manney, 1983). This assay can detect as little as approximately 1 ng of the enzyme.

The natural substrate for Bar proteinase is the peptide pheromone α-factor (available from Sigma) with the following sequence:

Trp-His-Trp-Leu-Gln-Leu-Lys-Pro-Gly-Gln-Pro-Met-Tyr.

The proteinase cleaves α-factor between Leu_6 and Lys_7, as determined by sequencing of cleavage products isolated by HPLC (Figure 6B). Although this site may be considered unusual for cleavage by an aspartyl protease, it should be noted that pepsin can cleave α-factor at one of several sites, one of which yields cleavage products with the same retention times as those from Bar cleavage (Figure 6C). Using α-factor cleavage as detected by HPLC as an assay, we have determined reaction conditions for the purified proteinase. The enyzme has a pH optimum of approximately 5.0 - 5.3 but retains >50% activity from *ca.* pH 2.6 to 6.8, and slight activity (1 - 2%) can be detected at pH 1.1. It does not require Ca^{+2}, other divalent cations, or any other cofactors.

A variety of natural and synthetic peptides have been assayed for cleavage by Bar proteinase, as shown in Table 1. Bar proteinase apparently has quite strict substrate specificity for α-factor (substrate 1) and only closely related peptides (although Arg can substitute for Lys at the scissile bond, substrate 10), as it does not cleave a similar sequence

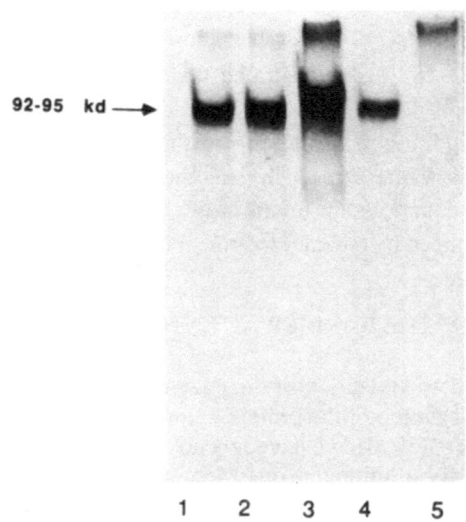

92-95 kd →

1   2   3   4   5

Figure 5. Immunoblot of secreted Bar proteinase from wild-type and mutant yeast strains. Lanes 1, 2, and 4, protein secreted by *mnn1 mnn9* mutant strains (see Figure 4); lane 3, protein secreted by a *mnn9* mutant; lane 5, protein secreted by a wild-type strain. Bar proteinase was concentrated from cell-free culture supernatants by ethanol precipitation (MacKay *et al.*, 1988) and detected with a rabbit polyclonal antibody raised against Bar polypeptide produced in *E. coli*.

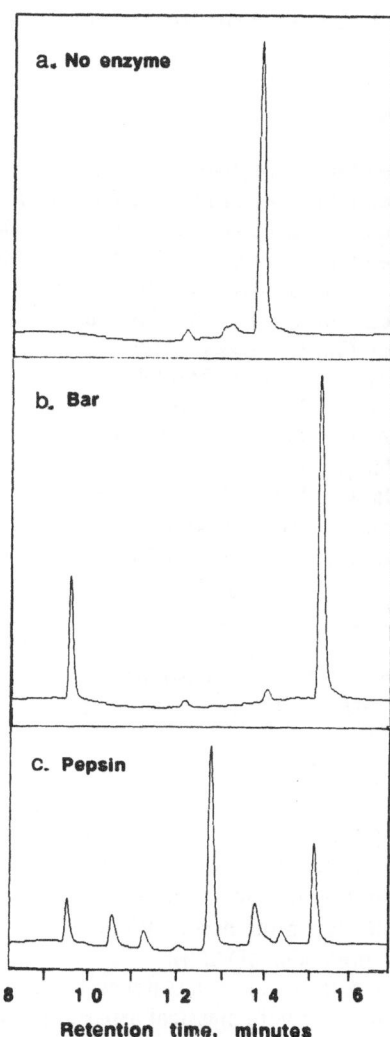

Figure 6. Detection of α-factor cleavage by HPLC. Standard reaction conditions (100 μl) are 50 mM sodium citrate, pH 5.0, 5 μg α-factor (*ca.* 30 μM), and ≤50 ng purified enzyme for 15 minutes at 30°C. Reactions are terminated by the addition of 1.5 μl of concentrated sulfuric acid and 90 μl was subjected to HPLC on a C18 column. Under the gradient conditions used, substrate eluted at approximately 14 minutes. With Bar proteinase, the C-terminal and N-terminal cleavage products eluted at approximately 9.4 and 15.3 minutes, respectively. a) No enzyme; b) Bar proteinase; c) porcine pepsin.

in platelet-derived growth factor (peptide 11) or in a synthetic peptide (14) provided by Ben Dunn. It is likewise inactive on two peptides generally cleaved by aspartyl proteinases (12 and 13). Like other aspartyl proteinases, Bar requires four amino acids (P_4 - P_1) on the amino terminal side of the scissile bond (peptides 2-5), but is inactive with even four residues (P_1'- P_4') on the carboxyl-terminal side of the bond (peptides 6-9), unlike other enzymes that require only three (Dunn *et al.*, 1987).

We have been unable to find any reversible inhibitors of Bar proteinase. As expected, the enzyme is resistant to EDTA, ε-amino caproic acid, and a number of serine protease inhibitors, including phenylmethyl sulfonylfluoride, aprotinin, leupeptin, tosyl arginyl methyl ester, and tosyllysyl methyl ester. However, it is also completely resistant to

Table 1. Substrate specificity of Bar proteinase[a]

	Peptide	Cleavage
1	Trp-His-Trp-Leu-Gln-*Leu-Lys*-Pro-Gly-Gln-Pro-Met-Tyr	+
2	His-Trp-Leu-Gln-*Leu-Lys*-Pro-Gly-Gln-Pro-Met-Tyr	+
3	Trp-Leu-Gln-*Leu-Lys*-Pro-Gly-Gln-Pro-Met-Tyr	+
4	Leu-Gln-Leu-Lys-Pro-Gly-Gln-Pro-Met-Tyr	-
5	Gln-Leu-Lys-Pro-Gly-Gln-Pro-Met-Tyr	-
6	His- Trp-Leu-Gln-Leu-Lys-Pro-Gly-Gln	-
7	Trp-Leu-Gln-Leu-Lys-Pro-Gly-Gln	-
8	Trp-Leu-Gln-Leu-Lys-Pro-Gly	-
9	Trp-Leu-Gln-Leu-Lys-Pro	-
10	Trp-His-Trp-Leu-Gln-*Leu-Arg*-Pro-Gly-Gln-Pro-Met-Tyr	+
11	· · · -Pro-Thr-Val-Gln-Leu-Arg-Pro-Val-Gln-Val-Arg-Lys-· · ·[b]	-
12	Lys-Pro-Ile-Glu-Phe-Nph-Arg-Leu[c]	-
13	Ala-Pro-Ala-Lys-Phe-Nph-Arg-Leu[c]	-
14	Arg-Phe-Leu-Glu-Nph-Lys-Pro-Gly-Gln-Pro[c,d]	-
15	Lys-Pro-Glu-Ile-Nph-Lys-Ser-Glu-Lys-Ile[c,e]	-

[a]Scissile bond indicated by italics. [b]Internal sequence of the B chain of platelet-derived growth factor, which was reduced and alkylated prior to incubation with Bar proteinase. [c]Synthetic peptides kindly provided by Dr. Ben Dunn. [d]Sequence based on α-factor. [e]Sequence based on a proposed internal cleavage site within Bar proteinase.

pepstatin A, although pepsin cleavage of α-factor under identical conditions was inhibited by 1 mM pepstatin. None of the uncleaved peptides in Table 1 inhibit Bar cleavage of α-factor. From its strict substrate specificity, it seems unlikely that any of the general reversible protease inhibitors will inhibit Bar proteinase. (We have not yet investigated irreversible inhibitors such as diazoacetyl norleucine [DAN]).

In summary, from the data available, it is not certain if the Bar enzyme is an aspartyl proteinase. Its sequence homology with classical aspartyl proteinases, proposed structure, inactivation by mutation of a putative active site aspartic acid, and its acid pH optimum are all consistent with this classification. However, its unique and strict substrate specificity and resistance to pepstatin A argue that this enzyme could differ significantly from aspartyl proteinases, possibly in structure or in mechanism of action.

ROLE OF THE THIRD DOMAIN IN PROMOTING EXPORT OF BAR PROTEINASE

In addition to the two presumably catalytic domains that are homologous to typical aspartyl proteinases, the Bar enzyme has a third domain of approximately 191 amino acids with a unique sequence. The domain contains three cysteines, four potential asparagine-linked glycosylation sites (see Figure 1), and 33% serine + threonine. SDS gel electrophoresis of nonglycosylated Bar polypeptide expressed in *E. coli* vs. the *mnn1 mnn9* yeast protein ± deglycosylation with endoglycosidase H (which removes N-linked, but not fungal O-linked carbohydrate) indicates that the polypeptide contains N-linked glycosylation as well as other substantial post-translational modification (S. Welch, unpublished). Based on analysis of expressed fragments of the polypeptide, a large part of the additional modification appears to occur in the third domain and probably arises from O-glycosylation

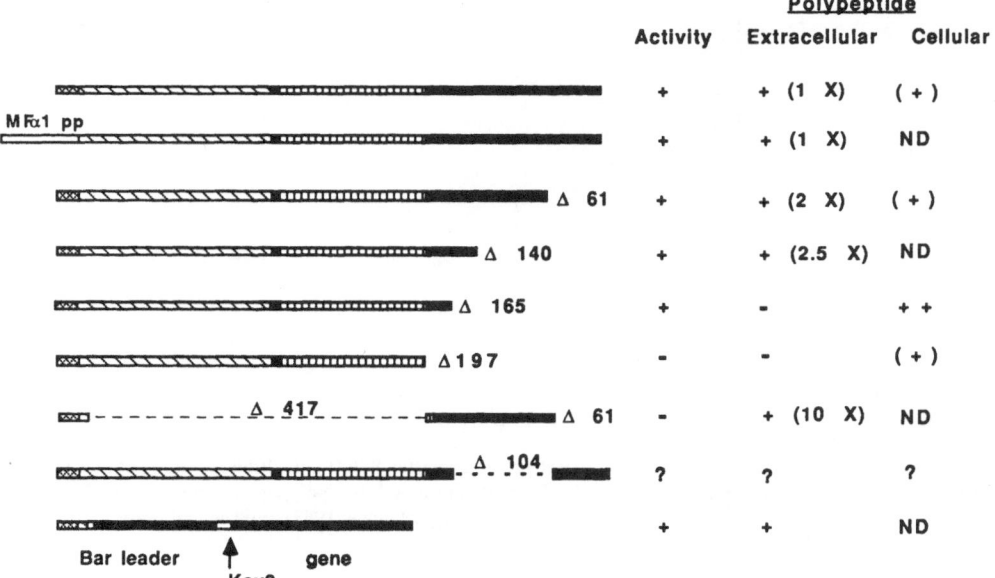

| | Activity | Polypeptide | |
		Extracellular	Cellular
	+	+ (1 X)	(+)
	+	+ (1 X)	ND
Δ 61	+	+ (2 X)	(+)
Δ 140	+	+ (2.5 X)	ND
Δ 165	+	-	+ +
Δ197	-	-	(+)
Δ 61	-	+ (10 X)	ND
	?	?	?
	+	+	ND

Figure 7. Effect of deletions in the third domain on protein export. See Figure 1 for explanation of the schematic representation of Bar polypeptide. Top figure, full length polypeptide; second figure, replacement of the Bar signal peptide with the preprosequence from the *S. cerevisiae MFα1* gene. See text for explanation of other figures.

of this serine/threonine-rich region, although other modifications cannot be ruled out at this time. It should be noted that many extracellular fungal enzymes contain O-glycosylated serine/threonine-rich domains (e.g., Salovuori *et al.*, 1987; Tomme *et al.*, 1988).

C-terminal deletion analysis indicated that at least part of the third domain is responsible for export of Bar proteinase. With the full length protein, at least 95% of the polypeptide is found in the culture medium under steady-state conditions. As shown in Figure 7, deletion of up to 140 amino acids from the carboxyl terminus did not decrease either activity or export, but removing an additional 25 amino acids eliminated nearly all exported activity, although significant cell-associated activity and protein could be detected. Removing 197 amino acids including six from the proposed C-terminus of the second domain led to loss of both activity and export. However, deletion of nearly all of the first two domains while retaining the third domain abolished proteinase activity but not export of the polypeptide fragment. These results suggest that the 25 amino acids between the endpoints of Δ140 and Δ165 may contain a specific sequence necessary for protein export or stability of the extracellular enzyme. Alternatively, perhaps a minimal size of the third domain (or amount of O-glycosylation?) is required for export and is relatively independent of the precise sequence. Bar expression plasmids containing internal deletions within the third domain are under construction to test these possibilities.

The properties of the third domain have been exploited in the design and construction of a novel leader for the secretion of heterologous recombinant proteins (Figure 7). A DNA restriction fragment encoding the proposed Bar signal peptide and the first 10 amino acids of the first domain was joined with another fragment representing the last 6 (approximately) amino acids of the second domain plus most of the third domain and an oligonucleotide encoding five amino acids of a Kex2 cleavage site. (The yeast Kex2 proteinase is a calcium-dependent serine protease that normally processes the precursors for α-factor and for killer toxin [Fuller *et al.*, 1989].) This leader has been compared to the one commonly used for the secretion of heterologous proteins (MacKay, 1986), that is, the preprosequence from the

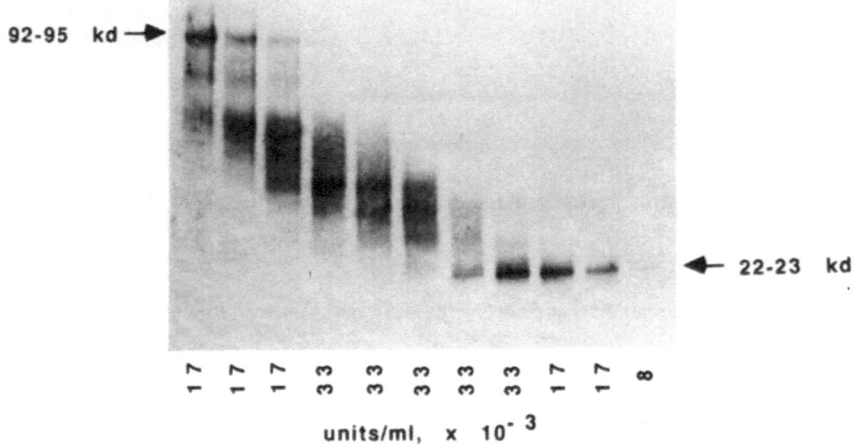

Figure 8. Immunoblot of active fractions of Bar proteinase after S-200 gel filtration. Polypeptide was detected with a rabbit polyclonal antiserum raised against full length Bar polypeptide produced in *E. coli*. Activity in each fraction was determined with a biological assay (Manney, 1983); units are arbitrary.

MFα1 gene that encodes the precursor for α-factor pheromone (Kurjan & Herskowitz, 1982). The Bar leader has been at least as effective as the *MFα1* signal for the secretion and export of human insulin precursor, platelet-derived growth factor, epidermal growth factor, transforming growth factor-α, and porcine urokinase.

SPECULATION ABOUT ACTIVE FRAGMENTS OF BAR PROTEINASE

As shown in Figure 5, Bar proteinase is secreted from *mnn1 mnn9* mutant cells as an apparent 92-95 kDa protein. However, during purification the protein undergoes extensive proteolysis with little or no loss of activity. S-200 gel filtration of pooled fractions from mono-Q chromatography resulted in several overlapping protein peaks with Bar proteinase activity in all fractions. Western blot analysis of these fractions demonstrated the existence of multiple, fairly discrete fragments, at least some of which appeared to retain activity (Figure 8). We were particularly intrigued by the small fragment (*ca.* 22-23 kDa; last 3-4 lanes), as this is the size predicted if a glycosylated fragment were released from the first domain by cleavage near the end of the domain. Perhaps such a monodomain fragment could form an active homodimer as shown for the HIV protease (Darke *et al.*, 1989; Meek *et al.*, 1989). It has also recently been reported (Bianchi *et al.*, 1990) that under controlled conditions pepsin can cleave itself near the end of its first domain, releasing a fragment that can be isolated as an active homodimer.[a] In any case, our results suggest that fragments of Bar proteinase retain proteolytic activity, although the precise sites and the enzyme(s) responsible for cleavages are unknown at this time.

---

[a] It should be noted however that, although we consistently observe cleavage of the purified enzyme, the small 22-23 kDa fragment is not always detected among the cleavage products. Western blots with an antibody specific for the third domain showed that some of the cleavages remove at least part of this domain. Moreover, cleavage patterns in purified samples of the mutant enzyme Bar:D287A indicate that some but probably not all of the cleavage is due to a contaminating protease.

ACKNOWLEDGMENTS

We wish to thank Teresa Gilbert and Peter Lockhart for last-minute DNA sequencing and oligonucleotide synthesis, respectively, and Ben Dunn for peptide substrates and useful discussion.

REFERENCES

Alber, T., & Kawasaki, G., 1982, Nucleotide sequence of the triose phosphate isomerase gene of *Saccharomyces cerevisiae*, *J. Mol. Appl. Genet.*, **1**:419.

Ballou, C. E., 1990, Isolation, characterization, and properties of *Saccharomyces cerevisiae mnn* mutants with nonconditional protein glycosylation defects, *Meth. Enzymol.*, **185**:440.

Bianchi, M., Boigegrain, R. A., Castro, B., & Coletti-Previero, M.-A., 1990, N-terminal domain of pepsin as a model for retroviral dimeric aspartyl protease, *Biochem. Biophys. Res. Comm.*, **167**:339.

Chan, R. K., & Otte, C. A., 1982, Isolation and genetic analysis of *Saccharomyces cerevisiae* mutants supersensitive to G1 arrest by a-factor and α-factor pheromones, *Mol. Cell. Biol.*, **2**:11.

Cross, R., Hartwell, L. H., Jackson, C., & Konopka, J. B., 1988, Conjugation in *Saccharomyces cerevisiae*, *Ann. Rev. Cell Biol.*, **4**:429.

Darke, P. L., Leu, C.-T., Davis, L. J., Heimbach, J. C., Diehl, R. E., Hill, W. S., Dixon, R. A. F., & Sigal, I. S., 1989, Human immunodeficiency virus protease: Bacterial expression and characterization of the purified aspartic protease, *J. Biol. Chem.*, **264**:2307.

Dunn, B. M., Jimenez, M., Weidner, J., Pennington, M., Carter, M., & Parten, B., 1987, Kinetic product analysis of aspartyl proteinases utilizing new synthetic substrates and reversed phase HPLC, *in*: "Proteins," J. J. L'Italien, ed., Plenum, New York.

Fuller, R. S., Brake, A., & Thorner, J., 1989, Yeast prohormone processing enzyme (KEX2 gene product) is a Ca^{2+}-dependent serine protease, *Proc. Natl. Acad. Sci., U.S.A.*, **86**:1434.

Herskowitz, I., 1989, A regulatory hierarchy for cell specialization in yeast, *Nature (London)*, **342**:749.

Hicks, J. B., & Herskowitz, I., 1976, Evidence for a new diffusible element of mating pheromones in yeast, *Nature (London)*, **260**:246.

Kurjan, J., & Herskowitz, I., 1982, Structure of a yeast pheromone gene: A putative α-factor precursor contains four tandem copies of mature α-factor, *Cell*, **30**:933.

Kronstad, J. W., Holly, J. A., & MacKay, V. L., 1987, A yeast operator overlaps an upstream activation site, *Cell*, **50**:369.

MacKay, V. L., 1986, Secretion of heterologous proteins in yeast, *in*: "The Biochemistry and Molecular Biology of Industrial Yeasts," G. G. Stewart, I. Russell, R. D. Klein, and R. R. Hiebsch, eds., CRC Press, Boca Raton, Florida.

MacKay, V. L., Welch, S. K., Insley, M. I., Manney, T. R., Holly, J., Saari, G. C., & Parker, M. L., 1988, *The Saccharomyces cerevisiae BAR1* gene encodes an exported protein with homology to pepsin, *Proc. Natl. Acad. Sci., U.S.A.*, **85**:55.

MacKay, V. L., Yip, C., Welch, S., Gilbert, T., Seidel, P., Grant, F., & O'Hara, P., 1990, Glycosylation and export of heterologous proteins expressed in yeast, *in*: "Recombinant Systems in Protein Expression," K. K. Alitalo, M.-L. Huhtala, J. Knowles, & A. Vaheri, eds., Elsevier, Amsterdam.

Manney, T. R., 1983, Expression of the *BAR1* gene in *Saccharomyces cerevisiae*: Induction by the α mating pheromone of an activity associated with a secreted protein, *J. Bacteriol.*, **155**:291.

Meek, T. D., Dayton, B. D., Metcalf, B. W., Dreyer, G. B., Strickler, J. E., Gorniak, J. G., Rosenberg, M., Moore, M. L., Magaard, V. W., & Debouck, C., 1989, Human immunodeficiency virus 1 protease expressed in *Escherichia coli* behaves as a dimeric aspartic protease, *Proc. Natl. Acad. Sci., U.S.A.*, **86**:1841.

Salovuori, I., Makarow, M., Rauvala, H., Knowles, J., & Kaariainen, L., 1987, Low molecular weight high-mannose type glycans in a secreted protein of the filamentous fungus *Trichoderma reesei*, *Bio/Technology*, **5**:152.

Sprague, G. F., Jr., & Herskowitz, I., 1981, Control of yeast cell type by the mating type locus. II. Identification and control of expression of the a-specific gene *BAR1*, *J. Mol. Biol.*, **153**:305.

Steden, M., Betz, R., & Duntze, W., 1989, Isolation and characterization of *Saccharomyces cerevisiae* mutants supersensitive to G1 arrest by the mating hormone a-factor, *Mol. Gen. Genet.*, **219**:439.

Tomme, P., Van Tilbeurgh, H., Pettersson, G., Van Damme, J., Vandekerckhove, J., Knowles, J., Teeri, T., & Claeyssens, M., 1988, Studies of the cellulolytic system of *Trichoderma reesei* QM 9414: Analysis of domain function in two cellobiohydrolases by limited proteolysis, *Eur. J. Biochem.*, **170**:575.

von Heinje, G., 1986, A new method for predicting signal sequence cleavage sites, *J. Mol. Biol.*, **184**:99.

CANDIDA ALBICANS ACID PROTEINASE: CHARACTERIZATION AND

ROLE IN CANDIDIASIS

Thomas L. Ray, Candia D. Payne and Brian J. Morrow+

The Marshall Dermatology Research Laboratories
Departments of Dermatology and +Biology
University of Iowa College of Medicine
Iowa City, Iowa   52242

INTRODUCTION

Candida albicans and related species are medically important yeast-like dimorphic fungi that are responsible, in part, for the rising incidence of serious, life-threatening opportunistic infections seen in immunocompromised and debilitated patients. In addition to minor localized infections of cutaneous, oropharyngeal and vaginal epithelium, *Candida* spp. produce transient or persistent fungemia that leads to systemic infections of virtually any organ (esp. liver, kidney and lung). Hematologic malignancy and transplant patients with prolonged neutropenia and antibiotic resistant fevers are particularly at risk. *C. albicans* and *C. tropicalis* are medically the most important, although other species do cause disease, but less frequently.

Identification of *Candida* virulence factors and traits has become increasingly important, but those identified to date have been of low potency. None compares in activity or action to classical bacterial exotoxins and endotoxins. Notably no single virulence factor has been deemed essential for the pathogenicity of *Candida* in any setting, though many appear to facilitate or contribute to the infection process. That a common commensal organism like *C. albicans* would possess highly active, potent toxins seems unlikely. The potential for opportunistic pathogenicity appears more likely to be derived from multiple low-potency factors, acting in concert, in the setting of altered and reduced host resistance. Among these factors, there is growing interest in secretory and membrane-associated hydrolytic enzymes.

Candida acid proteinase (CAP) is a putative *Candida* virulence factor. CAP is both a secreted (extracellular) and a cell wall-associated aspartyl proteinase that is predominantly expressed by the more pathogenic species of the *Candida* genus. Characterization of this enzyme, its role in the pathogenesis of candidiasis and expression by high-frequency "switch-phenotypes" will be reviewed.

Structure and Function of the Aspartic Proteinases
Edited by B.M. Dunn, Plenum Press, New York, 1991

The ability of *Candida* species to utilize keratin as a sole source of nitrogen[1] for growth suggested the possibility that an extracellular proteolytic activity was generated by these organisms. Staib[2] and co-workers[3] first demonstrated and purified an acidic proteinase (CAP) from *C. albicans*, the most medically important of the *Candida* species. Staib also demonstrated greater virulence of proteolytic strains than non-proteolytic strains of *C. albicans* in experimental murine *Candida* infections,[4] making the first association of this enzyme to strain virulence.

Several purification schemes and characterizations by different investigators[3,5,6,7,8] have revealed that the extracellular acid proteolytic activity of *C. albicans* is largely attributed to a carboxyl acid (aspartyl) proteinase (EC 3.4.23.6), possessing pepsin-like and cathepsin D-like properties. The secretory CAP is a single polypeptide chain mannoprotein with a molecular mass of 41 kilodaltons (range 40 to 45 kilodaltons), an optimal activity range of pH 2.2 to 4.5 (activity up to pH 5.5), an isoelectric point of 4.1 to 4.5 and a K_m for albumin of 70 mM.[3,5,7,8,9,10]

Generally inactive at neutral pH, CAP is irreversibly denatured by alkaline conditions (pH 7.5 to 8.5). It is inhibited by equimolar pepstatin, but 1,2-epoxy-3-(*p*-nitrophenoxy) propane (EPNP) and diazoacetylnorleucine methyl ester (DAN) have no effect. Chymostatin, but not leupeptin, partially inhibits the enzyme. It is resistant to phenylmethylsulfonyl fluoride (PMSF), soybean trypsin inhibitor (SBTI), N-ethylmaleimide (NEM), EDTA and EGTA.[5,7,8,9,10] The acid proteinase activity is partially inhibited by the serum protease inhibitors alpha-2-macroglobulin and alpha-1-antitrypsin[7,11] (summarized in Table 1).

Broad substrate susceptibility supports CAP as a general proteinase. Substrates include albumin, hemoglobin, transferrin, casein, IgA immunoglobulin heavy chains (but not light chains), keratin and denatured collagen.[3,5,7,8,10,12] Proteolysis of IgA may participate in Candida infections of mucosal surfaces, while keratinolytic activity may facilitate invasion across epithelial barriers. The collagenolytic activity might contribute to tissue invasion but remains speculative.

C-terminal leucine and N-terminal tryptophan residues have been reported,[5] but they do not agree with the deduced sequence derived from a proposed secretory CAP cDNA sequence.[13] Obtained by hybridization with a *S. cerevisiae* yeast *PrA* probe, this sequence clearly codes for an aspartyl proteinase, having high homology with porcine pepsin in the active site regions. Whether the clone codes for a cytoplasmic aspartyl proteinase (similar to non-secreted *S. cerevisiae* proteinase A) or secretory CAP is not resolved. Our preliminary studies of the N-terminal amino acid sequence (amino acids 1-24) from highly purified secretory CAP show no homology with the derived sequence of the above cDNA, nor is the terminal residue tryptophan. However, this sequence has partial homology (33% over a 21 residue overlap, plus 12 conservative replacements) with a reported N-terminal sequence for a secretory aspartic proteinase from the fungus *Rhizopus niveus*.[14] Registration is offset by only two residues from the N-terminus of the active *Rhizopus* enzyme (T.L. Ray, unpublished data). CAP enzyme is rich in aspartic acid, glutamic acid, serine and glycine.[5]

CAP is commonly secreted by the vast majority of *C. albicans* strains,[10] as well as most *C. stellatoidea* and *C. tropicalis* strains. Only occasional *C. parapsilosis* strains secrete low amounts of CAP, and it is rarely if all secreted by *C. guilliermondii*, *C. krusei*, *C. glabrata* and *C. pseudotropicalis*.[8,15,16] This correlates with the relative ranking of *Candida* species that cause medically important human infections, and exhibit virulence in experimental animal models.[17] Enzyme secretion is low when media contains simple nitrogen sources (amino acids or ammonium salts), but is elevated if the sole nitrogen source is protein.[18] The same enzyme is secreted independently of the protein source of nitrogen.[8]

Table 1. Biochemical and Physical Properties of *Candida* Acid Proteinase

Molecular weight	41,500 - 42,000
pH optimum	4.0 - 4.5
Isoelectric point (pI)	4.5
pH sensitivity	Inactivated at pH 7.5
Heat lability	100° C for 10 minutes
Heat stability	50° C
Amino acid composition	Asp, Ser, Glu, Gly ↑
PAS* and Con A† staining	+ (Mannoprotein)
Effect of divalent cations (1mM):	
Ca++, Mg++, Zn++, Cu++, Mn++	No effect
Effect of enzyme inhibitors:	
SBTI (10-100 µg/ml)	No effect
NEM (10-100 µg/ml)	No effect
PMSF (1-4 mM)	No effect
Pepstatin (10 ng-100 µg/ml)	5-93% inhibition
Chymostatin (10-100 µg/ml)	23-46% inhibition
Leupeptin (10-100 µg/ml)	No effect
EPNP (10-100 µg/ml)	No effect
DAN (10-100 µg/ml)	No effect
EDTA (1-10 mM)	No effect
EGTA (1-10 mM)	No effect
α-2-macroglobulin (10-100 µg/ml)	
pH 4.0	23-35% inhibition
pH 6.8	27-95% inhibition
Substrate susceptibility:	
BAEE (pH 6.2)	None
TAME (pH 8.1)	None
Collagen (denatured)	
pH 7.3	None
pH 4.5	Susceptible
Keratin azure (pH 3.5)	None
Bovine keratin	
pH 7.3	None
pH 4.5	Susceptible
Bovine serum albumin	
pH 7.3	None
pH 3.5 to 5.5	Susceptible
Hemoglobin	
pH 7.3	None
pH 4.5	Susceptible
Casein	
pH 7.3	None
pH 4.5	Susceptible

* PAS = Periodic acid-Schiff; † = Concanavalin A lectin

CAP is presumed to facilitate growth of *Candida* species in acidic environments that are poor in simple nitrogen sources, but rich in protein. Growth under these conditions does follow CAP dose-dependent kinetics. Glycolysis of carbon sources (sugars) provides sufficient acidification to generate the obligatory acid pH for enzyme activity.

ACID PROTEINASE AND *CANDIDA* VIRULENCE

Evidence for CAP as a virulence factor is mostly indirect or supportive in nature. *Candida* acid proteinase is secreted predominantly by pathogenic species, and not by non-pathogenic species of the *Candida* genus. Those species capable of invading the epidermis and dermis of the skin in experimental murine infections (*C. albicans, C. stellatoidea* and *C. tropicalis*) actively secrete CAP in culture and at sites of invasion in lesional skin, as demonstrated by direct immunofluorescence studies.[19,20,21]

Similarly, CAP antigen has been found in lesional tissue of systemic Candida infections in both humans and experimental animals.[6,22,23] Ruchel found CAP deposits in thrombosed cutaneous vessels of patients with disseminated candidiasis,[22] while Kondoh *et al.*[23] and Macdonald and Odds[6] found proteinase antigen in kidneys of experimental murine disseminated candidiasis.

Patients with *Candida* septicemia or local invasive infection have elevated circulating antibody to *Candida* acid proteinase as measured by enzyme immunoassay techniques. Macdonald and Odds[6,24] found 75% of patients with deep-seated *Candida* infections had elevated titers. Our studies found 13 of 14 *C. albicans* septicemia patients and 36 of 38 localized systemic *C. albicans* infection patients had IgG anti-CAP titers greater than 2 standard deviations above the mean of either controls or high risk patients without candidiasis.[25] Anti-CAP titers rose within one week of infection. Ruchel, in a study of 516 patients, found 11% of mucocutaneous candidiasis patients and 42% of visceral candidiasis patients had detectable CAP antigen.[26] Acid proteinase is undoubtably produced and secreted during the course of *Candida* infections.

Secretory acid proteinase activity of *Candida* strains has been correlated with more virulent behavior in experimental systemic infections in animals, beginning with Staib's correlation of proteolytic activity and increased lethality in genetically unrelated *C. albicans* strains.[4] More direct evidence has come from virulence studies using specific CAP-deficient *C. albicans* strains. Nitrosoguanidine induced mutants, deficient in CAP, are less virulent in mice and more susceptible to phagocytosis by human neutrophils than their CAP-sufficient parents.[27] Kwon-Chung *et al.* demonstrated reduced virulence in a nitrous acid induced CAP-deficient mutant. However, a CAP-sufficient spontaneous revertant was recovered by passage through mice which possessed virulence equal to that of the original parent.[28] This remains the most direct evidence for CAP as virulence factor to date. (Data summarized in Table 2.) A correlation has not been related to proteinase secretion in human clinical isolates.[29]

ROLE OF *CANDIDA* ACID PROTEINASE IN PATHOGENESIS

Adherence

A critical early step in the process of colonization and infection is the active adherence of blastoconidia (the yeast-like or spore forms of *Candida*) to oral, vaginal and skin epithelial cells. CAP-secreting strains of *C. albicans* possess enhanced adherence to buccal mucosal cells suggesting this enzyme may participate in the adherence phenomenon. A high correlation (r = 0.85) between proteinase secretion and adherence is reported[30] and the

Table 2. Virulence of CAP-Deficient Mutant Strains in Experimental Murine Systemic Candidiasis*

Strain	Strain Status	CAP Status	Average Survival Time (N=10)
C9	Original parent	CAP-sufficient	8.1 days
C9M1	Nitrous acid mutant	CAP-deficient	70% > 70 days
C9M1M	Spontaneous revertant	CAP-sufficient	8.6 days

*Data adapted from reference 28.

phenomenon is partially inhibited (89%) by pepstatin.[31] Scanning electron microscopy of *Candida* adherence to skin corneocytes reveals that adherent blastoconidia form circumjacent cavitations in the keratin-rich corneocytes. Embedded in these depressions, the organism becomes more firmly attached and initiates invasion steps. Cavitations form only about adherent CAP-secreting species and are blocked by pepstatin inhibition of CAP, implying a role for the enzyme in this process.[32] Similar cavitations are seen about adherent *Candida* blastoconidia adherent to buccal, gastrointestinal and vaginal epithelium.

Invasion

Proteolytic degradation of keratin and collagen[7,8,12,33] by CAP suggests a potential role in facilitating cell damage or tissue invasion by *Candida* hyphae, the fungal form of the organism. Only CAP-secreting species invade keratin-rich epidermis and collagen-rich dermis in experimental skin infections, while other species do not.[19,20,21] A study of *Candida* invasion in a chorioallantoic membrane model showed strong correlation between invasion and six CAP-sufficient, but not two CAP-deficient, *C. albicans* strains.[32] Invasion was associated with CAP antigen (demonstrated immunohistochemically) at invasion sites and the maintenance of acidity during culture. Both conditions may be important for successful tissue invasion. Similarly, CAP antigen has been demonstrated at invasion sites in experimental cutaneous murine candidiasis.[21]

Resistance to Phagocytosis

Conflicting studies associate CAP with both increased resistance[27] and increased susceptibility[33] to *Candida* uptake and killing by human polymorphonuclear cells. CAP-deficient mutants were less resistant than parents to killing in one study, suggesting a protective role for CAP. In the other study, unrelated strains, disparate for CAP, showed CAP secreting strains more susceptible to killing. It was speculated that CAP activated candidacidal factors in phagocytes. The potential for other variations between strains, in addition to CAP, obscures the interpretation of these studies, and the issue remains unresolved.

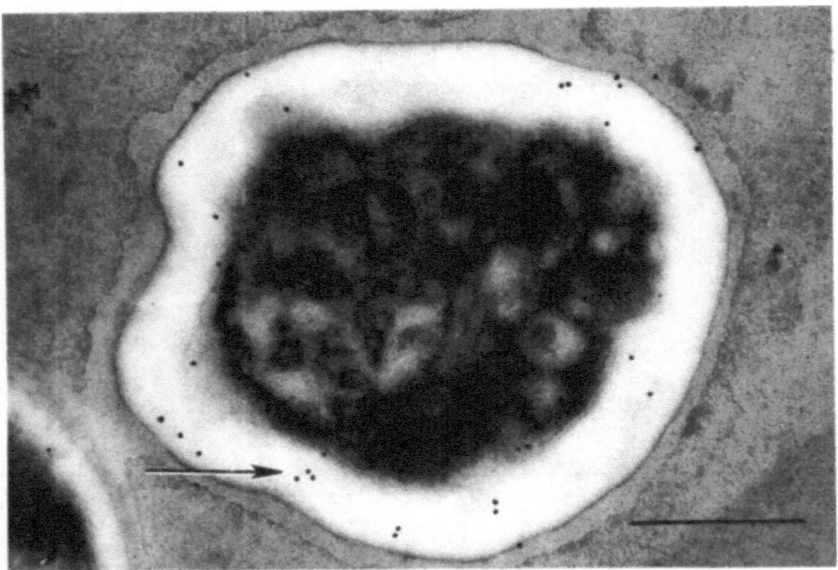

Figure 1. Immunoelectron micrograph of phenotype "opaque" blastoconidiae of *C. albicans* strain WO-1, stained with mouse monoclonal IgG antibody to CAP and visualized with 10 nm gold particle-conjugated goat anti-mouse IgG. CAP antigen is localized predominantly in the cell wall and peripheral cytoplasm. Focal columns of CAP (arrow) suggest specialized regions for enzyme secretion through the cell wall. Phenotype "white" of WO-1 lacks CAP and does not stain with anti-CAP antibody. (Bar = 1 micron)

Activation of Inflammatory Mediators

Recently, CAP has been shown to activate the plasma kinin-generating system, with production of vasoactive bradykinin in pharmacological amounts, by both *in vitro* and *in vivo* assays.[36] Activation is optimal at pH 5.2-5.5 and appears to be initiated by limited proteolysis of Factor XII (Hageman factor) by *Candida* proteinase, suggesting a procoagulant effect. A relationship to Factor X (Stewart-Prower factor) activation by *Candida* proteinase, as described by Ruchel,[37] is unclear. Whether intermediate zymogens of serine proteases are activated in this process is not resolved. However, CAP may potentially generate proinflammatory factors from the intrinsic coagulation and kallikrein-kinin systems by either direct or indirect mechanisms.

ACID PROTEINASE EXPRESSION BY HIGH-FREQUENCY "SWITCH-PHENOTYPES"

Individual strains of *C. albicans* and *C. tropicalis* spontaneously generate at high-frequency "switch phenotypes" that express variant colony and cellular morphologies.[38,39,40] With variants arising at rates of 10^{-2}, phenotype switching is not only frequent, but heritable and reversible. Strains express limited sets of colony morphologies (descriptively named, e.g. "star", "ring", "opaque", etc.) and each phenotype is capable of switching to any other variant and back, at similar frequencies. Restriction endonuclease digests of DNA confirm the strain relatedness of each phenotype. Similar phenotype switching has been shown in clinical isolates.[39,40] Because *Candida* is asexual and appears locked in a diploid state with balanced lethals, conventional recombinational events are unlikely. The genetic mechanism for this process is widely debated, but is unknown.

Table 3. Comparison of Experimental Systemic Murine Candidiasis with Phenotypes "White" and "Opaque" of *C. albicans* Strain WO-1

	"White" Infections	"Opaque" Infections
CAP Production	Low	High
Dose (# spores per tail vein)	5×10^6 (99% "White")	5×10^6 (99% "Opaque")
Survival (>30 d)	1/10 mice	3/10 mice
Mean Survival	5.8 days	16.7 days

Colony Forming Units at Death

Spleen	10^3	10^2
Kidney	10^6	10^6
Liver	10^3	10^2

% Phenotype recovered by organ over time

	"Opaque"	"White"
Spleen		
24 hrs	< 2	13
48 hrs	3	17
72 hrs	0	33
96 hrs	ND	41
120 hrs	ND	62
Kidney		
24 hrs	0	25
48 hrs	0	85
72 hrs	0	95
96 hrs	ND	100
120 hrs	ND	100
Liver		
24 hrs	6%	12
48 hrs	0	23
72 hrs	0	69
96 hrs	ND	69
120 hrs	ND	67

Switch phenotypes of laboratory strain 3153A and clinical isolate strain WO-1 variably express secretory CAP according to their specific phenotypes. Phenotype "wrinkled" of strain 3153A, but not "smooth" or "star", secretes moderate levels of functional CAP. Phenotype "opaque", but not "white", of strain WO-1 produces abundant levels of this enzyme[41] (Figure 1). The presence or absence of CAP is a constant trait of the specific phenotype expressed. The strong correlation of CAP secretion with specific phenotypes implies that regulation of this putative virulence factor is linked at least in part to the mechanism(s) of phenotype switching in *Candida* species.

The capability of *Candida* isolates to alter phenotypic, physiologic and secretory virulence traits at high frequency provides this eukaryotic organism with a greatly enhanced potential for adaptation. Phenotype switching may contribute to *Candida's* success as both a pathogen and a commensal by generating the diversity needed to adapt to the changing environment and defense mechanisms of the host.

To determine if switching and variable expression of CAP occured during murine systemic candidiasis, we studied the distribution and phenotype percentages of *Candida* colonies in visceral organs infected with *C. albicans* strain WO-1.[42, 43] Mice infected with phenotype "white" (low CAP production) had a mean survival of 5.8 days with predominantly "white" phenotype (>95%) being recovered from kidney, liver and spleen. Conversely, mice infected with "opaque" phenotype (high CAP production) had a mean survival of 16.7 days, but all survived a minimum of 11 days. At death, 62-100% of colonies recovered from internal organs were the "white" phenotype, rather than the inoculated "opaque" phenotype. Over the course of "opaque" infections, switching occurred in virtually one direction toward the "white" phenotype (Data summarized in Table 3.).

A single strain of *C. albicans* can express markedly different virulence patterns depending on the predominant phenotype present in the inoculum. "Opaque" was initially less virulent than "white", but mortality increased as phenotype switching increased the proportion and numbers of the "white" phenotype. "White" is the more lethal phenotype of strain WO-1 in systemic infections. However, "opaque" phenotype infections can exhibit similar but delayed virulence, during which conversion to the "white" phenotype occurs.

The role of CAP as a virulence factor in these studies is not apparent, nor critically tested. CAP is produced by "opaque" and not "white" phenotypes, but "opaque" is less virulent in this model. The results are indeed the reverse of those obtained by Kwon-Chung et al.[28] using the C9 and related strains of *C. albicans*. Because of multiple other differences between the two WO-1 phenotypes, besides CAP production, it is not possible to isolate the contribution of this single factor to the strain's virulent behavior. An invasive role for CAP is circumvented in this study by the intravenous route of inoculation and can not be assessed. However, CAP may contribute to virulence in some settings and not others.

CONCLUSIONS

A major extracellular proteolytic activity of medically important pathogenic species of the genus *Candida* can be attributed to an aspartic proteinase (CAP) with pepsin and cathepsin D-like properties. Its broad substrate specificity includes keratin and collagen, the major structural components of epithelial barriers. This secreted glycoprotein enzyme facilitates *Candida* growth by providing simple nitrogen sources through protein degradation. Circumstantial evidence supports CAP as a virulence factor that is capable of facilitating 1) adherence of the organism to epithelial cells during colonization, 2) invasion of epithelial tissues, 3) potential resistance to phagocytic cells and 4) generation of pro-inflammatory factors.

CAP is synthesized and secreted during human and experimental animal infections and deposited at sites of invasion and infection. Host antibody responses rise during such infections and CAP antigen can be detected, suggesting this enzyme might be a diagnostic marker of invasive or systemic infection.

Regulation of CAP synthesis/secretion is linked in part to the phenomenon of high frequency phenotype switching associated with some *Candida* species. Though CAP is neither a highly potent factor nor fully essential for some infection states, the cummulative studies support its participation in the pathogenesis of Candida infections. Rapid adaptation and variable expression of this putative virulence factor through phenotype switching may contribute to the dual behavior of this important opportunistic organism as a pathogen and a commensal. It also complicates the elucidation of the role of many traits and virulence factors controlled by switching mechanisms.

ACKNOWLEDGEMENTS

This work was supported in part by Public Health Service grants AI 18227 and AI 24344 from the National Institutes of Health.

REFERENCES

1. L. Kapica and F. Blank, Growth of *Candida albicans* on keratin as sole source of nitrogen, *Dermatologica* 115:81-105 (1957).
2. F. Staib, Serum-proteins as nitrogen source for yeast-like fungi, *Sabouraudia* 4:187-193 (1965).
3. H. Remold, H. Fasold and F. Staib, Purification and characterization of a proteolytic enzyme from *Candida albicans*, *Biochim. Biophys. Acta* 167:399-406 (1968).
4. F. Staib, Proteolysis and pathogenicity of *Candida albicans* strains, *Mycopathol Mycol Appl* 37:345-348 (1969).
5. R. Ruchel, Properties of purified proteinase from the yeast *Candida albicans*, *Biochim. Biophys. Acta* 659:99-113 (1981).
6. F. Macdonald and F.C. Odds, Inducible proteinase of *Candida albicans* in diagnostic serology and in the pathogenesis of systemic candidiasis, *J. Med. Microbiol.* 13:423-435 (1980).
7. M. Negi, R. Tsuboi, T. Matsui, and H. Ogawa, Isolation and characterization of proteinase from *Candida albicans*: Substrate specificity, *J. Invest. Dermatol.* 83:32-36 (1984).
8. T. L. Ray and C. D. Payne, Comparative production and rapid purification of *Candida* acid proteinase from protein-supplemented cultures, *Infect. Immun.* 58:508-514 (1990).
9. R. Ruchel, B. Boning and M. Borg, Characterization of a secretory proteinase of *Candida* parapsilosis and evidence for the absence of the enzyme during infection *in vitro*, *Infect. Immun.* 53:411-419 (1986).
10. R. Ruchel, R. Tegeler and M. Trost, A comparison of secretory proteinases from different strains of *Candida albicans*, *Sabouraudia* 20:233-244 (1982).
11. R. Ruchel and B. Boning, Detection of *Candida* proteinase by enzyme immunoassay and interaction of the enzyme with alpha-2-macroglobulin, *J. Immunol. Methods* 61:107-116 (1983).
12. M. Hattori, K. Yoshiura, M. Negi and H. Ogawa, Keratinolytic proteinase produced by *Candida albicans*, *Sabouraudia* 22:175-183 (1984).
13. T. J. Lott, L.S. Page, P. Boiron, J. Benson and E. Reiss, Nucleotide sequence of the *Candida albicans* aspartyl proteinase gene, *Nucleic Acid Res* 17:1779 (1989).
14. H. Horiuchi, K. Yanai, T Okazaki, M. Takagi and K Yano, Isolation and sequencing of a genomic clone encoding aspartic proteinase of *Rhizopus niveus*, *J. Bacteriol.* 170:272-278 (1988).
15. F. Macdonald, Secretion of inducible proteinase by pathogenic *Candida* species, *Sabouraudia* 22:79-82 (1984).
16. R. Ruchel, K. Uhlemann and B. Boning. Secretion of acid proteinases by different species of the genus Candida, *Zbl. Bakt. Hyg, I Abt. Orig. A* 255:537-548 (1983).

17. F.C. Odds, "*Candida* and Candidosis", 2nd ed. W.B. Saunders, Publ., Philadelphia, PA. (1988).

18. M. Crandall and J. E. Edwards, Jr., Segregation of proteinase-negative mutants from heterozygous *Candida albicans*. *J. Gen. Microbiol.* **133**:2817-2824 (1987).

19. T. L. Ray and K. D. Wuepper, Experimental cutaneous candidiasis in rodents, *J. Invest. Dermatol.* **66**:29-33 (1976).

20. T. L. Ray and K. D. Wuepper, Experimental cutaneous candidiasis in rodents. II. Role of the stratum corneum barrier and serum complement as a mediator of a protective inflammatory response, *Arch. Dermatol.* **114**:539-543 (1978).

21. T. L. Ray and C. D. Payne, Cytoplasmic Synthesis and Trans-Cell Wall Secretion of Acid Proteinase, a Virulence Factor, in Selected Switch Phenotypes of *Candida albicans, Clin. Res.* **38**:678A (1990).

22. R. Ruchel, On the role of proteinases from *Candida albicans* in the pathogenesis of acronecrosis, *Zbl . Bakt. Hyg., I Abt. Orig. A* **255**:524-536 (1983).

23. Y. Kondoh, K. Shimizu and K. Tanaka, Proteinase production and pathogenicity of *Candida albicans*. II. Virulence for mice of *C. albicans* strains of different proteinase activity, *Microbiol. Immunol.* **31**:1061-1069 (1987).

24. F. Macdonald and F. C. Odds, Purified *Candida albicans* proteinase in the serologic diagnosis of systemic candidosis, *J. Am. Med. Assoc.* **243**:2409-2411 (1980).

25. T. L. Ray and C. D. Payne, Detection of *Candida* acid proteinase (CAP) antibodies in systemic candidiasis by enzyme immunoassay, *Clin. Res.* **35**:711A (1987).

26. R. Ruchel, B. Boning-Stutzer and A. Mari, A synoptical approach to the diagnosis of candidosis, relying on serological antigen and antibody tests, on culture, and on evaluation of clinical data, *Mycoses* **31**:87-106 (1987).

27. F. Macdonald and F. C. Odds, Virulence for mice of a proteinase-secreting strain of *Candida albicans* and a proteinase-deficient mutant, *J. Gen. Microbiol.* **129**:431-438 (1983).

28. K. J. Kwon-Chung, D. Lehman, C. Good and P. T. Magee, Genetic evidence for role of extracellular proteinase in virulence of *Candida albicans, Infect. Immun.* **49**:571-575 (1985).

29. B. Schreiber, C. A. Lyman J. Gurevich, C. A. Needham, Proteolytic activity of *Candida albicans* and other yeasts, *Diagn. Microbiol. Infect. Dis.* **3**:1-5 (1985).

30. M. Ghannoum and K. Abu Elteen, Correlative relationship between proteinase production, adherence and pathogenicity of various strains of *Candida albicans J. Med. Vet. Mycol.* **24**:407-413 (1986).

31. M. Borg and R. Ruchel, Expression of extracellular acidproteinase by proteolytic *Candida* spp. during experimental infection of oral mucosa, *Infect. Immun.* **56**:626-631 (1988).

32. T. L. Ray and C. D. Payne, Scanning electron microscopy of epidermal adherence and cavitation in murine candidiasis: a role for *Candida* acid proteinase, *Infect. Immun.* **56**:1942-1949 (1988).

33. H. Kaminishi, Y. Hagihara, S. Hayashi and T. Cho, Isolation and characteristics of collagenolytic enzyme produced by *Candida albicans, Infect. Immun.* **53**:312-316 (1986).

34. K. Shimizu, Y. Kondoh and K. Tanaka, Proteinase production and pathogenicity of *Candida albicans*. I. Invasion into chorioallantoic membrane by *C. albicans* strains of different proteinase activity, *Microbiol. Immunol.* **31**:1045-1060 (1987).

35. T. Walther, M. Rytter, C. Schonborn and U.- F. Haustein, Differences in the intracellular killing of proteinase-positive and proteinase-negative *Candida albicans* strains by granulocytes, *Mykosen.* **29**:159-161 (1986).

36. H. Kaminishi, M. Tanaka, T. Cho, H. Maeda and Y. Hagihara, Activation of the plasma kallikrein-kinin system by *Candida albicans* proteinase, *Infect. Immun.* **58**:2139-2143 (1990).

37. R. Ruchel, On the renin-like activity of *Candida* proteinases and activation of blood coagulation *in vitro*, *Zbl. Bakt. Hyg., I Abt. Orig. A* **255**:368-379 (1983).

38. B. Slutsky, J. Boffo, and D. R. Soll, High frequency "switching" of colony morphology in *Candida albicans, Science* **230**:666-669 (1985).

39. B. Slutsky, M. Staebell, M. Pfaller and D. R. Soll. The "white-opaque transition": A second switching system in *Candida albicans, J. Bacteriol.* **169**:189-197 (1987).

40. D. R. Soll, R. Galask, S. Isley, T. V. G. Rao, D. Stone, J. Hicks, J. Schmid, K. Mac and C. Hanna, Switching of *Candida albicans* during successive episodes of recurrent vaginitis, *J. Clin. Microbiol.* **27**:681-690 (1989).

41. T. L. Ray, C. D. Payne and D. R. Soll, Variable expression of *Candida* acid proteinase by "switch-phenotypes" of individual *Candida albicans* strains, *Clin. Res.* **36**:687A (1988).

42. T. L. Ray and C. D. Payne, *Candida albicans* acid proteinase: a role in virulence, *in*: "Microbial Determinants of Virulence and Host Response," E. M. Ayoub, G. H.Cassell, W. C. Branche, Jr., and T. J. Henry, eds., Am Soc Microbiol, Washington, D.C. (1990).

43. T. L. Ray, C. D. Payne, B. J. Morrow and D. R. Soll, Switch phenotypes of *Candida albicans* strain WO-1 express diferent systemic and cutaneous virulence in murine models, *Clin. Res.* **37**:764A (1989).

PEPSTATIN-INSENSITIVE CARBOXYL PROTEINASES

Kohei Oda and Sawao Murao

Department of Agricultural Chemistry
University of Osaka Prefecture
Sakai, Osaka 591
Japan

INTRODUCTION

In 1960, when B. S. Hartley[1] classified proteinases based on the catalytic residue, no information was available concerning the catalytic residue of acid proteinases. The term "acid proteinase" was, therefore, given to the group of proteinases having an optimum pH in the acid region.

From such a background, we attempted to obtain a microbial inhibitor of acid proteinases, and obtained S-PI (Pepstatin Ac)[2] in 1970. Our discovery of S-PI and the development of active-site-directed reagents such as DAN (diazoacetyl-DL-norleucine methylester)[3] and EPNP [1,2-epoxy-3-(p-nitrophenoxy)propane][4] have resulted in remarkable progress in the study of acid proteinases. Now it is believed that two aspartic acid residues constitute the active site of the acid proteinases, and that the enzyme is specifically inactivated by pepstatin, DAN and EPNP.[5] Therefore, acid proteinases are now described as carboxyl or aspartic proteinases.

In this paper, we introduce the background of our research on pepstatin-insensitive carboxyl proteinase which should be classified in a new subclass,[6] and then we describe pepstatin-insensitive carboxyl proteinases isolated from *Pseudomonas* sp[7] and thermophilic *Bacillus* novosp.[8] The former was the first carboxyl proteinase isolated and characterized from a prokaryote, and the latter was the first carboxyl proteinase isolated from thermophilic bacteria. In addition, we describe a specific inhibitor of pepstatin-insensitive carboxyl proteinases, tyrostatin.[9]

THE DISCOVERY OF PEPSTATIN-INSENSITIVE CARBOXYL PROTEINASES AND THEIR PROPERTIES

The pepstatin-insensitive carboxyl proteinases which we describe here were found by testing with a carboxyl proteinase inhibitor, S-PI which we isolated in 1970.[2] The structure of S-PI is acetyl-valyl-valyl-AHMHA-Alanyl-AHMHA, as shown in Figure 1.[10] In the same

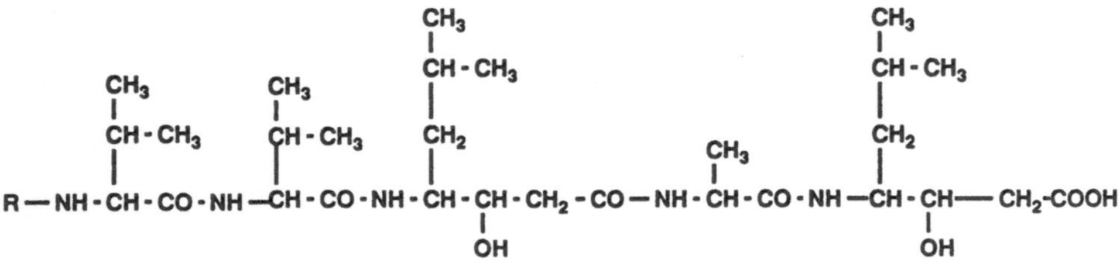

Acyl-Valyl-Valyl-AHMHA-Alanyl-AHMHA

Name	acyl group	organism	reference
S-PI	acetyl CH_3CO-	*St. naniwaensis*	2
Pepstatin	isovaleryl $(CH_3)_2CHCH_2CO-$	*St. testaceus*	11

Figure 1. Chemical structure of S-PI (Acetyl-pepstatin) and Pepstatin. AHMHA = 4-amino-3-hydroxy-6-methylheptanoic acid.

Table 1. Some Properties of *Scytalidium lignicolum* Carboxyl Proteinases

	S. lignicolum				Pepsin	*R. chinensis*
	A-l (A-2)		B	C		
Optimum pH	3.0 - 3.5		2.2	2.0	1.8 - 2.0	2.9 - 3.3
M W	40,000		21,969	360,000	34,500	35,000
pI	3.6 (3.8)		3.2	2.3	1.0	5.0 - 5.2
Amino acid composition	His; none		His; none	His; Yes	Ser; rich	His; none
Carbohydrate content	$\cong$ 10%		0	$\cong$ 30%	0	0
Inhibition with						
S-PI[a]	-		-	-	+	+
DAN	-		-	-	+	+
EPNP	-		+	-	+	+
Catalytic residues	-COOH[b]		-COOH[b] Glu$_{53}$ Asp$_{98}$	-COOH[b]	Asp$_{32}$ Asp$_{215}$	Asp$_{32}$ Asp$_{215}$

[a]S-PI; Acetyl-Pepstatin. [b]-COOH ; Zn(II)-PAD[31,32] and kinetics.[33]

Table 2. Hydrolysis of Z-X-Leu-Ala-Ala and Z-Phe-Y-Ala-Ala by *Scytalidium lignicolum* Carboxyl Proteinases

Carboxyl proteinase	k_{cat}/K_m (M^{-1} s^{-1})	Z—X—Leu—Ala—Ala Z—Phe—Y—Ala—Ala	
S. lignicolum A-1, A-2	8,000 - 11,000	↑	X=Phe, Ala, Tyr Y=Leu
S. lignicolum B	30 - 40	↑	X=Lys, Leu, Phe Y=Tyr, Lys
Pepsin and others*	200 - 1,000	↑	X=Phe, Leu Y=Tyr, Leu

Others* ; Carboxyl proteinases isolated from *Cladosporium* sp, *Rhizopus chinensis*, *Mucor miehei*, *Rhodotorula glutinus* and *Aspergillus saitoi*.

year H. Umezawa[11] isolated a pepstatin, containing isovaleryl as the acyl residue. Since 1970, we have examined the inhibitory spectrum of S-PI against all the carboxyl proteinases available. We considered that an S-PI-insensitive carboxyl proteinase, if there were any, should be distinct in the structure of the active site from any other known carboxyl proteinases.

From such a standpoint, we started screening for a particular microorganism producing an S-PI-insensitive carboxyl proteinase and succeeded in isolating *Scytalidium lignicolum* in 1972.[12,13] This strain produces four distinct carboxyl proteinases : A-1, A-2, B and C (Table 1).[14-17] None of them are inactivated by S-PI, DAN or EPNP, with the exception of carboxyl proteinase B, which is strongly inactivated by EPNP.

In addition to a unique behavior against inhibitors, carboxyl proteinases from *S. lignicolum* have unique substrate specificities.[18-20] When Z-tetrapeptides (Table 2)[18,19] such as Z-X-Leu-Ala-Ala and Z-Phe-Y-Ala-Ala are used as substrates, carboxyl proteinases A-1 and A-2 cleave these peptides at either the Leu-Ala or Y-Ala bond. The k_{cat}/K_m values are between 8,000 and 11,000 $M^{-1}sec^{-1}$. This is different from S-PI-sensitive carboxyl proteinases. Such S-PI-sensitive carboxyl proteinases as pepsin cleave the X-Leu and Phe-Y bonds. The k_{cat}/K_m values are between 200 and 1,000 $M^{-1}sec^{-1}$. Carboxyl proteinase B is even different from A-1 and A-2 because it cleaves the carboxyl terminal bonds, and its activity is very low compared with those of A-1 and A-2 enzymes. Thus the substrate specificities of carboxyl proteinases from *S. lignicolum* are distinguishable from those of other microbial carboxyl proteinases which are sensitive to pepstatin, DAN and EPNP.

We determined the amino acid sequence of carboxyl proteinase B (Figure 2).[21] The enzyme is a single polypeptide composed of 204 amino acid residues with a molecular weight of 21,969. The amino acid sequence of carboxyl proteinase B is quite different from those of pepstatin-sensitive carboxyl proteinases such as pepsin and penicillopepsin. Furthermore, unlike the other carboxyl proteinases, one of the catalytic residues of the enzyme is a glutamic acid. The active amino acid residue modified with EPNP was found to be Glu_{53}.[22,23] The other catalytic residue was found to be Asp_{98} by using a new inhibitor, 1-diazo-3-phenyl-2-propanone (DPP).[24]

The amino acid sequence around Glu_{53} shows a high homology with those around the active site Asp_{215} residue of porcine pepsin and other carboxyl proteinases (Figure 3). While

1		10		20

Thr-Val-Glu-Ser-Asn-Trp-Gly-Gly-Ala- Ile-Leu- Ile-Gly-Ser-Asp-Phe-Asp-Thr-Val-Ser-
21 30 40
Ala-Thr- Ala-Asn-Val-Pro-Ser-Ala-Thr-Gly- Ala-Ser-Gly-Gly-Ser-Ser- Ala- Ala-Trp-Val-
41 50 60
Gly- Ile-Asp-Gly-Asp-Thr-Cys-Gln-Thr- Ala- Ile-Leu-Glu-Thr-Gly-Phe-Asp-Trp-Tyr-Gly-
61 70 80
Asp-Gly-Thr-Tyr-Asp- Ala-Trp-Tyr-Glu-Trp-Tyr-Pro-Glu-Val-Ser-Asp-Asp-Phe-Ser-Gly-
81 90 100
Ile-Thr- Ile-Ser-Glu-Gly-Asp-Ser- Ile-Gln-Met-Ser-Val-Thr- Ala-Thr-Ser-Asp-Thr-Ser-
101 110 120
Gly-Ser- Ala-Thr-Leu-Glu-Asn-Leu-Thr-Thr-Gly-Gln-Lys-Val-Ser-Lys-Ser-Phe-Ser-Asn-
121 130 140
Glu-Ser-Ser-Gly-Leu-Cys-Arg-Thr-Asn- Ala-Glu-Phe- Ile- Ile-Glu-Asp-Phe-Glu-Glu-Cys-
141 150 160
Asn-Ser-Asp-Gly-Ser-Asp-Glu-Phe-Val-Pro-Phe- Ala-Ser-Phe-Ser-Pro- Ala-Val-Glu-Phe-
161 170 180
Thr-Asp-Cys-Ser-Val-Thr-Ser-Asp-Gly-Glu-Ser-Val-Ser-Leu-Asp-Asp- Ala-Gln- Ile-Thr-
181 190 200
Gln-Val- Ile- Ile-Asn-Asn-Gln-Asp-Val-Thr-Asp-Cys-Ser-Val-Ser-Gly-Thr-Thr-Val-Ser-
201 204
Cys-Ser-Tyr-Val

Figure 2. Amino acid sequence of *Scytalidium lignicolum* carboxyl proteinase B. Disulfide bonds connect Cys_{47}-Cys_{126}, Cys_{140}-Cys_{163} and Cys_{192}-Cys_{201}.

		53
Carboxyl proteinase B	-Cys-Gln-Thr-Ala-Ile-Leu-Glu*-Thr-Gly-Phe-	
		215
Porcine pepsin	-Cys-Gln-	-Ala-Ile-Val-Asp*-Thr-Gly-Thr-
Calf chymosin	-Cys-Gln-	-Ala-Ile-Leu-Asp*-Thr-Gly-Thr-
Penicillopepsin	-Ser-	-Gly-Ile-Ala-Asp*-Thr-Gly-Thr-
Rhizopuspepsin	-Ser-Phe-Asp-Gly-Ile-Leu-Asp*-Thr-Gly-Thr-	

	98	
Carboxyl proteinase B	-Thr-Ser-Asp*-Thr-Ser-Gly-Ser-Ala-Thr-Leu-	
	32	
Porcine pepsin	-Ile-Phe-Asp*-Thr-	-Gly-Ser-Ser-Asn-Leu-
Calf chymosin	-Leu-Phe-Asp*-Thr-	-Gly-Ser-Ser-Asp-Leu-
Penicillopepsin	-Asn-Phe-Asp*-Thr-	-Gly-Ser-Ser-Asp-Phe-
Rhizopuspepsin	-Asp-Phe-Asp*-Thr-	-Gly-Ser-Ala-Asp-Leu-

Figure 3. Comparison of amino acid sequences around Glu_{53} and Asp_{98} of *Scytalidium lignicolum* carboxyl proteinase B with those of active site aspartic acid residues of some other pepstatin-sensitive carboxyl proteinases.

Table 3. Classification of Carboxyl Proteinase

| Type | Inhibitor | | | Active center (-COOH) | Carboxyl proteinase |
	S-PI	DAN	EPNP		
Pepsin (S-PI-sensitive)	+	+	+	Asp_{32} Asp_{215}	Pepsin, Gastricsin, Chymosin Cathepsin D, Renin, Microbial carboxyl proteinases, (*Penicillium, Rhizopus, Aspergillus, Rhodotorula, Mucor, Cladosporium, Saccharomyces*, etc.)
	-	-	+	Glu_{53}, Asp_{98}	*S. lignicolum* B
Scytalidium (S-PI-insensitive)	-	-	-	Glu or Asp	*S. lignicolum* A and C *Lentinus edodes, Ganoderma lucidum, Pleurotus ostreatus, Flammulina velutipes, Pseudomonas* sp. No. 101, *Xanthomonas* sp. No. T-22, *Bacillus* novosp. MN-32(60°C), *Asp. niger* var. *macrosporus* A *Irpex lacteus.*

S-PI : acetyl-pepstatin, DAN :Diazoacetyl-DL-norleucine methylester, EPNP: 1-2-epoxy-3-(*p*-nitrophenoxy)propane

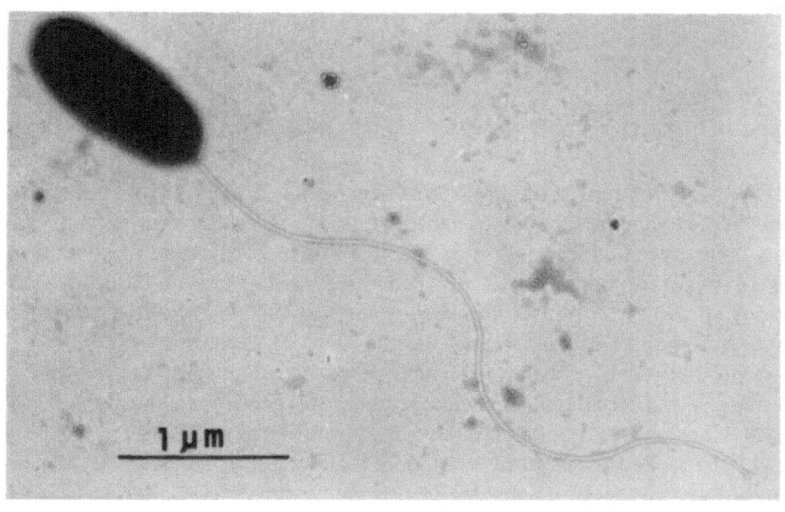

Figure 4 Electron microscopic photograph of *Pseudomonas* sp. No. 101.

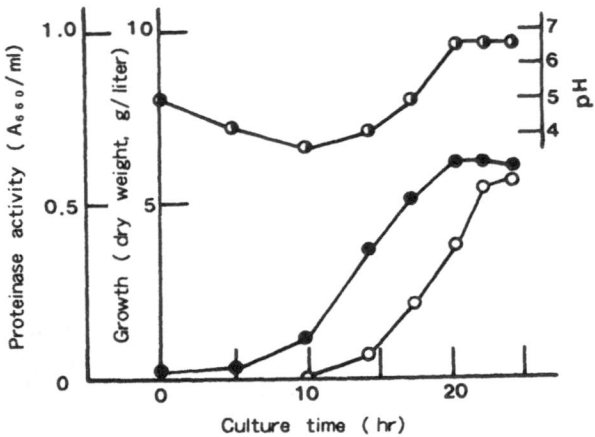

Figure 5. Time Course of *Pseudomonas* sp. carboxyl proteinase production. The *Pseudomonas* sp. No.101 was aerobically cultured in a 30-liter jar fermentor containing 20 liter medium (1% glucose, 0.5% polypeptone, and 0.5% meat extract, pH 5). The culture conditions were as follows: agitation at 300 rpm: aeration at 20 liters/min; temperture of 30°C. half-circles . pH; filled circles, growth(dry weight, g/liter); open circles, proteinase activity (A_{660}/ml).

the amino acid sequence around Asp_{98} also shows a high homology with those around the active site Asp_{32}, there is an insertion of serine residue between Thr and Gly. This was the first demonstration of a glutamic proteinase.

In our subsequent studies, we found that such enzymes having properties resembling *Scytalidium* enzymes were widely distributed among fungi,[25][26] basidiomycetes,[27-29] bacteria[7,30] and even in thermophilic bacteria as shown in Table 3. On the basis of the results described here and other investigations (studies on catalytic residues by means of Zn(II)-PAD[31][32] and kinetics[33]), we proposed that carboxyl proteinase should be classified into two groups: pepstatin-sensitive carboxyl proteinases and pepstatin-insensitive carboxyl proteinases.

PEPSTATIN-INSENSITIVE CARBOXYL PROTEINASE FROM *PSEUDOMONAS* SP.

In 1986, we started screening microorganisms, especially bacteria, for carboxyl proteinases. We were interested in looking for pepstatin-insensitive carboxyl proteinases, but at that time, there were not any studies on the occurrence of carboxyl proteinase in bacteria, regardless of pepstatin-sensitivity. In 1987, we succeeded in isolating two strains, *Pseudomonas* sp. No. 101[7] and *Xanthomonas* sp. No. T-22.[30] Figure 4 shows an electron microscopic photograph of *Pseudomonas* species.

The production of pepstatin-insensitive carboxyl proteinases by this strain was carried out by using a 30 liter Jar Fermentor under the conditions shown in Figure 5. The carboxyl proteinases began to accumulate at approximately 15 h after the beginning of cultivation and reached a maximum at 25 h. The culture fluid at 25 h was used for purification of the carboxyl proteinases.

Table 4. Summary of Purification of *Pseudomonas* sp. Carboxyl Proteinase

Procedure	Volume (ml)	Total protein (A_{280} x 10^{-3})	Total activity (PU x 10^{-3})	Specific activity (PU/A_{280})	Yield (%)
Culture filtrate	21600	66	533	8.1	100
$(NH_4)_2SO_4$ ppt	700	49	522	10.7	98
Acetone fractionation	520	18.6	496	26.7	93
DEAE-Sephadex A-50 1st	160	1.42	407	287	76
DEAE-Sephadex A-50 2nd	71	0.63	341	541	64
Sephadex G-75 1st	32	0.27	298	1103	56
Sephadex G-75 2nd	22	0.18	283	1572	53

Table 4 shows a summary of the purification of the carboxyl proteinase.[7] The purified enzyme represents approximately 190-fold purification over the original culture filtrate, with a 53 % recovery. From a 21 liter culture filtrate, about 105 mg of purified enzyme was obtained. We would like to emphasize that this enzyme was secreted in the culture fluid from gram-negative bacterium which has a double layer membrane around the cells. Figure 6 shows crystals of this enzyme. They are not large enough to study the three-dimensional structure properly, so we are attempting to obtain larger crystals.

The optimum pH for the action toward casein was around 3.0 and its molecular weight is 43,000 (Table 5).[7] This enzyme is an acidic protein having an isoelectric point at pH 3.2. From amino acid analysis, this enzyme contains one methionine residue. The absence of free thiol groups and the presence of 2 mol half-cystine indicates the presence of one disulfide bridge/mole enzyme.

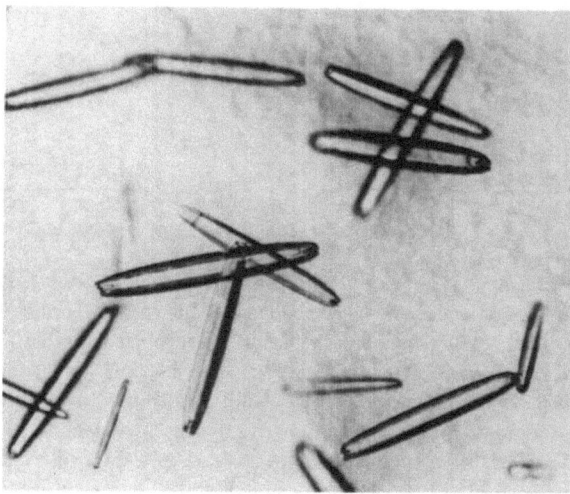

Figure 6. Crystals of *Pseudomonas* sp. carboxyl proteinase from $(NH_4)_2SO_4$ (0.2 saturated) at pH 4.3.

Table 5. Some Properties of *Pseudomonas* sp. Carboxyl Proteinase

Optimum pH (casein, 37°C, 15 min)	3.0
pH stability (37°C, 15 hr)	4.0 - 5.5
Inhibition with S-PI(acetyl-pepstatin)	none
DAN	none
EPNP	none
Molecular weight (SDS-PAGE)	43,000
Isoelectric point	3.2
Catalytic residues [Zn(II)-PAD]	-COOH

Amino acid composition

Asx 43.4 (40)[a]	Thr 30.9 (29)	Ser 50.5 (48)	Glx 35.1 (31)
Pro 11.4 (13)	Gly 50.7 (48)	Ala 40.2 (41)	1/2Cys 2.1 (2)
Val 19.9 (21)	Met 1.0 (1)	Ile 14.7 (16)	Leu 22.5 (23)
Tyr 17.8 (18)	Phe 10.4 (11)	Lys 6.0 (6)	His 2.1 (2)
Trp 6.8 (7)	Arg 5.7 (6)		

Total 363

[a] The numbers in parentheses are from the established amino acid sequence.

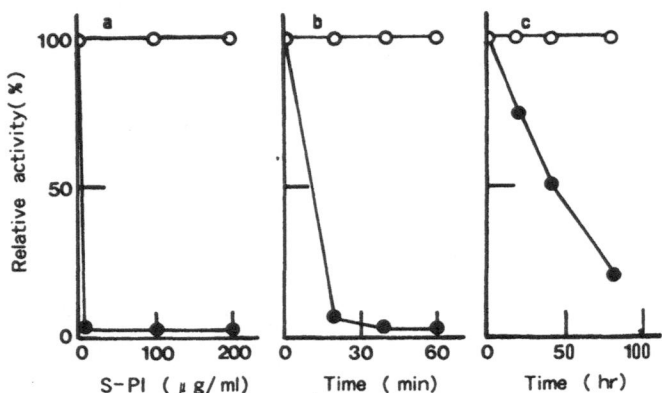

Figure 7. Inhibition of *Pseudomonas* sp. carboxyl proteinase with S-PI (Acetyl-Pepstatin), DAN and EPNP. (a) S-PI. 1 ml carboxyl proteinase dissolved in 50 mM sodium acetate buffer, pH 4 (10.4 µg/ml, 240 nM) was mixed with 0-40 µl S-PI (7.6 mM in methanol). After incubation at 37°C for 10 min, the remaining activities were assayed at pH 3.0 with casein as substrate. (b) DAN. 1 ml carboxyl proteinase dissolved in 50 mM sodium acetate buffer, pH 5 (610 µg/ml, 14 µM) were mixed with 83 µl CuSO4 (30 mM) and 50 µl DAN (60 mM)(final concentration: 12.5 µM; 2.19 mM; 2.65 mM, respectively. Enzyme/CuSO4/DAN = 1 :180 :210). A control was run without DAN. At the time indicated, samples were taken for testing the remaining activity. (c) EPNP. 3 ml carboxyl proteinase dissolved in 50 mM sodium acetate buffer, pH 5.0 (0.6 mg/ml, 14 µM) were incubated in the presence of 30 mg EPNP at 25°C with stirring. A control was run without EPNP. At the time indicated, samples were taken for testing the remaining activities. open circles . the *Pseudomonas* carboxyl proteinase; closed circles , porcine pepsin.

Table 6. Substrate Specificity of *Pseudomonas* sp. Carboxyl Proteinase

Substrate	$-P_3-P_2-P_1-P_1'-P_2'-P_3'-$	Hydrophobicity* Δgt (Kcal/mol)
Insulin B-chain		
Major	$-L-V-E_{13}-A_{14}-L-Y-$	-1.8—1.5—0.6—0.5—1.8—2.3-
	$-E-A-L_{15}-Y_{16}-L-V-$	-0.6—0.5—1.8—2.3—1.8—1.5-
	$-G-F-F_{25}-Y_{26}-T-P-$	0—2.5—2.5—2.3—0.4—?—
Minor	$-V-E-A_{14}-L_{15}-Y-L-$	-1.5—0.6—0.5—1.8—2.3—1.8-
	$-R-G-F_{24}-F_{25}-Y-T-$	—?—0—2.5—2.5—2.3—0.4-
Angiotensin I		
Major	$-R-V-Y_4-I_5-H-P-$	—?—1.5—2.3—3.0—0.5—?—
	$-H-P-F_8-H_9-L$	-0.5—?—2.5—0.5—1.8—1.8-
Minor	D_1-R_2-V-Y	0.5—?—1.5—2.3-

* Y. Nozaki and C. Tanford, J. Biol. Chem. **246**:2211 (1971).

S-PI, DAN and EPNP were examined for their effects on this enzyme with porcine pepsin as a control.[7] As shown in Figure 7-a, pepsin was completely inactivated by S-PI at a 1:1 ratio. On the other hand, the *Pseudomonas* carboxyl proteinase was not inhibited by S-PI even at a concentration of 200 μg/ml. DAN and EPNP also did not inhibit this enzyme as shown in Figure 7-b and 7-c, respectively.

We examined *Pseudomonas* carboxyl proteinase for substrate specificities using oxidized insulin B-chain and Angiotensin I.[34] When the oxidized insulin B-chain was used as the substrate, *Pseudomonas* carboxyl proteinase cleaved mainly the Glu-Ala, Leu-Tyr and Phe-Tyr bonds (Table 6). Angiotensin I was cleaved at first at the Tyr-Ile bond, and after that, the Phe-His bond was cleaved. These results suggest that *Pseudomonas* carboxyl proteinase has a preference for hydrophobic and bulky amino acid residues at both the P_1 and $P_{1'}$ positions. In addition, this enzyme has narrow substrate specificity compared with those of other enzymes.

Among the pepstatin-insensitive carboxyl proteinases we have studied, *Pseudomonas* carboxyl proteinase[7] was the first enzyme isolated from prokaryote cells regardless of its pepstatin-sensitivity. Therefore, *Pseudomonas* carboxyl proteinase is a good target for comparing the amino acid sequence as well as its active site structure with *Scytalidium lignicolum* B and also the pepstatin-sensitive enzymes reported so far.

The amino acid sequence of Pseudomonas carboxyl proteinase was elucidated as shown in Figure 8.[35] A portion, between Ser_{67} and Asp_{77}, has not been sequenced yet. The enzyme is a single polypeptide composed of 363 amino acid residues, containing one disulfide bridge. The enzyme contains one methionine residue. We attempted to cleave the protein at the methionine residue by cyanogen bromide but failed to obtain two fragments. This failure seems to be reasonable, since the sequence analysis showed that the methionine residue was oxidized to methionylsulfoxide.

Ala-Ala-Gly-Thr-Ala-Lys-Gly-His-Asn-Pro-Thr-Glu-Phe-Pro-Thr- Ile-Tyr-Asp- Ala-Ser-
21   30   40
Ser- Ala-Pro-Thr- Ala- Ala-Asn-Thr-Thr-Val-Gly- Ile- Ile-Thr- Ile-Gly-Gly-Val-Ser-Gln-
41   50   60
Thr-Leu-Gln-Asp-Leu-Gln-Gln-Phe-Thr-Ser- Ala-Asp-Gly-Leu- Ala-Ser-Val-Asn-Thr-Gln-
61   70   80
Thr- Ile-Gln-Thr-Gly-Ser-Ser-(Asx,Asx,Ser,Ser,Glx,Glx,Gly,Gly,Tyr)-Asp-Ser-Gln-Ser-
81   90   100
Ile-Val-Gly-Ser- Ala-Gly-Gly- Ala-Val-Gln-Gln-Leu-Leu-Phe-Tyr-Met- Ala-Asp-Gln-Ser
101   110   120
Ala-Ser-Gly-Asn-Gln-Gly-Leu-Thr-Ser- Ala-Phe-Asn-Gln- Ala-Val-Ser-Asp-Asn- Ala-Val-
121   130   140
Lys-Val- Ile-Asn-Val-Ser-Leu-Gly-Trp-Cys-Glu- Ala-Asp- Ala-Asn- Ala-Asp-Gly-Thr-Leu
141   150   160
Gln- Ala-Glu-Asp-Arg- Ile-Phe- Ala-Thr- Ala- Ala- Ala-Gln-Gly-Gln-Thr-Phe-Ser-Val-Ser-
161   170   180
Ser-Gly-Asp-Gln-Gly-Val-Tyr-Glu-Cys-Asn-Asn-Arg-Gly-Tyr-Pro-Asp-Gly-Ser-Thr-Tyr-
181   190   200
Ser- Val-Ser-Trp-Pro- Ala-Ser-Ser- Pro-Asn- Val- Ile- Ala- Val- Gly- Gly-Thr-Thr-Leu-Tyr-
201   210   220
Thr-Thr-Ser- Ala- Gly- Ala-Tyr-Ser-Asn-Glu-Thr-Val-Trp-Asn-Glu- Gly-Leu-Asp-Ser-Asp
221   230   240
Gly-Lys-Leu-Trp- Ala-Thr- Gly- Gly- Gly-Tyr-Ser-Val-Tyr- Glu-Ser-Lys-Pro-Ser-Trp-Gln-
241   250   260
Ser-Val-Val-Ser-Gly-Thr-Pro-Gly-Arg-Arg-Leu-Leu-Pro-Asp- Ile-Ser-Phe-Asp- Ala- Ala-
261   270   280
Gln-Gly-Thr-Gly- Ala-Leu- Ile-Tyr-Asn-Tyr-Gly-Gln-Leu-Gln-Gln- Ile-Gly-Gly-Thr-Ser-
281   290   300
Leu- Ala-Ser-Pro- Ile-Phe- Val- Gly-Leu-Leu- Ala-Arg-Leu- Gln-Ser- Ala-Asn-Ser-Asn-Ser-
301   310   320
Leu- Gly-Phe-Pro- Ala- Ala-Ser-Phe-Tyr-Ser- Ala- Ile- Ser- Ser- Thr- Pro- Ser-Leu- Val-His-
321   330   340
Asp- Val-Lys-Ser-Gly-Asn-Asn-Gly-Tyr- Gly- Gly-Tyr- Gly-Tyr-Asn-Ala-Gly-Thr-Gly-Trp-
341   350   360
Asp-Tyr-Pro-Thr- Gly-Trp- Gly-Ser-Leu-Asp- Ile- Ala-Lys-Leu-Ser- Ala-Tyr- Ile-Arg-Ser-
361   363
Asn-Gly-Phe

Figure 8. Amino acid sequence of *Pseudomonas* sp. No.101 carboxyl proteinase. A disulfide bond connnects Cys$_{130}$-Cys$_{169}$.

Interestingly, this enzyme has neither apparent sequence similarity to any pepstatin-sensitive carboxyl proteinases reported so far, nor to the pepstatin-insensitive carboxyl proteinase B of *Scytalidium lignicolum*.[21] Also absent is the characteristic active site aspartic sequence, Asp-Thr-Gly of pepstatin-sensitive carboxyl proteinases.

Recently we were able to obtain clones of this enzyme. Analysis of the nucleotide sequence is now under way.

PEPSTATIN-INSENSITIVE CARBOXYL PROTEINASE FROM THERMOPHILIC *BACILLUS* NOVOSP.

Our results reported here encouraged us to screen more extensively for new types of carboxyl proteinases. As the target, we chose thermostable enzymes. One reason is that, there has been nothing published on the occurrence of thermostable carboxyl proteinases, regardless of the pepstatin sensitivity. The second is that thermostable enzymes are applicable to industry.

In 1988, we succeeded in isolating a thermophilic strain[8] from acidic hot springs in Japan, which produces pepstatin-insensitive carboxyl proteinase in the culture filtrate. Optimum temperature and pH for the growth of the strain are 60°C and between pH 3.0 to 5.0, respectively. The morphological and physiological characteristics resemble those of *Bacillus acidocaldarius*. However, such characteristics as utilization of citric acid, the ability to reduce nitric acid, and so on were different. Therefore we tentatively identified this strain as *Bacillus* novosp.[8]

A summary of the purification of this enzyme is shown in Table 7. For the production of the thermostable enzyme, this strain was aerobically cultured at 60°C and pH 3.5, using 200 liter Jar Fermentor. The culture filtrate obtained after 46 h cultivation was used for this purification. After 6 purification steps, about 20 mg of purified enzyme was obtained with 13.8 % recovery from a 120 liter culture filtrate. The homogeneity of the enzyme was proven by electrophoresis and HPLC.

The optimum pH and temperature for the action of this enzyme was 3.2, and 70°C, respectively (Table 8).[8] The enzyme is fairly acid stable and thermostable, retaining 100% of initial activity between pH 2 to 5 after 24 h incubation at 50°C, and 60% activity after 10 min incubation at pH 4 and 80°C. The effects of various proteinase inhibitors on this enzyme were examined. The enzyme was not inhibited by serine, metallo, or thiol proteinase inhibitors. Such carboxyl proteinase inhibitors as S-PI, DAN or EPNP also did not show any inhibitory effects on this enzyme. The molecular weight of this enzyme was found to be 41,000 and the isoelectric point was found to be pH 3.5. The amino acid composition shows a high content of hydrophobic amino acids, with the mole percentage of the Gly, Ala and Pro residues 35% of the total.

Table 7. Summary of Purification of *Bacillus* novosp. Carboxyl Proteinase

Procedure		Volume (l)	Total activity (U X 10^{-3})	Total protein (mg x 10^{-3})	Specific activity (U/mg)	Yield %
Culture filtrate		120	400	1,330	0.30	100
(NH$_4$)$_2$SO$_4$ ppt		6.00	241	268	0.90	60.3
DEAE-Sephadex CL-6B		2.88	204	2.68	76.1	51.0
Sephadex G-100	1st	1.09	115	0.176	653	28.8
	2nd	0.90	82.8	0.092	900	20.7
TSK-DEAE 5PW	1st	0.25	65.0	0.029	2,240	16.3
	2nd	0.21	55.0	0.022	2,500	13.8

Table 8. Some Properties of *Bacillus* novosp. Carboxyl Proteinase

Optimum pH (casein, 70°C, 30 min)	3.2
pH stability (50°C, 24 hr)	2 - 5
Thermal stability (pH 4, 10 min)	70°C, 90%; 80°C, 60%; 90°C, 0%
Inhibition with S-Pl	none
DAN	none
EPNP	none
Molecular weight (SDS-PAGE)	41,000
Isoelectric point	3.5

Amino acid composition

Asx 40	Thr 25	Ser 26	Glx 34
Gly 51	Ala 55	1/2Cys 4	Val 30
Met 1	Ile 19	Leu 31	Tyr 11
Phe 10	Lys 5	His 5	Arg 12
Pro 35	Trp 7		

Total 401

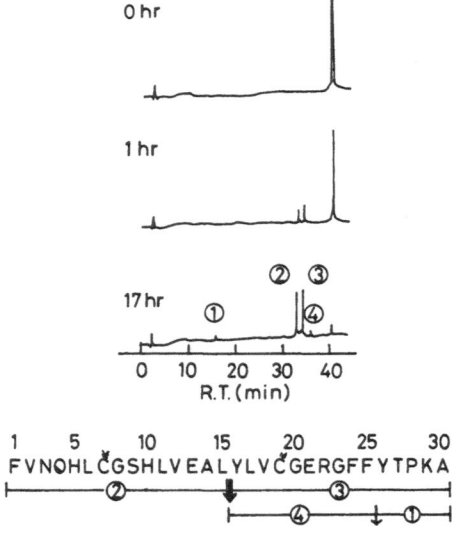

Figure 9. Hydrolysis of oxidized insulin B-chain by *Bacillus* novosp. carboxyl proteinase. Oxidized B-chain, 0.5 mg, in 20 mM KCl-HCl buffer, pH 3.0, was digested by incubation with 5 ng of *Bacillus* novosp. carboxyl proteinase at 20°C for indicated times (0 - 17 h). The digest (10 µl each) was then subjected to high performance liquid chromatography (HPLC) with a Nucleosile 5 C18 column. The peptides were eluted with an increasing linear gradient of acetonitrile, from 1 to 35 %, containing 0.1 % trifluoroacetic acid at flow rate of 0.8 ml per min at room temperature.

Table 9. Some Properties of *Bacillus* novosp. Carboxyl Proteinase and Thermopsin

Enzyme	*Bacillus* novosp. CP	Thermopsin
Microorganism	*Bacillus* novosp.	*Sulfolobus acidocardarius*
Growth	pH 3.0 - 5.0, 60°C	pH 2, 70°C
Optimum pH and Temp.	pH 3.0 and 70°C	pH 2.0 and 90°C
Thermostability	60 % (80°C, pH 4, 10 min)	100 % (80°C, pH 4.5, 2 days)
Molecular weight	41,000	51,000 (32,651)*
Inhibition with		
Pepstatin	none	Yes, $K_i = 2 \times 10^{-7}$
DAN	none	Yes, (non-specific)
EPNP	none	Yes, (non-specific)
Substrate specificity		
Oxidized B-chain	L_{15}—Y_{16}	L_{11}—V_{12}, L_{15}—Y_{16}, F_{24}—F_{25}
		F_{25}—Y_{26}, Y_{26}—T_{27}

*Based on the amino acid sequence

Substrate specificity was studied using oxidized insulin B-chain as a substrate. As shown in Figure 9, the Leu_{15}-Tyr_{16} bond was specificially cleaved. After overnight incubation, we detected a minor product from cleavage of the Phe_{25}-Tyr_{26} bond. These results suggest that this enzyme has a narrow specificity favoring large and hydrophobic residues on both sides of the scissile bond.

The pH dependency of the cleavage of the oxidized insulin B-chain was examined. An apparent pH optimum is around pH 2.6. At this pH and 30°C, k_{cat}/K_m value was determined to be 3.0×10^5 $M^{-1}sec^{-1}$.

In 1990, J. Tang[36,37] reported the presence of an extremely thermostable acid proteinase called thermopsin in *Sulfolobus acidocaldarius*, an archaebacterium. We compared the properties of *Bacillus* novosp. carboxyl proteinase with those of thermopsin (Table 9). Thermopsin is extremely stable at high temperatures and acidic pH regions. Thermopsin is inactivated by pepstatin, DAN or EPNP. Thermopsin cleaved the oxidized insulin B-chain at many points. Furthermore, the amino acid composition of thermopsin is very different from that of *Bacillus* novosp. carboxyl proteinase. For example, the mole percentage of Gly, Ala and Pro residues in the former is 21 %, while the mole percentage of those residues in the latter enzyme is 35 %. These observations suggest that *Bacillus* novosp. carboxyl proteinase has very different properties than that of thermopsin. Cloning of this enzyme is now under way.

A NOVEL PROTEINASE INHIBITOR, TYROSTATIN, INHIBITING SOME PEPSTATIN-INSENSITIVE CARBOXYL PROTEINASES

No specific inhibitors of pepstatin-insensitive carboxyl proteinase inhibitors were known previously, though such an inhibitor would seem to be useful in the study of the active site. Therefore, we attempted to obtain new peptide inhibitors of pepstatin-insensitive carboxyl proteinases by screening a number of microbes.

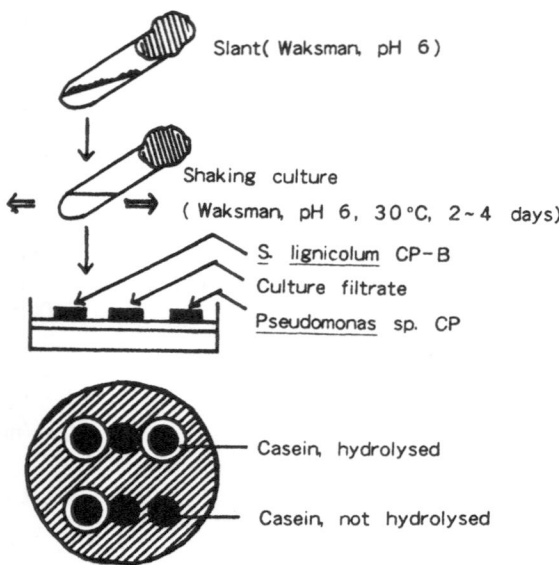

Slant(Waksman, pH 6)

Shaking culture

(Waksman, pH 6, 30 °C, 2~4 days)

S. lignicolum CP-B

Culture filtrate

Pseudomonas sp. CP

Casein, hydrolysed

Casein, not hydrolysed

Figure 10. Screening for pepstatin-insensitive carboxyl proteinase inhibitor. Casein agar plates were made as follows: first 15 ml of 1.5 agar solution (50 mM lactate buffer, pH 3.0) was poured into Petri-dish (φ 90 mm), and then 5 ml of 1.0 % agar and 1 % casein dissolved in the same buffer described above was overlaid.

Figure 10 shows schematically the screening method using casein agar plates. Paper disks were dipped in sample or target enzyme solutions and placed on casein agar plates. After overnight incubation, the occurrence of inhibitor is distinguished by the casein hydrolysis around the disks. By this simple method, we succeeded in isolating the strain, *Kitasatosporia* which produced an inhibitor. We named this inhibitor Tyrostatin.[9]

The chemical structure of tyrostatin was elucidated as shown in Figure 11.[9] The structure is N-isovaleryl-tyrosyl-leucyl-tyrosinal. One year later, we isolated the derivative of tyrostatin with the structure N-isovaleryl-tyrosyl-leucyl-tyrosinol. This derivative has no inhibitory activities against carboxyl proteinases, indicating that the C-terminal tyrosinal residue of tyrostatin is essential for inhibition.

The inhibition of various proteinases by tyrostatin was studied, and the results are summarized in Table 10.[9] Tyrostatin strongly inhibited pepstatin-insensitive carboxyl proteinases originating from *Pseudomonas* sp. No. 101 and *Xanthomonas* sp. No.T-22, but none of the other pepstatin-insensitive carboxyl proteinases from *Scytalidium lignicolum* or the thermostable carboxyl proteinase from *Bacillus* novosp. Furthermore, it inhibited all of the carboxyl proteinases tested from the pepstatin-sensitive groups, although the inhibitory activities were not very potent.

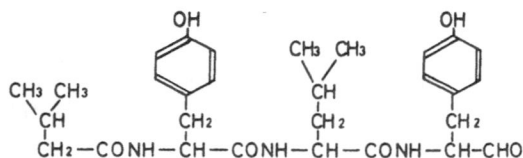

Figure 11. Chemical structure of tyrostatin.

Table 10. Inhibitory Spectrum of Tyrostatin on Carboxyl Proteinases. Inhibitory action of tyrostatin was estimated by measuring residual enzyme activity after incubation of each enzyme with tyrostatin (20 μg). Inhibitory ratings used are as follows : ++, more than 80 % inhibition; +, 20 - 80 % inhibition; ±, 10 - 20 % inhibition; -, less than 10 % inhibition of enzyme activity.

Carboxyl proteinase	Inhibition
Pepstatin-insensitive	
Pseudomonas sp. No. 101	++
Xanthomonas sp. No. T-22	++
Scytalidium lignicolum A	-
Scytalidium lignicolum B	-
Aspergillus niger var. *macrosporus* Type A	-
Bacillus novosp. (thermophile, 60°C)	-
Pepstatin-sensitive	
Porcine pepsin	+
Aspergillus niger var. *macrosporus* Type B	+
Rhodotorula glutinis	±
Cladosporium sp.	+
Rhizopus chinensis	+
Rhizopus nodosum	±

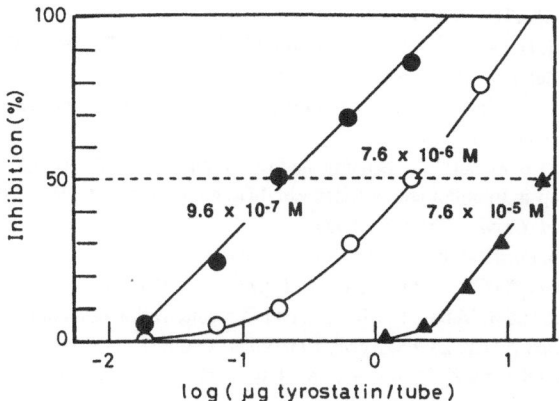

Figure 12. Effects of tyrostatin on carboxyl proteinases. Each tube contained 40 PU of enzyme. After incubation with various amounts of inhibitor at 37°C for 5 min, the remaining activity was assayed. filled circles , *Xanthomonas* sp. No. T-22 carboxyl proteinase (2.6 x 10^{-7} M, 40 PU); open circles , *Pseudomonas* sp. No. 101 carboxyl proteinase (2.2 x 10^{-7} M, 40 PU); filled triangles, pepsin (40 PU).

As shown in Figure 12, the dose of tyrostatin for 50% inhibition, ID_{50}, under the conditions used were 9.6×10^{-7} M, 7.6×10^{-6} M, and 7.6×10^{-5} M, for carboxyl proteinases from *Xanthomonas* sp. No. T-22, *Pseudomonas* sp. No. 101 and porcine pepsin, respectively. Recently the inhibition constant of tyrostatin for *Pseudomonas* carboxyl proteinase was found to be 2.5×10^{-9} M by using a synthetic peptide substrate.

Chemical synthesis is underway of tyrostatin derivatives capable of binding covalently with enzymes, in order that they may be used as probes for elucidating the catalytic residues of *Pseudomonas* sp. No. 101 carboxyl proteinase.

REFERENCES

1. B. S. Hartley, Proteolytic Enzymes, *Ann. Rev. Biochem.* **29**:45 (1960).

2. S. Murao and S. Satoi. New Pepsin Inhibitors (S-PI) from *Streptomyces* EF-44-201, *Agric. Biol. Chem.*, **34**:1265 (1970).

3. T. G. Rajagopalan, W. H. Stein and S. Moore, The Inactivation of Pepsin by Diazoacetyl-norleucine Methyl Ester, *J. Biol. Chem.*, **241**:4295 (1966).

4. J. Tang, Specific and Irreversible Inactivation of Pepsin by Substrate-like Epoxide, *J. Biol. Chem.*, **246**:4510 (1971).

5. J. Tang, Evolution in the Structure and Function of Carboxyl Proteases, *Molecular and Cellular Biochemistry*, **26**:93 (1979).

6. S. Murao and K. Oda, Pepstatin-Insensitive Acid Proteinases, *in*: "Aspartic Proteinases and Their Inhibitors," V. Kostka ed., Walter de Gruyter & Co., Berlin, New York (1985).

7. K. Oda, M. Sugitani, K. Fukuhara and S. Murao, Purification and Properties of a Pepstatin-insensitive Carboxyl Proteinase from a Gram-negative Bacterium, *Biochim. Biophys. Acta*, **923**:463 (1987).

8. S. Murao, K. Ohkuni, M. Nagao, K. Oda and T. Shin, A Novel Thermostable, S-PI(Pepstatin Ac)-insensitive Acid Proteinase from Thermophilic *Bacillus* novosp. Strain MN-32, *Agric. Biol. Chem.*, **52**:1629 (1988).

9. K. Oda, Y. Fukuda, S. Murao, K. Uchida and M. Kainosho, A Novel Proteinase Inhibitor, Tyrostatin, Inhibiting Some Pepstatin-insensitive Carboxyl Proteinases, *Agric. Biol. Chem.*, **53**:405 (1989).

10. M. Fukumura, S. Satoi. N. Kuwana and S. Murao, Structure Elucidation of New Pepsin Inhibitor (S-PI), *Agric. Biol. Chem.*, **35**:1310 (1971).

11. Umezawa, T. Aoyagi, H. Morishima, M. Matsuzaki, M. Hamada and T. Takeuchi, Pepstatin: A New Pepsin Inhibitor Produced by *Actinomycetes*, *J. Antibiot.*, **23**:259 (1970).

12. S. Murao, K. Oda and Y. Matsushita, New Acid Proteases from *Scytalidium lignicolum* M-133, *Agric. Biol. Chem.*, **36**:1647 (1972).

13. S. Murao, K. Oda and Y. Matsushita, Isolation and Identification of a Microorganism which Produces Non Streptomyces Pepsin Inhibitor and N-Diazoacetyl-DL-norleucine Methylester Sensitive Acid Proteases, *Agric. Biol. Chem.*, **37**:1417 (1973).

14. K. Oda and S. Murao, Purification and Some Enzymatic Properties of Acid Protease A and B of *Scytalidium lignicolum* ATCC 24568, *Agric. Biol. Chem.*, **38**:2435 (1974).

15. K. Oda, H. Torishima and S. Murao, Purification and Characterization of Acid Proteinase C of *Scytalidium lignicolum* ATCC 24568, *Agric. Biol. Chem.*, **50**:651 (1986).

16. K. Oda, S. Murao, T. Oka and K. Morihara, Some Physicochemical Properties and Substrate Specificities of Acid Proteases A-1 and A-2 of *Scytalidium lignicolum* ATCC 24568. *Agric. Biol. Chem.*, **40**:859 (1976).

17. K. Oda, 5. Murao, T. Oka and K. Morihara, Some Physicochemical Properties and Substrate Specificity of Acid Protease B of *Scytalidium lignicolum* ATCC 24568, *Agric. Biol. Chem.*, **39**:77 (1975).

18. K. Morihara, H. Tsuzuki, S. Murao and K. Oda, Pepstatin-Insensitive Acid Proteases from *Scytalidium lignicolum* -Kinetic Study with Synthetic Peptides, *J. Biochem.*, **85**:661 (1979).

19. K. Morihara, Comparative Specificity of Microbial Acid Proteinases, *in* : "Proteinases and Their Inhibitors," V. Turk and Lj. Vitale, ed., Mladinska Knjiga-Pergamon Press, Ljubljana, Oxford (1981).

20. K. Oda and S. Murao, Action of *Scytalidium lignicolum* Acid Proteases on Insulin B-chain, *Agric. Biol. Chem.*, **40**:1221 (1976).

21. T. Maita, S. Nagata, G. Matsuda, S. Maruta, K. Oda, S. Murao and D. Tsuru, Complete Amino Acid Sequence of *Scytalidium lignicolum* Acid Protease B, *J. Biochem.*, **95**:465 (1984).

22. D. Tsuru, S. Shimada, S. Maruta, T. Yoshimoto, K. Oda, S. Murao, T. Miyata and S. Iwanaga, Isolation and Amino Acid Sequence of a Peptide Containing an Epoxide-Reactive Residue from the Thermolysin-Digest of *Scytalidium lignicolum* Acid Protease B, *J. Biochem.*, **99**:1537 (1986).

23. D. Tsuru, A. Naotsuka, R. Kobayashi, T. Yoshimoto, K. Oda and S. Murao, Inactivation of *Scytalidium lignicolum* Acid Protease B with 1,2-Epoxy-3-(4-azido-2-nitrophenoxy)propane, *Agric. Biol. Chem.*, **53**:2751 (1989).

24. D. Tsuru, R. Kobayashi, N. Nakagawa and T. Yoshimoto, Inhibition of *Scytalidium lignicolum* Acid Protease B by l-Diazo-3-phenyl-2-propane, *Agric. Biol. Chem.*, **53**:1305 (1989).

25. W. H. Chang. S. Horiuchi, K. Takahashi, M. Yamasaki and Y. Yamada, The Structure and Function of Acid Proteases, VI. Effect of Acid Protease-Specific Inhibitors on the Acid Proteases from *Aspergillus niger* var. macrosporus, *J. Biochem.*, **80**:975 (1976).

26. H. Kobayashi, I. Kusakabe and K. Murakami, Purification and Characterization of Pepstatin-insensitive Carboxyl Proteinase from *Polyporus tulipiferae* (*Irpex lacteus*), *Agric. Biol. Chem.*, **49**:2393 (1985).

27 K. Oda, T. Terashita, M. Kono and S. Murao, Occurrence of *Streptomyces* Pepsin Inhibitor-insensitive Carboxyl Proteinase in *Basidiomycetes*, *Agric. Biol. Chem.*, **45**:2339 (1981).

28. T. Terashita, K. Oda, M. Kono and S. Murao, *Streptomyces* Pepsin Inhibitor-Insensitive Carboxyl Proteinase from *Lentinus edodes*, *Agric. Biol. Chem..* **45**:1937 (1981).

29. T. Terashita, K. Oda, M. Kono and S. Murao, *Streptomyces* Pepsin Inhibitor-Insensitive Carboxyl Proteinase from *Ganoderma lucidum*, *Agric. Biol. Chem.*, **48**:1029 (1984).

30. K. Oda, T. Nakazima, T. Terashita, K. Suzuki and S. Murao, Purification and Properties of an S-PI (Pepstatin Ac)-insensitive Carboxyl Proteinase from a *Xanthomonas* sp. Bacterium, *Agric. Biol. Chem.*, **51**:3073 (1987).

31. H. Nakatani, K. Hiromi, S. Satoi, K. Oda, S. Murao, and E. Ichishima, Studies on the Interaction between *Streptomyces* Pepsin Inhibitor and Several Acid Proteinases by Means of a Zinc(II)-Dye Complex as a Probe, *Biochim. Biophys. Acta*, **391**:415 (1975).

32. K. Oda and S. Murao, Additional Evidence for the Identity of *Scytalidium lignicolum* Acid Proteinases with the Carboxyl Proteinase Group: The Interaction between Angiotensin I and S-PI-Insensitive Acid Proteinases by Means of a Zinc(II)-Dye Complex as a Probe, *Agric. Biol. Chem.*, **50**:1995 (1986).

33. K. Oda and S. Murao, Kinetic Studies on S-PI(Pepstatin Ac)-insensitive Acid Proteinases of *Scytalidium lignicolum* ATCC 24568. Evidence Identifying Them with a Carboxyl Proteinase Group, *Agric. Biol. Chem..* **50**:659 (1986).

34. K. Oda and M. Sugitani. Substrate Specificity of Pepstatin-insensitive Carboxyl Proteinase isolated from *Pseudomonas* sp. No. 101, *Agric. Biol. Chem .*, in preparation.

35. K. Oda, et al., in preparation.

36. X-I. Lin and J. Tang, Purification, Characterization, and Gene Cloning of Thermopsin, a Thermostable Acid Protease from *Sulfolobus acidocaldarius*, *J. Biol. Chem.* **265**:1490 (1990).

37. M. Fusek, X-I. Lin and J. Tang, Enzymic Properties of Thermopsin, *J. Biol. Chem.* **265**:1496 (1990).

STRUCTURE AND FUNCTION OF A PEPSTATIN-INSENSITIVE ACID PROTEINASE

FROM *ASPERGILLUS NIGER* var. *MACROSPORUS*

Kenji Takahashi,[a] Masaru Tanokura,[a] Hideshi Inoue,[a] Masaki Kojima,[a]
Yutaka Muto,[a] Makoto Yamasaki,[b] Osamu Makabe,[c] Takao Kimura,[c]
Toshio Takizawa,[a] Toru Hamaya,[a] Eiichiro Suzuki[e] and Hiroshi Miyano[e]

[a]Department of Biophysics and Biochemistry
Faculty of Science and

[b]Department of Chemistry
College of Arts and Sciences
The University of Tokyo
Tokyo, Japan

[c]Pharmaceutical Research Center
Meiji Seika Kaisha, Ltd.
Yokohama, Japan

[d]Bioscience Laboratories
Meiji Seika Kaisha, Ltd.
Sakado-shi, Saitama, Japan

[e]Central Laboratories
Ajinomoto Co., Inc.
Kawasaki, Japan

INTRODUCTION

The fungus *Aspergillus niger* var. *macrosporus* produces two extracellular acid proteinases, proteinases A and B.[1-3] The acid proteinase B (M_r about 35 kDa) is sensitive to pepstatin, diazoacetyl-DL-norleucine methyl ester (DAN) and 1,2-epoxy-3-(p-nitrophenoxy)propane (EPNP) and thus belongs to the ordinary aspartic proteinase family.[4] On the other hand, the proteinase A (M_r about 22 kDa) is insensitive to pepstatin and also almost insensitive to DAN and EPNP,[4] and shows substrate specificity fairly different from that of pepsin-type aspartic proteinases.[5,6] These results indicate that the proteinase A belongs to a different acid proteinase family. This proteinase, therefore, seems to be an interesting object to investigate its structure/function relationships. Further, the study will contribute to a deeper understanding of the structure and function of the aspartic or acid proteinases in general.

As the first step toward this direction, we have determined the complete amino acid sequence of the enzyme. Further, we have cloned the genomic DNA and cDNA of the enzyme and analyzed their nucleotide sequences. Thus, the amino acid sequence of the precursor form of the enzyme has been deduced. These results revealed that the acid proteinase A is synthesized as a single chain precursor form of 282 amino acid residues and then processed to the two chain mature form composed of a light (L) chain of 39 residues and a heavy (H) chain of 173 residues. The amino acid sequence of the enzyme shows about 50% identity with that of *Scytalidium lignicolum* acid proteinase B.[7] In addition, we have investigated the secondary structure and pH dependent conformational change of the enzyme by circular dichroism (CD) and NMR spectroscopy. The results indicated that the enzyme is considerably rich in ß-sheet structure in the native state and that this native conformation is rapidly and irreversibly lost at pH around 6 and above with dissociation of the two chains and concomitant loss of activity. The two-dimensional NMR spectra of the enzyme also indicated a high content of ß-sheet structure.

EXPERIMENTAL

Acid proteinase A

The acid proteinase A was prepared using the crude enzyme mixture (Proctase) obtained from the culture medium of *Aspergillus niger* var. *macrosporus* by repeated chromatography on DEAE-Toyopearl columns.

Determination of the amino acid sequence

The reduced and S-pyridylethylated protein or the reduced and S-aminoethylated protein was separately digested by *Staphylococcus aureus V8* protease, trypsin, lysyl endopeptidase and α-chymotrypsin and the resulting peptides were fractionated by chromatography on a column (1.5 x 200 cm) of Sephadex G-50 (superfine) or a Mono Q (HR 5/5) column, and/or by HPLC on a Tosoh 120T reverse phase column (0.46 x 25 cm). Edman degradation and PTH-amino acid analysis were performed mainly using a protein sequencer model 477A/120A (Applied Biosystems, Inc.). The locations of the disulfide bonds were determined by isolation by HPLC and analysis of cystine-containing peptides from a chymotryptic digest of the unmodified enzyme.

Gene and cDNA cloning and sequencing

The specific probe for the acid proteinase A gene was PCR-amplified by using the total DNA of the fungus as a template and the synthetic primers corresponding to the amino acid sequences Trp-Tyr-Glu-Trp-Tyr-Pro (residues 41-46) and Met-Asp-Ile-Glu-Gln-Asp (residues 149-154) in the H chain. By using this probe, the genomic DNA of the enzyme was cloned from the λ gt10 gene library of the same fungus. The cDNA of the enzyme was selectively PCR-amplified by using the total cDNA of the fungus as a template and the synthetic oligonucleotides corresponding to the NH_2- and COOH-terminal portions of the precursor protein as primers. The nucleotide sequences were determined with a DNA sequencer model 370A (Applied Biosystems, Inc.).

CD spectra

CD spectra were measured in a Jasco J-600 CD spectropolarimeter with 4.7 - 22.6 μM protein solutions in 50 mM buffers, containing 0.2 M NaCl.

```
       1                          10                                           20
Glu-Glu-Tyr-Ser-Ser-Asn-Trp-Ala-Gly-Ala-Val-Leu-Ile-Gly-Asp-Gly-Tyr-Thr-Lys-Val-
                                  30                                           39
Thr-Gly-Glu-Phe-Thr-Val-Pro-Ser-Val-Ser-Ala-Gly-Ser-Ser-Gly-Ser-Ser-Gly-Tyr
```

H CHAIN

```
        1                          10                                          20
<Glu-Ser-Glu-Glu-Tyr-Cys-Ala-Ser-Ala-Trp-Val-Gly-Ile-Asp-Gly-Asp-Thr-Cys-Glu-Thr-
                                   30                                          40
 Ala-Ile-Leu-Gln-Thr-Gly-Val-Asp-Phe-Cys-Tyr-Glu-Asp-Gly-Gln-Thr-Ser-Tyr-Asp-Ala-
                                   50                                          60
 Trp-Tyr-Glu-Trp-Tyr-Pro-Asp-Tyr-Ala-Tyr-Asp-Phe-Ser-Asp-Ile-Thr-Ile-Ser-Glu-Gly-
                                   70                                          80
 Asp-Ser-Ile-Lys-Val-Thr-Val-Glu-Ala-Thr-Ser-Lys-Ser-Ser-Gly-Ser-Ala-Thr-Val-Glu-
                                   90                                         100
 Asn-Leu-Thr-Thr-Gly-Gln-Ser-Val-Thr-His-Thr-Phe-Ser-Gly-Asn-Val-Glu-Gly-Asp-Leu-
                                  110                                         120
 Cys-Glu-Thr-Asn-Ala-Glu-Trp-Ile-Val-Glu-Asp-Phe-Glu-Ser-Gly-Asp-Ser-Leu-Val-Ala-
                                  130                                         140
 Phe-Ala-Asp-Phe-Gly-Ser-Val-Thr-Phe-Thr-Asn-Ala-Glu-Ala-Thr-Ser-Gly-Gly-Ser-Thr-
                                  150                                         160
 Val-Gly-Pro-Ser-Asp-Ala-Thr-Val-Met-Asp-Ile-Glu-Gln-Asp-Gly-Ser-Val-Leu-Thr-Glu-
                                  170      173
 Thr-Ser-Val-Ser-Gly-Asp-Ser-Val-Thr-Val-Thr-Tyr-Val
```

Figure 1. The amino acid sequence of *Aspergillus niger* acid proteinase A.

NMR spectra

One-dimensional NMR spectra were measured in a Bruker AM-400 NMR spectrometer with 1 - 5 mM protein solutions in D_2O and 50 mM sodium acetate buffers-0.2 M NaCl. Two-dimensional NMR spectra (HOHAHA and DQF-COSY) were also measured in the same spectrometer at protein concentrations of 3 mM (in D_2O) and 5 mM (in H_2O) at pH 4.6 and 45°C.

HPLC gel filtration

A Tosoh G3000SW column was used with 50 mM buffers, containing 0.2 M NaCl as the solvent.

RESULTS AND DISCUSSION

Two-chain structure of the acid proteinase A

When the enzyme was examined by sodium dodecyl sulfate polyacrylamide gel electrophoresis, one major slow-moving band and one minor fast-moving band were obtained. Gel filtration on a Sephadex G-50 column at pH 7.5 also yielded two fractions, one major protein fraction and one minor peptide fraction (data not shown). Initially, the minor band (or fraction) was thought to be an autolysis product derived from the major band (or fraction). However, analysis of these fractions revealed that they are both components of the original protein. Indeed, the enzyme yielded a single protein fraction on Sephadex G-50 gel filtration at pH around 5 with an amino acid composition which was a sum of those of the two fractions separated at pH 7.5. Therefore, the enzyme was assumed to have a two-chain structure.

The amino acid sequence of the acid proteinase A

The complete amino acid sequence of the enzyme was determined by conventional methods of protein chemistry as shown in Figure 1. The enzyme was indeed shown to be a two-chain protein of 212 residues (M_r 22,265), composed of a 39-residue light (L) chain and a 173-residue heavy (H) chain bound non-covalently with each other. No peptide was obtained which connects the L and H chains. The heavy chain was blocked at the NH_2-terminus with a pyroglutamic acid and had two intra-chain disulfide bonds (H chain, residues 6-30 and 18-101). The enzyme is a fairly acidic protein (pI about 3.3) and contains only 4 basic amino acid residues (3 Lys at positions L 19, H 64, and H 72, 1 His at position H 90, and no Arg) while it has 37 free acidic amino acid residues (18 Asp and 19 Glu) per molecule of protein. The sequence of residues 38-52 in the H chain draws special attention since this 15-residue sequence includes 8 aromatic amino acid residues (5 Tyr, 2 Trp, and 1 Phe).

The amino acid sequence of the acid proteinase A has no homology with those of the typical aspartic or acid proteinases so far sequenced including thermopsin, another non-pepsin-type acid proteinase from *Sulfolobus acidocardarius*. As shown in Figure 2, sequence homology can be found only with *Scytalidium lignicolum* acid proteinase B,[7] which is also a pepstatin-insensitive acid proteinase.[9] The identity is about 50 %. Thus, these two proteinases clearly belong to the same acid proteinase family different from the ordinary aspartic proteinase family. However, there are interesting differences between these two enzymes . First, the *A. niger* enzyme has a two-chain structure, whereas the *S. lignicolum* enzyme has a single-chain structure. Secondly, the former has two disulfide bonds and the latter three disulfide bonds. Among these, only one disulfide bond (H chain, residues 18-101 in the case of the former) is common between them. Thirdly, two acidic residues, Glu_{53} and Asp_{98}, in the *S. lignicolum* enzyme were assumed to be the active-site residues from chemical modification studies.[10-12] In the *A. niger* enzyme, the residues at the corresponding positions are Gln (H chain, residue 24) and Lys (H chain, residue 72),

```
A. niger A:        EEYSSNWAGAVLIGDGYTKVTGEFTVPSVSAGSSGSSGY
                   ::: :: :::       :       :::     : :::
S. lignicolum B:   TVESNWGGAILIGSDFDTVSATANVPSATGASGGSS--


  <ESEEYCASAWWGIDGDTCETAILQTGVDFCYEDGQTSYDAWYEWYPDYAYDFSDITISE
    :::::::::::: :::: :: : : :: :::::::::  ::: :::::
  -------AAWVGIDGDTCQTAILETGFDW-YGDG--TYDAWYEWYPEVSDDFSGITISE
                   *

  GDSIKVTVEATSKSSGSATVENLTTGQSVTHTFSGNVEGDLCETNAEWIVEDFE-----SGD
  :::::  : :::  ::::: :::::::: : :: :   :: :::: : :::::    :
  GDSIQMSVTATSDTSGSATLENLTTGQKVSKSFS-NESSGLCRTNAEFIIEDFEECNSDGSD
        *

  SLVAFADFG-SVTFTNAEATSGGSTVGPSDATVMDIEQDGSVLTETSVSGDSVTVTYV
  : :: :   : ::    :: : :   ::       :  : :::: :    ::
  EFVPFASFSPAVEFTDCSVTSDGESVSLDDAQITQVIINNQDVTDCSVSGTTVSCSYV
```

Figure 2. Amino acid sequence comparison between *Aspergillus niger* acid proteinase A and *Scytalidium lignicolum* acid proteinase B. Double dots indicate identical residues. *, putative active site residue of *S. lignicolum* acid proteinase B. -, deletion.

```
┌──────────────────────────────────────────10───────────────────────────20─┐
│ Met-Lys-Phe-Ser-Thr-Ile-Leu-Thr-Gly-Ser-Leu-Phe-Ala-Thr-Ala-Ala-Leu-Ala-Ala-Pro-│
│                            30                                    40        │
│ Leu-Thr-Glu-Lys-Arg-Arg-Ala-Arg-Lys-Glu-Ala-Arg-Ala-Ala-Gly-Lys-Arg-His-Ser-Asn-│
│                            50                                    60        │
│ Pro-Pro-Tyr-Ile-Pro-Gly-Ser-Asp-Lys-Glu-Ile-Leu-Lys-Leu-Asn-Gly-Thr-Thr-Asn┤Glu-│
└────────────────────────────70───────────────────────────────────┘          80
  Glu-Tyr-Ser-Ser-Asn-Trp-Ala-Gly-Ala-Val-Leu-Ile-Gly-Asp-Gly-Tyr-Thr-Lys-Val-Thr-
                            90                                ┌───────100──┐
  Gly-Glu-Phe-Thr-Val-Pro-Ser-Val-Ser-Ala-Gly-Ser-Ser-Gly-Ser-Ser-Gly-Tyr┤Gly-Gly-│
  ┌─────────────────────────110─────────────────────────────────┘         120│
  │Gly-Tyr-Gly-Tyr-Trp-Lys-Asn-Lys-Arg├Gln-Ser-Glu-Glu-Tyr-Cys-Ala-Ser-Ala-Trp-Val-
  └───────────────────────130─────────┘                                  140
  Gly-Ile-Asp-Gly-Asp-Thr-Cys-Glu-Thr-Ala-Ile-Leu-Gln-Thr-Gly-Val-Asp-Phe-Cys-Tyr-
                            150                                    160
  Glu-Asp-Gly-Gln-Thr-Ser-Tyr-Asp-Ala-Trp-Tyr-Glu-Trp-Tyr-Pro-Asp-Tyr-Ala-Tyr-Asp-
                            170                                    180
  Phe-Ser-Asp-Ile-Thr-Ile-Ser-Glu-Gly-Asp-Ser-Ile-Lys-Val-Thr-Val-Glu-Ala-Thr-Ser-
                            190                                    200
  Lys-Ser-Ser-Gly-Ser-Ala-Thr-Val-Glu-Asn-Leu-Thr-Thr-Gly-Gln-Ser-Val-Thr-His-Thr-
                            210                                    220
  Phe-Ser-Gly-Asn-Val-Glu-Gly-Asp-Leu-Cys-Glu-Thr-Asn-Ala-Glu-Trp-Ile-Val-Glu-Asp-
                            230                                    240
  Phe-Glu-Ser-Gly-Asp-Ser-Leu-Val-Ala-Phe-Ala-Asp-Phe-Gly-Ser-Val-Thr-Phe-Thr-Asn-
                            250                                    260
  Ala-Glu-Ala-Thr-Ser-Gly-Gly-Ser-Thr-Val-Gly-Pro-Ser-Asp-Ala-Thr-Val-Met-Asp-Ile-
                            270                                    280
  Glu-Gln-Asp-Gly-Ser-Val-Leu-Thr-Glu-Thr-Ser-Val-Ser-Gly-Asp-Ser-Val-Thr-Val-Thr-
      282
  Tyr-Val
```

Figure 3. The deduced amino acid sequence of the precursor form of *Aspergillus niger* acid proteinase A. The prepropeptide and the intervening sequence are enclosed.

respectively, none of which appear to be the active site residues. These results suggest that both enzymes have diverged to a considerable extent from each other although they are thought to originate from the same ancestral protein. The *A. niger* enzyme, however, is also assumed to have catalytic carboxyl group(s) at the active site as judged from its pH-activity profile and our recent chemical modification studies including carboxyl group modification which are now under way.

The amino acid sequence of the prepro-form of the acid proteinase A deduced from the DNA sequences

Using the procedures described in the "MATERIALS AND METHODS" section, a genomic clone and a cDNA clone of the enzyme could be isolated and sequenced. The genomic clone had no intron in the coding region and the nucleotide sequence of this region coincided completely with that of the cDNA clone. The deduced amino acid sequence is shown in Figure 3. The encoded protein is a single chain precursor of 282 residues, and composed of the NH_2 terminal prepropeptide (59 residues), the L chain (39 residues), an intervening sequence (11 residues), and the H chain (173 residues), linked in this order. The deduced amino acid sequences of the L and H chains coincided completely with those obtained with the protein by conventional methods of protein chemistry except that the NH_2-terminal glutamine residue in the deduced sequence of the H chain was found as a pyroglutamic acid in the actual protein. Therefore, the enzyme is thought to be synthesized initially as a single chain preprotein and then processed to the mature two chain form.

The boundary between the signal sequence and the propeptide sequence has not been determined, but it is suggested to be most likely the Ala-Ala (residues 18-19) bond according to the method of von Heijne.[13] The propeptide sequence is fairly rich in basic residues and shows a slight homology with those of pepsin-type aspartic proteinases. This suggests the possibility that like pepsinogen, the proenzyme (i.e., zymogen) is activated autocatalytically to the active form. Indeed, the cleavage site between the propart and the L chain is the Asn_{59}-Glu_{60} bond, which is consistent with the substrate specificity of the enzyme. The Asn-X bond has been shown to be one of the most favorable bonds for the enzyme from studies on its substrate specificity toward proteins (K.Takahashi *et al.*, unpublished).

Further, the 11-residue intervening sequence, Gly-Gly-Gly-Tyr-Gly-Tyr-Trp-Lys-Asn-Lys-Arg, has a characteristic structure; rich in glycine, aromatic amino acids and basic amino acids. The occurrence of these basic residues (2 Lys and 1 Arg) in this segment draws special attention since the mature enzyme has only four basic residues (3 Lys and 1 His). At present, nothing is known about the proteinase(s) involved in the processing of this intervening sequence. The COOH-terminal dibasic amino acid sequence (Lys-Arg) may be utilized as one of the cleavage signals for the processing enzyme. So far we have failed to find in the culture medium the unprocessed form(s) of the enzyme which retains the intact intervening sequence. Presumably, this processing takes place before the secretion of the enzyme into the extracellular medium. According to the secondary structure prediction of Chou and Fasman,[14] the three processing sites, that is, the Asn_{59}-Glu_{60} bond between the propeptide and the L chain, the Tyr_{98}-Gly_{99} bond between the L chain and the intervening sequence, and the Arg_{109}-Gln_{110} bond between the intervening sequence and the H chain, appear to be included in the turn structures, and thus favorably located for the proteolytic attack.

pH-Dependent conformational change of the acid proteinase A

The acid proteinase A is extremely unstable above pH 6 and inactivated very rapidly. In order to know something about the conformation and its stability, the conformational change of the enzyme was analyzed as a function of pH values by combination of CD, one-dimensional NMR, gel filtration, and activity measurement. The results obtained by each of

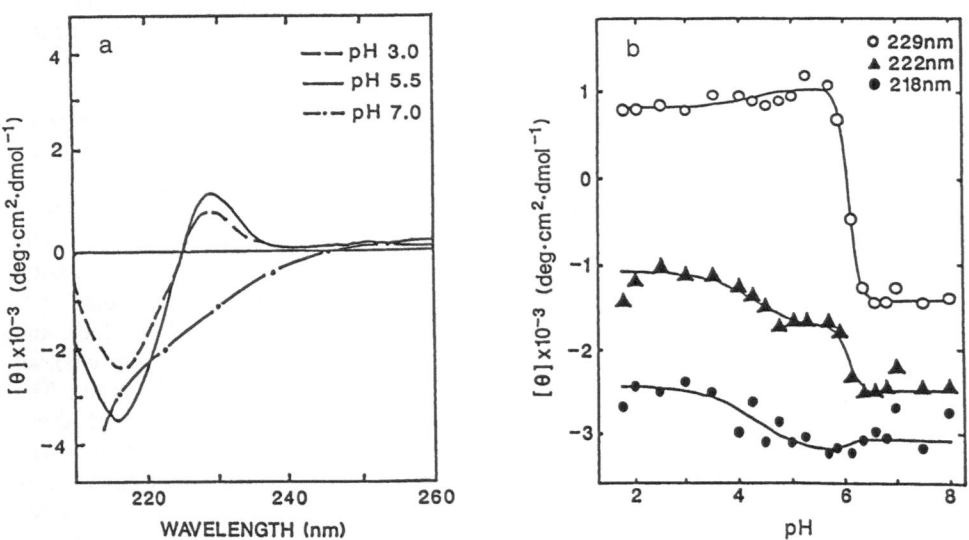

Figure 4. CD spectra of *Aspergillus niger* acid proteinase A (a) and the pH titration curves (b).

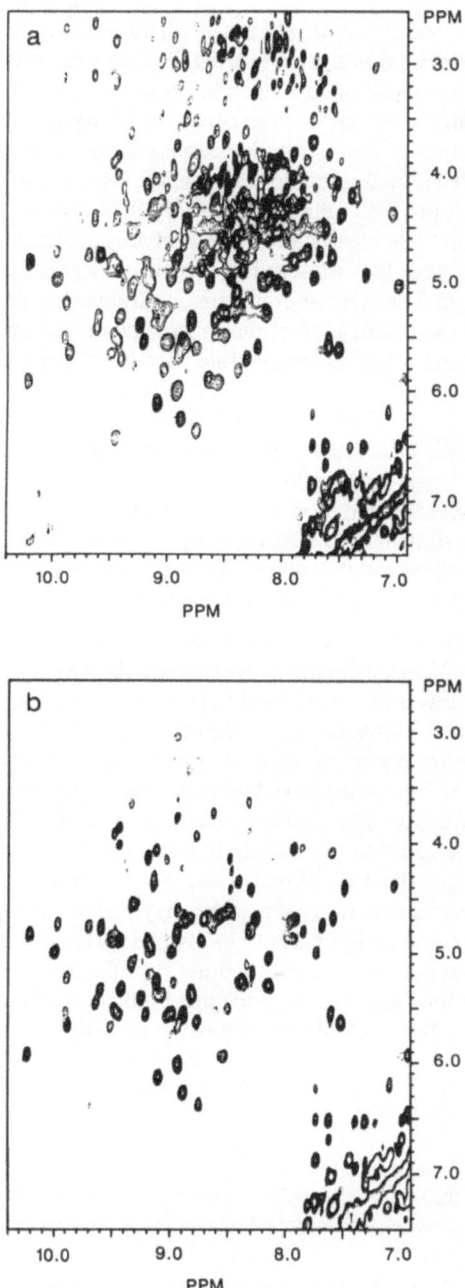

Figure 5. 2-Dimensional ¹H-NMR (HOHAHA) spectra (fingerprint region) of *Aspergillus niger* acid proteinase A. The spectra were measured at pH 4.6 and 45°C and at 400 MHz in H₂O (a) and D₂O (b)

these methods indicated that the enzyme is denatured rapidly and irreversibly in a narrow pH range between 6.0 and 7.0. Figure 4a shows some typical CD spectra of the enzyme at different pH values, and Figure 4b, pH titration curves of the CD spectra at three different wavelengths. The CD spectra at pH 3.0 and 5.5 in Figure 4a indicate that the enzyme has a high ß-sheet content in the weakly acidic pH range where the enzyme is stable. On the other hand, the CD spectrum at pH 7.0 indicates that the ß-sheet structure is largely lost. Almost the same results were obtained by other methods. In the case of CD spectra, an additional small conformational change was suggested to occur at pH around 4.0.

In addition, gel filtration analysis indicated that dissociation of the two chains occurs at pH around 6.0. The two chains could be isolated separately by Sephadex G-50 gel filtration after treatment at pH 8.5. One-dimensional NMR spectra of these isolated chains indicated that they have no ß-sheet structure. These chains could also be dissociated in 3 M guanidine-HCl at pH 5.5 with loss of activity. So far, however, we have been unable to regenerate any of the original activity or conformation by mixing these two isolated chains. These results indicate that the inactivation of the enzyme by denaturation at pH 6 and above or in 3 M guanidine-HCl at pH 5.5 is an irreversible process, accompanied by dissociation of the two chains .

Structural studies on the acid proteinase A using two-dimensional NMR spectroscopy

As a first step toward elucidation of the three-dimensional structure of the enzyme, we have measured its two-dimensional NMR spectra. Figure 5 shows the fingerprint regions of the HOHAHA spectra as typical examples. Among the many cross peaks observed in the HOHAHA spectrum (in H_2O), which contain those between amide protons and α– protons (Figure 5a), about 190 cross peaks were those between amide protons and α- or ß- protons as shown in the DQF-COSY spectrum (in H_2O) (data not shown). The fact that many of these cross peaks appear in the lower magnetic field region than the solvent signal suggests a high content of ß-sheet structure[15] in the molecule. Further, when the enzyme was kept in D_2O for over 24 h, about 100 amide protons remained unexchanged with deuterium (Figure 5b), suggesting that the enzyme has a tightly packed core structure in the molecule. The enzyme can be dissolved in a markedly high concentration (up to 10 mM at pH 4.5) in an aqueous solvent. A 5 mM enzyme solution gave a one dimensional NMR spectrum as sharp as a 1 mM enzyme solution, and the two-dimensional NMR spectra of both solutions were essentially the same. These results suggest that the enzyme has little tendency to aggregate at higher concentrations. Spectral assignments to individual residues are now under way.

In addition, we have recently succeeded in crystallization of the enzyme and started preliminary X-ray diffraction studies in the hope that both approaches would greatly facilitate the elucidation of the three-dimensional structure of the enzyme.

REFERENCES

1. Y. Koaze, H. Goi, K. Ezawa, Y. Yamada, and T. Hara, Fungal proteolytic enzymes. Part I. Isolation of two kinds of acid proteases excreted by *Aspergillus niger* var. *macrosporus, Agr. Biol. Chem.* **28**:216 (1964).

2. S. Horiuchi, M. Honjo, M. Yamasaki, and Y. Yamada, Studies on acid proteinases. I. Homogeneity and molecular weight of acid proteases from *Aspergillus niger* var. *macrosporus* (Proctase A and Proctase B), *Sci. Pap. Coll. Gen. Educ. Univ. Tokyo* **19**:127 (1969).

3. S. Horiuchi, M. Yamasaki, and Y. Yamada, Studies on acid proteases. II. Amino acid composition of acid proteases from *Aspergillus niger* var. *macrosporus* ("Proctases A and B"), *Sci. Pap. Coll. Gen. Educ., Univ. Tokyo* **19**:140 (1969)

4. W. -J. Chang, S. Horiuchi, K. Takahashi, M. Yamasaki, and Y. Yamada, The structure and function of acid proteases. IV. Effects of acid protease-specific inhibitors on the acid proteases from *Aspergillus niger* var. *macrosporus, J. Biochem.* **80**:975 (1976).

5. K. Iio and M. Yamasaki, Specificity of acid proteinase A from *Aspergillus niger* var. *macrosporus* towards B-chain of performic acid oxidized bovine insulin, *Biochim. Biophys. Acta* **429**:912 (1976).

6. E. Ido, T. Saito, and M. Yamasaki, Substrate specificity of acid proteinase A from *Aspergillus niger* var. *macrosporus, Agric. Biol. Chem.* **51**:2855 (1987).

7. T. Maita, S. Nagata, G. Matsuda, S. Maruta, K. Oda, and D. Tsuru, Complete amino acid sequence of *Scytalidium lignicolum* acid protease B, *J. Biochem.* **95**:465 (1984).

8. X. Lin and T. Tang, Purification, characterization, and gene cloning of thermopsin, a thermostable acid protease from *Sulfolobus acidocaldarius, J. Biol. Chem.* **265**:1490 (1990).

9. S. Murao and K. Oda, Pepstatin-insensitive acid proteinases, *in:* Aspartic Proteinases and Their Inhibitors, V. Kostka, ed., pp. 379-399, Walter de Gruyter, Berlin (1985).

10. D. Tsuru, S. Shimada, S. Maruta, T. Yoshimoto, K. Oda, S. Murao, T. Miyata, and S. Iwanaga, Isolation and amino acid sequence of a peptide containing an epoxide-reactive residue from the thermolysin digest of *Scytalidium lignicolum* acid protease B, *J. Biochem.* **99**:1537 (1986).

11. D. Tsuru, R. Kobayashi, N. Nakagawa, and T. Yoshimoto, Inhibition of *Scytalidium lignicolum* acid protease B by 1-diazo-3-phenyl-2-propanone, *Agric. Biol. Chem.* **53**:1305 (1989).

12. D. Tsuru, A. Naotsuka, R. Kobayashi, T. Yoshimoto, K. Oda, and S. Murao, Inactivation of *Scytalidium lignicolum* acid protease B with 1,2-epoxy-3-(4'-azido-2'-nitrophenoxy)propane, *Agric. Biol. Chem.* **53**:2751 (1989).

13. G. von Heijne, A new method for predicting signal sequence cleavage sites, *Nucleic Acids Res.* **14**:4683 (1986).

14. P. Y. Chou and G. D. Fasman, Empirical predictions of protein conformation, *Annu. Rev. Biochem.* **47**:251 (1978).

15. L. Szilagyi and O. Jardetzky, α-Proton chemical shifts and secondary structure in proteins, *J. Magn. Reson.* **83**:441 (1989).

INFECTION AND PATHOGENESIS OF CASH CROPS BY *BOTRYTIS CINEREA*:

PRIMARY ROLE OF AN ASPARTIC PROTEINASE

S. Movahedi,[1] C. G. Norey,[2] J. Kay[2] and J. B. Heale[1]

[1]Plant Cell and Molecular Sciences Research Group
King's College Kensington Campus
Campden Hill, London
W8 7AH, United Kingdom

[2]Department of Biochemistry
University of Wales College of Cardiff
P. O. Box 903, Cardiff, CF1 1ST
Wales, United Kingdom

Although extracellular enzymes are believed to be instrumental during invasion and subsequent infection in plant diseases caused by fungal necrotrophic pathogens, little direct evidence has been forthcoming to indicate a primary role for proteolytic enzymes in such infections. *Botrytis cinerea* is a ubiquitous necrotrophic fungus which affects fruit and vegetables both in the field and in post-harvest situations. It attacks the crop through wounds and accounts for a large proportion of the wastage that occurs with cash crops.[1]

A single spore isolate of this organism grown in liquid culture (in a medium supplemented with 0.4% gelatin) secreted increasing amounts of proteolytic activity into the medium with a maximum occurring after 48-72 hr at 24°C.[2] Indeed, such proteinase activity was also detected in leachates of *Botrytis* spore suspensions even before the initiation of germination. A similar time course of increasing proteolytic activity was determined in carrot slices that had been infected with *Botrytis*[2] and, after 2 days, grey mycelium was visible on the softened surface of the tissue. No such proteinase activity (measured with hemoglobin as substrate at pH 3.5) could be detected in healthy carrot tissue so that it would seem that the presence of proteinase is an early feature associated with fungal invasion.

The proteinase was purified to homogeneity (both from medium conditioned *in vitro* by fungal growth and from carrot tissue inoculated with *Botrytis*) by ammonium sulphate fractionation followed by FPLC chromatography on phenyl-Superose and Superose-12.[2] The enzyme thus obtained from both culture and infected tissue had a molecular weight of approximately 38 kDa and an isoelectric point of 4.0.[2] Its pH optimum for hemoglobin digestion was 3.5. Inhibitors of serine, cysteine and metalloproteinases (DFP, pCMB and EDTA, respectively) had no effect on the proteinase whereas isovaleryl pepstatin (at a concentration of 1 mM) blocked the activity of the purified enzyme completely. Similarly, all of the proteolytic activity (toward hemoglobin as substrate) in the culture filtrates was removed by inclusion of pepstatin in the assays. A similar activity with a pH optimum of 2.5 - 3.0 and which was inactivated by diazoacetyl-norleucine ester was reported earlier in culture

Table 1. Effect of statine-containing inhibitors on *B. cinerea* proteinase activity

Inhibitor	IC_{50}, (nM)
Iva-Val-Val-Sta-Ala-Sta	10
Lac-Val-Sta-Ala-Sta	1,000
Lac-Lys-Sta-Ala-Sta	1,000

Assays were carried out at $30°C$ in 0.17 M sodium formate buffer, pH 3.1. The substrate used was hemoglobin.

filtrates of an isolate of *B. cinerea* obtained from apples.[3] Thus, it would appear that an aspartic proteinase is solely responsible for the proteolytic activity that is present in the ungerminated *Botrytis* spores used and is produced during germination. The possibility of variability in the number and types of proteinase existing in other isolates of *Botrytis* cannot be excluded, however.

Potent inhibition of the aspartic proteinase activity was observed with isovaleryl-pepstatin (Iva-Val-Val-Sta-Ala-Sta; Table 1). By contrast, the water-soluble lactoyl- pepstatin (Lac-Val-Sta-Ala-Sta) and a lysine-containing analogue (Lac-Lys-Sta-Ala-Sta) were much less effective inhibitors (Table 1). These results are in agreement with earlier findings that fungal aspartic proteinases are more potently inhibited by the isovaleryl-variant of pepstatin.[4]

The inhibitor H-142 (Table 2) has also been shown previously to discriminate between aspartic proteinases isolated from different sources.[5] This compound is an efficient inhibitor of the proteinase purified from *Endothia parasitica* (Table 2) but has essentially no effect on penicillopepsin from *Penicillium janthinellum*. H-142 was equally without influence on the *Botrytis* proteinase activity, suggesting that this enzyme may resemble penicillopepsin more closely than the enzyme from *Endothia*.

Since proteinase production appears to be an early event in infection of fruit and vegetables, the possibility of reducing *Botrytis* infection by treatment with pepstatin was investigated. Eight separate isolates of the pathogen were obtained, one each from grapes, raspberries, strawberries, cabbage, broad bean, respectively and three from carrots. Spore suspensions were prepared in 0.1 M sodium acetate buffer, pH 3.5 in the presence or absence of 0.1 mM isovaleryl-pepstatin and inoculated into their respective host tissues,

Table 2. Effect of the compound, H-142 (Pro-His-Pro-Phe-His-ψ[Leu-CH_2-NH-Val]-Ile-His-Lys) on the proteinase activities of several fungi

Aspartic Proteinase from	K_i, (μM)
Endothia parasitica	0.15
Penicillium janthinellum	24
Botrytis cinerea	>20

Assays were carried out at pH 3.1

visible symptoms of infection being determined 48 hr later.[6] No effect of pepstatin was found on germination of the spores nor on the subsequent growth of the sporelings as determined through measurement of the length of the germ-tubes.[6] However, this treatment of *Botrytis* spores prior to inoculation generally strongly reduced and, in some cases, blocked completely the infection of the various hosts (Table 3). The extent of the observed reduction in infection seemed to be related to the earliness of the first appearance of proteinase activity by the different isolates; those which produced the aspartic proteinase as a very early event (e.g. the particularly virulent isolates 6 and 7 in which activity was readily detectable in ungerminated spores) were most susceptible to the effects of pepstatin.[6]

No proteinase activity was detected in healthy plant tissue so that the high levels measured upon *Botrytis* infection suggest a fungal origin and a primary role for this enzyme in pathogenesis, particularly since it is active in the pH range that develops in diseased tissue. The mechanism by which the proteinase mediates in pathogenesis is worthy of further investigation. It has been suggested that secretion of the enzyme into the host tissue is mandatory in advance of fungal growth. The purified enzyme has been shown to cause cell death in surface layers of treated carrot root slices, thus releasing nutrients available for further growth of the pathogen. Tests involving carrot protoplasts and suspension cells interacted with the enzyme suggest that the toxic effects originate from a degraded cell wall component.[6]

B. cinerea also poses a serious post-harvest problem because it frequently infects fruits and vegetables in cold stores. Since the purified aspartic proteinase still manifested approximately 15 % of its maximal activity when assayed at 0°C,[2] it may make a significant contribution to the degeneration of foodstuffs and loss of quality even at such cold-store temperatures.

Pathological roles for such fungal proteinases might be combatted by endogenous proteinase inhibitors positioned in the plant tissue. However, with the exception of an inhibitor present in potatoes [which has a strictly defined specificity toward (other) aspartic

Table 3. Effect of isovaleryl-pepstatin on the infection of fruits and vegetables by eight distinct isolates of *B. cinerea*

ISOLATE NUMBER AND HOST		REDUCTION IN INFECTION (%)
1	Grape	55
2	Raspberry	92
3	Strawberry	40
4	Cabbage	98
5	Broad bean (leaves)	95*
6	Carrot (root slices)	100
7	Carrot (")	100
8	Carrot (")	70

Spore suspensions from each of eight different *Botrytis* isolates were prepared in 0.1M sodium acetate buffer, pH 3.5 containing 0.1 mM isovaleryl-pepstatin and used to inoculate the respective hosts. Visible symptoms of infection were scored 48 hr after inoculation and the percentage reduction was calculated relative to control plants inoculated with spores prepared in buffer alone. *Buffer supplemented with 1% glucose and 1% yeast extract (in both control and pepstatin-treated preparations) since infection will only occur in broad bean leaves under these conditions.

proteinases[7,8] there is little indication that inhibitors of aspartic proteinases are present in the plant kingdom.[9]

Clearly then, the strategic deployment by the fungus of a hydrolytic activity belonging to the aspartic proteinase class is a tactic that permits relatively unimpeded access to the nutrients of the host tissue. Since blocking of the action of this enzyme results in the prevention of cell death so that nutrients are withheld, and is without effect on fungal germination or growth *per se*, the use of a specific proteinase inhibitor should not subject the organism to additional evolutionary pressure. Through the design and production of a synthetic inhibitor that is cheaper, more readily water-soluble and particularly, more specific than isovaleryl-pepstatin for use in the field and in post-harvest situations, intervention may be feasible to control the huge economic losses that are caused annually throughout the food-producing nations by this ubiquitous fungal pathogen.

ACKNOWLEDGEMENTS

We are most grateful to Drs. Steve Foundling, Theo Hofmann, Michael Szelke, Takaaki Aoyagi and Callum Campbell who generously provided us with samples of highly purified *Endothia* proteinase, penicillopepsin, H-142, lactoyl-pepstatin and its lysine analogue, respectively.

REFERENCES

1. J. R. Coley-Smith, K. Verhoeff and W. R. Jarvis, "The Biology of *Botrytis*," Academic Press, New York (1980).
2. S. Movahedi and J. B. Heale, *Physiol. Mol. Plant Pathol.* 36:289-302 (1990).
3. H. Urbanek and A. Kaczmarek, *Acta Biochim. Polonica* 32:101-109 (1985).
4. M. J. Valler, J. Kay, T. Aoyagi and B. M. Dunn, *J.Enzyme Inhib.* 1:77-82 (1985).
5. S. I. Foundling, J. Cooper, F. E. Watson, A. Cleasby, L. H. Pearl, B. L. Sibanda, A. Hemmings, S. P. Wood, T. L. Blundell, M. J. Valler, C. G. Norey, J. Kay, J. Boger, B. M. Dunn, B. J. Leckie, D. M. Jones, B. Atrash, A. Hallett and M. Szelke, *Nature* 327:349-352 (1987).
6. S. Movahedi and J. B. Heale, *Physiol. Mol. Plant. Pathol.* 36:303-324 (1990).
7. M. Mares, B. Meloun, M. Pavlik, V. Kostka and M. Baudys, *FEBS Lett.* 251:94-98 (1989).
8. A. Ritonja, I. Krizaj, P. Mesko, M. Kopitar, P. Lucovnik, G. Strukelj, J. Pungercar, D. J. Buttle, A. J. Barrett and V. Turk, *FEBS Lett.* 267:13-15 (1990).
9. J. Kay, *in*: "Aspartic Proteinases and their Inhibitors" (V. Kostka, ed.) pp 1-17 Walter de Gruyter, Berlin.(1985).

CRYSTAL STRUCTURES OF RHIZOPUSPEPSIN/INHIBITOR COMPLEXES

Kevin D. Parris,[†] Dennis J. Hoover[‡] and David R. Davies[†]

[†]Laboratory of Molecular Biology
NIDDK, NIH
Bethesda, Maryland 20892

[‡]Department of Medicinal Chemistry
Central Research Division
Pfizer Inc.
Groton, Connecticut 06340

INTRODUCTION

The crystal structures of the aspartic proteinases have been extensively studied over the past fifteen years (Hsu *et al.*, 1977; Subramanian *et al.*, 1977; reviewed by Davies, 1990). After the initial determinations of the native structures uncomplexed with inhibitors, a few complexes with pepstatin (Bott *et al.*, 1982) and with a fragment of pepstatin (James *et al.*, 1982) were reported. These studies showed that these inhibitors bound in the deep groove that separates the N- and C-terminal domains of the enzyme. Accompanying the binding of the inhibitor was a displacement of the "flap" region of the molecule, a hairpin loop that closes down on the inhibitor, the extent of the displacement depending on the initial location of the flap (James *et al.*, 1982; Bott *et al.*, 1982; Suguna *et al.*, 1987 and Cooper *et al.*, 1987). Since no other major conformational changes were observed, the crystals of these proteinases offered a convenient vehicle for examining a number of bound inhibitor conformations. Other factors facilitating the examination of these inhibitors were the availability of large numbers of renin inhibitors, and the fortunate ease of access to the combining site in several of the crystal forms, thus enabling the inhibitors to be soaked into the crystals.

In this paper we summarize some recent analyses of two isostere inhibitors, CP-69,799 and CP-82,218 (structures given below), with rhizopuspepsin and compare these results with previous observations on pepstatin (Bott *et al.*, 1985) and pepstatin-like (Suguna *et al.*, 1991) inhibitors and with a reduced inhibitor (Suguna *et al.*, 1987).

CP-69,799

CP-82,218

RESULTS

Except for pepstatin we have been unable to obtain co-crystals of the enzyme-inhibitor complex. Accordingly the data reported here concern crystals that were soaked in inhibitor. The inhibitor structure analyses are summarized in Table 1. Binding of the inhibitors and the accompanying flap closure had the effect of making the region around P_1 (notation of Schechter & Berger, 1967) inaccessible to solvent (Table 2).

CP-69,799

When this inhibitor was co-crystallized with endothiapepsin, a conformational change was observed in the enzyme which involved a rigid body rotation (Sali *et al.*, 1989) of the C-terminal domain by about 4 degrees relative to the N-terminal domain, together with a translation of 0.3Å. However, careful examination of the inhibited rhizopuspepsin showed no such conformational change. Excluding the flap, the rms deviation of the structures with and without inhibitor was 0.16Å for the alpha carbons and 0.15Å for all atoms. This change, which is within experimental error, together with the absence of any crystal cracking and the full substitution of the inhibitor on the enzyme, suggest that the presence of the inhibitor at the enzyme active site does not stress the crystal structure unduly. This, in turn implies that the energy driving any conformational change in the rhizopuspepsin is less than the energy of the crystal packing, which is generally regarded to be quite small.

Recent structure determinations of pepsin have revealed the flexibility of a C-terminal subdomain (Abad-Zapatero *et al.*, 1990; Sielecki *et al.*, 1990). It is possible that the observed changes in endothiapepsin bound to CP-69,799 are the result of interaction of this flexible subdomain with a different crystal environment.

Table 1. Summary of data collection and geometrical parameters for the rhizopuspepsin CP-69,799 and CP-82,218 inhibitor complexes. The target σ values used in refinement (PROFFT, Hendrickson, 1985; Finzel, 1987; Agarwal, 1978; Sheriff, 1987) are given in parentheses.

	CP-69,799	CP-82,218
Cell Parameters (P2$_1$2$_1$2$_1$)		
a, Å	60.55	60.48
b, Å	60.62	60.71
c, Å	107.44	106.93
Resolution range, Å	10 - 2.0	10 - 1.9
R Value, %	17.1	17.5
RMS deviation from ideality for		
Distances, Å		
Bond	0.012(.02)	0.011(.02)
Angle	0.038(.04)	0.033(.04)
Planar 1-4	0.041(.05)	0.036(.05)
Planar groups, Å	0.009(.02)	0.009(.02)
Nonbonded distances, Å		
Single torsion	0.18(.5)	0.18(.5)
Multiple torsion	0.21(.5)	0.19(.5)
Possible H-bonds	0.23(.5)	0.24(.5)
Torsion angles, deg		
Planar	1.4(2.0)	1.7(3.0)
Staggered	14.5(12.0)	13.4(15.0)
Orthonormal	29.7(20.0)	25.9(20.0)
Chiral Volume, Å^3	0.16(.15)	0.14(.15)
Thermal Restraints, Å^2		
Main chain bond	0.69(1.0)	0.62(1.0)
Main chain angle	1.14(1.5)	1.02(1.5)
Side chain bond	0.64(1.0)	0.67(1.0)
Side chain angle	1.04(1.5)	1.09(1.5)

Table 2. Inhibitor solvent accessibility changes upon binding to rhizopuspepsin. Surface areas are given in square angstroms. This solvent accessibility was assessed using the program MS of Connolly (1983) and programs developed by Sheriff *et al.* (1985). A radius of 1.7Å was assumed for the solvent probe; standard van der Waals radii (Case & Karplus, 1979) were used.

				RESIDUE			
CP-69,799	P_4	P_3	P_2	P_1	P_1'	P_2'	P_3'
	BOC	PHE	HIS	CHA	AHS	LYS	PHE
(ISOLATED)							
MAIN	21.28	28.59	27.40	30.63	36.91	23.93	31.80
SIDE	93.48	86.82	79.72	94.93	53.72	90.10	109.96
TOTAL	114.76	115.41	107.12	125.56	90.63	114.03	141.76
(BOUND)							
MAIN	22.35	9.96	0.00	0.00	0.00	6.19	32.20
SIDE	65.02	54.03	23.28	14.88	0.00	36.19	62.00
TOTAL	87.37	63.99	23.28	14.88	0.00	42.38	94.20

			RESIDUE		
CP-82,218	P_4	P_3	P_2	P_1	P_1'
	PIP	PHE	NLE	CHA	DFK
(ISOLATED)					
MAIN	12.99	29.05	22.42	21.88	19.63
SIDE	95.53	89.46	71.88	109.48	82.08
TOTAL	108.52	118.51	94.30	131.36	101.71
(BOUND)					
MAIN	17.71	7.05	0.00	0.00	5.13
SIDE	63.37	56.44	29.44	13.28	46.92
TOTAL	82.08	63.49	29.44	13.28	52.05

Abbreviations. BOC = N-tert-butoxycarbonyl; PIP = piperazine; CHA = cyclohexylalanine; NLE = norleucine; AHS = azahomostatine; DFK = difluoroketone.

The goals of the inhibitor studies on the aspartic proteinases have been two-fold. First, they would provide stereochemical data that would aid in the design of more potent inhibitors, and second, they could approximately define the conformation of the tetrahedral intermediate formed by the substrate during catalysis. This inhibitor has a dihydroxy structure at the active site and, for this region of the substrate, provides a very close approximation to the tetrahedral intermediate. The CF_2 group, however, although adopting a tetrahedral conformation, does not offer a good approximation in terms of hydrogen bonding to the substrate NH_2^+. Nevertheless, this inhibitor should, within fractions of an Å provide a reasonable approximation to the conformation of the tetrahedral intermediate.

Table 3 contains the distances between these and other inhibitors and the carboxylic acids groups of Asp_{35} and Asp_{218}. In the final models, both CP-69,799 and CP-82,218 are present in unit occupancies (Figure 1). They are bound in an extended conformation which is consistent with other inhibitors of rhizopuspepsin (Figure 2). Note the striking similarity in the conformations of the inhibitors from P_3 through the scissile bond. The inhibitors are held in position by a number of possible hydrogen bonds (Figure 3) between the backbone of the inhibitor and the surrounding protein atoms.

Table 3. Distances from atoms of the scissile bond surrogate to the oxygens of the catalytic aspartic acids. Column 3 indicates the atoms in the inhibitor which are equivalent to the atoms at the scissile bond of a normal peptide substrate (Column 2).

Inhibitor			D218		D35	
			OD2	OD1	OD2	OD1
CP-69,799	C	(CH(OH))	3.82	4.49	3.69	3.53
	O	(CH(OH))	2.70	2.87	2.53	3.28
	N	(CH$_2$)	3.76	5.02	4.89	4.88
CP-82,218*	C	(CH(OH))	3.71	4.30	3.60	3.50
	O1	(C(OH)$_2$)	4.66	4.81	3.28	2.49
	O2	(C(OH)$_2$)	2.59	2.92	2.89	3.68
	N	(CH$_2$)	3.59	4.85	4.82	4.79
Reduced	C	(CH$_2$)	3.45	3.57	3.39	3.27
	N		2.82	3.78	4.44	4.52
Pepstatin	C	(CH(OH))	3.50	4.29	3.72	3.42
	O	(CH(OH))	2.49	2.91	2.79	3.39
	N	(CH$_2$)	3.42	4.78	4.84	4.57
Inhibitor 2	C	(CH(OH))	3.45	4.28	3.68	3.46
(L-363,624)	O	(CH(OH))	2.58	2.90	2.57	3.21
	N	(CH$_2$)	3.15	4.65	4.79	4.73
Inhibitor 3	C	(CH(OH))	4.09	4.37	3.68	3.30
(L-363,889)	O	(CH(OH))	3.21	3.01	2.55	3.01
	N	(CH$_2$)	3.70	4.60	4.73	4.60

*O2 corresponds to the hydroxy which is present in other statine-like inhibitors (i.e., it is located between the catalytic aspartic acids).

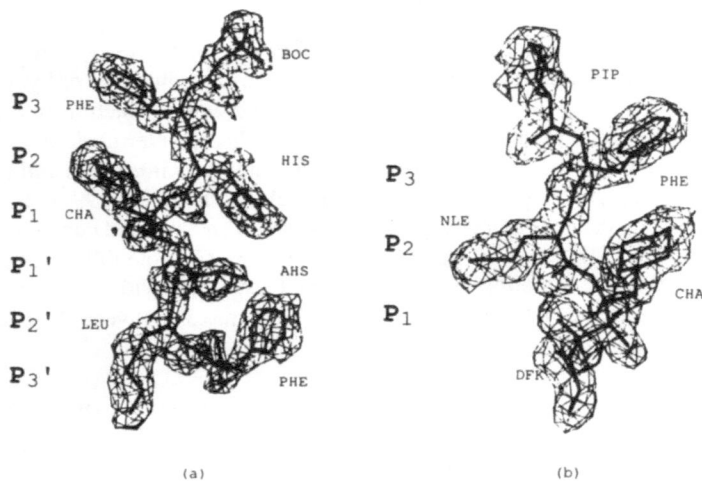

Figure 1. Final $2|F_o|-|F_c|$ electron density maps for inhibitors (a) CP-69,799 and (b) CP-82,218 with final models superimposed. Note that there is electron density for all residues of each inhibitor.

Figure 2. Superposition of inhibitors CP-82,218 (bold dash), CP-69,799 (bold line), pepstatin (dash dot), reduced peptide inhibitor (dash dot dot), inhibitor 1 (dash), and inhibitor 2 (line). The arrow indicates the carbonyl position of the scissile bond. The alignment of the inhibitors is described in the Discussion Section.

222

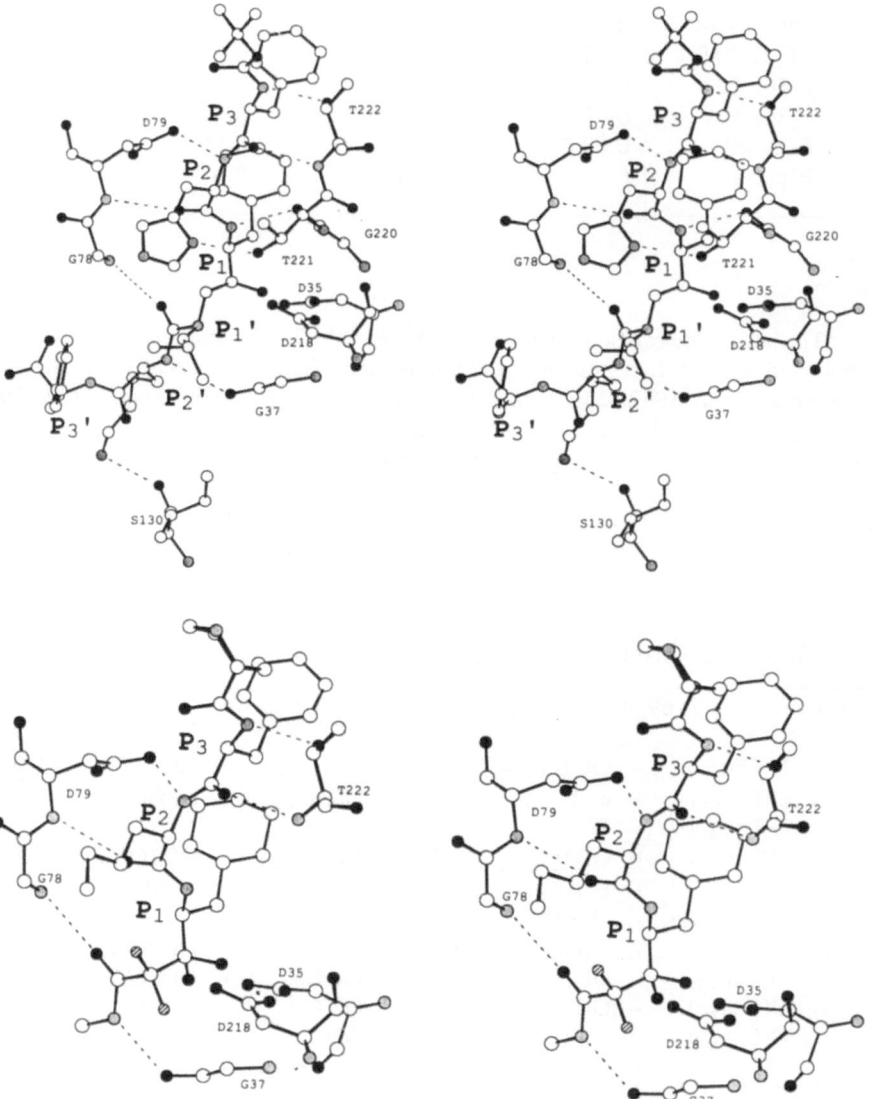

Figure 3. Stereo view of inhibitors (top) CP-69,799 and (bottom) CP-82,218 showing nearest contacts with the protein. The flap contains residues Asp78 and Gly79. Contacts short enough to be hydrogen bonds are shown dotted except for contacts between the hydroxyl group(s) of P_1 and the oxygens of Asp35 and Asp218.

Table 4. RMS deviation of the inhibitor structures from that of the native rhizopuspepsin

	RMS deviation (Å) of entire structure	RMS deviation (Å) for flap (residues 75-83)		
		All (59 atoms)	Main Chain (36)	Side Chain (23)
CP-69,799	0.16	0.40	0.43	0.36
CP-82,218	0.13	0.55	0.58	0.50
L-363,624	0.14	0.43	0.45	0.40
L-363,889	0.07	0.21	0.22	0.18
Reduced	0.08	0.54	0.59	0.48
Pepstatin	0.14	0.58	0.62	0.51

The active site region in the CP-82,218 and CP-69,799 complexes is quite similar to that observed in the native structure. In fact, the only change observed is the characteristic closure (James *et al.*, 1982; Bott *et al.*, 1982; Suguna *et al.*, 1987 and Cooper *et al.*, 1987) of the flap upon the inhibitor. For comparison, the movement in this region for the six rhizopuspepsin inhibitors studied thus far is tabulated in Table 4. This movement of the flap enhances the binding of the inhibitor by making several contacts, both hydrogen bond (Table 5) as well as van der Waals (Table 6), between the inhibitor and the flap.

Table 5. Potential hydrogen bonds between the inhibitors and rhizopuspepsin

Residue	CP-69,799		Å	CP-82,218		Å
P_3	N(Phe$_{401}$)	- OG1(Thr$_{222}$)	2.9	N(Phe$_{401}$)	- Wat$_{578}$	2.9
	O(Phe$_{401}$)	- N(Thr$_{222}$)	3.1	O(Phe$_{401}$)	- N(Thr$_{222}$)	3.0
	O(Phe$_{401}$)	- Wat$_{543}$	2.8			
P_2	N(His$_{402}$)	- OD1(Asp$_{79}$)	3.2	N(Asn$_{402}$)	- OD1(Asp$_{79}$)	3.1
	O(His$_{402}$)	- N(Asp$_{79}$)	3.1	O(Asn$_{402}$)	- N(Asp$_{79}$)	3.1
	NH(His$_{402}$)	- OG1(Thr$_{221}$)	2.8			
P_1	N(Cha$_{403}$)	- O(Gly$_{220}$)	2.9	O1(Cha$_{403}$)	- OD2(Asp$_{35}$)	3.3
	O(Cha$_{403}$)	- OD1(Asp$_{35}$)	3.3	O1(Cha$_{403}$)	- OD1(Asp$_{35}$)	2.6
	O(Cha$_{403}$)	- OD2(Asp$_{35}$)	2.5	O2(Cha$_{403}$)	- OD2(Asp$_{35}$)	2.8
	O(Cha$_{403}$)	- OD1(Asp$_{218}$)	2.9	O2(Cha$_{403}$)	- OD1(Asp$_{218}$)	2.9
	O(Cha$_{403}$)	- OD2(Asp$_{218}$)	2.7	O2(Cha$_{403}$)	- OD2(Asp$_{218}$)	2.6
P_1'	O(Cha$_{404}$)	- N(Gly$_{78}$)	3.0	O3(Dfk$_{403}$)	- N(Gly$_{78}$)	2.8
				N2(Dfk$_{403}$)	- O(Gly$_{37}$)	2.8
P_2'	N(Lys$_{405}$)	- O(G$_{37}$)	3.0			
	NZ(Lys$_{405}$)	- O(I$_{130}$)	2.8			
	NZ(Lys$_{405}$)	- Wat$_{506}$	3.2			
	NZ(Lys$_{405}$)	- Wat$_{777}$	3.2			
	O(Lys$_{405}$)	- Wat$_{666}$	3.4			
P_3'	OT(Phe$_{406}$)	- O(Ser$_{76}$)	3.0			

Table 6. Van der Waals contacts between the inhibitors and rhizopuspepsin

Subsite	CP-69,799			CP-82,218		
S_4	Boc_{401}:	Thr_{222}	Leu_{223}	Pip_{401}:	Thr_{222}	Phe_{278}
S_3	Phe_{401}:	Ile_{15}	Asp_{79}	Phe_{401}:	Ile_{15}	Asp_{79}
		Glu_{16}	Thr_{221}		Glu_{16}	Thr_{222}
S_2	His_{402}:	Gly_{78}	Thr_{221}	Nle_{402}:	Asp_{79}	Ile_{225}
		Asp_{79}	Trp_{294}		Gly_{220}	Trp_{294}
		Gly_{220}	Ile_{298}		Thr_{221}	Ile_{298}
S_1	Cha_{403}:	Asp_{33}	Ser_{81}	Cha_{403}:	Asp_{33}	Phe_{114}
		Asp_{35}	Phe_{114}		Asp_{35}	Leu_{122}
		Tyr_{77}	Asp_{218}		Tyr_{77}	Asp_{218}
		Asp_{79}	Gly_{220}		Ser_{81}	Gly_{220}
S_1'	Ahs_{404}:	Trp_{194}	Ile_{298}	Dfk_{403}:	Gly_{37}	Thr_{221}
		Ile_{216}			Asp_{218}	
S_2'	Lys_{405}:	Ser_{38}	Ile_{130}			
		Ile_{75}	Val_{133}			
		Ser_{76}	Trp_{194}			
S_3'	Phe_{406}:	Trp_{194}	Trp_{294}			

Another movement which is characteristic of inhibitor binding in rhizopuspepsin structures is the rearrangement of the side chain of Asn_{119}, a residue which is about 5Å removed from the binding pocket. In the native structure, this residue is hydrogen-bonded to Wat_{698} while in the inhibited structures, Wat_{698} has been displaced. This leads to a rearrangement of the Asn_{119} side chain in order that it may adopt a hydrogen-bonding interaction with Glu_{16}. This rearrangement is observed in the CP-82,218 complex, but in the CP-69,799 complex the residue is disordered. The final model of this side chain is such that it is evenly divided between the position observed in the native structure and that observed in other inhibited structures.

DISCUSSION

Although the mechanism of action for the aspartic proteinases is not fully known, there have been various mechanisms proposed which have their foundation in high resolution X-ray crystallographic studies of inhibited enzymes. The inhibitors generally have been short peptide chains which have been modified in the scissile bond region to prevent proteolysis. To date, various modifications have been studied using both NMR and X-ray crystallography (See Figure 4 for examples). Generally, these modifications include substitution of a reduced moiety at either the C or N positions of the scissile amide bond. For inhibitors which have been reduced at the amide carbonyl, two oxidation states have been studied. The first has been obtained from the substitution of the carbonyl with an alcohol (hydroxyethylene) while in the second, the carbonyl has been replaced with a methylene (reduced isostere). Substitution of a methylene for the amide N results in the ketomethylene and given proper predisposition, this ketone can be converted to a ketone hydrate yielding even more homology with the transition state. Another series of inhibitors has been based on the microbial inhibitor pepstatin which contains two residues of the rare amino acid statine. These pepstatin-like inhibitors have an insertion of two atoms in the backbone prior to the scissile bond which results in a frame shift that places the P_1' side chain in the S_2' pocket.

In the structural studies of inhibited complexes, the peptide inhibitors bind within a cleft that lies between the N- and C-terminal domains. This cleft is approximately 40Å long and a portion of the enzyme, a ß-hairpin loop between residues 76 and 83, (rhizopuspepsin numbering), projects over this binding groove. Due to the close interactions between the substrate and enzyme, both the amino acid composition (especially for renin; Green *et al.*, 1989) and length of the inhibitor are important mechanistic factors. Fruton (1976) and Hofmann *et al.* (1988) have demonstrated that the addition of residues to either P_3 or P_2' substantially increased k_{cat} without a corresponding increase in K_m. It was their conclusion that the additional binding energy was being used to stimulate catalysis through an induction of conformational change in either the peptide, the enzyme, or perhaps both. Pearl (1985) postulated that conformational changes in the substrate could result in distortion of the peptide bond from a planar conformation thereby inducing a tetrahedral conformation at the amide nitrogen which would facilitate proton acquisition. Supporting evidence for the theory of conformational change was provided by Sali *et al.* (1989) when they described an inhibitor of endothiapepsin, CP-69,799, which produced a domain rotation of 4 degrees between the N-terminal domain relative to the C-terminal domain. Previously, the only example of conformational change within a fungal aspartic proteinase upon complexation with a substrate was the closure of the flap down onto the bound substrate. Additional evidence for flexibility within the aspartic proteinase enzymes has been presented by both Abad-Zapatero *et al.*

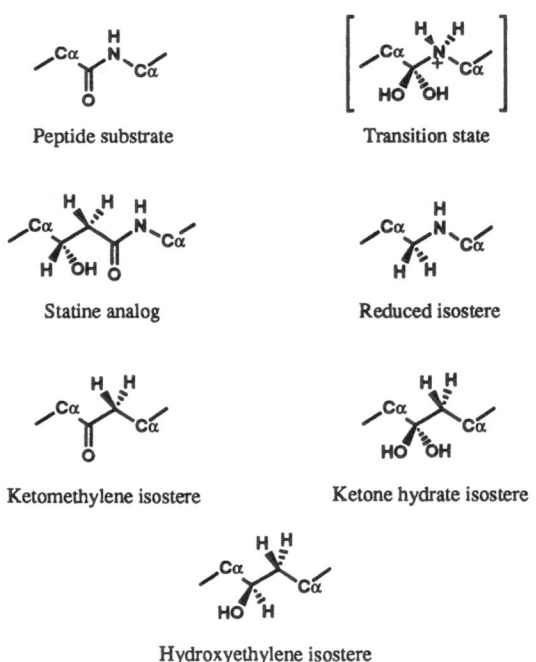

Peptide substrate   Transition state

Statine analog   Reduced isostere

Ketomethylene isostere   Ketone hydrate isostere

Hydroxyethylene isostere

Figure 4. Scissile bond modifications.

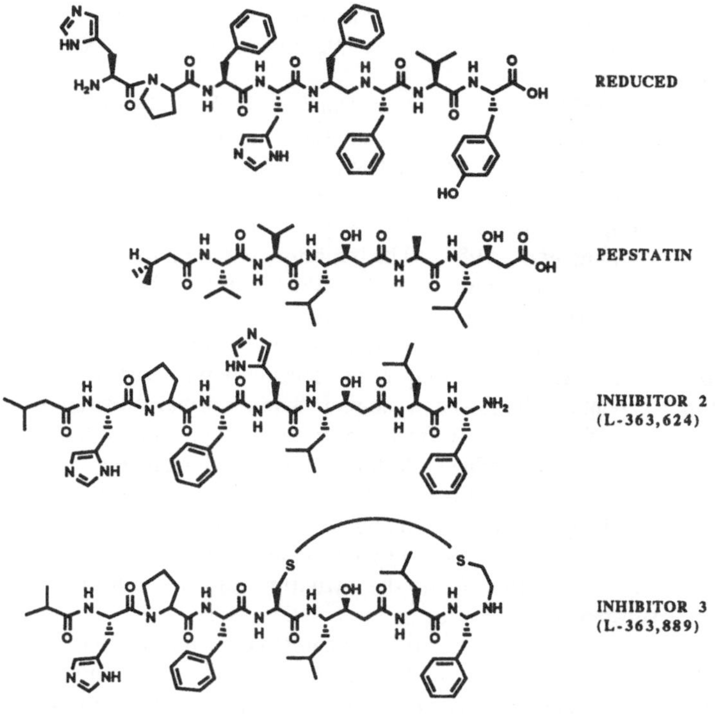

Figure 5. Additional inhibitors whose complexes with rhizopuspepsin have been studied crystallographically.

(1990) and Sielecki *et al.* (1990). In their analyses of pepsin (2.3Å and 1.8Å, respectively), they observed a flexible subdomain located in the C-terminal region when these structures were compared to the fungal proteases. This subdomain was characterized by a displacement through a screw rotation and translation relative to the remainder of the molecule.

To date, in addition to CP-69,799 and CP-82,218, four inhibitors of rhizopuspepsin (Figure 5) have been analyzed and a catalytic mechanism proposed which was based on the examination of rhizopuspepsin inhibited with the reduced peptide. These six complexes were analyzed to determine if any conformational changes, such as the domain rotation observed by Sali, *et al.* (1989) or any flexibility such as that discussed by Abad-Zapatero *et al.* (1990) and Sielecki *et al.*(1990) were present. The positions of the N-terminal domain (residues 1-194) and the C-terminal domain (residues 301 to 325) were considered to remain constant while the coordinates of the region considered variable (residues 195-300) were not constrained. The inhibitor structures were compared to the native enzyme using the program ALIGN (Satow *et al.*, 1986). The alignment was based on the position of the C_α of each residue in the fixed regions and the resulting rms deviations were 0.14 Å for Inhibitor 2, 0.07 Å for Inhibitor 3, 0.08 Å for the reduced inhibitor, 0.14 Å for pepstatin, 0.16 Å for CP-69,799, and 0.13 Å for CP-82,218. These data suggest neither domain rotation nor flexibility are present in any of these inhibitor complexes. When the aligned structures were superimposed and viewed using the Evans and Sutherland PS390 (FRODO; Jones 1978; Pflugrath *et al.*, 1984), the only mobility noted was in the flap region. This rms deviation of the flap region is tabulated in Table 4. In a final attempt to ascertain if there were any changes to the structure of the rhizopuspepsin upon binding CP-69,799, a plot of the rms shift for the main chain atoms of the enzyme versus residue number was made. Examination of Figure 6 once again shows that the only significant change between the CP-69,799 inhibited

227

rhizopuspepsin and the native enzyme is the movement of the flap (residues 75-83). Given these results, neither domain rotation nor translation is observed when CP-69,799 is bound to rhizopuspepsin.

Let us now consider the enzyme mechanism in the light of these data. Earlier proposals for the mechanism were based on the structure of pepstatin complexed to rhizopuspepsin (Bott *et al.*, 1982; Bott & Davies, 1983), the structure of penicillopepsin complexed to a pepstatin analog (James & Sielecki, 1985), and the structure of reduced peptide isosteres complexed to various fungal proteases (Pearl, 1985; Pearl & Blundell, 1984; Pearl, 1987; Suguna *et al.*, 1987). The inhibitors based on the scissile bond surrogate statine do not provide an unambiguous model for the binding of the substrate to the enzyme due to the frame shift resulting from the two additional atoms in the statine residue. The reduced peptide analyzed with rhizopuspepsin also does not provide an unambiguous model since this peptide lacks a hydroxyl group at the scissile bond. Due to the absence of this hydroxyl group, the peptide sits lower in the cleft than a normal substrate would. Thus in modelling the intermediate based on the reduced peptide, the position of the inhibitor must be adjusted and the hydroxyl group added. Although it is a statine analog, CP-82,218 provides the best substrate analog examined to date since it mimics both the transition state at the carbon as well as the nitrogen of the scissile bond. However, it is not the definitive substrate model for structural analysis since it lacks any residues beyond P_1'. Nevertheless, the information which is available from the structure of this inhibitor complexed to rhizopuspepsin is substantial.

There is general agreement that the mechanism does not involve the formation of a covalent bond between the substrate and the enzyme (Hofmann & Fink, 1984) but that it most likely proceeds via general base catalysis involving a molecule of water. James and Sielecki (1985) have proposed that this nucleophilic water is a water molecule bonded to the outside of Asp_{35}. Another mechanism, proposed by Suguna *et al.* (1987), suggested that the nucleophilic water was that water which was located between the catalytic aspartic acid residues in the native structure. In any case, a mechanism involving general base catalysis would involve a tetrahedral intermediate which would be characterized by two oxygens attached to the carbon of the scissile bond. In analysis of the CP-69,799 complex, the azahomostatine residue provides one oxygen of this tetrahedral intermediate and the position of the second can be deduced. However, analyzing a structure such as this can not

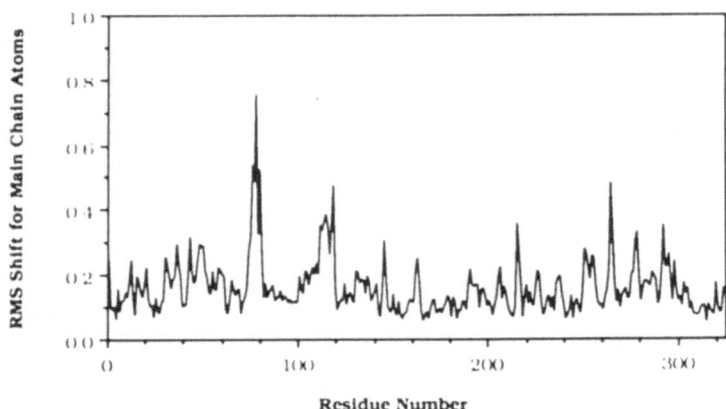

Figure 6. Graph showing the rms shift for the main chain atoms of rhizopuspepsin upon binding inhibitor CP-69,799.

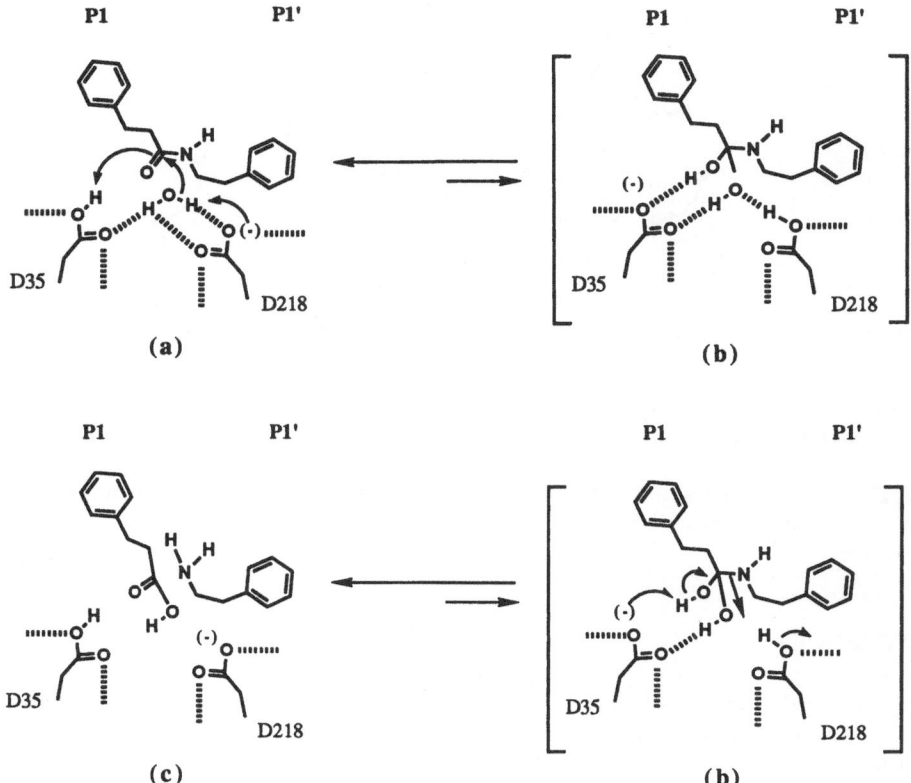

Figure 7. The catalytic mechanism proposed by Suguna et al. (1987).

unambiguously determine the identity of the nucleophilic water since the proposed tnucleophilic water molecules of James and Sielecki (1985) and of Suguna *et al.* (1987) are displaced upon inhibitor binding. Conversely, when complexed with rhizopuspepsin, inhibitor CP-82,218 possesses the dihydroxy intermediate; one of the oxygens in this structure does correspond to the nucleophilic water.

We conclude that in several respects the structures observed here fully support the mechanism proposed by Suguna *et al.* (1987). The central water molecule (Wat507) has sufficient room to swing out of the way of the approaching substrate and still maintain hydrogen bonds with the carboxylate oxygens. This allows the situation depicted in Figure 7a to occur. Subsequent attack of the water on the carbonyl carbon of the scissile bond and transfer of a hydrogen atom to Asp_{218} from the water gives the dihydroxy intermediate shown in Figure 7b. Comparison of this Figure with the CP-82,218 complex (Figure 8) shows good agreement, indicating that the modeling used in defining the substrate position from the reduced inhibitor must have been quite accurate. A similar model for the tetrahedral intermediate was also proposed by Blundell *et al.* (1987) based on inhibitor binding studies with endothiapepsin.

The CF_2 carbon atom is approximately placed so that the substrate can receive a proton from Asp_{218}. Whether this occurs in a concerted manner with nucleophilic attack on the carbonyl carbon atom or whether it occurs earlier, coincident with the strain induced in the

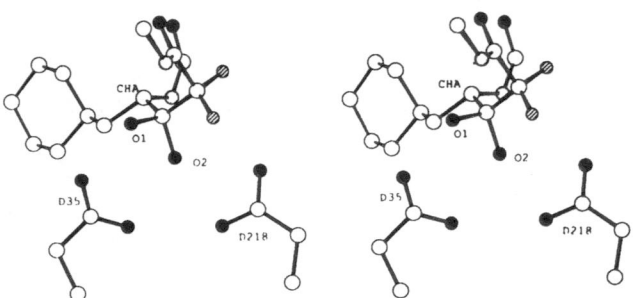

Figure 8. The active site as observed in the crystal structure of rhizopuspepsin complexed with CP-82,218.

substrate by its attachment to the enzyme, is beyond the scope of this investigation, and must await neutron crystallographic analysis.

NOTE: The coordinates of the complexes will be deposited in the Brookhaven Protein Data Bank. A manuscript describing the experimental details is in preparation and will be reported elsewhere.

REFERENCES

Abad-Zapatero, C., Rydel, T. J. & Erickson, J., 1990, Revised 2.3Å Structure of Porcine Pepsin: Evidence for a Flexible Subdomain, *Proteins, Structure, Function and Genetics*, **8**:62.

Agarwal, R. C., 1978, A new least squares refinement technique based on the fast Fourier transform algorithm, *Acta Crystallogr.*, Section A, **34**:791.

Blundell, T. L., Cooper, J., Foundling, S. I., Jones, D. M., Atrash, B. & Szelke, M., 1987, On the rational design of Renin inhibitors: X-ray studies of aspartic proteinases complexed with transition-state analogues, *Biochemistry*, **26**:5585.

Bott, R., Subramanian, E. & Davies D. R., 1982, Three-dimensional structure of the complex of the *Rhizopus chinensis* carboxyl proteinase and pepstatin at 2.5Å resolution, *Biochemistry*, **21**:6956.

Bott, R., Davies & D. R., 1983, Pepstatin binding to *Rhizopus chinensis* aspartyl proteinase, *in*: "Peptides: Structure and Function", V. J. Hruby and D. H. Rich, eds., Pierce Chemical Company.

Case, D. A. & Karplus, M., 1979, Dynamics of ligand binding to heme proteins, *J. Mol. Biol.*, **132**:343.

Connolly, M. L., 1983, Analytical molecular surface calculation. *J. Appl. Crystallogr.*, **16**:548.

Cooper, J., Foundling, S. I., Hemmings, A., Blundell, T. L., Jones, D. M., Hallett, A. & Szelke, M., 1987, X-ray studies of aspartic proteinases-statine complexes, *Eur. J. Biochemistry*, **169**:215.

Davies, D. R., 1990, The structure and function of the aspartic proteinases, *Ann. Rev. Biophys. Biophys. Chem.*, **19**:189.

Finzel, B. C., 1987, Incorporation of fast Fourier transforms to speed least squares refinement of protein structures, *J. Appl. Crystallogr.*, **20**:53.

Fruton, J. S., 1976, Mechanism of the catalytic action of pepsin and related acid proteinases, *Advances in Enzymology*, **44**:1.

Green, D. W., Aykent, S., Gierse, J. K. & Zupec, M. E., 1990, Substrate specificity of recombinant human renal renin: Effect of histidine in the P_2 subsite on pH dependence, *Biochemistry*, **29**:3126.

Hendrickson, W. A., 1985, Stereochemically restrained refinement of macromolecular structures, *in*: "Methods in Enzymology," H. W. Wyckoff, C. W. H. Hirs, S. N. Timasheff, eds., Academic Press, New York.

Hofmann, T., Allen, B., Bendiner, M., Blum, M. & Cunningham, A., 1988, Effect of secondary substrate binding in penicillopepsin: contribution of subsites S_3 and S_2' to k_{cat}, *Biochemistry*, **27**:1140.

Hsu, I.-N., Delbaere, L. T. J. & James, M. N. G., 1977, Penicillopepsin from *P. janthinellum*. Crystal structure at 2.8Å and sequence homology with porcine pepsin, *Nature* (London), **266**:140.

Hofmann, T., & Fink, A. L., 1984, Cryoenzymology of Penicillopepsin, *Biochemistry*, **23**:5247.

James, M. N. G. & Sielecki, A. R., 1985, Stereochemical analysis of peptide bond hydrolysis catalyzed by the aspartic proteinase penicillopepsin, *Biochemistry*, **24**:3701.

James, M. N. G., Sielecki, A., Salituro, F., Rich, D. H. & Hofmann, T., 1982, Conformational flexibility in the active sites of aspartyl proteinases revealed by a pepstatin fragment binding to penicillopepsin, *Proc. Natl. Acad. Sci.*, U.S.A., **79**:6137.

Jones, T. A.,1978, A graphics model building and refinement system for macromolecules, *J. Appl. Crystallogr*, **11**:268.

Pearl, L., 1985, The extended binding cleft of aspartic proteinases and its role in peptide hydrolysis, *in*: "Aspartic Proteinases and their Inhibitors," V. Kostka, ed., Walter de Gruyter, Berlin, New York.

Pearl, L., 1987, The catalytic mechanism of aspartic proteinases, *FEBS Letters*, **214**:8.

Pearl, L. H. & Blundell, T. L., 1985, The active site of aspartic proteinases, *FEBS Letters*, **174**:96.

Pflugrath, J. W., Saper, M. A. & Quiocho, F. A., 1984, *in*: "Methods and Applications in Crystallographic Computing," S. Hall and T. Ashida, eds., Clarendon, Oxford.

Sali, A., Veerapandian, B., Cooper, J. B., Foundling, S. I., Hoover, D. J. & Blundell, T. L., 1989, High-resolution X-ray diffraction study of the complex between endothiapepsin and an oligopeptide inhibitor: the analysis of the inhibitor binding and description of the rigid body shift in the enzyme, *EMBO J.*, **8**:2179.

Satow, Y., Cohen, G. H., Padlan, E. A. & Davies, D. R., 1986, Phosphocholine binding immunoglobulin Fab McPC603: An X-ray diffraction study at 2.7Å, *J. Mol. Biol.*, **190**:593.

Schechter, I. and Berger, A.,1967, On the size of the active site in proteases. I. Papain, *Biochem. Biophys. Res. Commun.*, **27**:157.

Sheriff, S., Hendrickson, W. A., Stenkamp, R. E., Sieker, L. C. & Jensen, L. H., 1985, Influence of solvent accessibility and intermolecular contacts on atomic mobilities in hemerythrin, *Proc. Natl. Acad. Sci. U.S.A.*, **82**:1104.

Sheriff, S., 1987, Addition of symmetry-related contact restraints to PROTIN and PROLSQ, *J. Appl. Crystallogr.*, **20**:53.

Sielecki, A. R., Fedorov, A. A., Boodhoo, A., Andreeva, N. S. & James, M. N. G., 1990, Molecular and Crystal Structures of Monoclinic Porcine Pepsin refined at 1.8Å Resolution, *J. Mol. Biol.*, **214**:143.

Subramanian, E., Swan, I. D. A., Liu, M., Davies, D. R., Jenkins, J. A., Tickle, I. J. & Blundell, T. L., 1977, Homology among acid proteases: Comparison of crystal structures at 3Å resolution of acid proteases from *Rhizopus chinensis* and *Endothia parasitica.*, *Proc. Natl. Acad. Sci.*, U.S.A., **74**:556.

Suguna, K., Padlan, E. A., Smith, C. W., Carlson, W. D. & Davies, D. R., 1987, Binding of a reduced peptide inhibitor to the aspartic proteinase from *Rhizopus chinensis*: implications for a mechanism of action. *Proc. Natl. Acad. Sci.*, U.S.A., **74**:556.

Suguna, K., Padlan, E. A., Bott, R., Boger, J. & Davies, D. R., 1991, Proteins: Structure, Function and Genetics, in press.

A YEAST EXPRESSION SYSTEM AND SITE-DIRECTED MUTAGENESIS OF A

FUNGAL ASPARTIC PROTEINASE, *MUCOR* RENNIN

Jun-ichi Aikawa, Ryuji Hiramatsu, Makoto Nishiyama,
Sueharu Horinouchi and Teruhiko Beppu

Department of Agricultural Chemistry
The University of Tokyo
Yayoi 1-1-1, Bunkyo-ku, Tokyo 113
Japan

INTRODUCTION

The cheese industry has long used a characteristic aspartic proteinase, chymosin (calf rennin) obtained from the calf stomach, as a milk coagulant for cheese making. A severe shortage of chymosin in 1950 stimulated efforts to find substituting microbial enzymes and Arima *et al.* succeeded in discovering such an enzyme from a fungal strain, *Mucor pusillus*.[1] A similar enzyme was subsequently found from a closely related species, *Mucor miehei*,[2] and these fungal aspartic proteinases are called *Mucor* rennin. Currently more than a half of cheese in the world is produced with *Mucor* rennin.

Milk clotting by the aspartic proteinases is due to cleavage of κ-casein, which functions as a stabilizer of milk micelle. Chymosin and *Mucor* rennin are discriminated from other members of the aspartic proteinases not only by high milk-clotting activity but also by low proteolytic activity, which assure high yields of the clotting material. On the other hand, alignment of their sequences with those of other aspartic proteinases indicates that chymosin and *Mucor* rennin may have fundamentally a three dimensional structure common to all the aspartic proteinases.

In the last several years we have constructed expression systems of chymosin and *M. pusillus* rennin (MPR) in *Escherichia coli*[3,4] and yeast,[5,6] respectively. Site-directed mutagenesis using these systems especially with chymosin indicated several residues such as Tyr$_{75}$ (in pepsin numbering) on the flap structure of the molecule as candidates for further analyses.[7,8] For this purpose, the yeast expression system which permits effective secretion of the correctly processed and active form of MPR has several advantages. In this paper, we describe characteristics of the yeast expression system as well as some results obtained by site-directed mutagenesis of MPR.

MATERIALS AND METHODS

Strains and Plasmids

Saccharomyces cerevisiae MC16(α, *leu2*, *his4*, *ade2*) was used as the host for production of MPR. JP1 is a plasmid in which the cloned MPR gene was placed under the control of the *GAL7* promoter.[4]

Mutagenesis

Site-directed mutagenesis was carried out using synthetic oligonucleotides by Kunkel's method[9] using a Bio-Rad MutaGene kit. All the mutations were checked by nucleotide sequencing using the M13-dideoxy chain termination method.[10]

Purification of yeast MPR

S. cerevisiae MC16 harboring JP1 was cultured aerobically at 30°C in YPD medium containing 2% Bacto-peptone (Difco), 1% Bacto-yeast extract (Difco) and 2% glucose. The growing cells were transferred into YPGal medium containing 3% galactose in place of glucose and cultured further aerobically. One liter of the supernatant from an early stationary culture was taken and the pH was adjusted to 5.5. It was then directly applied to a DEAE-Toyopearl 650M (Tosoh) column. The adsorbed protein was eluted with 50 mM phosphate (pH 5.5) or 50 mM acetate (pH 5.5) buffers containing 150 mM NaCl. The fractions containing MPR were concentrated by ultrafiltration, and the sample was applied to a Superose 12 gel filtration column on an FPLC system (Pharmacia Fine Chemicals). Milk-clotting activity was eluted coincidentally with a single protein peak. The peak was used as a purified yeast MPR. Commercial MPR from Meito Sangyo Co. was purified in essentially the same way.

```
MLFSKISSAILLTAASFALTSARPVSKQSDADDKLLALPLTSV    -24

NRKYSQTKHGQQAAEKLGGIKAFAEGDGSVDTPGLYDFDLEEY     20

AIPVSIGTPGQDFYLLFDTGSSDTWVPHKGCDNSEGCVGKRFF     63

DPSSSSTFKETDYNLNITYGTGGANGIYFRDSITVGGATVKQQ    106

TLAYVDNVSGPTAEQSPDSELFLDGIFGAAYPDNTAMEAEYGD    149

TYNTVHVNLYKQGLISSPVFSVYMNTNDGGGQVVFGGVNNTLL    192

GGDIQYTDVLKSRGGYFFWDAPVTGVKIDGSDAVSFDGAQAFT    235

IDTGTNFFIAPSSFAEKVVKAALPDATESQQGYTVPCSKYQDS    278

KTTFSLDLQKSGSSSDTIDVSVPISKMLLPVDKSGETCMFIVL    321

PDGGNQFIVGNLFLRFFVNVYDFGKNRIGFAPLASGYENN      361
```

Figure 1. Amino acid sequence of MPR. A closed triangle indicates the processing site of presequence and an open triangle indicates that of prosequence. Underlined sequences were confirmed by direct amino acid sequencing with proMPR and matured MPR secreted from the recombinant yeast. Boxed sequences indicate possible glycosylation sites.

EndoH treatment of yeast MPR

The yeast MPR was digested with endo-ß-N-acetylglucosaminidase H (endoH) by the method of Tarentino *et al.*[11] The endoH-digested yeast MPR was purified similarly by MonoQ and Superose 12 chromatographies.

Assay of Enzyme Activity

Milk clotting activity was measured and expressed according to Iwasaki *et al.*[12] Proteolytic activity was measured at pH 4.0 by a modified method of Anson *et al.*,[13] using acid-denatured hemoglobin as the substrate, as described.[7] Two chromogenic oligopeptides Leu-Ser-Phe(NO$_2$)*Nle-Ala-Leu-OMe (peptide I) and Lys-Pro-Ile-Glu-Phe*Phe(NO$_2$)-Arg-Leu-OH (peptide II), were used as the substrates to determine kinetic parameters, as described previously.[7]

RESULTS AND DISCUSSION

Expression and Secretion of Mucor *Rennin from Recombinant Yeast*

We have cloned the structural gene of *Mucor pusillus* rennin (MPR) from the chromosome of *M. pusillus* by using synthetic oligo-DNA probes and determined its nucleotide sequence.[14] The gene of MPR has no intron, and encodes the preproenzyme which contains 361 amino acids of mature MPR with an NH$_2$-terminal extension of 66 amino acids (Figure 1). In order to express the MPR gene, the promoter region of yeast *GAL7* was introduced upstream of the MPR coding sequence. The entire sequence containing the *GAL7* promoter and the MPR gene with its possible terminator region was subcloned on the *E. coli*-yeast shuttle vector pSS21, and the resulting plasmid designated JP1 was introduced into *S. cerevisiae* MC16 strains by using the alkali method. When the transformant was grown in YPGal medium, distinct milk clotting activity to form a halo on skim milk plates was detected by spotting the supernatant of the cultured medium. SDS-polyacrylamide gel electrophoresis showed a single major band of protein which was reactive with anti-MPR antibody.[15] The amount of the secreted MPR was estimated to be more than 200 mg/l and N-terminal amino acid sequencing of the protein produced by recombinant yeast (designated yeast MPR) was identical to that of the mature MPR produced by the original fungal strain (commercial MPR). The apparent molecular weight of the yeast MPR was 46 kDa. This was larger by 6000 Da than that of the commercial MPR due to glycosylation as described later.

During the early stage of induced culture, a larger protein of 51 kDa reactive with anti-MPR antibody was detected as a major extracellular protein.[6] We purified this 51-kDa protein by column chromatography. Its NH$_2$-terminal amino acid sequence was identical to the sequence starting from Arg$_{-44}$ in the prepropeptide region deduced from the nucleotide sequence of the cloned MPR gene (Figure 1). This result clearly demonstrated that the 51 kDa protein was proMPR which contained 44 additional amino acid residues at the NH$_2$-terminus of mature MPR as the prosequence and the NH$_2$-terminal 22 amino acids of preproMPR served as a signal peptide for the secretion of proMPR. This was the first observation on the secretion of a zymogen of fungal aspartic proteinase.

The mature MPR (200 µg/ml) caused quick clotting of casein (55 s) at pH 5.2, while no clotting was observed with the purified proMPR within 20 min. Since slow conversion of proMPR into mature MPR was observed to occur even at pH 5.2, we concluded that proMPR was actually inactive. The efficient conversion of proMPR to the mature MPR proceeded only at acidic pH. The conversion was inhibited by aspartic proteinase inhibitors,

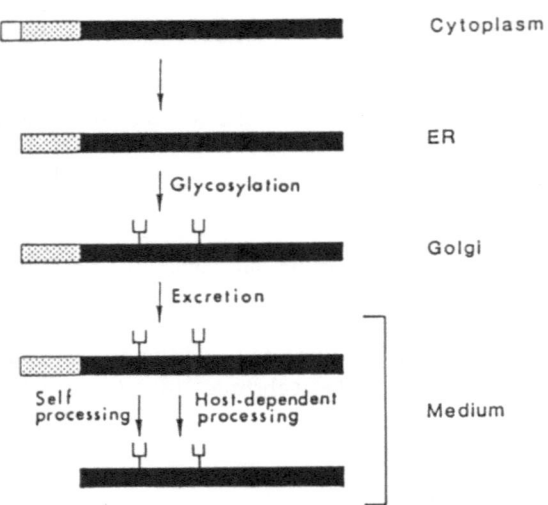

Figure 2. Process of MPR secretion from recombinant yeast cells. open box indicates presequence; dotted box indicates prosequence ; filled box indicates matured MPR.

such as diazoacetyl-DL-norleucine methyl ester and pepstatin. The conversion was highly dependent on the initial concentration of proMPR and addition of mature MPR accelerated the conversion. These results indicate that the conversion of yeast proMPR into mature MPR may be triggered by intramolecular processing of proMPR itself at acidic pH, which is accelerated by higher proteolytic activity of the processed product.

However, an inactive mutant, MPR D38G, in which one of the catalytic Asp residues, Asp_{38} (in MPR numbering), was exchanged to Gly by site-directed mutagenesis, was still secreted in the form of mature MPR by the yeast strain MC16. This result clearly showed that a certain proteinase(s) of the host cells played an alternative role in processing of proMPR. A yeast pep4-3 mutant strain, *S. cerevisiae* AB103-1, which is deficient in an aspartic proteinase, proteinase A, was found to excrete D38G mutant in the form of proMPR and no processing was observed even in prolonged cultivation.

From these results, we propose a model of the processing of preproMPR in the yeast secretion pathway along with possible glycosylation steps as shown in Figure 2. The presence of host-dependent processing in this system permits production of correctly processed mutant MPRs which have no proteolytic activity, which provides an advantage in utilization for protein engineering.

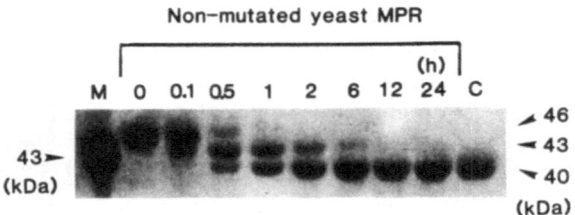

Figure 3. EndoH treatment of yeast MPR. Lane C; commercial MPR.

Table 1. Effects of EndoH treatment and mutations on milk clotting and proteolytic activities of yeast MPR

Enzyme	number of glycosylated residues	Clotting Activity (U/µg)	Proteolytic Activity (U/µg)	C/P	
Yeast MPR native (-endoH)	2	3.04 (45.1%)	3.85 (198%)	0.790	(22.8%)
native (+endoH)	0	5.61 (83.2%)	2.46 (127%)	2.28	(65.7%)
N79Q	1	4.39 (65.1%)	2.36 (122%)	1.86	(53.6%)
N188Q	1	4.08 (60.5%)	2.83 (146%)	1.44	(41.5%)
N79Q/N188Q	0	5.95 (88.3%)	2.18 (112%)	2.73	(78.7%)
commercial MPR	0	6.74 (100%)	1.94 (100%)	3.47	(100%)

Glycosylation of Mucor rennin by Yeast

The protein band of the active yeast MPR in SDS-polyacrylamide gel electrophoresis was stained with periodic acid-Schiff reagent which was used to detect sugar moieties. The color reaction of the yeast MPR was distinctly stronger than that of commercial MPR produced by the original fungal strain. Treatment of the yeast MPR with endoglycosidase H (endoH) caused conversion of the apparent molecular weight in two steps, first from 46 kDa to 43 kDa and then to 40 kDa which was identical to that of commercial MPR (Figure 3). This suggests that yeast MPR is glycosylated at two sites, although there are three possible Asn-linked glycosylation sites in the MPR sequence, Asn_{79}-Ile-Thr, Asn_{113}-Val-Ser, and Asn_{188}-Asn-Thr (in MPR numbering)(Figure 1).

We constructed the mutants, N79Q, N113Q and N188Q, in which one of these potentially glycosylated Asn was exchanged to Gln, and expressed them in the yeast host.[16] The supernatants after the cultivation in YPGal medium for 4 days were analyzed by SDS-polyacrylamide gel electrophoresis. Both N79Q and N188Q showed a reduced molecular size of about 43 kDa whereas that of N113Q was the same as that of non-mutated MPR (46 kDa). When both Asn_{79} and Asn_{188} were changed to Gln simultaneously (N79Q/N188Q), the size was decreased to that of the commercial MPR. All these data clearly showed that the yeast MPR was glycosylated at the two sites, Asn_{79} and Asn_{188}.

The yeast MPR showed distinctly lower milk clotting activity (C) along with relatively higher proteolytic activity (P) than those of the commercial MPR. EndoH treatment of the yeast MPR improved the C/P ratio to almost the same as that of commercial MPR (Table 1). Similar changes of the C/P ratio were observed with mutated MPR lacking one or both of the N-glycosylation sites. These results indicate that glycosylation of MPR causes distinct modulation of its enzymatic activity, probably due to a change in the specificity for scissile peptide bonds. However, no significant differences in the value of k_{cat} and K_m for the synthetic oligopeptides I and II were observed with these mutants. The difference in the lengths of substrates probably accounts for this inconsistency. A longer synthetic peptide with an appropriate sequence will be required for further analysis of the effect of glycosylation on the catalytic activity. Because of these possible effects of glycosylation on the catalytic properties, all the mutated MPRs produced by the yeast expression system were treated with endoH before examination of their enzymatic properties.

The secreted amounts of N79Q and N188Q were apparently reduced to one half and that of the double mutant N79Q/N188Q was reduced to about 1/20 of that of the non-mutated MPR. Distinct intracellular accumulation of the immunoreactive proteins was observed only with the mutations at the actual glycosylation sites. These results indicate that glycosylation at the two sites of MPR is required for efficient secretion from yeast cells.

Site-directed Mutagenesis of Mucor *rennin*

By using the yeast expression system, we initiated protein engineering of MPR. Since X-ray crystallographic analyses of several aspartic proteinases have revealed that the three dimensional structure of these enzymes are very similar to each other,[17] we may refer to the common structural feature hitherto observed among the aspartic proteinases as the primary basis to choose sites for mutagenesis. Various inhibitors specific to the aspartic proteinases by mimicking substrate peptides give enzyme-inhibitor complexes, whose crystallographic analyses provide information on the residues possibly involved in forming subsites for substrate recognition (Figure 4).[18,19,20]

On the other hand, our previous results by random and site-directed mutagenesis of chymosin have revealed several sites as candidates for further analyses.[7,8] Tyr_{75} in pepsin

Table 2. Milk clotting and proteolytic activities of mutant MPR's with substitution of Tyr_{75}

Enzyme	Clotting Activity (U/μg)	Proteolytic Activity (U/μg)	C/P
non-mutated	6.49 (100)	1.49 (100)	4.36 (100)
Y75F	0.63 (9.7)	0.35 (23.4)	1.80 (41.3)
Y75N	2.05 (31.6)	0.18 (11.9)	11.5 (264)
Y75W	-----(<0.01)		
Y75A	n. d.		
Y75C	n. d.		
Y75I	n. d.		
Y75L	n. d.		
Y75M	n. d.		
Y75P	n. d.		
Y75Q	n. d.		
Y75S	n. d.		
Y75V	n. d.		

n.d. = no measurable activity detected.

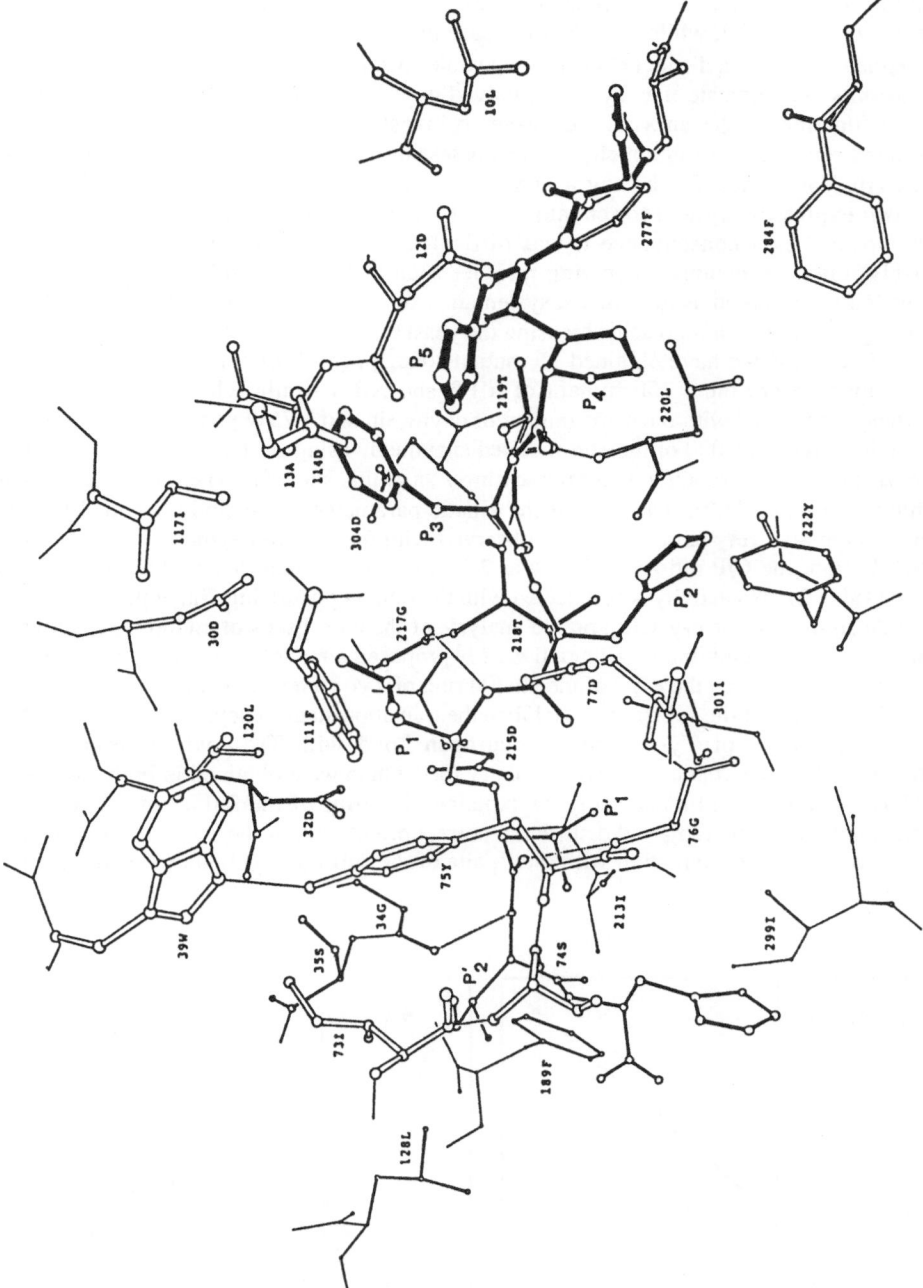

Figure 4. Enzyme-inhibitor complex of *Endothia* pepsin and inhibitor H261, illustrated according to the coordinates by T. Blundell's group.

numbering is such a residue, located on the flap probably participating in formation of the S_1 subsite and well conserved among all the aspartic proteinases. In our studies on chymosin, we found that the exchange of Tyr_{75} to Phe caused marked changes in kinetic parameters depending on substrates. A marked decrease in k_{cat} but almost no change of K_m was observed with peptide I, while a marked increase in K_m but no change of k_{cat} was observed with peptide II.[7] Thus a distinct change in substrate specificity was induced by the mutation. These results also indicate that Tyr_{75} is involved in not only substrate binding but also the catalytic function of the enzyme. In order to investigate the role of Tyr_{75} further, we constructed several other mutant chymosins possessing exchange of Tyr_{75} to other residues, such as Trp, Val, Ile and Thr, among which only Trp gave extremely low activity. However, the *E. coli* expression system for chymosin is dependent on the self processing activity of prochymosin. As a consequence it was difficult to obtain convincing results with the (probably) inactive mutants possessing residues such as Val, Ile and Thr at this position. This difficulty prompted us to initiate a systematic exchange of Tyr_{75} of MPR (Tyr_{82} in MPR numbering) to all other amino acids by using the yeast expression system.

Up to now, we have obtained 15 mutants (except for those with Gly, His, Lys and Arg). Among them the Y75F mutant of MPR showed a similar change in the kinetic properties as observed with the same mutant of chymosin and Y75W possessed a very weak proteolytic activity (1/1000 of the non-mutated enzyme by using skim milk as the substrate). However, to our surprise, the mutant possessing Asn at this site (Y75N) exhibited a distinct enzyme activity (Table 2). Changes of the kinetic parameters for peptides I and II by this mutation were also very similar to those observed with the mutant chymosin Y75F (Figure 5). In addition, the C/P ratio of Y75N was 2.5 times higher than that of the non-mutated enzyme (Table 2). No activity was detected with the mutant possessing Gln at this position.

According to the crystallographic analysis of the co-crystals of endothiapepsin with the statine-containing inhibitors, Blundell *et al.*[18] proposed a model for the transition state of the ES complex in which the edge of the phenol ring of Tyr_{75} interacts with a hydroxyl group of the scissile bond of the substrate to stabilize the transition state. They also postulated that the hydroxyl group of Tyr_{75} forms a hydrogen bond with Trp_{39} which assures this arrangement. The enzymatic properties of the Y75F, which were obtained in both chymosin and *Mucor* rennin, could most likely be explained in terms of the failure of Phe to form hydrogen bonding with Trp_{39} and disturb the correct orientation of the aromatic ring, which might be required for interaction with the P_1 site of the substrates (Figure 6A). In order to

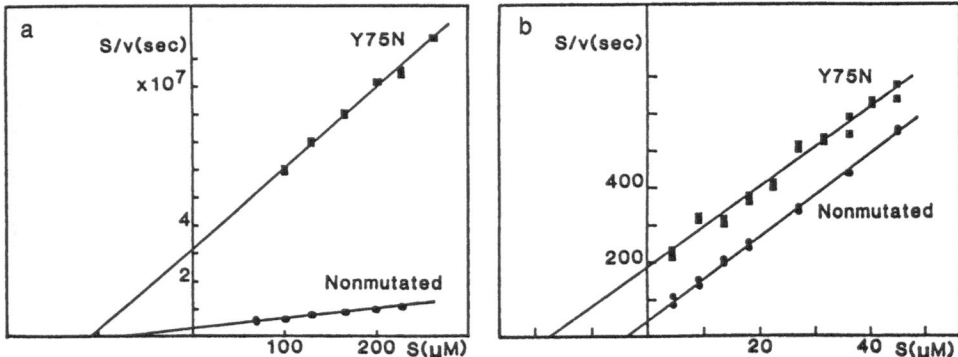

Figure 5. S/v vs. S plots of non-mutated and Y75N mutant MPRs for peptides I (a) and II (b) as substrates. filled circles, non-mutated; filled squares, Y75N.

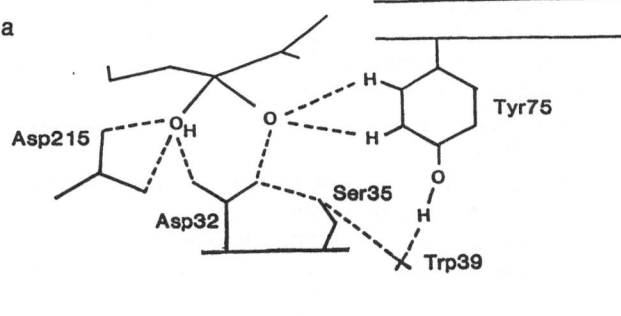

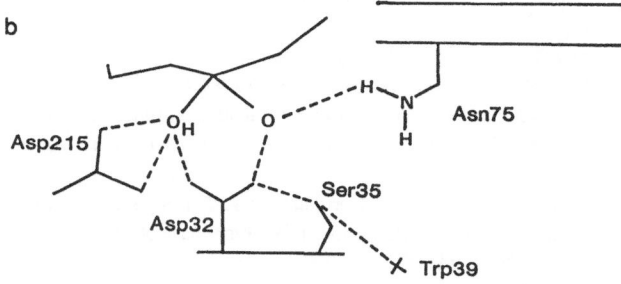

Figure 6. Model of transition state complexes of non-mutated (a) and Y75N (b) MPR.

explain the unexpected activity of the mutant MPR Y75N, we may assume that the amide group of Asn_{75} is pointed at almost the identical position to that of the aromatic ring of the tyrosine residue in the non-mutated enzyme and contributes to stabilize the transition state complex by forming a hydrogen bond as shown in Figure 6B.

Tyr_{75} on the flap is conserved among all the aspartic proteinases hitherto sequenced and is involved in their catalytic functions. In this sense the mutant MPR Y75N can be assumed as a new type aspartic proteinase, which will be useful for obtaining new information on this group of proteinases.

REFERENCES

1. K. Arima, S. Iwasaki, and G. Tamura, Milk clotting enzyme from microorganisms. Part I. Screening test and the identification of the potent fungus, *Agric. Biol. Chem.* **31**:540 (1967).
2. M. Ottensen, and W. Richert, The isolation and partial purification of an acid protease produced by *Mucor miehei*, *C. R. Trav. Lab. Carlsberg* **37**:301(1970).
3. Y. Kawaguchi, N. Shimizu, K. Nishimori, T. Uozumi, and T. Beppu, Renaturation and activation of calf prochymosin produced in an insoluble form in *Escherichia coli*, *J. Biotechnol.* **1**:307 (1984).
4. Y. Kawaguchi, S. Kosugi, K. Sasaki, T. Uozumi, and T. Beppu, Production of chymosin in *Escherichia coli* cells and its enzymatic properties, *Agric. Biol. Chem.* **51**:1871 (1987).
5. T. Yamashita, N. Tonouchi, T. Uozumi, and T, Beppu, Secretion of *Mucor* rennin, a fungal aspartic protease of *Mucor pusillus*, by recombinant yeast cells, *Mol. Gen. Genet.* **210**:462 (1987).
6. R. Hiramatsu, J. Aikawa, S. Horinouchi, and T. Beppu, Secretion by yeast of the zymogen form of *Mucor* rennin, an aspartic proteinase of *Mucor pusillus,* and its conversion to the mature form, *J. Biol. Chem.* **264**:16862 (1989).

7. J. Suzuki, K. Sasaki, Y. Sasao, A. Hamu, H. Kawasaki, M. Nishiyama, S. Horinouchi, and T. Beppu, Alteration of catalytic properties of chymosin by site-directed mutagenesis, *Prot. Eng.* **2**:563 (1989).

8. J. Suzuki, A. Hamu, N. Nishiyama, S. Horinouchi, and T. Beppu, Functional contribution of Thr$_{218}$, Lys$_{220}$ and Asp$_{304}$ in chymosin analyzed by site-directed mutagenesis, *Prot. Eng.* **4**:69 (1990).

9. T. A. Kunkel, Rapid and efficient site-specific mutagenesis without phenotype selection, *Proc. Natl. Acad. Sci. U.S.A.* **82**:488 (1985).

10. F. Sanger, A. R. Coulson, B. G. Barrelle, A. J. M. Smith, and B. A. Ros, Bacteriophage as an aid to rapid DNA sequencing, *J. Mol. Biol.* **143**:161 (1980).

11. A. L. Tarentino, T. H. Jr. Plummer, and F. Maley, The release of intact oligosaccharide from specific glycoprotein by endo-ß-*N*-acetylglucosaminidase H, *J. Biol. Chem.* **249**:818 (1974).

12. S. Iwasaki, G. Tamura, and K. Arima, Milk clotting enzyme from microorgamisms. Part II. The enzyme production and the properties of crude enzyme, *Agric. Biol. Chem.* **31**:546 (1967).

13. M. I. Anson, The estimation of pepsin, trypsin, papain, and cathepsin with hemoglobin, *J. Gen. Physiol.* **22**:79 (1938).

14. N. Tonouchi, H. Shoun, T. Uozumi, and T. Beppu, Cloning and sequencing of a gene for *Mucor* rennin, an aspartic protease from *Mucor pusillus*, *Nucleic Acids Res.* **14**:7557 (1986).

15. Y. Etoh, H. Shoun, T. Beppu, and K. Arima, Physiochemical and immunochemical studies on similarities of acid proteases *Mucor pusillus* rennin and *Mucor miehei* rennin," *Agric. Biol. Chem.* **43**:209 (1979).

16. J. Aikawa, T. Yamashita, M. Nishiyama, S. Horinouchi, and T. Beppu, Effects of glycosylation on the secretion and enzyme activity of *Mucor* rennin, an aspartic proteinase of *Mucor pusillus*, produced by recombinant yeast, *J. Biol. Chem.* **265**:13955 (1990).

17. D. R. Davies, The structure and function of the aspartic proteinases, *Annu. Rev. Biophys. Biophys. Chem.* **19**:189 (1990).

18. T. L. Blundell, J. Cooper, S. I. Foundling, D. M. Jones, B. Attrash, and M. Szelke, On the rational design of renin inhibitors: X-ray Studies of aspartic proteinases complexed with transition-state analogues, *Biochemistry* **26**:5585 (1987).

19. J. Cooper, S. Foundling, A. Hemmings, T. Blundell, D. M. Jones, A. Hallett, and M. Szelke, The structure of a synthetic pepsin inhibitor complexed with endothiapepsin, *Eur. J. Biochem.* **169**:215 (1987).

20. A. Sali, B. Veerapandian, J. B. Cooper, S. I. Foundling, D. J. Hoover, and T. L. Blundell, High-resolution X-ray diffraction study of the complex between endothiapepsin and an oligopeptide inhibitor: the analysis of the inhibitor binding and description of the rigid body shift in the enzyme, *EMBO J.* **8**:2179 (1989).

STUDIES ON THE MECHANISM OF ACTION OF PENICILLOPEPSIN

Theo Hofmann, Max Blum and Annie Cunningham

Dept. of Biochemistry
University of Toronto
Toronto, Ontario
Canada M5S 1A8

INTRODUCTION

Several high resolution structures of aspartic proteinases are now known, (see Gilliland et al.[1] for a recent list and earlier contributions in this volume) and detailed information has been obtained on the binding of substrate analogue inhibitors from the crystallographic analysis of their complexes with the enzymes. Several proposals for their mechanism of action have been made.[2-7] Yet we still do not fully understand how these enzymes function. The crystallographic data are insufficient to rationalize the extensive experimental information available from studies on specificity,[8] steady-state and presteady-state kinetics,[8] isotope exchange experiments,[9,10] transpeptidation reactions[11-13] and low-temperature kinetics.[14] Thus, we do not know much about the molecular events that are responsible for the so-called "secondary specificity",[8] that is, the very large increases in catalytic efficiency with increasing length of substrates, nor do we understand the structural details of transpeptidation reactions. It is also not clear what the rate-determining steps of the reactions are. In our recent studies we have attempted to throw some more light on these phenomena.

It is known from the work of Fruton[8] with pig pepsin that the elongation of substrates of pepsin from dipeptides to hexapeptides leads to very large increases in catalytic efficiency, but has no significant effect on K_m. This effect appeared to be incremental with each amino acid residue added. However, in a study with penicillopepsin we showed that the effect on k_{cat} is almost entirely due to the residues in positions P_3 and P_2' which bind into the enzyme's subsites S_3 and S_2' respectively.[15] Similar results have recently been obtained with rhizopuspepsin, endothiapepsin and to a lesser extent with pig pepsin (Hofmann, T., unpublished observations). We further showed that the occupation of these subsites has different effects on the temperature dependence of the catalysis.[16] Whereas two substrates of the series Ac-Ala$_m$-Lys*Nph-Ala$_n$-amide which have no amino acid in either P_3 or P_2' give linear Arrhenius plots, substrates with a residue in P_2', but not in P_3, give non-linear plots with a break at 10.5°C and relatively high energies of activation. Substrates with a residue in P_3 only or in P_3 and P_2' also give non-linear Arrhenius plots, but in these cases the transition temperature is at 14.2°C and the activation energies are considerably lower. Furthermore, we were able to show that pepstatin, which has a valine residue in P_3, induces a temperature dependent conformational change with a transition that corresponds to the kinetic transition observed with substrates with residues in P_3.[16] These results suggested that several of the different kinetic phenomena might be explained by conformational changes. However, the crystallographic structures of aspartic proteinases and their complexes with the substrate analog inhibitors known until fairly recently indicated that no major conformational changes in the enzymes accompanied binding, and, by implication, substrate binding and catalytic action.[2,17,18] The exception is the movement of the Tyr_{75} flap, a structural feature that is flexible in the free enzymes, but becomes more rigid as it assists in the tight binding of the ligands in the complexes. Recent studies with endothiapepsin have now shown that several inhibitors in which an amino acid occupies subsite S_3, induce a conformational change in the enzyme which consists of a small rotation of a domain consisting of residues 190-302 (pig pepsin numbering). This change has been described as a "rigid-body movement".[19,20] Furthermore, the data suggest that it is the occupation of the S_3 pocket that induces this conformational change, although other factors may also play a role. This structural domain also accounts for the main differences in the backbone structures among different aspartic proteinases.[20] It is therefore reasonable to infer that changes akin to those observed on ligand binding to endothiapepsin may also occur with other aspartic proteinases, and that such changes can account for at least some of the kinetic differences between substrates with S_3 occupancy and those without. Evidence for substrate dependent conformational changes has also been demonstrated with pig pepsin inhibitor complexes by Adab-Zapatero and colleagues and is described elsewhere in this volume.[36] These changes involve the same domain of the molecule as that found in endothiapepsin by Sali et al.[19,20]

One of the differences between short and long substrates could be due to differences in the rate-determining step. Information on rate-determining steps in hydrolytic reactions can be obtained from a study of solvent isotope effects. Three such studies have been carried out previously with pig pepsin, but they led to conflicting conclusions. No solvent isotope effects were observed with the substrates Ac-Phe*Tyr-OMe[22] and methyl phenyl sulfite,[23] whereas Hollands and Fruton[24] observed a solvent isotope effect of about two with Gly_3-Nph*Phe-OMe. The latter authors could not find a satisfactory explanation for this discrepancy and felt that further work was needed to resolve this problem. However, to the best of our knowledge no such additional study with any of the aspartic proteinases has been carried out. We therefore looked at the solvent isotope effect on the hydrolysis by penicillopepsin of the -Lys*Nph- bond of substrates of the series Ac-Ala_m-Lys*Nph-Ala_n-amide[29] and present some additional results in this paper.

Another unexplained phenomenon is the ability of aspartic proteinases, such as pepsin and penicillopepsin, to trap some amino acids and dipeptides non-covalently during transpeptidation reactions, but in such a manner that the trapped ligands cannot exchange with free ligands in solution.[11,25,26] The result of this trapping is that the major early products of the transpeptidation reactions are oligomers of the terminal amino acid of the substrate. In the study of the action of pig pepsin on Leu*Tyr-amide the first demonstrable products found were Leu_4 and Leu_3. Leu_2 and leucine appeared only as secondary products of the action of the enzyme on the higher oligomers.[27] Similarly, in a study of the action of penicillopepsin on Nph*Ala_2-amide, we found that the first product is a transient condensation product, Nph_2-Ala_2-amide, which is rapidly followed by Nph_3 and small amounts of Nph_4.[28] Again, the dimer and the free amino acids appear as products only much later. In both of these systems no free amino acid present during the reaction is incorporated into the transpeptidation products, as shown by experiments with radioactive compounds. The detailed analysis of these reactions, presented elsewhere,[28] led us to suggest that the transpeptidation substrate triggers a conformational change in the enzyme which allows it to trap the N-terminal free amino acid, or its dimer, apparently non-exchangeably, while releasing the C-terminal moiety of the substrate. Subsequently the trapped amino acid is condensed with another substrate molecule. We propose that this trapping is made possible by the presence of two strong electrostatic, as well as hydrogen-bond interactions between the ligand and the active site of the enzyme. Here we present results of a study of the pH-dependence of the transpeptidation reactions which add support to our proposal.

MATERIALS AND METHODS

Materials

Penicillopepsin was prepared as described.[30] The synthesis of peptides of the series Ac-Ala$_m$-Lys*Nph-Ala$_n$-amide, where m and n equal 0 - 3, and of other peptides, was described elsewhere.[28,31] Chemicals used for buffers were of the highest quality available. [^{18}O]-water (97 atom% ^{18}O) and D$_2$O (99.9 atom% D) were obtained from MSD Isotopes (Merck Frosst Canada Inc., Montreal).

Enzyme kinetics

All enzyme assays were carried out at 25°C in 20 mM formate (pH 3.5, pD 3.9), or acetate (pH 4 - 5.5, pD 4.4 - 5.9), adjusted to an ionic strength of 20 mM with NaCl. The aqueous solvent was H$_2$O, D$_2$O or [^{18}O]H$_2$O as appropriate. A Uvikon 820 spectrophotometer (Kontron AG, Zurich, Switzerland) was used with cells of 2 or 10 mm lightpaths and at wavelengths of 296, 306, 320 or 330 nm. The cells and wavelengths were chosen so that the initial absorbance of the substrate solution did not exceed 2.0.[15] The enzyme concentrations ranged between 0.5 and 300 µg/mL (15 nM and 9 µM), depending on the rate of the reaction, and were chosen to give initial rates in the first 2 - 10 min.

Analysis of ^{18}O incorporation

The details of the analysis of the incorporation of ^{18}O into the various products are described elsewhere.[28]

Transpeptidation reactions

Transpeptidation reactions were carried out in 20 mM acetate (pH 4.5 - 5.25) or formate (pH 2.0 - 3.5) at 30°C for periods up to 24 hours, and with enzyme concentrations of 35 µM and substrate concentrations of 1 mM. Samples were withdrawn at intervals and quenched with 1/4 volume of ammonia (5 M). Penicillopepsin is rapidly inactivated at pH > 7. The samples were stored at -20°C and centrifuged before analysis by HPLC on a C18 reverse phase Vydac 218TP104 column (250 x 4.6 mm) in 0.1% TFA, run at room temperature with a linear gradient from 0% to 50% acetonitrile over 25 min. Rates of disappearance of substrate and formation of products were calculated from the integrated peaks.

Model building

The conformational studies for the possible trapping of Nph and Nph$_2$ in the active site were carried out with the program MMS (version 1.3) which was obtained from Dr. S. Dempsey, San Diego, CA. We used the coordinates of the complex of the pepstatin analog Iva-Val-Val-Sta-OEt with penicillopepsin.[17] The Iva and OEt groups and the N- terminal valine were deleted. The side chains of Val$_2$ and Sta$_3$ were changed to tyrosine side chains. The latter were chosen to simulate p-nitrophenylalanine residues. The positions of the tyrosine side chains were then adjusted by rotation around the α - ß and ß -γ bonds in such a way that there were no non-bonded distances between them and atoms of the enzyme which were less than the sum of the corresponding van der Waals radii. The backbone atoms of the tyrosines were left in the positions which were occupied by those of Val_2 and Sta_3 of the inhibitor. A hydronium ion was then introduced in a position in which it was around 3 Å from one each of the oxygen atoms of the carboxyl groups of Asp_{215} and P_1 Tyr. The final model was checked for non-allowed contacts with the program INSIGHT (Biosym Technologies, Inc., San Diego, CA).

RESULTS AND DISCUSSION

Rate-determining steps

The results of the solvent isotope effect experiments are summarized in Table 1. The rate constants for the substrate Ac-Lys*Nph-amide which has no amino acid residues in P_3 and P_2' were independent of pH in the range 5.0 - 5.5 (pD 5.4 - 5.9) and showed no solvent isotope effect.[29] The K_m's, determined at pH 5.5 (pD 5.9) in separate experiments, were not significantly different (0.2 mM in H_2O and 0.25 mM in D_2O), and agreed with the previously determined value in H_2O of 0.22 ± 0.01.[32] The absence of a solvent isotope effect shows that the rate-determining step does not involve the bond breaking step. This step is now generally accepted to proceed via a general acid-base mechanism in which one of the active site aspartic acids activates a water molecule which then attacks the scissile bond and generates a tetrahedral intermediate. A second proton transfer to the nitrogen of the leaving carboxyl moiety completes the cleavage. Thus, there are two successive proton transfers either of which would be expected to show a solvent isotope effect if it were involved in the rate-determining step.

Earlier studies, such as the pre-steady state fluorescence measurements of Fruton[33] and the inhibitor binding experiments of Rich and Bernatowicz,[34] show that a minimal scheme for the action of aspartic proteinases is the following:

$$E + S \Leftrightarrow ES \Leftrightarrow (ES)^* \Leftrightarrow (ET)^* \Leftrightarrow (EP_1P_2)^* \Leftrightarrow EP_1P_2 \Rightarrow E + P_1 + P_2$$

where $(ES)^*$ represents an enzyme substrate complex in a conformation different from that of the ES complex, $(ET)^*$ a complex with the tetrahedral intermediate, $(EP)^*$ an enzyme product complex which undergoes a conformational change to EP that leads to the release of the

Table 1. Solvent isotope effects for penicillopepsin catalyzed hydrolyses. Assays were carried out in 20 mM formate or acetate buffers at pH 3.5 - 5.5 (pD 3.9 - 5.9), 25°C; substrate concentrations were 10 - 25 x K_m.

Substrate	k_{obs} (sec^{-1})[a]		k_H/k_D
	H_2O	D_2O	
Ac-Lys*Nph-amide[b]	0.013	0.012	1.09 ± 0.04
Ac-Ala$_2$-Lys*Nph-amide	1.12	0.77	1.45 ± 0.1
Ac-Ala-Lys*Nph-Ala$_2$-amide	1.15	0.7	1.64
Ac-Ala$_2$-Lys*Nph-Ala$_2$-amide[b]	40	19	2.11 ± 0.16

a) k_{obs} at pH 4.5 (pD 4.9), except for Ac-Lys*Nph-amide

b) from reference.[29]

246

product. The findings that demonstrable conformational changes are associated with the long substrates that occupy S_3, but not with the short substrates, suggests that the above scheme requires substrate-dependent adjustments. The absence of a solvent isotope effect with the substrate Ac-Lys*Nph-amide suggests that the rate-determining step lies either before or after the proton-transfer steps, that is in the binding events that lead to productive binding or in the product release step. We can eliminate the release of products as the rate-determining step, because, when the hydrolysis of Ac-Lys*Nph-amide is carried out in $[^{18}O]H_2O$, the product Ac-Lys contains only one atom of ^{18}O. This differs from the products of the acyl transpeptidation reactions, where some 90% of the carboxyl groups contain two atoms of ^{18}O and thereby provide evidence that the release of the N-terminal moiety of the peptide is rate-limiting.[28]

On the other hand the rate constant for Ac-Ala$_2$-Lys*Nph-Ala$_2$-amide, a substrate which binds into both S_3 and S_2', show a solvent isotope effect of 2.11 at pH 4.5 (Table 1). The same effect is also seen over the pH range 3.5 - 5.5 (pD 3.9 - 5.9).[29] In this case, too, the K_m values are unaffected. The solvent isotope effects for Ac-Ala$_2$-Lys*Nph-amide, which lacks a residue in P_2', and Ac-Ala-Lys*Nph-Ala$_2$-amide, which lacks a residue in P_3, are smaller than that for Ac-Ala$_2$-Lys*Nph-Ala$_2$-amide (Table 1), suggesting that separate and additive effects are induced by binding of amino acid residues in either S_3 or S_2'. A proton inventory was undertaken for Ac-Ala$_2$-Lys*Nph-Ala$_2$-amide (details are given by Cunningham et al.[29]), in order to estimate the number of protons involved in the isotope effect. The result was analyzed statistically according to Schowen[35] in terms of three possibilities, one involving a single proton, one with two protons and one with several protons in the rate-determining step. The analysis of the results showed a better fit for the two-proton and the multi-proton case, than for the single proton mechanism, but did not allow us to distinguish between the two former cases.

The explanation of the effect in molecular terms is not simple. The rate-determining step can be placed after the initial binding event and before product release, the latter because the analysis of Ac-Ala$_2$-Lys isolated from a digest carried out in $[^{18}O]$-water contained only one ^{18}O atom, as did Ac-Lys discussed above. This suggests that peptide bond cleavage had not been reversed. The actual peptide bond cleavage event would not be the rate-determining step, because this step involves transfers of only single protons in two successive steps. We suggest that the solvent isotope effect is due to an event preceding bond cleavage. Pearl[3] has proposed that binding to the enzyme induces the distortion of the scissile peptide bond towards a tetrahedral state and we suggested in turn[15] that this distortion is greatly assisted for "good" substrates by binding in subsites S_3 and S_2', and especially by the specific hydrogen bond formation in these sites. As we have shown in the Introduction, evidence has been obtained that binding in subsite S_3 is associated with conformational changes. We therefore propose that the reaction rate for the substrates with S_3 and S_2' occupancy is determined by the hydrogen bond formation in these two subsites and by the associated conformational changes, changes that probably involve protonic reorganizations. This proposal can account for a considerable solvent isotope effect. It would also explain why the solvent isotope effects for substrates which occupy only one of S_3 or S_2' are only about half of that observed when both subsites are occupied: binding in each subsite contributes separately to the effect. On the other hand with substrates that occupy neither of these subsites, no conformational changes have been observed. Peptide bond distortion is probably assisted only to a small degree, if at all, by the enzyme. This step would be slow because peptide bond distortion would depend on the probability that a small fraction of the bond was in the distorted state at any one time, due to intramolecular motion. We therefore propose that the latter event is the rate-determining step for these substrates. This leads us to suggest the following modified minimal scheme for the action of penicillopepsin, and probably other aspartic proteinases, where (a) represents the path for substrates without and (b) that for those with solvent isotope effects. ES^* is an enzyme-substrate complex in which only the substrate is distorted, E^* indicates that the enzyme is in an altered conformation. For

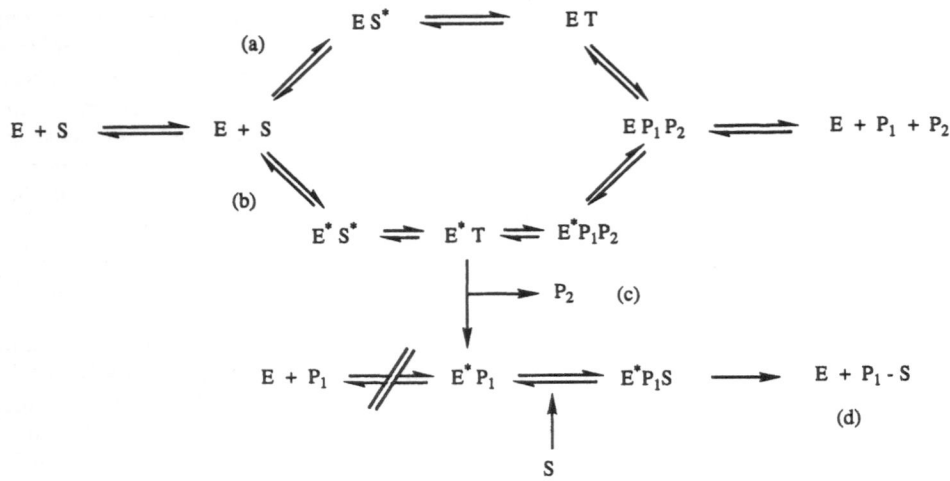

substrates without a solvent isotope effect the reaction ES $\Leftrightarrow$ ES* is the slowest step, for substrates which show a solvent isotope effect it is the reaction ES $\Leftrightarrow$ E*S*. (c) is a simplified scheme for transpeptidation, where the rate is limited by the release of the product P_1-S; the product P_1 (d) is not detectably released from the E$^*P_1$ complex[28] (see also below).

This scheme is still simplified. The conformational changes induced by the occupation of subsites S_3 and S_2' respectively are probably not the same, that is, the nature of E* differs for different substrates. Also, at present there is good evidence from X-ray analysis,[19,20] and CD and kinetic experiments[16] for a conformational change induced by binding in S_3, but the evidence for a conformational change following binding in S_2' comes only from the kinetics of temperature dependence,[16] as we have discussed in preceding paragraphs. Furthermore, substrates that undergo acyl transpeptidation probably also induce conformational changes (see below), although they need not have residues in P_3 and P_2', for example Leu-Tyr-amide. In this case it may be the positive charge on the α-amino group that is responsible for inducing a change. Acyl transpeptidation has been observed only with substrates with a free α-amino group.

The proposed schemes also account for the apparent discrepancy between the solvent isotope effect experiments with pig pepsin of Clement and Snyder[22] and Reid and Fahrney[23] on one hand, and Hollands and Fruton[24] on the other: only the latter used a substrate with a P_3 residue, and only that substrate showed a solvent isotope effect.

Trapping of intermediates during transpeptidation

Other reactions catalyzed by aspartic proteinases that involve conformational changes are acyl transpeptidations. These are characterized by the following phenomena. A) Trimers and tetramers of the N-terminal amino acid of such substrates as Leu-Tyr-amide with pig pepsin[27] or Nph-Ala$_2$-amide with penicillopepsin[28] are the first major products formed, the monomers and dimers appear only as late products of secondary reactions. B) If the transpeptidation reactions are carried out in the presence of an excess of that labelled amino acid which is at the N-terminus of the substrate no label appears in the oligomers formed, which means that the intermediate enzyme complexes like Enz.Nph and Enz.Nph$_2$ (Figure 1) apparently do not dissociate and also do not form from their free components to any measurable extent.

Our recent studies[28] have led to the formulation of the complex series of reactions shown in Figure 1. This scheme is based on the time course of the appearance of the products Nph$_2$-Ala$_2$-amide, Nph$_2$, Nph$_3$ and Nph$_4$ from the substrate Nph-Ala$_2$-amide; on

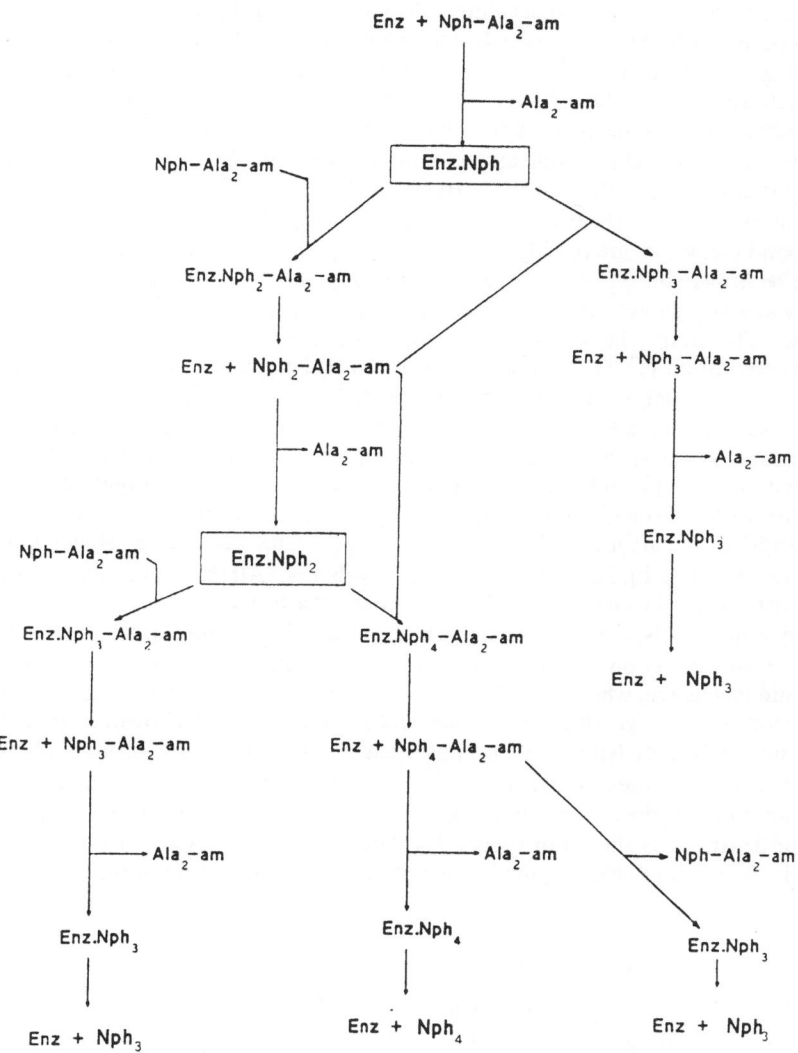

Figure 1. Scheme for the reactions leading to the formation of Nph₃ and Nph₄ by the action of penicillopepsin on Nph-Ala₂-amide. **Bold print:** products formed during the early phase of the reaction; boxed in: enzyme complexes of acceptor molecules for substrate and Nph₂-Ala₂-amide; am, amide.

the extent of incorporation of ^{18}O into those products; and on the study of the action of penicillopepsin on the intermediates Nph_2-Ala_2-amide and Nph_3-Ala_2-amide, and on Nph_2 and Nph_3, when these are used as substrates and as inhibitors.[28]. Figure 1 shows the main pathways by which the trimers and tetramers can be formed. The failure of Nph and Nph_2 to appear in the early stages of the reaction supports the postulate that the Enz.Nph and Enz.Nph_2 complexes cannot dissociate. Previously we proposed that the intermediates Nph_2-Ala_2-amide and Nph_3-Ala_2-amide underwent cleavage after internal translocation in the active site groove,[13] because we had not been able to detect them. However, we have *now* found small amounts of Nph_2-Ala_2-amide as the very first product[28] and therefore suggest that it is released from the enzyme before cleavage. Nph_3-Ala_2-amide cannot be detected, because it is a very good substrate and does not accumulate. When the reactions are carried out in $[^{18}O]$-water about 90 % of the carbonyl and carboxyl $[^{16}O]$-oxygens of all Nph-Nph bonds, but not of Nph-Ala bonds, are replaced with $[^{18}O]$-oxygen. This shows that the peptide bond cleavages are readily reversible, except for the Nph*Ala cleavage which was shown to be irreversible.[28] A study of the action of penicillopepsin on various intermediates in $[^{18}O]$-water also shows that the reverse reactions occur while the peptides are bound in the active site. This in turn indicates that the release of products, specifically of the N-terminal moiety, is rate limiting.[28]. The late products, Nph and Nph_2 are probably formed as follows. The time course analysis of the products by HPLC shows that oligomers longer than tetramers, such pentamers, hexamers and heptamers, form transiently in small amounts. They presumably arise by condensation of the trimers and tetramers. They are then hydrolyzed to give Nph and Nph_2 as secondary, or tertiary, products (Figure 2).

As we have mentioned in the Introduction there is little evidence for and much evidence against covalent trapping of Nph and Nph_2 in the active site. Hence the question arises as to how these ligands are trapped non-covalently. All the data available, including a very low affinity of the enzyme for Nph when the latter is used as an inhibitor ($K_i > 0.5$ M), indicate that the Enz.Nph complex cannot be formed from its free components, but can only form as a result of action of the enzyme on the substrate. We suggest that this induces a conformational change, which is followed by cleavage of the scissile bond, and by the release of the C-terminal moiety, Ala_2-amide. The conformational change, however, is stabilized by the presence of Nph or Nph_2. We propose that this stabilization is brought about by two electrostatic interactions, shown in Figure 3. One of them is between the N-terminal ammonium group of the trapped amino acid or dipeptide and the carboxylate of Asp_{77} (only the dipeptide ligand is shown in Figure 3). The side chain of Asp_{77} is sufficiently free to rotate to interact with the ammonium group of either the amino acid or the dipeptide ligand.

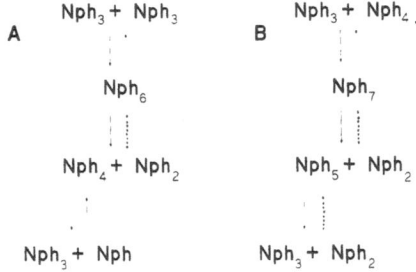

Figure 2. Secondary reactions catalyzed by penicillopepsin acting on Nph-Ala_2-amide during transpeptidation.

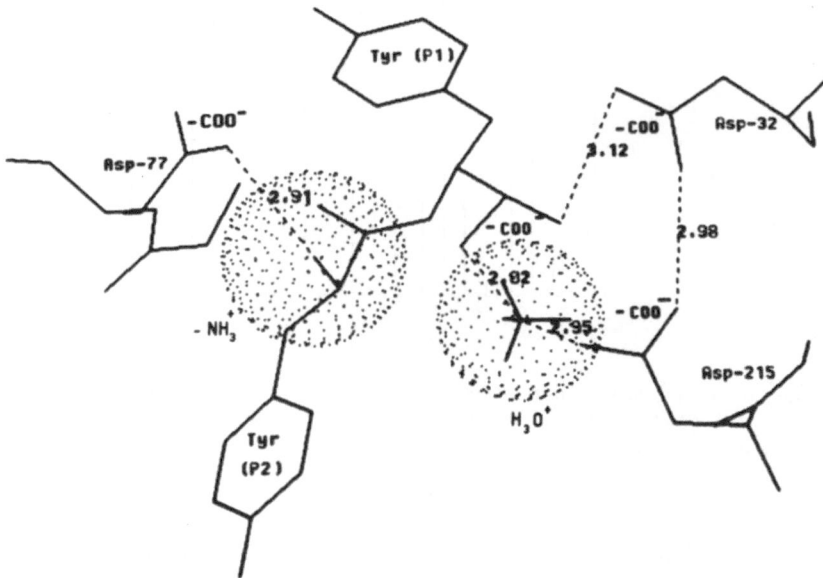

Figure 3. Proposal for the binding of Nph_2 in the active site of penicillopepsin. Nph residues are represented by Tyr residues. The dotted spheres represent the hydronium and ammonium ions respectively. The dotted lines and associated numbers represent possible hydrogen bonds

The other ion pair could form between the carboxylate of the ligand and a putative hydronium ion. This would prevent charge repulsion between the carboxylate of the ligand and the negative charge on the active site aspartic acid pair, by neutralizing one of them. Such a hydronium ion can readily be accommodated sterically and can also form potential hydrogen bonds with the carboxyls of the substrate and of the active site residues. We have postulated that these electrostatic, as well as additional hydrogen bond interactions, could provide sufficient binding energy to give an affinity of 10^{12} M^{-1}, an affinity sufficient to account for the apparent irreversible binding under the conditions used for the experiments.[28] We have now adduced supporting evidence for the proposed electrostatic stabilization by studying the effect of pH on the distribution of the major products of transpeptidation (Figure 4). Figure 4 shows the ratios of Nph_3, Nph_2 and Nph at a point where 50% of the substrate has been used up. The amount of Nph formed increases steadily from only trace amount at pH 5.25 until it is the major product at pH 2.0. Concurrently Nph_3 decreases and none is found at pH 2.0. The amount of Nph_2 increases initially with decreasing pH but then also falls off and makes up only a small proportion of the products at pH 2. A similar picture is seen when the initial rates of formation of the products relative to the rate of disappearance of Nph-Ala_2-amide are compared with each other (Table 2). At pH 5.25 no Nph or Nph_2 is formed initially, but their relative rates increase with decreasing pH to pH 2.75 for Nph_2 and pH 2.0 for Nph. In contrast the relative rates of Nph_3 decrease with decreasing pH and at pH 2.0 none is detectable.

This change in the relative amounts of products with decreasing pH is most readily explained as follows. As the pH drops the carboxylates, including that of Asp_{77} and the C-terminal carboxylates of the products Nph and Nph_2, become increasingly protonated, leading to a decrease in the affinity of these ligands. In other words, the release of Nph and Nph_2 from their enzyme complexes, which is not measurable at pH 5.25, increasingly

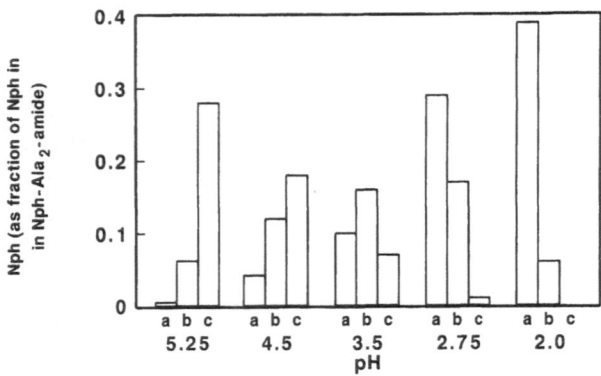

Figure 4. Effect of pH on penicillopepsin catalyzed transpeptidation. The bars represent the distribution of Nph (a), Nph_2 (b), and Nph_3 (c) at a point during the reaction when the substrate $Nph-Ala_2$-amide has decreased to 50% of the starting level.

competes with the condensation reaction of the complexes with acceptor molecules. The relative amount of Nph_2 increases down to pH 2.75 because Nph is probably more tightly held, since the pK_a of its carboxylate group is at least one pH unit lower than that of the dimer. Nph_2 can therefore still form readily by transpeptidation. The release of these ligands from their enzyme complexes occurs at the cost of the formation of Nph_3. The overall rate of the reactions as measured by the disappearance of the substrate shows a pH dependence with an optimum at pH 4.5 (Table 2). The formation of Nph_2 and Nph_3 also shows an optimum at pH 4.5, but falls off more sharply at lower pH.

From all the evidence available so far the transpeptidation reactions catalyzed by pig pepsin acting on substrates such as Leu-Tyr-amide[27] are strictly analogous to those catalyzed by penicillopepsin. This includes the fact that trapped Leu does not exchange with free Leu in solution.[26] However, the residue at position 77 in pepsin is a Thr and not an Asp, and therefore the N-terminal ammonium groups of the trapped ligands cannot form ion pairs with the side chain of that residue. An analysis by molecular graphics of the structure of a pepsin inhibitor complex will be needed in order to find out whether or not there is a negative charge in pepsin that could form an equivalent electrostatic interaction. Such an analysis will be done once the atomic coordinates for a pepsin-inhibitor complex become available.

Table 2. Effect of pH on rates of penicillopepsin catalyzed transpeptidation with $Nph*Ala_2$-amide as substrate. The experimental conditions were as described under Methods.

Peptide	rate constant ($sec^{-1} \times 10^3$) pH				
	5.25	4.5	3.5	2.75	2.0
$Nph-Ala_2$-amide	14	27	10	1.2	0.15
Nph	0	2.1	1.5	0.31	0.12
Nph_2	0	1.6	2.6	0.13	0.02
Nph_3	9	13.5	1.2	0.34	0
Nph_2-Ala_2-amide	6	7.8	2.0	~0	0

SUMMARY

We have summarized and discussed the evidence from solution and structural studies for the role of conformational changes in the catalytic action of penicillopepsin and pig pepsin, and by implication of other aspartic proteinases.

ACKNOWLEDGMENT

We thank Dr. Henrianna Pang for the ^{18}O analyses.

REFERENCES

1. G. L. Gilliland, E. L. Winborne, J. Nachman, and A. Wlodawer, The three-dimensional structure of recombinant chymosin at 2.3 Å resolution, *Proteins, Struct. Funct. Genet.* **8**:82-101 (1990).

2. K. Suguna, R. R. Bott, E. A. Padlan, E. Subramanian, S. Sherriff, G. Cohen, and D. R. Davies, Structure and refinement at 1.8 Å resolution of the aspartic proteinase from *Rhizopus chinensis, J. Mol. Biol.* **196**:877-900 (1987).

3. L. H. Pearl, The catalytic mechanism of aspartic proteinases, *FEBS Lett.* **214**:8-12 (1987).

4. L. Polgar, The mechanism of action of aspartic proteases involves "push-pull" catalysis, *FEBS Lett.* **219**:1-4 (1987).

5. G. Fischer, Acyl group transfer - aspartic proteinases, *in*: "Enzyme Mechanisms," M. I. Page, and A. Williams, eds. pp. 229-239, Roy. Soc. Chem., London 1987.

6. M. N. G. James, and A. R. Sielecki, Aspartic proteinases and their catalytic pathway, *in*: "Biological Macromolecules and Assemblies. Vol. 3: Active Sites of Enzymes", F. A. Jurnak and A. McPherson, eds. pp. 415-482, John Wiley and Sons, New York (1984).

7. T. Hofmann, B. M. Dunn, and A. L. Fink, Mechanism of action of aspartic proteinases, *Biochemistry* **23**:5253-5256 (1984).

8. J. F. Fruton, The mechanism of action of pepsin and related acid proteinases, *Adv. Enzymol.* **44**:1-36 (1976).

9. V. K. Antonov, L. M. Ginodman, Y. K Kapitannikov, T. N. Barshevskaya, A. G. Gurova, and L. D. Rumsh, Mechanism of pepsin catalysis: general base catalysis by the active site carboxylate ion, *FEBS Lett.* **88**:87-90 (1978).

10. V. K. Antonov, L. M. Ginodman, L. D. Rumsh, Y. K. Kapitannikov, T. N. Barshevskaya, L. P. Yavashev, A. G. Gurova, and L. I. Volkova, Studies on the mechanism of action of proteolytic enzymes using heavy oxygen exchange, *Eur. J. Biochem.* **117**:195-200 (1981).

11. T. T. Wang, and T. Hofmann, Acyl and amino intermediates in reactions catalyzed by pig pepsin, *Biochem. J.* **153**:691-699 (1976).

12. M. S. Silver and S. L. T. James, Enzyme catalysed condensation reactions which initiate rapid peptic cleavage of substrates. 2. Proof of mechanism for three examples, *Biochemistry* **20**:3183-3189 (1980).

13. M. Blum, A. Cunningham, M. Bendiner and T. Hofmann, Penicillopepsin,the aspartic proteinase from *Penicillium janthinellum*: Substrate binding and intermediates in transpeptidation reactions, *Biochem. Soc. Trans.* **13**:1044-1046 (1985).

14. T. Hofmann and A. L. Fink, Cryoenzymology of penicillopepsin, *Biochemistry* **23**:5247-5253 (1984).

15. T. Hofmann, B. Allen, M. Bendiner, M. Blum, and A. Cunningham, Effect of secondary substrate binding in penicillopepsin: contributions of subsites S_3 and S_2' to k_{cat}, *Biochemistry* **27**:1140-1146 (1988).

16. B. Allen, M. Blum, A. Cunningham, G. C. Tu, and T. Hofmann, A ligand-induced, temperature-dependent conformational change in penicillopepsin, evidence from nonlinear Arrhenius plots and from circular dichroism studies, *J. Biol. Chem.* **265**:5060-5065 (1990).

17. M. N. G. James, A. R. Sielecki, F. Salituro, D. H. Rich, and T. Hofmann, Conformational flexibility in the active sites of aspartyl proteinases revealed by a pepstatin fragment binding to penicillopepsin, *Proc. Natl. Acad. Sci., U.S.A.* **79**:6137-6141 (1982).

18. T. L. Blundell, J. Cooper, S. I. Foundling, D. M. Jones, B. Atrash, and M. Szelke, On the rational design of renin inhibitors: x-ray studies of aspartic proteinases complexed with transition-state analogues, *Biochemistry* **26**:5585-5590 (1987).

19. A. Sali, B. Veerapandian, J. B. Cooper, S. I Foundling, D. J. Hoover, and T. L. Blundell, High-resolution x-ray diffraction study of the complex between endothiapepsin and an oligopeptide inhibitor: the analysis of the inhibitor binding and description of the rigid body shift in the enzyme, *EMBO J.* **8**:2179-2188 (1989).

20. A. Sali, B. Veerapandian, J. B. Cooper, D. S. Moss, T. Hofmann, and T. L. Blundell, Rigid body movement and conformational differences in aspartic proteinases, *Proteins: Struct. Funct. Genet.* (submitted).

21. K. B. J. Schowen, Solvent hydrogen isotope effects, *in*: "Transition states of biochemical processes," R. D. Gandour and R. L. Schowen, eds., pp. 225-283, Plenum Press, New York (1978).

22. G. E. Clement and S. L. Snyder, The kinetics of the pepsin catalyzed hydrolysis of N-acetyl-L-phenyalanyl-L-tyrosine methyl ester, *J. Am. Chem. Soc.* **88**:5338-5339 (1966).

23. T. W. Reid and D. Fahrney, The pepsin catalyzed hydrolysis of sulfite esters, *J. Am. Chem. Soc.* **89**:5941-5943 (1967).

24. T. R. Hollands and J. S. Fruton, On the mechanism of pepsin action, *Proc. Natl. Acad. Sci., U.S.A.* **62**:1116-1120 (1969).

25. H. Neumann, Y. Levin, A. Berger and E. Katchalski, Pepsin catalyzed transpeptidation of the amino-transfer type, *Biochem. J.* **73**:33-41 (1959).

26. M. Takahashi, T. T. Wang, and T. Hofmann, Acyl intermediates in pepsin and penicillopepsin catalyzed reactions, *Biochem. Biophys. Res. Commun.* **57**:39-46 (1974).

27. M. K. Lutek, T. Hofmann, and C. M. Deber, Transpeptidation reactions of porcine pepsin: formation of tetrapeptides from dipeptide substrates, *J. Biol. Chem.* **263**:8011-8016 (1988).

28. M. Blum, A. Cunningham, H. Pang, and T. Hofmann, Mechanism and pathway of penicillopepsin catalyzed transpeptidation and evidence for non-covalent trapping of amino acid and peptide intermediates, *J. Biol. Chem.* **266**:9501-9507 (1991).

29. A. Cunningham, M. I. Hofmann, and T. Hofmann, Rate-determining steps in penicillopepsin catalyzed reactions, *FEBS Lett.* **276**:119-122 (1990).

30. T. Hofmann, Penicillopepsin, *Meth. Enzymol.* **45**:434-452 (1976).

31. T. Hofmann and R. S. Hodges, A new chromophoric substrate for penicillopepsin and other fungal aspartic proteinases, *Biochem. J.* **203**:603-610 (1982).

32. T. Hofmann, R. S. Hodges and M. N. G. James, Effect of pH on the activities of penicillopepsin and *Rhizopus* pepsin, and a proposal for the productive substrate binding mode in penicillopepsin, *Biochemistry* **23**:635-643 (1984).

33. J. S. Fruton, Fluorescence studies on the active site of proteinases, *Mol. Cell. Biol.* **32**:105-114 (1980).

34. D. H. Rich and M. S. Bernatowicz, Synthesis of analogues of the carboxyl protease inhibitor pepstatin. Effect of structure in subsite P_3 on inhibition of pepsin, *J. Med. Chem.* **25**:791-795 (1982).

35. K. L. Schowen, The proton inventory technique, *C.R.C. Crit. Rev. Biochem.* **17**:1-44 (1984).

36. C. Abad-Zapatero, D. N. Neidhart, T. J. Rydel, J. Luly, and J. Erickson, Refined crystal structures of porcine pepsin and inhibitor complexes: evidence for a flexible subdomain, (see earlier chapter in this volume)

THERMOPSIN, A THERMOSTABLE ACID PROTEASE FROM

SULFOLOBUS ACIDOCALDARIUS

Xinli Lin, Martin Fusek and Jordan Tang
Laboratory of Protein Studies
Oklahoma Medical Research Foundation
Oklahoma City, Oklahoma   73104

INTRODUCTION

Most of the well-studied aspartic proteases, including those derived from yeast, fungi, plants and animal sources, are stable in temperatures up to about 50° to 60°C. Aspartic proteases which can function at high temperature in the range of 80° to 100°C have not been reported so far. We searched for thermostable acid proteases in the thermoacidophilic archaebacteria, *Sulfolobus acidocaldarius, Sulfolobus solfataricus,* and *Thermoplasma acidophilum*, because these organisms grow best in acidic media in a pH near 2 and at temperature of 80°C. Using a highly sensitive radioassay (Lin *et al.*, 1989), we found proteolytic activities in the cultures of all three bacteria. The highest activity was found in *Sulfolobus acidocaldarius*. This protease was named thermopsin.

RESULTS AND DISCUSSION

The majority of thermopsin activity is tightly associated with the bacteria cells. The activity in the medium represents less than 10 % of total thermopsin. We tried various methods to release the enzyme from the cells, including sonication and incubations with acids, bases, salts, detergents, proteases, and organic solvents. None of these was very effective. Only the incubation of cells in 0.25 M sodium formate, pH 3.2, at 80°C for a long period of time (1 to 3 days) caused the partial release of bound thermopsin. Thermopsin is probably bound covalently through a rather stable linkage to the cell wall of the bacterium and only a small amount of the enzyme is released by the bacterium into the medium.

Thermopsin in the culture medium was purified to homogeneity with a five-step procedure including column chromatographies on DEAE-Sepharose CL-6B, phenyl-Sepharose CL-4B, Sephadex G-100, MonoQ FPLC and HPLC gel filtration (Lin & Tang, 1990). The purified thermopsin produces a single band on SDS-PAGE which also exhibited proteolytic activity.

The activity of thermopsin at different temperatures was investigated. The activity increases with temperature up to 90°C. At 100°C, the activity can still be measured although the enzyme inactivates slowly. The optimal activity of thermopsin is near pH 2. Studies on

Structure and Function of the Aspartic Proteinases
Edited by B.M. Dunn, Plenum Press, New York, 1991

proteolytic specificity showed that thermopsin is an endopeptidase with preference for large hydrophobic residues on both sides of the scissile bond, which is also a feature shared by many aspartic proteases, such as pepsin, gastricsin, chymosin and cathepsin D.

Thermopsin was inhibited by the aspartic protease inhibitors pepstatin, DAN and EPNP. Pepstatin inhibits thermopsin with a $K_i = 2 \times 10^{-7}$ M (Fusek *et al.*, 1990). This distinguishes thermopsin from the 'pepstatin-insensitive acid proteases' (Maita *et al.*, 1984) and further suggests that thermopsin is related to the aspartic proteases by catalytic mechanism. In addition, thermopsin is inhibited by mercuric acetate. Since *p*-chloromercuric benzoate and iodoacetic acid had no effect, the mercuric acetate inhibition is probably unrelated to modification of a thiol group. Other common protease inhibitors have no effect on thermopsin activity.

The temperature dependence of the kinetic parameters K_m, k_{cat} and k_{cat}/K_m was studied in the temperature range of 27-80°C (Fusek *et al.*, 1990). The results showed that K_m values only moderately increased between 26-60°C but sharply increased above 65°C. The k_{cat} values increased steadily over the entire temperature range. The values of k_{cat}/K_m, on the other hand, showed a plateau between 48 and 65°C, with a maximum at near 56°C. These results mean that at relatively low temperature (below 65°C), in spite of a favorable K_m, the hydrolytic activity of the enzyme would be restricted by a low V_{max} at these temperatures. This is understandable from the view point of bacterial physiology because *S. acidocaldarius* does not grow at these low temperatures. At higher temperatures, above 65°C, both k_{cat} and K_m increase rapidly with the increase of temperature. Thus at the high temperature, in spite of K_m becoming less favorable, a high rate of hydrolysis near that of V_{max} is achieved at high substrate concentration. The value of k_{cat}/K_m for thermopsin is highest in the temperature range 50-65°C. This means that in the temperature range where the organism could only marginally grow (around 60°C), thermopsin would respond to a lower substrate concentration. But at the favorable growing temperature (around 80°C), the enzyme would respond only to high substrate concentration. This property of thermopsin may contribute to the proper growth of the bacterium.

Thermopsin is extremely resistant to denaturation by detergent such as SDS. Even at high concentrations of SDS (up to 3 % w/v), thermopsin still has significant proteolytic activity towards various protein substrates at 80°C.

Cloning of Thermopsin

Thermopsin is a single-chain protein as indicated by gel electrophoresis and by a single N-terminal sequence. A genomic library of *S. acidocaldarius* was prepared and screened by an oligonucleotide probe designed from the N-terminal sequence of thermopsin. Five positive clones were isolated. From these clones the thermopsin gene was mapped and sequenced. The nucleotide sequence (Lin & Tang, 1990) showed that thermopsin structure was coded in 1020 bases. In the deduced protein sequence, there are 41 amino acid residues (including the initiation Met) preceding the N-terminal position of thermopsin. Most of these residues appear to be characteristic of a leader sequence. However, the presence in this region of a short pro sequence cannot be ruled out. Thermopsin has no apparent sequence similarity to aspartic proteases of the pepsin family nor to pepstatin-insensitive acid proteases and thus may represent a new class of acid proteases. There are 11 potential N-glycosylation sites on each thermopsin molecule. The molecular weight estimated from gel filtration (45,000) is larger than that calculated from the deduced sequence (32,651). In addition, Asn_{24} and Asn_{28} were undetected in N-terminal sequencing, suggesting that thermopsin is glycosylated at these positions and may be in some others of these 11 sites.

We have searched for sequence similarity in either the thermopsin gene or protein in the data bases of Genbank. The results showed that thermopsin gene and protein sequences have not been previously reported and are not significantly related to any known sequences in the Genbank.

In order to establish clearly the relationships of thermopsin in the cultural medium and that released by formic acid from the cells, the latter was also purified to homogeneity. The N-terminal sequence of the cell bound thermopsin was identical to that of enzyme purified from the culture medium. The enzymic properties are also similar in general. These results suggest that the medium thermopsin is likely derived from the cell bound enzyme and the thermopsin present at these two locations are the same enzyme.

Although our work has not established the physiological role of thermopsin in *S. acidocaldarius*, it seems reasonable to suggest that thermopsin is a digestive enzyme, which breaks down proteins in the culture medium of this archaebacteria to supply amino acids for bacterial growth and metabolism.

REFERENCES:

Fusek, M., Lin, X. & J. Tang, 1990, *J. Biol. Chem.* **265**:1496-1501.
Lin, X., Wong, R. N. S. & Tang, J., 1989, *J. Biol. Chem.* **264**:4482-4489.
Lin, X. & Tang J., 1990, *J. Biol. Chem.* **265**:1490-95.
Maita, T., Nagata, S., Matsuda, G., Oda, K., Murao, S. & Tsura, D., 1984, *J. Biochem.* **95**:465-475.

PURIFICATION OF AN ASPARTIC PROTEINASE FROM

ASPERGILLUS ACULEATUS

Steen B. Mortensen and Claus Dambmann

Bioscience and Industrial Biotechnology
Novo Nordisk A/S
2880 Bagsvaerd
Denmark

INTRODUCTION

Aspergilli are known to produce a number of extracellular enzymes (e.g. proteinases, carbohydrases, lipases). Several are utilized in commercial products, based on a main activity, but often containing minor side activities. Usually it is difficult to isolate small amounts of a side activity, but in such cases affinity chromatography can be a valuable tool, due to its specific action. We have succeeded in isolating a proteolytic side activity from the *Aspergillus aculeatus* derived carbohydrase product Viscozyme™ (Novo Nordisk A/S, Denmark). The proteinase was isolated by affinity chromatography on a bacitracin matrix, and was shown to be an aspartic proteinase.

MATERIALS AND METHODS

Reagents

Si-300 (30 m) polyol and Si 300-DEAE polyol were from Serva (Heidelberg, FRG). Tresyl chloride (2,2,2-trifluoroethanesulfonyl chloride), was from Fluka (Buchs, Switzerland). Bacitracin was from Sigma (St. Louis MO, U.S.A).

Buffers

A) 0.1 M sodium acetate (pH 4.5). B) 0.1 M sodium acetate/ 1 M NaCl/25 % 2-propanol (pH 4.5). C) 50 mM 2-morpholinoethanesulfonic acid (pH 6.1). D) 50 mM 2-morpholinoethanesulfonic acid/ 15 mM $CaCl_2$/ 10 mM NaCl (pH 6.1)

Activation of resin and coupling with bacitracin

The Si-300 material was activated by tresyl chloride and coupled with bacitracin as previously described (Mortensen *et al.*, 1989).

Structure and Function of the Aspartic Proteinases
Edited by B.M. Dunn, Plenum Press, New York, 1991

SDS-page

SDS-PAGE on Pharmacia Phastgel was performed according to the manufacturers recommendations (Pharmacia, Uppsala, Sweden).

Analysis of Aspergillus aculeatus *aspartic proteinase activity*

Proteolytic activity, determined as HPU (Hemoglobin Protease Units), was measured at 25°C, pH 5.0 and hemoglobin as the substrate. One HPU is defined as the amount of enzyme liberating 1 mM of primary amino groups (determined by comparison with a serine standard). Amino groups were determined with an OPA (*o*-phtaldialdehyde) reagent, giving a color that can be measured at 340 nm.

Clotting activity in column eluates was determined by a microtiter plate assay (Mortensen *et al.*, 1989). One hundred µl of a diluted sample from each fraction was pipetted into a microtiter plate well. One hundred µl (1 %) dried skim milk in Buffer D was added, and the milk-clotting activity was followed by measuring the change in absorbance at 540 nm in a Perkin Elmer (CT, U.S.A.) Lambda reader at one minute intervals.

HPLC

HPLC was performed with a Series 4 Perkin-Elmer HPLC pump, a LC95 detector, and a PE7700 data controller. Columns were an Si 300-DEAE polyol ion-exchanger (1.6x12 cm) and the bacitracin affinity column described in Materials and Methods. Proteins were detected at 280 nm.

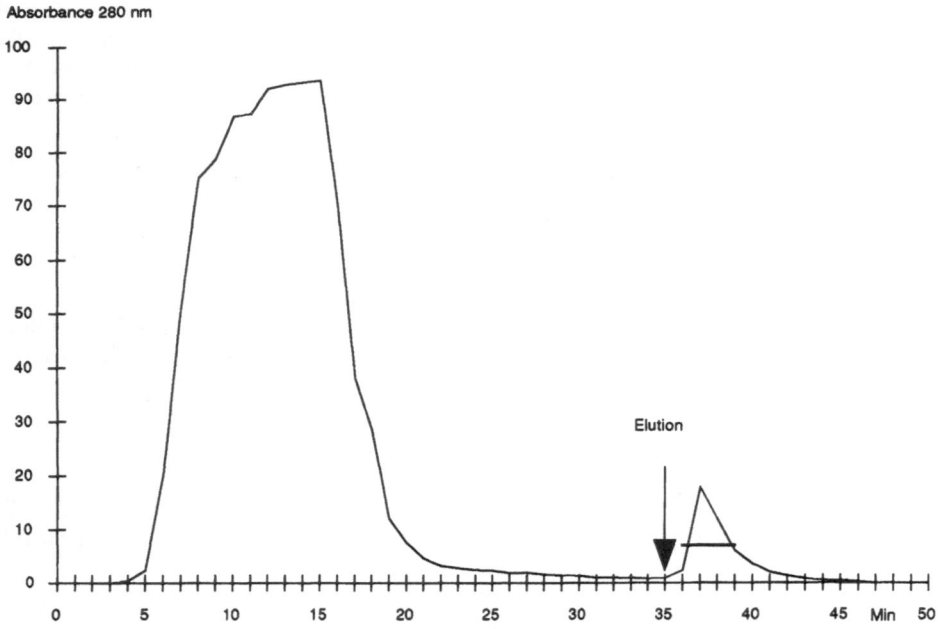

Figure 1. Bacitracin affinity chromatography of *A. aculeatus*. The sample (50 ml) was applied to the affinity column (1.6x12cm), equilibrated in 0.1 M sodium acetate (pH 4.5) at a flow-rate of 5 ml/min. Non-adsorbed material was passed through the column with 0.1 M sodium acetate (pH 4.5). Proteinase was eluted with 0.1 M sodium acetate/ 1 M NaCl/ 25 % 2-propanol (pH 4.5). Proteinase-active fractions are combined as indicated by the horizontal bar. Proteins were detected at 280 nm.

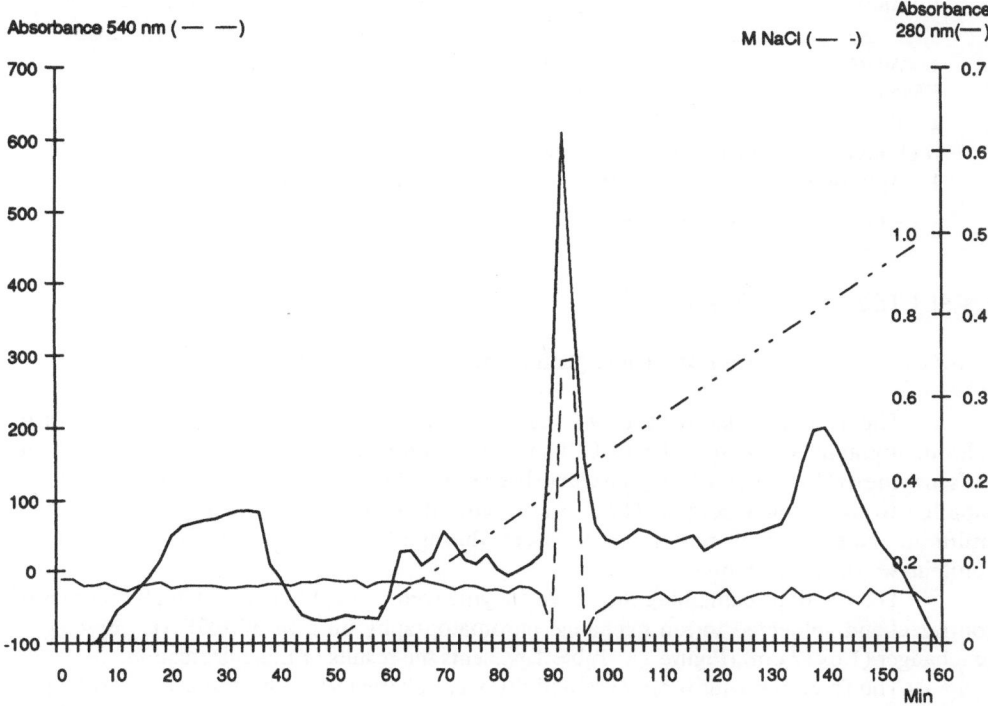

Figure 2. Separation of *A. aculeatus* aspartic proteinase from bacitracin affinity chromatograhy on a Si-300-DEAE column. The material from bacitracin affinity chromatography was added to 9 volumes of 50 mM MES(pH 6.1) (Running buffer C) and the *A. aculeatus* aspartic proteinase was purified by chromatography on the Si-300-DEAE column. The proteinase was eluted by a gradient from 0 to 1 M NaCl. Proteins were detected at 280 nm. Proteinase-active fractions (T = 88 - 96 min) were combined.

Table 1. Purification of *A. aculeatus* aspartic proteinase

Fraction	Volume, ml	Units	% recovery	Total protein	Specific activity	Purification
Starting material	50	5.33	100	5030	0.0011	1
Bacitracin pool	15	3.95	74	35.8	0.1103	100
DEAE pool	45	3.20	60	3.4	0.9412	856

Table 2. Effect of proteinase inhibitors on *A. aculeatus* aspartic proteinase

Inhibitor	Final concentration	Proteinase activity, %
Pepstatin	1.0 mM	0
PMSF	1 mg/ml	96
E-64	0.1 mM	87
p-choloro-mercurbenzoate	0.1 mM	106
Phosphoramidon	0.02 mM	93

RESULTS AND DISCUSSION

Purification and characterization of Aspergillus aculeatus *aspartic proteinase.*

The *Aspergillus aculeatus* aspartic proteinase was purified by affinity chromatography on a Si-300 HPLC-bacitracin column (Figure 1). The sample 25 ml Viscozyme™ (Novo Nordisk A/S product sheet B 456a) was added 25 ml buffer A, and was applied to the affinity column (12x1.6 cm), equilibrated in buffer A at a flow-rate of 5 ml/min. Nonadsorbed material was passed through the column with buffer A. Then proteinase was eluted with buffer B.

The fractions containing the active enzyme were pooled with an added 9 volumes of buffer C and subjected to ion exchange chromatography on a Si 300-DEAE polyol ion-exchanger (1.6x12 cm) (Figure 2). Table 1 presents the results of this purification.

The fractions with proteinase activity were characterized by the use of proteinase inhibitors and it was shown that the enzyme belonged to the group of aspartic proteinases (Table 2).

The apparent molecular weight of the aspartic proteinase determined by SDS-PAGE was 38 kDa (Figure 3). This value is comparable to molecular weights from other *aspergilli* , for example, *A.oryzae* which has been reported to have a molecular weight of 39.4 kDa (Davidson *et al.*, 1975).

Isoelectric focusing of the proteinase on a Pharmacia Phastgel apparatus revealed an isoelectric point of 5.8. The enzymatic activity was confirmed by an overlay consisting of 1% agarose and 0.5 % casein at pH 5.5. After incubation at 45°C for 20 h the proteinase activity was detected by clotting.

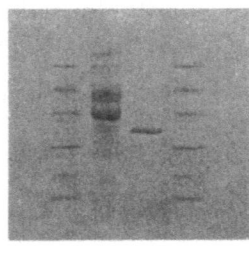

1 2 3 4

Figure 3. SDS-PAGE of *A. aculeatus* aspartic proteinase. Lanes 1 and 4 standard proteins: 92 kDa, 67 kDa, 43 kDa, 30 kDa, 20.1 kDa and 14.4 kDa. Lane 2, *A.aculeatus* supernatant starting material. Lane 3, Purified *A.aculeatus* aspartic proteinase.

The pH-dependent activity profile of the *A. aculeatus* aspartic proteinase determined using the HPU-method carried out at the appropriate pH values showed that the pH optimum was 5 and the temperature activity profile showed a maximum at 50°C using the HPU-method.

Aspartic proteinases have been purified by bacitracin affinity chromatography before (Stepanov *et al.*, 1981; Stepanov & Rudenskaya, 1983; Mortensen *et al.*, 1989). It seems that this method is generally applicable to isolation of small amounts of aspartic proteinases from complex mixtures.

REFERENCES

Davidson, R., Gertler, A. & Hofmann, T., 1975, *Biochem. J.* **147**:45-53.

Mortensen, S. B., Thim, L., Christensen, T., Wldike, H. F., Boel, E., Hjortsh, J. K. & Hansen, M. T., 1989, *J. Chromatogr.* **476**:227-233.

Stepanov, V. M. & Rudenskaya, G. N., 1983, *J. Appl. Biochm* **5**:420.

Stepanov, V. M., Rudenskaya, G. N., Garida, A. V. & Osterman, A. I., 1981, *J. Biochem. Biophys.* **5**:177.

EFFECT OF GROWTH CONDITIONS ON THE EXTRACELLULAR PRODUCTION

OF THE ASPARTIC PROTEINASE BY *CANDIDA ALBICANS*

M. Lam, V. Peterkin, E. Reiss and C. J. Morrison*

Centers for Disease Control
Atlanta, Georgia   30333

INTRODUCTION

There is increasing evidence that the aspartic proteinase (AP) of *C. albicans* (CA) is a pathogenic factor.[1-3] It is one of few fungal proteins known to be transported through the cell wall and, as a major antigenic protein of CA, allows for its use in the development of immunodiagnostic tests.[4] In order to facilitate this goal, increased production of AP is needed. Controversy exists as to which protein substrate allows for maximum AP production *in vitro*. Previous studies have used one strain of CA, one time point during AP production, and one cultivation temperature.[5,6] We compared the effect of temperature, substrate and strain on the kinetics of AP production *in vitro* over a 14 day period.

METHODS

Culture conditions

CA was grown for 48 hr at 25°C on Sabouraud's dextrose agar slants. Growth from slants was suspended in phosphate buffered saline, pH 7.2, washed by centrifugation, and suspended to a final concentration of 1×10^7 cells per ml in yeast carbon base (YCB) or yeast nitrogen base (YNB) medium supplemented with vitamins, 0.2 % glucose, and 0.2 % of one of the following: bovine serum albumin (BSA), bovine erthyrocyte hemoglobin (Hb), or bovine hoof keratin (K). Medium was adjusted to a pH of 5.6 with HCl prior to inoculation with CA. Liquid shake cultures were incubated at 140 rpm at 25°C or 37°C for a period of 14 days. CA strains tested were CBS 2730, 36B, and B311.

Proteinase assay.

An aliquot of culture medium was removed daily to assay for extracellular proteinase production. Two mls of 1 % BSA in 0.1 M sodium citrate buffer, pH 3.6, were added to 0.5 ml of culture supernatant, mixed by inversion, and incubated for 30 min at 37°C. The reaction was stopped by the addition of 5 % ice-cold trichloroacetic acid (TCA). Control samples contained 0.5 ml of culture supernatant but BSA was added after addition of TCA. Each sample was vortex-mixed, centrifuged, and the supernatant was passed through a millipore filter (0.45 μm pore size) attached to a 12 cc syringe. The absorbance of the filtrate was determined spectrophotometrically at a wavelength of 280 nm. In addition, culture pH and cell growth was monitored daily.

RESULTS

Extracellular proteinase production by CA strain CBS 2730 was measured during growth at 25°C in medium containing either BSA, Hb, or K as the protein substrate. BSA was found to allow for maximal AP production relative to the other substrates tested. The use of Hb or K resulted in intermediate or inapparent AP production, respectively. The order of substrate efficacy for AP production was maintained whether YCB or YNB medium was used (Figure 1).

When extracellular AP production by CBS 2730 was examined during growth at 37°C, BSA remained the best substrate for AP production. Overall AP production at 37°C was depressed relative to production at 25°C in all media where AP was detected (i.e. peak AP production was halved at 37°C relative to 25°C in YCB-BSA medium). The day on which peak AP production occurred (day 3) was earlier when CBS 2730 was grown at 37°C than at 25°C and the disappearance of detectable AP activity occurred earlier (day 11).

A precipitous decline in culture pH (from pH 5.6 to 3.5) occurred at day 4 for cultures grown at 25°C in YCB medium containing BSA, Hb, or K as protein substrate and in YNB medium containing K.

Cell growth was maximal in YCB medium containing either BSA or Hb as substrate. Intermediate growth occurred in YNB medium supplemented with BSA or Hb. Growth was inapparent in either YCB or YNB medium containing K as protein substrate.

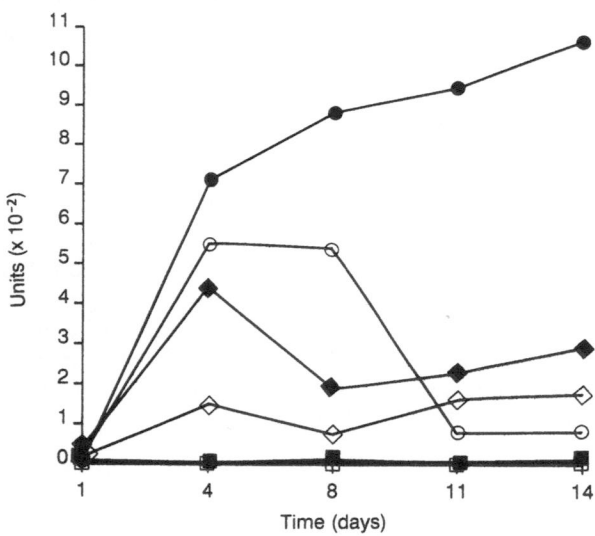

Figure 1. Effect of protein substrate and medium composition on the extracellular production of aspartic proteinase by *C. albicans* strain CBS 2730 during culture at 25°C. Symbols: solid circles, YCB-BSA medium; open circles, YNB-BSA medium; solid diamonds, YCB-Hb medium; open diamonds, YNB-Hb medium; solid squares, YCB-K medium; and open squares, YNB-K medium. Units of proteinase activity equal the absorbance reading at 280 nm obtained during proteinase assay for the test sample minus that for the respective control sample.

In order to determine whether these phenomena held true for other strains of CA, two other strains were tested. Both CA strains 36B and B311 gave maximal, intermediate and inapparent AP production when BSA, Hb, and K were used as substrates, respectively. These observations held whether YNB or YCB medium was used. Peak AP production in YCB-BSA medium at 25°C occurred earlier for strain 36 (day 2) than for CBS 2730 (see Figure 1) while strain B311 AP production peaked at day 12.

CONCLUSIONS

We found that for maximum AP production: YCB was superior to YNB, BSA was superior to Hb which was superior to K, and 25°C was superior to 37°C. These observations held for all three CA strains tested but the day of peak AP production differed between strains. Maximum cell growth occurred in YCB rather than YNB medium, BSA was equivalent to Hb as substrate for cell growth while K was a poor substrate, and cell growth was similar at 25°C and 37°C. Culture pH dropped most consistently in YCB medium relative to YNB medium regardless of substrate used. The decline in culture pH was independent of cell growth or proteinase production. These results indicate that YCB-BSA medium and a temperature of 25°C allows for maximum AP production and that CA strain and day of harvest affect AP yields.

REFERENCES

1. T. L. Ray and C. D. Payne, Scanning electron microscopy of epidermal adherence and cavitation in murine candidiasis: a role for *Candida* acid proteinase, *Infect. Immun.* **56**:1942-1949 (1988).
2. K. J. Kwon-Chung, D. Lehman, C. Good and P. T. Magee, Genetic evidence for role of extracellular proteinase in virulence of *Candida albicans*, *Infect. Immun.* **49**:571-575 (1985).
3. F. Odds, *Candida albicans* proteinase as a virulence factor in the pathogenesis of *Candida* infections, *Zbl. Bakt. Hyg. A* **260**:539-542 (1985).
4. R. Ruchel, B. Boning-Stutzer and A. Mari, A synoptical approach to the diagnosis of candidosis, relying on serological antigen and antibody tests, on culture, and on evaluation of clinical data, *Mycoses* **31**:87-106.
5. T. L. Ray and C. D. Payne, Comparative production and rapid purification of *Candida* acid proteinase from protein-supplemented cultures, *Infect. Immun.* **58**:508-514 (1990).
6. M. Negi, R. Tsuboi, T. Matsui, and H. Ogawa, Isolation and characterization of proteinase from *Candida albicans*: substrate specificity. *J. Invest. Dermatol.* **83**:32-36 (1984).

X-RAY ANALYSIS OF A DIFLUOROSTATONE RENIN INHIBITOR BOUND AS THE TETRAHEDRAL HYDRATE TO THE ASPARTIC PROTEASE ENDOTHIAPEPSIN

Dennis J. Hoover,[1]* Balusubramanian Veerapandian,[2] Jon B. Cooper,[2] David B. Damon,[1] Beryl W.Dominy,[1] Robert L. Rosati[1] and Tom L. Blundell[2]*

[1]Department of Medicinal Chemistry
Central Research Division, Pfizer, Inc.
Groton, CT 06340

and [2]Laboratory of Molecular Biology
Department of Crystallography
Birkbeck College
Malet Street, London WC1E 7HX

INTRODUCTION

In this study we report the X-ray analysis of a complex between the aspartic protease endothiapepsin (EC 3.4.23.6) and an inhibitor bound as a carbonyl hydrate (gem-diol) to the catalytic aspartates of this enzyme, in a manner closely resembling the putative tetrahedral intermediate in proteolytic cleavage of the peptide bond.[1] This study was undertaken in order to obtain a closer model of the interactions stabilizing this intermediate than those used in previous analyses, which were based on X-ray crystallographic data obtained from inhibitors lacking one or both of the hydroxyl residues of this species.[2,3]

MATERIALS AND METHODS

As renin inhibitors also may have high endothiapepsin affinity,[4] a potent renin inhibitor (CP-81,282) was chosen for this study which was derived by combining the high-affinity morpholino-4-carbonyl-Phe-Nle (P_4-P_2) sequence with the electrophilic, tight-binding difluorostatone[5,6] function. Synthesis of CP-81,282 with less than 3% epimerization at the cyclohexylmethyl-bearing carbon was accomplished by DEC-modified Pfitzner-Moffatt oxidation[7] of the corresponding (S)-hydroxy (non statine-like) peptidic alcohol (**1**). This substance was derived in standard fashion from ethyl 4(S)-(t-Boc-amino)-5-cyclohexyl-3(S)-hydroxy-2,2-difluoropentanoate,[6] whose synthesis together with the 3(R) isomer (1:1 ratio) was effected in high enantiomeric purity by addition of a mixture of the requisite aldehyde to $TiCl_4$ and zinc in THF.[8]

CP-81,282, Hydrate Form   Putative Proteolysis Intermediate

CP-81,282 proved a significantly more potent inhibitor of human renin and endothiapepsin (Table 1) than either **1** or the corresponding statine-like alcohol (**2**). This is doubtless due to its low barrier for formation of the tetrahedral hydrate ($K_{hydration}$ = 2.1 in 5:1 DMSO-water and 0.6 in wet chloroform, as measured by ^{19}F NMR), which binds tightly to these enzymes. CP-81,282 was stable to isomerization for weeks in these solvent systems but was found by ^{19}F NMR to equilibrate slowly with its presumably much less active P_1 isomer (final ratio 1:1.3, respectively) in buffered solution (pH 4.3 or 7.0).

Endothiapepsin was cocrystallized with a 10-fold excess of CP-81,282 from 0.1M NaOAc pH 4.6 buffer (2 mg/mL in enzyme, 2.2 M in ammonium sulfate, with a small quantity of acetone). The complex crystallized in a form isomorphous to the native enzyme. X-ray data were collected to 2.0 Å and the structure solved by difference Fourier methods and refined to R = 0.18.

RESULTS AND DISCUSSION

The inhibitor is bound in an extended form, with interactions generally similar to those of other aspartic protease-inhibitor complexes.[4] These include hydrogen bonds from P_3 NH to Thr_{219} OH, Thr_{219} NH to P_3 C=O, P_1 NH to Gly_{217} C=O and Thr_{218} OH, and interactions with the "flap" (P_2 NH to Asp_{77} carboxylate and Gly_{76} NH to both P_2 C=O and P_1' C=O).

Figure 1 shows interactions of the tetrahedral hydrate with catalytic aspartates (Asp_{32} and Asp_{215}). The pro-(R) (statine-like) hydroxyl oxygen occupies an equivalent position to conserved water in the native enzymes, and to the statine-like hydroxyl oxygen in complexes of inhibitors containing statine and related structures.[4] The second hydroxyl oxygen of the hydrate is located within hydrogen-bonding distance of the outer Asp_{32} carboxyl oxygen (2.6 Å) and more remotely, Gly_{34} C=O (3.2 Å, not shown).

Table 1. Enzyme inhibition by CP-81,282 and the related difluorostatines

Compound		IC_{50}, (nM), Human Plasma Renin[a]	K_i, (nM) Endothiapepsin[b]
1	3(S) (non statine-like) alcohol	420	6200
2	3(R) (statine-like alcohol)	10	920
CP-81,282	ketone/hydrate	1	11

[a] pH 7.4 (n = 6), [b] pH 3.1 (n = 6).

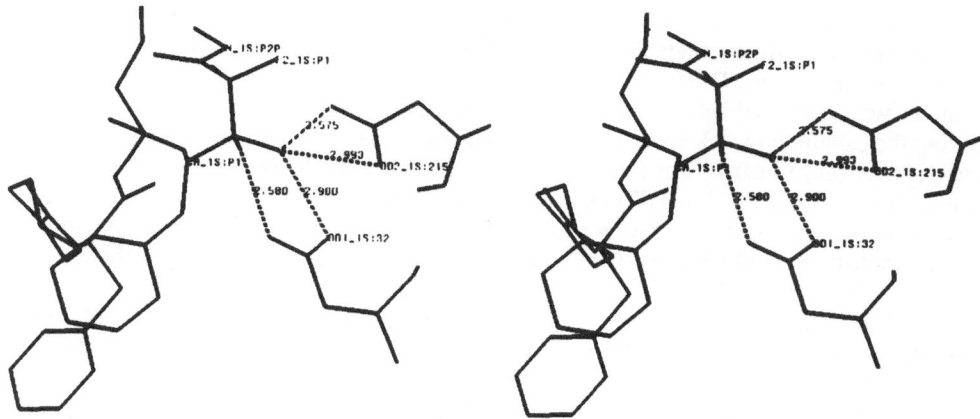

Figure 1. Stereo view of the hydrate-aspartate interactions of the complex with selected interoxygen distances in Å.

The probable locations of the three protons relevant to catalysis (two from the hydrate hydroxyls and one from the aspartates, whose positions are of course not defined by this X-ray analysis) are nevertheless apparent from this data. The short interoxygen distances of 2.6 Å, and attendant favorable hydrogen bond geometries, point strongly to the locations of two of these protons, one between the outer Asp_{215} carboxyl oxygen and the statine-like hydroxyl oxygen, and the other between outer Asp_{32} carboxyl oxygen and the second hydrate oxygen. The third proton is probably located between the statine-like hydroxyl oxygen and the inner Asp_{32} carboxyl oxygen (2.9 Å). Though the interoxygen distance from the statine-like hydroxyl to Asp_{215} carboxylate inner oxygen is nearly equivalent (3.0 Å), such a hydrogen bond is clearly improbable by comparison, because of the unacceptable geometry involved. Two complexes are consistent with these proton locations (Figure 2). These differ only in the proximity of each of the three protons to one or the other of the pairs of interacting oxygens. The first (2a) appears the more likely, as the ionized Asp_{32} is stabilized by two hydrogen bonds from the hydrate, and additionally, from the enzyme (Ser_{35} γ-OH, Gly_{34} NH, not shown). In 2b, the ionized Asp_{215} receives but one hydrogen bond from the hydrate and two hydrogen bonds from the enzyme (from Thr_{218} γ-OH, Gly_{217} NH, not shown), but only one of these is possible in human renin which has Ala at 218. In 2b, the O-H bond of the statine-like hydroxyl is nearly eclipsed with the hydrate carbon-CF_2 bond (33° dihedral angle), which is also less favorable than in 2a, where these bonds are staggered (148° dihedral).

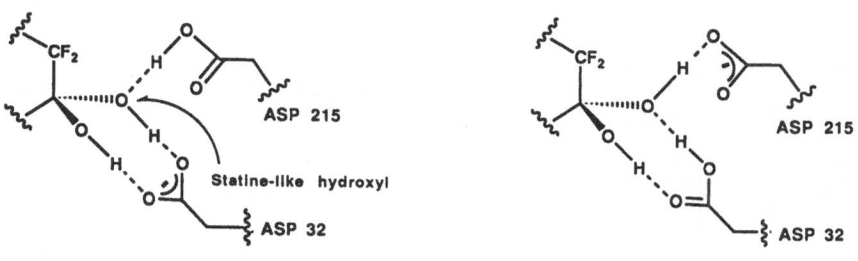

(a) Proposed Arrangement   (b) Alternative Arrangement

Figure 2. Alternative possible arrangements of the hydrate-aspartate protons.

Molecular orbital calculations were performed on simple models of 2a and 2b constructed from difluoroacetaldehyde hydrate ($H\text{-}C(OH)_2\text{-}CF_2\text{-}H$) and formic acid/formate fragments whose coordinates were extracted from the X-ray data (minus the requisite protons, which were added). Only the three protons bound to oxygen were allowed to move during the calculations, which were performed using the Gaussian 86 program and STO-4G basis set. Arrangement 2a was found substantially (16 kcal/mole) more stable than 2b. The calculation performed with the STO 6-31G basis set also indicated a large stability difference (8 kcal/mole in favor of 2a). These assessments are qualitative, since these simple models ignore contributions of other enzyme residues, which have some influence on the aspartyl ionization state in more detailed theoretical models of native aspartic proteases.[9]

The interactions of Figure 2a provide a potential model for those stabilizing a tetrahedral intermediate in proteolysis. This arrangement (Figure 3b) would smoothly follow Asp_{32}-mediated protonation of the substrate carbonyl and concomitant addition of the conserved water molecule, the latter negatively polarized by hydrogen-bonding to the Asp_{215} carboxylate outer oxygen. The process is doubtless facilitated by distortion of the substrate, possibly by rigid body movement[4] in the complex. These events leave the lone pair on pyramidalized nitrogen in an antiperiplanar relationship to the newly formed C-O bond[10], an orientation from which a proton might conceivably be obtained from solvent, either directly, or perhaps via a highly ordered water which is also observed, resulting in product formation. Modelling studies with inhibitors bearing a scissile bond surrogate of the correct dipeptide length as well as C-terminal residues indicate that both of these possibilities would require some movement in the complex and could be potentially difficult.

Alternatively, nitrogen inversion (for dimethylamine requiring ca. 5 kcal/mole[11]) followed by rotation of the $C(OH)_2\text{-}N$ bond (Figure 3c), would leave the lone pair approximately positioned for protonation by the Asp_{215} proton H-bonded to the statine-like hydroxyl oxygen. Either Asp_{32} oxygen might participate in deprotonation of the gem-diol during product formation. Additionally (not shown), a shift in the hydrogen bond from the statine-like hydroxyl from Asp_{32} inner oxygen to Asp_{215} inner oxygen during nitrogen protonation would also contribute to breakdown of the tetrahedral intermediate. With this event, both the formation, then the breakdown of the tetrahedral intermediate are mediated by carboxylic acid dimer-like structures where first Asp_{32}, then Asp_{215} are equally important partners.[12] A proton shuttled between the inner oxygens would restore charge to Asp_{215}.

In summary, the structure of the CP-81,282/endothiapepsin complex provides a firm basis for refining existing proposals on the mechanism of aspartic proteases. No donors sufficient to stabilize an oxyanion in the substrate are present, as in the serine proteases. The data instead point strongly toward a gem-diol intermediate formed by Asp_{32}-mediated protonation of the substrate carbonyl function, which is stabilized by dual hydrogen bonds to a buried Asp_{32} carboxylate. The nucleophile is most likely the Asp-bound water conserved in the native aspartic protease X-ray structures. A number of these events have been previously proposed.[3] Further insight into aspartic protease action may come from future X-ray studies of yet closer mimics[7] of the tetrahedral proteolysis intermediate.

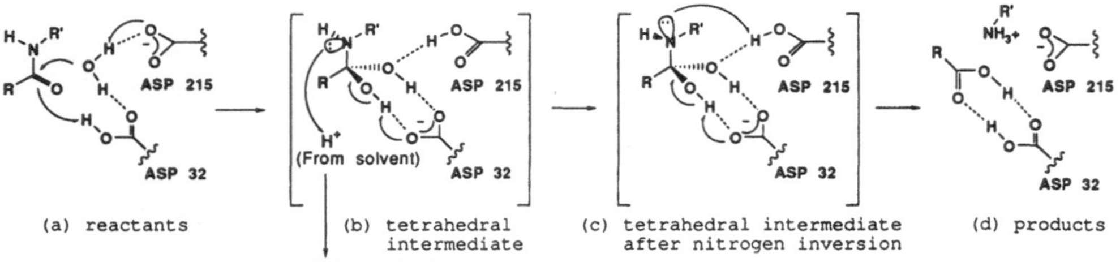

(a) reactants (b) tetrahedral intermediate (c) tetrahedral intermediate after nitrogen inversion (d) products

(d) products

Figure 3. Possible proteolysis events.

ACKNOWLEDGEMENT

We wish to thank A. Sali for stimulating discussions and I. M. Purcell, K. A. Simpson, A. Rauch, and W. R. Murphy for the inhibition constants.

REFERENCES

1. Full experimental details will be reported elsewhere. The coordinates of the complex will be deposited in the Brookhaven Protein Data Bank.

2. M. N. G. James and A. R. Sielecki, Stereochemical analysis of peptide bond hydrolysis catalyzed by the aspartic proteinase penicillopepsin, *Biochemistry* **24**:3701 (1985).

3. K. Suguna, E. A. Padlan, C. W. Smith, W. D. Carlson, and D. R. Davies, Binding of a reduced peptide inhibitor to the aspartic proteinase from *rhizopus chinensis*: implications for a mechanism of action, *Proc. Natl. Acad. Sci. U.S.A.* **84**:7009 (1987).

4. A. Sali, B. Veerapandian, J. B. Cooper, S. I. Foundling, D. J. Hoover, and T. L. Blundell, High-resolution X-ray diffraction study of the complex between endothiapepsin and an oligopeptide inhibitor: the analysis of the inhibitor binding and description of the rigid body shift in the enzyme, *Embo J.* **8**:2179 (1989), and refs. therein.

5. M. H. Gelb, J. P. Svaren, and R. H. Abeles, Fluoro-ketone inhibitors of hydrolytic enzymes, *Biochemistry* **24**:1813 (1985).

6. S. Thaisrivongs, D. T. Pals, W. M. Kati, S. R. Turner, L. M. Thomasco, and W. Watt, Design and synthesis of potent and specific renin inhibitors containing difluorostatine, difluorostatone, and related analogs, *J. Med. Chem.* **29**:2080 (1986), and references therein.

7. D. B. Damon and D. J. Hoover, Synthesis of the ketodifluoromethylene dipeptide isostere, *J. Am. Chem. Soc.* **112**:6439 (1990).

8. T. Ishihara, T. Yamanaka, and T. Ando, New low-valent titanium catalyzed reaction of chlorodifluoromethyl ketones leading to difluorinated ß-hydroxy ketones, *Chem. Lett.* 1165 (1984).

9. A. Goldblum, Theoretical calculations on the acidity of the active site in aspartic proteinases, *Biochemistry* **27**:1653 (1988).

10. P. Deslongchamps, Stereoelectronic control in the cleavage of tetrahedral intermediates in the hydrolysis of esters and amides, *Tetrahedron* **31**:2463 (1975).

11. R.A. Eades, D. A. Well, D. A. Dixon, and C. H. Douglass, Jr., Inversion barriers in methyl-substituted amines, *J. Phys. Chem.* **85**:976 (1981).

12. L. Polgar, The mechanism of action of aspartic proteases involves 'push-pull' catalysis, *FEBS Lett.* **219**:1 (1987).

SUBSTRATE SPECIFICITY STUDY OF RECOMBINANT *RHIZOPUS CHINENSIS* ASPARTIC PROTEINASE

W. Todd Lowther,[#] Zhong Chen,[*] Xin-li Lin,[*] Jordan Tang[*] and Ben M. Dunn[#]

[#]Department of Biochemistry and Molecular Biology
J. Hillis Miller Health Center, Box J-245
University of Florida
Gainesville, Florida   32610

[*]Protein Studies Laboratory
Oklahoma Medical Research Foundation
Oklahoma City, Oklahoma   73104

INTRODUCTION

Rhizopuspepsin, a model aspartic proteinase from the fungus *Rhizopus chinensis*, has recently been cloned and expressed by Chen *et al.* (1991). High resolution crystallographic analysis of rhizopuspepsin and complexes with active site ligands has been reported by Davies' group (Parris *et al.*, this volume). Our initial characterization of the substrate specificity of the active site is described in this report. This study will enable future comparisons between kinetic and crystallographic data from other aspartic proteinases as well as for use in planning and analyzing site-directed mutagenesis studies.

Kinetic parameters were determined for a series of chromogenic substrates with systematic substitutions in the substrate residues P_4, P_5 and P_1 to study possible hydrophobic, electrostatic and hydrogen bonding interactions of the active site. Peptides with various multiple substitutions were also analyzed.

MATERIALS AND METHODS

Recombinant rhizopuspepsin was obtained through the expression of a recombinant clone as described in Lin *et al.*, this volume. The enzyme was dissolved in distilled water and aliquots stored at -20°C.

Octapeptide substrates were generated by the solid phase method with an Applied Biosystems Model 430A Synthesizer. All peptides were shown to be greater than 85 % pure by reverse phase HPLC. Stock solutions of the peptides were made in distilled water and quantitated by amino acid analysis.

Substrate hydrolysis was followed by the decrease in the average absorbance from 284-324 nm using a Hewlett Packard 8452A diode array spectrophotometer. All reactions were performed at 37°C in 0.1 M sodium formate buffer, pH 3.5. Amino acid analysis of HPLC purified reaction products showed that all peptides were cleaved specifically at the Phe-Nph bond (Nph = p-Nitrophenylalanine) (Dunn $et\ al.$, 1986). The K_m and V_{max} values were determined from the initial rates of at least six different peptide substrate concentrations using Marquardt analysis. The amount of active enzyme was determined by fitting the curve generated by the competitive titration with the inhibitor V-S-Q-N-Lψ[CH(OH)CH$_2$]V-I-V (U85548E, Sawyer $et\ al$., 1990) (K_i = 0.042 ± 0.014 nM) with the Henderson equation for tight binding inhibitors (Henderson, 1972). The standard deviations of the k_{cat} and k_{cat}/K_m values were propagated using equations derived by standard procedures for non-independent or correlated errors as outlined by Meyer (1975).

RESULTS AND DISCUSSION

Rhizopuspepsin seems to be able to bind and cleave a wide variety of substrates. The kinetic parameters k_{cat}, K_m and k_{cat}/K_m, derived from the observations of six analogs of the P$_5$ position, are given in Table 1. The set of amino acids (Lys, Ser, Asp, Arg, Ala and Leu) provide a range of potential electrostatic, hydrophobic and hydrogen bonding interactions. The changes in this position appear to be easily accomodated with little or no effect on the overall specificity of the enzyme for substrate. The lysine substituted peptide may not be bound as tightly as the others, but it does show a two to three fold increase in substrate turnover.

Table 2 shows the effects of substitutions made in P$_4$. The proline in the parent or control peptide has been substituted with Ser, Asp, Arg, Ala, and Leu exploring a range of functionalities. The arginine substitution leads to both a decrease in k_{cat} as well as a small increase in K_m. The severe decrease in the specificity for this substitution and the acceptance of an aspartic acid points to a possible electrostatic interaction in the S$_4$ binding pocket and its preference for a negative charge. The serine substitution also reduces the k_{cat} to a value in the

Table 1. Kinetic parameters of peptide substrates with substitutions in P$_5$

P$_5$	P$_4$	P$_3$	P$_2$	P$_1$	P$_1$'	P$_2$'	P$_3$'	k_{cat} (sec^{-1})	K_m (μM)	k_{cat}/K_m (μM^{-1}s^{-1})
K	P	A	K	F	X	R	L	18.1 ± 2.0	14.4 ± 3.3	1.25 ± 0.32
S	P	A	K	F	X	R	L	4.9 ± 0.5	4.0 ± 1.0	1.21 ± 0.32
D	P	A	K	F	X	R	L	5.3 ± 0.7	7.0 ± 2.5	0.76 ± 0.29
R	P	A	K	F	X	R	L	5.7 ± 0.6	3.3 ± 1.4	1.74 ± 0.76
A	P	A	K	F	X	R	L	6.7 ± 0.6	5.8 ± 0.7	1.14 ± 0.17
L	P	A	K	F	X	R	L	5.3 ± 0.7	6.5 ± 2.0	0.82 ± 0.21

X = p-Nitrophenylalanine

Table 2. Kinetic parameters of peptide sustrates with substitutions in P_4

P_5	P_4	P_3	P_2	P_1	$P_1{}'$	$P_2{}'$	$P_3{}'$	k_{cat} (sec^{-1})	K_m (μM)	k_{cat}/K_m (μM^{-1}s^{-1})
K	P	A	K	F	X	R	L	18.1 ±2.0	14.4 ± 3.3	1.25 ± 0.32
K	S	A	K	F	X	R	L	9.3 ± 1.1	22.3 ± 4.1	0.42 ± 0.09
K	D	A	K	F	X	R	L	21.7 ± 2.7	29.7 ± 6.3	0.73 ± 0.18
K	R	A	K	F	X	R	L	2.7 ± 0.3	43.3 ± 9.9	0.06 ± 0.02
K	A	A	K	F	X	R	L	14.6 ± 1.4	19.4 ± 1.0	0.75 ± 0.08
K	L	A	K	F	X	R	L	21.5 ± 2.0	13.7 ± 0.7	1.57 ± 0.17

X = *p*-Nitrophenylalanine

range of the P_5 variation peptides (ca. 5 sec^{-1}) while still exhibiting a higher K_m.

The substitutions in P_1, as shown in Table 3, suggest possible steric requirements. The ready clevage of the leucine and phenylalanine derivatives by rhizopuspepsin illustrates a requirement for a γ–branched amino acid or a similarly branched, large ring structure for productive binding and hydrolysis of the substrate. The alanine and ß-branched valine substitutions, however, show a marked decrease or no cleavage under standard assay conditions with higher enzyme concentrations. This suggests that the S_1 binding pocket may be tighter and smaller around the ß-carbon and can not accomodate the bulky valine side chain.

Table 3. Kinetic parameters of peptide substrates with substitutions in P_1

P_5	P_4	P_3	P_2	P_1	$P_1{}'$	$P_2{}'$	$P_3{}'$	k_{cat} (sec^{-1})	K_m (μM)	k_{cat}/K_m (μM^{-1}s^{-1})
K	P	A	K	A	X	R	L	4.2 ± 0.4	24.3 ± 3.6	0.17 ± 0.03
K	P	A	K	V	X	R	L	PC	PC	PC
K	P	A	K	L	X	R	L	16.1 ± 2.2	18.8 ± 5.4	0.86 ± 0.27
K	P	A	K	F	X	R	L	18.1 ± 2.0	14.4 ± 3.3	1.25 ± 0.32

X = *p*-Nitrophenylalanine. PC = Poorly cleaved under standard assay conditions. A ten-fold excess of enzyme (8.9 nM) did cleave the substrate (36 μM), but at an extremely slow rate (4.6 * 10^{-6} AU/s) when compared to the cleavage of K-P-A-K-F-X-R-L (38 μM, 17 * 10^{-6} AU/s, 0.76 nM enzyme).

Table 4. Kinetic parameters of substrates with multiple substitution

P_5	P_4	P_3	P_2	P_1	$P_1{}'$	$P_2{}'$	$P_3{}'$	k_{cat} (sec^{-1})	K_m (μM)	k_{cat}/K_m (μM^{-1}s^{-1})
K	P	I	E	F	X	R	L	21.8 ± 1.9	5.7 ± 1.6	3.84 ± 1.16
K	P	N	Q	F	X	R	L	45.1 ± 4.2	10.8 ± 0.9	4.18 ± 0.53
K	P	V	S	Y	X	R	L	13.9 ± 2.0	10.2 ± 4.3	1.36 ± 0.60
P	P	T	I	F	X	R	L	53.5 ± 5.2	14.4 ± 2.8	3.73 ± 0.81
K	P	T	V	F	X	R	L	13.0 ± 2.5	22.9 ± 9.5	0.57 ± 0.26

X = *p*-Nitrophenylalanine

The observations from the substrates with multiple substitutions (Table 4) are harder to compare, but do point to some important interactions. The best specificity (highest values of k_{cat}/K_m) for all of the substrates in this study is seen when the P_3 and P_2 positions do not contain alanine or lysine, respectively. One perhaps suprising observation was the facile cleavage of the peptide containing two conformation restraining prolines in P_4 and P_5. This evidence lends additional support for the unimportance of the P_5 position in influencing substrate binding and cleavage. This rapid hydrolysis may, however, be due to the overriding positive effects caused by the changes in P_2 and P_3. If proline in P_5 does not make any contribution to binding and subsequent cleavage, the Ile in P_2 may be preferred over Valine causing a four fold increase in k_{cat}.

From this preliminary study, the interactions of substrates seen in the active site of rhizopuspepsin corroborate trends already seen in pepsin by Rich and Bernatowicz (1982), Dunn *et al.*, (1986 and 1987) in mammalian and fungal enzymes, as well as by Hofman (1988) in penicillopepsin. The two most critical subsites appear to be S_2 and S_3. The binding of the substrates in this study are relatively the same while the efficiency of the enzyme of turnover of the substrate varies greatly with k_{cat} values ranging from 2.7 to 53.5 sec^{-1}. Further systematic substitutions in the P_2 and P_3 positions will hopefully yield a better understanding of the possible electrostatic interactions in S_2 subsite and the size and hydrophobicity requirements of S_3 subsite. Finally, additional experiments are underway to explore interactions on the "prime" or C-terminal side of the active site cleft. Our aim is to correlate these observations with the high-resolution crystallography.

REFERENCES

Chen, Z., Koelsch, G., Han, H., Wang, X., Lin. X., Hartsuck, J. & Tang, J., 1991, Recombinant rhizopuspepsinogen; expression, purification, and activation properties of recombinant rhizopuspepsinogens, *J. Biol. Chem.* in press.

Dunn, B. M., Jimenez, M., Parten, B. F., Valler, M. J., Rolph, C. E. & Kay, J., 1986, A systematic series of synthetic chromophoric substrates for aspartic proteinases, *Biochem. J.* 237:899-906.

Dunn, B. M., Valler, M. J., Rolph, C. E., Foundling, S. I., Jimenez, M. & Kay, K., 1987, The pH dependence of the hydrolysis of chromogenic substrates of the type, Lys-Pro-Xaa-Yaa-Phe-

(NO$_2$)Phe-Arg-Leu, by selected aspartic proteinases: evidence for specific interactions in subsites S$_3$ and S$_2$, *Biochim. Biophys. Acta* **913**:122-130.

Henderson, P., 1972, A linear equation that describes the steady-state kinetics of enzymes and subcellular particles interacting with tightly bound inhibitors, *Biochem. J.* **127**:321-333.

Hofman, T., Allen, B., Bendiner, M., Blum, M. & Cunningham, A., 1988, *Biochemistry* **27**:1140-1146.

Meyer, S. L., 1975, "Data Analysis for Scientists and Engineers", John Wiley and Sons, Inc., New York.

Parris, K. D. , Hoover, D. J. & Davies, D. R., 1991, Crystal structures of rhizopuspepsin/inhibitor complexes, this volume.

Rich, D. & Bernatowicz, M., 1982, Synthesis of analogues of the carboxyl proteinase inhibitor pepstatin. Effect of structure in subsite P$_3$ on the inhibition of pepsin, *J. Med. Chem.* **25**:791-795.

Sawyer, T. K., Tomasselli, A. G., Poorman, R. A., Hui, J. O., Hinzmann, J., Staples, D. J., Maggiora, L. L., Smith, C. W. & Heinrikson, R., 1990, Design, structure-activity and specificity of highly potent P$_1$-P$_1$'-modified pseudopeptidyl inhibitors of HIV-1 aspartyl protease, *in* : "Peptides: Chemistry, Structure and Biology, Proceedings of the Eleventh American Peptide Symposium", J. E. River and G. R. Marshall, eds., ESCOM Science Publishers B.V., Leiden, The Netherlands.

LOCALIZATION OF CATHEPSIN D IN ENDOSOMES: CHARACTERIZATION

AND BIOLOGICAL IMPORTANCE

Janice S. Blum,[1] Maria L. Fiani[2] and Philip D. Stahl[1]

[1]Department of Cell Biology and Physiology
Washington University School of Medicine
Saint Louis, Missouri   63110

[2]Laboratorio di Biologia Cellular
Istituto Superiore di Sanita
Rome, Italy   00161

INTRODUCTION

Endosomal Proteolysis

Proteases were initially identified in endosomes through studies of receptor-ligand transport.[1] During receptor-mediated endocytosis, cell surface receptors bind exogenous ligands (Figure 1). These receptor-ligand complexes are internalized by clathrin-coated vesicles, which give rise to endosomes. Shortly after endosome formation, the internal pH of these vesicles drops to between pH 5 - 6.[2] Many internalized receptor-ligand complexes dissociate upon endosome acidification, with the released receptors recycling back to the cell surface. Ligands delivered into endosomes undergo a variety of fates including transport back to the cell surface[3] or sorting to different intracellular compartments such as lysosomes and the Golgi.[4,5] Susceptible protein ligands are cleaved in endosomes indicating that these vesicles also serve as a processing compartment.[1,6-8]

In macrophages, comparing the transport and accumulation of two distinct ligands in endosomes provided the first insight that proteases are active in these vesicles.[5] Both ß-glucuronidase and bovine serum albumin derivatized with mannose residues (mannose BSA) are ligands recognized by the macrophage specific mannose receptor.[9] Each of these ligands are delivered via this receptor into the endocytic pathway. ß-Glucuronidase is a lysosomal enzyme and naturally resistant to low pH and proteases. Thus, this ligand is internalized by cells and accumulates intracellularly. In contrast, mannose BSA is taken up by macrophages and rapidly proteolyzed.[5] Proteolytic fragments of this radiolabeled ligand are detected as early as 2 - 5 min within cells. Cleavage fragments of mannose BSA are released from cells after 5 - 10 min and can be detected in the medium as trichloroacetic acid soluble material (Figure 2). Pretreatment of cells with pepstatin A, an inhibitor of aspartic proteases and leupeptin, an inhibitor of cysteine proteases diminishes mannose BSA proteolysis and

Structure and Function of the Aspartic Proteinases
Edited by B.M. Dunn, Plenum Press, New York, 1991

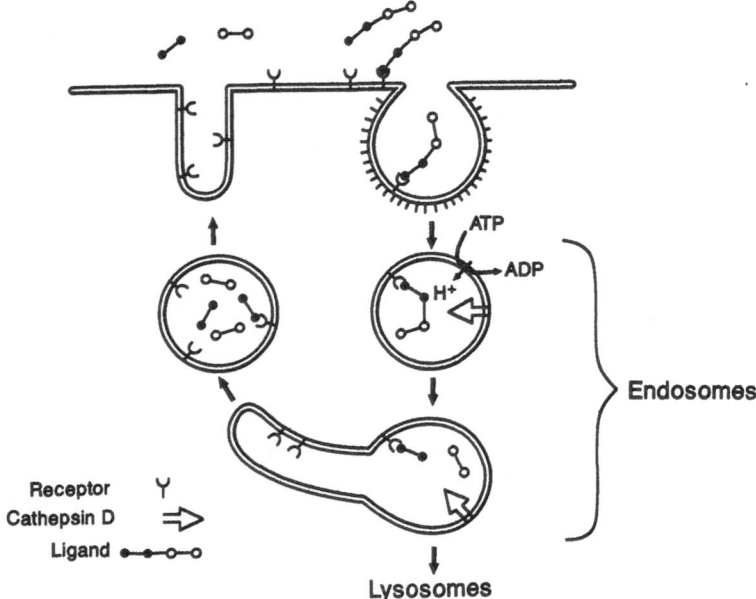

Figure 1. Ligand transport into proteolytic endosomes. Receptor-ligand complexes are formed at the cell surface and internalized via coated pits. These coated pits pinch off to form coated vesicles which subsequently shed their clathrin coats to become endosomes. Internalized ligands are released from receptors as endosomes acidify due to the action of proton pumping ATPases. The released receptors cycle back to the cell surface to be reutilized in endocytosis. Susceptible ligands are cleaved in endosomes with proteolytic fragments being retained or released from the cell.

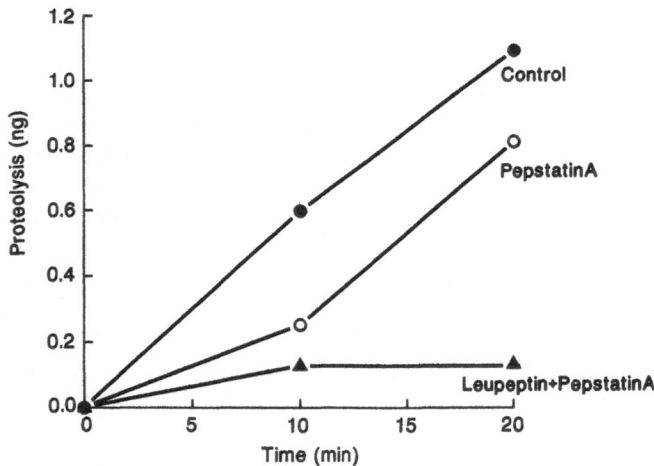

Figure 2. Proteolysis of mannose BSA in J774 macrophages. [^{125}I]-radiolabeled mannose BSA is internalized by cells and proteolyzed following its transport into endosomes. Fragments of [^{125}I]-mannose BSA are released from cells into the medium. Proteolysis was monitored by following the release of acid soluble fragments of ligand into the medium. Pretreatment of cells with pepstatin A (0.5 mg/ml) and leupeptin (0.5 mM) diminished proteolysis and resulted in an intracellular accumulation of [^{125}I]-mannose BSA.

release. Cleavage of mannose BSA is therefore catalyzed by both aspartic and cysteine proteases.

These results demonstrate that mannose BSA is rapidly delivered into an intracellular compartment containing proteases. Subcellular fractionation studies using macrophages show that ligands are transported from the plasma membrane into endosomes after 2 - 5 min.[10] These studies confirm that ß-glucuronidase initially accumulates in endosomes and is subsequently sorted as an intact molecule to lysosomes. Mannose BSA is delivered into endosomes and rapidly proteolyzed with little of this ligand being detected in lysosomes.[5] Inhibition of mannose BSA cleavage with pepstatin A, results in ligand accumulation in both endosomes and lysosomes. Thus, preventing the initial cleavage of mannose BSA in endosomes permits its delivery and accumulation in lysosomes.

To confirm that endosomal vesicles were indeed capable of ligand proteolysis, endosomes containing mannose BSA were isolated and incubated *in vitro*. Proteolysis of mannose BSA in isolated endosomes is stimulated by ATP, which drives vesicle acidification. Ionophores such as nigericin or weak bases like NH_4Cl block proteolysis by dissipating the intravesicular pH gradient (Table 1). In the absence of ATP, degradation of mannose BSA is observed when vesicles are incubated in acidic medium below pH 6. Preliminary experiments using other ligands indicate that proteolysis is observed at neutral pH using endosomal vesicles.[11] For example, proteolysis of the toxin ricin A chain in endosomes is observed over a broad pH range. Thus, proteolysis of this ligand may occur in very early endosomes prior to a significant drop in vesicle pH. Differences in the pH optima observed for proteolysis may reflect differences in ligand susceptibility or the presence of several different proteases in endosomes. Using isolated endosomes we have confirmed that both aspartic and cysteine proteases are active in macrophage endosomes (Table 2). Endosomal vesicles gradually acidify as they are transported through cells so that proteases with different pH optima may function in these vesicles. Aspartic protease activity in endosomes is observed predominately at acidic pH while cysteine proteases are active over a broader pH range. Changes in endosomal pH may provide a means of regulating the action of different proteases and their cleavage specificity.

IDENTIFICATION OF CATHEPSIN D IN ENDOSOMES

The protease cathepsin D is active in macrophage endosomes and lysosomes. Cathepsin D is found as a 46 kDa protein in both rabbit and mouse macrophage endosomes.[5,8] The contents of endosomal vesicles can be radiolabeled by incubating endosomes containing lactoperoxidase with ^{125}I and peroxide.[5] When endosomal proteins are labeled by this method and passed over a pepstatin agarose affinity column, a single 46 kDa protease is isolated. This macrophage protease could be immunoprecipitated only with antibodies recognizing cathepsin D. Amino terminal sequence analysis also confirmed that this endosomal protease is cathepsin D.

Table 1. Requirements for Endosomal Proteolysis

Treatment	Control	Nigericin	NH_4Cl	-ATP
% Proteolysis	100	32.0	43.0	58.7

Endosomal vesicles containing radiolabeled mannose BSA were isolated and incubated *in vitro* in isotonic buffer pH 7.2 with an ATP regenerating system.[5] Nigericin (10 µM) or NH_4Cl (10 mM) inhibit ligand cleavage. The dependence on ATP reflects a need for endosome acidification during the proteolysis of mannose BSA.

Table 2. Identification of proteinases responsible for ligand cleavage in endosomes

Addition	% Inhibition	Specificity
Pepstatin A (145 µM)	54.9	aspartic proteases
Leupeptin (500 µM)	52.9	cysteine proteases
o-Phenanthroline (1 mM)	0.0	metalloproteases
aprotinin (1 unit/ml)	1.1	serine proteases

Isolated endosomes containing radiolabeled mannose BSA were incubated *in vitro* in isotonic buffer at pH 4.5 at 37°C. Protease inhibitors were added to the assay and the cleavage of mannose BSA monitored.

Cathepsin D is synthesized as a 53 kDa inactive precursor protein and converted to the active 46 kDa protease following transport through Golgi compartments.[12] In macrophages a significant proportion of the newly synthesized cathepsin D is tightly associated with membranes. Approximately half of the 53 kDa precursor is present in a soluble form and is secreted directly into the medium without further processing. The membrane bound 53 kDa protease is converted to a soluble form intracellularly following its transport beyond the Golgi into endosomes.[13] This conversion temporally follows the processing of cathepsin D to its 46 kDa active form (Figure 3). The transformation from membrane bound to soluble protease is slowed in cells treated with monensin, an ionophore which disrupts protein transport beyond the Golgi. The association of cathepsin D with membranes could be important in sorting or retaining this protease in endosomes. Lysosomal cathepsin D, also a 46 kDa protease in rodents, is found as a soluble enzyme.[13] Endosomal cathepsin D may be a precursor to the lysosomal protease, with release of membrane bound protease serving as a signal for its sorting or transport to lysosomes. The molecular mechanisms responsible for protease association with membranes have not been definitively identified. Mannose-6-phosphate residues are present on the N-linked oligosaccharides of cathepsin D and promote protease binding to mannose-6-phosphate receptors.[14] These receptors are responsible for shuttling lysosomal enzymes from the Golgi to prelysosomal

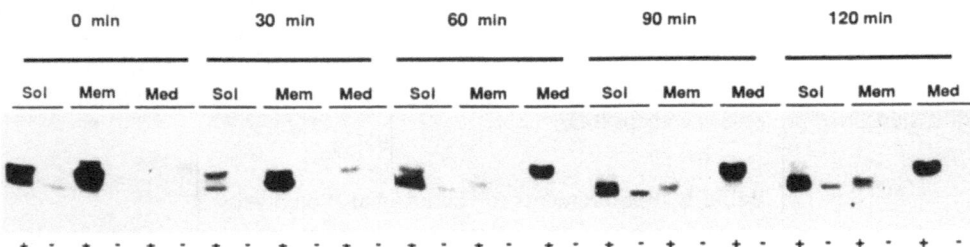

Figure 3. Association of cathepsin D with membranes during protease biosynthesis. Mouse peritoneal macrophages were pulse radiolabeled with ^{35}S-methionine for 15 min then chased for 0, 30, 60, 90 or 120 min in the absence of radiolabel. Cell lysates were prepared at each time point and fractionated into soluble (sol) and membrane (mem) samples.[13] Aliquots from soluble and membrane fractions as well as the medium (med) were immunoprecipitated with an antibody specific for cathepsin D (+) or control serum (-) and then analyzed by gel electrophoresis. The protease is initially synthesized as a 53 kDa precursor present in both a soluble and membrane bound form. The soluble precursor is secreted by cells into the medium after 30 min. Membrane-associated cathepsin D is trimmed to a 46 kDa protease which shifts to soluble fractions during this processing over the course of 60 - 90 min.

vesicles or late endosomes.[15] However, cathepsin D binding to mannose-6-phosphate receptors is not entirely responsible for the association of this protease with membranes. Conditions which release lysosomal enzymes from the receptor such as exposure to low pH and mannose-6-phosphate do not trigger cathepsin D release from membranes. Treating fibroblasts with weak bases causes secretion of intracellular proteases as they are dissociated from the mannose-6-phosphate receptor. However a similar treatment does not affect protease retention to membranes, again indicating that the membane association of the protease is independent of the mannose-6-phosphate receptor. Studies in human macrophages using weak bases suggest there is an alternate mechanism aside from the mannose-6-phosphate receptor for protease sorting to lysosomes in these cells.[16] Protease association with membranes may be related to this alternate sorting or transport process. In agreement with this hypothesis, we have observed that the cysteine protease, cathepsin L is also transiently associated with membranes.

The localization of cathepsin D in endosomes has been investigated using immunocytochemistry at the level of electron microscopy.[17] Endosomes are identified by incubating macrophages with mannose BSA coated with 20 nm gold for 5 min (Figure 4). Cathepsin D is detected in these vesicles using an antibody specific for this protease tagged with 10 nm gold. The protease is concentrated in lysosomes, however a significant proportion is found in endosomes and small vesicles in the periphery of cells. Similar small vesicles have also been identified in B-cells and were found to contain both cathepsin D and B.[18] These small vesicles may serve to shuttle proteases from the Golgi to endosomes or from endosome to endosome. Immunoreactive cathepsin D is not detected on the plasma membrane of macrophages, nor could protease activity be detected on the cell surface.

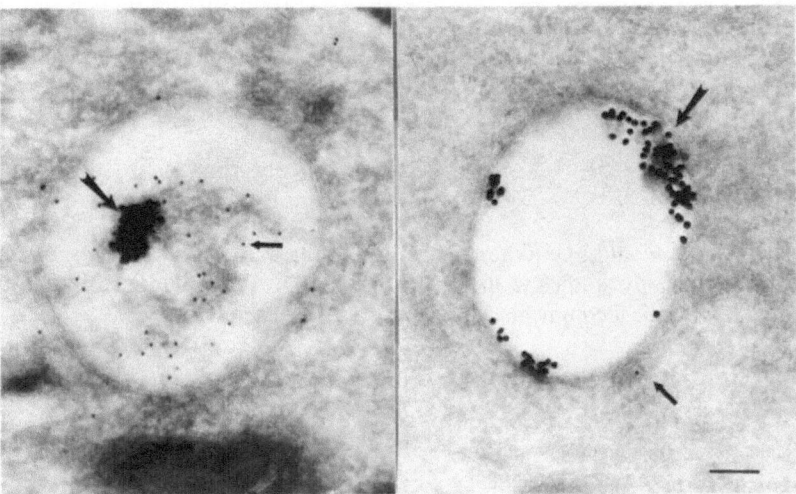

Figure 4. Immunolocalization of cathepsin D in endosomes. Mouse J774 macrophages were incubated for 5 min with mannose BSA conjugated with 20 nm gold particles. During this incubation the mannose BSA is delivered into endosomes. The cells were then fixed, cryosectioned and stained with an antibody recognizing cathepsin D which was tagged with 10 nm protein A gold. Cathepsin D (10 nm gold, small arrow) is shown here in endosomes with mannose BSA (20 nm gold, large arrow). The protease was also observed in small vesicles located near endosomes or throughout the cytoplasm. The magnification is indicated with 1 cm representing 100 nm.

FUNCTION OF ENDOSOMAL CATHEPSIN D

The biological importance of endosomal proteases is suggested by the cleavage of a number of specific proteins within these vesicles. Cathepsin D has been implicated in the endosomal proteolysis of hormones, toxins and immunologically important antigens. Macrophages internalize bovine parathyroid hormone and cleave this protein to its bioactive fragment in endosomes.[19] Proteolysis of this hormone occurs in low pH endosomes and is catalyzed by cathepsin D. Active fragments of parathyroid hormone generated in endosomes are subsequently released from cells and may function in bone resorption and calcium homeostatis. The toxic component of ricin, A chain, is also proteolyzed in macrophage endosomes by cathepsin D.[11] Ricin A chain is transported from endosomes into the cytoplasm where it inactivates protein synthesis and causes cell death. Inhibition of toxin cleavage with protease inhibitors diminishes ricin A chain cytotoxicity. Endosomal proteolysis catalyzed by cathepsin D may be required for toxin transport into the cytoplasm or binding to ribosomes. Understanding the mechanism of toxin transport, and the role of endosomal proteases in this process, may provide a novel strategy for drug delivery into cells. Finally, endosomal proteases such as cathepsin D appear to play an important role in the processing of foreign antigens for immune recognition. Macrophages and B cells internalize and proteolytically process antigens in endosomes.[20,21] These fragments of antigen bind class II histocompatibility proteins and are displayed on the cell surface for recognition by T cells. Antigen processing is inhibited by compounds which block both cysteine and aspartic proteases.[22,23] Studies to correlate cathepsin D cleavage specificity with known peptide antigens further suggest a role for this protease in antigen processing.[24] Cathepsin D has been co-localized in endosomes with class II histocompatibility proteins, and these vesicles may represent the compartment where antigen is processed and where the resulting peptides bind class II proteins.[18]

The identification of proteases such as cathepsin D in endosomes provides new insights into the complex function of these intracellular vesicles. Future studies must address the molecular mechanisms responsible for targetting and retaining these proteases in endosomes. Such processes may also provide important insight into the evolution of protease structure. Whether endosomal proteases are present at high levels in all cells also remains to be determined. The role of these proteases in hormone and antigen processing may indicate that the expression of individual endosomal proteases will vary with cell function.

ACKNOWLEDGEMENTS

The authors would like to acknowledge and thank Marilyn A. Levy for conducting the electron microscopy studies to localize cathepsin D. We would also like to thank Dr. Micheal Koval for critical comments concerning this manuscript.

RERENCES

1. S. Diment and P. Stahl, Macrophage endosomes contain proteases which degrade endocytosed protein ligands, *J. Biol. Chem.* **260**:15311 (1985).
2. B. Tycko and F. R. Maxfield, Rapid acidification of endocytic vesicles containing α_2-macroglobulin, *Cell* **28**:643 (1982).
3. C. Tietze, P. Schelesinger and P. Stahl, Mannose-specific endocytosis receptor of alveolar macrophages: Demonstration of two functionally distinct intracellular pools of receptor and their role in receptor recycling, *J. Cell Biol.* **92**:417 (1982).

4. J. L. Goldstein, R. G. W. Anderson and M. S. Brown, Coated pits, coated vesicles and receptor-mediated endocytosis, *Nature* **279**:679 (1979).

5. M. D. Snider and O. C. Rogers, Intracellular movement of cell surface receptors after endocytosis: Resialylation of asialo-transferrin receptor in human erythroleukemia cells, *J. Cell Biol.* **100**:826 (1985).

6. M. Roederer, R. Bowser and R. F. Murphy, Kinetics and temperature dependence of exposure of endocytosed material to proteolytic enzymes of low pH: Evidence for a maturation model for the formation of lysosomes, *J. Cell Physiol.* **131**:200 (1987).

7. F. G. Hamel, B. I. Posner, J. J. M. Bergeron, B. H. Frank and W. C. Duckworth, Isolation of insulin degradation products from endosomes derived from intact rat liver, *J. Biol. Chem.* **263**:6703 (1988).

8. J. S. Blum, R. Diaz, S. Diment, M. Fiani, L. Mayorga, J. S. Rodman, P. D. Stahl, Proteolytic processing in endosomal vesicles, *Cold Spring Harbor Symposia on Quantitative Biology* **54**:287 (1989).

9. P. Stahl, P. Schlesinger, E. Sigardson, J. S. Rodman and Y. C. Lee, Receptor-mediated pinocytosis of mannose glycoconjugates by macrophages: Characterization and evidence for receptor recycling, *Cell* **19**:207 (1980).

10. T. Wileman, R. L. Boshans, P. Schlesinger and P. Stahl, Monensin inhibits recycling of macrophages and ligand delivery to lysosomes, *Biochem. J.* **220**:665 (1984).

11. J. S. Blum, M. L. Fiani and P. D. Stahl, Characterization of neutral and acidic proteases in endosomal vesicles, *J. Cell Biol.* **109**:188a (1989).

12. A. Hasilik and E. F. Neufeld, Biosynthesis of lysosomal enzymes in fibroblasts, *J. Biol. Chem.* **255**:4937 (1980).

13. S. Diment, M. Leech and P. Stahl, Cathepsin D is membrane-associated in macrophage endosomes, *J. Biol. Chem.* **263**:6901 (1988).

14. S. Kornfeld and I. Mellman, The biogenesis of lysosomes, *Annu. Rev. Cell Biol.* **5**:483 (1989).

15. G. Griffiths, B. Hoflack, K. Simons, I. Mellman and S. Kornfeld, The mannose-6-phosphate receptor and the biogenesis of lysosomes, *Cell* **52**:329 (1988).

16. T. Braulke, H. J. Geuze, J. W. Slot, A. Hasilik and K. von Figura, On the effects of weak bases and monensin on sorting and processing of lysosomal enzymes in human cells, *Eur. J. Cell Biol.* **43**:316 (1987).

17. J. S. Rodman, M. A. Levy, S. Diment and P. D. Stahl, Immunolocalization of endosomal cathepsin D in rabbit aveolar macrophages, *J. Leuk. Biol.* **48**:116 (1990).

18. L. E. Guagliardi, B. Koppelman, J. S. Blum, M. S. Marks, P. Cresswell and F. M. Brodsky, Co-localization of molecules involved in antigen processing and presentation in an early endocytic compartment, *Nature* **343**:133 (1990).

19. S. Diment, K. Martin and P. Stahl, Cleavage of parathyroid hormone in macrophage endosomes illustrates a novel pathway for intracellular processing of proteins, *J. Biol. Chem.* **264**:13403 (1989).

20. K. L. McCoy and R. H. Schwartz, The role of intracellular acidification in antigen processing, *Immunol. Rev.* **106**:129 (1988).

21. E. R. Unanue and P. M. Allen, The basis for the immunoregulatory role of macrophages and other accessory cells, *Science* **236**:551 (1987).

22. J. Puri and Y. Factorovich, Selective inhibition of antigen presentation to cloned T cells by protease inhibitors, *J. Immunol.* **141**:3313 (1988).

23. H. Takahashi, K. B. Clase and J. A. Berzofsky, Identification of proteases that process distinct epitopes on the protein, *J. Immunol.* **142**:2221 (1989).

24. J. M. van Noort and A. C. M. van der Drift, The selectivity of cathepsin D suggests an involvement of the enzyme in the generation of T-cell epitopes, *J. Biol. Chem.* **264**:14159 (1989).

PROTEOLYTIC ACTIVATION OF HUMAN PROCATHEPSIN D

Gary Richo and Gregory E. Conner

Department of Cell Biology and Anatomy
University of Miami School of Medicine
Miami, Florida

INTRODUCTION

The occurrence of multiple proteolytic processing steps is typical of lysosomal enzyme biosynthesis, but not characteristic of the other aspartyl proteases. Cathepsin D, a lysosomal aspartyl protease, is proteolytically processed several times during biosynthesis. A summary of these processing steps is shown in Figure 1. Although the activity and structure of cathepsin D has been studied in great detail[1,2,3,4] and the activation of procathepsin D is known to be dependent upon this proteolytic processing, the proteases and the exact cleavage sites which the proteases use to accomplish the processing of cathepsin D are, for the most part, unknown. The activation of the aspartyl protease proenzymes pepsinogen and prorenin has been explored in depth.[4,5] In this article we have summarized our recent studies on the proteolytic activation of human procathepsin D.

ISOLATION OF PROCATHEPSIN D

After co-translational removal of the presequence, procathepsin D (392 aa) is rapidly converted to single chain cathepsin D (366 aa) during transport to or immediately following arrival in the lysosome. Due to the rapid processing, procathepsin D does not accumulate inside cells. In order to study the activation of procathepsin D, it is necessary to purify the proenzyme from activated forms. We have accomplished this by two separate methods. First, pepstatin affinity chromatography was used to isolate procathepsin D synthesized and secreted by cultured cells. Expression of human procathepsin D in bacteria was the second method employed to isolate the proenzyme.

Pepstatin Affinity Chromatography

Tang and coworkers have shown that pepsinogen and pepsin can be differentiated by their pH dependent affinity for pepstatin.[6] We have demonstrated that procathepsin D and cathepsin D can also be separated by their differential affinity to pepstatin columns.[7]

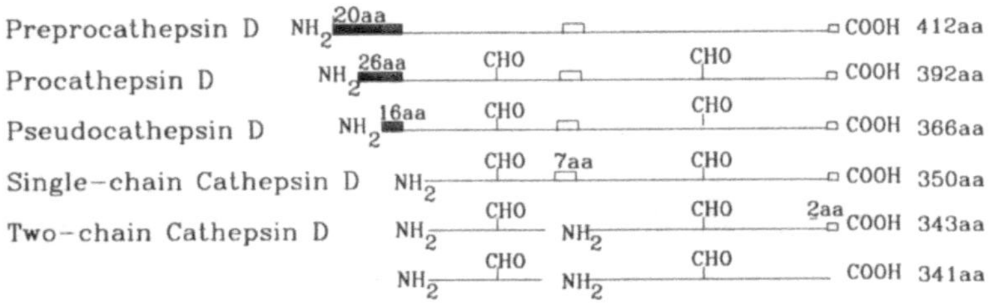

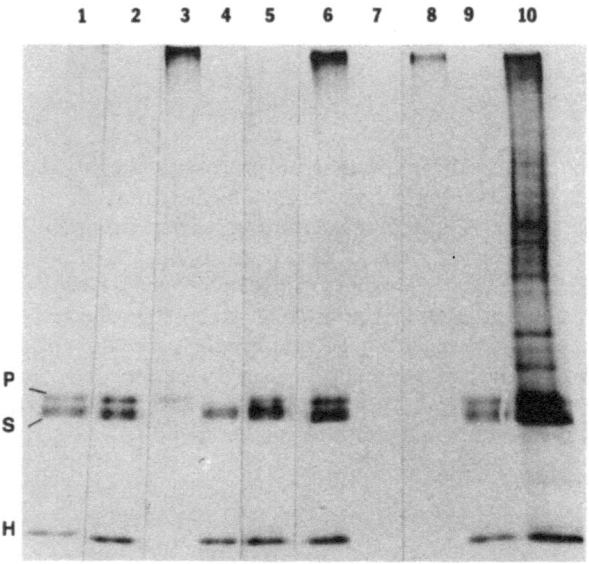

Figure 1. A summary of the processing steps during the biosynthesis of human cathepsin D. CHO refers to asparagine-linked carbohydrate. Boxed areas indicate amino acid residues removed during processing.

Figure 2. Pepstatin binding characteristics of procathepsin D and mature cathepsin D. Cultured porcine (PK15) cells were grown in [^{35}S]-methionine so that all forms of the enzyme would be labeled. After detergent extraction, equivalent volumes of the extract were either incubated at 4°C for the duration of the experiment in buffers of different pH or were mixed with pepstatinyl-agarose beads at 4°C in buffers at different pH in the presence or absence of soluble pepstatin. Beads were washed extensively in the binding buffers and eluted with pH 8.3 buffer. Samples were then immunoprecipitated, analyzed on a polyacrylamide SDS gel, and fluorographed. Lane 1, no treatment before immunoprecipitation; Lanes 2 and 5, incubation at pH 5.3 and 3.5 respectively for the duration of the experiment and not applied to pepstatinyl-agarose columns to control for endogenous proteolysis. Lanes 3 and 4, unbound and bound fractions of extracts applied to pepstatinyl-agarose in pH 5.3 buffer, respectively. Lanes 6 and 7, unbound and bound fractions respectively, of extracts preincubated in 0.1 M soluble pepstatin in pH 3.5 buffer before addition of pepstatinyl-agarose. Lanes 8 and 9 are unbound and bound fractions of extracts applied to pepstatinyl-agarose at pH 3.5. Lane 10 is a non-immunoprecipitated sample equivalent to lane 9. P indicates procathepsin D, S indicates single chain cathepsin D, and H indicates heavy chain cathepsin D. (Taken from reference 8, courtesy of The Biochemical Society, London)

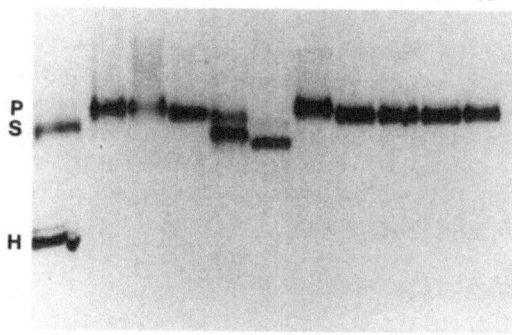

Figure 3. Autocatalytic proteolysis of purified procathepsin D. Culture media, conditioned by cells growing in [³⁵S]-methionine, was subjected to two cycles of pepstatinyl-agarose chromatography. The unbound fraction after the first cycle of chromatography at pH 5.3 was applied to a second column at pH 3.5. Material which bound to the second column was eluted in pH 8.3 buffer, analyzed on a 12 % polyacrylamide SDS gel and fluorographed. Lane 1, cathepsin D isolated by pepstatinyl-agarose chromatography at pH 5.3 from cells which were pulsed and chased so that only mature forms of the enzyme would be labeled. Lane 2, the procathepsin D obtained from the second cycle of pepstatinyl-agarose chromatography without any post column treatment. Lane 3 is procathepsin D incubated in pH 8.3 buffer. Lane 4 is procathepsin D incubated in pH 3.5 buffer. Lanes 5 and 6 are identical to lanes 3 and 4 but received endogylcosidase treatment following incubation. Lanes 7-11 are procathepsin D incubated in pH 3.5 buffer with various protease inhibitors: lane 7, 0.1 μg/ml (1 μM) pepstatin; lane 8, 1 μg/ml chymostatin; lane 9, 1 μg/ml leupeptin; lane 10, 0.04 % DMSO. P indicates procathepsin D, S indicates single chain cathepsin D, and H indicates heavy chain cathepsin D. (Taken from reference 8, courtesy of The Biochemical Society, London)

Following metabolic incorporation of [³⁵S]-methionine by porcine PK15 cells and extraction of cultures with nonionic detergent, procathepsin D can be separated from activated forms of cathepsin D by applying the extract to pepstatinyl agarose at pH 5.3. Activated forms of cathepsin D bound to the immobilized pepstatin while procathepsin D flowed through the column. Procathepsin D was then purified and concentrated by reapplication to another pepstatin affinity column at pH 3.5 followed by elution at slightly alkaline pH (Figure 2). The specificity of this separation was demonstrated by the addition of soluble pepstatin to extracts before application to the affinity support.

Cultured cells secrete small amounts of lysosomal enzymes into the culture media. In most cases, including cathepsin D, the majority of the secreted enzyme is unprocessed proenzyme. We have taken advantage of this property of cultured cells to enrich for procathepsin D in the starting material used for pepstatin affinity columns. Using media from cell cultures which have grown in the presence of [35S]-methionine, we have purified procathepsin D by the two step pepstatin affinity chromatography process described above. This highly purified procathepsin D was then used to examine the ability of the proenzyme to autocatalytically activate at acid pH in a fashion similar to pepsinogen autoactivation.

Autoactivation of Purified Procathepsin D

After incubation of procathepsin D at pH 3.5 for 30 min at 37°C, a change in apparent size can be detected by SDS gel electrophoresis (Figure 3).[7,8] This change is inhibited by the addition of pepstatin but not by the addition of other protease inhibitors. Endoglycosidase H treatment demonstrated that the change in mobility on SDS gels was not due to processing of carbohydrate. Thus it appears that procathepsin D undergoes autoproteolysis *in vitro*. A change in activity and pepstatin binding affinity accompanied the change in mobility.

The processed enzyme was not identical in size to the single chain enzyme isolated from cells. It also did not correlate to known intermediates which could be detected during radioactive amino acid pulse-chase studies of cultured cells. It is possible that this *in vitro* intermediate corresponds to the pseudopepsin and pseudochymosin intermediates found during activation of pepsinogen and prochymosin. Conditions which might generate a fully mature amino-terminus could not be explored because of the exceedingly small quantities of procathepsin D which are isolated from the cell cultures and media. It is also possible that the presence of contaminating but catalytic amounts of activated cathepsin D in the purified proenzyme was responsible for this cleavage.

Bacterial Expression of Human Procathepsin D

To increase the amount of procathepsin D available for study, human fibroblast procathepsin D cDNA was transferred to the bacterial plasmid pET3a[9] and expressed in *E. coli*.[10] A methionine codon for initiation of protein synthesis was appended to the region encoding procathepsin D. After induction of transcription by the *T7* RNA polymerase, large inclusion bodies were evident in bacteria containing the procathepsin D plasmid but not in bacteria containing a control plasmid (Figure 4). Analysis of isolated inclusion bodies by SDS gel electrophoresis and Western blots demonstrated that they contained a protein of the expected size for non-glycosylated procathepsin D. This protein reacted with anti-porcine cathepsin D antiserum. After solubilization in urea and DEAE chromatography, the amino-terminal amino acid sequence of the purified protein was determined to be that of methionyl-procathepsin D. As synthesized in the bacteria, the procathepsin D was not active using either the hemoglobin assay[11] or a chromogenic peptide assay.[12]

Refolding of Bacterial Procathepsin D

Procathepsin D inclusion bodies were solubilized in 7 M urea, 50 mM mercaptoethanol, and 50 mM CAPS buffer at pH 10.7. Soluble material was diluted into 100 volumes of water. After adjusting the pH to 8.7 with HCl, the protein was held at room temperature for 2 hours. Following acidification to pH 3.5, no activity could be detected using a peptide substrate,[12] Lys-Pro-Ile-Glu-Phe-NO$_2$Phe-Arg-Leu. Proteolytic activity could be detected after longer incubations at room temperature in acid pH. Figure 5 shows the time dependent increase in cathepsin D activity during incubation in 20 mM Na formate

pET3a   pTCPSD

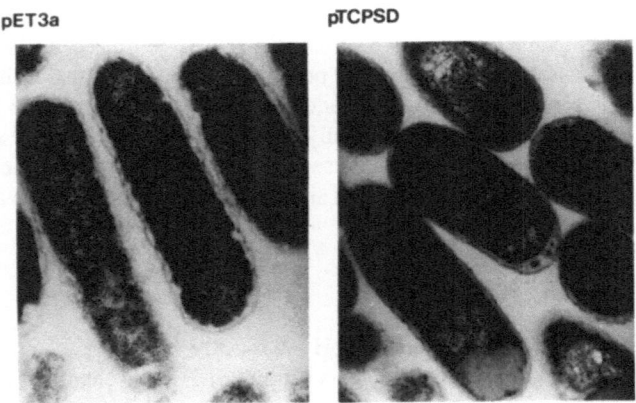

Figure 4. Procathepsin D expressed in bacteria. Transmission electron micrographs of cells that contain the expression vector without insert (pET3a) or with the procathepsin D cDNA (pTCPSD1). (Taken from reference 10, courtesy of Mary Ann Liebert, Inc.)

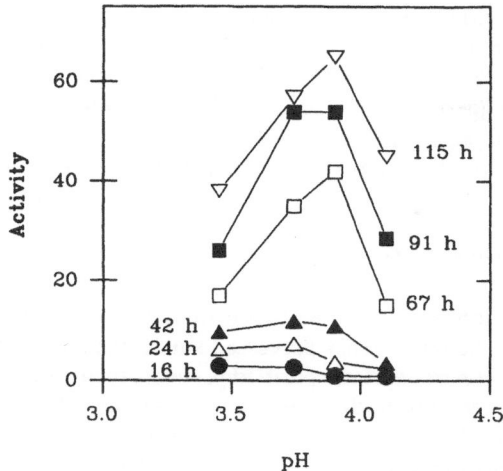

Figure 5. Effect of pH on the formation of cathepsin D activity. After refolding urea-solubilized cathepsin D as described in the text, enzyme was incubated in 20 mM Na formate buffers of different pH values. Proteolytic activity was measured at the indicated times using a peptide substrate assay.[10,12] Activity is expressed as the change in absorbance units at 300 nm x 10^{-6} per sec.

buffers of different pH values. Since activation of procathepsin D was shown to occur rapidly using proenzyme isolated from culture media, the slow appearance of activity after acidification was a surprising result.

Addition of glutathione in different ratios of oxidized and reduced forms to the procathepsin D after dilution from urea did not abrogate the lag in formation of activity. Addition of glutathione did increase the rate of appearance of activity in the acidified mixtures (Figure 6). This glutathione dependence suggested that the slow formation of activity was due to an acid dependent slow refolding reaction.

Characterization and Purification of Bacterial Cathepsin D

The existence of an acid-dependent refolding reaction was confirmed by application of the active renaturation mixtures to a pepstatin affinity column. Since procathepsin D isolated from cell cultures was shown to bind to pepstatin, any correctly folded proenzyme in the acidified renaturation mixture should also bind to the pepstatinyl agarose. No proenzyme was bound and eluted from the pepstatin column in a specific fashion. A processed form of cathepsin D in the renaturation mixture did however bind specifically to the affinity column. Apparently procathepsin D undergoes autocatalytic proteolysis immediately after it folds to the correct conformation. It is possible that the high concentration of denatured or incorrectly folded procathepsin D in the mixtures competes strongly for the active site of the renatured proenzyme. If this is the case, correctly folded but unprocessed enzyme would not be detected by the pepstatin affinity technique. Another purification method which is not dependent on binding to the active site must be used to rule out this possibility.

Amino acid sequence analysis of the amino-terminus of this molecule demonstrated that cleavage had occurred after Leu_{26p}. Thus, this molecule probably corresponds to the intermediate found after *in vitro* activation of procathepsin D isolated from cell culture media. The cleavage site found in this partially processed cathepsin D resembles that found in pseudopepsin[13,14] and pseudochymosin[15] and consequently this form has been termed pseudocathepsin D.

293

The pH optima and K_m of pseudocathepsin D and human placental cathepsin D were determined using the peptide substrate[12] and the values for both enzymes were indistinguishable. In addition, the intrinsic fluorescence spectra of the two enzymes were identical. These data suggested that the pseudocathepsin D was correctly folded and that the 16 amino acids of the propeptide remaining on the molecule had little effect on the activity or structure of the molecule when compared to fully mature two-chain enzyme from human placenta.

Significance of Pseudocathepsin D

Longer incubations of the renaturation mixtures resulted in steadily increasing quantities of proteolytic activity. Fully processed single chain enzyme was not detected in these mixtures. Incubation of pepstatin purified pseudocathepsin at different pH values did not result in further processing to the single chain enzyme. The *in vivo* significance of pseudocathepsin D is not clear at this time. Preliminary experiments using protease inhibitors suggested that activation of procathepsin D occurs through the activity of a cysteinyl protease rather than an aspartic protease. Using mouse L cells and leupeptin, intermediates in activation could be detected on gels which could possibly correspond to pseudocathepsin D. Taken together the preliminary data suggested that activation may occur in one step by a cysteinyl protease or via a two step processing pathway in which procathepsin D is first converted autocatalytically to pseudocathepsin D and then to the single chain enzyme by the action of a different protease.

The various pathways by which procathepsin D can be activated is important for several reasons. First the conversion of procathepsin D to single-chain enzyme has been assumed to be autocatalytic[7,8] and to coincide with arrival of procathepsin D in an acidic compartment. The possibility that processing can occur by mechanisms other than autocatalysis suggests that this may not be a valid assumption. Second, some breast tumors secrete large quantities of procathepsin D. Previously it has not been apparent how this procathepsin D could participate in invasion of the tumors. Activation of procathepsin D by other extracellular proteases could allow participation of cathepsin D in proteolysis of extracellular matrix in local regions of low pH.

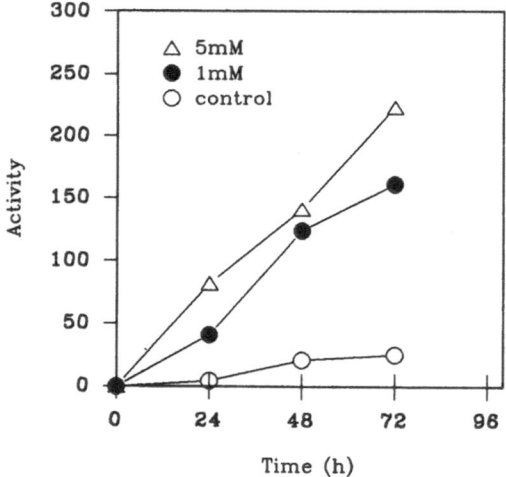

Figure 6. Effect of Glutathione on the formation of cathepsin D activity. After refolding urea-solubilized cathepsin D as described in the text, enzyme was incubated at pH 3.7 in the absence or presence of 1 mM or 5 mM glutathione. Proteolytic activity was measured using a peptide substrate assay[10,12] and activity is expressed as the change in absorbance units at 300 nm x 10^{-6} per sec.

Some aspartyl protease zymogens such as pepsinogen are capable of autoproteolytic activation at two different sites.[16] The structural features which are involved in these cleavages do not appear to exist in prorenin as it is not able to undergo any autocatalytic cleavage.[5] Procathepsin D on the other hand appears to be capable of one of the autoproteolytic cleavages seen during the activation of pepsinogen. Examination of procathepsin D structure and comparison to prorenin and pepsinogen may demonstrate which features are necessary for either or both of these processing sites.

SUMMARY

Procathepsin D is a short-lived inactive precursor of the lysosomal aspartyl protease, cathepsin D. Pulse-chase analysis using radiolabeled amino acids demonstrated the existence of several biosynthetic intermediates during formation of mature cathepsin D (summarized in Figure 1). Procathepsin D is capable of autocatalytic cleavage to pseudocathepsin D. This was demonstrated using small quantities of procathepsin D isolated from cell culture media as well as using a non-glycosylated form of procathepsin D synthesized in a bacterial expression system. Complete conversion to the single-chain cathepsin D appears to require a second enzyme which is inhibited by leupeptin. This conclusion was drawn from the inability to produce single-chain enzyme from either procathepsin D or pseudocathepsin D *in vitro* as well as observations from addition of protease inhibitors to cell cultures. It appears that the conversion of procathepsin D to active single-chain enzyme falls between the paradigms of pepsinogen autoactivation and prorenin conversion by a separate enzyme.

ACKNOWLEDGEMENTS

The authors thank Dr. Ben Dunn and Paula Scarborough for peptide substrate, many helpful discussions, use of their laboratory where refolding experiments were initiated. This work was supported by grants from the NIH, GM35812 to G. C. and BRSG funds to the University of Miami School of Medicine.

REFERENCES

1. A. J. Barrett, in: "Proteinases in Mammalian Cells and Tissues," A. J. Barrett, ed., North Holland Publishing Co., New York, pp. 209-248 (1977).
2. G. E. Conner, G. Blobel and A. H. Erickson, in: "Lysosomes: their role in protein breakdown," H. Glaumann and J. Ballard, eds., Academic Press, London, pp. 151-162 (1987).
3. J. G. Shewale, T. Takahashi and J. Tang, in: "Aspartic Proteinases and Their Inhibitors," V. Kostka, ed., Walter de Gruyter, Berlin, pp. 101-116 (1985).
4. J. Tang and R. N. S. Wong, J. Cell. Biol. 33:53-63 (1987)
5. T. Inagami, K. Misono, J. -J. Chang, Y. Takii and C. Dykes, in: "Aspartic Proteinases and Their Inhibitors," V. Kostka, ed., Walter de Gruyter, Berlin, pp. 319-337 (1985).
6. J. Mariciniszyn, J. S. Huang, J. A. Hartsuck and J. Tang, J. Biol. Chem. 251:7095-7102 (1976).
7. G. E. Conner, Biochem. J. 263:601-604 (1989).
8. A. Hasilik, K. von Figura, E. Conzelmann, H. Nehrkorn and K. Sandhoff, Eur. J. Biochem. 125:317-321 (1982).
9. A. H. Rosenberg, B. N. Lade, D. -S. Chui, S. W. Linl, J. J. Dunn and F. W. Studier, Gene 56:125-135 (1987).
10. G. E. Conner and J. A. Udey, DNA and Cell Biology 9:1-9 (1990).

11. J. S. Huang, S. S. Huang and J. Tang, *J. Biol. Chem.* **254**:11405-11417 (1979).

12. B. M. Dunn, M. Jimenez, B. F. Parten, M. J. Valler, C. E. Rolph and J. Kay, *Biochem. J.* **237**:899-906 (1986).

13. C. W. Dykes and J. Kay, *Biochem J.* **153**:141-144 (1976).

14. K. Asbaek Christensen, V. Barkholt Pedersen and B. Foltmann, *F.E.B.S. Letters* **76**:214-218 (1977).

15. V. Barkholt Pedersen, K. Asbaek Christensen and B. Foltmann, *Eur. J. Biochem.* **94**:573-580 (1979).

16. T. Kageyama and K. Takahashi, *in*: "Aspartic Proteinases and Their Inhibitors," V. Kostka, ed., Walter de Gruyter, Berlin, pp. 265-282 (1985).

BIOLOGICAL SIGNIFICANCE AND ACTIVITY CONTROL OF CATHEPSIN E

COMPARED WITH CATHEPSIN D

Kenji Yamamoto, Hideaki Sakai,[§] Eiko Ueno[§] and Yuzo Kato[§]

Department of Pharmacology
Kyushu University
Faculty of Dentistry
Fukuoka 812
Japan

and

[§]Department of Pharmacology
Nagasaki University School of Dentistry
Nagasaki 852
Japan

INTRODUCTION

Protein degradation in mammalian cells is thought to occur via two major pathways: a lysosomal and a non-lysosomal pathway. The former may participate in degradation of the majority of cellular proteins nonspecifically and the latter may preferentially degrade abnormal and short-lived proteins.[1] Recently, a number of non-lysosomal proteinases, such as cytosolic proteinases and plasma membrane-associated proteinases, have been identified in mammalian tissues. Most of the non-lysosomal enzymes have been shown to exhibit optimal activity at neutral pH. The aspartic proteinases are one of the four known main classes of proteinases and catalyze the hydrolysis of a variety of protein substrates below pH 5. The aspartic proteinases in mammalian cells are tentatively classified into two groups. One is a secretory group consisting of enzymes that function in extracellular spaces (pepsin, gastricsin etc.). The other is a non-secretory group consisting enzymes that function primarily within the cell. Cathepsins D and E are the two main non-secretory aspartic proteinases. Cathepsin D is a typical and well characterized lysosomal enzyme that is identified in almost all the mammalian cells. The wide distribution of cathepsin D throughout most tissues suggests its general role in proteolysis of cellular proteins. Besides its lysosomal role, cathepsin D has been suggested to be involved in a variety of physiological and pathological processes, for example in the proteolytic processing of lysosomal enzymes[2-4] and inflammatory and neoplastic disease states.[5-7] By contrast, cathepsin E is a relatively poorly characterized enzyme. Recent immunochemical studies have demonstrated that cathepsin E is a non-lysosomal protein, a part of which is present in the cytosol.[8-10] However, the endogenous substrates for cathepsin E are not known and the cellular function of this enzyme is therefore unclear.

Although cathepsin E has been clearly distinguished from cathepsin D on the basis of immunochemical and physicochemical properties,[11-16] they are very similar in many biochemical and catalytic features, such as substrate specificity, susceptibility to various proteinase inhibitors and pH optimum.[16-20] The results suggest that the two enzymes may share some roles in proteolysis of cellular proteins. However, the functional partnership of cathepsins D and E still remains to be answered. As part of our ongoing investigation for cellular functions of cathepsin E, we have clarified, using discriminative antibodies specific for each enzyme, that the distribution of cathepsins D and E in various rat tissues and cells are clearly different.[8] Cathepsin D is found in all of the tissues tested, while cathepsin E is limitedly distributed in certain cell types such as lymphoid tissues, gastrointestinal tracts and urinary organs. Of the cell types tested, erythrocytes have been shown to contain cathepsin E as the only aspartic proteinase. These results appear to be of particular importance, since such differences may reflect different physiological roles for the two enzymes.

SYNTHESIS AND ACCUMULATION OF CATHEPSINS E AND D IN MURINE FRIEND ERYTHROLEUKEMIA (FEL) CELLS DURING DMSO-INDUCED DIFFERENTIATION

In mature erythrocytes cathepsin E is normally associated with the cytoplasmic surface of the membrane in a latent form.[19] Cathepsin D is not found in mature erythrocytes. We have previously shown that the latent cathepsin E can be readily activated by a variety of procedures, such as cell aging *in vitro*, oxidant challenge, heating, and exposure to non-lytic concentrations of phospholipase C (from *Bacillus cereus*)[21, 22] and that the activation of the enzyme is accompanied by its dissociation from the membrane. Since its activation is closely correlated to the increased degradation of the membrane proteins, the results suggested the

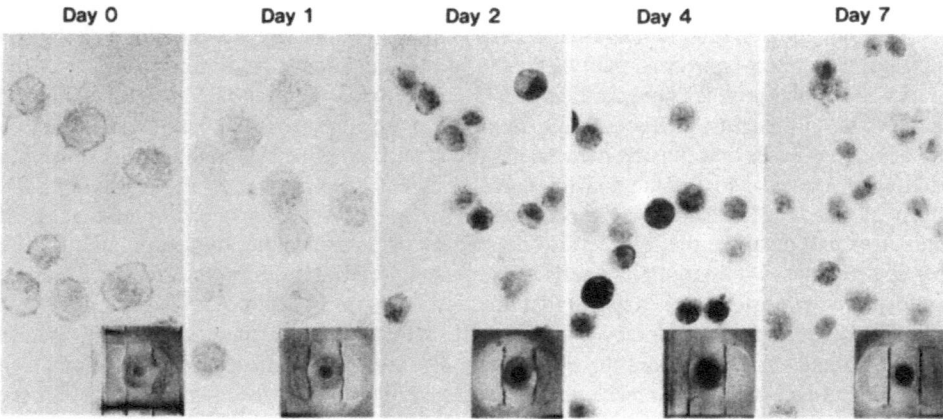

Figure 1. Photomicrographs of Wright-Giemsa stained FEL cells. Cells that were grown in DME supplemented with 13 % inactivated FBS, 5 % BSA and 1.8 mM iron dextran for 12 h, were plated on fibronectin-coated dishes and incubated for 1 h at 37°C. Erythroid differentiation was induced by incubating the attached cells in the medium containing 1.8 % DMSO at 37°C. After 4 days of induction, the culture medium was replaced with fresh DMSO-free medium and incubation was continued for an additional 3 days. Cells were collected at the indicated time points, stained with Wright-Giemsa and analyzed with a light microscope.

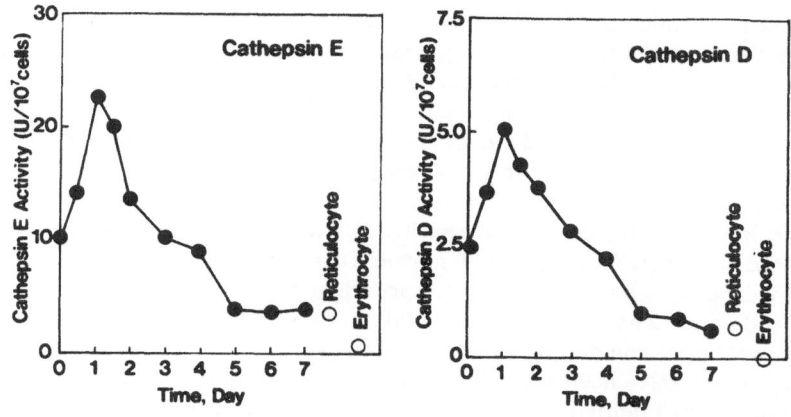

Figure 2. Changes in the levels of cathepsins E and D in FEL cells during DMSO-induced differentiation, as compared with their levels in murine reticulocytes and erythrocytes. Cell extracts were subjected to immunoprecipitation using discriminative antibodies specific for each enzyme. The levels of the two enzymes were calculated from the amounts of enzyme activity immunoprecipitated.

possible role of cathepsin E in the removal of senescent or damaged erythrocytes from blood circulation.[22]

The erythrocytes are the end product of a complex differentiation process which requires several days to be completed. During this process, proliferating multipotential stem cells respond to erythropoietin and differentiate through a succession of developmental stages (proerythroblasts, basophilic erythroblasts, polychromatophilic erythroblasts) to the normoblast stage. Then, upon extrusion of the nucleus, reticulocytes were formed which enter into the blood circulation, where they complete their maturation to the final erythrocyte stages. Murine Friend erythroleukemia (FEL) cells, which have been used extensively as an *in vitro* model for erythroid differentiation, are blocked at early stages of their normal differentiation pathway, presumably between the burst forming unit-erythroid stage and the colony forming unit-erythroid stage. FEL cells have been shown to differentiate into enucleating cells and reticulocytes when treated with dimethyl sulfoxide (DMSO) on fibronectin-coated dishes for 7 days.[23] In an attempt to study the biosynthesis and subcellular localization of cathepsin E during erythroid differentiation, we have used this system. Initially, we examined the morphology of Wright-Giemsa stained cells during DMSO-induced FEL cell differentiation in order to characterize the differentiation process with this system. As shown in Figure 1, uninduced cultures consisted predominantly of large cells with large nuclei. After DMSO induction for 4 days, most cells were small, with a condensed nucleus, and accumulated hemoglobin. After 7 days of DMSO treatment, cells that became distinctly smaller in size were mostly enucleating or had enucleated. The morphological changes of FEL cells during erythroid differentiation were consistent with the data of Patel and Lodish.[23]

As the second step in examining the synthesis and subcellular location of cathepsin E in comparison with those of cathepsin D during DMSO-induced FEL differentiation, we determined whether murine cathepsins E and D were recognized by rabbit antibodies against rat cathepsins E and D. Immunoprecipitation and Western immunoblotting analyses showed that each antibody immuno-reacted selectively with its murine counterpart. Thus, we used these antibodies to examine the levels of cathepsins E and D and their subcellular localization in FEL cells during DMSO-induced differentiation. The cells were collected at various stages of this process and ultrasonicated in the presence of 1 % Triton X-100. After centrifugation, the cell extracts, in which more than 90 % of the total hemoglobin-hydrolyzing activity at pH 3.5 was recovered, were subjected to immunoprecipitation and Western immunoblotting using the antibodies specific for cathepsins E and D. As shown in Figure 2, no significant difference was observed concerning their accumulating profiles in cells during 7 day differentiation. In uninduced cells, both cathepsins E and D were found at considerable levels and their synthesis was distinctly induced by treatment with DMSO. The levels of cathepsins E and D, which were calculated from the amounts of enzyme activity immunoprecipitated on the basis of the specific activities of purified enzymes, were nearly equal in induced cells (E, 0.15 µg/mg cell protein; D, 0.16 µg/mg cell protein). Within 1 day of DMSO-induced differentiation, both cathepsins E and D levels in cells were dramatically increased and corresponded to 220 % and 200 % of the respective enzymes present in an equivalent number of uninduced cells. An additional incubation of the cells with DMSO resulted in a progressive decrease in the cellular levels of both enzymes. After 4 days of DMSO-induced differentiation, when most of the cells accumulated high levels of hemoglobin and exhibited morphology typical of the normoblast, cathepsins E and D lowered to the levels in uninduced cells. The levels of these enzymes further declined over the next 3 days. Reticulocytes that were prepared from phenylhydrazine-treated mice showed the cathepsins E and D levels equivalent to those in FEL cells after 7 days of differentiation. Importantly, murine mature erythrocytes exhibited a complete loss of cathepsin D. A small part of cathepsin E remained in the mature cells in the membrane-associated form. It is worth noting that the ratio of cathepsins E and D contents in FEL cells had not significantly changed during DMSO-induced differentiation for 7 days.

SUBCELLULAR LOCALIZATION OF CATHEPSINS E AND D IN FEL CELLS DURING DMSO-INDUCED DIFFERENTIATION

To determine the cellular localization of cathepsin E and D in DMSO-induced FEL cells, the cells were subjected to separation into a soluble cytosol and an insoluble particulate fraction by centrifugation at 200,000 x g for 60 min after disruption of cells by nitrogen cavitation. The levels of cathepsins E and D were analyzed for both fractions by immunoprecipitation. In induced cells, cathepsin E was present both in the cytosol fraction and in the particulate fraction at an equal level, whereas cathepsin D was found with about 70 % in the particulate fraction and with about 30 % in the cytosol fraction (Figure 3). No significant change was observed in the distribution of cathepsin D into the cytosol and

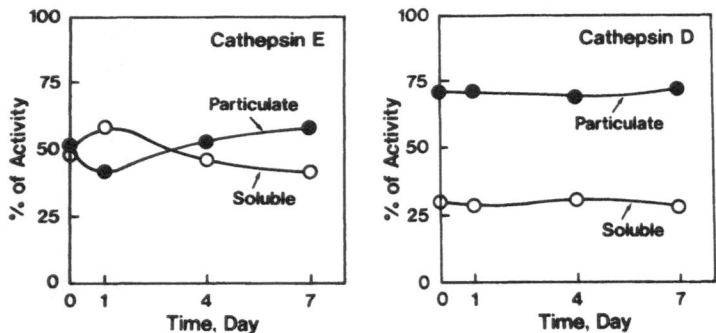

Figure 3. Relative abundance of cathepsins E and D in the particulate and cytosol fractions of FEL cells during differentiation.

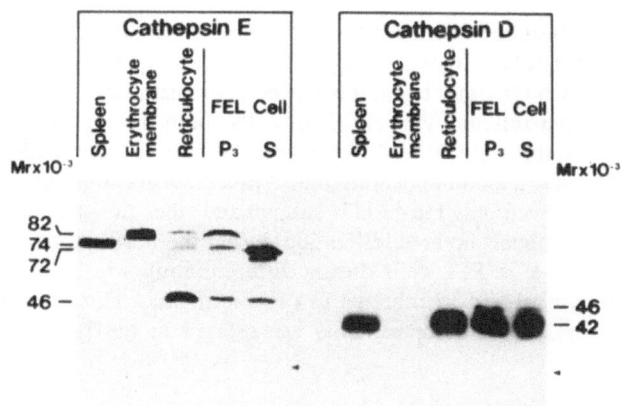

Figure 4. Immunoblots of cathepsins E and D from SDS-polyacrylamide gels. Each sample was separated on 8 % polyacrylamide gels containing 0.2 % SDS and then transferred to nitrocellulose membranes followed by immunoreaction with the antibodies specific for each enzyme.

particulate fractions throughout erythroid differentiating period. By contrast, the distribution of cathepsin E into the two fractions was distinctly changed during this period. The soluble/sedimentable cathepsin E in DMSO-induced FEL cells was 1.4 on one day of differentiation, 0.9 on 4 days of differentiation, and 0.7 on 7 days of differentiation.

To gain further insight into the intracellular localization of cathepsins E and D in FEL cells during DMSO-induced differentiation, the lysed cells were further separated by differential centrifugation into subcellular fractions which sedimented at 480 x g for 10 min (P1), 9,750 x g for 10 min (P2), and 200,000 x g for 60 min (P3). It was found, from the distribution of marker enzymes, that P1 fraction contained cell debris, nuclei, and some membraneous materials; P2 fraction was enriched in lysosomes and other membraneous organelles. P3 fraction consisted of microsomes and some membraneous materials. The final supernatant (S) was enriched in cytoplasmic matrix and contained some materials from other membraneous organelles. Upon one day of DMSO-induced differentiation, the most dramatic increase in the levels of cathepsins E and D was observed in the S and P3 fractions, respectively, and their levels corresponded to about 300 % and 400 % of those present in uninduced cells. Within 4 days of DMSO-induced differentiation, the level of cathepsin E in S fraction lowered to the levels observed in uninduced cells, but its level in P1 fraction remained at a high level which corresponded to about 230 % of that in uninduced cells. Similarly, the cathepsin D level in P3 fraction distinctly decreased, but its level in P2 fraction remained relatively constant after 4 days of DMSO-induced differentiation. Importantly, cathepsin D was completely eliminated in mature erythrocytes. The maturation of reticulocytes into erythrocytes at a final stage of erythroid differentiation was accompanied by complete loss of cytosolic cathepsin E. A significant amount of the membrane-associated cathepsin E only remained in erythrocytes. The early and abundant appearance of cathepsin E in the cytosol and its exclusive disappearance during erythroid differentiation suggests some roles of this enzyme in erythropoiesis.

Western immunoblotting analysis revealed that cathepsin E associated with murine erythrocyte membrane had an apparent MW of 80 kDa, whereas the enzyme from murine spleen, where it appeared to be present as the soluble protein,[24] had a MW of 74 kDa (Figure 4). In reticulocytes, both 80 kDa and 74 kDa subunits were observed. A subunit of MW approximately 46 kDa observed in reticulocytes appeared to be the monomeric form of the enzyme. In DMSO-induced FEL cells for one day, the 80 kDa subunit was found exclusively in the particulate fraction, and the 74 kDa subunit appeared in the soluble cytoplasmic fraction. On the other hand, it has been demonstrated, by SDS-PAGE followed by immunoblotting, that reticulocytes and FEL cells (both uninduced and induced cells) contained a major subunit with MW 42 kDa and a minor subunit with MW 46 kDa of cathepsin D. The 46 kDa subunit appeared to be a precursor form of cathepsin D, since the purified cathepsin D showed only the 42 kDa subunit and since the 46 kDa subunit tended to convert to the 42 kDa subunit upon acidification during the purification. Our experiments indicated that cathepsin E in FEL cells during differentiation appeared at least in the two compartments, the cytosol and membraneous compartments. However, it remains to be answered whether these two compartments are related in the process of cathepsin E biosynthesis.

POSSIBLE CONTROL OF ERYTHROCYTE CATHEPSIN E ACTIVITY

Our previous work has showed that cathepsin E is a non-lysosomal protein.[7, 8] Thus, we consider that cathepsin E must participate in the non-lysosomal proteolysis in cells. Since, however, the endogenous substrates are not known, the precise role of this enzyme is unclear. It has been demonstrated that cathepsin E in human erythrocytes is responsible for autodegradation of the membrane proteins observed in the pH range of 3.5 to 5.0.[19] At pH values above 5.5, however, only very minor degradation of the membrane proteins was observed. Cathepsin E purified from the erythrocyte membrane was also essentially inactive at pH values above 5.5 by itself (Figure 5). The results raised the question whether cathepsin E might not participate in cellular proteolysis under physiological conditions. If cathepsin E plays a role in normal workings of the mature erythrocyte, it is essential to investigate which mechanisms are responsible for the regulation of cathepsin E activity.

Recently, it has been demonstrated that cathepsin E from human erythrocyte membranes exhibits virtually full activity on two synthetic chromogenic substrates and casein at pH 5.8 in the presence of ATP.[25] Hence it has been suggested that ATP might stabilize the

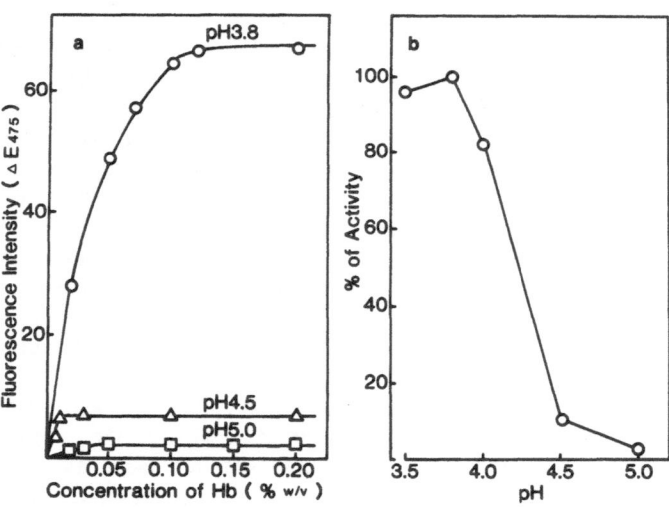

Figure 5. Effects of substrate concentration (a) and pH (b) on the cathepsin E activity. Buffer used was 0.1 M sodium acetate. In (b), hemoglobin concentration was 0.1 % and the maximal activity is taken as 100 %.

Table 1. Effect of various phosphorous compounds on hydrolysis of hemoglobin by Cathepsin E from human erythrocyte membranes

Compound	Concentration (mM)	Relative Activity (%)
None	-	100
Na_3PO_4	6.25	133
$Na_4P_2O_7$	6.25	184
$Na_5P_3O_{10}$	6.25	245
GTP	6.25	359
CTP	6.25	253
ATP	6.25	331
ADP	6.25	259
AMP	6.25	149
AMP-PCP	6.25	266

The reaction mixture contained in a total volume of 550 μl of sodium phosphate buffer (0.1 M), hemoglobin (0.1 %) and other additions as specified. Reactions were started by addition of the enzyme solution. Incubation was carried out at 40°C for 40 min. The reaction was stopped by addition of 550 μl of 5 % of trichloroacetic acid and acid soluble products were determined by fluorescamine method.

enzyme molecule dramatically in such a way to maintain its active conformation. We thus extended this work to better understand the activation of cathepsin E by ATP and its analogues. For this, we used hemoglobin as a substrate for cathepsin E assay because it was most specific for cathepsin E. The assay was carried out by the fluorescamine method.[26]

As expected from the previous study,[25] cathepsin E was significantly activated by both inorganic phosphates and nucleotides (Table 1). The maximum activation effect was observed with GTP and ATP that corresponded to about 360 % and 330 % of the enzyme activity measured in the absence of these ligands, respectively. Although ADP and AMP exhibited a significant activation effect, their potency was distinctly lower than that of ATP. The data indicated that the activation rate became greater with increasing number of phosphate groups. The non-hydrolyzable AMP-PCP analogue produced the noticeable activation of cathepsin E which was equivalent to that obtained with ADP, suggesting that ATP hydrolysis may not be required for activation. Pepstatin, the specific inhibitor of aspartic proteinases, completely abolished all of the enzyme activity. Since three phosphate groups and adenine or guanine base are necessary for the maximum activation, cathepsin E appears to have at least two specific sites for binding the phosphate group and the nucleoside. The catalytic activity of cathepsin E was clearly decreased at pH values above 5.5 even in the presence of ATP, but the activation rate was approximately equivalent to that observed at pH 5.5. Similarly, the activation rate was roughly invariant over the pH values between 4.5 and 5.0, although the catalytic activity of cathepsin E both in the presence and in the absence of ATP significantly increased as the pH was lowered. The activation effect of ATP was not observed below pH 4.0 because of precipitation of hemoglobin by ATP. Watabe *et al.*[27] reported that cathepsin D from bovine spleen is greatly activated by polyphosphates, including ATP when bovine serum albumin is used as a substrate at pH 4.6. Similar results were obtained with cathepsin D from bovine kidney and spleen.[28] However, the cathepsin D activation by ATP was not observed with hemoglobin as a substrate in those studies. As shown in Figure 6, the effect

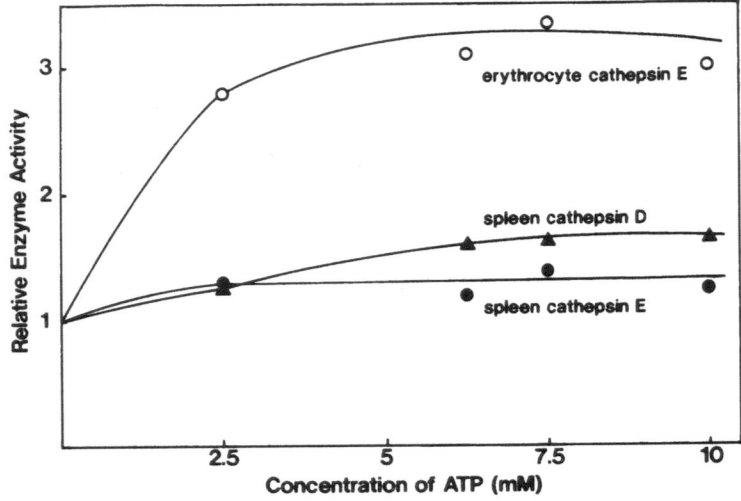

Figure 6. Comparison of the ATP activation effect on Hb-hydrolysis by cathepsin E purified from human erythrocyte membranes (open circles) and from rat spleen (filled circles) and by cathepsin D from rat spleen (filled triangles).

of ATP on rat spleen cathepsins D and E, as compared with that observed with cathepsin E from human erythrocyte membranes, is very small.

The data in Table 2 show the stabilization effect of ATP on heat inactivation of cathepsin E from erythrocyte membranes. In the presence of ATP, 85 % of the initial enzyme activity remained after incubation at 50°C for 20 min. However, the incubation without ATP resulted in a faster loss of the activity. This thermostabilizing effect of ATP was also observed at 60°C. It is thus considered that ATP is useful not only for maintaining the enzyme in the active conformation at around neutral pH, but also for protecting the enzyme from heat inactivation.

Table 2. Effect of ATP on thermostability of cathepsin E from human erythrocyte membranes

Temperature (°C)	Preincubation Time (min)	% of Activity Remaining	
		-ATP	+ATP
50	0	100	100
50	10	60	89
50	20	45	85
60	5	5	50
60	10	0	22

The enzyme was preincubated for 5 min at 0°C in the absence or presence of 6.25 mM ATP and then assayed with hemoglobin as a substrate at pH 5.5.

Table 3. Effect of nucleic acids on hydrolysis of hemoglobin by Cathepsin E from human erythrocyte membranes

Nucleic Acid	Concentration (μg/ml)	Relative Enzyme Activity (%)
None	-	100
DNA (calf thymus)	5	113
DNA (calf thymus, heat-denatured)	5	170
plasmid pUC18	5	125
plasmid pUC18 (double-strand linear)	5	139
plasmid pUC18 (single-strand linear)	5	208
poly(dA)	5	271
tRNA (brewer's yeast)	5	274
mRNA (rabbit globin)	5	252
rRNA 5S (*E. coli* MRE600)	5	319
rRNA 16S & 23S (*E. coli* MRE600)	5	215
poly(A)	5	132
poly(U)	5	287
poly(C)	5	141

Reactions were carried out under the same conditions as described in Table 1.

It seems of significance that cathepsin E from erythrocyte membranes can be activated by physiological concentrations of ATP and can exhibit its catalytic activity even at around neutral pH. Since old erythrocytes have elevated cellular levels of ATP[29] and since the membrane-associated cathepsin E is activated in aged erythrocytes,[22] its activity may be regulated by ATP. This implies that the physiological significance of cathepsin E in erythrocytes may be related to the removal of senescent erythrocytes from the blood circulation. Besides erythrocytes, we have recently demonstrated that cathepsin E localizes in microvilli of various cell types, such as renal proximal tubule cells, gastric parietal cells and hepatic cell borders.[24] Although it remains to be answered whether cathepsin E in these cells is present in a latent form and converts to an active form under physiological conditions, the present study suggests the possibility that ATP may be a important factor in regulating its catalytic activity in such cells. Several lines of evidence have also shown that cathepsin E localizes in the cytosolic compartment in such cell types as immature erythroid cells, lymphocytes and gastric mucosa.[8-10] Hence, to gain further insight into the general regulation mechanism of proteolysis by cathepsin E, we examined the direct activation of the enzyme by various nucleic acids. As indicated in Table 3, Hb-hydrolysis by cathepsin E at pH 5.5 was distinctly activated by low concentrations of RNAs (nM levels). The rate of this activation appeared to be dependent on molecular sizes of RNAs. The activated proteolytic activity by RNAs was totally inhibited by pepstatin. DNAs were less effective in stimulating the proteolytic activity than RNAs. However, a denatured or a lower molecular size DNA significantly activated the proteolytic activity. The data presented here suggest the possibility that cathepsin E may be regulated by a part of RNAs, as well as by ATP, to play a part in intracellular extralysosomal proteolysis. Further studies on the regulation of this interesting enzyme in cells are in progress.

REFERENCES

1. B. Grinde and P. D. Seglen, *Biochim. Biophys. Acta* **632**:73-86 (1980).
2. Y. Nishimura and K. Kato, *Arch. Biochem. Biophys.* **260**:712-718 (1988).
3. Y. Nishimura and K. Kato, *Arch. Biochem. Biophys.* **261**:64-71 (1988).
4. Y. Nishimura and T. Kawabata, K. Furuno and K. Kato, *Arch. Biochem. Biophys.* **271**:400-406 (1989).
5. A. R. Poole, *in*: "Dynamics of Connective Tissue Macromolecules" P. M. C. Burleigh and A. R. Poole, eds., North-Holland Publishing Company, Amsterdam (1975).
6. F. Capony, M. Morisset, A. J. Barrett, J. P. Capony, P. Broquet, F. Vignon, M. Chambon, P. Louisot and H. Rochefort, *J. Cell Biol.* **104**:253-262 (1987).
7. T. Saku, H. Sakai, N. Tsuda, H. Okabe, Y. Kato and K. Yamamoto, *Gut* **31**:1250-1255 (1990).
8. H. Sakai, T. Saku, Y. Kato and K. Yamamoto, *Biochim. Biophys. Acta* **991**:367-375 (1989).
9. S. Yonezawa, K. Fujii, Y. Maejima, K. Tamoto, Y. Mori and N. Muto, *Arch. Biochem. Biophys.* **267**:176-183 (1988).
10. E. Ichimaru, H. Sakai, T. Saku, K. Kunimatsu, Y. Kato, I. Kato and K. Yamamoto, *J. Biochem.* **108**:1009-1015 (1990).
11. K. Yamamoto, 0. Kamata, N. Katsuda and K. Kato, *J. Biochem.* **87**:511-516 (1980).
12. C. Lapresle, V. Puizdar, C. Porchon-Bertolotto, E. Joukoff and V. Turk, *Biol. Chem. Hoppe-Seyler* **367**:523-526 (1986).
13. R. A. Jupp, A. D. Richards, J. Kay, B. M. Dunn, J. B. Wyckoff, I. M. Samloff and K. Yamamoto, *Biochem. J.* **254**:895-898 (1988).
14. K. Yamamoto, E. Ueno, H. Uemura and Y. Kato, *Biochem. Biophvs. Res. Commun.* **148**:267-272 (1987).
15. I. M. Samloff, R. T. Taggart, T. Shiraishi, T. Branch, W. A. Reid, R. Heath, R. W. Lewis, M. J. Valler and J. Kay, *Gastroenterology* **93**:77-84 (1987).
16. M. Takeda, E. Ueno, Y. Kato and K. Yamamoto, *J. Biochem.* **100**:1269-1277 (1986).
17. K. Yamamoto, N. Katsuda and K. Kato, *Eur. J. Biochem.* **92**:499-508 (1978).
18. K. Yamamoto, N. Katsuda, M. Himeno and K. Kato, *Eur. J. Biochem.* **95**:459-467 (1979).
19. K. Yamamoto, M. Takeda, H. Yamamoto, M. Tatsumi and Y. Kato, *J. Biochem.* **97**:821-830 (1985).
20. S. Yonezawa, T. Tanaka and T. Miyauchi, *Arch. Biochem. Biophys.* **256**:499-508 (1987).
21. K. Yamamoto, H. Yamamoto, M. Takeda and Y. Kato, *Biol. Chem. Hoppe-Seyler* **369**:315-322 (1988).
22. K. Yamamoto, M. Yamada and Y. Kato, *J. Biochem.* **105**:114-119 (1989).
23. V. P. Patel and H. F. Lodish, *J. Cell Biol.* **105**:3105-3118 (1987).
24. T. Saku, H. Sakai, Y. Shibata, Y. Kato and K. Yamamoto, *J. Biochem.*, in press.
25. D. J. Thomas, A. D. Richards, R. A. Jupp, E. Ueno, K. Yamamoto, I. M. Samloff, B. M. Dunn and J. Kay, *FEBS Lett.* **243**:145-148 (1989).
26. C. Schwabe, *Anal. Biochem.* **53**:484-490 (1973).
27. S. Watabe, A. Terada, T. Ikeda, H. Kouyama, S. Taguchi and N. Yago, *Biochem. Biophvs. Res. Commun.* **89**:1161-1167 (1979).
28. S. Pillai, R. Botti and J. E. Zull, *J. Biol. Chem.* **258**:9724-9728 (1983).
29. G. L. Dale, S. L. Norenberg, T. Suzuki and L. Forman, *in*: "The Red Cell," G. J. Brewer, ed., Alan R. Liss, Inc., New York (1989).

EXPLOITING THE MOLECULAR TEMPLATE OF ANGIOTENSINOGEN IN THE DISCOVERY AND DESIGN OF PEPTIDYL, PSEUDOPEPTIDYL AND PEPTIDEMIMETIC INHIBITORS OF HUMAN RENIN: A STRUCTURE-ACTIVITY PERSPECTIVE

Tomi K. Sawyer, Jackson B. Hester, Heinrich J. Schostarez,
S. Thaisrivongs, Gordon L. Bundy, Li Liu, V. Susan Bradford,
Anne E. DeVaux, Douglas J. Staples, Linda L. Maggiora,
Ruth E. TenBrink, John H. Kinner, Clark W. Smith, Donald T. Pals,
Sally J. Couch, Jessica S. Hinzmann, Roger A. Poorman,
Howard M. Einspahr, Barry C. Finzel, Keith D. Watenpaugh,
Boryeu Mao, Dennis E. Epps, Ferenc J. Kezdy and Robert L. Heinrikson

Upjohn Laboratories
The Upjohn Company
Kalamazoo, Michigan   49001

INTRODUCTION

The design of potent and pharmacologically effective, substrate-related inhibitors of renin has been the subject of intensive pharmaceutical discovery research for about one decade.   Milestone achievements in synthetic tailoring of fragment analogs of angiotensinogen (ANG; Figure 1) have been documented in terms of identifying renin inhibitors of subnanomolar potency, sustained *in vivo* hypotensive activity, stability towards proteolytic degradation, and, more recently, oral bioavailability and decreased systemic clearance.[1] By chemical modification of ANG-based derivatives, structure-activity analysis, and computer-assisted molecular modeling of peptidyl, pseudopeptidyl and peptidemimetic inhibitors using 3-D structural models of human renin, there currently exists a rather sophisticated wealth of information of relevance to the "rational" design of prototypic renin-targeted cardiovascular therapeutic agents.   Such efforts have bridged biochemistry, medicinal chemistry, computational and biophysical chemistry, and *in vivo* pharmacology including, in a few cases, clinical evaluation in humans.

In retrospect, the renin substrate (ANG) has provided the key chemical entity to design and investigate promising renin inhibitors by the aforementioned interdisciplinary research strategies. For this purpose we have focused considerable attention on the ANG molecular template and the target enzyme to advance the "logic" of systematic chemical tailoring of ANG-based renin inhibitors. In this report we detail our efforts in the structure-activity analysis of ANG-based renin inhibitors as well as our progress in deciphering the structure-function properties of the target enzyme active site by x-ray crystallography and computer-assisted molecular modeling methods. Furthermore, we highlight both past and recent structure-activity studies of renin inhibitors within the framework of a simple ANG-based, renin inhibitor classification system.

RENIN INHIBITOR STRUCTURE-ACTIVITY RELATIONSHIPS

A recent review by Greenlee[1] provides an excellent basis to evaluate a major bulk of the literature describing the design and structure-activity relationships of chemically-modified peptidyl, pseudopeptidyl and peptidemimetic inhibitors of renin. Although the diversity of chemical modifications is pronounced, the primary focus of design strategies reported throughout recent years has been on the P_3-P_1' tetrapeptidyl component of ANG. In retrospect, it is important to point out that the first reported discoveries of high affinity (IC_{50} or K_i values < 10 nM) renin inhibitors included compounds having dipeptidyl (or greater) functionalization at both the N- and C-terminii (relative to the P_1-P_1' Leu-Val sites of ANG). These early studies described the fundamental contribution of the N-terminal ANG_{6-9} sequence to renin inhibitor binding affinity and emphasized, in particular, the significance of the P_3-P_2 substructure of ANG in this regard. Altogether, these structure-activity data contributed to an apparent "conceptual bias" for renin inhibitor design strategies which have been primarily focused on the Phe_8-His_9-Leu_{10}-Val_{11} fragment of the ANG template. Nevertheless, some exceptional deviations from this trend have been disclosed, and our own recent contributions (*vide infra*) also advance less "traditional" approaches and provide a new perspective of the structure-activity relationships of ANG-based renin inhibitors.

A PROPOSED RENIN INHIBITOR CLASSIFICATION SYSTEM BASED ON THE SUBSTRATE MOLECULAR TEMPLATE

To compare the structure-activity relationships of lead compounds and prototypic therapeutic agents which have been discovered in this intense search for pharmacologically-effective inhibitors of renin over the past decade, we have implemented a simple classification system which focuses on the ANG template and ANG binding subsite interactions (Figure 2). The nomenclature of the proposed classification is as follows: **Class-I** renin inhibitors structurally embody both N- and C-terminal aminoacyl extension beyond the P_1-P_1' site (i.e., P_2-P_5 and P_2'-P_3', respectively), including structural mimetics thereof; **Class-II** renin inhibitors structurally embody only N-terminal aminoacyl extension beyond the P_1-P_1' site; and **Class-III** renin inhibitors structurally embody only C-terminal aminoacyl extension beyond the P_1-P_1' site.

This classification system generically encompasses all documented P_1-P_1' "transition state" modifications. Furthermore, this classification system may provide a basis for future structure-activity advancements of yet chemically novel ANG-derived inhibitors of renin

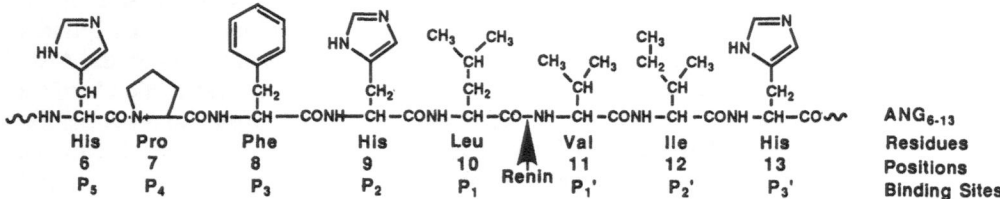

Figure. 1. ANG chemical structure and binding site nomenclature.

Class-I

W-His-Pro-Phe-His-X-Y-Ile-His-Z
W-Pro-Phe-His-X-Y-Ile-His-Z
W-Pro-Phe-His-X-Y-Ile-Z
W-Phe-His-X-Y-Ile-Z
W-His-X-Y-Ile-Z

Class-II

W-His-Pro-Phe-His-X-Y-Z
W-Pro-Phe-His-X-Y-Z
W-Phe-His-X-Y-Z
W-His-X-Y-Z

Class-III

W-X-Y-Ile-His-Z
W-X-Y-Ile-Z

Figure 2. A proposed classification system for ANG-based inhibitors of renin. Specifically, W and Z may include diverse functionalities except that of $C\alpha$- or $N\alpha$-linked amino acids, respectively. Furthermore, X-Y generically includes all reported dipeptidyl, pseudodipeptidyl and related P_1-P_1' functionalities. Finally, the P_5-P_2 His-Pro-Phe-His and P_2'-P_3' Ile-His sequences of candidate Class I-III inhibitors may be modified by other amino acids, or mimics thereof, which simulate ANG-like binding to the renin active site S_5-S_2 and S_2'-S_3' subsites (see text for details).

which, for example, may be devoid of both N- and C-terminal aminoacyl extension (proposed as Class-IV) or devoid of P_1-P_1' functionalization (proposed as Class-V).

Specific examples of ANG-based renin inhibitors which are designated as Class-I, Class-II and Class-III are given below. Furthermore, within the framework of this classification system we detail the structure-activity relationships of a series of Upjohn compounds which, to some extent, compromises the aforementioned "conceptual bias" of design strategies that have been primarily dedicated to ANG-based derivatives of the Class-I or Class-II type as well as a "transition-state" mimicry approach which has almost exclusively focused on $CH(OH)$-CH_2 type synthetic isosteres of the P_1-P_1' scissile amide.

Class-I Renin Inhibitors

In retrospect, the discovery of ANG-based renin inhibitors effecting potency in the micromolar range and *in vivo* activity was first realized with P_1-P_1' Phe-Phe substituted ANG derivatives exemplified by the decapeptide **RIP**.[2] Subsequent synthetic tailoring of **RIP** by P_1-P_1' Pheψ[CH_2NH]Phe substitution as well as N- and C-terminal aminoacyl modifications[3-5] resulted in 10- to 1,000-fold superior binding affinities (cf., **H-110** and **U-70714E**, Table 1A; *vide infra*). P_1-P_1' Leuψ[CH_2NH]Val, statine (Sta) and Leuψ[$CH(OH)CH_2$]Val substitutions were also found[3,6-12] to yield potent renin inhibitors (cf. **H-142**, **ISCRIP** and **U-70504E**, Table 1A; and **H-261** and **U-71038**, Table 1B; Figure 3). In the specific case of **ISCRIP** and **H-261**, we have advanced the structure-activity relationships of these prototypic renin inhibitors to further explore the P_5-P_2' and P_2'-P_3' sequences and identify potent Class-I analogs of reduced molecular size (Table 1).

Relative to the **ISCRIP** we determined[8] that the P_3' Phe contributed only slightly to renin inhibitory potency, whereas substitution of Ftr, Trp, Tyr or Phe for His at the P_5 site resulted in significantly improved potency (cf., **U-70504E**, **U-71941E**, **U72294E**, and **U-72234E** relative to **1** and **2**, in Table 1A). Further exploration of the structure-activity relationships of the N-terminal ANG_{6-9} sequence of **U-70504E** in terms of D-amino acid substitutions (cf. **3-6**) which were shown to significantly compromise the renin inhibitory potency by 100- to 10,000-fold. Finally, we determined[9] the effects of N_α-methyl-His (NMeHis) at the P_2 site of **U-70504E** as well as related derivatives having hydrophobic

Table 1A. Representative Class-I Renin Inhibitor Structure-Activity Relationships

Entry	Structure	IC_{50}, nM (K_i, nM)	Ref.
RIP	Pro-His-Pro-Phe-His-Phe-Phe-Val-Tyr-Lys	(2000)	2
H-108	Pro-His-Pro-Phe-His-Pheψ[CH$_2$NH]Phe-Val-Tyr-Lys	40	3
H-142	Pro-His-Pro-Phe-His-Leuψ[CH$_2$NH]Val-Ile-His-Lys	10	3
H-113	His-Pro-Phe-His-Leuψ[CH$_2$NH]Val-Ile-His	119	3
H-76	His-Pro-Phe-His-Leuψ[CH$_2$NH]Leu-Val-Tyr	1000	3
H-110	His-Pro-Phe-His-Pheψ[CH$_2$NH]Phe-Val-Tyr	820	3
U-70531E	D-His-Pro-Phe-His-Pheψ[CH$_2$NH]Phe-Val-Tyr	500	4
U-70714E	Ac-Ftr-Pro-Phe-His-Pheψ[CH$_2$NH]Phe-Val-Tyr-NH$_2$	3	4,5
ISCRIP	Iva-His-Pro-Phe-His-Sta-Ile-Phe-NH$_2$	1.9	7
U-70504E	Ac-Ftr-Pro-Phe-His-Sta-Ile-NH$_2$	0.1	8
U-71941E	Ac-Trp-Pro-Phe-His-Sta-Ile-NH$_2$	1.6	9
U-72294E	Ac-Tyr-Pro-Phe-His-Sta-Ile-NH$_2$	0.4	9
U-72234E	Ac-Phe-Pro-Phe-His-Sta-Ile-NH$_2$	0.4	9
1	Ac-His-Pro-Phe-His-Sta-Ile-NH$_2$	6.8	8
2	Ac-Pro-Phe-His-Sta-Ile-NH$_2$	31	
3	Ac-D-Ftr-Pro-Phe-His-Sta-Ile-NH$_2$	14	9
4	Ac-Ftr-D-Pro-Phe-His-Sta-Ile-NH$_2$	76	9
5	Ac-Ftr-Pro-D-Phe-His-Sta-Ile-NH$_2$	1700	9
6	Ac-Ftr-Pro-Phe-D-His-Sta-Ile-NH$_2$	260	9
7	Ac-Pro-Phe-NMeHis-Sta-Ile-NH$_2$	80	9
8	Ac-Trp-Pro-Phe-NMeHis-Sta-Ile-NH$_2$	12	9
9	Ac-Tyr-Pro-Phe-NMeHis-Sta-Ile-NH$_2$	21	9
10	Ac-Phe-Pro-Phe-NMeHis-Sta-Ile-NH$_2$	290	9
U-71613E	Ac-Ftr-Pro-Phe-NMeHis-Sta-Ile-NH$_2$	0.2	9

Abbreviations: Sta, statine; Iva, isovaleryl, Ftr, N^{in}-For-Trp; NMeHis, N_α-methyl-His.

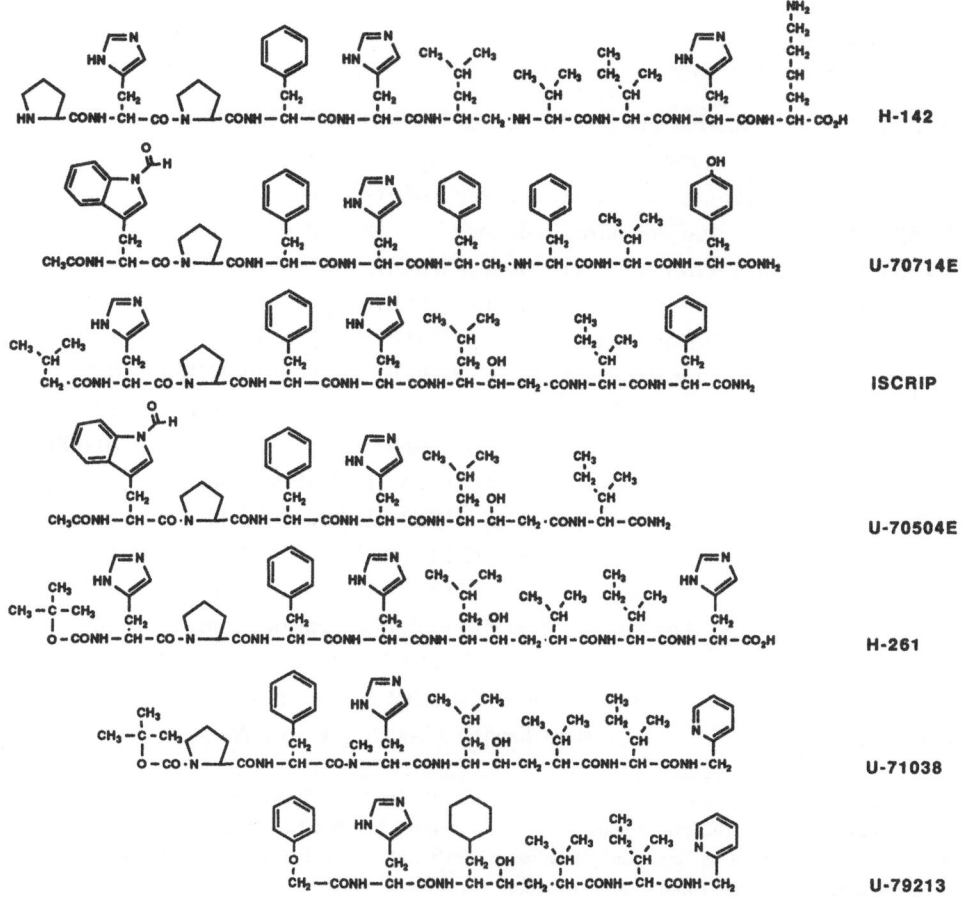

Figure 3. Representative Class-I renin inhibitors. The IC_{50} (or K_i) values for these compounds are given in Table 1).

aminoacyl substitution (i.e., Phe, Tyr and Trp) and found noteworthy potency differences in the resultant renin inhibitors (cf., **7-9** and **U-71613E**). In this series Ac-Ftr-Pro-Phe-NMeHis-Sta-Ile-NH$_2$ (**5**) was identified to the most potent and proteolytically-stable (data not shown) inhibitor of renin.

As compared to the P_1-P_1' Leuψ[CH(OH)CH$_2$]Val-substituted ANG$_{6-13}$-based renin inhibitor **H-261** we advanced[10-12] the structure-activity relationships of a series of analogs which incorporated P_2 NMeHis (cf., **U-71038**, **U-77436** and **U-78707**; Table 1B). These compounds differ chemically only in terms of N- and/or C-terminal functionalization. From a biological standpoint, the comparative *in vitro* potencies and *in vivo* pharmacological properties (data not shown) of these compounds inspired a significant effort to exploit the molecular template of the shared P_4-P_2' Pro-Phe-NMeHis-Leuψ[CH(OH)CH$_2$]Val-Ile sequence as a key intermediate to explore other polar and H$_2$O-solubilizing functionalization at the N- and/or C-terminus. Nevertheless, such compounds have been handicapped by their low (< 1-10 %) absolute or relative bioavailabilities in terms of oral administration. To simplify the chemical structure of this series of renin inhibitors in terms of peptidic character and/or molecular size we focused synthetic efforts on P_3 Phe replacements since this was a key site in the ANG molecular template for binding interactions with the target enzyme as

Table 1B. Representative Class-I Renin Inhibitor Structure-Activity Relationships (continued)

H-261	Boc-His-Pro-Phe-His-Leuψ[CH(OH)CH$_2$]Val-Ile-His	0.7	6
U-71038	Boc-Pro-Phe-NMeHis-Leuψ[CH(OH)CH$_2$]Val-Ile-Amp	0.3	10
U-77436	Thmac-Pro-Phe-NMeHis-Leuψ[CH(OH)CH$_2$]Val-Ile-AmpO	0.6	11
U-82146	Pro-Phe-NMeHis-Leuψ[CH(OH)CH$_2$]Val-Ile-AmpO	4.5	11
U-85197	Thmac-Pro-Phe-NMeHis-Leuψ[CH(OH)CH$_2$]Val-Ile	46	
U-71436E	Ac-Ftr-Pro-Phe-NMeHis-Leuψ[CH(OH)CH$_2$]Val-Ile-NH$_2$	0.2	12
U-78707E	Ac-Pro-Phe-NMeHis-Leuψ[CH(OH)CH$_2$]Val-Ile-NH$_2$	1.1	12
11	Ac-Pro-Phe-His-Leuψ[CH(OH)CH$_2$]Val-Ile-NH$_2$	0.5	12
12	Ac-Phe-His-Leuψ[CH(OH)CH$_2$]Val-Ile-NH$_2$	10	12
13	Ac-His-Leuψ[CH(OH)CH$_2$]Val-Ile-NH$_2$	>10,000	12
14	Poa-His-Leuψ[CH(OH)CH$_2$]Val-Ile-Amp	10	13
15	Cbz-His-Leuψ[CH(OH)CH$_2$]Val-Ile-Amp	180	
16	Ind-His-Leuψ[CH(OH)CH$_2$]Val-Ile-Amp	1.3	
17	Noa-His-Leuψ[CH(OH)CH$_2$]Val-Ile-Amp	2.4	
18	Poa-NMeHis-Leuψ[CH(OH)CH$_2$]Val-Ile-Amp	20	
19	Poa-His-Leuψ[CH(OH)CH$_2$]Leu-Ile-Amp	36	13
20	Poa-His-Leuψ[CH(OH)CH$_2$]Cha-Ile-Amp	4,000	13
U-79213	Poa-His-Chaψ[CH(OH)CH$_2$]Val-Ile-Amp	0.4	13

Abbreviations: Sta, statine; Iva, isovaleryl; Ftr, N^{in}-For-Trp; Boc, tert-butyloxycarbonyl; Poa, phenoxyacetyl; Cbz, carbobenzoxycarbonyl; Ind, 2-carboxy-indolyl; Noa, 1-naphthyloxyacetyl; Amp, 2-aminomethylpyridine; Cha, cyclohexylalanine; NMeHis, Nα-methyl-His; Thmac, tris(hydroxymethyl)-methylaminocarbonyl; and AmpO, Amp(N-oxide).

exemplified by systematic N-terminal aminoacyl deletion studies of Ac-Pro-Phe-His-Leuψ[CH(OH)CH$_2$]Val-Ile-NH$_2$ (cf., **11-13**; Table 1B). Noteworthy was the detailed structure-activity investigation of Class-I tetrapeptidemimetic derivatives of the generic formula W-His-Xaaψ[CH(OH)CH$_2$]Yaa-Ile-Z (cf., **14-20** and **U-79213**; Table 1B). Specifically, this series of potent ANG$_{8-13}$-mimetic derivatives incorporate only two bona fide amino acid residues; namely, His at the P$_2$ site and Ile at the P$_2$' site. Most importantly, this series of renin inhibitors illustrate the effects of P$_3$, P$_1$, and P$_1$' side-chain (or side-chain mimetic) chemical modifications on binding affinity as highlighted by the discovery[13] of Poa-His-Chaψ[CH(OH)CH$_2$]Val-Ile-Amp (**U-79213**; Figure 3), a subnanomolar potent derivative.

Class-II Renin Inhibitors

As defined above, Class-II compounds emphasize N-terminal aminoacyl elaboration beyond the P$_1$-P$_1$' site, and are defined to incorporate (optional) only limited C-terminal functionalization such as alkylamidation (e.g., Mba or εLys). This category of ANG-based derivatives has undoubtedly been the most intensively studied throughout recent years

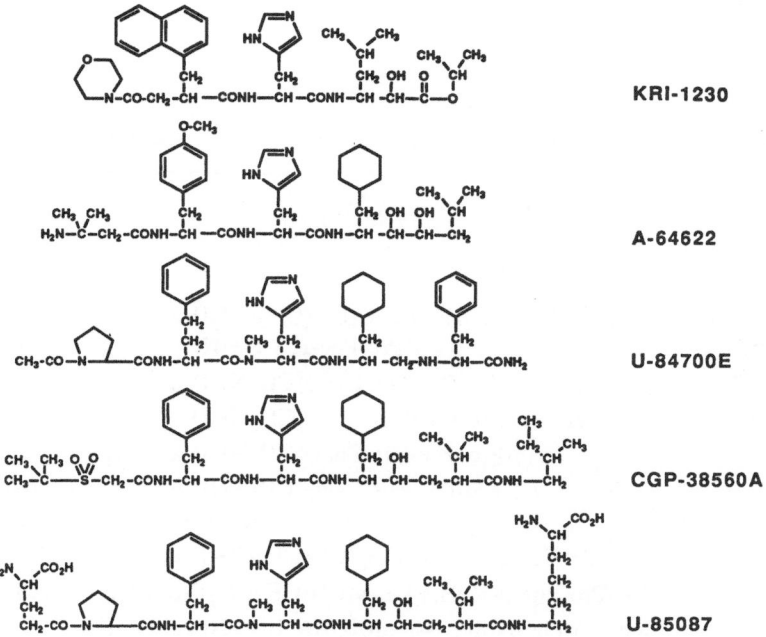

Figure 4. Representative Class-II renin inhibitors. The reported IC$_{50}$ values of KRI-1230[14], A-64662[15] and CGP-38560A[16] are 4.7 nM, 14 nM, and 2 nM, respectively. The IC$_{50}$ (or K$_i$) values of U-85087 and U-84700E (**36**) are given in Table 2

(Figure 4), and several Class-II renin inhibitors have been investigated *in vivo* in animal models as well as in human clinical testing (e.g., **A-64662**, **KRI-1230** and **CGP-38560A**). Indeed, impressive discoveries have been made in the search for potent, pharmacologically-effective, low molecular weight, and orally active "second generation" renin inhibitors of the Class-II type.

As shown in Table 2, **U-73083E** was a high affinity inhibitor (K$_i$ = 0.3 nM) of renin, and as compared to other P$_1$-P$_1$' pseudodipeptide-modified ANG$_{6-11}$ derivatives the Leuψ[CH(OH)CH$_2$]Val functionality was superior to Sta, Leuψ[CH$_2$NH]Val and Chaψ[CH$_2$NH]Val in terms of *in vitro* potency.[8,17] Again, the contribution of the hydrophobic aminoacyl functionality at the P$_5$ site (Ftr) of **U-73083E** to renin inhibitory potency was observed relative to the compound **24**. Interestingly, in the ANG$_{7-11}$ template a more dramatic difference in potencies was realized for this series of P$_1$-P$_1$' pseudodipeptidyl modified inhibitors of renin.[18] Compound **25** was particularly informative as it may be considered a C-truncated analog of **U-71038**, and it was determined that the Mba moiety provided a good replacement for the Ile-Amp group with respect to renin inhibitory potency. In contrast, the εLys moiety of **26** was found to effect decreased potency, but with a second modification by P$_1$-P$_1$' Chaψ[CH(OH)CH$_2$]Val a nanomolar potent ANG$_{7-11}$ derivative (**27**) was identified.[19] Relative to the latter compound (**27**) a series of analogs sharing a common P$_4$-P$_1$' Pro-Phe-NMeHis-Chaψ[CH(OH)CH]Val sequence were explored in terms of structure-activity (cf., **28-30** and **U-85087**; Table 2). Noteworthy was the finding that both N- and C-terminal modifications effected only slight deviations to the *in vitro* potency.

Table 2. Representative Class-II Renin Inhibitor Structure-Activity Relationships

Entry	Structure	IC$_{50}$, nM (K$_i$, nM)	Ref.
U-73083E	Ac-Ftr-Pro-Phe-His-Leuψ[CH(OH)CH$_2$]Val-NH$_2$	0.3	8,17
U-71852E	Ac-Ftr-Pro-Phe-His-Sta-NH$_2$	3.8	8,17
U-75867E	Ac-Ftr-Pro-Phe-His-Leuψ[CH$_2$NH]Val-NH$_2$	21.0	8,17
U-78918E	Ac-Ftr-Pro-Phe-His-Chaψ[CH$_2$NH]Val-NH$_2$	2.0	17
21	Ac-Pro-Phe-His-Chaψ[CH$_2$NH]Val-NH$_2$	81	12,18
22	Ac-Pro-Phe-His-Leuψ[CH$_2$NH]Val-NH$_2$	10,000	12,18
23	Ac-Pro-Phe-His-Sta-NH$_2$	1,000	12,18
24	Ac-Pro-Phe-His-Leuψ[CH(OH)CH$_2$]Val-NH$_2$	370	
25	Boc-Pro-Phe-NMeHis-Leuψ[CH(OH)CH$_2$]Val-Mba	0.8	
26	Boc-Pro-Phe-NMeHis-Leuψ[CH(OH)CH$_2$]Val-εLys	23	
27	Boc-Pro-Phe-NMeHis-Chaψ[CH(OH)CH$_2$]Val-εLys	1.9	
28	Thmac-Pro-Phe-NMeHis-Chaψ[CH(OH)CH$_2$]Val-εLys	1.6	
29	Thmac-Pro-Phe-NMeHis-Chaψ[CH(OH)CH$_2$]Val-NH$_2$	0.8	
30	γGlu-Pro-Phe-NMeHis-Chaψ[CH(OH)CH$_2$]Val-Mba	0.3	
U-85087	γGlu-Pro-Phe-NMeHis-Chaψ[CH(OH)CH$_2$]Val-εLys	1.0	19
U-71909E	Ac-Pro-Phe-His-Pheψ[CH$_2$NH]Phe-NH$_2$	13	12,18
31	Ac-D-Pro-Phe-His-Pheψ[CH$_2$NH]Phe-NH$_2$	31	12,18
32	Ac-Pro-D-Phe-His-Pheψ[CH$_2$NH]Phe-NH$_2$	>10,000	12,18
33	Ac-Pro-Phe-D-His-Pheψ[CH$_2$NH]Phe-NH$_2$	>10,000	12,18
34	Ac-Pro-Phe-NMeHis-Pheψ[CH$_2$NH]Phe-NH$_2$	110	12,18
35	Ac-Pro-Hph-NMeHis-Pheψ[CH$_2$NH]Phe-NH$_2$	4.0	
36	Ac-Pro-Hph-NMeHis-Chaψ[CH$_2$NH]Phe-NH$_2$	(0.1)	12,18
37	Tric-Pro-Hph-NMeHis-Chaψ[CH$_2$NH]Phe-NH$_2$	(0.1)	
38	Ac-Pro-Hph-NMeHis-Chaψ[CH$_2$NH]Phe-εLys	2.3	
39	Ac-Pro-Phe-NMeHis-Chaψ[CH$_2$NH]Phe-NH$_2$	10	12,18
40	Ac-Pro-Hph-His-Chaψ[CH$_2$NH]Phe-NH$_2$	(0.1)	12,18
U-79465E	Ac-Pro-Phe-His-Chaψ[CH$_2$NH]Phe-NH$_2$	(0.1)	12,18

Abbreviations (also refer to Table 1 footnote): Mba, 2-methyl-butylamine; γGlu, γ-carboxy-linked Glu; εLys, ε-amino-linked Lys; Hph, homophenylalanine; Tric, tricine (or N,N,N-tris(hydroxymethyl)-Gly).

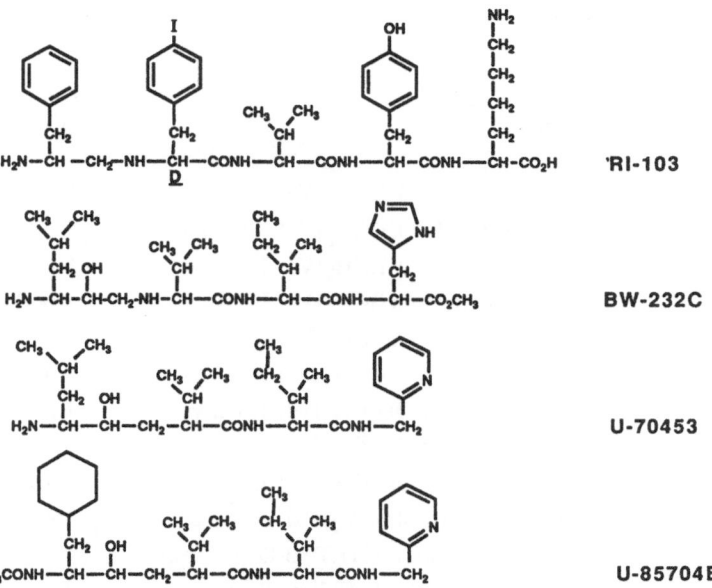

Figure 5. Representative Class-III renin inhibitors. The reported[21-23] IC_{50} values of compounds **RI-103**, **BW-232C** and **U-70453E** are 2000 μM, 400 μM, and 10 μM, respectively. The IC_{50} of **U-85704E** is given in Table 3.

However, the *in vivo* pharmacological properties of these Class-II renin inhibitors clearly showed the contribution of polar functionalization at the N- and C-terminii as best exemplified by the potent and sustained hypotensive activities of **U-85087** (*in vivo* data not shown).

We have also recently advanced[12,18] a series of Class-II compounds which clearly indicate the relative strength of P_1-P_1' scissile bond modification of the CH_2-NH type (versus CH(OH)-CH_2) as summarized in Table 2 (cf., **U-71909E**, **18-22** and **U-79465E**). Specifically, the ANG_{7-11} derivative **U-71909E** was systematically modified throughout its P_4-P_2 Pro-Phe-His sequence and it was determined that P_4 D-Pro or P_3-P_2 Hph-NMeHis substitutions were best in terms of *in vitro* potency and stability towards proteolytic degradation (data not shown). Relative to these compounds further modification by P_1-P_1' Chaψ[CH_2NH]Phe advance a yet more potent analogs (cf., **36-38**) in which polar N- or C-terminal functionalization was explored by structure-activity. Interestingly, it was determined that P_3 Hph itself did not contribute to improved potency per se, but in combination with P_2 NMeHis substitution a noteworthy difference between P_3-P_2 Phe-NMeHis and Hph-NMeHis modifications of **U-79465E** was identified (cf., **36, 39** and **40**). From a historical standpoint, it is also interesting to point out that the Class-I P_1-P_1' Leuψ[CH_2NH]Val-modified renin inhibitor **H-142** was one of the first compounds evaluated[20] in the human being. Relative to **H-142** the above series of P_1-P_1' Chaψ[CH_2NH]Phe-modified compounds were designed[12,18] by systematic chemical elaboration of the prototypic fragment derivative Ac-Pro-Phe-His-Leuψ[CH_2NH]Val-NH$_2$ (**22**) and exemplify noteworthy improvement in potency (> 10,000-fold) as well as the stability to proteolytic degradation (i.e., P_4-P_3 or P_3-P_2 cleavage).

Table 3. Representative Class-III Renin Inhibitor Structure-Activity Relationships

Entry	Structure	IC_{50}, nM (K_i, nM)	Ref.
U-70453	Leuψ[CH(OH)CH$_2$]Val-Ile-Amp	10,000	13
41	Chaψ[CH(OH)CH$_2$]Val-Ile-Amp	(250)	13
42	Boc-Chaψ[CH(OH)CH$_2$]Val-Ile-Amp	5,600	13
43	Tba-Chaψ[CH(OH)CH$_2$]Val-Ile-Amp	7,600	13
44	Poa-Chaψ[CH(OH)CH$_2$]Val-Ile-Amp	330	13
45	γGlu-Chaψ[CH(OH)CH$_2$]Val-Ile-εLys	>10,000	
46	Ac-Chaψ[CH(OH)CH$_2$]Val-Ile-εLys	1,200	
47	Ac-Chaψ[CH(OH)CH$_2$]Val-Ile-NH$_2$	730	
48	Ac-Chaψ[CH(OH)CH$_2$]Val-Ile-His-NH$_2$	1,000	
49	Ac-Chaψ[CH(OH)CH$_2$]Val-Ile-AmpO	150	
50	Ac-Chaψ[CH(OH)CH$_2$]Val-D-Ile-Amp	>10,000	
U-85704E	Ac-Chaψ[CH(OH)CH$_2$]Val-Ile-Amp	37	

Abbreviations: Tba, tert-butyloxycarbonyl (refer also to Tables I and II footnotes).

Class-III Renin Inhibitors

This general category of ANG-based renin inhibitors is, perhaps, the most obscure in the sense that although the earliest studies in this field of research were apparently targeted on the ANG_{10-13} template no major breakthroughs as related to the identification of potent (i.e., nanomolar) compounds have occurred. Thus, a paucity of reported structure-activity data exists for such Class-III renin inhibitors. Nevertheless, studies which have been reported include disclosures of a variety of structurally-diverse compounds (Figure 5).

Our identification[13] of the tripeptidemimetic analog 41 (U-82159E) as a competitive inhibitor of renin prompted our interests to further advance the structure-activity analysis of a series of its derivatives in an attempt to discover more potent Class-III compounds. Indeed, as summarized in Table 3 N-terminal modification of the P_1 amino functionality by Boc, Tba, Poa, and Ac groups (cf., 42-44, and U-85704E) resulted in considerable potency changes of which the Ac moiety was optimal. At the P_2-P_3' sites it was determined that the Ile-Amp moiety of U-85704E was superior to a number of modifications including Ile-εLys, Ile-NH$_2$, Ile-His-NH$_2$, Ile-AmpO and D-Ile-Amp (cf., 46-50). Finally, it is very timely to note that recent structure-activity studies[23,24] on the retroviral aspartyl protease of human immunodeficiency virus have showed the promising therapeutic (AIDS) potential of Class-III type compounds such as 43 (U-81749E) and 44 (U-82222E) in terms of their antiviral properties *in vitro*.

THREE-DIMENSIONAL STRUCTURAL STUDIES OF RENIN AND RENIN-INHIBITOR COMPLEXES

Biophysical and computational analysis of human renin and/or renin-inhibitor complexes by x-ray crystallography and related molecular modeling strategies remains to be fully exploited to advance our detailed understanding of the target enzyme in terms of both its 3-D architecture and active site chemistry. Relative to other aspartyl proteases (Table 4) the

Table 4. Crystallographic Structures of Renin and Related Aspartyl Proteases

Aspartyl Protease		Crystal Structure, (Resolution, Å)	
Name	Residues	Native Enzyme	Enzyme-Inhibitor Complex
Renin (Human)	339	Yes (2.5[a], 2.5[b])	U-79465E (2.5[b])
Renin (Mouse)	336	Yes (2.8[c])	No*
Pepsinogen (Porcine)	371	Yes (1.8[d])	No
Rhizopuspepsin	325	Yes (1.8[e])	Pepstatin (2.5[e])
			U-70531E (1.8[f])
			CP-69799 (2.0[g])
			CP-82218 (1.95[g])
Endothiapepsin	330	Yes (2.1[h])	H-261 (1.6[i])
			L-363564 (2.2[j])
			H-142 (2.1[j])
Penicillopepsin	323	Yes (2.1[k])	Pepstatin fragment (1.8[l])
Pepsin (Porcine)	299	Yes (2.3[m])	A-63218 (1.8[m])
			A-62702 (1.8[m])

*Note: Unsolved crystal structures of renin-inhibitor complexes are excluded from this literature synopsis.
References: [a]Sielecki *et al.*[25]; [b]Watenpaugh *et al.*[26]; [c]Navia *et al.*[27]; [d]James and Sielecki[28]; [e]Bott *et al.*[29]; [f]Suguna *et al.*[30]; [g]Parris *et al.*[31]; [h]Pearl and Blundell[32]; [i]Veerapandian *et al.*[33]; [j]Blundell *et al*[34]; [k]James and Sielecki[35]; [l]James *et al.*[36]; [m]Abad-Zapatero *et al.*[37].
Inhibitors: See references for chemical structures.

availability of crystallographic data on renin and/or renin-inhibitor complexes is only very recently been described.[25,26] However, there has been success in the crystallographic study of complexes of ANG-based renin inhibitors with other aspartyl proteases (e.g., endothiapepsin, rhizopuspepsin and porcine pepsin).

Attempts to incorporate knowledge of the structure of the target enzyme, human renin, into inhibitor design efforts have generally taken two independent paths. The first approach has combined knowledge of the renin sequence and the structures of several related aspartyl proteases (e.g., rhizopuspepsin, endothiapepsin, pepsinogen) within the framework of known physico-chemical principles to develop computationally-derived 3-D models (e.g., CKH-RENIN[38]) of the human renin. The CKH-RENIN model has been available for some time now and we have implemented[4,5,8,12,17,18,39] its use in inhibitor design efforts. The second approach has implemented conventional techniques of protein crystallography in the quest for experimentally-derived models of renin and/or renin-inhibitor complexes. Crystallographic studies of human renin have used protein expressed by recombinant means. Heterologous expression has been achieved in a variety of cells, but the most successful production of active and properly folded human renin has been developed from expression of the cloned gene in Chinese hamster ovary (CHO) cells and purified by affinity chromatography.[40,41] The protein is naturally glycosylated by post-translational modifications at two Asn side-chains. We have crystallized the protein in two different forms: a glycosylated form in which attachement of carbohydrate is under the control of the CHO cell, and a deglycosylated form in which most of the attached carbohydrates has been removed enzymatically. Our structure determination efforts have concentrated exclusively on the deglycosylated form because crystals of that form are significantly better for crystallographic analysis (higher resolution overall and fewer molecules per asymmetric unit). Crystals of deglycosylated renin are in the tetragonal space group I4 with a = b = 130 Å, c = 41 Å and one molecule per asymmetric unit. This form appears to be identical to that

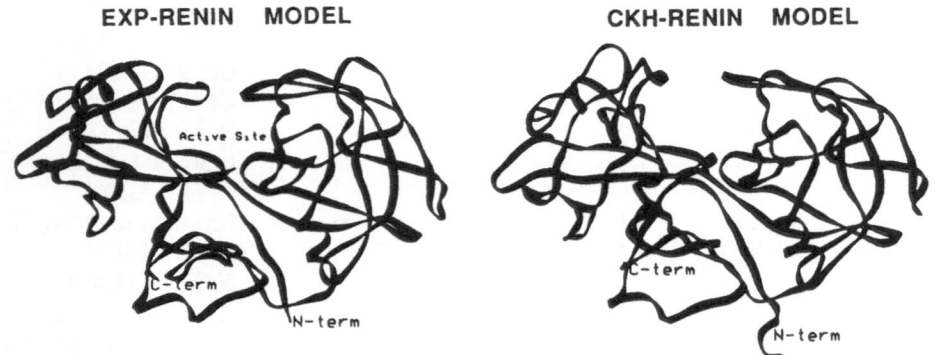

EXP-RENIN MODEL   **CKH-RENIN MODEL**

Figure 6. Ribbon drawings of the human renin cystallographic model (EXP-RENIN, left) and computational model (CKH-RENIN, right).

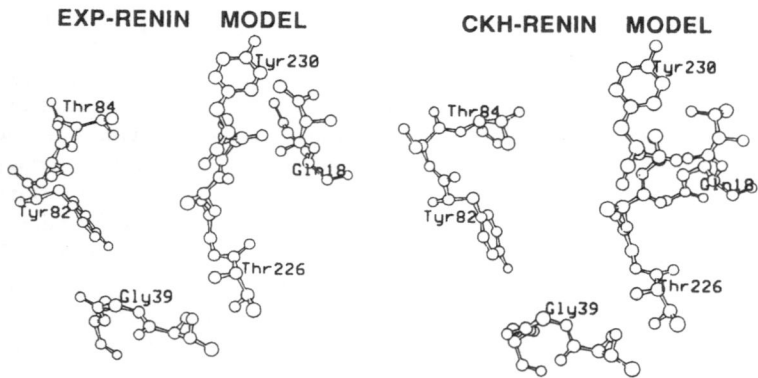

EXP-RENIN MODEL   **CKH-RENIN MODEL**

Figure 7. A portion of the active site of human renin in atomic detail as derived from the EXP-RENIN (left) and CKH-RENIN (right) models.

318

previously reported and used for successful structure determination by Sielecki and co-workers.[25] Several data sets were collected from crystals of deglycosylated renin with synchrotron radiation at an X-ray diffraction data-collection facility[42] at the Photon Factory in Tsukuba, Japan.

The structure of the Upjohn deglycosylated renin has been solved by application of molecular replacement methods. The model used for structure solution was composed of those regions common to the crystal structures of pepsinogen,[43] rhizopuspepsin[30] and penicillopepsin.[35] This structurally conserved framework amounted to approximately 60 % of the renin molecule. The majority of the polypeptide chain has been modeled and is being refined against a 2.5 Å resolution data set. The current value of R, a measure of agreement between observed and calculated data, is 0.24. The crystallographic refinement and model building effort on renin is an iterative process and continues, but is nearing completion in the near future. A ribbon drawing (Figure 6) of the main chain of our x-ray crystallographic model (EXP-RENIN) may be compared to a similar rendering of the homology-based model (CKH-RENIN). There remain a few regions of the EXP-RENIN model in which structure is in question. For example, portions of the loops arrayed above the active active site are not well defined. Regions where computational and crystallographic models diverge by more than 3 Å (rms) for significant stretches correspond to sequence numbers 50-60, 105-120, 250-260 and 275-310, all in the loop regions noted above. A more detailed comparison of a portion of the active site in the two models shows many features in common (Figure 7), although there are some significant differences. Specifically, while segments consisting of residues 38-40 and 226-230 are very similar in the two models, the loops around residues 83-84 and the side chains of residue 18 are configured differently. As refinement of the EXP-RENIN nears completion, it is hoped that differences such as these uncovered between the two models will aid in further refinements in inhibitor design and in homology-based modeling techniques.

As noted in Table 4, our very recent x-ray crystallographic determination[26] of a U-79465E-human renin complex provides a noteworthy opportunity to investigate the target enzyme active site more accurately and advance molecular modeling studies of inhibitors sharing substructural features with U-79465E, a subnanomolar potent class-II inhibitor incorporating P_1-$P_{1'}$ Chaψ[CH_2NH]Phe within its ANG_{7-11} template. Significant deviations between our crystallographically-derived, active site structure of U-79465E (data not shown) and that previously determined for this inhibitor using CKH-RENIN-based molecular modeling were observed. Albeit somewhat disppointing, these results do provide us impetus to re-evaluate the structure-conformation-activity relationships of renin inhibitors within the scope of our EXP-RENIN (i.e., native and inhibitor-bound) models. The details of this work will be published elsewhere.

Finally, it remains to be clearly understood what specific 3-D molecular interactions, whether intermolecular or intramolecular, are of major impact to renin binding to inhibitors. Based on structure-activity studies (*vide supra*), it is apparent that the substrate template modifications point towards the P_3-P_1 and P_1-$P_{2'}$ as critical substructural entities in the design of Class-I, Class-II, and Class-III inhibitors of renin. Recently, we have reported[44] the thermodynamic binding parameters for a series of ANG-based renin inhibitors which suggested that an important determinant of ligand interaction with the target enzyme is the solution-phase conformation of the inhibitor itself. Nevertheless, hydrophobic interactions appear to be the primary mode of inhibitor binding to the renin active site for those compounds surveyed in this investigation, and it is noted that the P_5, P_3, P_1 and $P_{1'}$ sites of ANG-based renin inhibitors reported to date have illustrated the successful incorporation of lipophilic side-chain functionalities (i.e., indolylmethyl, naphthylmethyl, cyclohexylmethyl, benzyl) as substitutions resulting in significantly increased binding affinity. Again, it is expected that crystallographic analysis of renin-inhibitor complexes will provide insights to further address these fundamental questions in this area of protein biophysical chemistry and structure-function.

CONCLUSIONS

This report summarizes a portion of our research efforts at The Upjohn Company which has been directed towards the design and identification of renin-targeted cardiovascular therapeutic agents. From a synthetic standpoint the design of renin inhibitors has almost exclusively been derived from chemical modification of ANG. In terms of structure-activity relationships, many potent ANG-based inhibitors of renin have been discovered which incorporate a variety of P_1-P_1' substitutions as well as significant modification of P_4-P_2 sequence. The C-terminal P_2'-P_3' sequence has been essentially abandoned in recent years as exemplified by an increasing amount of literature of Class-II compounds. Nevertheless, we document in this report the discovery of promising Class-III tripeptidemimetic inhibitors of renin. Overall, the design of renin inhibitors has been tedious to date as primarily due to the paucity of a crystallographically-defined 3-D structures of the target enzyme and/or, more importantly, an inhibitor bound complex. Future research efforts may take advantage of such information to provide a greater degree of rational drug design to renin inhibitor discovery strategies using computer-assisted molecular modeling techniques and programs to probe the exquisite topography of the renin active site. Specifically, such future work based on EXP-RENIN (native and inhibitor-bound) models is expected to provide a much more accurate starting point for computer-assisted 3-D molecular modeling strategies aimed at the design of yet novel peptidemimetic inhibitors of human renin which might be further chemically tailored to, perhaps, overcome key liabilities of such compounds tested to date as related to oral absorption, clearance, and pharmacologic duration of action.

REFERENCES

1. W. J. Greenlee, Renin inhibitors, *Med. Res. Rev.* **10**:173 (1990).
2. J. Burton, R. J. Cody, J. A. Herd and E. Haber, Specific inhibition of renin by an angiotensinogen analog: Studies in sodium depletion and renin-dependent hypertension, *Proc. Natl. Acad. Sci. U.S.A.* **77**:5476 (1980).
3. M. Szelke, B. Leckie, A. Hallett, D. M. Jones, J. Sueiras, B. Atrash and A. F. Lever, Potent new inhibitors of human renin, *Nature* **299**:555 (1982).
4. T. K. Sawyer, D. T. Pals, C. W. Smith, H. S. Saneii, D. E. Epps, D. J. Duchamp, J. B. Hester, R. E. TenBrink, D. J. Staples, A. E. DeVaux, J. A. Affholter, G. F. Skala, W. M. Kati, J. A. Lawson, M. R. Schuette, B. V. Kamdar and D. E. Emmert, Transition state-substituted renin inhibitory peptides: structure-conformation-activity studies on N^{in}-formyl-Trp and Trp modified congeners, *in*: "Peptides, Structure and Function (Proceedings of the Ninth American Peptide Symposium)", C. M. Deber, V. J. Hruby and K. D. Kopple, eds., Pierce Chemical Co., Rockford, p. 729 (1985).
5. D. E. Epps, B. Mao, D. J. Staples and T. K. Sawyer, Structure-conformation-activity relationships of renin inhibitory peptides studied by resonance energy transfer coupled with molecular modeling, *Int. J. Peptide Protein Res.* **31**:22 (1988).
6. M. Szelke, D. M. Jones, B. Atrash, A. Hallett and B. J. Leckie, Novel transition-state inhibitors of renin, *in*: "Peptides, Structure and Function (Proceedings of the Eighth American Peptide Symposium)", V. J. Hruby and D. H. Rich, eds., Pierce Chemical Co., Rockford, p. 579 (1983).
7. J. Boger, N. S. Lohr, E. H. Ulm, M. Poe, E. H. Blaine, G. M. Fanelli, T.-Y. Lin, L. S. Payne, T. W. Schorn, B. I. LaMont, T. C. Vassil, I. I. Stailito, D. F. Veber, D. H. Rich and A. S. Boparai, Novel renin inhibitors containing the amino acid statine, *Nature* **303**:81 (1983).
8. T. K. Sawyer, D. T. Pals, B. Mao, D. J. Staples, A. E. DeVaux, L. L. Maggiora, J. A. Affholter, W. Kati, D. Duchamp, J. B. Hester, C. W. Smith, H. H. Saneii, J. H. Kinner, M. Handschumacher and W. Carlson, Design, structure-activity and molecular modeling studies of potent renin inhibitory peptides

having N-terminal N^{in}-For-Trp (Ftr): Angiotensinogen congeners modified by P_1-$P_{1'}$ Phe-Phe, Sta, Leuψ[CH(OH)CH$_2$]Val or Leuψ[CH$_2$NH]Val substitutions, *J. Med. Chem.* **31**:18 (1988).

9. T. K. Sawyer, A. E. DeVaux, D. J. Staples, D. T. Pals, W. Kati, B. Mao and D. Duchamp, Structure-conformation-activity relationships of a highly potent transition state substituted peptide inhibitor of human renin, Abstracts of the 18th Central Regional Meeting of the American Chemical Society, Bowling Green, OH (June 1-5), No. 217, p. 95 (1986).

10. S. Thaisrivongs, D. T. Pals, D. W. Harris, W. M. Kati and S. R. Turner, Design and synthesis of a potent and specific renin inhibitor with a prolonged duration of action *in vivo*, *J. Med. Chem.* **29**:2088 (1986).

11. G. Bundy, D. T. Pals, J. A. Lawson, S. J. Couch, M. F. Lipton and M. A. Mauragis, Potent renin inhibitory peptides containing hydrophilic end groups, *J. Med. Chem.* **33**:2276 (1990).

12. T. K. Sawyer and B. Mao, Molecular modeling logic in the design of renin-targeted cardiovascular therapeutics, *Chemistry Today* **8**:53 (1990).

13. T. K. Sawyer, H. Schostarez, J. Hester, L. Liu, V. S. Bradford, D. Staples, R. TenBrink, A. Tomasselli, J. Hui, T. McQuade, W. G. Tarpley, D. Pals, J. Hinzmann, R. Poorman, J. Moon, W. J. Howe, R. Heinrikson, A. Wlodawer, M. Jaskolski, C. Craik, D. DeCamp and B. Dunn, Design and structure-activity-selectivity of HIV protease inhibitors: Molecular modeling studies based on a 2.5 Å crystallographic structure of a subnanomolar affinity ψ[CH(OH)CH$_2$] modified inhibitor complexed to the target enzyme, Poster Abstracts of the 22nd National Medicinal Chemistry Symposium, Austin, TX (July 29-August 2), American Chemical Society, p. 3 (1990).

14. K. Iizuka, T. Kamijo, T. Kubota, K. Akahane, H. Harada, I. Shimoka, H. Umeyama and Y. Kiso, New potent renin inhibitors, *in*: "Peptide Chemistry 1987", T. Shiba and S. Sakakikibara, eds., Protein Research Foundation, Osaka, , p. 649 (1988).

15. H. D. Kleinert, J. R. Luly, P. A. Marcotte, T. J. Perun, J. J. Plattner and H. Stein, Improvements in the stability and biological activity of small peptides containing novel Leu-Val replacements, *FEBS Lett.* **230**:38 (1988).

16. H. Rueger, P. Buhlmayer, W. Fuhrer, R. Goschke, V. Rasetti, J. Stanton and J. Wood, Orally-active renin inhibitors" Abstracts and Slides of the 21st National Medicinal Chemistry Symposium, Minneapolis (June 19-23), American Chemical Society, p. 69 (1988).

17. T. K. Sawyer, D. T. Pals, B. Mao, L. L. Maggiora, D. J. Staples, A. E. DeVaux, H. J. Schostarez, J. H. Kinner and C. W. Smith, Structure-conformation-activity relationships of renin inhibitory peptides having P_1-$P_{1'}$ Xaaψ[CH$_2$NH]Yaa substitutions: molecular modeling and crystallography studies, *Tetrahedron* **44**:661 (1988).

18. T. K. Sawyer, L. L. Maggiora, L. Liu, D. J. Staples, V. S. Bradford, B. Mao, D. T. Pals, B. M. Dunn, R. Poorman, J. Hinzmann, A. E. DeVaux, J. A. Affholter and C. W. Smith, Highly potent ψ[CH$_2$NH]-modified psuedopeptidyl inhibitors of renin: molecular modeling and aspartyl protease selectivity studies, *in*: "Peptides, Chemistry and Biology (Proceedings of the Eleventh American Peptides Symposium)", G. R. Marshall and J. Rivier, eds., Escom Science Publishers, Ae Leiden, p. 46 (1990).

19. S. Thaisrivongs, D. T. Pals, D. W. DuCharme, S. R. Turner, G. L. DeGraaf, J. A. Lawson, S. J. Couch and M. V. Williams, Renin inhibitory peptides. Incorporation of polar, hydrophilic end groups into an active renin inhibitory peptide template and their evaluation in a human renin-infused rat model and in concious sodium-depleted monkeys, *J. Med. Chem.* **34**:633 (1991).

20. D. J. Webb, P. J. O. Manhem, S. G. Ball, G. Inglis, B. J. Leckie, A. F. Lever, J. J. Morton, J. I. S., Robertson, G. D. Murray, J. Menard, A. Hallet, D. M. Jones and M. Szelke, *J. Cardiovasc. Pharmacol.*, **10** (Suppl. 7):S69 (1987).

21. J. Burton, H. Hyun and R. E. TenBrink, The design of substrate analog renin inhibitors, *in*: "Peptides, Structure and Function", Proceedings of the Eighth American Peptide Symposium, V. J. Hruby and D. H. Rich, eds., Pierce Chemical Co., Rockford, IL, p. 559 (1983).

22. J. G. Dann, D. K. Stammers, C. J. Harris, R. J. Arrowsmith, D. E. Davies, G. W. Hardy and J. A. Morton, Human renin: A new class of inhibitors, *Biochem. Biophys. Res. Commun.* **134**:71 (1986).

23. T. J. McQuade, A. G. Tomasselli, L. Liu, V. Karacostas, B., Moss, T. K. Sawyer, R. L. Heinrikson and W. G. Tarpley, A synthetic human immunodeficiency virus protease inhibitor with potent antiviral activity arrests HIV-like particle maturation, *Science* **247**:454 (1990).

24. T. K.Sawyer, J. B. Hester, S. Thaisrivongs, J. Fisher, A. G. Tomasselli, W. G. Tarpley, W. J. Howe and R. L. Heinrikson, Advances in HIV protease inhibitor design, structure-activity and active site molecular modeling, 200th National Meeting of the American Chemical Society, Washington, D.C. (1990).

25. A. R. Sielecki, K. Hayakawa, M. Fujinaga, M. E. P. Murphy, M. Fraser, A. K. Muir, C. T. Carilli, J. A., Lewicki, J. D., Baxter and M. N. G. James, Structure of recombinant human renin, a target for cardiovascular-active drugs, at 2.5 Å resolution, *Science* **243**:1346 (1989).

26. K. D. Watenpaugh, H. M. Einspahr, B. C. Finzel, L. L. Clancy, A. M. Mulichak, D. R. Holland, R. A. Poorman, J. O. Hui, R. L. Heinrikson, K. Murakami, A. Shoda, L. L. Maggiora and T. K. Sawyer, Crystallographic studies of a renin-renin inhibitor complex and comparison with other aspartyl proteinase-inhibitor complexes, Abstracts of the 12th American Peptide Symposium (June 16-21), Boston, MA (1991).

27. M. A. Navia, J. P. Springer, M. Poe, J. Boger and K. Hoogsteen, Preliminary X-ray crystallographic data on mouse submaxillary gland renin and renin-inhibitor complexes, *J. Biol. Chem.*, **259**: 12714 (1984).

28. M. N. G. James and A. R. Sielecki, Molecular structure of an aspartic proteinase zymogen, porcine pepsinogen, at 1.8 Å resolution, *Nature* **319**:33 (1986).

29. R. Bott, E. Subramanian and D. R. Davies, Three-dimensional structure of the complex of *rhizopus chinensis* carboxyl proteinase at 2.5 Å resolution, *Biochemistry* **21**:6956 (1982).

30. K. Suguna, E. A. Padlan, C. W. Smith, W. D. Carlson and D. R. Davies, Binding of a reduced bond peptide inhibitor to the aspartic proteinase from *rhizopus chinensis*: Implications for a mechanism of action, *Proc. Natl. Acad. Sci. U.S.A.* **84**:7009 (1987).

31. K. Parris, D. Hoover and D. Davies, Crystal structures of rhizopuspepsin/inhibitor complexes, *in*: "Structure and Function of Aspartic Proteinases: Genetics, Structures, Mechanisms", Proceedings of the 1990 Aspartic Proteinase Conference, Sonoma, CA (B. M. Dunn, ed.), Plenum Press, New York, in press (1991).

32. L. H. Pearl and T. L. Blundell, The active site of aspartic proteinases, *FEBS Lett.* **174**:96 (1984).

33. B. Veerapandian, J. B. Cooper, A. Sali and T. L. Blundell, X-Ray analyses of aspartic proteinases. III. Three-dimensional structure of endothiapepsin complexed with a transition-state isoster inhibitor of renin at 1.6 Å resolution, *J. Mol. Biol.* **216**:1017 (1990).

34. T. L. Blundell, J. Cooper, S. I. Foundling, D. M. Jones, B. Atrash and M. Szelke, On the rational design of renin inhibitors: X-Ray studies of aspartyl proteinases complexed with transition-state analogues, *Biochemistry* **26**:5587 (1987).

35. M. N. G. James and A. R. Sielecki, Structure and refinement of penicillopepsin at 1.8 Å resolution, *J. Molec. Biol.* **163**:299 (1983).

36. M. N. G. James, A. R. Sielecki, F. Salituro, D. H. Rich and T. Hofmann, Conformational flexibility in the active sites of aspartyl proteinases revealed by a pepstatin fragment binding to penicillopepsin, *Proc. Natl. Acad. Sci. USA* **79**:6137 (1982).

37. C. Abad-Zapareto, T. J. Rydel, D. Neidhart, J. Luly and J. W. Erickson, Inhibitor binding induces structural changes in porcine pepsin, *in*: "Structure and Function of Aspartic Proteinases: Genetics, Structure, Mechanisms", Proceedings of the 1990 Aspartic Proteinase Conference, Sonoma, CA (B. M. Dunn, ed.), Plenum Press, New York, in press (1991).

38. W. Carlson, M. Karplus and E. Haber, Construction of a model for the three-dimensional structure of human renal renin, *Hypertension* **7**:13 (1985).

39. S. Thaisrivongs, B. Mao, D. T. Pals, S. R. Turner and L. T. Kroll, Renin inhibitory peptides. A ß-aspartyl residue as a replacement for the histidyl residue at the P-2 site, *J. Med. Chem.* **33**:1337 (1990).

40. T. Imai, H. Miyazaki, S. Hirose, H. Hori, T. Hayashi, R. Kageyama, H. Ohkubo, S. Nakanishi and K. Murakami, Cloning and sequence analysis of cDNA for human renin precursor, *Proc. Natl. Acad. Sci. U.S.A* .**80**:7405 (1983).

41. R. A. Poorman, D. P. Palermo, L. E. Post, K. Murakami, J. H. Kinner, C. W. Smith, I. Reardon and R. L. Heinrikson, Isolation and characterization of native human renin derived from the Chinese hamster ovary cells, *Proteins: Structure, Function and Genetics* **1**:139 (1986).

42. N. Sakabe, A focusing Weissenberg camera with multi-layer-line screens for macromolecular crystallography, *J. Appl. Cryst.* **16**:542 (1983).

43. J. Hartsuck, S. J. Remington, private communication (1989).

44. D. E. Epps, J. Cheneey, H. Schostarez, T. K. Sawyer, M. Prairie, W. C. Krueger and F. Mandel, Thermodynamics of the interaction of inhibitors with the binding site of recombinant human renin, *J. Med. Chem.* **33**:2080 (1990).

DESIGN OF RENIN INHIBITORS CONTAINING CONFORMATIONALLY

RESTRICTED MIMETICS OF THE P_1-P_1' AND P_1 THROUGH P_2' SITES

Peter D. Williams, Linda S. Payne, Debra S. Perlow,
M. Katharine Holloway, Peter K. S. Siegl, Terry W. Schorn,
Robert J. Lynch, John J. Doyle, John F. Strouse, George P. Vlasuk,
Karst Hoogsteen[+], James P. Springer[+], Bruce L. Bush[+],
Thomas A. Halgren[+], Jan tenBroeke[+], William J. Greenlee[+],
Anthony D. Richards[++], John Kay[++] and Daniel F. Veber

Merck, Sharp and Dohme Research Laboratories
West Point, Pennsylvania, 19486

[+]Merck, Sharp and Dohme Research Laboratories
Rahway, New Jersey, 07065

[++]Department of Biochemistry
University of Wales College of Cardiff
P. O. Box 903, Cardiff, CF1 1ST, Wales
United Kingdom

INTRODUCTION

The clinical efficacy of converting enzyme inhibitors[1] for reducing blood pressure in a large percentage of hypertensive patients has aroused considerable interest in developing agents that interrupt the renin-angiotensin system at other points, for example by blockade of the angiotensin II receptor[2] or by inhibition of the aspartic proteinase, renin. Substrate based design of renin inhibitors in which the scissile P_1-P_1' dipeptide is replaced with a non-hydrolyzable group, often a mimetic of a tetrahedral transition state for amide bond hydrolysis, has provided a useful approach for obtaining a variety of inhibitor structure types,[3] many with K_i's of better than 10^{-9} M. A clinically useful renin inhibitor, however, has not yet emerged due to poor pharmacokinetics (i.e., metabolism or rapid clearance) or poor oral absorption, problems often encountered with peptidic drug targets. An example of this is seen with **1**, a "tetrapeptide" inhibitor[4] that spans the P_4 through P_3' sites of the renin substrate and which utilizes the statine analog, ACHPA[5], as a P_1-P_1' dipeptide replacement (Figure 1). Although **1** is quite potent *in vitro*, very low levels of drug are found in the blood after oral administration to the rhesus monkey at 50 mg/kg, and the half life after intravenous administration is short (< 1 h). Rapid biliary excretion of intact drug has been demonstrated for a number of other renin inhibitors.[6]

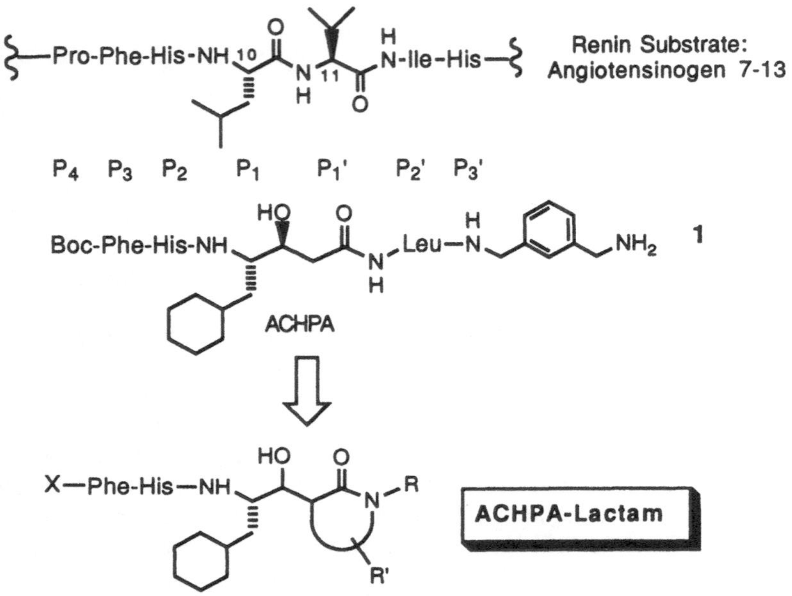

Figure 1. Conformationally constrained ACHPA as a P_1-P_1' mimetic.

As an approach to overcome these problems, we sought to reduce the molecular size and "peptide character" of inhibitors related to **1**, consistent with maintaining good levels of *in vitro* potency. Our approach was to position a P_1' side chain mimetic on the ACHPA backbone, a binding element that is absent in the statine/ACHPA design, such that a peptidic binding element elsewhere in the molecule could be removed. From previous work in our laboratories we were aware that potency is maintained with alkyl substitution on C2 of statine,[7] and also with certain ACHPA tertiary amides. From this we reasoned that cyclization between C2 and the ACHPA amide nitrogen to give an "ACHPA-lactam" (Figure 1) might provide a useful variation in that the lactam ring can serve as a rigid template for positioning hydrophobic groups to probe the enzyme active site in a spatially defined manner, especially with regard to the location of the S_1' subsite. Such ACHPA-lactams should be synthetically accessible via aldol condensation of a lactam enolate with Boc-L-cyclohexylalaninal.[5]

RESULTS AND DISCUSSION: *IN VITRO* RENIN INHIBITION

Aldol condensation of the lithium enolate of 1-methyl-2-pyrrolidinone with Boc-L-cyclohexylalaninal gave a chromatographically separable mixture of four diastereomers in roughly equal proportion. Conversion of each of these isomers to its Boc-L-phenylalanyl-L-histidyl derivative provided inhibitors **2a-d** for evaluating the the enzyme's diastereomer preference. As can be seen from the results given in Table 1 using human plasma renin, there is a marked preference for the 2S, 3S diastereomer (assignment of stereochemistry in **2a** comes from single crystal X-ray analysis of this isomer at the aldol stage). Quite remarkably, **2b**, the isomer which differs from **2a** only in its configuration at C2, loses greater than 3 orders of magnitude of potency, which gives an indication of the enzyme's sensitivity to the spatial arrangement of functionality in this part of the inhibitor structure. Both 3R epimers **2c** and **2d** are greater than an order of magnitude less potent than **2a**. The potency of **2a** is improved relative to the comparably substituted ACHPA amide **3**, suggesting that there may

Table 1. Diastereomer and ring size preference for ACHPA-lactam inhibitors

	n	C2	C3	Renin [a]	Cathepsin D [b]	Cathepsin E [b]
		Configuration		K_I^* or IC_{50} (nM)		
2a	1	S	S	10	600*	>10,000
2b	1	R	S	>20,000	>10,000	>10,000
2c	1	?[c]	R	280	>10,000	>>10,000
2d	1	?[c]	R	470	>10,000	>>10,000
4	2	S	S	47	-	-
5	3	S	S	200	135*	170*

[a] human plasma renin assay, pH 7.4; [b] human enzyme, pH 3.1
[c] configuration not determined

be some inherent benefit derived from the lactam conformational constraint. Having established the preferred ACHPA-lactam diastereomer, we next examined the effect of lactam ring size. As shown in Table 1, the 6- and 7-membered 2S, 3S diastereomers **4** and **5** are both poorer inhibitors than **2a**, with potency decreasing upon increasing lactam ring size. The 5-membered ACHPA-lactam **2a** was thus chosen for further studies in which substituents are incorporated on the lactam ring to probe the enzyme active site.

To aid in the design/discovery process, a computer model of human renin was constructed for use in assessing various inhibitor-enzyme interactions. The enzyme model was constructed by aligning the human renin sequence with the superimposed X-ray structures of the fungal enzymes endothiapepsin,[8] rhizopuspepsin[9] and penicillopepsin.[10] Structurally conserved regions were taken from the rhizopuspepsin X-ray structure, and variable regions were taken from the proteinase whose sequence best aligned with the human sequence in that region. X-ray structures of several fungal enzyme-inhibitor complexes[11] provided starting geometries for docking the ACHPA-lactam inhibitors in the renin enzyme model. Docked structures were then energy minimized in the static active site using OPTIMOL, a modified MM2 force field program created at Merck. A schematic representation of the hydrogen bonding arrangement for the **2a**-enzyme model complex is shown in Figure 2.

Boc-Phe-His-NH **3** (IC_{50} = 45 nM)

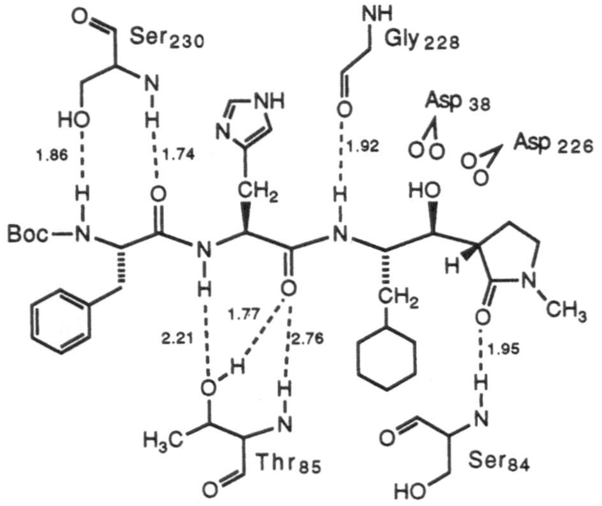

Figure 2. Schematic representation of the hydrogen bonding network for **2a** docked in the human renin model. Distances are given in angstroms.

Of note in the lactam portion of the inhibitor is the hydrogen bond formed between the lactam carbonyl oxygen and the backbone NH of Ser_{84} in the flap region of the enzyme. This interaction provides a rationalization for the low potency of **2b**, which in our enzyme model cannot form this hydrogen bond without the rest of the lactam ring encountering unfavorable van der Waals contacts with the enzyme. Other lactam ring rotamers of **2b** also result in a poor fit due to unfavorable contacts with the enzyme. Comparison of the proposed enzyme-bound forms of **2a** and the renin partial substrate, angiotensinogen$_{7-13}$ in which the scissile amide bond is present as its tetrahedral hydrate, suggested that small substituents on the 4- or 5-positions of the γ-lactam ring in **2a** can occupy the same space as the P_1' side chain of substrate, while substituents on the lactam nitrogen, depending on their size and flexibility, can be oriented out towards solvent or towards the P_2' and P_3' positions. The experimental results for some of these substitutions are presented in the following discussion.

The effect of substitution on the lactam nitrogen was investigated with compounds **6a-d** (Table 2). The data indicate that although these changes do not offer substantial improvements to inhibitor potency, the wide range of substituent sizes and polarities tolerated at this position could be useful for varying physical properties that affect drug absorption and distribution (e.g., log P, aqueous solubility, ionic state). The N-cyclohexyl ACHPA-lactam derivative **6a** gave crystals from methanol solution that were suitable for X-ray analysis. In the solid state, **6a** was found to adopt an extended conformation (pseudo-ß-sheet), similar to its proposed bound conformation in our enzyme model (Figure 3). Rotations of 60° and 95° about the Phe ψ and ACHPA-lactam C2-C3 bonds in the solid state conformer produce a backbone conformation very similar to the modeled bound conformer. To the extent that the solid state conformation of **6a** resembles its solution conformation, our model of the enzyme-inhibitor complex suggests that only minimal torsional reorganization of the inhibitor is required for productive binding.

The availability of homochiral 5-substituted 2-pyrrolidinones from D- or L-pyro-glutamic acid provided a convenient entry to diastereomerically pure 5-substituted ACHPA-lactam inhibitors (Table 2). From the series of inhibitors obtained using L-pyro-glutamic acid (**7a-d**), it can be seen that potency decreases with increasing substituent size. Even with the smallest substituent, no benefit is derived (compare **7a** with **2a**).

Table 2. ACHPA-Lactam substituent effects

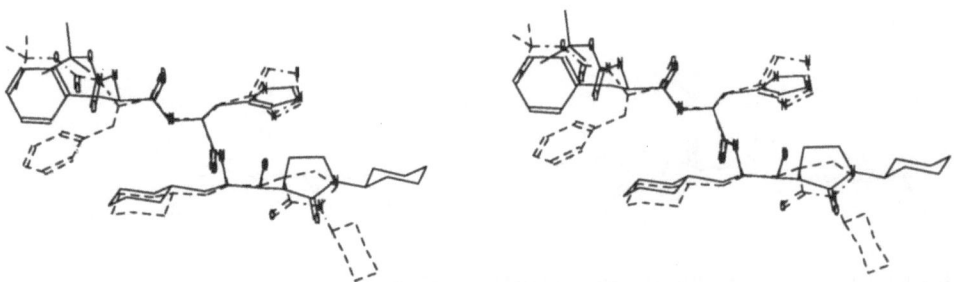

Boc-Phe-His-NH ... (structure with R^1, R^2, R^3 on lactam ring)

	R^1	R^2	R^3	K$_i$* or IC$_{50}$ (nM) Renin [a]	Cathepsin D [b]	Cathepsin E [b]
2a	CH$_3$	H	H	10	600*	>10,000
6a	c-C$_6$H$_{11}$	H	H	10	500*	240*
6b	-CH$_2$CH$_2$OH	H	H	7.7	-	-
6c	-CH$_2$CH$_2$OCH$_2$Ph	H	H	24	-	-
6d	-(CH$_2$)$_3$NHCH$_2$Ph	H	H	14	-	-
7a	CH$_3$	-CH=CH$_2$	H	8.0	175*	360*
7b	CH$_3$	-CH$_2$OH	H	14	-	-
7c	CH$_3$	-CH$_2$OCH$_3$	H	71	-	-
7d	CH$_3$	-CH$_2$OCH$_2$Ph	H	4,400	115*	ca. 10,000
8	CH$_3$	H	-CH=CH$_2$	2.5	250*	165*
9a	CH$_3$	CH$_3$	CH$_3$	1.3	150*	>>10,000
9b	n-C$_4$H$_9$	CH$_3$	CH$_3$	2.4	-	-

[a] human plasma renin assay, pH 7.4; [b] human enzyme, pH 3.1

On the other hand, a small hydrophobic group on the other face of the lactam ring, i.e., **8**, the isomer of **7a** obtained from D-pyro-glutamic acid, proved to be a beneficial substitution. Optimal potency in the 5-substituted series was obtained with the geminally substituted analog **9a**, the potency of which is similar to larger ACHPA-containing inhibitors such as **1**. The ACHPA-lactam terminus in **9a** thus represents an advance in that it provides an equipotent yet significantly lower molecular weight replacement for the peptidic C-

Figure 3. Comparison of solid state conformer (solid) and proposed enzyme-bound conformer (dashed) of **6a**.

terminus of ACHPA-containing inhibitors. Using our renin model, the increased potency of
9a relative to **2a** is consistent with having located the enzyme S_1' binding site, as
superposition of the bound forms of **9a** and hydrated angiotensinogen[7-13] shows a good
spatial overlap of the ACHPA-lactam gem-dimethyl group and the substrate P_1' side chain. It
is also interesting to note the structural similarity of the ACHPA-lactam terminus in **9a** with
equipotent inhibitors containing hydroxyethylene (**10**) and alkyl diol[12] (**11**) non-peptide C-
termini (Figure 4). In addition to the cyclohexylmethyl P_1 side chain and "transition state"
hydroxyl group, each contains a second polar group capable of accepting a hydrogen bond
and another hydrophobic group to mimic the P_1' side chain. When these structures are
docked in our renin model, the aforementioned functionality overlap nicely. Each of these
non-peptide C-termini thus represents a unique structural array for producing the same set of
interactions with the enzyme.

The N-butyl group in **9b** (Table 2) did not increase potency relative to the N-methyl
analog **9a**, although our model of the inhibitor-enzyme complex suggested that the flexible
butyl group can be oriented into the site occupied by the P_2' side chain of hydrated
angiotensinogen[7-13]. We therefore became interested in the possibility of further enhancing
potency by constructing constrained analogs in which the lactam N-substituent is forced to
occupy the S_2' site. Using our renin model, several bicyclic lactam templates of the general
type shown in Figure 5 were examined. Oxa-indolizidinone **13** emerged as a reasonable
target from both modeling and synthetic considerations, with the R group serving as a
mimetic of the P_2' side chain. An enantio- and stereospecific preparation of the oxa-
indolizidinone template was developed that allows for variations of the R group (Figure 5).
Acetate **12** was obtained from L-pyro-glutamic acid in 5 steps. Allylsilane trapping of the

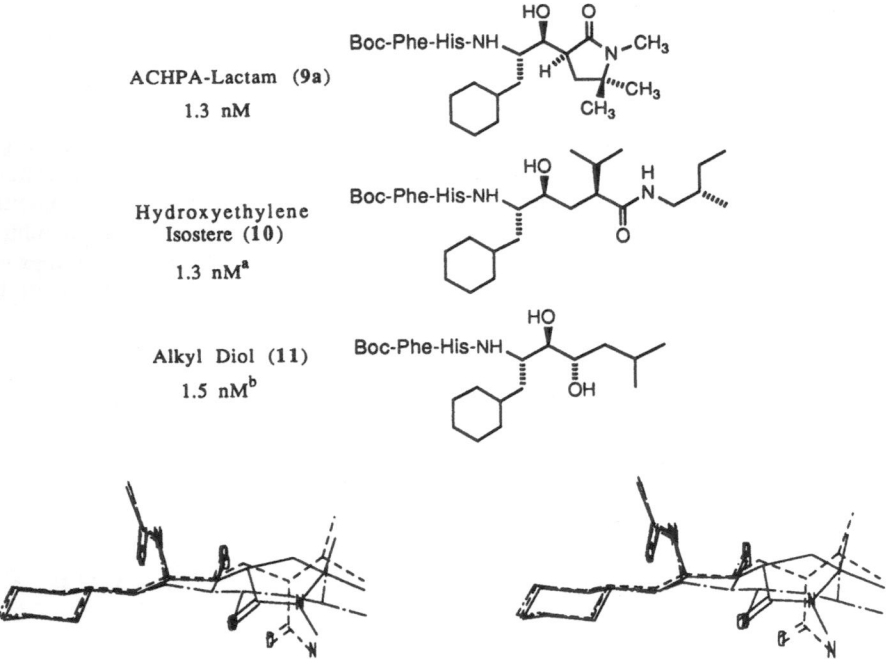

Figure 4. Comparison of three types of non-peptide C-termini. Stereoplot shows overlap of C-termini (9a
solid; 10 dot-dashed; 11 dashed) for inhibitors docked in the renin enzyme model. [a]W. J. Greenlee, S.
deLaszlo, Merck Sharp and Dohme Research Laboratories, Rahway, New Jersey, unpublished results;
[b]reference 12.

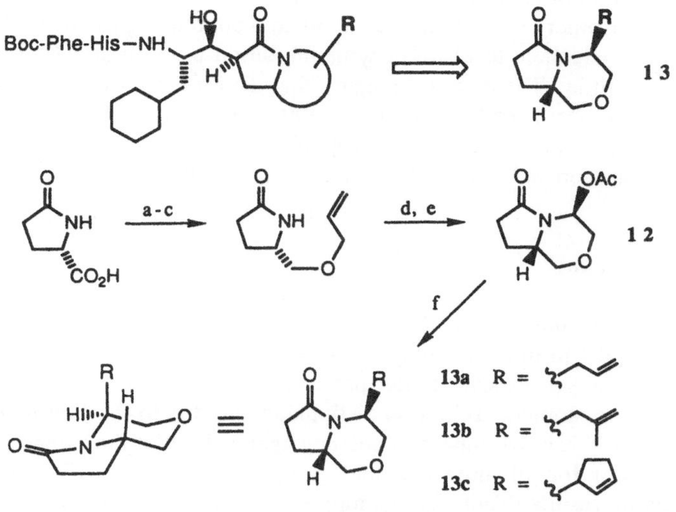

Figure 5. Design and synthesis of a conformationally restricted template for probing the $S_{2'}$ site. [a]$SOCl_2$, MeOH; [b]$NaBH_4$; [c]n-BuLi, THF, allyl bromide; [d]O_3, -78°C, Me_2S; [e]Ac_2O, DMAP; [f]BF_3OEt_2, allyltrimethylsilane derivative, CH_2Cl_2, 0°C.

Table 3. Substituent effects in the bicyclic ACHPA-lactam series

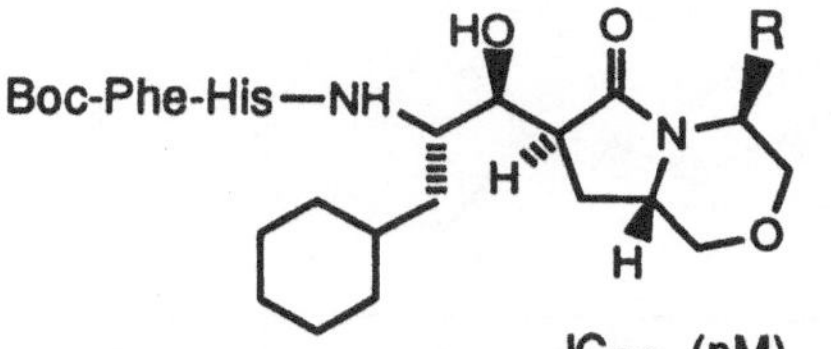

	R	IC$_{50}$ (nM)	
		PRA [a]	purified renin [b]
14a		1.3	0.74
14b		1.0	0.67
14c	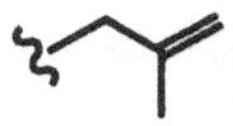	0.97	0.73
14d	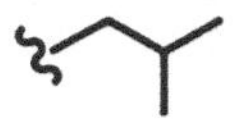	2.0	0.82
14e	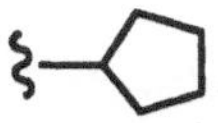	1.0	0.98

[a]human plasma renin assay, pH 7.4

[b]purified human kidney renin assay, pH 7.4

acyliminium species derived from acetate **12** by treatment with a Lewis acid gave exclusively the axially-substituted products, **13a-c**. Aldol condensation of these lactams with Boc-L-cyclohexylalaninal, isomer separation, and amino acid couplings provided inhibitors **14a-e** (Table 3). Although excellent levels of enzyme inhibition were observed for these analogs, potency varied very little. These results suggest that the different hydrophobic groups may not be involved in a very specific interaction with the enzyme. Indeed, a more precise assessment of the contribution of these hydrophobic groups to the overall binding energy must await the preparation and testing of the inhibitor where R is hydrogen.

Enzyme Selectivity Studies

Several of the monocyclic ACHPA-lactam inhibitors were examined for their selectivity with respect to human cathepsin D and E. From the results given in Table 1, cathepsin D shows the same diastereomer preference as renin, i.e., for the 2S, 3S isomer. With cathepsin E, the isomers are not as distinguishable due to the low potency. The 7-membered ring analog **5** shows enhanced potency against both of the cathepsins such that a non-specific inhibitor for all three enzymes results. The effect of lactam substitution produced interesting results (Table 2). Compared to **2a**, substitution with a hydrophobic group enhances potency against cathepsin D in all of the examples studied, whereas cathepsin

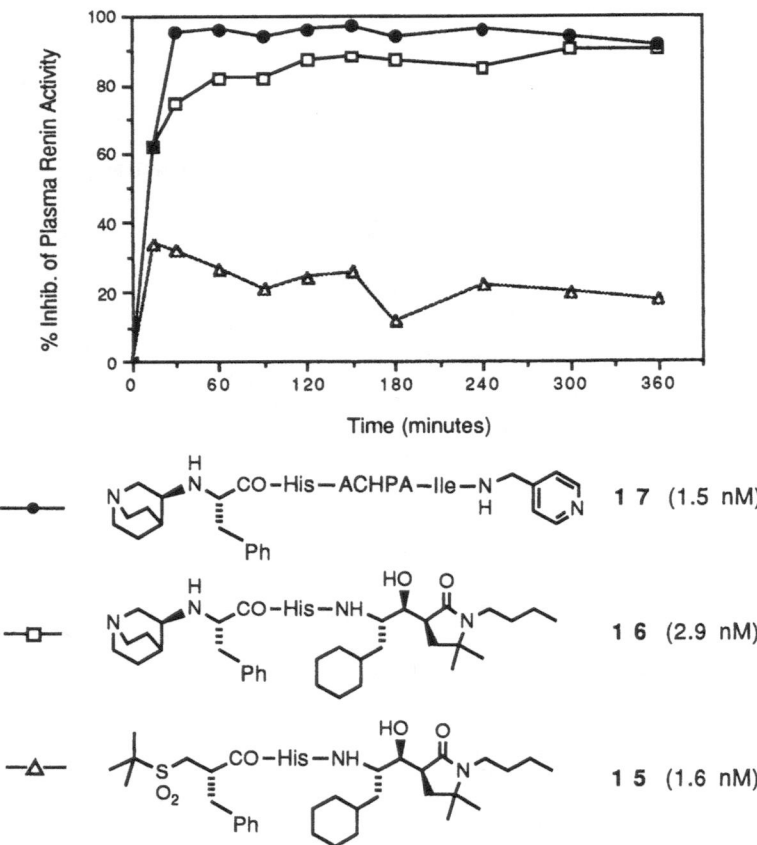

Figure 6. Inhibition of plasma renin activity after oral administration to sodium-deficient rhesus monkeys (12 μmol/kg; n=2).

E responds favorably to only selected substitutions (**6a, 7a, 8**). The benzyloxymethyl substituent (**7d**) substantially reduces potency against both renin and cathepsin E, resulting in a selective cathepsin D inhibitor. The optimal substitution pattern for renin inhibition (**9a**) provides greater than 100-fold selectivity against both cathepsins.

In vivo *studies*

In order to assess the ACHPA-lactam peptidomimetic strategy as it relates to *in vivo* properties, three nearly equipotent inhibitors were administered orally to conscious, sodium-depleted rhesus monkeys at a dose of 12 μmol/kg (Figure 6). These inhibitors incorporate 2S-(t-butylsulfonylmethyl)-3-phenylpropionyl[13] or N^α-(3S-quinuclidyl)-L-phenylalanyl[14] replacements for the Boc-Phe group in order to avoid the possible chemical instability of the Boc group in the gut. The plot of % inhibition of monkey plasma renin versus time provides a measure of duration of action in the blood. In comparing the two ACHPA-lactam inhibitors **15** and **16**, it is evident that structural changes at the N-terminus can greatly affect the amount of drug present in the blood. Interestingly, the comparison between **16** and the ACHPA-containing inhibitor **17** indicates that the latter has a somewhat better duration of action, a result not anticipated based on the higher molecular weight and increased "peptide character" of **17** relative to **16**. The p.o./i.v. ratio as determined by measuring drug levels in the blood by bioassay indicates that the oral bioavailability for each of these inhibitors is too low (i.e., <5%) to be considered for further investigation as a clinical candidate. Thus, additional variations in molecular size and physical properties are needed to define parameters that improve absorption for these inhibitors.

CONCLUSIONS

Two new series of renin inhibitors that incorporate mono- or bicyclic lactam-constrained ACHPA groups (ACHPA-lactams) encompassing the P_1-P_1' or P_1-P_1'-P_2' sites were developed. These lactams offer special advantage in that their conformational rigidity allows structural modifications which provide new insights about the active site of renin. Using a computer model of human renin, docking experiments with the most potent of these inhibitors suggests that hydrophobic substituents can be positioned on the lactam templates to mimic the P_1' or P_2' side chains of angiotensinogen. The conformation of ACHPA-lactam **6a** was determined by X-ray crystallographic analysis and compares very favorably to the proposed enzyme-bound form in our renin model. Variations in lactam ring size and substitution pattern can be used to alter enzyme selectivity with respect to cathepsin D and E, which suggests a broader application of the ACHPA-lactam motif for obtaining selective inhibitors of other aspartic acid proteinases. Compared to ACHPA-containing inhibitors of similar potency, **9a** represents a significant reduction in molecular size and peptide character. Improved pharmacokinetic properties and oral bioavailability in the rhesus monkey, however, did not follow from these changes. The ACHPA-lactam variation thus represents one iteration in a stepwise path to designing true peptidomimetics of therapeutic utility.

REFERENCES

1. a) M. A. Ondetti and D. W. Cushman, *Ann. Rev. Biochem* **51**:283 (1982). b) C. S. Sweet and E. H. Blaine, *in*: "Cardiovascular Pharmacology," M. J. Antonaccio, ed., Raven Press, New York, (1984) 119. c) M. J. Wyvratt and A. A. Patchett, *Med. Res. Rev.* **5**:483 (1985).

2. a) P. C. Wong, A. T. Chiu, W. A. Price, M. J. M.C. Thoolen, D. J. Carini, A. L. Johnson, R. I. Taber and P. B. M. W. M. Timmermans, *J. Pharmacol. Exp. Ther.* **247**:1 (1988). b) J. V. Duncia, A. T. Chiu,

D. J. Carini, G. B. Gregory, A. L. Johnson, W. A. Price, G. J. Wells, P. C. Wong, J. C. Calabrese and P. B. M. W. M. Timmermans, *J. Med. Chem.* **33**:1312 (1990). c) D. J. Carini, J. V. Duncia, A. L. Johnson, A. T. Chiu, W. A. Price, P. C. Wong and P. B. M. W. M. Timmermans, *J. Med. Chem.* **33**:1330 (1990).

3. a) W. J. Greenlee, *Pharm. Res.* **4**:364 (1987). b) W. J. Greenlee, *Med. Res. Rev.* **10**:173 (1990).

4. M. G. Bock, R. M. DiPardo, B. E. Evans, R. M. Freidinger, K. E. Rittle, L. S. Payne, J. Boger, W. L. Whitter, B. I. LaMont, E. H. Ulm, E. H. Blaine, T. W. Schorn and D. F. Veber, *J. Med. Chem.* **31**:1918 (1988).

5. J. Boger, L. S. Payne, D. S. Perlow, N. S. Lohr, M. Poe, E. H. Blaine, E. H. Ulm, T. W. Schorn, B. I. LaMont, T. -Y. Lin, M. Kawai, D. H. Rich and D. F. Veber, *J. Med. Chem.* **28**:1779 (1985).

6. a) J. Boger, C. D. Bennett, L. S. Payne, E. H. Ulm, E. H. Blaine, C. F. Homnick, T. W. Schorn, B. I. LaMont and D. F. Veber, *Regul. Pept. (Suppl. 4)*, 8 (1985). b) J. J. Plattner, P. A. Marcotte, H. D. Kleinert, H. H. Stein, J. Greer, G. Bolis, A. K. L. Fung, B. A. Bopp, J. R. Luly, H. L. Sham, D. J. Kempf, S. H. Rosenberg, J. F. Dellaria, B. De, I. Merits and T. J. Perun, *J. Med. Chem.* **31**:2277 (1988).

7.a) D. F. Veber, M. G. Bock, S. F. Brady, E. H. Ulm, D. W. Cochran, G. M. Smith, B. I. LaMont, R. M. DiPardo, M. Poe, R. M. Freidinger, B. E. Evans and J. Boger, *Biochem. Soc. Trans.* **12**:956 (1984). b) M. G. Bock, R. M. DiPardo, K. E. Rittle, J. Boger, R. M. Freidinger and D. F. Veber, *J. Chem. Soc., Chem. Comm.* 109 (1985).

8. L. H. Pearl and T. L. Blundell, *FEBS Lett.* **174**:96 (1984).

9. K. Suguna, E. A. Padlan, C. W. Smith, W. D. Carlson and D. R. Davies, *Proc. Nat. Acad. Sci. USA*, **84**:7009 (1987).

10. M. N. G. James and A. R. Sielecki, *J. Mol. Biol.* **163**:299 (1983).

11. a) S. I. Foundling, J. Cooper, F. E. Watson, A. Cleasby, L. H. Pearl, B. L. Sibanda, A. Hemmings, S. P. Wood, T. L. Blundell, M. J. Valler, C. G. Norey, J. Kay, J. Boger, B. M. Dunn, B. J. Leckie, D. M. Jones, B. Atrash, A. Hallet, M. Szelke, *Nature* **327**:349 (1987). b) D. R. Davies and K. Suguna, Preliminary coordinate sets for enzyme-inhibitor complexes of rhizopuspepsin were graciously provided.

12. J. R. Luly, N. BaMaung, J. Soderquist, A. K. L. Fung, H. Stein, H. D. Kleinert, P. A. Marcotte, D. A. Egan, B. Bopp, I. Merits, G. Bolis, J. Green, T. J. Perun and J. J. Plattner, *J. Med. Chem.* **31**:2264 (1988).

13. P. Buehlmayer, A. Caselli, W. Fuhrer, R. Goeschke, V. Rasetti, H. Rueger, J. L. Stanton, L. Criscione and J. M. Wood, *J. Med. Chem.* **31**:1839 (1989).

14. W. J. Greenlee, J. tenBroeke, S. de Laszlo, P. K. Chakravarty, V. J. Camara. D. G. Hangauer, K. J. Fitch, C. S. Sarnella, A. A. Patchett, P. D. Williams, D. S. Perlow, D. V. Veber, R. J. Lynch, J. J. Doyle, T. W. Schorn, J. F. Strouse and P. K. S. Siegl, *in*: "Peptides: Chemistry, Structure, and Biology. Proceedings of the Eleventh American Peptide Symposium," J. E. Rivier and G. R. Marshall, eds., Escom Leiden, 411, 1990.

EFFICIENT MUTAGENESIS, EXPRESSION AND PURIFICATION OF

PROCATHEPSIN D

Deepali Sachdev, Jeff Schorey and John Chirgwin

Department of Medicine, Division of Endocrinology
and Department of Biochemistry
University of Texas Health Science Center
7703 Floyd Curl Drive
San Antonio, Texas   78284-7877

INTRODUCTION

High level expression of proteins in mammalian cells and the ability to distinguish an expressed protein from its endogenous counterpart are necessary when examining the structure/function relationships of many ubiquitously expressed proteins, such as the lysosomal aspartyl protease procathepsin D. We have achieved a high level of protein expression by transient transfection into 293 cells (a human embryonal kidney cell line transformed with adenovirus). Distinction between endogenous and introduced protein has been accomplished by fusing a 13 amino acid segment derived from the c-*myc* sequence to the C-terminus of procathepsin D. This fusion was devised by H.R.B. Pelham (1988). The *myc* extension permits the isolation and purification of the expressed protein by immunoaffinity chromatography with an available monoclonal antibody, 9E10, directed against 11 amino acids of the peptide. Investigation of structure/function relationships requires repeated rounds of mutagenesis *in vitro* followed by expression and assay in mammalian cells. We have facilitated this process by constructing a vector that can be shuttled between bacterial and mammalian hosts without any intermediate subclonings.

METHODS AND RESULTS

Vector

The expression vector used is called pSKCMVPCDMf1 and is 5.6 kb in size. It was constructed by placing the complete coding sequence (Faust *et al.*, 1985) of procathepsin D downstream of the cytomegalovirus (CMV) immediate early promoter. The termination codon of cathepsin D was mutated by Pelham (1988) allowing read-through into the added *myc* extension. The prokaryotic region of the vector is derived from bluescript (Stratagene Corporation, San Diego, CA), which gives high yields of DNA. An f1 segment was added to permit preparation of single-stranded phagemid DNA. The vector also carries an SV40 origin of replication, an *E. coli* origin of replication and ampicillin resistance sequences. DNA manipulations were by standard means. Plasmid DNA was purified by the simple polyethylene glycol precipitation protocol described in Sambrook *et al.* (1989). DNA so prepared requires no cesium banding or column purification and efficiently and reproducibly transfects mammalian cells.

Transient expression

The 293 cell line constitutively expresses adenoviral early proteins, enhance stable transcription from a number of viral promoters, including CMVIE. Efficient translation of RNA transcripts is enhanced by cotransfection with pAdVA (viral associated) DNA. This system has been shown to give very high levels of secretion of growth hormone by Gorman *et al.* (1990). 293 cells are grown to subconfluency, then DNA-calcium phosphate precipitate is added to the cells. Three hours later, the cells are given a 15 % glycerol shock. After 48 hours the protein secreted into the medium is collected. In the case of labeled protein, the cells are changed into medium containing ^{35}S-methionine for the final 12 hours. The procathepsin D-*myc* fusion protein is secreted into the medium via the constitutive secretory pathway, probably by saturating the lysosomal mannose 6-phosphate receptor pathway. The cells secrete well in synthetic media such as PC-1 (Ventrex), which contains less than 1 mg/ml protein, thus facilitating purification of recombinant proteins. Labeled proteins were analyzed by immunoprecipitation using immobilized protein A in place of a second antibody. Immunoprecipitates were analyzed by denaturing polyacrylamide gel electrophoresis in the presence of SDS and mercaptoethanol. Labeled proteins were visualized by fluorography.

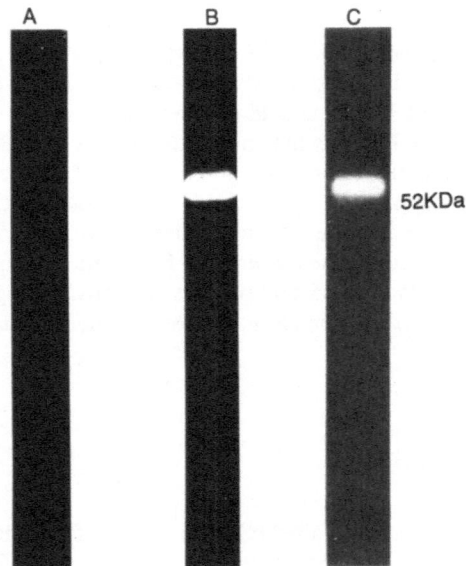

Figure 1. Characterization of 293 cell conditioned medium following CaPO$_4$ transfection and overnight labeling with ^{35}S methionine. Lane A: Immunoprecipitation of mock transfected medium with cathepsin D antibody. Lane B: pSKCMVPCDMf1 transfected medium immunoprecipitated with cathepsin D antibody. Lane C: Procathepsin D-*myc* fusion protein immunoaffinity purified with immobilized 9E10 monoclonal. Same starting material as lane B, not immunoprecipitated.

General procedures have been described by Faust *et al.* (1987). A rabbit polyclonal antibody was raised against human cathepsin D purified from placentas by affinity chromatography on pepstatinyl agarose.

Characterization of expressed procathepsin D-myc fusion protein

The 9E10 mouse monoclonal IgG (Evan *et al.*, 1985) was purified from ascites fluid by binding to protein A Sepharose. The antibody was covalently crosslinked *in situ* to the resin with dimethyl pimelimidate using a commercial kit (Pierce Chemicals, Rockford, IL). The fusion protein was purified from the 293 conditioned medium by immunoaffinity chromatography on this column. Following washing of the column, the purified fusion protein could be eluted with glycine/HCl pH 2.8, glycine/NaOH pH 11, or 1 M urea. The last is the preferred method since it avoids the autoactivation that occurs at low pH. We have observed loss of enzymatic activity following high pH elution.

Enzymatic activity

The procathepsin D-*myc* fusion protein exhibits proteolytic activity upon acid activation. The pSKCMVPCDMf1 transfected 293 conditioned medium has approximately 20 times more activity relative to that in estrogen-induced MCF_7 human breast tumor cell line conditioned medium, as assayed by ability to release 3% TCA soluble peptides from ^{14}C-labeled hemoglobin at pH 4.5. The assay was a micromodification of the classical procedure of Barrett (1977).

Endocytosis

The secreted procathepsin D-*myc* fusion protein can be endocytosed by mammalian cells, such as the CHO line. Labeled fusion protein was taken up via the plasma membrane mannose 6-phosphate receptor system (Kornfeld, 1987) and delivered to the lysosome. Ligand specificity for the receptor was demonstrated by competitive inhibition with mannose 6-phosphate but not galactose 6-phosphate. Delivery to the lysosome was confirmed by the proteolytic conversion of the 52 kDa pro form to the two-chain mature enzyme. This step occurs in the lysosome and is seen as 34+14 kDa bands on denaturing gels (data not shown).

CONCLUSION AND DISCUSSION

Using the above methods, we have shown: 1) High level of expression of procathepsin D-*myc* fusion protein; 2) Purification of the fusion protein from the conditioned medium by immunoaffinity chromatography of the *myc* extension; 3) Enzymatic activity and mannose 6-phosphorylation of the oligosaccharide side chains of the secreted recombinant proteins.

A variety of single amino acid changes in the cathepsin D sequence have been introduced by oligonucleotide-directed mutagenesis *in vitro* by the method of Kunkel *et al.*, 1987. Most of these mutants are efficiently expressed in mammalian cells and show proteolytic activity upon acid activation. Since no recloning steps are required when shuttling between bacterial and mammalian host cells, it is practical to make and test a given mutation in a few weeks, and a number of mutants can be generated simultaneously. It is also practical to accumulate multiple mutations by sequential rounds of mutagenesis.

The 293 cell system gives usefully high levels recombinant protein in conditioned medium. These cells are very hardy and express well in serum-free conditions. The 293 cells give substantially higher production of recombinant protein than the commonly used monkey kidney *cos* cell line (Gorman *et al.*, 1990). We presume that the high level of expressed procathepsin D within the golgi apparatus of the transfected cells exceeds the capacity of the mannose 6-phosphate receptor pathway leading to the lysosome and that this overloading results in secretion of the protein into the medium via the constitutive secretory default pathway. This general approach to exploration of procathepsin D structure/function relationships via rapid mutagenesis, re-expression, and purification should be applicable to other members of the aspartyl protease gene family, provided that C-terminal extension does not interfere with folding, stability, or activity.

ACKNOWLEDGMENTS

We thank Dr. Hugh Pelham for generously providing the CDM cathepsin D-*myc* fusion DNA, Sun Pil Brown for maintaining the mammalian cells and Gloria Peché for preparing the manuscript. The work described was supported by grants from the Veterans Administration and the Welch Foundation for Chemical Research (grant AQ1123) to JMC, who is an Associate Research Career Scientist of the VA.

REFERENCES

Barrett, A. J., 1977, Cathepsin D and other carboxyl proteases, *in*: "Proteinases in Mammalian Cells and Tissues", A. J. Barrett, ed., Elsevier/North-Holland Press, New York, pp. 209-248.

Evan, G. I., Lewis, G. K., Ramsay, G. & Bishop, J. M., 1985, Isolation of monoclonal antibodies specific for human c-*myc* proto-oncogene products, *Mol. Cell. Biol.* **5**:3610-3616.

Faust, P. F., Kornfeld, S. & Chirgwin, J. M., 1985, Cloning and sequence analysis of cDNA for human cathepsin D, *Proc. Natl. Acad. Sci. U.S.A.* **82**:4910-4914.

Faust, P. F., Chirgwin, J. M. & Kornfeld, S., 1987, Renin, a secretory glycoprotein, acquires phosphomannosyl residues, *J. Cell Biol.* **105**:1947-1955.

Gorman, C. M., Gies, D. R. & McCray, G., 1990, Transient production of proteins using an adenovirus transformed cell line, *DNA and Protein Engineering Techniques* **2**:3-10.

Kornfeld, S., 1987, Trafficking of lysosomal enzymes, *FASEB J.* **1**:462-468.

Kunkel, T. A., Roberts, J. D. & Zakour, R. A., 1987, Efficient mutagenesis without selection, *Methods Enzymol.* **154**:367-382.

Pelham, H. R. B., 1988, Evidence that luminal ER proteins are sorted from secreted proteins in a post-ER compartment, *EMBO J.* **7**:913-918.

Sambrook, J., Fritsch, E. F. & Maniatis, T., 1989, "Molecular Cloning", 2nd ed., Cold Spring Harbor Laboratory Press, New York.

MAPPING OF LYSOSOMAL TARGETING DETERMINANTS OF CATHEPSIN D

Jeff Schorey and John Chirgwin

Department of Medicine
Division of Endocrinology
and Department of Biochemistry
University of Texas Health Science Center at San Antonio
7703 Floyd Curl Drive
San Antonio, Texas   78284-7877

INTRODUCTION

The precursors of soluble mammalian lysosomal hydrolases are sorted from other, secretory glycoproteins as they exit the golgi apparatus, by virtue of binding to the mannose 6-phosphate glycoprotein receptor. This recycling receptor is responsible for the delivery to the lysosome of precursors carrying mannose 6-phosphate tags. It remains unclear how the cell distinguishes between lysosomally and non-lysosomally destined proteins, such that only the former are efficiently phosphorylated at the 6 position of mannose residues on their oligosaccharide side chains. Current evidence suggests that this discrimination is based on recognition of a stable domain in the folded structure of the lysosomal precursors and not of a primary sequence signal (Kornfeld, 1987). Cathepsin D is the only member of the aspartic protease gene family known to be lysosomal. Most of the other lysosomal cathepsins are thiol proteases. Cathepsin D shows a high degree of sequence identity to other aspartic proteases (Faust *et al.*, 1985) and was therefore expected to have a 3-dimensional structure similar to that of pepsin, for which high resolution structures have been described. Thus, cathepsin D is a model of choice for identifying those residues which form the recognition domain necessary for phosphorylation and lysosomal targeting. To aid in prediction of the recognition domain, models of human, porcine and bovine cathepsins D were generated. Potentially important residues on cathepsin D were identified by comparison of the cathepsin D models to other aspartic protease structures. Residues conserved among the cathepsins D but not between them and renin and pepsin were changed by site-directed mutagenesis and the corresponding proteins expressed. The mutated cathepsins D were tested for the presence of phosphorylated mannose residues.

Structure and Function of the Aspartic Proteinases
Edited by B.M. Dunn, Plenum Press, New York, 1991

METHODS AND RESULTS

Cathepsin D model

The graphics program FRODO and the X-ray crystal structure data for porcine pepsinogen were used in making the human cathepsin D structural model in collaboration with H. Lueke and F. Quiocho at the Howard Hughes Medical Institute at Baylor College of Medicine. Atomic coordinates were obtained from Protein Data Base files. The residues on the pepsinogen structure were substituted with the corresponding human cathepsin D residues (161 substitutions total). One deletion and three insertions were also made. The crude structure was refined by the energy minimization program, Amber. At present, the model lacks the pro sequence and the 11 amino acid insertion following residue 91 in the pepsin structure. Models of the porcine and bovine cathepsin D were made in a similar manner. The cathepsin D models were compared to a renin model (Sibanda *et al*, 1984) and to the X-ray structures of porcine pepsinogen, endothiapepsin and penicillopepsin. As expected, the cathepsin D models showed a high degree of conformational conservation with the known aspartic protease structures. A striking feature is the high number of surface lysine residues in cathepsin D relative to other aspartic proteases. For example, human cathepsin D has 19 lysines, bovine chymosin 6, mouse renin 12 and porcine pepsin 1.

Mutagenesis

The procathepsin D expression vector pSKCMVPCDMf1 (described by Sachdev *et al*., in this volume) was propagated in the *E. coli* host CJ236 in the presence of uridine and single stranded phagemid DNA rescued with M13 helper phage. Purified DNA was converted to double stranded form *in vitro* as described by Kunkel *et al*. (1987). Mutagenic 17-base oligonucleotides contained central single-base mismatches, changing a lysine codon to a glutamic by substituting G for A at the first base of the codon. Double stranded DNA was transformed into a standard *E. coli* host which preferentially degrades the unmutated, dU-containing first strand. The colonies were screened by stringent hybridization to ^{32}P end-labeled elongation primers. Parental colony hybrids melt out at significantly lower temperature than do the successful mutants, which no longer have any mismatches to the mutagenic oligonucleotides. The positives were picked for characterization and in certain cases, used for subsequent rounds of mutagenesis. The mutations produced and tested include: the single mutations: LYS-172-GLU, LYS-206-GLU, and LYS-276-GLU (LYS 173, 207 and 278 respectively using pepsin numbering): the double mutations: LYS-172,206-GLU, LYS-172,276-GLU, LYS-206,276-GLU; and the triple mutation LYS-172,206,276-GLU.

Transfection and expression

Parental and mutant DNA derivatives of pSKCMVPCDMf1 were grown in *E. coli* and purified by the polyethylene glycol selective precipitation protocol described in Sambrook *et al*. (1989). DNAs were transiently transfected into mammalian 293 cells by CaPO$_4$ coprecipitation as described by Gorman *et al*. (1990).

All of the expressed proteins carry a 13aa extension of their C-termini, which includes an 11aa peptide of the human c-*myc* protein that is efficiently recognized by the 9E10 monoclonal antibody. *Myc* fusion proteins expressed by the transformed 293 cells and secreted into the culture medium were isolated by immunoaffinity chromatography using this monoclonal. All of the mutants were found to be expressed and secreted as 52 kDa procathepsin D-*myc* fusion proteins (see figure).

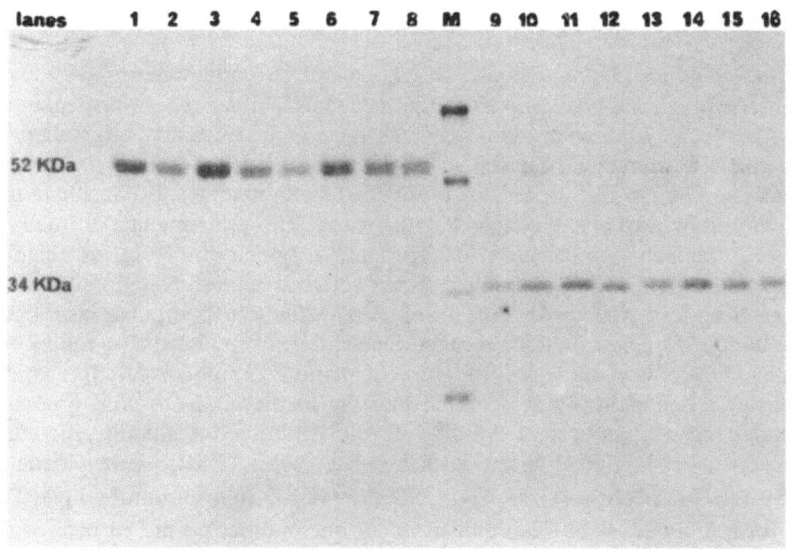

Figure 1. Immunoprecipitation of human cathepsin D-*myc* fusion protein using a rabbit polyclonal antiserum to human placental cathepsin D. Lanes 1-8: Immunoprecipitation of 293 cell conditioned medium containing ^{35}S methionine labeled mutated and parental procathepsin D-*myc* fusion proteins. Lanes 9-16: Immunoprecipitated mature cathepsin D following endocytosis by CHO cells incubated with 293 conditioned medium containing labeled procathepsin D-*myc* protein. Lanes 1, 9: LYS-172-GLU; lanes 2, 10: LYS-206-GLU; lanes 3, 11: LYS-276-GLU; lanes 4, 12: LYS-172, 206-GLU; lanes 5, 13: LYS-172, 276-GLU; lanes 6, 14: LYS-206, 276-GLU; lanes 7, 15: LYS-172, 206, 276-GLU; lanes 8, 16: parental cathepsin D; M: molecular weight marker.

Activity

The lysine mutant procathepsin D fusion proteins were found to have acid activatable enzymatic activity. Release of ^{14}C-hemoglobin peptides in 3 % TCA soluble form was determined following incubation at pH 4.5 and 37°C. This assay can detect enzymatic activity in 50 nl of medium conditioned by MCF_7 human breast tumor cells, an abundant source of the proenzyme. Activity in medium of 293 cells transfected by pSKCMVPCDMf1 was 20 fold higher. No difference between the acid activatable proteolytic activities of the parental and triple lysine mutant procathepsin D-*myc* fusion proteins was detected.

Endocytosis

Conditioned media from transfected 293 cells labeled with ^{35}S-methionine, containing procathepsin D fusion proteins, were incubated with CHO (chinese hamster ovary) cells to determine if the mutated procathepsin D-*myc* proteins were competent for endocytosis via the mannose 6-phosphate receptor. As can be seen in Figure 1 (lanes 9-16), the expressed procathepsin D-*myc* fusion proteins were endocytosed via the mannose 6-phosphate receptor of CHO cells.

Endocytosis was blocked by 5 mM mannose 6-phosphate but not by 5 mM galactose 6-phosphate (data not shown). The receptor delivers endocytosed ligands to the lysosomal compartment, where they are proteolytically matured. This results in conversion of 52 kDa procathepsin D into its mature two-chain form, which is seen as a 36 kDa heavy chain band on denaturing gels. The 14 kDa light chain has only one labeled methionine residue and is not seen in the figure. The *myc* peptide determinant is lost in the lysosome (Pelham, 1988).

DISCUSSION

The computer graphics models of cathepsin D from human, pig and cow revealed many conserved surface lysine residues that occur on neither renin nor pepsin structures, suggesting that some of these residues may contribute to the lysosomal targeting determinant. We have begun systematically to replace the surface lysines of human procathepsin D by mutagenesis *in vitro* to test the role of these residues in mannose 6-phosphate tagging, by conversion of individual lysine codons to glutamates. This experiment is formally similar to a controlled chemical modification experiment, in which single lysine residue can be individually succinylated. We reasoned that if certain lysine side chains were involved in specific binding to a lysosomal recognition site on the phosphotransferase, then binding would be inhibited by changing the local surface charge from plus one to minus one. When the mutant cDNAs are transiently expressed in the 293 human cell line, much of the procathepsin D is constitutively secreted into the medium. Both wild type and mutant proteins are expressed and secreted at high level. A number of mutants, including a third generation one carrying three Lys to Glu changes, have been tested. The triple mutant carries LYS 173→GLU LYS 207→GLU, and LYS 278→GLU (pepsin numbering). These three residues form a cluster within ten angstroms of one another on the surface of the human cathepsin D model. All the members of this set of mutants (three single, three double, and one triple) are stably folded and possess acid activatable proteolytic activity. None of the mutations to date has ablated phosphorylation and mannose 6-phosphate receptor binding activity.

ACKNOWLEDGEMENTS

We thank Dr. H. R. B. Pelham for the *myc* fusion construct, Dr. Tom Blundell for the human renin coordinates, Sun Pil Brown for culturing cells and Gloria Peché for preparing the manuscript. The work was supported by grants from the Veterans Administration and the Welch Foundation for Chemical Research (AQ1123) to JMC, who is an Associate Research Career Scientist of the VA.

REFERENCES

Faust, P. F., Kornfeld, S. & Chirgwin, J. M., 1985, Cloning and sequence analysis of cDNA for human cathepsin D, *Proc. Natl. Acad. Sci. U.S.A.* **82**:4910-4914.

Gorman, C. M., Gies, D. R. & McCray, G., 1990, Transient production of proteins using an adenovirus transformed cell line, *DNA and Protein Engineering Techniques* **2**:3-10.

Kornfeld, S., 1987, Trafficking of lysosomal enzymes, *FASEB J.* **1**:462-468.

Kunkel, T. A., Roberts, J. D. & Zakour, R. A., 1987, Efficient mutagenesis without selection, *Methods Enzymol.* **154**:367-382.

Pelham, H. R. B., 1988, Evidence that luminal ER proteins are sorted from secreted proteins in a post-ER compartment, *EMBO J.* **7**:913-918.

Sibanda, B. L., Blundell, T., Hobart, P. M., Fogliano, M., Bindra, J. S., Dominy, B. W. & Chirgwin, J. M., 1984, Computer graphics modelling of human renin, *FEBS Lett.* **174**:102-111.

Sambrook, J., Fritsch, E. F. & Maniatis, T., 1989, "Molecular Cloning", 2nd ed., Cold Spring Harbor Laboratory Press, New York.

COMPARISON OF KINETIC PROPERTIES OF NATIVE AND RECOMBINANT

HUMAN CATHEPSIN D

Paula E. Scarborough,[1] Gary R. Richo,[2] John Kay,[3]
Gregory E. Conner[2] and Ben M. Dunn[1]

[1]Department of Biochemistry and Molecular Biology
University of Florida
Gainesville, Florida   32610

[2]Department of Anatomy and Cell Biology
University of Miami
Miami, Florida   33101

[3]Department of Biochemistry
University of Wales College of Cardiff
P.O. Box 903
Cardiff, CF1 1ST
Wales
United Kingdom

INTRODUCTION

The aspartic proteinase cathepsin D has been shown to function primarily in the degradation of proteins in the lysosomes of most higher eukaryotic cells and, recently, has been evaluated as a prognostic factor in node-negative breast cancer. Synthesized as a preproenzyme and cleaved cotranslationally to proenzyme, procathepsin D is then further processed to the single-chain and finally two-chain active forms of cathepsin D which reside in the lysosome. The rapidity of the initial cleavage of procathepsin D makes it difficult to isolate the zymogen form. In order that the processing[1] and active site requirements of cathepsin D may be examined, human fibroblast procathepsin D has been expressed in *E. coli*, refolded from solubilized inclusion bodies, and purified using a pepstatin affinity column. To confirm that the recombinant enzyme is an acceptable model for studying substrate specificity and the processing mechanism *in vitro*, kinetic properties of recombinant cathepsin D were compared to those of native cathepsin D isolated from human placenta and spleen. Kinetic parameters were determined for a series of synthetic chromogenic peptide substrates derived from a parent substrate with the following sequence.

$$\text{Lys-Pro-Ile-Glu-Phe*Nph-Arg-Leu}$$

Replacements in substrate subsites P_1, P_2, P_3, P_4, and P_5 were examined.

Structure and Function of the Aspartic Proteinases
Edited by B.M. Dunn, Plenum Press, New York, 1991

Based on earlier studies showing that prosequences of porcine pepsin[2] and human renin[3] have inhibitory effects on the corresponding enzyme in a pH-dependent fashion, synthetic prosequences of human procathepsin D, procathepsin E, progastricsin and pepsinogen along with calf prochymosin were used in preliminary inhibition studies to further examine active site requirements. Again the data from recombinant cathepsin D were compared to the native enzymes.

MATERIALS AND METHODS

Human fibroblast procathepsin D was expressed in and isolated from *E. coli* cells in the form of insoluble inclusion bodies which were stored in 1 mg aliquots at -70°C. In order to recover active enzyme, inclusion bodies were solubilized and reduced at a concentration of 2 mg/ml in 50 mM CAPS, 50 mM ß-ME, 8 M Urea, pH 10.7 and incubated at room temperature for 30 minutes. After being spun in a Speed Vac for 10 minutes, the sample was added dropwise to water to a concentration of 0.5 nmol/ml and the pH was adjusted to 8.7. After a two hour incubation, glutathione was added at concentrations of 10 mM reduced and 1 mM oxidized glutathione. After two more hours the pH was lowered to 3.7 and after 48 hours, the refolding mixture was applied onto a pepstatinyl-agarose affinity column equilibrated with 0.1 M sodium formate buffer pH 3.5, then eluted with 0.1 M sodium phosphate buffer pH 8.0. Column fractions containing hydrolytic activity were pooled and stored at -70°C.

For the purposes of purifying placental cathepsin D, frozen human placental tissue was thawed and homogenized in 10 mM Tris-HCl pH 7.4, 0.5 % Brij 35 in a Waring blender. The homogenate was centrifuged at 10,000 rpm for 30 minutes in a Sorvall GSA rotor. The pH of the supernatant was adjusted to 3.7 with 5.7 N HCl, and sodium acetate was added to a concentration of 0.1 M. Following a 30 minute incubation at 0°C, the precipitate formed was removed by a 30 minute centrifugation at 10,000 rpm. The acidic supernatant was applied to a pepstatinyl-agarose column equilibrated with 0.1 M sodium acetate pH 3.5, 0.1 % Brij 35, 1 M NaCl, and washed until the OD_{280} of the effluent returned to baseline. The column was eluted with 50 mM Tris-HCl pH 8.6, 0.1 % Brij 35, 1 M NaCl. The eluent was dialyzed against 10 mM sodium phosphate pH 7.0, 0.1% Brij 35, 0.1 M NaCl then applied to a 1 ml DEAE-Sephacel column equilibrated with the same buffer. Cathepsin D was recovered in the breakthrough fraction.

Human spleen cathepsin D was prepared as described by Afting and Becker.[4] All peptide substrates and inhibitors were synthesized and purified by HPLC as described previously.[5,6] Concentrations of substrate solutions were determined by amino acid analysis.

Activity assays were performed on a Hewlett-Packard Diode array Spectrophotometer monitoring the average decrease in absorbance from 284-324 nm. All K_m and V_{max} values were determined at 37°C in 0.1 M sodium formate pH 3.5. Propart inhibition studies were done in 0.1 M sodium acetate buffer pH 4.5 based on earlier reports of pH dependence of propart inhibition. All enzyme preparations were titrated with a tightly binding inhibitor which was a gift from the Upjohn Company.

OBSERVATIONS AND DISCUSSION

The peptide sequences examined in substrate studies and the kinetic parameters obtained are detailed in Table 1. Peptide 1, of sequence **Lys-Pro-Ile-Glu-Phe*Nph-Arg-Leu**, was a far superior substrate, yielding a specificity constant which was one to two orders of magnitude greater than the other peptides examined thus far, providing a sensitive assay for cathepsin D activity. Specifically, the kinetic parameters obtained for each enzyme

Table 1. Substrate Specificity of Human Cathepsin D

Peptide Sequence	Substrate Specificity x 10^3 (V_{max}/K_m)[a]		
	Placenta	Spleen	Recombinant
1 Lys-Pro- Ile-Glu-Phe*Nph-Arg-Leu	77.5 ± 2.86	53.6 ± 4.09	40.8 ± 3.77
2 Lys-Pro-*Ala-Asp*-Phe*Nph-Arg-Leu	0.91 ± 0.02	0.78 ± 0.08	0.53 ± 0.01
3 Lys-Pro-*Asn-Gln*-Phe*Nph-Arg-Leu	0.83 ± 0.01	0.47 ± 0.04	0.73 ± 0.01
4 Lys-Pro-*Thr-Val*-Phe*Nph-Arg-Leu	4.70 ± 0.51	1.34 ± 0.12	2.98 ± 0.10
5 Lys-Pro-*Val-Ser-Tyr**Nph-Arg-Leu	0.37 ± 0.03	1.37 ± 0.01	0.30 ± 0.01
6 *Ala*-Pro-*Ala-Lys*-Phe*Nph-Arg-Leu	0.67 ± 0.04	0.22 ± 0.02	0.33 ± 0.001
7 *Asp*-Pro-*Ala-Lys*-Phe*Nph-Arg-Leu	1.27 ± 0.11	0.52 ± 0.05	0.92 ± 0.03
8 *Leu*-Pro-*Ala-Lys*-Phe*Nph-Arg-Leu	1.05 ± 0.07	0.40 ± 0.04	0.44 ± 0.01
9 *Arg*-Pro-*Ala-Lys*-Phe*Nph-Arg-Leu	0.18 ± 0.01	0.09 ± 0.01	0.11 ± 0.004
10 *Ser*-Pro-*Ala-Lys*-Phe*Nph-Arg-Leu	0.49 ± 0.02	0.16 ± 0.02	0.23 ± 0.01
11 Lys-Pro-*Ala-Lys*-Phe*Nph-Arg-Leu	NC[b]	NC	NC
12 Lys-Pro-*Ala-Lys-Ala**Nph-Arg-Leu	NC	NC	NC
13 Lys-Pro-*Ala-Lys-Leu**Nph-Arg-Leu	NC	NC	NC
14 Lys-Pro-*Ala-Lys-Val**Nph-Arg-Leu	NC	NC	NC
15 Lys-*Alà-Ala-Lys*-Phe*Nph-Arg-Leu	NC	NC	NC
16 Lys-*Asp-Ala-Lys*-Phe*Nph-Arg-Leu	NC	NC	NC
17 Lys-*Leu-Ala-Lys*-Phe*Nph-Arg-Leu	NC	NC	NC
18 Lys-*Arg-Ala-Lys*-Phe*Nph-Arg-Leu	NC	NC	NC
19 Lys-*Ser-Ala-Lys*-Phe*Nph-Arg-Leu	NC	NC	NC

[a]Velocities were measured in units of M/sec/μg of enzyme; substrate concentrations were 5 μM to 300μM; enzyme quantities required for assays were in the range of 0.015 μg to 0.33 μg.

[b]In reactions containing up to 5 μg enzyme, 400 μM peptide, and monitored for up to 30 min at 37°C, no measurable cleavage was observed.

Table 2. Inhibition of Cathepsin D by Aspartic Proteinase Prosequences

	p-2	p-1	p1	p2	p3	p4	p5	p6	p7	p8	p9	p10	p11	p12	p13	plac	spl	rec
						Prosequence Peptide[a]											% Inhibition	
hpd	S	A	L	V	R	I	P	L	H	K	F	T	S	I	R	45	23	82
hpg		A	V	V	K	V	P	L	K	K	F	K	S	I	R	49	34	79
hpp		I	M	Y	K	V	P	L	I	R	K	K	S	L	R	20	33	32
hpe	G	S	L	H	R	V	P	L	R	R	H	P	S	L	K	7	4	0
bpc	A	E	I	T	R	I	P	L	Y	K	G	K	S	L	R	23	21	27

[a]Position numbering based on porcine pepsinogen sequence; synthetic peptides are derived from the prosequences of the human proenzymes procathepsin D (hpd), progastricsin (hpg), pepsinogen (hpp), procathepsin E (hpe) and calf prochymosin (bpc).

preparation were as follows: for human placenta cathepsin D, $K_m = 50 \pm 4$ μM and $V_{max} = 4$ x 10^{-6} M/sec/μg of enzyme; for human spleen cathepsin D, $K_m = 36 \pm 7$ μM and $V_{max} = 2$ x 10^{-6} M/sec/μg of enzyme; and for recombinant human fibroblast cathepsin D, $K_m = 40 \pm 8$ μM and $V_{max} = 2$ x 10^{-6} M/sec/μg of enzyme. Generally speaking peptides 2 through 5 listed in Table 1 gave appreciable rates of cleavage. However Michaelis binding constants were apparently at least an order of magnitude larger than those observed for peptide 1, contributing to the poor specificity constants. Most notably, substituting Ala-Asp into P_3-P_2, thereby conserving the hydrophobic-negative charge combination, was extremely detrimental to binding and catalysis.

The dramatic results observed in the series of peptides with lysyl residues in both P_5 and P_2 (peptides 11 through 19) suggested that the combination of positively charged residues in these two positions is unacceptable to cathepsin D. This result was not seen with human cathepsin E, porcine pepsin or rhizopuspepsin which all cleaved substrates of this nature.[7,8] Interestingly, upon replacement of the positive charge in P_5 (peptides 6 through 10), minimal recovery of cleavage was observed. Results from peptide 9 with Arg in P_5 served to further support the conclusions drawn concerning the combination of positive charges in both P_5 and P_2. Preliminary inhibition data (Table 2) showed that not only was cathepsin D inhibited by the corresponding zymogen prosequence, but also by the highly homologous prosequences of other aspartic proteinases.

Overall the data demonstrated the suitability of the recombinant enzyme for further studies which will include detailing active site requirements, processing studies and mutagenesis of residues thought to be critical. The emphasis on the next series of synthetic peptide substrates will be on improving both binding and catalysis in order to further examine active site requirements. The results demand study of synthetic peptides with single substitutions in P_2 and P_3 in order to determine whether the preference of the Ile-Glu combination in peptide 1 is a question of charge, hydrophobicity, size or geometry. Single systematic replacements in the P_5 and P_4 positions while retaining the Ile-Glu in P_3-P_2 should yield more information on the importance of the positive charge in P_5 and the possible role of the Pro in P_4. Determining K_i values and pH-dependence for all of the propart peptides examined should also reveal information on the active site requirements of cathepsin D as well as the other mammalian aspartic proteinases.

REFERENCES

1. G. Richo and G. E. Conner, *in*: "Structure and Function of the Aspartic Proteinases: Genetics, Structures and Mechanisms," Ben M. Dunn, ed., Plenum Press, New York.

2. B. M. Dunn, C. Deyrup, W. G. Moesching, W. A. Gilbert, R. J. Nolan and M. L. Trach, (1978) *J. Biol. Chem.* **253**:7269-7275.

3. A. D. Richards, J. Kay, B. M. Dunn, C. M. Bessant and P. A. Charlton, (1991) *FEBS Lett.* in press.

4. E.-G. Afting and M.-L. Becker. (1981) *Biochem. J.* **197**:519-522.

5. A. D. Richards, L. H. Phylip, W. G. Farmerie, P. E. Scarborough, A. H. Alvarez, B. M. Dunn, Ph-H. Hirel, J. Konvalinka, P. Strop, L. Pavlickova, V. Kostka and J. Kay. (1990) *J. Biol. Chem.* **265**: 7733-7736.

6. J. Konvalinka, P. Strop, J. Velek, V. Cerna, V. Kostka, L. H. Phylip, A. D. Richards, B. M. Dunn and J. Kay. (1990) *FEBS Lett.* **268**:35-38.

7. C. R. Rao, P. E. Scarborough, W. T. Lowther, J. Kay, B. Batley, S. Rapundalo, S. Klutchko, M. D. Taylor and B. M. Dunn, *in*: "Structure and Function of the Aspartic Proteinases: Genetics, Structures and Mechanisms," Ben M. Dunn, ed., Plenum Press, New York.

8. W. T. Lowther, Z. Chen, X.-L. Lin, J. Tang and B. M. Dunn, *in*: "Structure and Function of the Aspartic Proteinases: Genetics, Structures and Mechanisms," Ben M. Dunn, ed., Plenum Press, New York.

CATHEPSIN D INHIBITOR FROM POTATO TUBERS (*Solanum tuberosum* L.)

M. Mareš, M. Fusek, V. Kostka, and M. Baudyš

Institute of Organic Chemistry and Biochemistry
Czechoslovak Academy of Science
Flemingovo 2, CS-166 10
Prague, Czechoslovakia

INTRODUCTION

The interest in naturally occuring inhibitors of proteolytic enzymes has been increasing mainly because of their possible use as therapeutic agents. Very little is known, however, about inhibitors of aspartic proteinases of protein character; their number include a pepsin and gastricsin inhibitor isolated from the roundworm *Ascaris suum* and *A. hominis*,[1,2] intact activation (propart) peptides[3] and the IA$_3$ inhibitor of yeast proteinase A.[4] So far, no cathepsin D inhibitor of tissue origin has been described in spite of the fact that this major lysosomal aspartic proteinase[5] plays an important role in many physiological or pathophysiological processes. We have been able to characterize a cathepsin D and trypsin inhibitor of protein character (PDI), first isolated from potato tubers in our laboratory in 1976,[6] to determine its complete amino acid sequence and to align it with the sequences of the family of soybean trypsin inhibitor.[7] In this study we report some of the basic structural characteristics of PDI homologs found in potatoes.

MATERIALS AND METHODS

The isolation of PDI homologs from potatoes of commercial origin was achieved in the final FPLC step by ion-exhange chromatography on a Mono Q column.[7] The inhibition activities were tested using bovine spleen cathepsin D[6] and trypsin.[8] The homogeneity of inhibitors was examined by SDS PAGE in 12.5 % gels.[9] The tryptic glycopeptide was prepared according to Mares, *et al.*[7] Amino sugars, neutral sugars and fucose were quantified as described elsewhere.[10] The C-terminal sequences were determined by carboxypeptidase Y digestion.[11] Crystallization of PDI-III was performed as described by Baudys *et al.*,[23] and PDI-IV was crystallized from 20 % LiCl.

RESULTS

The separation of PDI homologs is shown in Figure 1. By using our standard procedure[7] we were able to obtain three new components: PDI-I, II and IV, in addition to the standard peak of PDI-III.

PDI-III

The complete amino acid sequence of PDI-III is given in Figure 2. A sugar moiety linked to Asn_{19} has the following composition: glucosamine, 1.9 residues; neutral sugars, 4-5 residues (mannose equivalents); fucose, 1.0 residue. Similar values were obtained for the corresponding 9-residue glycopeptide isolated from the tryptic digest of PDI-III. This shows that Asn_{19} is the only glycosylation site in the PDI-III molecule. The composition of the oligosaccharide moiety corresponds to the so-called high mannose-type.[12]

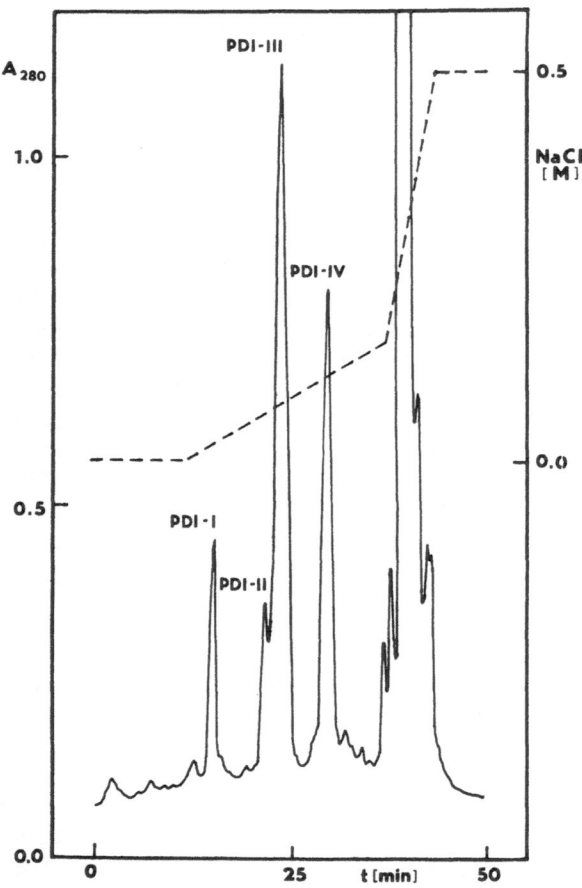

Figure 1. FPLC chromatography of crude PDI on Mono Q 5/5 column with separation of homologs PDI-I to -IV. A solution of 0.1 M ethanolamine-HCl, pH 10.0, was used as starting buffer. The lyophilizate of crude PDI (10 mg) dissolved in 2 ml of starting buffer was applied and eluted by the NaCl gradient shown at a flow rate of 1 ml/min.

```
              10              +  20                  30                  40
PDI  E S P L P K P V L D T N G K E L N P N S S Y R I I S I G R G A L G G D V Y L G K S P N
STI          D F V L D N E G N P L E N G G T Y Y I L S D - - I T A F G G I R A A P T G N
WTI          E P L L D S E G E L V R N G G T Y Y I L P Q - R W A L G G G I E A A A T G T
ETI            V L L D G N G E V V Q N G G T Y Y L L P Q - V W A Q G G G V Q L A K T G E
BTI          R E L L D V D G N F L R N G G S Y Y I V P A - F R G K G G G L E L A R T G S
WAI      D P P P V H D T D G N E L R A D A N Y Y V L P A - N R A H G G L T M A P G H T R
WBA      A D D P V Y D A E G N K L V N R G K Y T I V S F - - - S D G A G I D V V A T G N

              50                  60              *   70                  80
PDI  S D A P C P D G V F R Y N S - - - - - D V G P S G T P V P F I P L S G G I F E D Q L L
STI  E - - R C P L T V V Q S R N E L D K G I G T I I I S S P Y R I R F I A E - G H P L S L K
WTI  E - - T C P L T V V R S P N - - - - - - E V S V G E P L R I S S Q L R S G F I P D Y S
ETI  E - - T C P L T V V Q S P N - - - - - - E L S D G K P I R I E S R L R S T F I P D D D
BTI  E - - T C P R T V V Q A P A - - - - - - E Q S R G L P A R L S T P P R I R Y I G P E F
WAI  - - - R C P L F V S Q E A D - - - - - - G Q R D G L P V P I A P H G G A P S D K I I R
WBA  E N P E D P L S I V K S T R N I M Y A T S I S S E D K T - - - - - - P P Q P R N I L

              90                  100                 110
PDI  N I - - - Q F N I A T V K L C V S - - Y T I W K V G N L N A Y F R T M L L E T G G T I
STI  F D - - - S F A V I M - - L C V G - I P T E W S V V E D L P E G P A V K I G E N K D A
WTI  L V - - - R I G F A N P P K C A P - - S P W W T V V E D Q P Q Q P S V K L S E L K S T
ETI  E V - - - R I G F A Y A P K C A P - - S P W W T V V E D E Q E G L S V K L S E D E S T
BTI  Y L T - I E F F E E Q K P P S C L R D S N L Q W K V E E E S Q I - - - V K I A S K E E E
WAI  L S T D V R I S F R A Y T T C V Q - S T E W H I D S S E L V S G P R H V I T G P N R D
WBA  E N - - M R L K I N F A T D P H K - - G D V W S V V D F Q P D G Q Q L K L A G R Y P N

      120                 130                 140                 150
PDI  G Q - - A D S S Y F K I V K L S - - - N F G Y N L L Y C P I T P P F L C P F C R D D N
STI  M - - - - D G W F R L E R V S D D E F N N Y K L V F C P Q Q A - - - - - - - - E D D
WTI  K F - - - - D Y L F K F E K V T S K - F S S Y K L K Y C A - - - - - - - - - K R D
ETI  Q F - - - - D Y P F K F E Q V S D K - L H S Y K L L Y C E G - - - - - - - - - K H E
BTI  Q L - - - - F G S F Q I K P Y - - - R D D Y K L V Y C E P Q Q G G R X X X X X X X L
WAI  P S P S G R E N A F R I E K Y S G A E V H E Y K L V F C G - - - - - - - - - - D
WBA  Q V - - - - K G A F T I Q K G S N - T P R T Y K L L F C P V G S P - - - - - - - -

      160                 170                 180
PDI  F C A K V G V V I Q N - - G K R R L A L V - N E N P L D V L F Q E V
STI  K C G D I G I S I D H D D G T R R L V V S - K N K P L V V Q F Q K L D K E S L
WTI  T C K D I G I Y P D Q K - G Y A R L V V T - D E N P L V V I F K K V E S S
ETI  K C A S I G I N R D Q K - G Y R R L V V T - E D N P L T V V L K K D E S S
BTI  E C K D L G I S I D D D - N N R R L A V K - E G D P L V V Q F V N A D R E G N
WAI  S C Q D L G V F R D L K G G A W F L G A T - - E P Y H V V V F K K A P P A
WBA  - C K N I G I S T D P E - G K K P L V V S Y Q S D P L V V K F H R H E P E
```

Figure 2. Comparison of amino acid sequences of PDI-III and Kunitz-type structures. PDI, inhibitor PDI-III of cathepsin D from potato;[7] STI, soybean trypsin inhibitor;[18] WTI, winged bean trypsin inhibitor;[19] ETI, trypsin inhibitor DE-3 from *E. Latissima*;[20] BTI, trypsin inhibitor DE5 from *A. Pavonina*;[21] WAI, wheat α-amylase/subtilisin inhibitor;[14] WBA, winged bean albumin-1.[15] The resiudues are numbered according to PDI. Glycosylated Asn$_{19}$ of PDI is marked by a cross, the putative active site for trypsin inhibition (Arg$_{67}$) by an asterisk. The sequences were aligned with preservation of the intact ß-sheet segments of STI identified in X-ray diffraction studies.[22]

PDI-I

The molecular weight of PDI-I is (from SDS PAGE) close to that of PDI-III. From N-terminal sequence analysis this form is truncated with N-terminal Glu$_1$ missing in comparison to PDI-III. The identity of the C-terminal sequence of 8 residues of PDI-I and III was confirmed by carboxypeptidase Y digestion. PDI-I has a higher content of mannose equivalents (6-7 residues) compared to PDI-III.

PDI-IV

The molecular weight is 2 kDa lower than that of PDI-III and PDI-I. This difference can be accounted for by the absence of the oligosaccharide moiety. The length of the polypeptide chain of PDI-IV corresponds to that of PDI-III as follows from N- and C-terminal sequencing. Asn_{19} of PDI-III is replaced by Asp in the same position of the PDI-IV molecule. This amino acid replacement resulting in the absence of the glycosylation signal can be either genetically encoded (PDI-IV would then be an isoinhibitor of PDI-III) or may be ascribed to a difference in post-translational modification. In view of the recently published potato cDNA sequence[13] differring from the PDI-III amino acid sequence in two residues only (Asp_{19} and Pro_{88}) the former explanation seems more plausible.

PDI-II

From the determination of the N-terminal 21-residue sequence PDI-II is a homolog with both N-terminal Glu_1 missing and with Asp in position 19. This finding is in accordance with the molecular weight of PDI-II (similar to that of PDI-I) and its chromatographic behavior on a Mono Q column.

DISCUSSION

The above data and earlier results[7,13] seem to point to two closely related isoinhibitors: PDI-III and PDI-IV. Different N-terminal processing of these sequences may then generate further isoforms such as PDI-I and PDI-II, respectively. Sequencing of the internal part of the molecules is in progress in our laboratory to confirm this hypothesis.

The sequence of PDI clearly indicates that this protein belongs to the Kunitz soybean trypsin inhibitor family. A new alignment of these structures (homologies varying between 17 and 57 %) is presented in Figure 2. Also shown are the bifunctional α-amylase/subtilisin inhibitor from wheat,[14] the winged bean albumin-1 with no inhibitory activity[15] (the region of reactive loop is deleted in the alignment) and the inhibitors of serine proteinases. Partial sequence data on cysteine proteinase inhibitors from potatoes[16] provide evidence of a broad funtional divergence of the Kunitz structural type.

It seems probable that homologs of the Kunitz family are encoded by a whole multigenic family. Our results obtained with the PDI isoinhibitors are in agreement with this statement. This is also supported by data on the NID ("novel cathepsin D inhibitor") from potatoes,[17] another isoinhibitor of PDI-III and PDI-IV.

The possible physiological role of differences found to exist in the structures of the PDI homologs described here is not clear at present. Preliminary inhibition data obtained with cathepsin D and trypsin have not revealed any significant difference in the activity of PDI-I, III and IV. The question of the presence or absence of the glycosylation site in these strongly homologous protein in relation to PDI targeting and localization deserves special interest. PDI-III and PDI-IV were successfully crystallized and the determination of their three-dimensional structures[23] should cast more light on this inhibitor type.

REFERENCES

1. G. M. Abu-Erreish and R. J. Peanasky, Pepsin inhibitors from *Ascaris lumbricoides*, *J. Biol. Chem.* 249:1558 (1974).
2. M. R. Martzen, B. A. McMullen, K. Fujikawa and R. J. Peanasky, Aspartic protease inhibitors from the parasitic nematode *Ascaris*, *in*: "The Aspartic Proteinases: Genetics, Structures and Mechanisms," B. M. Dunn, ed., Plenum Press, New York, (1991).
3. M. Baudyš, I. Pichova, J. Pohl and V. Kostka, Chicken pepsin - activation peptide (p1-p42) complex isolated and artificially formed: a comparison, *in*: "Aspartic Proteinases and Their Inhibitors," V. Kostka, ed., W. de Gruyter, Berlin (1985).

4. T. Dreyer, M. J. Valler, J. Kay, P. Charlton and B. M. Dunn, The selectivity of action of the aspartic proteinase IA$_3$ from yeast (*Saccharomyces cerevisiae*), *Biochem. J.* **231**:777 (1975).

5. A. J. Barrett, Cathepsin D and other carboxyl proteinases, *in*: "Proteinases of Mammalian Cells and Tissues," A. J. Barrett, ed., Elsevier, Amsterdam (1977).

6. H. Keilova and V. Tomašek, Isolation and some properties of cathepsin D inhibitor from potatoes, *Collect. Czech. Chem. Commun.* **41**:489 (1975).

7. M. Mareš, B. Meloun, M. Pavlik, V. Kostka and M. Baudyš, Primary structure of cathepsin D inhibitor from potatoes and its structure relationship to soybean trypsin inhibitor family, *FEBS Lett.* **251**:94 (1989).

8. H. Keilova and V. Tomasek, Further characterization of cathepsin D inhibitor from potatoes, *Collect. Czech Chem. Commun.* **41**:2440 (1976).

9. U. K. Laemmli, Cleavage of structural proteins during the assembly of the head of bacteriophage T4, *Nature* **227**:680 (1970).

10. M. Baudyš and V. Kostka, Covalent structure of chicken pepsinogen, *Eur. J. Biochem.* **136**:89 (1983).

11. R. Hayashi, Carboxypeptidase Y in sequence determination of peptides, *Methods Enzymol.* **47**:84 (1977).

12. R. Kornfeld and S. Kornfeld, Assembly of asparagine-linked oligosaccharides, *Annu. Rev. Biochem.* **54**:631 (1985).

13. G. Štrukelj, J. Pungerčar, A. Ritonja, I. Krizaj, F. Gubenšek, I. Kregar and V. Turk, Nucleotide and deduced amino acid sequence of an aspartic proteinase inhibitor homologue from potato tubers (*Solanum tuberosum* L.), *Nucleic Acids Res.* **18**:4605 (1990).

14. K. Maeda, The complete amino-acid sequence of the endogeneous α-amylase inhibitor in wheat, *Biochim. Biophys. Acta* **871**:250 (1986).

15. A. A. Kortt, P. M. Strike and J. De Jersey, Amino acid sequence of a crystalline seed albumin (winged bean albumin-1) from *Psophocarpus tetragonolobus* (L.) DC, *Eur. J. Biochem.* **181**:403 (1989).

16. T. Popovic, J. Brzin, M. Drobnic-Kosorok and V. Turk, Specificity of cysteine proteinase inhibitors from potato tubers, 14th International Congress of Biochemistry, Prague, Czechoslovakia, abstr. MO 002 (1988).

17. A. Ritonja, I. Krizaj, P. Mesko, M. Kopitar, P. Lucovnik, B. Strukelj, J. Pungerčar, D. J. Buttle, A. J. Barrett and V. Turk, The amino acid sequence of a novel inhibitor of cathepsin D from potato, *FEBS Lett.* **267**:13 (1990).

18. S. H. Kim, S. Hara, S. Hase, T. Ikenaka, H. Toda, K. Kitamura and N. Kaizuma, Comparative study on amino acid sequences of Kunitz-type soybean trypsin inhibitors, Ti^a, Ti^b, and Ti^c, *J. Biochem.* **98**:435 (1985).

19. M. Yamamoto, S. Hara and T. Ikenaka, Amino acid sequence of two trypsin inhibitors from winged bean seeds (*Psophocarpus tetragonolobus* [L] DC), *J. Biochem.* **94**:849 (1983).

20. F. J. Joubert, c. Heussen and E. B. D. Dowle, The complete amino acid sequence of trypsin inhibitor DE-3 from *Erithrina Latissima* seeds, *J. Biol. Chem.* **260**:12948 (1985).

21. M. Richardson, F. A. P. Campos, J. Xavier-Filho, M. L. R. Macedo, G. M. C. Maia and A. Yarwood, The amino acid sequence and reactive (inhibitory) site of the major trypsin isoinhibitor (DE5) isolated from seeds of the Brasilian Carolina tree (*Adenanthera pavonina* L.), *Biochim. Biophys. Acta* **872**:134 (1986).

22. A. D. McLachlan, Three-fold structural pattern in the soybean trypsin inhibitor (Kunitz), *J. Mol. Biol.* **133**:557 (1979).

23. M. Baudyš, M. Ghosh, K. Harlos, M. Mareš, V. Kostka and C. C. F. Blake, Crystallization and preliminary crystallographic study of cathepsin D inhibitor from potatoes, *J. Mol. Biol.* in preparation.

ASPARTIC PROTEINASE FROM BARLEY SEEDS IS RELATED TO

ANIMAL CATHEPSIN D

K. Törmäkangas,[1] P. Runeberg-Roos,[1] A. Östman,[1], C. Tilgmann,[1]
P. Sarkkinen,[2] J. Kervinen,[1,] L. Mikola,[2] and N. Kalkkinen[1]

[1]Institute of Biotechnology
University of Helsinki
Karvaamokuja 3
SF-00380 Helsinki, Finland

[2]Department of Cell Biology
University of Jyvaskyla
Vapaudenkatu 4
SF-40100 Jyvaskyla, Finland

INTRODUCTION

In contrast to the well-characterized mammalian aspartic proteinases, plant aspartic proteinases have received little attention so far. Aspartic proteinase activity has been detected, for example, in resting seeds of scots pine (Salmia *et al.*, 1978), soybean (Bond & Bowles, 1983), barley and wheat (Morris *et al.*, 1985) as well as in leaves of orange (Garcia-Martinez & Moreno, 1986) and barley (Kervinen *et al.*, 1990). Aspartic proteinases have been purified from the seeds of rice (Doi *et al.*, 1980), cucumber, squash (Polanowski *et al* 1985) and wheat (Dunaevsky *et al.*, 1989) as well as from the leaves of tomato (Rodrigo *et al.*, 1989). The plant aspartic proteinases have been reported to enhance the hydrolysis of at least wheat (Belozersky *et al.*, 1989) and cocoa (Heinrichs *et al.*, 1990) storage proteins. Rodrigo *et al.* (1989) have also suggested that the biological action of the pathogenesis related proteins in tomato leaves could be regulated by aspartic proteinases. Taken together, the specific functions of plant aspartic proteinases remain largely unknown.

Here we report the complete nucleotide sequence and experimental data on the partial amino acid sequences of an aspartic proteinase from barley (*Hordeum vulgare* L. cv Kustaa) seeds. A putative processing scheme is also suggested for the enzyme. To our best knowledge this is the first plant aspartic proteinase cloned and sequenced.

Purification and characterization of the enzyme

The aspartic proteinase activity (hydrolysis of hemoglobin at pH 3.7) was purified from extracts of barley seeds by affinity chromatography on a pepstatin-Sepharose column.

With affinity purified enzyme preparation the optimal pH for the hydrolysis of hemoglobin was from 3.5 to 3.9. The enzymatic activity was completely inhibited by pepstatin, whereas other selective proteinase inhibitors known to inhibit serine, cysteine and metalloproteinases did not affect the activity (Sarkkinen *et al.*, submitted).

SDS-PAGE of the purified enzyme preparations under nonreductive conditions revealed three bands: 43 kDa, 29 kDa and 11 kDa. On reduction with 2-mercaptoethanol the 43 kDa band was absent but new bands appeared at 32 kDa and 16 kDa. Affinity purified enzyme preparations were further purified by FPLC on a Mono Q column and the subunits were then separated by reversed phase chromatography. Upon partial sequencing of tryptic peptides from the purified subunits, peptides with the same amino acid sequences were found in the 32 kDa and 29 kDa subunits as well as in the 16 kDa and 11 kDa subunits. The N-terminal amino acid sequences of the 32 kDa and 29 kDa subunits were identical, but those of the 16 kDa and 11 kDa differed from each other (Sarkkinen *et al.*, submitted).

Figure 1. cDNA and amino acid sequences of a barley seed aspartic proteinase. The N-termini of the subunits are marked with arrows. The tryptic peptides sequenced are underlined. Potential active sites are marked with asterisks, potential signal sequence cleavage sites with Δ and a potential glycosylation site with +.

356

The 43 kDa protein appears to be a heterodimer where two subunits (32+16 kDa) are covalently bound by disulfide bridges. The plant aspartic proteinases purified so far have all been monomeric, the size varying between 37 kDa (Rodrigo *et al.*, 1989) and 65 kDa (Doi *et al.*, 1980). Thus, the barley aspartic proteinase is similar in size to the other plant aspartic proteinases, but differs from them by being heterodimeric. In native PAGE assays, however, two active aspartic proteinases were present, suggesting that also the 29 kDa and 11 kDa proteins might form a complex held together by noncovalent bridges (data not shown).

Construction and screening of the cDNA library

A cDNA library was constructed from developing barley embryo mRNA (16 days post anthesis) and screened using synthetic oligonucleotide mixtures encoding amino acid sequences Lys-Asn-Tyr-Met-Asn-Ala and Tyr-Cys-Gln-Phe-Asp-Met-Gly. The initial screening yielded a 1.6 kb cDNA clone lacking the N-terminal part of the enzyme. In the second screening the 5'-part of the 1.6 kb clone was used as a probe and it identified a 1.8 kb full length cDNA containing a putative initiation methionine, a signal sequence, a prosequence, the N-terminal sequences of 32 kDa, 29 kDa, 16 kDa and 11 kDa subunits and the polyadenylation signal (Runeberg-Roos *et al.*, submitted).

Primary structure and putative processing of the enzyme

The 1.8 kb cDNA was sequenced with the dideoxy method from both strands. Computer analysis (PCGENE, PSIGNAL) of the cDNA sequence data (Figure 1) predicted a hydrophobic signal sequence (presequence) of about 20 amino acids to be cleaved from the 508 residue polypeptide, but the exact location of the cleavage site remains to be determined. The N-termini of both the 32 kDa and 29 kDa subunits start from the serine residue at position 67. This gives a putative prosequence of about 45 amino acids, which is equal in length to the prosequences of other aspartic proteinases such as porcine cathepsin D and chymosin (Faust *et al*, 1985; Tang & Wong, 1987). The potential active sites are located at Asp_{101}-Thr_{102}-Gly_{103} and Asp_{238}-Ser_{289}-Gly_{290}, similar to the other aspartic proteinases.

According to the amino acid and cDNA sequence data the following hypothetical processing scheme of the barley aspartic proteinase was constructed (Figure 2).

Our earlier protein analyses suggested that the larger (32+16 kDa) enzyme is an intermediate precursor of the smaller (29+11 kDa) enzyme. The presence of the N-termini of all subunits (32, 29, 16 and 11 kDa) in the same transcript as well as the presence of a single 2.0 kb mRNA in the Northern blots (Runeberg-Roos *et al.*, submitted) confirms this hypothesis. In addition, during the processing, a disulfide bridge in the cleaved polypeptide is removed and the 29 kDa and 11 kDa subunits remain held together by noncovalent bonds.

In comparison with the mammalian aspartic proteinases the barley enzyme has an extra 104 amino acids inserted approximately 317 amino acids from the initiation methionine, and containing the N-terminal sequence of the 16 kDa subunit. The N-terminus of the 11 kDa subunit is located immediately after the insert. The insert is located at approximately the same position as intron 7 in the human renin gene (Miyazaki *et al.*, 1984), the human prochymosin pseudogene (Örd *et al.* 1990), and the human pepsinogen A and C genes (Sogawa *et al.*, 1983; Hayano *et al.*, 1988). Interestingly, the 104 amino acid insert has certain homology with the CaMV genome. However, the origin of the 104 amino acid insert as well as its evolutionary significance remains to be elucidated.

According to the amino acid sequence data barley aspartic proteinase is homologous to porcine cathepsin D, human cathepsin D and yeast proteinase A. The homology is split between two regions of the barley enzyme, leaving 104 nonhomologous amino acids in between. In the N-terminal region there is a 52 % identity over 248 amino acids between the

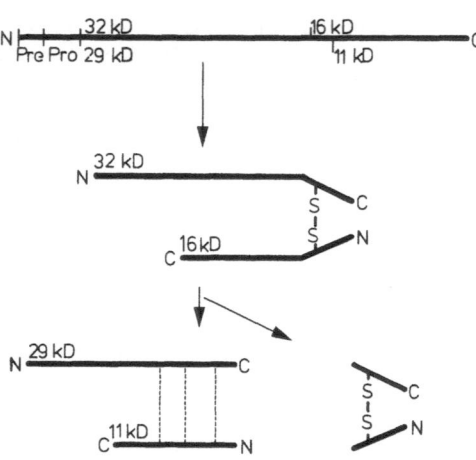

Figure 2. Processing scheme of the barley aspartic proteinase. The 40 kDa protein is formed by proteolytic processing of the 43 kDa precursor form.

porcine cathepsin D and the barley enzyme. With the C-terminal 88 amino acids, an identity of 53 % is found (Runeberg-Roos *et al.*, submitted).

CONCLUSIONS

While all the other plant aspartic proteinases purified so far are single chain proteins, the barley enzyme resembles mammalian cathepsin D in being heterodimeric. However, the barley enzyme is processed at a different site than the mammalian cathepsin D. This is due to a barley-specific 104 amino acid insert which contains the barley enzyme processing site; this insert is not present in mammalian cathepsin Ds. The processing step in this area seems not to affect the activity of the barley aspartic proteinase, since both the 32+16 kDa and 29+11 kDa forms of the enzyme are active. The processing site of mammalian cathepsin D differs from the sequence at the corresponding location in the barley aspartic proteinase, which may be the reason for the latter not to be processed at this site.

REFERENCES

Belozersky, M. A., Sarbakanova, Sh. T., & Dunaevsky, Ya. E., 1989, Aspartic proteinase from wheat seeds: isolation, properties and action on gliadin, *Planta* **177**:321-326.

Bond, H. M., & Bowles, D. J., 1983, Characterization of soybean endopeptidase activity using exogenous and endogenous substrates, *Plant Physiol.* **72**:345-350.

Doi, E., Shibata, D., Matoba, T., & Yonezawa, D., 1980, Characterization of pepstatin-sensitive acid protease in resting rice seeds, *Agric. Biol. Chem.* **44**:741-747.

Dunaevsky, Y., Sarbakanova, S. T., & Belozersky, M. A., 1989, Wheat seed carboxypeptidase and joint action on gliadin of proteases from dry and germinating seeds, *J. Exp. Botany* **40**:1323-1329.

Faust, P. L., Kornfeld, S., & Chirgwin, J. M., 1985, Cloning and sequence analysis of cDNA for human cathepsin D, *Proc. Natl. Acad. Sci. U.S.A.* **82**:4910-4914.

Garcia-Martinez, J. L., & Moreno, J., 1986, Proteolysis of ribulose-1,5-bisphosphate carboxylase/oxygenase in Citrus leaf extracts, *Physiol. Plant.* **66**:377-383.

Hayano, T., Sogawa, K., Ichihara, Y., Fujii-Kuriyama, Y., & Takahashi, K., 1988, Primary structure of human pepsinogen C gene, *J. Biol. Chem.* **263**:1382-1385.

Heinrichs, H., Xiong, Q., Voigt, J., Kirchhoff, P., & Biehl, B., 1990, An aspartic endoprotease and a serine exoprotease are involved in storage protein degradation in cocoa seeds, *Physiol. Plant.* **79**: abstract A15.81.

Kervinen, J., Kontturi, M., & Mikola, J., 1990, Changes in the proteinase composition of barley leaves during senescence in field conditions, *Cereal Res. Comm.* **18**:191-197.

Miyazaki, H., Fukamizu, A., Hirose, S., Hayashi, T., Hori, H., Ohkubo, H., Nakanishi, S., & Murakami, K., 1984, Structure of the human renin gene, *Proc. Natl. Acad. Sci. U.S.A.* **81**:5999-6003.

Morris, P. C., Miller, R. C., & Bowles, D. J., 1985, Endopeptidase activity in dry harvest-ripe wheat and barley grains, *Plant. Sci.* **39**:121-124.

Polanowski, A., Wilusz, T., Kolaczkowska, M. K., Wieczorek, M., & Wilimowska-Pelc, A., 1985, Purification and characterization of aspartic proteinases from *Cucumis sativus* and *Cucurbita maxima* seeds, *in*: "Aspartic proteinases and their inhibitors", pp 49-52. V. Kostka ed., Walter de Gruyter and Co., New York.

Rodrigo, I., Vera, P., & Conejero, V., 1989, Degradation of tomato pathogenesis-related proteins by an endogenous 37 kDa aspartyl endoproteinase, *Eur. J. Biochem.* **184**:663-669.

Runeberg-Roos, P., Törmäkängas, K., & Östman, A., Primary structure of a barley grain aspartic proteinase - a plant aspartic proteinase resembling mammalian cathepsin D, submitted.

Salmia, M. A., Nyman, S. A., & Mikola, J. J., 1978, Characterization of the proteinases present in germinating seeds of Scots pine, *Pinus sylvestris*, *Physiol. Plant.* **42**:252-256.

Sarkkinen, P., Kalkkinen, N., Tilgmann, C., Siuro, J., Kervinen, J. & Mikola, L., 1990, Aspartic proteinase from barley grains is related to mammalian lysosomal cathepsin D, Submitted.

Sogawa, K., Fujii-Kuriyama, Y., Mizukami, Y., Ichihara, Y., & Takahashi, K., 1983, Primary structure of human pepsinogen gene, *J. Biol. Chem.* **258**:5306-5311.

Tang, J., & Wong, R. N. S., 1987, Evolution in the structure and function of aspartic proteinases, *J. Cell. Biochem.* **33**:53-63.

Örd, T., Kolmer, M., Villems, R., & Saarma, M., 1990, Structure of the human genomic region homologous to the bovine prochymosin-encoding gene, *Gene* **91**:241-246.

IMMUNOHISTOCHEMICAL AND IMMUNOCYTOCHEMICAL LOCALIZATION OF

CATHEPSIN E COMPARED WITH CATHEPSIN D

Hideaki Sakai, Takashi Saku,[a] Yuzo Kato and Kenji Yamamoto[b]

Department of Pharmacology
[a]Department of Pathology
Nagasaki University School of Dentistry
Nagasaki 852, Japan

[b]Department of Pharmacology
Kyushu University Faculty of Dentistry
Fukuoka 812, Japan

INTRODUCTION

Recent reports have suggested that cathepsin E is a non-lysosomal aspartic proteinase and that this enzyme participates in extralysosomal proteolysis.[1-3] Although its enzymatic and structural properties have been demonstrated,[4-7] its physiological function is still unknown. Although cathepsin E can be clearly distinguished from another intracellular aspartic proteinase cathepsin D,[8-10] the functional relationship between the two enzymes remains to be answered. Recently, the distribution of cathepsins E and D has been shown to be markedly different in various rat tissues and cells by immunochemical analyses employing discriminative antibodies specific for each enzyme. The results indicated that cathepsin D has a ubiquitous distribution, while cathepsin E has a relatively limited distribution. Of the cell types tested, gastrointestinal tracts, lymphoid tissues, urinary tracts and blood cells contained high levels of cathepsin E. The other rat tissues had little or no detectable cathepsin E. The stomach is one of the most interesting organs, because the greatest accumulation of cathepsin E is observed in the gastric mucosa[1,11] and because this tissue contains not only a non-secretory type of aspartic proteinases (cathepsins E and D) but also a secretory type of the enzymes (pepsin, gastricsin etc.). In order to clarify the functional characteristics of cathepsin E in human stomach, we have examined the intracellular localization of cathepsin E in a variety of physiological and pathological processes, as compared with that of cathepsin D, by immunohistochemical and immunocytochemical analyses using the antibodies specific for each enzyme.

Structure and Function of the Aspartic Proteinases
Edited by B.M. Dunn, Plenum Press, New York, 1991

RESULTS AND DISCUSSION

As shown in Figure la, cathepsin E was diffusely and intensively demonstrated in the subvacuolar region of foveolar epithelial cells. Parietal cells gave fine granular stainings for cathepsin E, while the mucous neck cells and the chief cells showed no positive staining for cathepsin E. When immunohistochemical localization of cathepsin D was also examined in normal human gastric tissues, coarse granules positive for this enzyme were found on the free surface of foveolar epithelial cells, in parietal cells and in chief cells. The results indicated that the distributions of cathepsins E and D were distinct in the normal gastric mucosa and they were consistent with those observed in rat gastric mucosa.[1] Since parietal cells have been shown to be positive for both cathepsins E and D, the subcellular localization of these two enzymes was examined at the electron microscopic level by using immunogold techniques. As shown in Figure lb, cathepsin E was scattered in the cytoplasmic matrix. Gold particles for cathepsin E were also concentrated in cistic spaces of tubulovesicular system and around microvilli of intracellular canaliculi. By contrast, the immunolocalization of cathepsin D was confined to lysosomes. The foveolar epithelial cells presented similar distribution patterns of cathepsins E and D to those of the parietal cells. The density of gold particles for cathepsin D was higher in lysosomes of the parietal cells than of any other type of cells in the stomach. In the previous studies, we have also shown the extralysosomal localization of cathepsin E in rat peritoneal elicited neutrophils and macrophages.[3,12]

The differential distribution and subcellular localization of cathepsins E and D were of particular importance, since such differences must reflect different physiological roles of the two enzymes. Cathepsin D is exclusively associated with the lysosomal proteolysis, while cathepsin E appears to be involved in the extralysosomal proteolysis. The high levels of cathepsin E in the cytoplasm in foveolar epithelial cells and parietal cells suggest the possibility that the enzyme may be related to characteristics of these cells, such as with high absorptive and secretive activities and with a rapid regenerating activity. It is worth noting that cathepsin E localizes the intramicrovilli of parietal cells. A similar intramicrovilluos localization was observed in renal proximal tuble cells, intestinal and tracheobronchial epithelial cells.[1,12] These results suggest that cathepsin E is involved in the specialized function of these cells, such as absorption, secretion and digestion of various cellular components. It was also of interest that cathepsin E was greatly accumulated in the cisternae of rER and in the dilated nuclear envelope of the gastric cells. The localization in the rER suggests that the membrane-bound ribosomes may be a major site for the biosynthesis of cathepsin E, although it remains to be answered how the enzyme segregated into the cisternae is transported into the cytosol.

To gain further insight into the cellular function of cathepsin E in human gastric cells, we examined the subcellular localization of the enzyme, as compared with that of cathepsin D, in cells with pathological disease states. Chronically inflamed gastric mucosa showed basically the same immunohistochemical distribution of cathepsins E and D as normal mucosa, and erosive or ulcerative changes did not influence the basic staining patterns of either enzyme. Metaplastic foveolar or grandular epithelium tended to lose immunoreactivity for each enzyme. There were no immunoreactions to cathepsins E and D in the epithelium when dysplastic changes were seen. No immunoreaction for cathepsin E was observed in cells of well differentiated tubular or papillary adenocarcinoma (Figure lc). Similarly, cathepsin D disappeared in these specimens, except that stromal infiltrated macrophages in the carcinoma foci were distinctly positive for cathepsin D. In moderately differentiated adenocarcinoma, cathepsins E and D were found in some of cells which formed bizarre ductal structures. The immunostaining for cathepsin E was diffusely found in the cytoplasm, while the immunoreaction products for cathepsin D were always in coarse granules which were evenly distributed in the cytoplasm but not in the apical surface. In poorly differentiated adenocarcinoma, intensive and diffuse cytoplasmic stainings for both cathepsins E and D

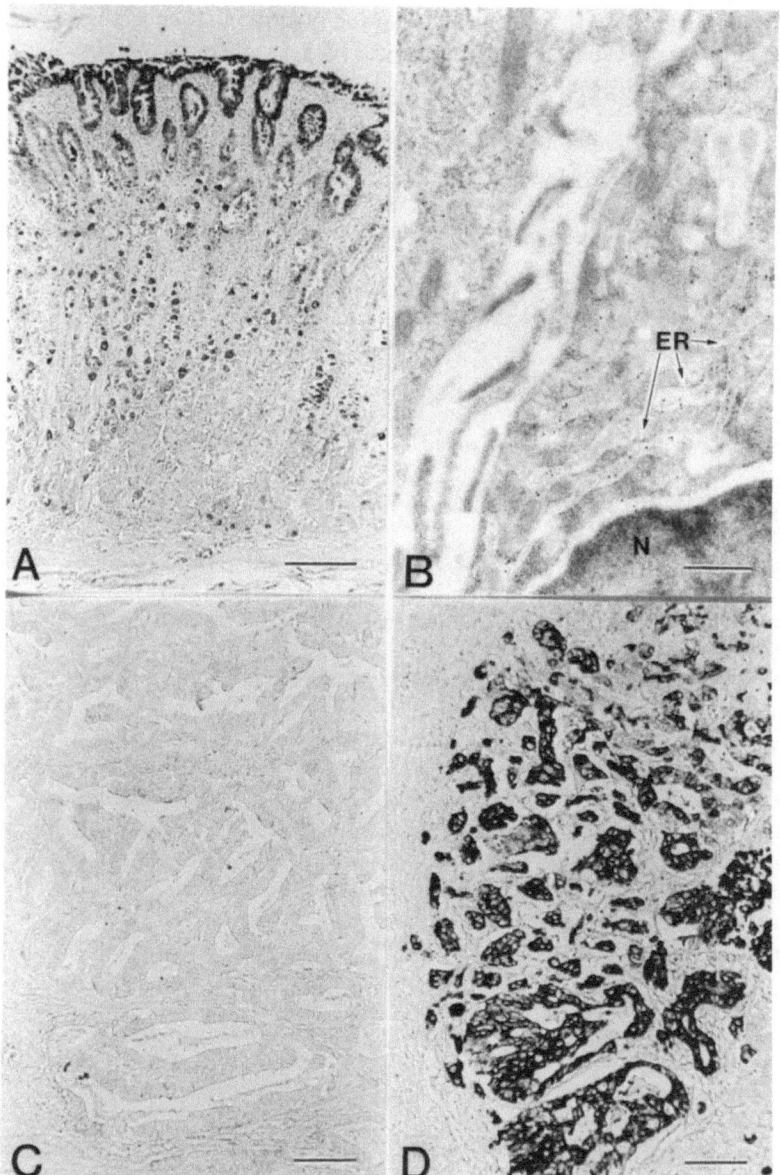

Figure 1. Immunohistochemical and immunocytochemical localization of cathepsin E in normal mucosa and cancerous lesions of the human stomach. A) Immunoperoxidase staining for cathepsin E in normal human gastric mucosa, counterstained with hematoxylin. Cathepsin E localized intensively in the cytoplasm of foveolar epithelium. Parietal cells also show positive staining for cathepsin E. Bar = 200 μm. B) Ultrastructural immunogold staining for cathepsin E in parietal cells of normal gastric mucosa. Cathepsin E is localized in the cytosol and rER. A small amount of labeling also is present along the microvilli facing the intercellular space. N, nucleus. Bar = 0.5 μm. C) and D) are immunoperoxidase stainings for cathepsin E in well differentiated and poorly differentiated adenocarcinoma, respectively, counterstained with hematoxylin. No positive reaction was observed in the tumor cells of well-differentiated adenocaricinoma, while prominent immunoreaction for cathepsin E was found in the cytoplasm of poorly differentiated adenocarcinoma cells. Bars = 100 μm.

were observed in most of the tumor cells (Figure 4d). A similar observation has been reported by Shiraishi et al.,[13] who showed that the positive immunostaining for the slow moving proteinase (now called as cathepsin E) was higher in diffuse cell types (including our poorly differentiated cell types) than in intestinal cell types (including our well differentiated cell types).

It is difficult to explain why both cathepsins E and D are absent in precancerous lesions and well differentiated adenocarcinoma. However, since it is known that intracellular processing of some proteins is at least suppressed during the early stages of malignant transformation, the dysfunctioning of cathepsins E and D may play a part in the development of carcinoma in the gastric mucosa.

REFERENCES

1. H. Sakai, T. Saku, Y. Kato and K. Yamamoto, *Biochim. Biophys. Acta* **991**:367-375 (1989).

2. S. Yonezawa, K. Fujii, Y. Maejima, K. Tamoto, Y. Mori and N. Muto, *Arch. Biochem. Biophys.* **267**:176-183 (1988).

3. E. Ichimaru, H. Sakai, T. Saku, K. Kunimatsu, Y. Kato, I. Kato and K. Yamamoto, *J. Biochem.* **108**:1009-1015 (1990).

4. K. Yamamoto, N. Katsuda and Y. Kato, *Eur. J. Biochem.* **92**:499-508 (1978).

5. M. Takeda, E. Ueno, Y. Kato and K. Yamamoto, *J. Biochem.* **100**:1269-1277 (1986).

6. T. Azuma, G. Pals, T. K. Mohandas, J. M. Couvreur and R. T. Taggart, *J. Biol. Chem.* **264**:16748-16753 (1989).

7. S. Yonezawa, T. Takahashi, M. Ichinose, K. Miki, J. Tanaka and S. Gasa, *Biochem. Biophys. Res. Commun.* **166**:1032-1038 (1990).

8. K. Yamamoto, 0. Kamata, N. Katsuda and K. Kato, *J. Biochem.* **87**:511-516 (1980).

9. C. Lapresle, V. Puizdar, C. Porchon-Bertolotto, E. Joukoff and V. Turk, *Biol. Chem. Hoppe-Seyler* **367**:523-526 (1986).

10. R. A. Jupp, A. D. Richards, J. Kay, B. M. Dunn, J. B. Wyckoff, I. M. Samloff and K. Yamamoto, *Biochem. J.* **254**:895-898 (1988).

11. N. Muto, M. Yamamoto, S. Tani and S. Yonezawa, *J. Biochem.* **103**:629-632 (1988).

12. T. Saku, H. Sakai, Y. Shibata, Y. Kato and K. Yamamoto, *J. Biochem.* in press.

13. T. Shiraishi, I. M. Samloff, R. T. Taggart and G. N. Stemmermann, *Dig. Dis. Sci.* **33**:1466-1472 (1988).

ORIGINS OF THE MULTIPLE CATHEPSIN E TRANSCRIPTS OBSERVED

IN HUMAN GASTRIC MUCOSA AND GASTRIC ADENOCARCINOMA

Takeshi Azuma,[1,2,3] Keiichi Kawai[1] and R. Thomas Taggart [2,3]

[1]Department of Preventive Medicine
 Kyoto Prefectural University of Medicine
 Kyoto, Japan

[2]Department of Molecular Biology and Genetics
 Wayne State University Medical School
 Detroit, Michigan

[3]Center for Molecular Biology
 Wayne State University
 Detroit, Michigan

INTRODUCTION

We previously reported the cloning and sequence analysis of human gastric cathepsin E (CTSE) cDNA clones. The cDNA clones were isolated from a gastric adenocarcinoma recombinant library using a set of complementary nondegenerate 18mer oligonucleotide probes specific for a 6 residue sequence surrounding the first active site region of all previously characterized human aspartic proteinases.[1] Northern analysis of poly (A+) RNA isolated from CTSE producing gastric adenocarcinoma cell line revealed three transcripts (3.6, 2.6 and 2.1 kb). The multiple CTSE transcripts are a unique finding among aspartic proteinases and may result from several potential causes including; multiple genes, rearranged genes, alternative initiation of transcription, alternative splicing or alternative polyadenylation. In the present study, we examined additional CTSE cDNA clones in an attempt to determine the origin of multiple transcripts observed in gastric adenocarcinoma. We present here evidence that two major CTSE transcripts (2.6 and 2.1 kb) result from alternative polyadenylation of the primary CTSE transcript.

MATERIALS AND METHODS

Isolation and characterization of cDNA recombinant clones

Double stranded complementary DNA was prepared following the procedure of Grubler and Hoffman.[2] After methylation of internal *Eco*RI sites and the addition of *Eco*RI linkers, the cDNA was ligated into the *Eco*RI site of ZAPII (Strategene). The phage were packaged and recombinants were selected by plating on *E. coli* strain XL-1 (Strategene). Duplicate nitrocellulose filters containing denatured recombinant phage were screened in parallel with each of the 18 base probes corresponding to the coding and anticoding sequence of the six residue sequence of the first active site aspartyl group of pepsinogen A and pepsinogen C;

Act1 SENSE	5'-GACACCGGCTCC TCCAAC
Act1 ANTISENSE	CTGTGGCCGAGGAGGTTG-5'

Filters were prehybridzed in 4X SSC containing 0.5 % SDS, 2X Denhardts, 250 μg/ml sonicated salmon testes DNA at 37°C for 3 h and then hybridized overnight with 5' end labeled oligonucleotide probe under the same conditions. Hybridizing phage were identified after washing in 4X SSC at 47-55°C and autoradiography. Phage were purified by successive rounds of plating and hybridization. The purified phage were converted to the corresponding plasmid form (Bluescript II, Strategene) utilizing the plasmid excision procedure provided by the manufacturer. Dideoxy chain termination sequencing procedures employing [^{35}S]-dATP were performed on both strands of denatured plasmid cDNA inserts using T7 DNA polymerase (Pharmacia) and oligonucleotides specific for regions flanking the *Eco*RI cloning site and CTSE coding sequences spaced at approximately 400 bp intervals.

Northern analysis

RNA was purified from human gastric fundic mucosa following the procedure of Rall *et al.*[3] One gram of pulverized tissue was homogenized for one min at room temperature in 10 ml of lysis buffer containing 5 M guanidinium thiocyanate, 100 mM Tris-HCl (pH 7.5), 10 mM EDTA (pH 8.0) and 14 % (v/v) ß-mercaptoethanol. After dissolution, the homogenate was clarified by centrifugation (10,000 X g, 4°C, 30 min). RNA was precipitated (16-24 h, 4°C) by adding 5.5 volumes of 4 M LiCl. The RNA and residual protein and DNA were resuspended in 5 ml of 3 M LiCl containing 4 M urea by vortexing vigorously for one min. The volume of the suspension was adjusted to 30 ml and the precipitate was collected by centrifugation as described above. The RNA pellet was dissolved in 2.5 ml of 1 % SDS and further purified by sequential extraction of the aqueous phase with a phenol:chloroform-isoamyl alcohol solution (25:24:1, v/v), and chloroform-isoamyl alcohol (24:1, v/v), followed by precipitation with ethanol. Poly (A$^+$) RNA was purified by chromatography on oligo (dT) cellulose (Pharmacia).

Human gastric fundic mucosal poly (A$^+$) RNA was analyzed by electrophoresis. RNA was resuspended in 10 mM potassium phosphate (pH 7.5), 1 M glyoxal, and 50 % dimethyl sulfoxide and denatured by heating to 60°C for 10 min. After electrophoresis through a 1 % agarose gel, denatured RNA and DNA size standards were transferred to nitrocellulose and hybridized to random primer [^{32}P]-dCTP labeled CTSE cDNA probe (AGS402) which contained the entire predicted CTSE coding sequence.[1]

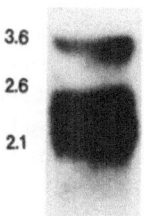

Figure 1. Northern analysis of poly (A+) RNA isolated from human gastric fundic mucosa after electrophoresis of denatured RNA in 1 % agarose. The sizes of hybridizing transcripts to cDNA probe for CTSE are indicated in kilobases.

RESULTS AND DISCUSSION

In our initial report we described a CTSE composite sequence (2158 bp) derived from cDNA clones that included an open reading frame of 1188 bases along with portion of the 5'-untranslated (49 bp) and 3'-untranslated (921 bp) regions.[1] The predicted amino acid sequence included a 396 residue molecule (M_r = 42,745) that contained two regions characteristic of the two conserved active site regions of all known aspartic proteinases and exhibited greater than 53 % identity to human pepsinogen A. A polyadenylation signal sequence, AATAAA, was noted 337 nucleotides downstream from the stop codon. Northern analysis of poly (A+) RNA isolated from CTSE producing gastric adenocarcinoma cell line revealed three transcripts (3.6, 2.6, and 2.1 kb). In the present study, the same CTSE transcript pattern as that seen in poly (A+) RNA from the cell line was observed in poly (A+)

```
TAAGGAGGGGCCTTGTGTCTGTGCCTGCCTGTCTGACAGACCTTGAATATGTTAGGCTGG
GGCATTCTTTACACCTACAAAAAGTTATTTTCCAGAGAATGTAGCTGTTTCCAGGGTTGC
AACTTGAATTAAGACCAAACAGAACATGAGAATACACACACACACACACATATACACACA
CACACACTTCACACATACACACCACTCCCACCACCGTCATGATGGAGGAATTACGTTATA
CATTCATATTTTGTATTGATTTTTGATTATGAAAATCAAAAATTTTCACATTTGATTATG
AAAATCTCCAAACATATGCACAAGCAGAGATCATGGTATAATAAATCCCTTTGCAACTCC
ACTCAGCCCTGACAACCCATCCACACACGGCCAGGCCTGTTTATCTACACTGCTGCCCAC
TCCTCTCTCCAGCTCCACATGCTGTACCTGGATCATTCTGAAGCAAATTCCGAGCATTAC
ATCATTTTGTCCATAAATATTTCTAACATCCTTAAATATACAATCGGAATTCAAGCATCT
CCCATTGTCCCACAAATGTTTGGCTGTTTTTGTAGTTGGATTGTTTGTATTAGGATTCAA
GCAAGGCCCATATATTGCATTTATTTGAAATGTCTGTAAGTCTCTTTCCATCTACAGAGT
TTAGCACATTTGAACGTTGCTGGTTGAAATCCCGAGGTGTCATTTGACATGGTTCTCTGA
ACTTATCTTTCCTATAAAATGGTAGTTAGATCTGGAGGTCTGATTTTGTGGCAAAAATAC
TTCCTAGGTGGTGCTGGGTACTTCTTGTTGCATCCTGTCAGGAGGCAGATAATGCTGGTG
CCTCTCTATTGGTAATGTTAAGACTGCTGGGTGGGTTTGGAGTTCTTGGCTTTAATCATT
CATTACAAAGTTCAGCATTTT - (A)
                    - ACCTGATCGTTTCAGTGGTACTTGATGATCATTGCTGA
GATCCACACTATAATTAGGGGTGGCAGAACAGGTGTTTTCTAATTCTGCTATCCCTTTGG
CATTTGTTAGTTGGAATTCTTCTATAAAAAACATAGGCCGGGTACAGTGCTCACGCCTGT
AATCCTAGCACTTTCGGAGGCCAAGGCAGGCAGATCACGAGGTCAAGAGATGGAGACTAT
CCTGGCCAACATGGTTAAACCCCTTCTCTACTAAAAGTACAAAAATTAGCCAGGCATGGT
GGCACACGCCTGTAGTCCCAGCTACCCAGGAAGCTGAGGCAGGAGAATCGCTTGAACCCA
GGAGACAGAGGCTGCAGTGAGCCAAGATCACGCCACTGCACTCCAGCCTGGCAACAGAGC
GAGACTCCTTCTC - (A)
```

Figure 2. The sequence of 3'-untranslated region of CTSE. The positions of stop codon and the polyadenylation signal sequence, AATAAA, are underlined. The location of the poly(A) tail is indicated by (A).

RNA from human gastric fundic mucosa (Figure 1). The CTSE transcripts are somewhat larger than might otherwise be expected for a precursor protein of $M_r=42,745$. The composite 2.1 kb CTSE sequence contained a relatively large 3'-untranslated region. It is interesting to note that the cathepsin D transcript also contains an extended 3'-untranslated flanking region similar in length to that which we have identified for CTSE.[4] It has been speculated that extended 3'-untranslated sequences may be a common feature of intracellular proteinases.

Further analysis of the gastric adenocarcinoma CTSE cDNA clones provided evidence for alternative polyadenylation of the primary transcript. Two different positions of the poly (A) tail, located 577 and 988 nucleotides downstream from AATAAA, have been identified (Figure 2).

We have reported a restriction fragment length polymorphism of CTSE.[5] The inheritance was examined in 4 families and is consistent with co-dominant expression of two alleles at probably a single locus. From these findings it therefore seems likely that the two major CTSE transcripts (2.1 and 2.6 kb) result from alternative polyadenylation. The characterization of the CTSE gene is in progress, and these studies will assist in the identification of the origins of the multiple transcripts.

ACKNOWLEDGEMENTS

This work was funded by a grant from the Center for Molecular Biology at Wayne State University and a grant in aid for encouragement of young scientists from the Ministry of Education, Science and Culture (02857109) Japan.

REFERENCES

1. T. Azuma, G. Pals, T. K. Mohandas, J. M. Couvreur and R. T. Taggart, Human gastric cathepsin E; predicted sequence, localization to chromosome 1, and sequence homology with other aspartic proteinases, J. Biol. Chem. 264:16748-16753 (1989).
2. U. Grubler and B. J. Hoffman, A simple and very efficient method for generating cDNA libraries, Gene 25:263-269 (1983).
3. L. B. Rall, J. Scott and G. I. Bell, Human insulin-like growth factor I and II mRNA: isolation of cDNAs and analysis of expression, in: "Methods of Enzymology," 147A, D. Barnes and D. A. Sirbasku, eds. p239-248, Academic Press, New York (1987).
4. P. L. Faust, S. Kornfeld and J. M. Chirgwin, Cloning and sequence analysis of cDNA for human cathepsin D, Proc. Natl. Acad. Sci. U.S.A. 82:4910-4914 (1985).
5. M. P. Johnson, T. Azuma, F. A. Boudi and R. T. Taggart, RFLP for human cathepsin E (CTSE), Nucl. Acids Res. 17:10147 (1989).

HUMAN STOMACH CATHEPSIN E ACTION ON HUMAN IMMUNOGLOBULINS

A. F. Kisseljov, S. V. Gulnik and N. I. Tarasova

Moscow State University
Department of Chemistry
Moscow 119899
USSR

INTRODUCTION

Human stomach mucosa contains four immunologically distinct aspartic proteinases - pepsin A, gastricsin or pepsin C (in the form of zymogens), cathepsin D and cathepsin E. Enzymatic properties, primary structure, cellular origin, activation mechanism and physiological role of pepsin, gastricsin and cathepsin D are well characterized. Much less information is available about cathepsin E although the enzyme has drawn much attention in recent years.

For comparison of enzymatic properties of stomach proteinases in hydrolysis of proteins we have undertaken a study of their action on human antibodies. Immunoglobulins were chosen as substrates because their primary and tertiary structures are well characterized and information about action of numerous proteinases on them is available. In addition, cathepsin E was found in lymphoid associated tissues and thus is believed to be involved in the immune response. Thus, information about the ability of cathepsin E to hydrolyze antibodies can contribute to our understanding of the physiological role of that enzyme as well.

METHODS

Purified human immunoglobulins were incubated with pepsin, gastricsin and cathepsin E. Samples of reaction mixture were removed in definite time intervals, boiled with SDS-PAGE sample buffer and analyzed by SDS-PAGE. The concentration of three enzymes in the reaction mixtures were adjusted so that they had equal milk-clotting activities.

Structure and Function of the Aspartic Proteinases
Edited by B.M. Dunn, Plenum Press, New York, 1991

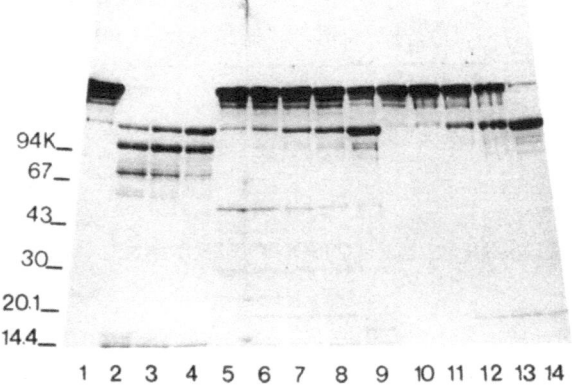

Figure 1. Digestion of human IgG with gastric proteinases at pH 4.0. SDS-PAGE under non-reducing conditions. Lane 1 - control sample; lanes 2-4 - pepsin digestion, incubation times 1, 2, 4 hours respectively; lanes 5-9 - cathepsin E digestion, incubation times 1, 2, 4, 6, 22 hours; lanes 10-14 - gastricsin digestion, incubation times 1, 2, 4, 6, 22 hours. 100 μl of 1.4 mg/ml immunoglobulin G solution was incubated with 15 μl of the enzyme in 0.1 M acetate buffer pH 4.0 at 37°C. Pepsin concentration in stock solution was 0.1 mg/ml (checked by amino-acid analysis). Solutions of gastricsin and cathepsin E were diluted so that they had milk-clotting activity exactly equal to that of 0.1 mg/ml solution of pepsin (0.1 mg/ml of gastricsin and 0.05 mg/ml of cathepsin E). 20 μl of reaction mixtures were removed at appropriate time, divided into 2 aliquots. Each aliquot was boiled with 10 μl 2x concentrated Laemli sample buffer, containing ß-mercaptoetanol and without it, and analysed by SDS-PAGE in 6-20 % gradient polyacrylamide gel. Pepsin and gastricsin were obtained by activation of corresponding zymogens. Pepsinogen and progastricsin were purified from human stomach mucosa extract exactly as described[11] Partially purified cathepsin E was obtained by the method of Tarasova *et al.*[12] Final purification was achieved by preparative polyacrylamide gel electrophoresis in Davies system in a gradient gel. The band, corresponding to the enzyme, was localized with the help of caseogram[13] and cut out. The enzyme was eluted in a Biotrap unit (Schliescher and Schuell) in 15 mM NH_4HCO_3.

RESULTS

At pH 5.0 none of three proteinases degraded immunoglobulins A and G. At pH 4.0 pepsin cleaved antibodies of both classes rapidly, while cathepsin E action was very weak (Figure 1). The velocities of human immunoglobulin G hydrolysis by pepsin, gastricsin and cathepsin E related approximately as 100:4:1. For human immunoglobulins class A the same ratio was 5:5:1. The molecular weights of pepsin hydrolysis products were in a good agreement with published results of other authors.[1-4] Gastricsin and cathepsin E cleaved immunoglobulins G almost in the same way as pepsin, that is, in a hinge region predominantly yielding (Fab)$_2$ and F$_c$ fragments (100 kDa and 21 kDa in size). In the case of pepsin further hydrolysis of F$_c$ fragment was detectable after 1 hour of incubation, meanwhile gastricsin showed the same extent of cleavage in 24 hours while cathepsin E required two days to achieve similar cleavage. Traces of F$_c$ fragment splitting by cathepsin E were observed after 2 days of incubation. The site of hydrolysis differed from that of pepsin, as the resulting parts were 19 kDa and 17.5 kDa, while pepsin gave several 14-16 kDa fragments.

In the case of immunoglobulin A the pattern of cleavage by all three enzymes was less clear, probably due to lower specificity of the proteases action. The characterization of the fragments of immunoglobulin A hydrolysis by gastric proteases is in progress now.

Cathepsin E is known to be activated by ATP in reactions of peptide hydrolysis at pH > 5.0.[5] No influence of the triphosphate was detected in the case of immunoglobulins cleavage at pH 4-5.5. This observation is consistent with the proposal that pH influences the reaction under investigation through the changes in immunoglobulin conformation mainly.

We have not detected any alteration of milk clotting activity of cathepsin E at pH 5.3 in the presence of ATP either. So it is probable that cathepsin E displays ATP-dependence of action only in the case of low molecular weight substrates.

DISCUSSION

Thus, cathepsin E is the least active gastric protease in immunoglobulin cleavage. At the same time the specific milk-clotting activity of cathepsin E was found to be approximately two times higher than that of pepsin and gastricsin. Kinetic parameters of peptide hydrolysis do not differ much for pepsin[6,7] and cathepsin E.[8,9] The main structural difference of cathepsin E from other mammalian aspartic proteases is that cathepsin E exists in a dimeric form. Elucidation of cDNA structure of the enzyme[10] and the data on molecular weight determination permit the suggestion that cathepsin E exists predominantly as a dimer consisting of two identical and catalytically active subunits. Nothing is known about the spatial arraignment of subunits, but one can assume that the dimerization leads to sterical hindrance in binding of large protein substrates. This may be a reason for lower catalytic activity against immunoglobulins, compared to monomeric aspartic proteases. However, this same argument does not yield a satisfactory explanation for the high milk-clotting activity of cathepsin E. The other reason for low rates of immunoglobulin hydrolysis by cathepsin E can be an unfavorable primary structure of the molecule that is a target for proteolytic action, i.e., of the hinge region of immunoglobulins. Unlike pepsins, the amino acid preference of cathepsin E is poorly characterized and the enzyme specificity awaits more careful study. Precise localization of cathepsin E hydrolysis sites in immunoglobulins is in progress now.

REFERENCES

1. R. Heimer, S. S. Schnoll and A. Primack, *Biochemistry* **6**:127-134 (1967).
2. H. Bennich and M. W. Turner, *Biochim. Biophys. Acta* **175**:388-395 (1969).
3. T. E. Michaelsen, B. Fangione and E. C. Franklin, *J.I mmunol.* **119**:558-563 (1977).
4. A. G. Pardo, E. S. Rosenwaser and B. Fangione, *J. Immunol.* **121**:1040-1044 (1978).
5. D. J. Thomas, A. D. Richards, R. A. Jupp, E. Ueno, K. Yamamoto, I. M. Samloff, B. M. Dunn and J. Kay, *FEBS Lett.* **243**:145-148 (1989).
6. B. M. Dunn, B. Parten, M. Jimenez, C. E. Rolph, M. J. Valler and J. Kay, *in*: "Aspartic Proteinases and their Inhibitors," V. Kostka, ed., 221-243, W. de Gruyter, Berlin (1985).
7. B. M. Dunn, M. J. Valler, C. E. Rolph, S. I. Foundling, M. Jimenez and J. Kay, *Biochim. Biophys. Acta* **913**:122-130 (1987).
8. I. M. Samloff, R. T. Taggart, T. Shiraishi, T. Branch, W. A. Reid, R. Heath, R. W. Lewis, M. J. Valler and J. Kay, *Gastroenterology* **93**:77-84 (1987).
9. R. A. Jupp, A. D. Richards, J. Kay, B. M. Dunn, J. B. Wyckoff, I. M. Samloff and K. Yamamoto, *Biochem. J.* **254**:895-898 (1988).
10. T. Azuma, G. Pals, T. K. Mohandas, J. M. Couvreur and R. T. Taggart, *J. Biol. Chem.* **264**:16748-16753 (1989).
11. P. K. Ivanov, M. M. Chernaya, A. E. Gustchina, I. V. Pechik, S. V. Nikonov and N. I. Tarasova, *Biochim. Biophys. Acta* (in press) (1990).
12. N. I. Tarasova, B. Foltmann and P. Szecsi, *Biochim. Biophys. Acta* **869**:96-100 (1986).
13. B. Foltmann, P. B. Szecsi and N. I. Tarasova, *Anal. Biochem.* **146**:353-360 (1985).

THE ENGINEERING OF RECOMBINANT ACTIVE HUMAN PRERENIN AND

ITS EXPRESSION IN MAMMALIAN AND INSECT CELLS

J. A. Norman, R. Baska, O. Hadjilambris, D. Youngsharp and R. Kumar

Departments of Cardiovascular Biochemistry and Molecular Biology
The Bristol-Myers Squibb Pharmaceutical Research Institute
Princeton, New Jersey

INTRODUCTION

Human renin, an aspartic proteinase, plays an important role in the regulation of the renin angiotensin cascade that regulates blood pressure and electrolyte balance. The cDNA encoding human renin has been cloned and sequenced[1] and is known to encode a 406 amino acid preprorenin protein that is processed by signal peptidase during secretion to release prorenin as a 386 amino acid zymogen. The 46 amino acid "pro" domain can then be removed by a renin processing enzyme to produce enzymatically active renin by cleavage at an Arg-Leu bond. Many investigators have mimicked the effects of the renin processing enzyme by trypsin activation *in vitro* where high concentrations of trypsin are incubated with prorenin for brief periods of time followed by excess trypsin inhibitor to stop secondary proteolytic processing by trypsin.[2] We have constructed prerenin by deleting the "pro" segment of the preprorenin cDNA and expressed this construct transiently and stably in mammalian cells and in the baculovirus insect cell expression system.

METHODS

Plasmid Construction

All plasmids were constructed using the techniques and guidelines described by Sambrook *et al.*[3] The cDNA encoding all but the first 10 nucleotides of the 5' untranslated region, the preprorenin coding region and all of the 3' untranslated region was removed from a pIBI76 plasmid by cutting with *Xba* I and *Bam* HI restriction enzymes and inserted into a pREX II expression plasmid. The subsequent pREN IIΔ expression plasmid contained the cDNA encoding human preprorenin downstream from an RSV LTR promotor. The pREN IIΔ plasmid also contains an SV 40 origin of replication making it suitable for COS cell expression. Figure 1 illustrates how pRENIII, the expression plasmid encoding prerenin, was constructed from 4 different parts to encode active renin with a Thr-Met substitution for Leu-Thr-Thr at the N-terminus. The DNA sequence corresponding to the 5' untranslated region, including the restriction enzyme sites *Sal* I, *Not* I and *Xba* I, all of the 20 amino acids of the "pre" region and the first two amino acids, Thr-Met, of renin were synthesized by polymerase chain reaction (PCR) using two synthetic deoxyoligonucleotide primers of 47 (3') and 49 (5') nucleotides according to the method of Saiki *et al.*[4] The amplification of this segment of pRENII∆ resulted in a 116 bp fragment that was digested with *Sal* I and *Nco* I restriction enzymes to produce a 104 bp fragment that was purified by gel electrophoresis. A segment of the renin cDNA corresponding to the 3rd to the 20th amino acid was produced by chemically synthesizing 2 DNA strands of 53 and 49 nucleotides. These DNA fragments were annealed such that a 5' CATG was extended from the double stranded product and would anneal with the 5' extended GTAC *Nco* I restriction site from the PCR amplified DNA. The DNA sequence encoding amino acids 21 through the termination codon (TGA) of the 3' untranslated region was obtained by cleavage of pRENII∆ plasmid DNA with *Sca* I and *Bam* HI restriction enzymes. The expression plasmid vector for cloning was obtained by cleavage of pREXIII with *Sal* I and *Bam* HI. The renin cDNA segment *Sca* I-*Bam* HI and expression vector pREXIII (*Sal* I-*Bam* HI) were purified by agarose gel electrophoresis. A 4 fragment ligation reaction was set up to assemble the complete prerenin expression plasmid. The transfer vector for baculovirus expression, pBacRen, was constructed by cutting the "prerenin" cDNA from pREN III using *Xba* I and *Bam* HI restriction enzymes and inserting it into a pVL 1392 transfer vector.

Transfection Protocols

Transient transfection of COS1 cells was used to examine the expression and secretion of both prorenin and renin. Transfection of COS 1 cells was carried out with pRENII∆ or pREN III (20 µg) using lipofectin to mediate the uptake of DNA. After the overnight transfection step, the cells were maintained in DMEM + 10% FBS and the media was sampled for renin secretion every 24 hours for the following 10 days. A permanent cell line secreting active renin was produced by co-transfection of NRP cells with pREN III and pHyg (selectable marker) using electroporation. After 24 hours, the cells were maintained in media containing 150 µg/ml Hygromycin B. Resistant clones were selected after 14 days and grown to confluency in 60 mm petri dishes then assayed for renin production. The positive clone B/1 was subsequently subcloned and assayed again for renin production. The production of a recombinant baculovirus containing the cDNA encoding prerenin was done by co-transfection of Sf9 insect cells with wild type baculovirus DNA and the pBacRen transfer vector. Three days after co-transfection of a 25 cm^2 flask of Sf9 cells, infection was visually evident by the presence of multiple occlusion bodies. The media was sampled and assayed for the production of active renin. The recombinant virus was subsequently purified in a plaque identification assay. The pure recombinant viral stock was then used to infect a 30 ml suspension of Sf9 cells and renin secretion was monitored with respect to time after infection.

Renin Assay

Immunoreactive renin was quantitated by a direct immunoassay[5] utilizing two monoclonal antibodies directed toward different epitopes on renin (Diagnostics Pasteur, Paris).

Enzymatically active renin was measured by the production of angiotensin I (AI) using human angiotensinogen as a substrate. Angiotensin I was quantitated by RIA using a kit produced by Biotecx Laboratories, Friendswood, TX.

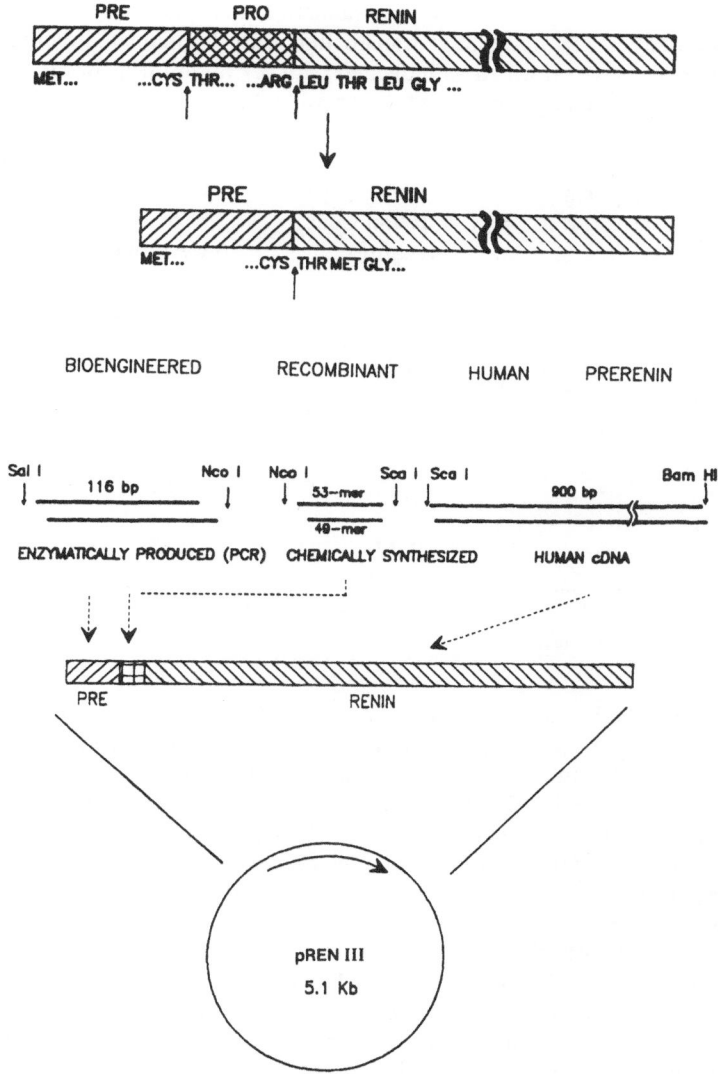

Figure 1. Strategy for the engineering of "prerenin".

RESULTS AND DISCUSSION

The pRENIII construct was tested by transient expression in COS 1 cells and compared with pRENIIΔ. Expression of active renin and prorenin was optimal after 3 days and remained surprisingly high for up to 14 days. This long duration of expression suggests that the mRNA encoding both prerenin and preprorenin have long half-lives in COS 1 cells. Since both the immunoassay and the enzymatic assay were performed on the culture media from these transfected COS 1 cells, it was possible to estimate the specific activity of renin in these samples. As shown in Table 1, the levels of immunoreactive renin are higher in the media from COS 1 cells transfected with pRENIIΔ than pRENIII but the specific activity of

Table 1. Expression of two forms of recombinant human renin in COS-1 cells. Culture medium was assayed 114 hours after transfection with the indicated expression plasmid. Trypsin treatment was performed as indicated with 0.1 mg/ml trypsin for 15 min at 20°C followed by the addition of 2 mg/ml soybean trypsin inhibitor.

Expression Plasmid	Trypsin Treatment	Active Renin Conc (ng/ml)	Renin Activity (ng AI/min/ml)	Specific Activity* (μg AI/min/mg)
pREN IIΔ (preprorenin)	-	0.07	0.13	2810 ± 770
n = 5	+	2.9	2.38	818 ± 67
pREN III (prerenin)	-	0.65	0.84	$1360 + 140$
n = 4	+	0.77	0.67	$870 + 60$

* Calculated specific activity of purified human renin (1500 Int. Units/mg) = 900 μg AI/min/mg (pH 7.0, 37°C, 0.5 μM human angiotensinogen)

active renin is higher than trypsin-activated prorenin. This indicates that the trypsin activation step leads to secondary proteolysis of renin with a loss of activity and that the expressed active renin is more homogeneous. This conclusion is confirmed by the lower specific activity of trypsin-treated active renin which is similar to that seen with trypsin-activated prorenin. The substitution of Thr-Met for Leu-Thr-Leu at the amino-terminus of renin does not appear to have any detrimental effect on the folding or activity of renin since the activity of this enzyme is preserved and the antibodies to the different epitopes used in the immunoassay recognize the active conformation of renin. This amino terminal substitution permits the Cys-Thr scissile bond present in the prepro junction of preprorenin to be preserved. It was reasoned that this Cys-Thr bond may be necessary for signal peptidase to recognize and cleave the signal sequence from prerenin. Additionally, the introduction of Met as the penultimate amino-terminal amino acid introduces a novel *Nco* I restriction site into the cDNA encoding prerenin. This restriction site will allow other cDNAs to be readily substituted for renin if they begin with a Methionine and take advantage of the prerenin signal sequence to permit their secretion. Previously, Harrison *et al.*,[6] have demonstrated that active renin can be expressed in mouse myeloma cells when transfected with a plasmid vector containing a targeted deletion of the prepro domain of preprorenin that has been replaced with a nucleotide sequence encoding a 19 amino acid heavy chain immunoglobulin leader sequence. It is not clear whether this construct is generally useful in many different cell types that do not express immunoglobulin genes. Additionally, the approach used in the present study to target the deletion of the "pro" region utilized the most current methodology in molecular biology.

The transient expression of active renin was also attempted in Sf9 insect cells using a baculovirus expression vector. Recombinant baculovirus containing the cDNA encoding human prerenin was purified by a plaque identification assay after co-transfection of Sf9 cells with wild type baculovirus DNA and the transfer vector pBacRen. The recombinant viral stock was used to infect a 30 ml suspension of Sf9 cells at a density of 10^6 cells/ml. Enzymatically and immunologically active renin secretion was monitored as a function of time after infection by measuring the levels of renin present in the culture media every 24 hours. Optimal renin secretion was evident 4 to 5 days after infection with the highest level of renin secretion reaching greater than 1 mg/L at day 5. Renin secretion decreased dramatically after day 5 when cell lysis was visually evident. These results indicate that insect cells are capable of correctly processing the prerenin protein allowing the active form of renin to be secreted at very high levels. This expression system may be optimized for the overproduction of active renin for subsequent purification and physical studies.

NRP cells, an NIH 3T3 cell line immortalized by polyoma virus, was used as a host for the establishment of a permanent cell line secreting active human renin. This cell line was identified as a surviving clone, B/1, after transfection of NRP cells with linearized pRENIII and pHyg. The B/1 clone was subcloned and one of the subclones, B10, was used for subsequent growth studies. The B10 subclone secretes approximately 100 pg renin/10^6 cells when grown in the presence of both G418 and Hygromycin. This cell line will secrete renin when kept in 2 % fetal bovine serum and has been maintained for approximately 6 months. This cell line may be useful for future studies to examine the processing and secretion of active renin.

In summary, active renin has been expressed in three different cell lines using a novel DNA construct employing 2 different expression vectors. It is evident that mammalian and insect cells have a signal peptidase capable of identifying and cleaving the 20 amino acid signal peptide, "pre", sequence in prerenin to permit the secretion of the active form of this enzyme.

REFERENCES

1. T. Imai, H. Miyazaki, S. Hirose, H. Hori, T. Hayashi, R. Kageyama, H. Ohkubu, S. Nakanishi and K. Murakami, Cloning and sequence analysis of cDNA for human renin precursor, *Proc. Natl. Acad. Sci. U.S.A.* **80**:7405 (1983).
2. S. A. Atlas, T. E. Hesson, J. E. Sealey, B. Dharmgrongartama, J. E. Laragh, M. C. Ruddy and M. Aurell, Characterization of inactive renin ("prorenin") from renin-secreting tumors of nonrenal origin, *J. Clin. Invest.* **73**:437 (1984).
3. J. Sambrook, E. F. Fritsch and T. Maniatis, "Molecular Cloning: A Laboratory Manual," 2nd Ed. Cold Spring Harbor Laboratory, Cold Spring Harbor, New York (1989).
4. R. K. Saiki, Enzymatic amplification of β-globin genomic sequences restriction site analysis for diagnosis of sickle cell anemia, *Science* **230**:1350 (1985).
5. J. Menard, T.-T. Guyenne, P. Corvol, B. Pau, D. Simon and R. Roncucci, Direct immunometric assay of active renin in human plasma, *J. Hypertension* **3(Suppl. 3)**:S275 (1985).
6. T. M. Harrison, The pro-peptide is not necessary for active renin secretion from transfected mammalian cells, *Proteins: Structure, Function, and Genetics* **5**:259 (1989).

SIMPLE PROCEDURE FOR RECOVERY OF CRYSTALLIZABLE HUMAN

RECOMBINANT RENIN FROM MAMMALIAN CELL-CONDITIONED MEDIUM

K. F. Geoghegan, A. J. Lanzetti, M. J. Ammirati, D. E. Danley,
B. A. O'Connor and P. M. Hobart

Central Research Division
Pfizer, Inc.
Groton, Connecticut   06340

INTRODUCTION

The structural analysis of human renin's active site is a critical component to understanding the enzyme's exquisite specificity. Broad interest in the molecular basis of substrate recognition by renin is due in large part to its pivotal role in blood pressure regulation. It is assumed that a precise elucidation of renin-inhibitor complexes will guide the design of novel agents representing a new generation of antihypertensive drugs.

Structural studies of renin, especially human renin, were limited because of the low abundance of the enzyme in both kidney tissue and serum. However, recombinant DNA cloning of human renin coding sequences has furnished, in addition to the complete sequence of the human renin protein and its precursor, a means to express human renin heterologously. In most cases, including the present work, the human renin precursor (preprorenin) coding sequence has been expressed in mammalian cells where its protein sequence signals the processing and release of prorenin into the culture medium. This release into culture media provides an obvious advantage for further purification. Several methods have been employed to isolate either prorenin or renin from cultured cell-conditioned medium using either conventional chromatography (Carilli *et al.*, 1988), active site-directed affinity chromatography (Heinrikson, *et al.*, 1989), or immunoaffinity chromatography (Higashimori, *et al.*, 1989).

This report describes a relatively simple purification strategy for renin based in large part on a classical observation that prorenin binds to immobilized Cibacron Blue (CB) resin under conditions where renin does not. Moreover, it is a relatively fast procedure requiring only commercially available reagents and yields an enzyme of high specific activity which is sufficiently pure to give diffraction-quality crystals.

EXPRESSION OF RECOMBINANT PRORENIN

Recombinant human prorenin was obtained from a cloned mouse cell line stably transformed with the human renin gene (Hobart *et al.*, 1984). Beginning with a 12.8 kb

Structure and Function of the Aspartic Proteinases
Edited by B.M. Dunn, Plenum Press, New York, 1991

human genomic DNA fragment encoding preprorenin, a plasmid expression vector was constructed by inserting a large mouse metallothionein enhancer/promoter DNA fragment (Kpnl^{-730} to Bgl2^{-64})(Glanville et al., 1981) into a unique Kpnl site located approximately 145 bases 5' of the presumptive transcriptional initiation site of the human renin gene. This vector, together with pSV2neo, was then integrated into genomic DNA of cultured mouse fibroblast cells (L929) using the method of $CaPO_4$ precipitation and G418 selection. Conditioned media from transformed cells were screened by immunodetection (Western blots) and renin enzyme assays. A clonal cell line (L929-2234) releasing more than 1 μg of human prorenin (per 10^6 cells per 24 hour period) into the growth media was isolated and used as a source of recombinant protein for subsequent purification.

PURIFICATION OF RECOMBINANT RENIN USING CIBACRON BLUE CHROMATOGRAPHY

Large scale mammalian cell cultures (performed at Invitron, Inc., St. Louis, MO) were used to prepare 100 liter volumes of conditioned medium containing low level serum (0.1%). The conditioned medium was concentrated 100-fold to provide a starting material for renin recovery having a protein concentration of 15 mg/ml. The major protein component of this material was identified by SDS-PAGE to be the bovine serum albumin.

Early recovery experiments revealed a difficulty derived from the use of L929 cells. This fibroblast-like cell line releases substantial quantities of collagen precursors into the medium along with prorenin. Trypsin-catalysed activation of the prorenin in crude media samples was accompanied by the formation of a heterogeneous population of collagen fragments which were found to adhere tenaciously to active renin. Renin, which appeared to be highly purified on silver stained SDS-PAGE, was shown by amino acid and protein sequence analyses to be severely contaminated with collagen-derived peptides. To correct this, ammonium sulfate treatment (40 % saturation) was introduced as an initial step to precipitate the collagen precursors, leaving prorenin in solution.

Following the ammonium sulfate step, the prorenin-containing sample (0.2 M NaCl; 0.02 M Tris-HCl, pH 8.0) was loaded onto a column of Blue Sepharose Fast Flow (Pharmacia). After the column had been washed, the bound protein, including prorenin and BSA, was eluted using 1.4 M NaCl, 20 mM Tris-HCl, pH 8.0. Following dialysis against

Table 1. Summary of Renin Purification

	Volume ml	Protein mg/ml	Activity Gu/ml	total(GU)	Sp Act Gu/mg	Yield %	Fold Purification
Medium Conc.	600	13.0	777	390,000	52	100	
(NH$_4$)$_2$SO$_4$ Sup.	970	3.7	575	560,000	155	143	3
Peak CB I	800	0.8	784	630,000	1018	162	20
Activated/diluted	2500	0.2	126	320,000	630	82	12
CB II/conc.	730	0.3	467	340,000	1567	87	30
Peak SSFF	452	0.2	335	150,000	1675	39	32

Activity values for the first three steps were derived by assaying trypsin-activated samples of these prorenin containing fractions; the radioimmunoassay method of renin assay was employed. Abbreviations: GU, Goldblatt unit of renin activity; CB, Cibacron Blue; SSFF, S Sepharose Fast Flow.

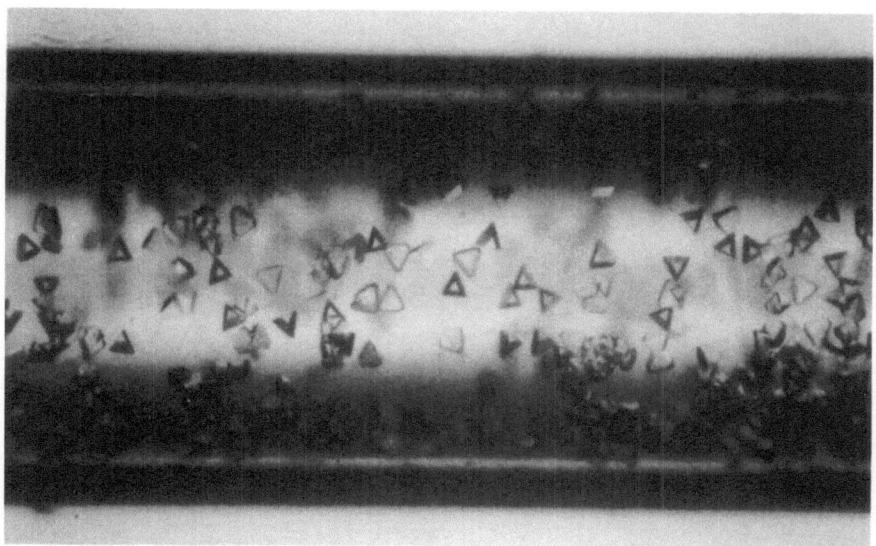

Figure 1. Microcrystals of recombinant human renin (shown in 1 mm capillary tube).

0.2 M NaCl, 0.02 M Tris, pH 7.4, the crude prorenin fraction was treated with immobilized trypsin (Trypsin-30 ENZYGEL, Boehringer Mannheim) at a level that effected full conversion of prorenin to renin in 3-5 h at 20°C. Pilot studies to establish the required quantity of insoluble trypsin were performed using SDS-PAGE (PhastSystem, Pharmacia) to monitor the conversion. Immobilized trypsin was subsequently removed by centrifugation and the supernatant fraction was reloaded onto a CB column (previously equilibrated with the pH 7.4 buffer). With the basic propeptide now absent, renin passed through the column without binding while a large majority of the contaminating proteins were once again retained. The flow through fraction was equilibrated with 0.005 M NaCl, 0.01 M sodium acetate pH 5.0, and the renin was concentrated and freed of residual impurities by cation exchange chromatography performed on S Sepharose Fast Flow (Pharmacia).

REFERENCES

Carilli, C., Wallace, L., Smith, L., Wong, M., and Lewicki, J., 1988, Semi-preparative purification of recombinant human renin and prorenin, *J. Chromatog.* **444**:203-208.

Glanville, N., Durham, D., Palmiter, R., 1981, Structure of mouse metallothionein-I gene and its mRNA, *Nature* (London) **292**: 267-269.

Heinrikson, R., Hui, J., Zurcher-Neely, H., and Poorman, R., 1989, A structural model to explain the partial catalytic activity of human prorenin, *Am. J. Hypertension* **2**:367-380.

Higashimori, K., Mizuno, K., Nakajo, S., Boehm, F., Marcotte, P., Egan, D., Holleman, W., Heusser, C., Poisner, A., and Inagami, T., 1989, Pure human inactive renin. Evidence that native inactive renin is prorenin, *J. Biol. Chem.* **264**:14662-14667.

Hobart, P., Fogliano, M., O'Connor, B., Schaeffer, I. and Chirgwin, J. 1984, Human renin gene: structure and sequence analysis, *Proc. Natl. Acad. Sci U.S.A.* **81**:5026-5030.

Lim, L., Stegeman, R., Leimgruber, N., Gierse, J., and Abdel-Mefuid, S., 1989, Preliminary crystallographic study of glycosylated recombinant human renin, *J. Mol. Biol.* **210**:239-240.

SUBSTRATE SPECIFICITY OF HUMAN RENIN: THE EFFECT OF SUBSTITUTIONS AT THE AMINO TERMINUS AND P_3 POSITION OF THE SUBSTRATE

David W. Green and Meheryar N. Rivetna

Department of Biological Sciences
Corporate Research and Development, Monsanto Company
700 Chesterfield Village Parkway
Chesterfield, Missouri   63198

INTRODUCTION

Renin (E.C. 3.4.23.15) regulates the initial step in the generation of the potent pressor octapeptide angiotensin II by cleaving the decapeptide angiotensin I from the amino terminus of angiotensinogen (Ondetti & Cushman, 1982).   In contrast to structurally homologous cellular aspartic proteases, renin specificity is restricted to angiotensinogen and little variation from the substrate sequence is tolerated. Inspection of the crystal structure (Sielecki *et al.*, 1989) indicates substrate residues P_4 to P_3' (nomenclature of Schechter & Berger, 1967, where P_1-P_1' is the scissile peptide bond) interact directly with renin. We have determined the steady-state kinetic parameters of human renin with substrates (Table 1) that differ in the amino terminal P_7 to P_{10} residues (PTDP and KIHPFHLLVYS) and P_3 residue (KIHPXHLLVYS). Our results are consistent with the structural predictions; that is, substitutions outside of the P_4 to P_3' sequence have no significant effect on renin activity but substituting *p*-nitro-phenylalanine (*p*NF, X) for phenylalanine at the P_3 position decreases the reactivity of renin.

MATERIALS AND METHODS

Peptide Synthesis

The peptides KIHPFHLLVYS and KIHPXHLLVYS were synthesized according to the procedure of Baranay and Merrifield (1979) and PTDP as reported previously (Green *et al.*, 1990). Peptide composition was verified by amino acid analysis (J. F. Zobel, Monsanto Co.) and fast-atom bombardment mass spectroscopy (P. C. Toren, Monsanto Co.).

Kinetic Measurements

Initial velocity measurements of renin activity were performed essentially as previously described (Green *et al.*, 1990). Stock solutions of peptides were prepared in

Table 1. Substrates used with human renin

Substrate	P_{10}	P_9	P_8	P_7	P_6	P_5	P_4	P_3	P_2	P_1	P_1'	P_2'	P_3'	P_4'
PTDP	Asp	Arg	Val	Tyr	Ile	His	Pro	Phe	His	Leu	Leu	Val	Tyr	Ser
KIHPFHLLVYS				*Lys*	Ile	His	Pro	Phe	His	Leu	Leu	Val	Tyr	Ser
KIHPXHLLVYS				*Lys*	Ile	His	Pro	p*NF*	His	Leu	Leu	Val	Tyr	Ser

Abbreviations: PTDP, porcine tetradecapeptide, Substitutions from PTDP are italicized; *p*NF is *p*-nitro-phenylalanine.

reaction buffer (100 mM MES pH 6.5 and 10 % weight/volume PEG) except for PTDP, which was initially dissolved in DMSO and diluted in reaction buffer so that the final DMSO concentration was less than 0.1%. Peptide concentrations were determined from the extinction coefficient of tyrosine except for KIHPXHLLVYS, which was determined by comparing the HPLC peak area (see below) to that of a known amount of KIHPFHLLVYS. Renin (32 nN final concentration) was incubated with 5 to 300 µM PTDP or KIHPFHLLVYS for 5 to 20 min and 320 nN enzyme with KIHPXHLLVYS for 60 to 240 min. The reaction was injected onto a Vydac C-18 column (10 x 250 mm) and the substrate and two product peaks separated by running a linear 15 to 70 % acetonitrile gradient with 0.1 % TFA over 25 min (2.0 ml min^{-1}). Peak area was determined using a Spectra-Physics SP4270 computing integrator coupled to a Kratos Spectraflow 783 detector (216 nm, 0.5 AUFS) or a Hewlett Packard 1040A diode array detector. The amount of product (LVYS) formed was determined by comparison of the product peak area to an LVYS standard curve. Kinetic constants were determined from fitting the initial velocities to the Michaelis-Menten equation using the non-linear regression analysis program of R. J. Leatherbarrow ("ENZFITTER", Elsevier-BIOSOFT).

RESULTS AND DISCUSSION

Renin cleaved PTDP, KIHPFHLLVYS and KIHPXHLLVYS (Table 1) exclusively between the Leu-Leu peptide bond, which allowed the rate of cleavage to be compared by measuring the amount of LVYS generated. The kinetic constants K_m (Michaelis constant), k_{cat} (turnover number) and k_{cat}/K_m (catalytic efficiency) obtained with the three substrates are listed in Table 2.

As reported by Poorman *et al.* (1987), substituting Lys for Asp-Arg-Val-Tyr on the amino terminus dramatically enhanced the solubility of the substrate and the substitution had no significant effect on the kinetic parameters (Table 2). Replacing Asp-Arg-Val-Tyr (PTDP) with Lys (KIHPFHLLVYS) increased K_m 3-fold and k_{cat} 2-fold, resulting in a 2-fold decrease in k_{cat}/K_m. These results are consistent with the structural prediction that substrate residues outside of the P_4 to P_3' sequence should not directly interact with renin (Sielecki *et al.*, 1989) and hence have little influence on catalysis.

The substitution of Phe with pNF at position P_3 did not change K_m, but k_{cat} and k_{cat}/K_m with KIHPXHLLVYS were decreased 100-fold relative to KIHPFHLLVYS (Table 2). Since turnover and catalytic efficiency are both decreased by the same order of magnitude, the result cannot be attributed to nonproductive substrate binding.

It was surprising that a substitution two peptide bonds from the scissile bond (P_1-P_1') would influence k_{cat} but not K_m. However, this is consistent with Fruton's (1976) observations of pepsin substrate specificity, where substrate substitutions 2 to 3 peptide bonds from the scissile bond affected k_{cat} but not K_m. Fruton suggested substrate binding in pepsin may be a step-wise process where the P_1 and P_1' residues serve as the initial "nucleation site" and the binding of flanking substrate residues continues until a complex is formed that can proceed through catalysis. The species specificity of renin is at least in part due to a Leu to Val P_1' substitution that affects K_m (Burton & Quinn, 1988). Thus, it seems reasonable that the nucleation model of substrate binding in the structurally homologous pepsin could be applied to renin to explain the effect of a P_3 substitution on K_m.

The nucleation model could also explain the effect on k_{cat}/K_m observed with the P_3 substitution. In a Michaelis-Menten mechanism (which renin follows), k_{cat}/K_m can not only describe the formation of the initial enzyme-substrate complex, but also subsequent isomerizations of the enzyme-substrate complex. It appears renin undergoes an isomerization following substrate binding based on structural comparison to aspartic protease-inhibitor complexes. Renin has a peptide loop or "flap" that in homologous cellular aspartic proteases folds over the active site after inhibitor binding (Sielecki et $al.$, 1989). In addition, Sali et $al.$ (1989) have shown that a P_3 residue can bind in different orientations in an endothiapepsin-inhibitor complex. Thus, the effect of the Phe to pNF P_3 substitution on k_{cat}/K_m with renin may be to interfere with an isomerization of the enzyme-substrate complex that is required for catalysis.

If isomerization of the renin-substrate complex is hindered by the P_3 substitution, k_{cat} could be affected by the ability of the renin-substrate complex to assume an orientation that is favorable for hydrolysis of the scissile bond. Crystal structures of aspartic protease-inhibitor complexes indicate that the scissile bond may undergo torsional strain during catalysis (Suguna et $al.$, 1987), which may require "anchoring" of the substrate to the enzyme by residues flanking the scissile bond. If the P_3 substitution hinders the anchoring, it is possible to see how a residue two peptide bonds from the scissile bond could decrease the rate of peptide bond hydrolysis. Whether the Phe to pNF substitution could do this by disrupting hydrophobic, steric or solvation interactions cannot be determined from the present analysis.

Table 2. Kinetic constants for human renin with PTDP,[a] KIHPFHLLVYS[b] and KIHPXHLLVYS[b]

Substrate	K_m μM	k_{cat} nmoles LVYS min^{-1} ml^{-1} (μg renin)$^{-1}$	k_{cat}/K_m
PTDP	34 ± 5.8	35 ± 2.3	1.1
KIHPFHLLVYS	110 ± 20	59 ± 5.0	0.52
KIHPXHLLVYS	92 ± 14	0.52 ± 0.040	0.0056

[a]100 mM MES, pH 6.5, 0.1 % DMSO, 10% PEG, 37°C; [b]100 mM MES, pH 6.5, 10 % PEG, 37°C

REFERENCES

Baranay, G. & Merrifield, R. B., 1979, *in*: "The Peptides", E. Gross & J. Meienhofer, eds., Vol. 2, pp. 1-284, Academic Press, New York.

Bradford, M., 1976, *Anal. Biochem.* **72**:248-254.

Burton, J. & Quinn, T., 1988, *Biochim. Biophys. Acta* **952**:8-12.

Fruton, J. S., 1976, *Adv. Enzymology* **44**:1-36.

Green, D. W., Aykent, S., Gierse, J. K. & Zupec, M. E., 1990, *Biochemistry* **29**:3126-3133.

Ondetti, M. A. & Cushman, D. W., 1982, *Ann. Rev. Biochem.* **51**:283-308.

Poorman, R. A., Palmero, D. P., Post, L. E., Murakami, K., Kinner, J. H., Smith, C. W., Reardon, H. & Heinrickson, R. L., 1987, *Proteins: Struct. Funct. Genet.* **1**:139-145.

Sali, A., Veerapandian, B., Cooper, J. B., Foundling, S. I., Hoover, D. J. & Blundell, T. L., 1989, *EMBO J.* **8**:2179-2188.

Schechter, I. & Berger, A., 1967, *Biochem. Biophys. Res. Commun.* **27**:157-162.

Sielecki, A. R., Hayakawa, K., Fujiaga, M., Murphy, M. E. P., Fraser, M., Muir, A. K., Carilli, C. T., Lewicki, J. A., Baxter, J. D. & James, M. N. G., 1989, *Science* **243**:1346-1351.

Suguna, K., Padlan, E. A., Smith, C. W., Carlson, W. D. & Davies, D. R., 1987, *Proc. Natl. Acad. Sci. U.S.A.* **84**:7009-7013.

SUBSTRATE ANALOGUE RENIN INHIBITORS CONTAINING REPLACEMENTS OF

HISTIDINE IN P_2 OR ISOSTERES OF THE AMIDE BOND BETWEEN P_3 AND P_2 SITES

P. Raddatz, A. Jonczyk, C. J. Schmitges and J. Sombroek

E. Merck
Pharmaforschung
D-6100 Darmstadt
Germany

INTRODUCTION

The search for orally active renin inhibitors as therapeutic agents continues to represent a challenging target for medicinal chemists.[1] Analogues of the angiotensinogen region flanking the bond split by renin have turned out to be very potent and specific inhibitors of renin. However, the high affinity of these angiotensinogen analogues for human renin is often associated with fast hydrolysis between P_3 and P_2 sites by the intestinal serine protease chymotrypsin.[2]

Since stability against proteolytic attack in the digestive tract is a requirement for orally active peptides,[3] we focused our synthetic efforts on angiotensinogen analogues that are resistant to chymotrypsin and that retain a high specificity and high inibitory potency for human renin.

In our synthetic strategy we chose to prepare modified compounds based on the potent renin inhibitors I and II reported by J. Boger[4] and M. Szelke,[5], respectively.

CHEMISTRY

The P_2-P_3 isosteres and the P_1-P_1' isosteres (Boc-ACHP-OH)[6] and Boc-Leuψ[CHOHCH$_2$]Val)[7] were synthesized by standard procedures.

BIOLOGY

In Vitro *Enzyme Inhibition.*

The IC_{50} values of the compounds against renin and the related aspartic proteases pepsin and cathepsin D are summarized in Table 1. The renin IC_{50} data were obtained by incubating human EDTA-plasma at pH 5.5 in the presence of angiotensinase inhibitors with different concentrations of the compounds to be tested for 2 h at 37°C. The angiotensin I generated during the incubation was quantified with a commercial radioimmunoassay kit (NEN).

The pepsin and cathepsin D IC_{50} values were determined by incubating hemoglobin with 20 U of porcine pepsin at pH 1.8 for 10 min. at 35.5°C and with 100 mU of bovine cathepsin D at pH 3.2 for 20 min. at 37°C, respectively. Hemoglobin is degraded by these enzymes to liberate peptides soluble in trichloroacetic acid. The concentration of the peptides is determined by their absorbance at 280 nm. The concentration of the inhibitor that inhibits peptide liberation (= pepsin or cathepsin D activity) by 50 % is calculated.

Degradation by Chymotrypsin

The enzymatic degradation of the synthetic peptides was performed at room temperature (24°C) with bovine α-chymotrypsin (45 mU/mg). The 0.05 M Tris-buffer contained 0.02 M CaCl$_2$ and was adjusted to pH 7.4 with HCl. Because of poor solubility all peptides were dissolved in formamide. Aliquots of the solutions of peptides in organic solvent and enzyme in aqueous buffer were mixed in autosampler vials to give final concentrations of 0.5 mg/ml peptide, 0.375 mg/ml chymotrypsin and 25 % (vol/vol) formamide. The content of each vial was acidified with 5 % TFA in 80 % isopropanol at desired stop times and analyzed by HPLC. The vials for time t_0 contained no enzyme.

The stability of the synthetic compounds towards chymotrypsin was examined by HPLC at 254/220 nm on a reversed phase column (Lichrosorb RP-8, 7 μm, 250 x 4 mm, E. Merck) in 0.3 % trifluoroacetic acid at 1 ml/min with a gradient of isopropanol (1 - 80 %) for 60 min. The remaining amount of undegraded peptides was expressed in % remaining HPLC area at 254 nm and plotted against time. $T_{1/2}$ (min) for degradation by chymotrypsin was calculated (Table 1, 'stable' means no detectable degradation at 8 h).

RESULTS

The replacement of histidine in P_2 of ACHP based tetrapeptides (Table 1) by glycine (III), ß-alanine (IV) and L-alanine (VI) led to compounds with renin inhibitory potency and specificity comparable to the standard peptide (I). Peptide V with γ-amino butyric acid in P_2 was about 10-fold less potent. Incubation of these peptides with the digestive enzyme chymotrypsin revealed a rapid degradation of histidine (I), glycine (III) and L-alanine (VI) containing compounds. Surprisingly the ß-alanine (IV) and the γ-amino butyric acid (V)

Table 1. Structures of renin inhibitors, inhibition parameters and stability to attack by chymotrypsin

	P_3	P_2		Human Renin	IC$_{50}$ [nM] Cathepsin D	Pepsin	$t_{1/2}$ (min) for Degradation by Chymotrypsin
I	Boc-Phe-	His	-ACHP-Ile-N〔pyridylmethyl〕	2.9	5800	> 10000	115
II	Boc-Phe-	His	-Leuψ[CHOHCH$_2$]Val-Ile-His-OH	2.4	n. d.	n. d.	75
III	Boc-Phe-	NHCH$_2$CO	-ACHP-Ile-N〔aminopyrimidinylmethyl〕 H$_2$N	2.5	13000	38500	118
IV	Boc-Phe-	NH(CH$_2$)$_2$CO	-ACHP-Ile-N〔aminopyrimidinylmethyl〕 H$_2$N	1.5	7900	18500	stable
V	Boc-Phe-	NH(CH$_2$)$_3$CO	-ACHP-Ile-N〔aminopyrimidinylmethyl〕 H$_2$N	28.5	3200	> 10000	stable
VI	Boc-Phe-	Ala	-ACHP-Ile-N〔aminopyrimidinylmethyl〕 H$_2$N	0.53	28	950	55
VII	Boc-	Pheψ[COCH$_2$]-Gly	-ACHP-Ile-N〔aminopyrimidinylmethyl〕 H$_2$N	8.5	> 10000	> 10000	n. d.
VIII	Boc-	Pheψ[CHOHCH$_2$]-Gly (S)	-ACHP-Ile-N〔aminopyrimidinylmethyl〕 H$_2$N	4.2	> 10000	> 10000	stable
IX	Boc-	Pheψ[CHOHCH$_2$]-Gly (R)	-ACHP-Ile-N〔aminopyrimidinylmethyl〕 H$_2$N	1.2	> 10000	10000	stable
X	Boc-	Pheψ[COCH$_2$S]-Gly	-ACHP-Ile-N〔aminopyrimidinylmethyl〕 H$_2$N	2.1	28000	53000	stable
XI	Boc-Phe-	NHCH$_2$CO	-Leuψ[CHOHCH$_2$]-Val-Ile-N〔aminopyrimidinylmethyl〕 H$_2$N	5.4	3900	1400	83
XII	Boc-Phe-	NH(CH$_2$)$_2$CO	-Leuψ[CHOHCH$_2$]-Val-Ile-N〔aminopyrimidinylmethyl〕 H$_2$N	4.1	4700	830	stable
XIII	Boc-	Pheψ[COCH$_2$S]Gly	-Leuψ[CHOHCH$_2$]Val-Ile- N〔aminopyrimidinylmethyl〕 H$_2$N	5.4	8800	6000	stable

containing inhibitors displayed a remarkable resistance against chymotrypsin under these conditions.

These findings were confirmed by compounds XI and XII representing a class of inhibitors with the Leuψ[CHOHCH$_2$]Val-isostere in position P$_1$-P$_1$'. In comparison to peptide III, which has a glycine in P$_2$, several isostere-containing compounds (VII - X) display comparable potencies, with the range varying from 3.4-fold less potent (VIII) to 2-fold more potent (IX).

Some representative examples of the isostere-containing compounds (VIII - X) were tested for sensitivity to chymotryptic attack. As expected these derivatives showed enhanced resistance to chymotrypsin when compared to the glycine containing standards III and XI.

CONCLUSION

1. Incorporation of ß-alanine in position P$_2$ of ACHP or Leuψ[CHOHCH$_2$]Val based tetrapeptides gave highly active renin inhibitors (compounds IV and XII). These inhibitors show a high specificity for renin and a remarkable stability against chymotrypsin.

2. The incorporation of isosteres of the amide bond (ketomethylenes, hydroxyethylenes and the corresponding prolonged thio-analogues) between P$_3$ and P$_2$ yielded compounds (VII-X and XIII) with comparable renin inhibitory activity as the glycine-containing standard peptides III and XI.

REFERENCES

1. W. J. Greenlee, *Med. Res. Rev.* 10:173 (1990); W. J. Greenlee, *Pharm. Res.* 4:364 (1987); J. M. Wood, J. L. Stanton and K. G. Hofbauer, *J. Enzyme Inhibition* 1:169 (1987); H. D. Kleinert, *Am. J. Hypertens.* 2:800 (1989).

2. Nomenclature as described by I. Schechter and A. Berger, *Biochem. Biophys. Res. Comm.* 27:157 (1967).

3. M. J. Humphrey and P. S. Ringrose, *Drug Metab. Rev.* 17:283 (1986).

4. J. Boger, L. S. Payne, D. S. Perlow, M. Poe, E. H. Blaine, E. H. Ulm, T. W. Schorn, B. LaMont, M. G. Bock, R. M. Freidinger, B. E. Evans and D. F. Veber, Ninth American Peptide Symposium, Toronto, (1985).

5. M. Szelke, D. H. Jones, A. Hallett and B. Atrash, WO 84/03044.

6. P. Raddatz, H. E. Radunz, G. Schneider and H. Schwarz, *Angew. Chem.* 100:414 (1988); For a review see: H.-J. Altenbach, *Nachr. Chem. Tech. Lab.* 36:756 (1988).

7. For a review see: R. Herming, *Nachr. Chem. Tech. Lab.* 38:460 (1990).

MOLECULAR MODELING OF RENIN INHIBITOR P_2 SUBSTITUENTS

E. A. Lunney,* C. C. Humblet,* J. T. Repine,* T. L. Blundell,**
J. B. Cooper** and B. L. Sibanda**

*Parke-Davis Pharmaceutical Research
Ann Arbor, Michigan 48105

**Laboratory of Molecular Biology
Department of Crystallography
Birkbeck College
London
United Kingdom

INTRODUCTION

In the absence of an accurate three-dimensional structure of human renin, the crystal structures of fungal enzymes have provided reasonable templates for the elaboration of a renin model.[1] In addition, the increasing number of bound inhibitor structures determined from co-crystallizations with endothiapepsin,[2] have considerably enhanced our understanding of the multiple binding modes available to the side chains. These experimental data have formed the bases for molecular modeling studies[3] applied to a limited series of highly flexible thiourea P_2 side chains. The results provide evidence for an inter-sidechain dependency that will ultimately govern the particular binding affinity. These findings are in agreement with an independent study recently reported in the literature.[4]

RESULTS

Various renin inhibitors have now been cocrystallized with endothiapepsin. Although the inhibitor-bound structures identify the availability of multiple binding modes, in particular for the P_5, P_2 and the P_3' subsites, they consistently retain a common extended backbone conformation for the P_3 to P_1 residues with sidechains alternating north-south. A knowledge-based human renin structural model, combined with the experimentally derived inhibitor binding mode was used to manually dock the backbone frame of a limited series of renin inhibitors (Table 1).

Table 1. Structures of renin inhibitors used in this work with IC_{50} values obtained with monkey renin

COMPOUND IC_{50} RENIN (MONKEY) nM

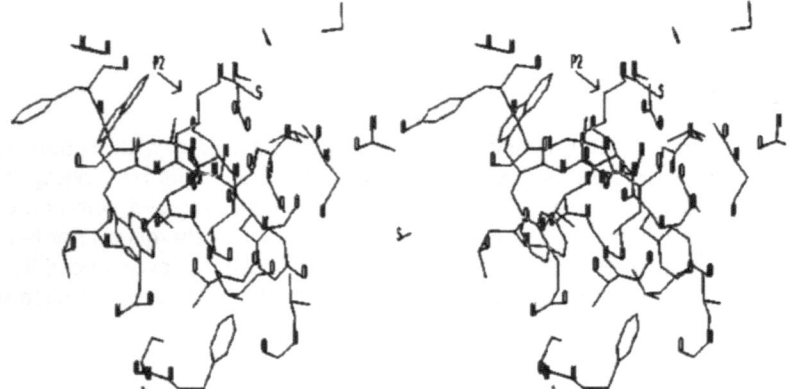

Compound	IC_{50} RENIN (MONKEY) nM
1	23
2	10,560
3	62

Figure 1. Stereoview of favorable P_2 conformation.

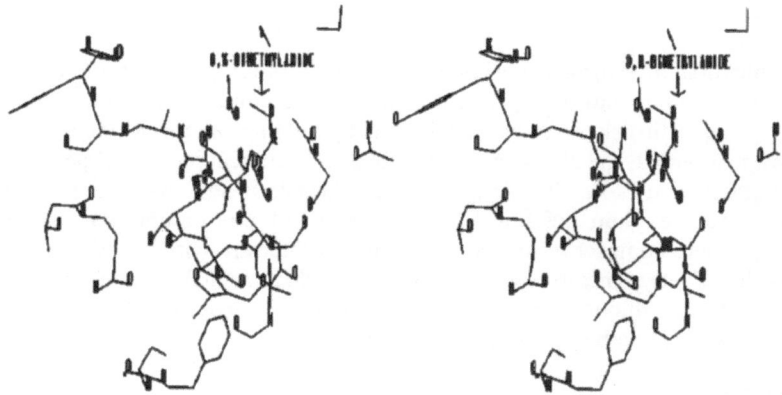

Figure 2. Stereoview showing a favorable conformation for binding of N,O-dimethylamide into the S_1' region.

Conformational analyses were applied to the P_4 to P_2 substructure common to **1** and **2** to determine accessible binding sites for the highly flexible thiourea P_2 sidechain. The results identified a favorable P_2 conformation (Figure 1) that extends into the S_1' region.

Looking further into the P_1' and P_2' substructures of these two analogs, it was surprising that the single substitution of the N,O-dimethylamide moiety in **2** for the methylbutylamide would lead to an essentially inactive analog. Beginning with a crystal structure from the Cambridge Structural Database[5] for the N,O-dimethylamide fragment, molecular modeling experiments indicated compatibility with the cleft and that certain low energy conformers occupied the S_1' binding site (Figure 2). These would interfere with the binding of the long P_2 side chain as shown in Figure 3. Furthermore, modeling indicated that the methylbutylamide can not significantly occupy the S_1' site when aligned for hydrogen bonding interactions.

To substantiate this hypothesis of the interference between the flexible P_2 side chain and a group in the S_1' site, **3**, an analog of **2** in which the thiourea substituent was replaced with the more compact histidine residue was synthesized. This analog exhibited strong binding affinity with an $IC_{50} = 62$ nM (Table 1).

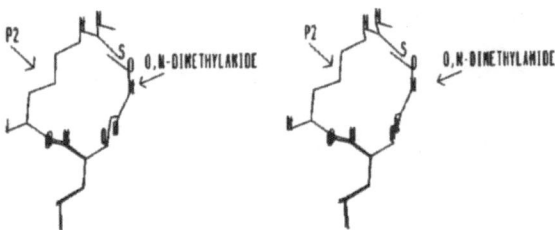

Figure 3. Stereoview showing an unfavorable interaction between P_2 thiourea substituent and N, O-dimethylamide in the P_1' region.

DISCUSSION

Molecular modeling experiments indicate a binding pocket for the flexible P_2 sidechain that extends into the S_1' site. Compound **1** contains a statine and so no P_1' sidechain occupies this site to interfere with this P_2 binding mode. Modeling studies have also shown that with **2**, the N,O-dimethylamide at P_2' can occupy the S_1' pocket left vacant upon having statine at P_1-P_1'. The long thiourea sidechain is no longer free to bind in the S_1' site. This could explain the low binding affinity of this compound. Based on this hypothesis, a more compact residue at P_2 would not interfere with the N,O-dimethylamide occupying the S_1' site. Therefore an analog of **2**, in which a smaller, standard P_2 substituent is substituted for the thiourea residue, should be a more potent inhibitor. This was shown to be the case with **3** exhibiting strong binding affinity.

CONCLUSION

Based on molecular modeling studies, upon binding, the long, flexible P_2 side chains of renin inhibitors can extend into the S_1' binding site. Analogs in which these P_2 groups are not competing with a P_1' or a P_2' residue for this binding site, have proven to be potent inhibitors, as in the case of **1**. Conversely, with **2**, the N,O-dimethylamide can interfere with occupation of the S_1' site by the long P_2 side chain, resulting in an essentially inactive compound.

REFERENCES

1. T. L. Blundell, B. L. Sibanda, A. M. Hemmings, S. I. Foundling, I. J. Tickle, L. H. Pearl and S. P. Wood, A rational approach to the design of renin inhibitors, *in*: "Molecular Graphics and Drug Design," Elsevier Science, B.V. (Biomedical Division), Amsterdam (1986).

2. T. L. Blundell, J. Cooper and S. I. Foundling, On the rational design of renin inhibitors: x-ray studies of aspartic proteinases complexed with transition-state analogues, *Biochem.* **26**:5585 (1987).

3. The molecular modeling studies were carried out using the Sybyl software package (Version 3.5), Tripos Associates, Inc., a subsidiary of Evans and Sutherland, 1699 S. Hanley Rd., Suite 303, St. Louis, Missouri 63144.

4. D. E. Epps, J. Cheney, H. Schostarez, T. K. Sawyer, M. Prairie, W. C. Krueger and F. Mandel, Thermodynamics of the interaction of inhibitors with the binding site of recombinant human renin, *J. Med. Chem.* **33**:2080 (1990).

5. (a) F. H. Allen, O. Kennard and R. Taylor, Systematic analysis of structural data as a research technique in organic chemistry, *Acc. Chem. Res.* **16**:146 (1983). (b) F. H. Allen, S. Bellard, M. D. Brice, B. A. Cartwright, A. Doubleday, H. Higgs, T. Hummelink, B. G. Hummelink-Peters, O. Kennard, W. D. S. Motherwell, J. R. Rodgers and D. G. Watson, *Acta. Crystallogr.* **B35**:2331 (1979).

HUMAN IMMUNODEFICIENCY VIRUS PROTEINASE:

NOW, THEN, WHAT'S NEXT?

Mary C. Graves

Department of Molecular Genetics
Roche Research Center
Hoffmann-La Roche Inc.
Nutley, New Jersey   07110

INTRODUCTION

"Art is I; science is we."   Claude Bernard (1813-1878)

Unlike in art, progress in the field of science comes as a result of the efforts of many people: "we" scientists. It emerges through the accumulation of work in countless laboratories, from those working for years on basic model systems, reaching all the way to those studying applied aspects of a specific system. This interdependent continuum of basic and applied aspects is exemplified in the evolving story of the human immunodeficiency virus protease (HIV PR).

The speed with which information accumulated about HIV -- from basic virology, to enzymology of key proteins, to clinical aspects -- remains unrivaled in the history of infectious disease research. Much of the success can be credited to timing: the emergence of the virus coincided with the developing fields of immunology and molecular biology. However, the speed with which many viral enzymes were identified and characterized may be attributed directly to the groundwork laid in basic mammalian and avian retroviral research and, for the viral protease in particular, to the wealth of information gained in the aspartic proteinase arena.   In this chapter, I would like to share one person's perspective on one of the fastest moving areas of science today: HIV-1 PR.

In considering the HIV PR story, I tend to recall it as a series of major accomplishments. This chapter, therefore, will be organized with these highlights in mind and will be presented in rough chronological order. A cautionary note: the goal of this chapter is to illuminate specific milestones in this brief story, and a direct acknowledgement of everyone's contributions is not possible or appropriate. Also, although data are beginning to accumulate on HIV-2 and SIV PR, the comments here will be restricted to HIV-1 PR.

The first descriptions of the "AIDS viruses," later renamed HIV, provided early circumstantial evidence for several viral enzymes, including the viral protease. In 1985, molecular cloning and sequence determination of these viruses revealed putative *gag* and *pol* open reading frames that were homologous to those of other known retroviruses. In retroviruses, these open reading frames are first translated into the gag and the gag-pol fusion precursor proteins that are later processed to mature species, at least in part by a viral protease[1,2] (Figure 1). In the *pol* open reading frame, note was taken of the now famous "Asp-Thr-Gly" region and of its homology to the active site of known retroviral proteases.[3] Toh *et al*.[4] also suggested that the retroviral proteases were related to the cellular aspartic proteinases based on this sequence conservation. Amino terminal sequencing of major gag proteins p24 CA[5,6] (see reference 7 for nomenclature) as well as NC[5] permitted the assignment of two cleavage sites. Later, amino terminal sequence determinations of two pol proteins, RT[8,9] and IN,[9] provided additional processing sites. How many of these were processed by the still putative viral protease was a matter of much speculation. Interestingly, only two of these sites (Figure 1 and Figure 2: site A, Tyr-Pro and site F, Phe-Pro) displayed sequence conservation, and they became the focus of those turning to the study of HIV PR.

The first specific HIV PR milestone was described in a hallmark paper by Kramer, *et al*. [10] in March of 1986. That work demonstrated that the PR coding region was in the 5' end of the *pol* open reading frame as predicted, but more important, it proved that active PR could be produced by recombinant means. The gag-pol fusion protein was synthesized in yeast and PR-dependent processing of gag was observed. This proved to be an excellent start to the field; however, no new major data appeared for about 1.5 years. This period was "the calm before the storm" that has yet to subside.

The second major milestone was the observation of PR autoprocessing and the demonstration that a gene encoding only the mature domain of PR (99 amino acids) could direct the synthesis of active PR. Several groups had observed a region of homology to the two aromatic-Pro cleavage sites (see above) that was located in pol just downstream of the gag-pol overlap site. Processing at this point (site E) would produce a 99 amino acid protein spanning the putative conserved active site. Work by Debouck, *et al*. [11] at the end of 1987 revealed that when a subsegment of the *pol* open reading frame was expressed in *E. coli*, the product underwent proteolysis to yield a 10 kDa species. Co-expression of the abbreviated pol precursor and the gag precursor resulted in accurate processing of gag. These results

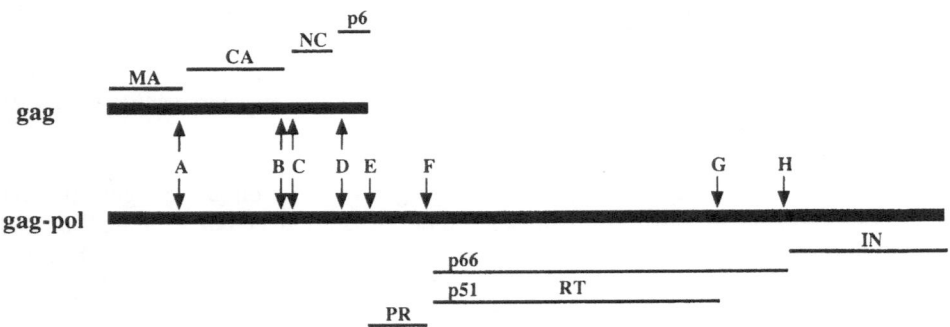

Figure 1. Location of HIV PR processing sites in the viral gag and gag-pol precursor molecules. Thick bars represent precursor polyproteins; thin bars represent mature products. Letters A through H designate the site of PR processing. Mature product abbreviation is as recommended.[7] Note that RT exists as a heterodimer of p51 and p66 subunits as shown. Other sites of processing have been noted in gag, see Figure 2.

TYPE	SEQUENCE	LOCATION
CLASS I	Ser-Gln-Asn-Tyr-:-Pro-Ile-Val-Gln	A
aromatic-Pro	Ser-Phe-Asn-Phe-:-Pro-Gln-Ile-Thr	E
	Thr-Leu-Asn-Phe-:-Pro-Ile-Ser-Pro	F
CLASS II	Ala-Arg-Val-Leu-:-Ala-Glu-Ala-Met	B
Glx	Ala-Thr-Ile-Met-:-Met-Gln-Arg-Gly	C
	Pro-Gly-Asn-Phe-:-Leu-Gln-Ser-Arg	D
OTHER	Ala-Glu-Thr-Phe-:-Tyr-Val-Asp-Gly	G
	Arg-Lys-Ile-Leu-:-Phe-Leu-Asp-Gly	H

Figure 2. Sequences of the HIV PR cleavage sites in the viral gag and gag-pol precursor molecules. The sequences are grouped into classes as described in the text. The -:- symbol indicates site of hydrolysis. The location is given by letter A through H (Figure 1). Three additional sites in gag[20] have not been well studied: site between B and C (class II : Ala-Glu-Ala-Met-:-Ser-Gln-Val), site internal to NC (class "other": Arg-Gln-Ala-Asn-:-Phe-Leu-Gly), site internal to p66 (class II: Phe-Arg-Ser-Gly-:-Val-Glu-Thr).

were soon confirmed and extended in several labs.[12-17] In early 1988, we went on to demonstrate that the 10-11 kDa species itself was active, and that a gene encoding only the 99 residue portion (Pro69-Phe167) was required for activity.[12] That is to say, correct PR folding does not require either the processing of termini or precursor sequences beyond that found in the mature domain. Thus, the single domain alone could be expressed to yield active PR. In quick succession, a variety of labs reported the production of recombinant active enzyme in *E. coli* (see below).

"Trust in Allah, but tie your camel"   Arabian Proverb

Many scientists trusted the concepts that HIV PR was 1) an aspartic PR due to sequence homology and 2) crucial for the life cycle of the virus. But while the data were consistent with these hypotheses, definite proof was lacking. Thus, the appearance of another hallmark paper was greeted with much enthusiasm (and relief!). Kohl and colleagues[18] created an mutant enzyme by altering the putative active site Asp_{25} to Asn. The mutated gene was incorporated back into the virus and the effect of this single change on virus function was analyzed. No gag processing was observed in the resulting virus particles and the particles produced were noninfectious. Later, the effect of PR mutants, which displayed immature particle morphology, was demonstrated visually by electron microscopy.[19] Thus, we have an elegant proof of the hypothesis that PR is crucial for virus viability, a result that also strengthened the interest of the pharmaceutical industry in HIV PR as a therapeutic target.

Concomitant with these results, investigations on the substrate specificities of PR were underway. First, a substrate "data base" was built by elucidating the remaining maturation sites in the gag and pol precursors. The sites previously identified within gag were confirmed and more were identified in the Oroszlan laboratory,[20] where they bravely purified proteins directly from virus particles for amino and carboxy-terminal analyses. Subsequently, they also identified analogous gag and pol cleavage sites of HIV-2 and SIV.[21] Determinations of the amino terminus of autoprocessed PR[14,22,23] and the RT p51/p66 site[24] completed the list of known pol processing sites. A cumulative list of the typically recognized sites is given in Figure 2. Second, many groups demonstrated the general ability of purified HIV PR to accurately cleave peptides containing these sites, albeit with quite disparate efficiencies (see reference 25 and references therein). These sequences have been grouped in several ways, none of which is entirely satisfying since no single structural theme is obvious from their primary sequence. The first group is the aromatic-Pro class, which was mentioned earlier, and is exemplified by the two sites that bound PR as well as the site between gag MA and CA proteins. In addition to the aromatic-Pro at the scissile bond, other residues appear to be conserved: Asn at P_2, Ser or Thr at P_4, and a long-chain hydrophobic residue at P_2'. The second class can be termed "Glx" (Gln or Glu), which simply has hydrophobic residues at P_1 and P_1' and the conserved Gln or Glu at P_2'. The catch-all category "other" encompasses the sites that do not have any features in common with the first two. These additional data allowed cleavage sites other than "aromatic-Pro" to be used as models for transition-state inhibitor design.

THE PROTEIN IS CHARACTERIZED

It almost seems unfair to group a tremendous body of work by a legion of scientists into the single milestone of "biochemical characterization" of the PR enzyme. First, it has been proven to be an aspartic protease. One can pick and choose among the ways this point has been approached: sequence homology,[4,26] three-dimensional structural homology,[27-29] active site aspartate mutagenesis[18,30-32] and inhibition by pepstatin[33] (K_i ~1 μM), a general inhibitor of this class of proteases. In early reports, isovaleryl-pepstatin appeared to inhibit HIV PR rather poorly (0.1-1 mM) when compared to cellular aspartyl proteinases.[16,17,23,30] [These early results may be explained by the recent observation that pepstatin forms visible polymers at mM concentrations, thus reducing the effective free pepstatin concentration (see Shoeman, et al., this volume).] But work by Richards and colleagues[34] demonstrated that a P_4 substituted pepstatin, acetyl-pepstatin, was indeed a potent inhibitor of HIV-1 PR (K_i = 20 nM).

The enzyme has now been purified by a number of groups[23,33,35-37] and shown to be active as a dimeric species[33,38] as predicted.[26] Kinetic parameters for several peptide substrates have been determined as summarized in a recent review.[25] It has been difficult to directly compare results from the different labs since there is not an accepted standard for assay conditions. Substrate K_m values vary but are typically in the mM to high nM range, although K_m appears to be improved by higher ionic strength.[34, 39] The turnover numbers are in the range of 2-70 per sec[33,36,38-41] and probably vary with peptide substrate as well as assay conditions (V_{max} appears to be reduced by high ionic strength). The pH optimum ranges from 4 to 6,[33, 34, 36, 42] which also varies as a function of ionic strength (concentrations of 0.05 M to 1 M monovalent ion have been used). pH appears to influence substrate K_m rather than k_{cat}.[39]

While recombinant PR was being produced by a variety of labs, ambitious peptide chemists undertook the herculean task of synthesizing the 99 residue peptide. At least two groups, at Cal Tech[43] and Merck,[44] were successful in chemically synthesizing the peptide and refolding at least a portion of it to an active species. It was an impressive feat for peptide chemistry as well as for protein purification when the active species was later isolated and used for crystallization studies.[28]

It is possible that the PR gene has been mutagenized more than any other gene to date. Some reports describe only single site changes, but the most thorough study by far was from the Swanstrom group which produced a family of mutants that spanned the entire coding region.[32,45] Their results agreed with crystallographic data (below) in distinguishing the parts of the protein that were most likely involved in critical functions, such as dimer contacts and substrate binding, from those that were less crucial for activity.

Another tremendous milestone was achieved in 1989: the three dimensional crystal structure solution of the native enzyme. It was first solved by the Merck group[27] and soon modified in the dimer contact region by the Wlodawer group[28] using synthetic PR. A third report from the Blundell group[29] using recombinant material confirmed the modified structure. These papers gave us an exquisite picture of the enzyme, showing that the two identical subunits were related by an exact two-fold axis of symmetry. Structural comparison to the larger cellular and bacterial monomeric aspartic proteinases reveals how Nature has economized and streamlined this class of enzymes. While the achievement of structure determination itself was outstanding, perhaps even more so was the willingness of the Wlodawer group to share this vital information as quickly as possible with the scientific community at large. That quick dissemination of information was crucial to the rapid progress made by other labs in solving later crystal structures of PR:inhibitor complexes[46] (also, see other chapters in this volume).

INHIBITORS ARE DEVELOPED

While the enzyme itself was being well characterized and studied, a flurry of industrial and academic activity culminated in the synthesis of many substrate-based transition-state analog compounds that are effective inhibitors.[46-51] The best of these display quite potent activity not only against the enzyme *in vitro* but also against the virus in cultured cells. Of importance is the fact that several different classes of inhibitors have been described as shown by representatives in Figure 3. The chemical classes represented range from the hydroxyethylene (A, B) to hydroxyethylamine (D) to the statine isostere (E). Efficacy of reduced bond inhibitors described thus far have not approached those of other classes. Inhibitors span regions encompassing subsites P_4-P_4' (E), P_2-P_3' (A), and P_3-P_2' (D). The inhibitors shown also demonstrate that the peptide chain can be in register (A, B) or out of register either by the addition of a methylene group, as shown by Roche compound (D), or short, as in acetyl-pepstatin and the symmetrical inhibitor (E, C). All these classes have yielded potent inhibitors whose K_i varies from subnanomolar to about 70 nM, with more inhibitors continually being synthesized (see other chapters in this volume).

Thus, with the information obtained from the native crystal structure and the production of potent inhibitors, descriptions of three dimensional structures of PR:inhibitor complexes were anxiously anticipated. Four structures of complexes have been published to date[46,50-52] and others appear in this volume. The inhibitor peptide chains are all in extended conformations, making efficient use of backbone hydrogen bond capability. It appears that the enzyme binding cleft can accommodate residues spanning P_3-P_3', with some variation at P_4: a P_4 acetyl group being exposed to solvent [52] but a P_4 Ser being protected.[51] These or related inhibitors are under development in several companies, the best of which will soon be tested in man. Moreover, the structures of enzyme:inhibitor complexes, as well as ones sure to follow, will provide the basis for further rational design of inhibitors. Hope remains high that some PR inhibitor may be successfully developed in the clinic as an anti-HIV drug.

WHAT'S NEXT?

When considering these milestones, the HIV PR field can easily be described as one of the most rapidly evolving in biological science today. In less than 5 years, we have gone from the first description of the target in the literature, to producing and characterizing recombinant and synthetic enzyme, to having three-dimensional structural information, to having potent inhibitors produced and tested in man.

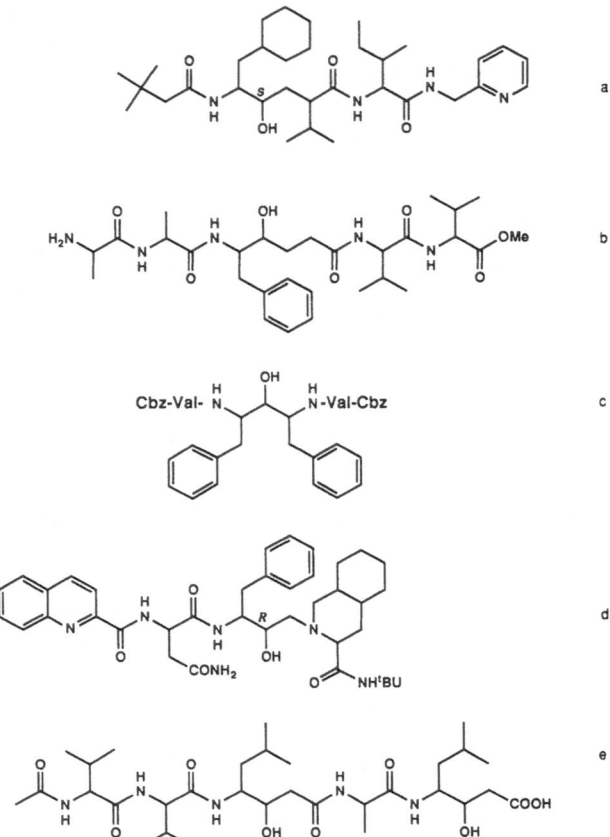

Figure 3. Examples of potent substrate-based inhibitors of HIV PR. a: hydroxyethylene inhibitor U-81749,[48] K_i = 70 nM. b: hydroxyethylene #8 in reference 47, K_i = 18 nM. c: C2 symmetrical inhibitor A-74704,[46] K_i = 4.5 nM. d: hydroxyethylamine Ro 31-8959,[49] K_i < 0.4 nM. e: acetyl-pepstatin,[34] K_i = 20 nM. When known, the configuration at the transition state isostere is given.

Yet, in the midst of this "applied" research directed towards PR inhibitor development, one can easily find intriguing "fundamental" questions regarding PR that will be the basis of future research. Surely, this fertile field of research will continue to hold the interest of both academic and industrial scientists. While one cannot presume to know all directions the studies will take, many areas currently are being pursued.

Several obvious mechanistic questions remain regarding the viral precursor activation. For example, is there an ordered maturation of gag or gag-pol precursors? Experiments using gag precursors synthesized *in vitro* or recombinant precursors synthesized *in vivo* as substrate seem to suggest so.[53,54] If so, then do the different "classes" of cleavage sites play a role in the ordered process? Is it significant that the aromatic-Pro sites bound PR? Evidence appears to indicate that local secondary/tertiary structure surrounding the cleavage site has a greater impact than the exact sequence at the cleavage site, at least for some of the sites.[54] Curiously, the genetic order of the *gag* coding domains mirrors their spatial arrangement in the viral particle. Is this order maintained by a specific processing regimen?

Perhaps most intriguing is the question: what is the control mechanism by which the PR domain (zymogen) becomes activated in the budding virus? First, all gag and gag-pol precursors transit the cell cytoplasm to become membrane associated via the amino-terminal myristic acid moiety, where gag protein begins to oligomerize. Since PR is active only as a dimer, and since the concentration of gag-pol is ~20- to 50-fold lower than gag precursor, then on a numerical basis, gag-pol:gag-pol dimer formation would not be favored over gag:gag-pol dimer/oligomer. Therefore, are simply membrane compartmentalization and low concentration of the PR precursor species the controlling factors, or are there additional mechanisms?

Turning to the enzyme mechanism itself, a dichotomy clearly exists in the molecular requirements for a PR cleavage site (and presumably an inhibitor). On the one hand, the enzyme is highly specific for certain sites and does not indiscriminately hydrolyze bulk protein. But on the other hand, many of the recognized sites have little or nothing in common. For example, the two peptide sequences that appear to be hydrolyzed the most efficiently are: S-F-N-F-:-P-Q-I and A-T-I-M-:-M-Q-R.[23,55] Perhaps there is an underlying truth that Nature has hidden in the clues that we have yet to discern. However, recent work has given us additional clues as to molecular requirements. Two non-viral proteins are substrates for PR and are hydrolyzed at the sites S-L-N-L-:-R-E-T (vimentin[40]) and G-D-A-L-:-L-E-R (*pseudomonas* exotoxin[56]). Peptides containing these sites are hydrolyzed with similar efficiencies as those spanning cleavage sites in the viral precursors. The presence of the positively charged Arg in the P_1' position of vimentin was unexpected, but is clearly permissible. The vimentin site resembles a hybrid of the aromatic-Pro (P_4 Ser and P_2 Asn) and Glx ($P_{2'}$ Glu) classes. Recent results clearly demonstrate the importance of the P_2' residue within the context of gag cleavage site A.[57] Others have found that the nature of the P_2 residue impacts on the k_{cat} but not K_m value.[41] Also, HIV-1 PR appears to accurately process HIV-2 precursors even though some sites differ at the P_2 and P_1 positions.[58] Are rules emerging from all the examples to point out preferences in positions P_4, P_2, and P_2' for substrate recognition and hydrolysis?

Is the relative importance of a residue's position context dependent? For example, does the importance of the P_2 residue differ depending on the rest of the sequence? Is the flexibility more in the enzyme or the substrate? Analysis of the crystal structures of PR complexed to a variety of inhibitors demonstrates very little difference in the PR binding pockets,[46,50-52] suggesting that the substrate, or inhibitor, must assume a certain conformation to effectively occupy the subsites. Can we learn more about this flexibility by studying the bound inhibitors? Somehow, inhibitors whose peptide backbone is in register or longer or shorter by one atom are all effectively bound. How is this difference in length accommodated?

Different types of questions arise when considering PR as a target for antiviral therapy. Perhaps the most important is: will inhibitors of PR succeed as an effective treatment for HIV infections? Certainly many pharmaceutical firms believe so and are working intensely for verification. Will the promise of *de novo* rational drug design based on structural information come to fruition in the case of HIV PR? The number of labs that have independently obtained crystal structure information on the native PR or a PR:inhibitor complex probably exceeds that for any other protein system. And the clear goal is to use this information for improving inhibitor design for drug development. One recent report demonstrates the potential of this approach,[59] as well as several contributions to this book. Are we standing on the threshold of the future of drug design?

These and other questions will continue to be the focus of research. The PR saga as outlined here is the culmination of basic groundwork laid in the fields of retrovirology and aspartic proteinases. Without this body of information, it would have been impossible to achieve these impressive milestones in such a short period of time. Just as advances in the HIV PR arena relied upon past successes in other fields, may the contributions described here fuel related areas of research. The science that "we" practice and share now can only further expand future success in both basic and applied research.

ACKNOWLEDGEMENTS

I am especially thankful to the scientists from the aspartic proteinase field who were enthusiastic teachers to us "newcomers". Recognition is also given to my colleagues and competitors who were remarkably open with their data and ideas. I also thank Louis Todaro for his help in assembling the figure of inhibitors.

REFERENCES

1. C. Dickson, R. Eisenman, H. Fan, E. Hunter and N. Teich, Protein biosynthesis and assembly, in: "RNA Tumor Viruses," R. Weiss, N. Teich, H. Varmus and J. Coffin, eds., 2nd Ed., Cold Spring Harbor Lab., Cold Spring Harbor, NY (1984).

2. A. M. Skalka, Retroviral proteases: first glimpses at the anatomy of a processing machine, *Cell* **56**:911 (1989).

3. L. Ratner, W. Haseltine, R. Patarca, K. J. Livak, B. Starcich, S. F. Josephs, E. R. Doran, J. A. Rafalski, E. A. Whitehorn, K. Baumeister, L. Ivanoff, S. R. Petteway Jr, M. L. Pearson, J. A. Lautenberger, T. S. Papas, J. Ghrayeb, N. T. Chang, R. C. Gallo and F. Wong-Staal, Complete nucleotide sequence of the AIDS virus, HTLV-III, *Nature* **313**:277 (1985).

4. H. Toh, M. Ono, K. Saigo and T. Miyata, Retroviral protease-like sequence in the yeast transposon TY 1, *Nature* **315**:691 (1985).

5. R. Sanchez-Pescador, M. D. Power, P. J. Barr, K. S. Steimer, M. M. Stempien, S. L. Brown-Shimer, W. W. Gee, A. Renard, A. Randolph, J. A. Levy, D. Dina and P. A. Luciw, Nucleotide sequence and expression of an AIDS-associated retrovirus (ARV-2), *Science* **227**:484 (1985).

6. S. Wain-Hobson, P. Sonigo, O. Danos, S. Cole and M. Alizon, Nucleotide sequence of the AIDS virus, LAV, *Cell* **40**:9 (1985).

7. J. Leis, D. Baltimore, J. M. Bishop, J. Coffin, E. Fleissner, S. P. Goff, S. Oroszlan, H. Robinson, A. M. Skalka, H. M. Temin and V. Vogt, Standardized and simplified nomenclature for proteins common to all retroviruses, *J. Virol.* **62**:1808 (1988).

8. F. dM. Veronese, T. D. Copeland, A. L. DeVico, R. Rahman, S. Oroszlan, R. C. Gallo and M. G. Sarngadharan, Characterization of highly immunogenic p66/p51 as the reverse transcriptase of HTLV-III/LAV, *Science* **231**:1289 (1986).

9. M. M. Lightfoote, J. E. Coligan, T. M. Folks, A. S. Fauci, M. A. Martin and S. Venkatesan, Structural characterization of reverse transcriptase and endonuclease polypeptides of the acquired immunodeficiency syndrome retrovirus, *J. Virol.* **60**:771 (1986).

10. R. A. Kramer, M. D. Schaber, A. M. Skalka, K. Ganguly, F. Wong-Staal and E. P. Reddy, HTLV-III gag protein is processed in yeast cells by the virus *pol*-protease, *Science* **231**:1580 (1986).

11. C. Debouck, J. G. Gorniak, J. E. Strickler, T. D. Meek, B. W. Metcalf and M. Rosenberg, Human immunodeficiency virus protease expressed in *Escherichia coli* exhibits autoprocessing and specific maturation of the gag precursor, *Proc. Natl. Acad. Sci. U.S.A.* **84**:8903 (1987).

12. M. C. Graves, J. J. Lim, E. P. Heimer and R. A. Kramer, An 11-kDa form of human immunodeficiency virus protease expressed in *Escherichia coli* is sufficient for enzymatic activity, *Proc. Natl. Acad. Sci. U.S.A.* **85**:2449 (1988).

13. J. Mous, E. P. Heimer and S. F. J. Le Grice, Processing protease and reverse transcriptase from human immunodeficiency virus type I polyprotein in *Escherichia coli*, *J. Virol.* **62**:1433 (1988).

14. E. P. Lillehoj, F. H. R. Salazar, R. J. Mervis, M. G. Raum, H. W. Chan, N. Ahmad and S. Venkatesan, Purification and structural characterization of the putative gag-pol protease of human immunodeficiency virus, *J. Virol.* **62**:3053 (1988).

15. H.-G. Kräusslich, H. Schneider, G. Zybarth, C. A. Carter and E. Wimmer, Processing of *in vitro*-synthesized gag precursor proteins of human immunodeficiency virus (HIV) type 1 by HIV proteinase generated in *Escherichia coli*, *J. Virol.* **62**:4393 (1988).

16. C.-Z. Giam and I. Boros, *In vivo* and *in vitro* autoprocessing of human immunodeficiency virus protease expressed in *Escherichia coli*, *J. Biol. Chem.* **263**:14617 (1988).

17. J. Hansen, S. Billich, T. Schulze, S. Sukrow and K. Moelling, Partial purification and substrate analysis of bacterially expressed HIV protease by means of monoclonal antibody, *EMBO J.* **7**:1785 (1988).

18. N. E. Kohl, E. A. Emini, W. A. Schleif, L. J. Davis, J. C. Heimbach, R. A. F. Dixon, E. M. Scolnick and I. S. Sigal, Active human immunodeficiency virus protease is required for viral infectivity, *Proc. Natl. Acad. Sci. U.S.A.* **85**:4686 (1988).

19. C. Peng, B. K. Ho, T. W. Chang and N. T. Chang, Role of human immunodeficiency virus type 1-specific protease in core protein maturation and viral infectivity, *J. Virol.* **63**:2550 (1989).

20. L. E. Henderson, T. D. Copeland, R. C. Sowder, A. M. Schultz and S. Oroszlan, Analysis of proteins and peptides purified from sucrose gradient banded HTLV-III, *in*: "Human retroviruses, cancer, and AIDS: approaches to prevention and therapy," D. Bolognesi, ed., Alan R. Liss, Inc., New York (1988).

21. L. E. Henderson, R. E. Benveniste, R. Sowder, T. D. Copeland, A. M. Schultz and S. Oroszlan, Molecular characterization of gag proteins from simian immunodeficiency virus (SIV_{Mne}), *J. Virol.* **62**:2587 (1988).

22. M. C. Graves, J. J. Lim, M. A. Zicopoulos, T. J. Stoller, M. C. Miedel, Y.-C. E. Pan, W. Danho and C. M. Nalin, Expression and characterization of human immunodeficiency virus-1 protease, *in*: "Proteases of Retroviruses," V. Kostka, ed., Walter de Gruyter & Co., Berlin (1989).

23. H.-G. Kräusslich, R. H. Ingraham, M. T. Skoog, E. Wimmer, P. V. Pallai and C. A. Carter, Activity of purified biosynthetic proteinase of human immunodeficiency virus on natural substrates and synthetic peptides, *Proc. Natl. Acad. Sci. U.S.A.* **86**:807 (1989).

24. M. C. Graves, M. C. Meidel, Y.-C. E. Pan, M. Manneberg, H.-W. Lahm and F. Grüninger-Leitch, Identification of a human immunodeficiency virus-1 protease cleavage site within the 66,000 dalton subunit of reverse transcriptase, *Biochem. Biophys. Res. Comm.* **168**:30 (1990).

25. J. Kay and B. M. Dunn, Viral proteinases: weakness in strength, *Biochim. Biophys. Acta* **1048**:1 (1990).

26. L. H. Pearl and W. R. Taylor, A structural model for the retroviral proteases, *Nature* **329**:351 (1987).

27. M. A. Navia, P. M. D. Fitzgerald, B. M. McKeever, C.-T. Leu, J. C. Heimbach, W. K. Herber, I. S. Sigal, P. L. Darke and J. P. Springer, Three-dimensional structure of aspartyl protease from human immunodeficiency virus HIV-1, *Nature* **337**:615 (1989).

28. A. Wlodawer, M. Miller, M. Jaskólski, B. K. Sathyanarayana, E. Baldwin, I. T. Weber, L. M. Selk, L. Clawson, J. Schneider and S. B. H. Kent, Conserved folding in retroviral proteases: crystal structure of a synthetic HIV-1 protease, *Science* **245**:616 (1989).

29. R. Lapatto, T. Blundell, A. Hemmings, J. Overington, A. Wilderspin, S. Wood, J. R. Merson, P. J. Whittle, D. E. Danley, K. F. Geoghegan, S. J. Hawrylik, S. E. Lee, K. G. Scheld and P. M. Hobart, X-ray analysis of HIV-1 proteinase at 2.7 Å resolution confirms structural homology among retroviral enzymes, *Nature* **342**:299 (1989).

30. S. Seelmeier, H. Schmidt, V. Turk and K. von der Helm, Human immunodeficiency virus has an aspartic-type protease that can be inhibited by pepstatin A, *Proc. Natl. Acad. Sci. U.S.A.* **85**:6612 (1988).

31. S. F. J. Le Grice, J. Mills and J. Mous, Active site mutagenesis of the AIDS virus protease and its alleviation by *trans* complementation, *EMBO J.* **7**:2547 (1988).

32. D. D. Loeb, C. A. Hutchison III, M. H. Edgell, W. G. Farmerie and R. Swanstrom, Mutational analysis of human immunodeficiency virus type 1 protease suggests functional homology with aspartic proteinases, *J. Virol.* **63**:111 (1989).

33. P. L. Darke, C.-T. Leu, L. J. Davis, J. C. Heimbach, R. E. Diehl, W. S. Hill, R. A. F. Dixon and I. S. Sigal, Human immunodeficiency virus protease: bacterial expression and characterization of the purified aspartic protease, *J. Biol. Chem.* **264**:2307 (1989).

34. A. D. Richards, R. Roberts, B. M. Dunn, M. C. Graves and J. Kay, Effective blocking of HIV-1 proteinase activity by characteristic inhibitors of aspartic proteinases, *FEBS Lett.* **247**:113 (1989).

35. J. E. Strickler, J. Gorniak, B. Dayton, T. Meek, M. Moore, V. Magaard, J. Malinowski and C. Debouck, Characterization and autoprocessing of precursor and mature forms of human immunodeficiency virus type 1 (HIV 1) protease purified from *Escherichia coli*, *Proteins* **6**:139 (1989).

36. A. G. Tomasselli, M. K. Olsen, J. O. Hui, D.J. Staples, T. K. Sawyer, R. L. Heinrikson and C.-S. C. Tomich, Substrate analogue inhibition and active site titration of purified recombinant HIV-1 protease, *Biochemistry* **29**:264 (1990).

37. J. Rittenhouse, M. C. Turon, R. J. Helfrich, K. S. Albrecht, D. Weigl, R. L. Simmer, F. Mordini, J. Erickson and W. E. Kohnbrenner, Affinity purification of HIV-1 and HIV-2 proteases from recombinant *E. coli* strains using pepstatin-agarose, *Biochem. Biophys. Res. Comm.* **171**:60 (1990).

38. T. D. Meek, B. D. Dayton, B. W. Metcalf, G. B. Dreyer, J. E. Strickler, J. G. Gorniak, M. Rosenberg, M. L. Moore, V. W. Magaard and C. Debouck, Human immunodeficiency virus 1 protease expressed in *Escherichia coli* behaves as a dimeric aspartic protease, *Proc. Natl. Acad. Sci. U.S.A.* **86**:1841 (1989).

39. A. D. Richards, L. H. Phylip, W. G. Farmerie, P. E. Scarborough, A. Alvarez, B. M. Dunn, Ph.-H. Hirel, J. Konvalinka, P. Strop, L. Pavlickova, V. Kostka and J. Kay, Sensitive, soluble chromogenic substrates for HIV-1 proteinase, *J. Biol. Chem.* **265**:7733 (1990).

40. R. L. Shoeman, B. Höner, T. J. Stoller, C. Kesselmeier, M. C. Miedel, P. Traub and M. C. Graves, Human immunodeficiency virus type 1 protease cleaves the intermediate filament proteins vimentin, desmin and glial fibrillary acidic protein, *Proc. Natl. Acad. Sci. U.S.A.* **87**:6336 (1990).

41. J. Konvalinka, P. Strop, J. Velek, V. Cerna, V. Kostka, L. H. Phylip, A. D. Richards, B. M. Dunn and J. Kay, Sub-site preferences of the aspartic proteinase from the human immunodeficiency virus, HIV-1, *FEBS Lett.* **268**:35 (1990).

42. S. Billich, M.-T. Knoop, J. Hansen, P. Strop, J. Sedlacek, R. Mertz and K. Moelling, Synthetic peptides as substrates and inhibitors of human immune deficiency virus-1 protease, *J. Biol. Chem.* **263**:17905 (1988).

43. J. Schneider and S. B. H. Kent, Enzymatic activity of a synthetic 99 residue protein corresponding to the putative HIV-1 protease, *Cell* **54**:363 (1988).

44. R. F. Nutt, S. F. Brady, P. L. Darke, T. M. Ciccarone, C. D. Colton, E. M. Nutt, J. A. Rodkey, C. D. Bennett, L. H. Waxman, I. S. Sigal, P. S. Anderson and D. F. Veber, Chemical synthesis and enzymatic activity of a 99-residue peptide with a sequence proposed for the human immunodeficiency virus protease, *Proc. Natl. Acad. Sci. U.S.A.* **85**:7129 (1988).

45. D. D. Loeb, R. Swanstrom, L. Everitt, M. Manchester, S. E. Stamper and C. A. Hutchison III, Complete mutagenesis of the HIV-1 protease, *Nature* **340**:397 (1989).

46. J. Erickson, D. J. Neidhart, J. VanDrie, D. J. Kempf, X. C. Wang, D. W. Norbeck, J. J. Plattner, J. W. Rittenhouse, M. Turon, N. Wideburg, W. E. Kohlbrenner, R. Simmer, R. Helfrich, D. A. Paul and M. Knigge, Design, activity, and 2.8 Å crystal structure of a C2 symmetric inhibitor complexed to HIV-1 protease, *Science* **249**:527 (1990).

47. G. B. Dreyer, B. W. Metcalf, T. A. Tomaszek, Jr., T. J. Carr, A. C. Chandler, III, L. Hyland, S. A. Fakhoury, V. W. Magaard, M. L. Moore, J. E. Strickler, C. Debouck and T. D. Meek, Inhibition of human immunodeficiency virus 1 protease *in vitro*: rational design of substrate analogue inhibitors, *Proc. Natl. Acad. Sci. U.S.A.* **86**:9752 (1989).

48. T. J. McQuade, A. G. Tomasselli, L. Liu, V. Karacostas, B. Moss, T. K. Sawyer, R. L. Heinrikson and W. G. Tarpley, A synthetic HIV-1 protease inhibitor with antiviral activity arrests HIV-like particle maturation, *Science* **247**:454 (1990).

49. N. A. Roberts, J. A. Martin, D. Kinchington, A. V. Broadhurst, J. C. Craig, I. B. Duncan, S. A. Galpin, B. K. Handa, J. Kay, A. Kröhn, R. W. Lambert, J. H. Merrett, J. S. Mills, K. E. B. Parkes, S. Redshaw, A. J. Ritchie, D. L. Taylor, G. J. Thomas and P. J. Machin, Rational design of peptide-based HIV proteinase inhibitors, *Science* **248**:358 (1990).

50. M. Miller, J. Schneider, B. K. Sathyanarayana, M. V. Toth, G. R. Marshall, L. Clawson, L. Selk, S. B. H. Kent and A. Wlodawer, Structure of complex of synthetic HIV-1 protease with a substrate-based inhibitor at 2.3 Å resolution, *Science* **246**:1149 (1989).

51. A. L. Swain, M. M. Miller, J. Green, D. H. Rich, J. Schneider, S. B. H. Kent and A. Wlodawer, X-ray crystallographic structure of a complex between a synthetic protease of human immunodeficiency virus 1 and a substrate-based hydroxyethylamine inhibitor, *Proc. Nat. Acad. Sci., U.S.A.* **87**:8805 (1990).

52. P. M. D. Fitzgerald, B. M. McKeever, J. F. VanMiddlesworth, J. P. Springer, J. C. Heimbach, C.-T. Leu, W. K. Herber, R. A. F. Dixon and P. L. Darke, Crystallographic analysis of a complex between HIV-1 protease and acetyl-pepstatin at 2.0 Å resolution, *J. Biol. Chem.* **265**:14209 (1990).

53. S. Erickson-Viitanen, J. Manfredi, P. Viitanen, D. E. Tribe, R. Tritch, C. A. Hutchison III, D. D. Loeb and R. Swanstrom, Cleavage of HIV-1 gag polyprotein synthesized *in vitro*: sequential cleavage by the viral protease, *AIDS Res. Human Retroviruses* **5**:577 (1989).

54. K. Partin, H.-G. Kräusslich, L. Ehrlich, E. Wimmer and C. Carter, Mutational analysis of a native substrate of the human immunodeficiency virus type 1 proteinase, *J. Virol.* **64**:3938 (1990).

55. P. L. Darke, R. F. Nutt, S. F. Brady, V. M. Garsky, T. M. Ciccarone, C.-T. Leu, P. K. Lumma, R. M. Freidinger, D. F. Veber and I. S. Sigal, HIV-1 protease specificity of peptide cleavage is sufficient for processing of gag and pol polyproteins, *Biochem. Biophys. Res. Comm.* **156**:297 (1988).

56. A. G. Tomasselli, J. O. Hui, T. K. Sawyer, D. J. Staples, D. J. FitzGerald, V. K. Chaudhary, I. Pastan and R. L. Heinrikson, Interdomain hydrolysis of a truncated *Pseudomonas* exotoxin by the human immunodeficiency virus-1 protease, *J. Biol. Chem.* **265**:408 (1990).

57. N. Margolin, W. Heath, E. Osborne, M. Lai and C. Vlahos, Substitutions at the $P_{2'}$ site of gag p17/p24 affect cleavage efficiency by HIV-1 protease, *Biochem. Biophys. Res. Comm.* **167**:554 (1990).

58. S. F. J. Le Grice, R. Ette, J. Mills and J. Mous, Comparison of the human immunodeficiency virus type 1 and 2 proteases by hybrid gene construction and *trans*-complementation, *J. Biol. Chem.* **264**:14902 (1989).

59. R. L. DesJarlais, G. L. Seibel, I. D. Kuntz, P. S. Furth, J. C. Alvarez, R. R. O. de Montellano, D. L. DeCamp, L. M. Babé and C. S. Craik, Structure-based design of nonpeptide inhibitors specific for the human immunodeficiency virus 1 protease, *Proc. Natl. Acad. Sci. U.S.A.* **87**:6644 (1990).

SUBSTRATE SPECIFICITY OF THE HUMAN (TYPE 1) AND

SIMIAN IMMUNODEFICIENCY VIRUS PROTEASES

Christine Debouck

Department of Molecular Genetics
SmithKline Beecham Pharmaceuticals
King of Prussia, Pennsylvania

INTRODUCTION

All retroviruses have a small (ca. 10 kilobases) RNA genome that contains three major open reading frames typically organized in the order 5'-LTR-*gag-pol-env*-LTR-3'. The retroviral genome also encodes a protease of the aspartyl class that is responsible for the processing of the gag and gag-pol polyprotein precursors, both primary products of translation (see Dickson *et al.*, 1984 for a review). Because they are essential for viral maturation and infectivity (Crawford & Goff, 1985; Katoh *et al.*, 1985; Kohl *et al.*, 1988; Peng *et al.*, 1989), retroviral proteases have been the subject of intense investigation, particularly as a therapeutic target for the treatment of HIV infections (see Davies, 1990; Debouck & Metcalf, 1990; Hellen *et al.*, 1989; Kay & Dunn, 1990, for reviews). In this context, the design of specific inhibitors of retroviral proteases requires the characterization of the mechanism of action and the substrate preferences of these enzymes. We discuss here a genetic approach towards understanding the substrate recognition by the HIV-1 protease and its closely related congener from the simian immunodeficiency virus, the SIV_{MAC} protease.

PROCESSING OF NATURAL SUBSTRATES BY THE HIV-1 PROTEASE IN BACTERIA

The processing of gag-pol in HIV-1 infected cells appears to be limited to the sites shown in Figure 1 (Copeland & Oroszlan, 1988; Henderson *et al.*, 1988; Lillehoj *et al.*, 1988). We have used recombinant DNA technology to produce the HIV-1 protease in the bacterium *Escherichia coli* as described in detail elsewhere (Debouck *et al.*, 1987). The HIV-1 protease was expressed in its mature and active form through "auto"-processing of a larger precursor containing both upstream and downstream flanking sequences from the *pol* open reading frame. Primary sequence determination showed that the recombinant HIV-1 protease is 99 amino acid-long, starting with Pro_{69} and ending with Phe_{167} from the *pol* open reading frame (Debouck *et al.*, 1987; Strickler *et al.*, 1989). This sequence is identical to the one reported for the protease isolated directly from HIV-1 virions (Copeland & Oroszlan, 1988; Lillehoj *et al.*, 1988).

We have shown that the recombinant HIV-1 protease properly processes HIV-1gag and pol precursor polyproteins when these are co-expressed in the same bacterial cells with the protease. The protease accurately processed an HIV-1 Pr55^{gag} precursor in *E. coli* yielding mature, authentic recombinant p24 gag protein beginning with Pro$_{133}$ and ending with Leu$_{363}$ from the *gag* open reading frame (Debouck *et al.*, 1987; Hassell *et al.*, in preparation) which correspond to the termini of p24 gag isolated from viral particles (Casey *et al.*, 1985; Henderson *et al.*, 1988). Similarly, the protease processed an HIV-1 pol precursor in *E. coli* to yield mature, authentic recombinant reverse transcriptase that consisted of an equimolar mixture of two polypeptides, p66 and p51, that have also been described for the enzyme isolated from HIV-1 virions. The amino-terminus of both chains was shown to be Pro$_{168}$ from the *pol* open reading frame (Mizrahi *et al.*, 1989) and was identical to that of genuine viral reverse transcriptase (diMarzo Veronese *et al.*, 1986; Lightfoote *et al.*, 1986). The carboxyl-terminus of the recombinant p51 chain was shown to correspond to Phe$_{607}$ of the *pol* open reading frame (Mizrahi *et al.*, 1989; Becerra *et al.*, 1990; Graves *et al.*, 1990), but cannot be compared to the carboxyl-terminus of the viral p51 chain since the latter has not been determined to date.

CONSTRUCTION OF A HETEROLOGOUS SUBSTRATE FOR THE HIV-1 PROTEASE

Since the processing of the HIV-1 gag-pol polyprotein by the HIV-1 protease is apparently limited to the sites shown in Figure 1, the cleavage specificity of the enzyme must be highly stringent. However, the lack of sequence conservation among these sites indicates that the determinants of cleavage specificity do not reside entirely within the primary sequence of the cleavage sites. Exceptions are sites 1, 5, and 6, which fit the consensus sequence, Ser/Thr.Xaa.Yaa.Phe/Tyr*Pro (with cleavage at *), also found in cleavage sites for other retroviral proteases (Pearl & Taylor, 1987). In addition, all sites share an overall hydrophobicity (Figure 1). Residues in the P$_1$ and P$_1$' positions are strictly hydrophobic. Residues in P$_2$ and P$_2$'consist of hydrophobic or uncharged polar amino acids (with one exception in site 2). Charged residues are scarce in the P$_4$ to P$_4$' positions and are strictly

SITE	JUNCTION	PRIMARY SEQUENCE OF CLEAVAGE SITE		
1	p17*p24	SER.GLN.ASN.**TYR**	*	**PRO.ILE.VAL**.GLN
2	p24*X	**ALA**.ARG.**VAL.LEU**	*	**ALA**.GLU.**ALA.MET**
3	X*p7	**ALA**.THR.**ILE.MET**	*	**MET**.GLN.ARG.GLY
4	p7*p6	**PRO**.GLY.ASN.**PHE**	*	**LEU**.GLN.SER.ARG
5	*PR	SER.**PHE**.ASN.**PHE**	*	**PRO**.GLN.**ILE**.THR
6	PR*RT	THR.**LEU**.ASN.**PHE**	*	**PRO.ILE**.SER.**PRO**
7	p51*RN	**ALA**.GLU.THR.**PHE**	*	**TYR.VAL**.ASP.GLY
8	p66*IN	ARG.LYS.**ILE.LEU**	*	**PHE.LEU**.ASP.GLY

Figure 1. Cleavage sites for the HIV-1 protease within the HIV-1 gag-pol polyprotein precursor. The primary sequence of the cleavage sites is shown with the three-letter designation for amino acid residues. Boldfaced residues represent non polar amino acids, underlined residues correspond to charged polar amino acids and the remaining residues are uncharged polar amino acids. The genetic sequences flanking the cleavage sites (*) are shown using the nomenclature recommended by Leis *et al.* (1988).

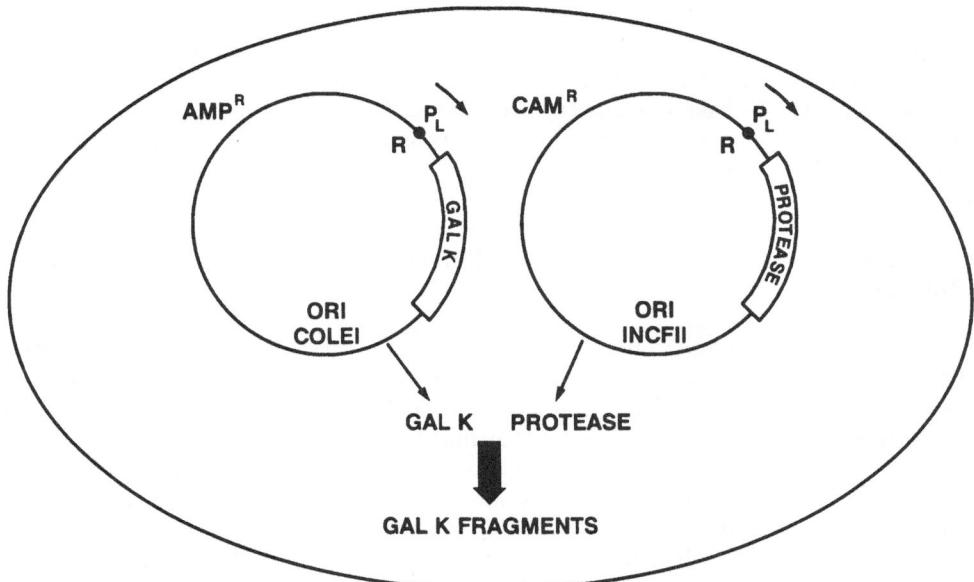

Figure 2. Proteolytic processing of heterologous polyprotein in bacteria. The double-plasmid system used to demonstrate retroviral protease activity in *E. coli* is depicted. A single bacterial cell is shown containing the two compatible expression vectors. One plasmid (at the right) directs the expression of the protease (incFII origin-based plasmid bearing the chloramphenicol-resistance marker, CAM[R]). The other plasmid (at the left) directs the expression of the engineered galactokinase (colE1 origin-based vector bearing the ampicillin-resistance marker, AMP[R]). In both plasmids, the expression is directed by the strong and regulatable Lambda P_L promoter and can be turned on (induction) by temperature shift or by addition of nalidixic acid (Shatzman & Rosenberg, 1987).

(with one exception) excluded from the P_2 to P_2' positions. It is also possible that features of the secondary structure of these cleavage sites contribute to their recognition by the enzyme.

In order to determine whether the HIV-1 protease exhibits cleavage preference towards certain sites and to assist in the design of specific inhibitors of the enzyme, several laboratories have synthesized small peptides corresponding to the cleavage sites shown in Figure 1 (Billich *et al.*, 1988; Darke *et al.*, 1988; Schneider & Kent, 1988; Krausslich *et al.*, 1989; Moore *et al.*, 1989). Results from these peptidolysis experiments have sometimes been difficult to interpret because of the inherent hydrophobicity and subsequent poor solubility of these peptides. In an effort to circumvent these technical difficulties and to obtain more accurate information on the cleavage site preferences of the HIV-1 protease, we chose to engineer each gag-pol cleavage site individually within a heterologous protein. For this purpose, we selected the *E. coli* galactokinase gene product (galK) which has convenient indicator properties. A [galK]+ phenotype results in red colonies on MacConkey galactose indicator plates, whereas a [galK]- phenotype gives rise to white colonies on these plates (Miller, 1972). We initially inserted the p17-p24 gag cleavage site (site 1, Tyr*Pro) in the middle of the galactokinase gene, away from the presumed active sites of the enzyme (Debouck *et al.*, 1985). Upon co-expression of this engineered galactokinase together with the HIV-1 protease in our double-plasmid expression system (Figure 2), we observed that galactokinase was completely processed in the two expected halves, indicating that it was a competent substrate for the HIV-1 protease in this system. Interestingly, this processing also resulted in the enzymatic inactivation of galactokinase, thereby providing us with a colorimetric plate assay for HIV-1 protease activity. Indeed, in the absence of protease or in the presence of mutated, inactive protease, the bacterial colonies turn red on the indicator plates, whereas in the presence of active protease, the colonies remain white on the same plates (Figure 3). We have similarly engineered the other gag-pol cleavage sites (sites 2-8) at the same position within galactokinase in order to compare and quantify the HIV-1 proteolytic activity towards these various sites.

PROCESSING OF ENGINEERED SUBSTRATES BY THE HIV-1 PROTEASE IN BACTERIA

To assess the relative efficiency of cleavage at the various HIV-1 gag-pol sites by the HIV-1 protease, galactokinase engineered with each individual cleavage site was co-expressed with active HIV-1 protease using our double-plasmid expression system. The processing of galactokinase was monitored by immunoblot analysis using a rabbit polyclonal antibody raised against galactokinase. In all cases, efficient cleavage was observed upon induction of gene expression from the plasmids. By examining the extent of processing after shorter periods of induction and calculating the ratio of processed versus unprocessed galactokinase, we were able to rank the efficiency of cleavage by the HIV-1 protease at these various sites. As shown in Table 1, the HIV-1 protease cleaves all sites with similarly high efficiency, with the following order of preferences: Phe*Pro > Met*Met > Tyr*Pro > Leu*Phe >Phe*Tyr = Phe*Leu > Leu*Ala. (Cleavage at site 6, Phe*Pro was not determined in this set of experiments). The difference between the best and worst cleavage efficiency was less than 3-fold (compare Phe*Pro (site 5), processed at 90 % to Leu*Ala (site 2), processed at 32 %). The ranking obtained by this approach was similar to the one deduced from peptidolytic assays carried out by some (Darke *et al.*, 1988; Krausslich *et al.*, 1989), but not other laboratories (Billich *et al.*, 1988; Schneider & Kent, 1988). The best agreement between our assay and these various peptidolytic assays was observed for site 3 (Met*Met) and site 5 (Phe*Pro) which are the most efficiently cleaved by the HIV-1 protease. The

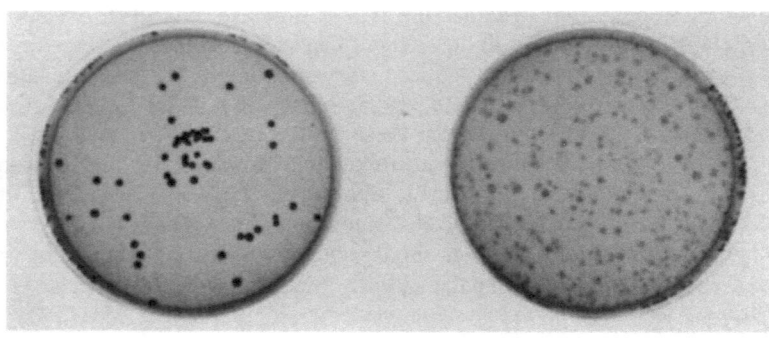

| NO or INACTIVE | ACTIVE |
| PROTEASE | PROTEASE |

Figure 3. Colorimetric plate assay for HIV-1 protease activity. Bacteria containing two compatible plasmids expressing HIV-1 protease and galactokinase engineered with the p17-p24 gag cleavage site, respectively, were plated on MacConkey galactose agar plates and incubated at 32°C overnight. The plates were then switched to 39°C to induce gene expression from the plasmids. In the absence of HIV-1 protease activity, full-length galactokinase is produced and the colonies turn red upon heat induction (left plate). In the presence of HIV-1 protease activity, galactokinase is proteolytically processed and the colonies remain white (right plate).

Table 1. Cleavage of engineered galactokinase by HIV-1 and SIV proteases

		% cleavage	
Site	Scissile bond	HIV-1 PR[a]	SIV PR[a]
1	Tyr * Pro	81.6	31.4
2	Leu * Ala	32.1	18.5
3	Met * Met	83.4	52.4
4	Phe * Leu	50.7	26.7
5	Phe * Pro	90.0	45.6
6	Phe * Pro	ND	ND
7	Phe * Tyr	51.8	0.0
8	Leu * Phe	58.2	44.4

[a]% cleavage is calculated by (amount of galactokinase processed/total amount of full-length and processed galactokinase) x 100 as determined by immunoblot analysis and autoradiogram densitometry.
ND: not determined.

discrepancies between the various assays and laboratories most probably result from the poor solubility of most peptide substrates, problem that is alleviated in our assay by the use of a full-length soluble molecule as the backbone for the presentation of the cleavage sites to the protease.

EXPRESSION AND ANALYSIS OF SIMIAN IMMUNODEFICIENCY VIRUS PROTEINASE

We have expressed mature, active simian immunodeficiency virus (SIV_{MAC}) protease in *E. coli* as a first step towards the evaluation of SIV infection in macaques as a model for the preclinical assessment of AIDS therapeutics targeted to the HIV-1 protease. The SIV protease was produced in *E. coli* using a recombinant expression system in which the mature protease"auto"-processed from a precursor form as described for the HIV-1 enzyme (Deckman *et al.*, in preparation). A full biochemical characterization of the enzyme and its comparison to the HIV-1 protease are described elsewhere (Grant *et al.*, 1991a,b). The SIV protease is 99-amino acids long with 49 % or 73 % of its amino acid residues being identical or similar, respectively, to the corresponding residues in the HIV-1 enzyme. In an effort to gain information on how primary structure differences affect protease activity and cleavage specificity, we examined the ability of the SIV protease to process galactokinase engineered with the various HIV-1 gag-pol cleavage sites. The results in Table 1 show that the SIV protease is capable of cleaving most HIV-1 protease cleavage sites with about half the efficiency of cleavage exhibited by the HIV-1 enzyme. Since the level of expression of the two enzymes in *E. coli* is identical (data not shown), this observation suggests that the SIV protease is intrinsically less active than its HIV-1 congener. Site 8 (Leu*Phe) was cleaved by the HIV-1 and SIV proteases with very similar efficiency. In contrast, site 7 (Phe*Tyr) was cleaved very poorly, if at all, by the SIV protease. This site, which corresponds to the cleavage site within the HIV-1 reverse transcriptase, was also cleaved very poorly when a peptide, Ac-Ala-Glu-Thr-Phe*Tyr-Val-Asp-NH2, was used as substrate (Grant *et al*, 1990a). In an effort to uncover the amino acid residue(s) responsible for this disparity in substrate preference, we positioned the amino acid substitutions found in the SIV protease within the

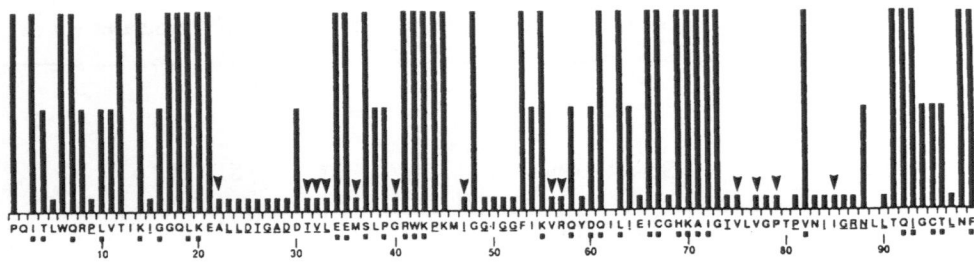

Figure 4. Schematic representation of the of structure-function information on the HIV-1 protease. This representation is a modification of the one published by Loeb *et al*, 1989. The HIV-1 protease primary sequence is shown with the one-letter designation for amino acid residues. The vertical bars summarize the data published by Loeb *et al*, 1989 on the phenotype of HIV-1 protease mutants (tall bar: fully active; medium-size bar: partially active; short bar: inactive). The amino acids that are different in the primary sequence of the SIV$_{MAC}$ protease (Kornfeld *et al*, 1987; Myers *et al*, 1990) are shown as either solid squares (position where a nonconservative substitution does not inactivate the protease) or downward solid arrow (position where a nonconservative substitution inactivates the protease).

primary sequence of the HIV-1 enzyme in relation to the phenotype of substitutions at all positions (Figure 4). Although most substitutions found in the SIV enzyme affect residues of HIV-1 protease that tolerate mutations, several substitutions were found to affect residues that are absolutely required for the HIV-1 protease activity. When these changes were located within the three-dimensional structure of the HIV-1 protease (not shown), three of them were found within the active site cleft of the protease, Ile$_{32}$ (Val in HIV-1 protease), Val$_{47}$ (Ile) and Ile$_{82}$ (Val). Interestingly, these correspond to residues in HIV-1 proteasethat have been shown to be part of the S$_1$/S$_1$' (Val$_{82}$) and S$_2$/S$_2$' Val$_{32}$, Ile$_{47}$) substrate binding pockets (Miller *et al.*, 1989). In two cases, valine residues in the HIV-1 protease are replaced by the larger and more branched amino acid isoleucine in the SIV protease. Although this is a conservative substitution, the difference in size between valine and isoleucine should change the size of these substrate binding pockets so that larger amino acid side chains in the substrate, such as phenylalanine and tyrosine, cannot be well accommodated within the SIV protease active site cleft. We are currently investigating the effect of these three substitutions, separately or in combination, on the activity and substrate preference of the HIV-1 protease.

We have also tested the sensitivity of the SIV protease towards a few potent peptide analog inhibitors that were developed against the HIV-1 enzyme and contain the hydroxyethylene isostere replacement for the scissile bond. We found that these peptide analog inhibitors are equipotent *in vitro* towards these two retroviral enzymes. These results demonstrate that, despite its primary structure and substrate preference differences as compared to the HIV-1 protease, the SIV enzyme can be used to evaluate inhibitors directed towards the HIV-1 protease activity. It also gives further support for the usefulness of SIV infection as a model for the preclinical assessment of HIV-1 protease inhibitors as anti-AIDS drugs.

CONCLUSIONS

We have expressed authentic, active HIV-1 and SIV retroviral proteases in the bacterium *Escherichia coli*. A genetically engineered system was developed for the determination of the substrate preference and proteolytic cleavage efficiency by these two enzymes. The results obtained for the HIV-1 enzyme were consistent with data published using classical enzymology and peptidolytic assays. This "genetic enzymology" approach is particularly useful for the analysis of processing at hydrophobic cleavage sites that are not easily amenable to *in vitro* peptidolytic assays and can be applied to the characterization of other proteases.

ACKNOWLEDGEMENTS

We acknowledge Robert Craig, Ingrid Deckman and Stephan Grant for their contribution to the work described in this chapter. We also thank L. Altland and C. Del Tito for their assistance in preparing the manuscript. This work was supported in part by NIH grants AI 24845 and GM 39526.

REFERENCES

Becerra, S. P., Clore, G. M., Gronenborn, A. M., Karlstrom, A. R., Stahl, S. J.,Wilson, S. H. & Wingfield, P. T., 1990, Purification and characterization of the RNase H domain of HIV-1 reverse transcriptase expressed in recombinant *Escherichia coli*, *FEBS Letters* **270**:76.

Billich, S., Knoop, M-T., Hansen, J., Strop, P., Sedlacek, J., Mertz, R. & Moelling, K., 1988, Synthetic peptides as substrates and inhibitors of human immunodeficiency virus-1 protease, *J. Biol. Chem.* **263**:17905.

Casey, J. M., Kim, Y., Andersen, P. R., Watson, K. F., Fox, J. L. & Devare,S. G., 1985, Human T-cell lymphotropic virus type III: immunologic characterization and primary structure analysis of the major internal protein, p24, *J. Virol.* **55**:417.

Copeland, T. D. & Oroszlan, S., 1988, Genetic locus, primary structure, and chemical synthesis of human immunodeficiency virus protease, *Gene Anal. Technol.* **5**:109.

Crawford, S. & Goff, S. P., 1985, A deletion in the 5' part of the *pol* gene of Moloney murine leukemia virus blocks proteolytic processing of the gag and pol polyproteins, *J. Virol.* **53**:899.

Darke, P. L., Nutt, R. F., Brady, S. F., Garsky, V. M., Ciccarone, T. M., Leu, C-T., Lumma, P. K., Freidinger, R. M., Veber, D. F. & Sigal, I. S., 1988, HIV-1 protease specificity of peptide cleavage is sufficient for processing of gag and pol polyproteins, *Biochem. Biophys. Res. Commun.* **156**:297.

Davies, D.R., 1990, The structure and function of the aspartic proteinases, *Ann. Rev. Biophys. Biophys. Chem.* **19**:189.

Debouck, C., Riccio, A., Schumperli, D., McKenney, K., Jeffers, J., Hughes, C., Rosenberg, M., Heustersbreute, M., Brunel, F. & Davison, J., 1985, Structure of the galactokinase gene of *Escherichia coli*, the last (?) gene of the *gal* operon, *Nucl. Acids Res.* **13**:1841.

Debouck, C., Gorniak, J. G., Strickler, J. E., Meek, T. D., Metcalf, B. W. & Rosenberg, M., 1987, Human immunodeficiency virus protease expressed in *Escherichia coli* exhibits autoprocessing and specific maturation of the gag precursor, *Proc. Natl. Acad. Sci. U.S.A.* **84**:8903.

Debouck, C. & Metcalf, B. W., 1990, Human immunodeficiency virus protease: a target for AIDS therapy, *Drug Dev. Res.* **21**:1.

Deckman, I. C., *et al.*, in preparation.

Dickson, C., Eisenman, R., Fan, H., Hunter, E. & Teich, N., 1984, Protein biosynthesis and assembly, *in*: "Molecular biology of tumor viruses", R. Weiss, N. Teich, H. Varmus & J. Coffin, eds., Cold Spring Harbor Laboratory, Cold Spring Harbor, NY, p.513.

diMarzo Veronese, F., Copeland, T., Vico, A., Rahman, R., Oroszlan, S., Gallo, R. C. & Sarngadharan, M. G., 1986, Characterization of highly immunogenic p66/p51 as the reverse transcriptase of HTLV-III/LAV, *Science* **231**:1289.

Grant, S. K., Deckman, I. C., Minnich, M., Culp, J., Franklin, S., Dreyer, G. B., Tomaszek, T. A., Debouck, C. & Meek, T. D., 1991a, Purification and biochemical characterization of recombinant SIV protease and comparison to HIV-1 protease, Biochemistry **30**:8424.

Grant, S. K., Deckman, I. C., Brooks, I., Metcalf, B. W., Debouck, C., Meek, T. D. & Hensley, P., 1991b, Conformational stability and dimer dissociation of recombinant HIV-1 and SIV retroviral proteases, submitted for publication.

Graves, M. C., Meidel, M. C., Pan, Y-C. E., Manneberg, M., Lahm, H-W. & Gruninger-Leitch, F., 1990, Identification of a human immunodeficiency virus-1 protease cleavage site within the 66,000 dalton subunit of reverse transcriptase, *Biochem. Biophys. Res. Commun.* **168**:30.

Hassell, A. M., *et al.*, in preparation.

Hellen, C. U. T., Krausslich, H-G. & Wimmer, E., Proteolytic processing of polyproteins in the replication of RNA viruses, *Biochemistry* **28**:9881.

Henderson, L. E., Copeland, T. D., Sowder, R. C., Schultz, A. M. & Oroszlan, S., 1988, Analysis of proteins and peptides purified from sucrose gradient banded HTLV-III, *in*: "Human retroviruses, cancer, and AIDS: Approaches to Prevention and Therapy", D. Bolognesi, ed., Alan R. Liss, Inc., p. 135.

Katoh, I., Yoshinaka, Y., Rein, A., Shibuya, M., Odaka, T. & Oroszlan, S., 1985, Murine leukemia virus maturation: protease region for conversion from "immature" to "mature" core forms and for virus infectivity, *Virology* **145**:280.

Kay, J. & Dunn, B. M.,1990, Viral proteinases: weakness in strength, *Bioch.Biophys. Acta* **1048**:1.

Kohl, N. E., Emini, E. A., Schleif, W. A., Davis, L. J., Heimbach, J. C., Dixon, R. A. F., Scolnick, E. M. & Sigal, I. S., 1988, Active human immunodeficiency virus protease is required for viral infectivity, *Proc.Natl. Acad. Sci. U.S.A.* **85**:4686.

Kornfeld, H., Riedel, N., Viglianti, G. A., Hirsch, V. & Mullins, J. I., 1987, Cloning of HTLV-4 and its relation to simian and human immunodeficiency viruses, *Nature* **326**:610.

Krausslich, H-G., Ingraham, R. H., Skoog, M. T., Wimmer, E., Pallai, P. V. & Carter, C. A., 1989, Activity of purified biosynthetic proteinase of human immunodeficiency virus on natural substrates and synthetic peptides, *Proc.Natl. Acad. Sci. U.S.A.* **86**:807.

Leis, J., Baltimore, D., Bishop, J. M., Coffin, J., Fleissner, E., Goff, S. P., Robinson, H., Skalka, A. M., Temin, H. M. & Vogt, V., 1988, Standardized and simplified nomenclature for proteins common to all retroviruses, *J. Virol.* **62**:1808.

Lightfoote, M. M., Coligan, J. E., Folks, T. M., Fauci, A. S., Martin, M. A. & Venkatesan, S., 1986, Structural characterization of reverse transcriptase and endonuclease polypeptides of the acquired immunodeficiency syndrome retrovirus, *J. Virol.* **60**:771.

Lillehoj, E. P., Salazar, F. H. R., Mervis, R. J., Raum, M. G., Chan, H. W., Ahmad, N. & Venkatesan, S., 1988, *J. Virol.* **62**:3053.

Loeb, D. D., Swanstrom, R., Everitt, L., Manchester, M., Stamper, S. E. & Hutchinson III, C. A., 1989, Complete mutagenesis of the HIV-1 protease, *Nature* **340**:397.

Miller, J. H., (ed), 1972, "Experiments in molecular genetics", Cold Spring Harbor Laboratory, Cold Spring Harbor, NY.

Miller, M., Schneider, J., Sathyanarayana, B. K., Toth, M. V., Marshall, G. R., Clawson, L., Selk, L., Kent, S. B. H. & Wlodawer, A., 1989, Structure of complex of synthetic HIV-1 protease with a substrate-based inhibitor at 2.3 Å resolution, *Science* **246**:1149.

Mizrahi, V., Lazarus, G. M., Miles, L. M., Meyers, C. A. & Debouck, C., 1989, Recombinant HIV-1 reverse transcriptase: purification, primary structure, and polymerase/ribonuclease activities, *Arch. Biochem. Biophys.* **273**:347.

Moore, M. L., Bryan, W. M., Fakhoury, S. A., Magaard, V. W., Huffman, W. F., Dayton, B. D., Meek, T. D., Hyland, L., Dreyer, G. B., Metcalf, B. W., Strickler, J. E., Gorniak, J. G. & Debouck, C., Peptide substrates and inhibitors of the HIV-1 protease, *Biochem. Biophys. Res. Commun.* **159**:420.

Myers, G., Rabson, A. B., Berzofsky, J. A., Smith, T. F. & Wong-Staal, F., (eds), 1990, "Human retroviruses and AIDS: a compilation and analysis of nucleic acid and amino acid sequences", Los Alamos National Laboratory, Los Alamos, NM.

Pearl, L. & Taylor, W. R., 1987, Sequence specificity of retroviral proteases, *Nature* **328**:482.

Peng, C., Ho, B. K., Chang, T. W. & Chang, N. T., 1989, Role of human immunodeficiency virus type 1-specific protease in core protein maturation and viral infectivity, *J. Virol.* **63**:2550.

Schneider, J. & Kent, S., 1988, Enzymatic activity of a synthetic 99 residue protein corresponding to the putative HIV-1 protease, *Cell* **54**:363.

Shatzman, A. R. & Rosenberg, M., 1987, Expression, identification, and characterization of recombinant gene products in *E. coli, Meth. Enzymol.* **152**:661.

Strickler, J. E., Gorniak, J., Dayton, B., Meek, T., Moore, M., Magaard, V., Malinowski, J. & Debouck, C., 1989, Characterization and auto processing of precursor and mature forms of human immunodeficiency virus type 1 (HIV1) protease purified from *Escherichia coli, Proteins* **6**:139.

EXPRESSION AND CHARACTERIZATION OF GENETICALLY LINKED HOMO-

AND HETERO-DIMERS OF HIV PROTEINASE

Hans-Georg Kräusslich,[1] Anke-Mareil Traenckner[1] and
Friedrich Rippmann[2]

[1]Institut für Virusforschung/ATV
Deutsches Krebsforschungszentrum
D-6900 Heidelberg

[2]Laboratory of Mathematical Biology
The National Institute for Medical Research
The Ridgeway, Mill Hill
London NW7 1AA
United Kingdom

INTRODUCTION

Infectious retroviral particles are composed of an inner core structure enclosed in a host-derived plasma membrane that contains the viral glycoproteins. In the case of HIV, this inner core consists of a ribonucleoprotein complex (two identical molecules of genomic RNA associated with the viral nucleocapsid [NC] protein and probably also with the viral enzymes reverse transcriptase [RT], integrase [IN] and proteinase [PR]) encased in a capsid [CA] shell (Gelderblom *et al.*, 1989). All structural components of the viral core (derived from the viral *gag* gene) as well as the replication enzymes (derived from the *pol* gene) are synthesized and assembled as polyprotein precursors. Proteolytic processing by the virus-encoded, virion-associated proteinase takes place only during and after budding of the viral particle from the plasma membrane and processing is not required for the release of immature, non-infectious particles but is necessary for maturation of infectious virions (reviewed in Kräusslich & Wimmer, 1988). This elaborate mechanism of synthesizing different stable polyproteins at defined rates enables the virus to target many components of the viral particle to the site of assembly using only a single targeting signal. It requires, however, that a proteolytic enzyme is packaged into the virion that is capable of separating the different functional domains, thus allowing viral replication to occur. The activity of such an enzyme has to be tightly regulated since premature processing would dissociate the components of the virion from the site of assembly and incomplete or unfaithful processing should interfere with viral uncoating and replication. Controlled limited proteolysis can be achieved by synthesizing an inactive form of the viral proteinase as part of the polyprotein which is activated upon assembly of the viral particle. At the site of assembly, cellular or viral factors may be encountered that lead to activation of an inactive zymogen possibly by post-translational modification. Alternatively there might not be activation of a zymogen but rather relief of inactivation by dissociating a cellular or viral inhibitor bound to the polyprotein. Conceivably, retroviral polyproteins may require binding of cellular proteins for targeting to the appropriate subcellular localization (e.g. the plasma membrane as site of assembly). Such polyprotein complexes may prevent proteolytic activity by local or general unfolding or by binding and obliterating the active site or substrate binding pocket. A third non-exclusive possibility would be that initiation of the proteolytic cascade requires cooperative action of multiple molecules of the viral proteinase and, thus, a critical concentration can only be achieved in the budding particle.

Retroviral proteinases have been assigned to the family of aspartic proteinases on the basis of sequence homology and biochemical and mutational analysis (reviewed in Hellen *et al.*, 1989; Kay & Dunn, 1990). Determination of the three-dimensional structure of the proteinases of Rous sarcoma virus (Jaskólski *et al.*, 1990) and HIV (Wlodawer *et al.*, 1989) confirmed this assignment and revealed that the active enzymes are dimers of identical polypeptide chains, with a fold similar to other aspartic proteinases. Each polypeptide chain is less than half the size of cellular aspartic proteinases (99 amino acids in the case of HIV-1) and contains only a single active-site Asp residue. Analysis of the crystal structure of HIV PR (Wlodawer *et al.*, 1989) revealed that the subunits of the dimer are related by a nearly perfect twofold axis of symmetry which passes between the flaps, the active-site Asp residues and the two C-terminal strands of the dimer interphase. The terminal strands from each monomer form a common four-stranded antiparallel ß-sheet (the dimer interface) which therefore differs from the six-stranded interphase region of pepsin-like proteinases. The arrangement of the N- and C-termini of the two monomers in this terminal sheet suggests that it should be possible to covalently link the two monomers by a short hinge region without distorting the overall fold of the enzyme.

METHODS

We therefore wanted to generate a bacterial expression vector containing two copies of HIV PR linked by a short hinge region. To this end, two mutagenesis vectors were constructed containing the coding region of HIV PR flanked by either a methionine codon at

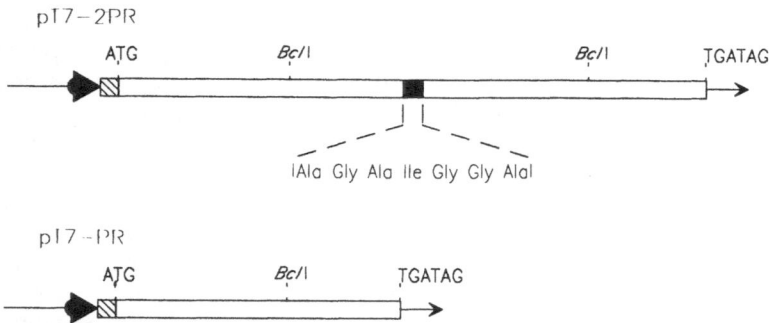

Figure 1. Structure of the HIV PR gene in the expression plasmids *pT7*-PR and *pT7*-2PR. The coding sequence for HIV PR (derived from strain BH10; Ratner *et al.*, 1985) is shown as open box, the hinge region as filled box and the non-translated region from *T7* gene10 (Rosenberg *et al.*, 1987) as hatched box. The *T7* promoter (black triangle), initiation (ATG) and termination (TGATAG) codons are indicated. *pT7*-PR was generated by deleting the small *Bcl*I fragment from *pT7*-2PR. For clarity this diagram is not to scale. [modified from: Kräusslich, submitted for publication]

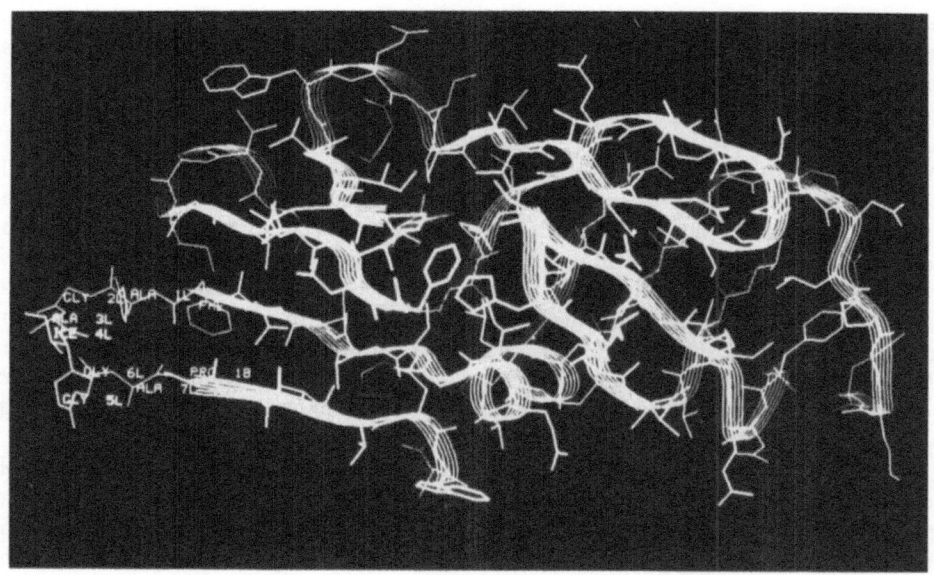

Figure 2. Molecular modeling of the single chain dimer of HIV PR. The hinge region was placed on the PR dimer using the coordinates of Wlodawer *et al.* (1989) and the program INSIGHT on an IRIS workstation.

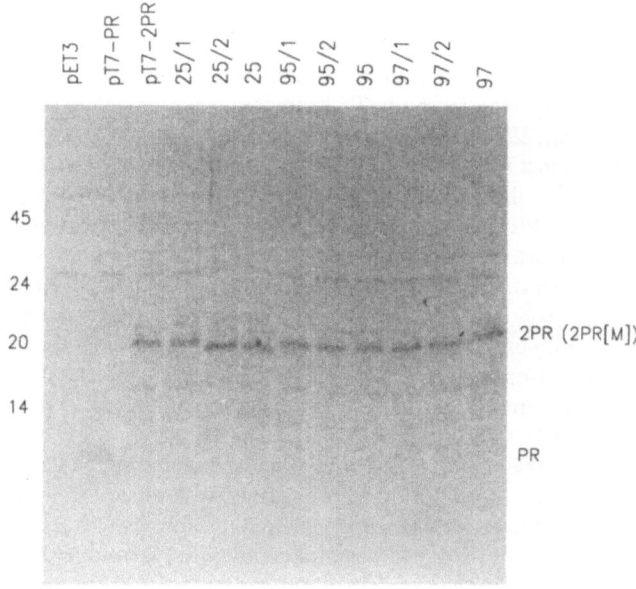

Figure 3. Western blot analysis showing the expression products from plasmids *pET3*, *pT7*-PR, *pT7*-2PR and mutants thereof after transformation and induction of *E.coli* BL21(DE3)pLysS (Moffatt & Studier, 1987). Induced bacteria were disrupted by sonication, centrifuged 15 min at 5000 x g and equal amounts were loaded on a 17.5 % polyacrylamide gel. Proteins were transferred to nitrocellulose and stained with an antibody against HIV PR (gift of P. Barr). On the top, the plasmids used are indicated. Molecular weight standards are indicated on the left, migration of the monomer (PR), the single chain dimer (2PR) and mutants thereof is identified on the right.

419

the 5' end or by two stop codons at the 3' end. At the 3' end of the first molecule and at the 5' end of the second molecule we inserted short sequences encoding a hinge region. These sequences share a common restriction site, thus allowing easy cloning of a genetically linked dimer of HIV PR. Each of the two mutagenesis vectors could be used to introduce specific mutations into the PR sequence by oligonucleotide-directed mutagenesis. Thus, both heterodimers and the homodimer for a specific mutation could be generated using a single oligonucleotide. The final expression vector contained, under the control of promoter and untranslated region of phage *T7* gene 10, two copies of HIV PR preceded by a Met codon and linked by a sequence encoding a 7 amino acids hinge region (*pT7-2PR*; Figure 1; Kräusslich, submitted for publication). By removal of the small *Bcl*I fragment (Figure 1) we also constructed a plasmid containing only a single copy of HIV PR without flanking sequences (*pT7-PR*; Figure 1). This plasmid is similar to the previously described pHIVproPII (Kräusslich *et al.*, 1988) but does not contain the first 68 amino acids of the *pol* open reading frame preceding the PR sequence. Expression of HIV PR from *pT7-PR* was several fold higher than the expression observed with pHIVproPII (not shown).

The hinge region linking the two PR domains was designed to be structurally flexible and not to be a substrate for HIV PR. In order to optimize the sequence of this hinge, molecular modeling of the genetically linked dimer was performed on an IRIS workstation using the coordinates of HIV PR (Wlodawer *et al.*, 1989). Figure 2 shows a view of the terminal sheet of the dimer with the hinge region covalently linking the two PR polypeptide chains. The terminal Phe_{99} of the first PR copy and the Pro_1 of the second copy as well as the amino acids of the hinge are identified.

RESULTS AND DISCUSSION

The different constructs were transformed into *E.coli* BL21(DE3)pLysS (Moffatt & Studier, 1987) and expression from the *T7* promoter was induced essentially as described before (Kräusslich *et al.*, 1989). The expression products were analyzed by immunoblotting using an antiserum against PR (gift of P. Barr) and were shown to have the expected sizes of approximately 11 kDa for the monomer and 23 kDa for the covalently linked dimer (Figure 3, lanes *pT7-PR* and *pT7-2PR*, respectively; Kräusslich, submitted for publication). The single chain dimer of HIV PR was stable on prolonged incubation and no immunoreactive band at or close to the migration of the monomer could ever be detected. We consistently found a minor immunoreactive band of approximately 18 kDa (Figure 3) which appeared independent of the proteolytic activity of the expression product and may be due to internal initiation or to cleavage by a bacterial enzyme. Activity of the expression products was analyzed by incubation of bacterial extracts with an *in vitro* translated gag precursor (pr55^{gag}) as described (Kräusslich *et al.*, 1988). The *in vitro* translated pr55 was entirely stable to incubation without addition of bacterial extracts and with extracts containing the vector *pET3* (Figure 4; Rosenberg *et al.*, 1987). In addition, no processing was observed upon incubation with extracts from uninduced bacteria containing the expression vectors *pT7-PR* and *pT7-2PR*. Incubation with extracts from induced bacteria containing plasmid *pT7-PR* or *pT7-2PR*, on the other hand, gave complete processing of the gag precursor (Figure 4). The processing pattern was identical to that observed after incubation with purified HIV PR (Kräusslich *et al.*, 1989) confirming the specificity of this reaction (Kräusslich, submitted for publication). In dilution experiments we consistently observed a 2- to 5-fold higher activity for extracts containing the genetically linked dimer (Table 1). In addition, in some experiments we also observed some processing after incubation with uninduced bacterial extracts containing plasmid *pT7-2PR*.

In previous experiments, we had shown that expression of HIV PR is toxic to bacterial cells and that plasmids expressing an active enzyme will lead to very slow bacterial

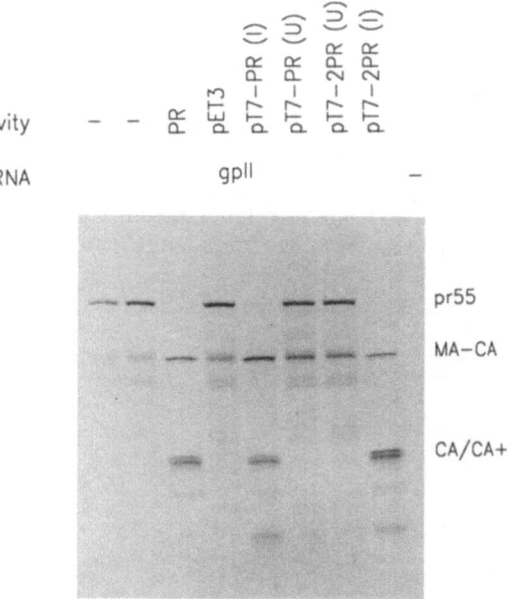

Figure 4. Activity of expression products on *in vitro* synthesized gag substrate. *In vitro* transcription and translation of pr55^{gag} from plasmid pHIVgpII has been described previously (Kräusslich *et al.*, 1988). Translation products were incubated with bacterial extracts that had been clarified by centrifugation at 5000 x g for 15 min. At the top of the panel the plasmids used for transformation of BL21(DE3)pLysS are indicated [(U): uninduced, (I): induced]. Incubation was for 60 min at 37°C using essentially the same conditions as described previously (Kräusslich *et al.*, 1989). The reactions were separated on a 17.5 % polyacrylamide gel and were analyzed by autoradiography. The lane on the right indicates a mock translation without added RNA. On the right, the positions of relevant HIV structural proteins are indicated.

homodimer showed that activity was either totally abolished or reduced by >2 x 10³ (Kräusslich, submitted for publication, see also Table 1). Activity could not be restored by mixing of the two heterodimers indicating that only the single chain dimer and not multimers thereof constitute the active moiety (Table 1; Kräusslich, submitted for publication).

These experiments underscore the catalytic role of the Asp$_{25}$ and Asp$_{25}$' residues in the single chain dimer but they also indicate the potential of this system to generate stable heterodimers of HIV PR and to study the effects of asymmetric mutations on substrate recognition and activity of the enzyme. To this end, we designed a first set of mutations that was expected to yield similar or different phenotypes when present in the homodimer and in the heterodimer. Substitution of the active site Asp$_{25}$ was expected to abolish the enzyme's catalytic activity even when present in only one PR copy and this was shown to be the case (see above). Substitution of Cys$_{95}$ by Met$_{95}$, on the other hand, was predicted to be tolerated when expressed as homo- and heterodimer. This prediction was based on the results of molecular modeling analysis and on previous mutagenesis experiments. Both cysteines of HIV PR were replaced by L-α-amino-n-butyric acid (Aba) in the chemically synthesized enzyme without apparent loss of activity (Schneider & Kent, 1988). Moreover, a Cys95Met Transformation of mutant forms of the single chain dimer containing an active site substitution (Asp25Ala) in either the first or the second or both copies of the PR coding sequence had no effect on bacterial growth rates (Figure 5) indicating that toxicity was due to the enzyme's activity and that the Asp$_{25}$ and Asp$_{25}$' residues from both copies of the dimer are required to constitute a competent active site. Accordingly, incubation of the *in vitro* translated substrate with bacterial extracts containing the mutant heterodimers and the

Table 1. Relative activities of expression products from *pT7*-2PR and derivatives on pr55^{gag}. The constructs indicated were transformed into BL21(DE3)pLysS and extracts of induced bacteria were incubated with *in vitro* synthesized pr55^{gag} as described in the legend to Figure 4. The mutations present in the constructs are identified in the middle column (25/1 has an Asp/Ala substitution of residue 25 in the first PR copy). For assaying relative activities, at least three dilutions of each bacterial extract were used. Samples of extracts containing *pT7*-2PR were used in each experiment to standardize the amount of processing. Relative activities were calculated from the dilution factors required to give approximately equal levels of processing. The activity of extracts containing *pT7*-2PR was arbitrarily set as 100.

Construct	Mutation	Activity
PR	--	> 20
2PR	--	100
25/1	D/A	0
25/2	D/A	< 0.05
25/1+	D/A+	
25/2	D/A	< 0.05
25	D/A/ D/A	0
95/1	C/M	>20
95/2	C/M	> 20
95	C/M C/M	>20
97/1	L/F	> 20
97/2	L/F	>10
97	L/F L/F	< 0.5
26/1	T/S	> 20
26	T/S T/S	> 20

Dilution range assayed, 5 x 10$^{-4.}$

growth when transformed into *E. coli* BL21(DE3) where basal level expression from the *T7* promoter is possible. Conversely, transformation of constructs expressing active PR into the strain BL21(DE3)pLysS which gives much tighter control of expression (Moffatt & Studier, 1987) does not alter the generation time of the bacteria (Kräusslich *et al.*, 1989). We wanted to make use of this toxicity of HIV PR for the development of a simple initial screening assay for active PR. BL21(DE3) and BL21(DE3)pLysS were both transformed with either the vector alone or with plasmids expressing the monomeric and dimeric forms of PR. Figure 5 shows that all three plasmids gave similar numbers of resistant colonies when transformed into the tightly regulated strain whereas only the vector plasmid yielded colonies when transformed into BL21(DE3). On closer inspection tiny colonies could be found on the plates containing transformants of pT7-PR whereas no colonies at all were observed for pT7-2PR. This observation is in agreement with the higher activity of the 2PR expression product against an *in vitro* translated substrate (see above). In some instances we observed very few colonies of normal size after transformation with *pT7*-PR and these colonies may contain mutated inactive forms of PR as has been described recently (Baum *et al.*, 1990). mutation has been reported to yield active PR on expression in *E. coli* (Z. Hostomsky & K. Appelt, personal communication) , whereas the conservative Cys95Ser exchange abolished activity (Debouck *et al.*, 1989). We therefore introduced a Cys95Met mutation into each of the two mutagenesis vectors described above by using oligonucleotide-directed mutagenesis on uridine-substituted single-stranded DNA templates (Kunkel, 1985). The mutation was verified by sequence analysis and fortuitous second site mutations were ruled out by

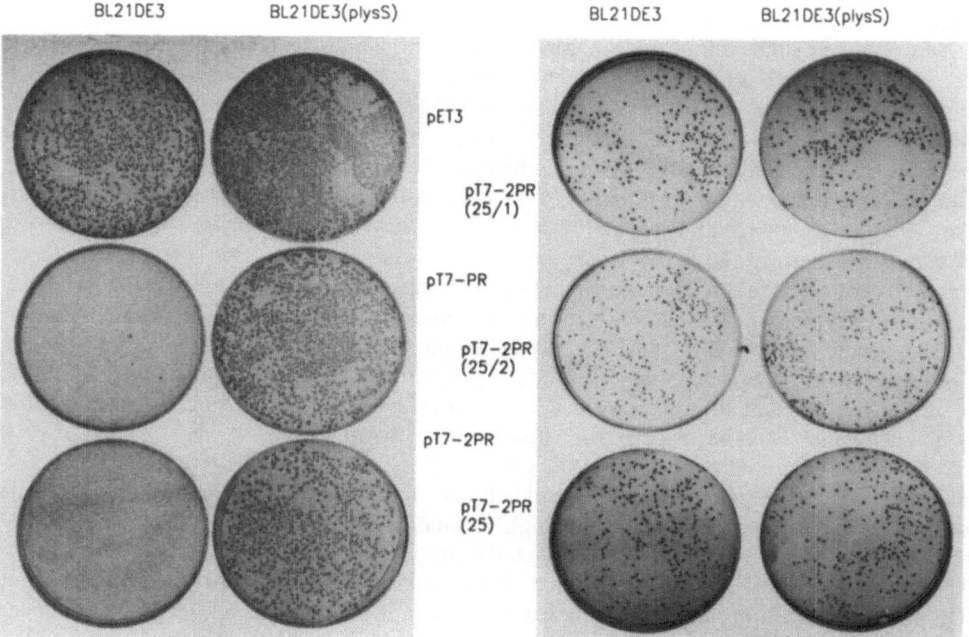

Figure 5. Expression of HIV PR and mutants thereof in *E. coli* B L21(DE3) and BL21(DE3)pLysS (Moffatt & Studier, 1987). The plasmids indicated were transformed into the two expression strains and equal amounts were plated.

sequencing the entire PR coding region for each mutant. Subsequently, the vectors expressing the two heterodimers and the homodimer (Cys95Met) were generated by joining a mutated first copy with a wild-type second copy and so on. Expression of all three constructs in BL21(DE3) did not yield any colonies indicating that the mutant enzymes were active. This was confirmed by incubation of bacterial extracts with *in vitro* synthesized pr55 which gave processing similar to the wild-type single chain dimer in all cases (Table 1). Expression levels were also comparable (Figure 3) indicating that the specific activities of the mutant homo- and heterodimers were similar to the wild-type single chain dimer.

One of our major research interests focusses on the dimerization of HIV PR and the stability of the dimer interface and we wanted to make use of the genetically linked dimer to analyze the contributions of specific amino acids in the terminal ß-strands. The formation of ß-sheet structures on the "surface" of a protein, i.e. with one side exposed to the solvent and the other side pointing to the interior of the protein seems to depend on certain patterns in the amino acid composition. A systematic survey of ß-sheet structures (Lifson & Sander, 1980) shows that hydrophilic and hydrophobic residues alternate, the hydrophilic side chains pointing to the solvent and the hydrophobic ones cooperating in the inside of the protein. This arrangement of alternating hydrophilic and hydrophobic side chains is also found in the terminal ß-sheets of the retroviral proteinases (Wlodawer *et al.*, 1989; Jaskólski *et al.*, 1990). The central residues of the ß-sheet in HIV PR are Leu_{97}, $Leu_{97'}$ flanked by Ile_3, $Ile_{3'}$. Loeb *et al.* (1989) have shown that the Ile3Leu substitution is tolerated in HIV PR as would be expected from this conservative exchange. The Ile3Ser substitution, on the other hand, abolished the enzyme's activity. Moreover, an Ile3Asn substitution which retained activity in their experimental system was subsequently shown to be completely inactive when assayed for *E. coli* toxicity (Baum *et al.*, 1990). Substitution of Leu_{97} in the C-terminal ß-strand with serine also abolished the activity of the mutant enzyme (Loeb *et al.*, 1989), again supporting the importance of the central hydrophobic interaction. The Leu97Phe substitution, on the other hand, although it retained the potential for hydrophobic interaction led to an inactive phenotype (Loeb *et al.*, 1989). Visual inspection of the three-dimensional structure of HIV PR shows that the reason for the inactive phenotype may be a distortion of the geometry of the active site (especially the fireman's grip holding the catalytic Asp_{25}, Asp_{25}' in place) rather than an interference with the formation of the ß sheet. Therefore the possible interaction of Phe_{97}, Phe_{97}' with the residues of the fireman's grip was investigated further. The volume of the cavity formed by the ß-sheet and the lower residues of the active site seems not to accomodate the two relatively large phenyl rings, compared to the two Leu side chains of the wild-type enzyme, whereas accomodation of one phenyl ring seems not impossible (Figure 6). The genetically linked PR dimer offers the unique possibility to create such heterodimers and test for their biological activity. Moreover, generating the two heterodimers should allow us to analyze whether the mutation had an equal effect when introduced into the first or second chain (as would be expected for a quasi-symmetric molecule).

We therefore introduced the Leu97Phe mutation into both copies of PR and generated vectors for the expression of the homo- and hetero-dimers. Expression levels of the different constructs were similar to the wild-type single chain dimer (Figure 3). Transformation of the three constructs into *E.coli* BL21(DE3) gave large colonies in the case of the plasmid expressing the homodimer indicating that this mutant single chain dimer is inactive. The heterodimers containing the Leu97Phe substitution in the first or second copy of PR gave intermediate phenotypes with much smaller colonies in the case of 97/1 (not shown). The results of this toxicity screen correlated well with the activities of the mutant enzymes against the viral gag polyprotein. Activity of the Leu97Phe homodimer was reduced by a factor of

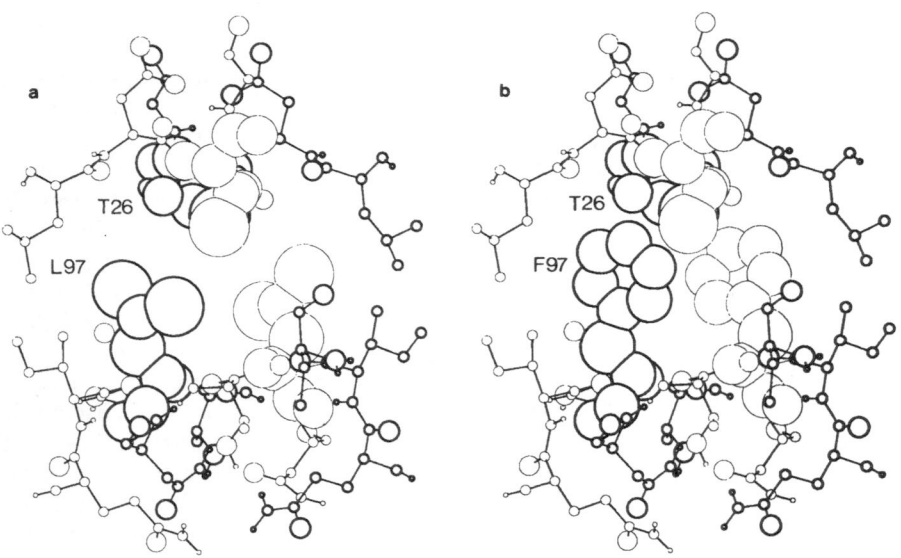

Figure 6. a) View of the active site residues Leu_{24}, Asp_{25}, Thr_{26}, and Gly_{27} of both wild-type molecules (forming the fireman's grip, top) and their interactions with Leu_{97}, Leu_{97}' pointing up from the ß-sheet which is formed by the N- and C- terminal ends of both molecules. Spheres of 50 %-Van der Waals radius are drawn around Leu_{97}, Leu_{97}' and Thr_{26}, Thr_{26}'. One molecule is drawn in thicker lines. b) View of the active site residues of the Leu97Phe mutant homodimer. The phenyl rings collide with Thr_{26}, Thr_{26}'. One molecule is drawn in thicker lines.

>2 x 10² when compared to the wild-type single chain dimer (Table 1). It is likely that a similarly low activity would not have been detected in the experimental system used by Loeb *et al.* (1989), therefore giving an inactive phenotype. Both heterodimers containing the Leu97Phe substitution, on the other hand, retained proteolytic activity and induced correct processing of the synthetic HIV polyprotein. Activity of the heterodimer containing the mutation in the second PR copy was considerably lower than the wild type enzyme whereas the same mutation in the first chain had only a minor effect on activity (Table 1). These different *in vitro* activities of the two heterodimers correlate well with the observed toxicities of the respective enzymes. The reason for the differences is presently unknown. Taken together these results confirm the predictions that were based on molecular modeling of the mutant enzymes.

A particular feature of the active site of retroviral PRs is the side chain to backbone interaction (fireman's grip) of the Thr hydroxyl group (Thr$_{26}$ in the case of HIV PR) accepting a hydrogen from the amide group of Thr in the other molecule and donating its hydrogen to the carbonyl O of Leu (Leu$_{24}$ in the case of HIV PR) in the other molecule (Figure 7). The two loops holding the two Asp-residues of the active site in place are thus effectively stabilized by 4 hydrogen bonds. This leaves only a very restricted choice for a rational mutation of Thr$_{26}$ in HIV PR. The only other small amino acid which is able to form this specific side chain to backbone interaction is serine (Figure 7), and an Asp, Ser, Gly triplet is in fact found in the active site of the proteinase of Rous sarcoma virus (Toh *et al.*, 1985), in contrast to the Asp, Thr, Gly triplet in the other PRs. The Thr26Ser mutation was therefore introduced into one and both PR copies of the single chain dimer. Expression of both constructs in BL21(DE3) failed to give any colonies at all indicating that both mutant enzymes were active. This was confirmed by incubation of the respective bacterial extracts with the synthetic HIV polyprotein. Both enzymes were active and induced specific processing of the precursor protein comparable to the wild type single chain dimer (Table 1).

Taken together the described experiments suggest that an HIV PR homodimer (Phe$_{97}$, $_{97}$' and Ser$_{26}$, Ser$_{26}$') and possibly also the heterodimer (Phe$_{97, 97}$' and Ser$_{26}$ Thr$_{26}$') may have restored proteinase activity. The most hindering interaction of Leu$_{97}$, Leu$_{97}$' appears to be with Thr$_{26}$, Thr$_{26}$' (Figure 6) and, if this hypothesis were correct, generating more space for the aromatic rings may relieve the distortion of the active site geometry. Homo- and hetero-dimers with a Thr26Ser substitution should clear the cavity from one and two methyl groups, respectively, thus allowing more space for the accomodation of the phenyl rings. We are currently testing this hypothesis by generating the respective homo- and hetero-dimers and testing their proteolytic activity. Additional experiments are designed to analyze the contributions of the other residues of the terminal ß-strands towards stability of the dimer and activity of the enzyme.

We have shown that a stable active single chain dimer of HIV PR can be generated by covalently linking two copies of PR by a short flexible hinge region. Since the two PR copies are cloned in two different mutagenesis vectors, any mutation can easily be introduced into each of the two copies and the two heterodimers and the homodimer can be generated in a single experiment. Using a simple screening assay based on the toxicity of PR in *E.coli*, mutagenesis and initial screening can be performed in a very short period of time. This system should therefore lend itself to a detailed evaluation of the contribution of specific amino acids towards the stability and activity of PR. Asymmetric mutagenesis appears particularly interesting in the flap region of PR (to assay for the role of the flap in the determination of substrate specificity) and in the terminal strands to assay for the contribution of specific amino acids to dimer stabilization.

In addition, this system should allow the introduction of compensatory mutations, based on molecular modeling, that should restore PR activity by relieving distortions. We are currently pursuing this approach by trying to restore activity of the Leu97Phe homodimer and by testing the effects of other mutations in the terminal sheets. Besides the potential to

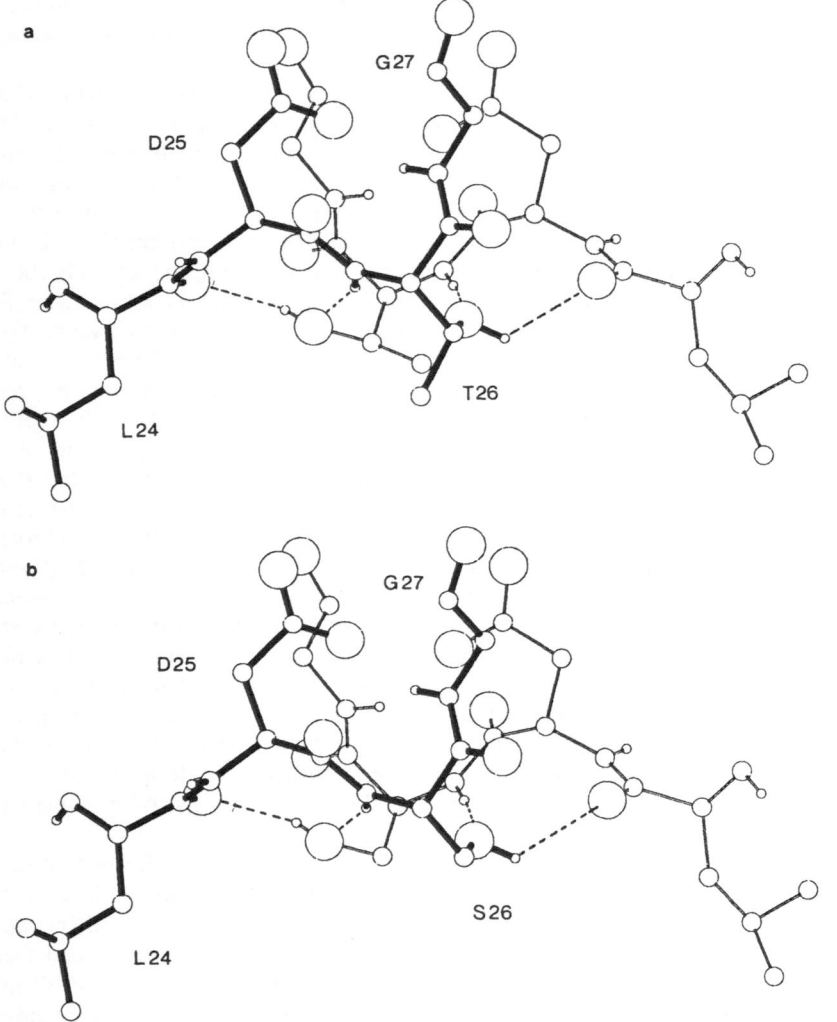

Figure 7. Hydrogen bond stabilization of the fireman's grip. a) wild type fireman's grip (Leu$_{24}$, Asp$_{25}$, Thr$_{26}$, Gly$_{27}$). The four hydrogen bonds are drawn as dashed lines. One molecule is drawn thicker. b) mutant fireman's grip (Leu$_{24}$, Asp$_{25}$, Ser$_{26}$, Gly$_{27}$). The interactions are identical to those of the wild type.

generate heterodimers of HIV PR, the single chain dimer enables us to study the role of PR during the assembly process. As discussed in the beginning, synthesis of a monomeric inactive form of PR as a component of the polyprotein may allow the targeting of unprocessed polyproteins to the plasma membrane requiring only a single targeting signal. The observation that the expression of a fully competent dimeric proteinase leads to premature processing, thus preventing particle assembly, suggests that concentration of the polyprotein and subsequent dimerization may in fact be the rate limiting step for processing to occur (Kräusslich, submitted for publication). Tight regulation of PR activity may therefore play a crucial role in the viral replication cycle.

ACKNOWLEDGEMENTS

We thank J. J. Dunn for plasmid *pET3* and P. Barr for antiserum against PR. We are grateful to H. zur Hausen for continued support and interest and to G. Sczakiel, M. Fäcke, K. Mergener, V. Bosch and M. Pawlita for discussions and suggestions. We also thank W. Weinig for oligonucleotide synthesis. This work was supported in part by a grant from the Bundesgesundheitsamt to HGK.

REFERENCES

Baum, E. Z., Bebernitz, G. A. & Gluzman, Y., 1990, Isolation of mutants of human immunodeficiency virus protease based on toxicity of the enzyme in *Escherichia coli*, *Proc. Natl. Acad. Sci. U.S.A.* **87**:5573-5577.

Debouck, C., Dreyer, G. B., Gorniak, J. G., Malinowski, J., Meek, T. D., Moore, M. L. & Strickler, J. E., 1989, Expression and structure-function characterization of HIV-1 proteinase, *in*: "Viral proteinases as targets for antiviral chemotherapy," Kräusslich, H.-G., Oroszlan, S. and Wimmer, E., eds., Cold Spring Harbor Laboratory Press, Cold Spring Harbor.

Gelderblom, H. R., Özel, M. & Pauli, G., 1989, Morphogenesis and morphology of HIV structure-function relations, *Arch. Virol.* **106**:1-13

Hellen, C. U. T., Kräusslich, H.-G. & Wimmer, E., 1989, Proteolytic processing in the replication of RNA viruses, *Biochemistry* **28**:9881-9890.

Jaskólski, M., Miller, M., Mohana Rao, J. K., Leis, J. & Wlodawer, A., 1990, Structure of the aspartic protease from Rous sarcoma retrovirus refined at 2 Å resolution, *Biochemistry* **29**:5889-5898.

Kay, J. & Dunn, B. M., 1990, Viral proteinases: weakness in strength, *Biochim. Biophys. Acta* **1048**:1-18.

Kräusslich, H.-G., Ingraham, R. H., Skoog, M. T., Wimmer, E., Pallai, P. V. & Carter, C. A., 1989, Activity of purified biosynthetic proteinase of human immunodeficiency virus on natural substrates and synthetic peptides, *Proc. Natl. Acad. Sci. U.S.A.* **86**:807-811.

Kräusslich, H.-G., Schneider, H., Zybarth, G., Carter, C. A. & Wimmer, E., 1988, Processing of *in vitro*-synthesized gag precursor proteins of human immunodeficiency virus (HIV) type 1 by HIV proteinase generated in *Escherichia coli*, *J. Virol.* **62**:4393-4397.

Kräusslich, H.-G. & Wimmer, E., 1988, Viral proteinases, *Annu. Rev. Biochem.* **57**:701-754.

Kunkel, T., 1985, Rapid and efficient site-specific mutagenesis without phenotypic selection, *Proc. Natl. Acad. Sci. U.S.A.* **82**:488-492.

Lifson, S. & Sander, C., 1980, Composition, cooperativity and recognition in proteins, *in*: "Protein Folding," Jaenicke, R., ed., Elsevier/ North Holland Biomedical Press.

Loeb, D. D., Swanstrom, R., Everitt, L., Manchester, M., Stamper, S. E. & Hutchison III, C. A., 1989, Complete mutagenesis of the HIV-1 protease, *Nature* **340**:397-400.

Moffatt, B. A. & Studier, F. W., 1987, *T7* lysozyme inhibits transcription by *T7* RNA polymerase, *Cell* **49**:221-227.

Ratner, L., Haseltine, W., Patarca, R., Livak, K. J., Starcich, B., Josephs, S. F., Doran, E. R., Rafalski, J. A., Whitehorn, E. A., Baumeister, K., Ivanoff, L., Petteway, S. R., Pearson, M. L., Lautenberger, J. A., Papas, T. S., Ghrayeb, J., Chang, N. T., Gallo, R. C. & Wong-Staal, F., 1985, Complete nucleotide sequence of the AIDS virus HTLV-III, *Nature* **316**:277-284.

Rosenberg, A. H., Lade, B. N., Chui, D., Lin, S., Dunn, J. J. & Studier, F. W., 1987, Vectors for selective expression of cloned DNAs by *T7* RNA polymerase, *Gene* **56**:125-135.

Schneider, J. & Kent, S. B. H., 1988, Enzymatic activity of a synthetic 99 residue protein corresponding to the putative HIV-1 protease *Cell* **54**:363-368.

Toh, H., Ono, M., Saigo, K. & Miyata, T., 1985, Retroviral protease-like sequence in the yeast transposon *Ty1*, *Nature* **315**:691-692.

Wlodawer, A., Miller, M., Jaskólski, M., Sathyanarayana, B. K., Baldwin, E., Weber, I. T., Selk, L. M., Clawson, L., Schneider, J. & Kent, S. B. H., 1989, Conserved folding in retroviral proteases: crystal structure of a synthetic HIV-1 protease, *Science* **245**:616-621.

EXPRESSION OF THE HIV ASPARTIC PROTEASE FUSED TO A BACTERIAL

PHENOTYPIC MARKER

Bruce D. Korant and Chris J. Rizzo

Central Research & Development Department
Du Pont Experimental Station, E328
Wilmington, Delaware   19880-0328

INTRODUCTION

The retroviruses encode a processing protease which is essential for virus replication. The enzyme is an aspartic protease, and cleaves both the *gag* and *gag-pol* gene products to yield mature virus containing active reverse transcriptase. In the absence of active protease only non-infectious viral particles are produced, so the enzyme has become an important target for drug designers, with the aim of benefiting people infected with the human immunodeficiency viruses, HIV-1 and HIV-2, and related retroviruses, which are agents of other serious diseases in man and domestic animals.

The present study reports the construction of a simple polyprotein, in which the amino terminal portion of the HIV-1 (IIIB) *pol* gene, including the viral protease (pro) and the first twenty-four amino acids of the viral reverse transcriptase are fused in proper translational reading frame with the ß-lactamase gene of *Bacillus licheniformis*,[1] designated BLA. Two viral cleavage sites are present in the resulting fusion protein, located at either end of the protease, and a third, bacterial cleavage site is located at the junction of the signal peptide and first amino acid of mature BLA. The plan in designing the fusion protein was that cleavage would be largely under control of the viral protease, and that cleavage should result in appearance of enzymatically active BLA which could be scored as penicillin resistance in bacteria expressing the fusion protein.

RESULTS

The fusion protein[2] was expressed in *E. coli* K12, strain BL-21 (DE-3), a strain which was designed by W. Studier to express the *T7* phage RNA polymerase, under *Lac* operator control, if exposed to a chemical such as IPTG, which inactivates the Lac repressor.

Others[3,4] have demonstrated a toxic effect of the HIV-1 protease in *E. coli*. We believe this is due to degradation of certain essential *E. coli* proteins when HIV-1 protease is expressed in the cytosol.

Table 1. Recovery of HIV protease after expression as a protease-BLA fusion protein in *E. coli* cell compartments, measured by Western blotting

Cell Compartment	HIV Protease % Recovery	HIV Protease % Soluble
Periplasm	36	91
Cytoplasm	54	8

Comparing the cloning efficiency in BL-21 of the pro-BLA fusion versus protease alone showed at least a 100-fold enhancement, and restriction digests and sequence analysis showed the correct nature of the transformed DNA.

Our favored explanation for our observed higher transformation frequency is that the fusion of HIV protease to BLA still permits the normal migration of BLA to the periplasm to take place, effectively moving the HIV protease to a cell compartment which is less prone to lethal damage by a protein hydrolase. We investigated this using Western blotting of both cell compartments; the cytoplasm and the periplasmic space, 45 min after induction of the biosynthesis of the protease-BLA fusion protein. The results, summarized in Table 1, confirm that almost one-half of the expressed HIV protease is detected in the periplasmic compartment, and that most of the viral protease remaining in the cytoplasm was aggregated.

By comparison, after expressing the HIV-1 protease directly, we found that less than 3 % was detected in the periplasmic space, and that of the bulk found in the cytoplasm, less than 5 % was soluble.[5]

The specific enzymatic activity of HIV protease was measured on a peptide substrate, CSQNYPVV, corresponding to the HIV p17/p24 cleavage site in the viral gag polypeptide. The soluble form of the enzyme recovered from the periplasmic space was compared with disaggregated, refolded HIV protease from pellet fraction[5] and the results indicated a six-fold greater specific activity of the soluble enzyme. This is likely due to some degree of mis-folded enzyme present in the preparation originating with the insoluble form of the enzyme.

The specific activity of the BLA produced from the fusion protein was not measured, but it was sufficiently active to permit the cells to grow on 150 μg/ml of penicillin, which was the highest concentration tested. The HIV protease and BLA polypeptides, which were overproduced in the cells to about 5 %-10 % of total cell protein, were recovered and the amino-terminal sequences were determined and found to be correct for mature protease and BLA,[6] see Table 2.

Table 2. Amino-terminal sequences of polypeptides expressed from the HIV protease-BLA fusion protein

Polypeptide	Sequence Predicted	Sequence Found (Picomoles)
HIV-protease (11 kDa)	P-Q-V-T-L	P (95) Q (70) V (80) T (30) L (20)
BLA (mature)	S-Q-P-A-E-K-N	S (120) Q (125) P (90) A(40) E (20)

Table 3. Expression level of the HIV protease in cells containing the *pET*-11 plasmid as percent of total bacterial protein

Generation Number	HIV Protease Expression (% Total)
1	11
10^6	9
10^{12}	10

As an alternate system, the plasmid *pET*-11 was used to express the HIV protease directly, unfused to BLA. In this plasmid, constructed by W. Studier (personal communication), the expression of the protease was under double control, with the *T7* promoter site under *lac* repressor control. The plasmid also contains a gene coding for the *lac* repressor, which is produced constitutively. Adding IPTG to *E. coli* BL-21 cultures containing the *pET*-11 plasmid inactivates the *lac* repressor, releasing its suppression of *T7* polymerase synthesis and allowing transcription of the HIV protease gene. Results summarized in Table 3 show the expression of the protease is maintained over at least 10^{12} generations, in small-scale fermentations (up to one liter).

DISCUSSION

HIV protease is not easily expressed in *E. coli*, because of its cytotoxic properties[3,4] and because of its insolubility. Two new HIV PR expression systems are described in this paper. One involves the plasmid *pET*-11, which contains a sophisticated double set of controls dependent on the constitutive production of the *lac* repressor in the culture. Transformants carrying the gene for HIV protease are easily obtained, and are extremely stable genetically.

A disadvantage of direct expression via *pET*-11 is that the HIV protease is largely insoluble, and must be solubilized by a strong denaturant, such as 8 molar urea, and then refolded. Therefore, as an alternate approach, we constructed a gene expressing HIV protease fused to a bacterial reporter, namely BLA. The lactamase detoxifies the protease, we suggest by transporting it to the periplasmic space. Once there, cleavage events under HIV protease control initiate the release of the mature viral enzyme from the polyprotein. Also released is the mature BLA protein, which is enzymatically active and produces a penicillin-resistance phenotype. The latter effect is easily scored, and should be a simple way to assay the action of the HIV protease and perhaps other proteases of interest within viable *E. coli*. The HIV protease found in the periplasm is mostly soluble, and has a higher specific enzymatic activity than refolded HIV protease obtained from intracellular aggregates. Since our system generates several percent of total cell protein as soluble HIV protease, it would seem to offer a useful new option for expression of HIV protease and related enzymes.

ACKNOWLEDGEMENT

Thanks to Professor J. O. Lampen for providing us the clone of *B. licheniformis* lactamase.

REFERENCES

1. P. Mezes, W. Wang, E. Yeh and J. Lampen, *J. Biol. Chem.* **258**:11211 (1983).
2. B. Korant and C. Rizzo, *Biol. Chem. Hoppe.-Seyler.* **371**:271 (1990).
3. Z. Hostomsky, K. Appelt and R. Ogden, *Biochem. Biophys. Res. Commun.* **161**:1056 (1989).
4. E. Baum, G. Bebernitz and Y. Gluzman, *Proc. Nat'l. Acad. Sci. U.S.A.* **87**:5573 (1990).
5. Y.-S. Cheng, M. McGowan, C. Kettner, J. Schloss, S. Erickson-Viitanen and F. Yin, *Gene* **87**:243 (1990).
6. K. Neugebauer, K. Sprengel and H. Schaller, *Nucleic Acids Res.* **9**:2577 (1981).

COMPARISON OF THREE INHIBITOR COMPLEXES OF

HUMAN IMMUNODEFICIENCY VIRUS PROTEASE

Amy L. Swain, Alla Gustchina,* and Alexander Wlodawer

Macromolecular Structure Laboratory
NCI-Frederick Cancer Research and Development Center
ABL-Basic Research Program
Frederick, Maryland   21702-1201
*On leave from the V.A. Engelhardt Institute of Molecular Biology
Academy of Sciences of the USSR, Moscow, USSR

INTRODUCTION

The protease encoded by human immunodeficiency virus (HIV) presents one of the most promising targets for designing new classes of drugs which could be therapeutically beneficial in cases of acquired immune deficiency syndrome (AIDS). A rational approach to drug design begins with characterization of the target enzyme and of enzyme-inhibitor complexes. In addition to kinetic data, structural information can elucidate precise interactions between inhibitors and the enzyme that may be responsible for effective inhibition. To aid in this drug design effort a large number of complexes of HIV-1 protease with substrate-based inhibitors are being studied, and several crystal structures of these complexes have been solved. Five structures of HIV-1 protease-inhibitor complexes have already been published (Miller *et al.*, 1989; Erickson *et al.* 1990; Fitzgerald *et al.*, 1990; Swain *et al.*, 1990; Jaskólski *et al.*, 1991). It is necessary to compare these structures in order to draw general conclusions about the nature of protein-inhibitor interactions.

The coordinate sets for three protease-inhibitor complex structures solved in this laboratory are used as the basis of comparisons described here. All structures were solved in the same orthorhombic crystal form ($P2_12_12_1$), thus the influence of crystal packing on the structure of the complex must be the same in each case. The sequences of the inhibitors bound in the complexes are given in Table 1, along with other relevant information. These inhibitors are hexa-, hepta- and octapeptides with different sequences, each one based on characteristics of a known substrate sequence. Each of the inhibitors has a different non-hydrolyzable group replacing the otherwise scissile peptide bond. They all bind to the protease in the same general conformation making similar contacts with the enzyme (Gustchina & Weber, 1990). The hexapeptide, MVT101, has a reduced peptide bond in place of the normal scissile bond. The heptapeptide, JG365, contains a hydroxyethylamine linkage, and the octapeptide, U85548e, has a hydroxyethylene linkage. The hydroxy-derived linkages of the latter two inhibitors are believed to mimic a tetrahedral intermediate of the proteolysis reaction.

Structure and Function of the Aspartic Proteinases
Edited by B.M. Dunn, Plenum Press, New York, 1991

Table 1. Sequences of inhibitors and relevant information on their properties and characteristics of the complexes with HIV PR

Inhibitor	P5	P4	P3	P2	P1	Sequence Linkage	P1'	P2'	P3'	K_i	Resol. (Å)	R
MVT-101			Ac-Thr	Ile	Nle	CH_2N	Nle	Gln	Arg	760 nM	2.3 Å	0.176
JG-365		Ac-Ser	Leu	Asn	Phe	$CH(OH)CH_2N$	Pro	Ile	Val-OMe	0.24 nM	2.4 Å	0.146
U-85548e	Val	Ser	Gln	Asn	Leu	$CH(OH)CH_2$	Val	Ile	Val	< 1 nM	2.5 Å	0.138

Table 2. Residues of HIV PR involved in polar contacts with bound inhibitors

	JG-365	U85548e	MVT-101
P_5		Val_{201}	
P_4	Ser_{101}	Ser_{202}	$Acetyl$
	Asp_{29} Asp_{30} Gly_{48}	Asp_{30} Gly_{48}	Gly_{48}
P_3	Leu_{202}	Gln_{203}	Thr_{201}
	Asp_{29}	Asp_{29}	Asp_{29}
P_2	Asn_{203}	Asn_{204}	Ile_{202}
	Asp_{29} Asp_{30} Gly_{48}	Asp_{29} Asp_{30} Gly_{48}	Gly_{48}
P_1	Phe_{204}	Leu_{205}	Nle_{203}
	Asp_{25} Gly_{27} Asp_{125}	Asp_{25} Gly_{27} Asp_{125}	Gly_{27}
P_1'	Pro_{205}	Val_{206}	Nle_{204}
P_2'	Ile_{206}	Ile_{207}	Gln_{205}
	Gly_{127} Asp_{129}	Gly_{127} Asp_{129}	Gly_{127} Asp_{129} Gly_{148}
P_3'	Val_{207}	Val_{208}	Arg_{206}
	Gly_{148}	Gly_{148}	Asp_{129} Gly_{148}

Table 3. Residues of HIV PR involved in nonpolar contacts with bound inhibitors

	JG-365	U85548e	MVT-101
P_5		Val_{201}	
P_4	Ser_{101}	Ser_{202}	$Acetyl$
P_3	Leu_{202}	Gln_{203}	Thr_{201}
	Arg_{108}	Gly_{48}	Gly_{48}
		Val_{182}	
P_2	Asn_{203}	Asn_{204}	Ile_{202}
	Ala_{28}	Ala_{28}	Ala_{28}
			Asp_{30}
	Ile_{47}	Ile_{47}	Ile_{47}
	Gly_{49}		
	Ile_{84}		Ile_{84}
	Ile_{150}		Ile_{150}
P_1	Phe_{204}	Leu_{205}	Nle_{203}
	Gly_{49}		
	Ile_{50}		Ile_{50}
		Leu_{123}	Leu_{123}
			Asp_{125}
	Pro_{181}		Pro_{181}
	Val_{182}	Val_{182}	Val_{182}
	Ile_{184}	Ile_{184}	Ile_{184}
P_1'	Pro_{205}	Val_{206}	Nle_{204}
		Leu_{23}	
		Asp_{25}	
	Pro_{81}	Pro_{81}	Pro_{81}
	Val_{82}		Val_{82}
	Ile_{84}	Ile_{84}	Ile_{84}
			Gly_{127}
	Gly_{149}	Gly_{149}	
	Ile_{150}		
P_2'	Ile_{206}	Ile_{207}	Gln_{205}
	Ile_{50}	Ile_{50}	Ile_{50}
	Ala_{128}	Ala_{128}	Ala_{128}
	Val_{132}		Val_{132}
	Ile_{147}	Ile_{147}	Ile_{147}
		Gly_{148}	
	Ile_{184}		
P_3'	Val_{207}	Val_{208}	Arg_{206}
	Arg_8		Arg_8
		Pro_{81}	
	Asp_{129}		

METHODS

The interactions between an inhibitor and protease were identified by a function CONTACT within the molecular graphics program FRODO (Jones, 1985). Polar contacts (Table 2) between protease and inhibitor are those between oxygen atoms, nitrogen atoms or oxygen and nitrogen atoms which have distances less than or equal to 3.5 Å. Nonpolar contacts (Table 3) are defined as distances between carbon atoms which are less than or equal to 4.5 Å. In cases where there were both a polar and nonpolar contact to a particular residue, the polar contact took priority and was listed in Table 2.

Protein Asymmetry and Inhibitor Disorder

The native enzyme is a homodimer composed of two identical molecules related by crystallographic symmetry (Wlodawer *et al.*, 1989), but the two molecules can be distinguished in the crystals of the complexes. The clearest means to distinguish one monomer from the other is with respect to the orientation of the mobile flap regions (Figure 1). The peptide bond between residues 50 and 51 is turned approximately 180° in molecule 1 relative to its position in molecule 2. The torsion angles identifying monomer 1 are ψ_{50} in the range of -20° to -45° and ϕ_{51} between -65° and -90°. In monomer 2, ψ_{150} is in the range of 130° to 140° and ϕ_{151} is between 90° and 115°. It is in this context that the interactions between each inhibitor and the protease will be described.

The question of whether the inhibitors are bound in a unique orientation within the crystal has been the subject of certain disagreement. Our original report of the structure of MVT101 described only one orientation of the inhibitor (Miller *et al.*, 1989), but that conclusion has now been modified as new refinement with 2 Å resolution data indicated conformational disorder (M. Miller, personal communication). Similar disorder has been reported by Fitzgerald *et al.* (1990) for the 2 Å structure of acetyl-pepstatin crystallized as a complex with the HIV-1 protease in a different crystal form. Disorder has not been observed for JG365 nor for U85548e, but Jaskólski *et al.*, (1990) made calculations which indicated that up to 30 % disorder would not have been detectable due to limitations of the available data. Other research groups have also reported that disorder is present for some inhibitors,

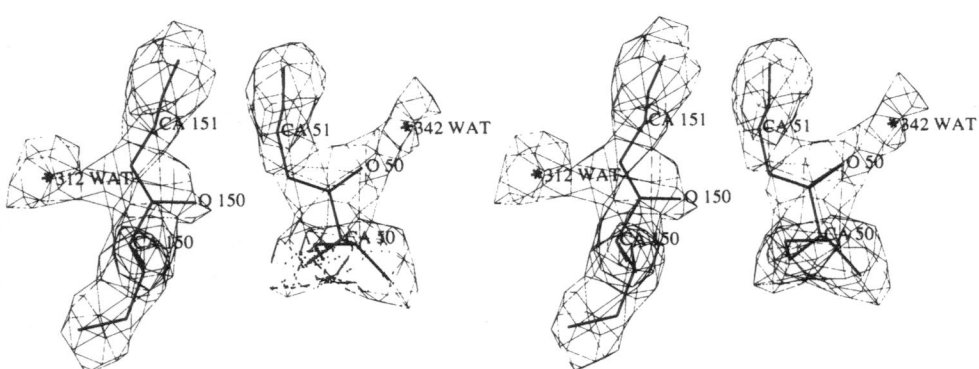

Figure 1. Stereoview of the Ile-Gly tips of the flaps of peptide-inhibited protease superimposed on $(2|F_o|-|F_c|)\alpha_c$ electron density. The peptide bond of monomer 2 (150-151) is oriented differently than that of monomer 1 (50-51) (see text). There is a hydrogen bond between O150 and N51, and water molecules 312 and 342 may assist in stabilizing this conformation.

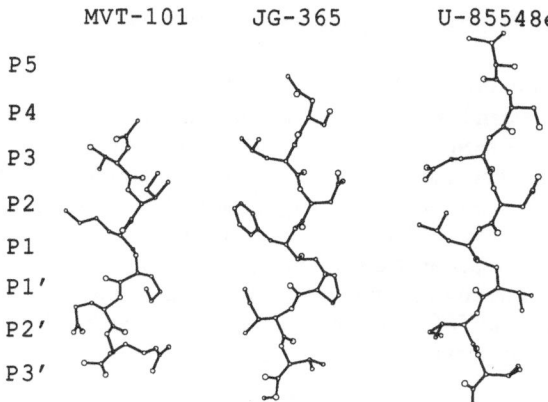

Figure 2. Structures of MVT101, JG365 and U85548e.

but not for the others, and this may be dependent on the nature of the ligand. Since the interactions between the inhibitor bound in two directions and the enzyme are similar (M. Miller, personal communication) and the refinement is still in progress, only the original structure of the MVT101 complex will be discussed here. The structures of the three inhibitors are shown in Figure 2.

Differences Among the Protease in the Structures of Three Complexes

The structure of the protease is very similar among the three protease-inhibitor complexes. Table 4 lists the results of alignment of each protease molecule with the others. As the structures are visually compared using computer graphics, the notable differences among the three models are side chain positions at the surface of the protein, and a different orientation of a peptide bond in the ß-bend 115-118 of U85548e. This latter difference places two atoms of U85548e in sufficiently different position from the analogous atoms of JG365 and MVT101 that they are not included in the alignment procedure. This ß-bend at the outer edge of the protease is involved in crystal lattice contacts and has high temperature factors, but has no apparent bearing on inhibitor binding. Therefore, with relevance to protease-inhibitor interactions, the three protease structures can be considered identical.

Table 4. Results of alignment of each proteinase-inhibitor complex with the other complexes

Alignment	pairs of atoms	max. dist. between atoms	rms deviation
U on J	196	0.936	0.358
M on J	198	1.010	0.373
U on M	196	1.128	0.433

U = U85548e; M = MVT-101; J = JG-365.

All three inhibitors are bound in the protease active site with torsion angles (ϕ,ψ) corresponding to an extended conformation. The only exceptions are torsion angles in the non-scissile groups. As seen in Table 2, the contacts between the main chain of the inhibitor and the protease main chain are the same for the three complexes. The nonhydrolyzable scissile bond analog of each inhibitor is aligned with the carboxyl groups of the active site aspartates. In the cases of JG365 and U85548e, the hydroxyl group at the P_1-P_1' junction is positioned between the aspartate carboxyl groups, within hydrogen bonding distance to each carboxylate oxygen. The configuration of the tetrahedral carbon of both of those inhibitors is *S*. Other similarities and differences will be addressed specifically according to inhibitor residue/protease binding pocket.

Subsites P_1-P_1'

The side chains of the residues in the protease comprising these pockets are mostly hydrophobic, although there are polar contacts between the side chains of the active site aspartates and the main chain hydroxyl groups of JG365 and U85548e. Residues occupying positions P_1 and P_1' of the inhibitor are usually hydrophobic and can be moderately large. Each of these three inhibitors has hydrophobic moieties at P_1 and P_1', with Phe-Pro, the most asymmetric dipeptide, found at this junction in JG365. At the Leu-Val junction of U85548e there is a size difference of only one carbon in the side chain and in MVT101 the junction is symmetric, with sequence Nle-Nle. However, the conformation of the norleucine side chains within the binding cleft makes this bound inhibitor topologically asymmetric. In all three cases, the side chain which extends furthest from the main chain of the inhibitor resides in the S_1 pocket. The norleucine in the S_1' pocket is compressed and the smaller of the side chains of the other two inhibitors are in this pocket. This indirect evidence implies that, as seen in these crystal structures, the S_1' pocket is smaller than the S_1 pocket.

An important hydrophobic contact of P_1 and P_1' is with $Pro_{181(81)}$. In the more compact pocket, S_1', each inhibitor contacts Pro_{81} (Table 3). In the pocket S_1, MVT101 and JG365 make contact with Pro_{181}. However, the Leu_{205} side chain in S_1 of U85548e is not long enough to make that contact. The position of the protease loop in which Pro_{81} (monomer 1) resides is different from the loop in monomer 2 (containing Pro_{181}). Each of these loops makes different contacts with a symmetry-related molecule through $Trp_{6(106)}$. The contacts between this loop of monomer 1 and its symmetry neighbor are more extensive in comparison with those in monomer 2. Therefore, the loop containing Pro_{81} is closer to the body of the protease than that containing Pro_{181}, thus making the S_1' binding pocket more compact and allowing a water-mediated contact between this loop and the flap of monomer 2. These are the structural features that contribute to the observed asymmetry between the S_1 and S_1' binding pockets in this crystal form.

Subsites P_2-P_2'

Although the side chains of the residues forming the S_2 and S_2' pockets are mostly hydrophobic, both hydrophilic and hydrophobic residues belonging to the inhibitor can occupy these sites. JG365 and U85548e have asparagine at P_2 while MVT101 has the similar glutamine at P_2'. MVT101 has isoleucine in P_2 while JG365 and U85548e have isoleucine in P_2'. The isoleucine side chains are observed in three different orientations for the three different inhibitors, forming hydrophobic contacts with different groups in the binding pocket. In the case of the amides, however, the side chain points toward the extremity of the inhibitor, avoiding the hydrophobic contacts of the pockets, and is apparently held in place by hydrogen bond interactions. The amide group of asparagine in U85548e is directed toward the OG of serine in the S_4 pocket, with the distance between OD1 of asparagine and OG of serine of 3.62 Å. The asparagine of JG365 also points in this general direction. The P_4 serine OG of JG365 is positioned differently than in U85548e, but since no density was present for this side chain (Swain *et al.*, 1990) it could well have the same conformation as the P_4 serine in U85548e and provide stabilizing force for the P_2 Asn. In the case of glutamine in P_2' of MVT101, the amide group is again pointing toward the extremity of the inhibitor, approaching the C terminal group of MVT101. The P_2' amide nitrogen - C-terminal nitrogen distance is 2.82 Å. Since the positions of the nitrogen and oxygen of an amide can only be inferred indirectly, this could still be a reasonable H-bond. This side chain makes a polar contact with the main chain of Gly_{48}. For both the P_2 and P_2' sites, the amide side chains are also stabilized by polar contact with the carbonyl oxygens of the P_3 and P_3' inhibitor residues. There are also polar contacts between the P_2 amide groups of JG365 and U85548e and protease main chain atoms of Asp_{29} and Asp_{30}.

Subsites P_3-P_3'

Aliphatic, polar, and ionic side chains are represented in these inhibitors at the P_3 and P_3' sites. However, regardless of which type of side chain is present, S_3 is distinct from S_3' and the S_3 pocket is similar among the three complexes. The outer "wall" of the S_3 and S_3' pockets is formed by the juxtaposition of protease loops containing $Phe_{53(153)}$ and $Pro_{181(81)}$. The difference in the positions of those residues may be due to the presence of a tryptophan side chain from a symmetry related molecule inserted between these loops. In S_3' Trp_{106} is inserted between Phe_{153} and Pro_{81} so that the Arg_{206} of MVT101 must assume a strained and compact configuration. In S_3 Trp_6 is too far to directly influence the binding of residue P_3. In all three structures we observe a large positional difference between the loop 78-82 of monomer 1 and 178-182 of monomer 2, as well as between Trp_6 of monomer 1 and Trp_{106} of momomer 2. This observation lends credence to the assumption that the crystal lattice contacts can be partially responsible for the observed asymmetry of the protease dimer. Two side chains at P_3, glutamine in U85548e and threonine in MVT101, have potential for polar interactions, yet do not make such contacts with the protease. The OG of threonine is in hydrogen bonding distance of the adjacent carbonyl oxygen of the acetyl group. The leucine of JG365 points toward the outer edge of the pocket while the glutamine of U85548e points toward the interior of the pocket, seemingly doing no more than avoiding repulsive interactions. However, the OE1 of glutamine is 3.6-3.75 Å from all three nitrogens of the Arg_{108} amide group. Though these distances are longer than acceptable hydrogen bonds, there is some attractive force between these atoms.

For the protease-MVT101 complex, the pocket S_3' is slightly different from that found in the other complexes due to a different position of the Arg_8 side chain. Residues in these inhibitors at P_3' are either valine or arginine. When valine is present, Arg_8 extends into the S_3' pocket. When arginine is present at P_3', Arg_8 is rotated by a single torsion angle away from the pocket. Even so, the Arg_{206} side chain of MVT101 curves around, removing the charged end from the otherwise hydrophobic pocket, and forming a salt bridge with the side chain of Asp_{129}.

Subsites P_4

Only two inhibitors, JG365 and U85548e, have residues with side chains at the P_4 site. In both cases this residue is serine. S_4 can barely be referred to as a binding pocket because only the portion of the P_4 side chain which is oriented toward the carboxyl end of the inhibitor is surrounded by atoms of the protease. As was mentioned previously, the serine side chains are positioned differently in the structures of JG365 and U85548e. The most reliable model for this side chain is protease-U85548e, because in this complex the electron density for the serine OG was obvious (Jaskólski *et al.*, 1991). Also supporting the orientation of serine in U85548e is a 3.62 Å contact, just beyond normal H-bonding distance, between the P_4 serine OG and the OD1 of P_2 asparagine. In the complex of protease with JG365, density for the serine OG was absent so the side chain was positioned to maximize polar contacts (Swain *et al.*, 1990). Serine OG of JG365 hydrogen bonds to Asp_{29}, Asp_{30} and to Wat318. Other water molecules help to anchor this residue in JG365; Wat368 hydrogen bonds to the main chain nitrogen while Wat378 hydrogen bonds to the carbonyl oxygen of the C-terminal ester group. These two waters are not present in U85548e, probably in part because U85548e has a P_5 residue, but there are water molecules in similar positions in MVT101, which has no P_5 residue.

Subsite P_5

U85548e is the only inhibitor with a residue at P_5. This really is not an S_5 pocket, since the valine residue extends out of the protease binding groove. The side chain makes no contact with the protease and its temperature factor is quite high, indicating probable motion. It is not expected that this residue contributes to the inhibitory properties of U85548e.

Conserved Waters

A number of water molecules were observed in the inhibitor complexes and some of them are conserved among the three structures. The most intriguing water molecule resides between each inhibitor and the tips of the flaps. In JG365 and U85548e it is labeled 301, in MVT101 it is listed as Wat511. It is tetrahedrally coordinated to main chain nitrogen atoms from residues 50 and 150 of the flaps, and to carbonyl oxygens from substituents P_2 and $P_{1'}$ of the inhibitor.

In all three structures two water molecules are observed in the S_3 site. In U85548e and MVT101 the closest water mediates contacts between P_3 and the side chain of Asp_{29}, the carbonyl of Gly_{27} and the other water molecule. In JG365, though this water is in the same position, it is slightly beyond hydrogen bonding distance of the inhibitor. Only in JG365 the same configuration exists in the S_3' site. Here, water 361 mediates contacts among the carbonyl oxygen of the $P_{2'}$ residue and the side chain of Asp_{129}, the carbonyl oxygen of Gly_{127} and water 304. Only in U85548e is the water analogous to Wat361 of JG365 conserved, and only in MVT101 is the water analogous to Wat304 of JG365 conserved.

DISCUSSION

The structures of the protease observed in all three complexes are strikingly similar. The protease binds the inhibitors with a complementary conformational adjustment of both the protease and the inhibitor. The adjustment of the protease seems to be primarily the closing of the flaps over the protease along with a hinge-like shift in the two monomers toward the center of the molecule. The inhibitor adjustment is particularly observable in the protease-MVT101 complex where the P_1' norleucine and the P_3' Arg side chains are bent to fit into the pockets. The inhibitors also seem to aid in stabilization of their conformation through the interactions between their own side chains of P_2 and P_4 (or the N-terminus in the case of MVT101).

There are only a few regions of major positional differences between monomer 1 and monomer 2 among the three proteases. Those differences which are obviously due to crystal contacts are loops 15-18 (115-118) and 37-40 (137-140). These loops are distant from the active site and seem to have no effect on inhibitor binding. The more subtle differences of the loop containing $Pro_{81(181)}$ and $Trp_{6(106)}$ must have a direct impact on inhibitor binding as they help to define the S_1, S_1' and S_3, S_3' pockets. There still is no direct evidence for the cause of these differences. Most puzzling remains the orientation of the peptide bonds in the tips of the flaps. However, their orientation does make structural sense in providing an attractive force between the flaps helping to maintain their closure around the inhibitor. The orientation of this peptide bond has no bearing on contacts between the protease side chains and inhibitor, as disorder is observed in some protease-inhibitor complex structures (Miller, unpublished; Fitzgerald *et al.*, 1990).

The important result of this comparison is that while the interactions between inhibitors and the protease are very similar, subtle differences are observed. The polar interactions between the extended inhibitor main chain and protease main chain are the same in the three structures. The polar interactions between the active site aspartates of the two inhibitors and the hydroxyl groups at the scissile bond analogs are consistent with their tighter binding. It is clear that more than three structures should be compared to draw general conclusions about the mechanism of inhibitor binding, and in particular the structures solved in other crystal forms should be brought into this comparison. This should enable the design of side chain modifications which would provide optimal binding and redesign of the main chain, to change the peptidic nature of the inhibitors. Such steps would be required to transform protease inhibitors into anti-AIDS drugs.

ACKNOWLEDGEMENT

This study was sponsored by the National Cancer Institute, DHHS, under contract number N01-CO74101 with ABL.

REFERENCES

Erickson, J., Neidhart, D. J., VanDrie, J., Kempf, D. J., Wang, X. C., Norbeck, D., Plattner, J. J., Rittenhouse, J., Turon, M., Wideburg, N., Kohlbrenner, W. E., Simmer, R., Helfrich, R., Paul, D. & Knigge, M., 1990, *Science* **249**:527-533.

Fitzgerald, P. M. D., McKeever, B. M., VanMiddlesworth, J. F., Springer, J. P., Heimbach, J. C., Leu, C.-T., Herber, W. K., Dixon, R. A. F. & Darke, P. L., 1990, *J. Biol. Chem.* **265**:14209-14219.

Gustchina, A. & Weber, I. T., 1990, *FEBS Lett.* **269**:269-272.

Jaskólski, M., Tomasselli, A. G., Sawyer, T. K., Staples, D. G., Heinrikson, R. L., Schneider, J., Kent, S. B. H. & Wlodawer, A., 1991, *Biochemistry,***30**:1600-1609.

Jones, A., 1985, *Methods Enzymol.* **115**:157-171.

Miller, M., Schneider, J., Sathyanarayana, B. K., Toth, M. V., Marshall, G. R., Clawson, L., Selk, L., Kent, S. B. H. & Wlodawer, A., 1989, *Science* **246**:1149-1152.

Swain, A. L., Miller, M., Green, J., Rich, D. H., Schneider, J., Kent, S. B. H. & Wlodawer, A., 1990, *Proc. Natl. Acad. Sci. U.S.A.* **87**:8805-8809.

COMPARISONS OF THE SEQUENCES, 3-D STRUCTURES AND MECHANISMS OF

PEPSIN-LIKE AND RETROVIRAL ASPARTIC PROTEINASES

Tom L. Blundell, Jon B. Cooper, Andrej Šali and Zhan-yang Zhu

Imperial Cancer Research Fund Unit of Structural Molecular Biology
Department of Crystallography
Birkbeck College
University of London
Malet Street, London WC1E 7HX
United Kingdom

INTRODUCTION

The discovery of pseudo-two fold symmetry in the 3-D structures of the pepsin-like aspartic proteinases by Tang, James, Blundell and coworkers (Tang *et al.*, 1978) has led to a productive series of hypotheses in subsequent years. These hypotheses concerned a dimeric ancestor of the aspartic proteinases (Tang *et al.*, 1978), the close equivalence of the two active site aspartates (Pearl & Blundell, 1984), the similarity of the specificity sites on either side of the scissile bond (Blundell *et al.*, 1983) and the structure of the retroviral proteinases as dimeric homologues of the pepsins (Pearl & Taylor, 1987; Blundell *et al.*, 1988). Now that the crystal structures of several pepsins, retroviral proteinases and their inhibitors have been defined by X-ray analysis (see other chapters in this volume), these ideas can be reassessed more precisely and usefully. In this paper we provide short descriptions of the relationships between the secondary and tertiary structures of the two lobes of aspartic proteinases and the subunits of retroviral proteinases. We use these to provide an optimal alignment of sequences, an identification of residues that are important to the "aspartic proteinase fold" and a phylogenetic tree of structures. We describe the analogies between inhibitor binding of the retroviral and cellular proteinases and the rigid group rotations that are consequential upon occupation of the specificity pockets. Finally we discuss a mechanism of hydrolysis that is consistent with the structures of this family of enzymes.

3-D STRUCTURES

Figure 1 shows representative 3-D structures of monomeric pepsin-like (endothiapepsin; Blundell *et al.*, 1990) and dimeric retroviral proteinases (HIV proteinase; Wlodawer *et al.*, 1989; Lapatto *et al.*, 1989) viewed from equivalent directions perpendicular to the 2-fold axis. The similarity in general shape can be seen although the monomeric pepsin is roughly 60 % larger than the retroviral dimer. The close relationship between their folds is best seen by considering the arrangements of the ß-strands and α-helices shown schematically in Figure 2 for the motif that is common between each lobe of a pepsin and a subunit of a retroviral enzyme.

Most of the structures of the retroviral and pepsin-like proteinases are comprised of ß-strands arranged in a simple, symmetrical way. Each subunit of the retroviral proteinase and each lobe of a pepsin-like proteinase comprises two similar motifs formed from anti-parallel strands: a, b, c and d for the first and a', b', c' and d' for the second (Blundell *et al.*. 1985; 1990; Miller *et al.*, 1989a; Lapatto *et al.*, 1989) as shown in Figure 2. These are organised together in a distorted sheet (sheet 1). Strands c and d' and strands c' and d form two pairs of parallel strands. Strands b and c and strands b' and c' form anti-parallel ß-hairpins that are folded over sheet 1 and hydrogen-bonded together around intra-domain two-fold axis to give a second sheet (sheet 2), which is orthogonal to the first. In the retroviral

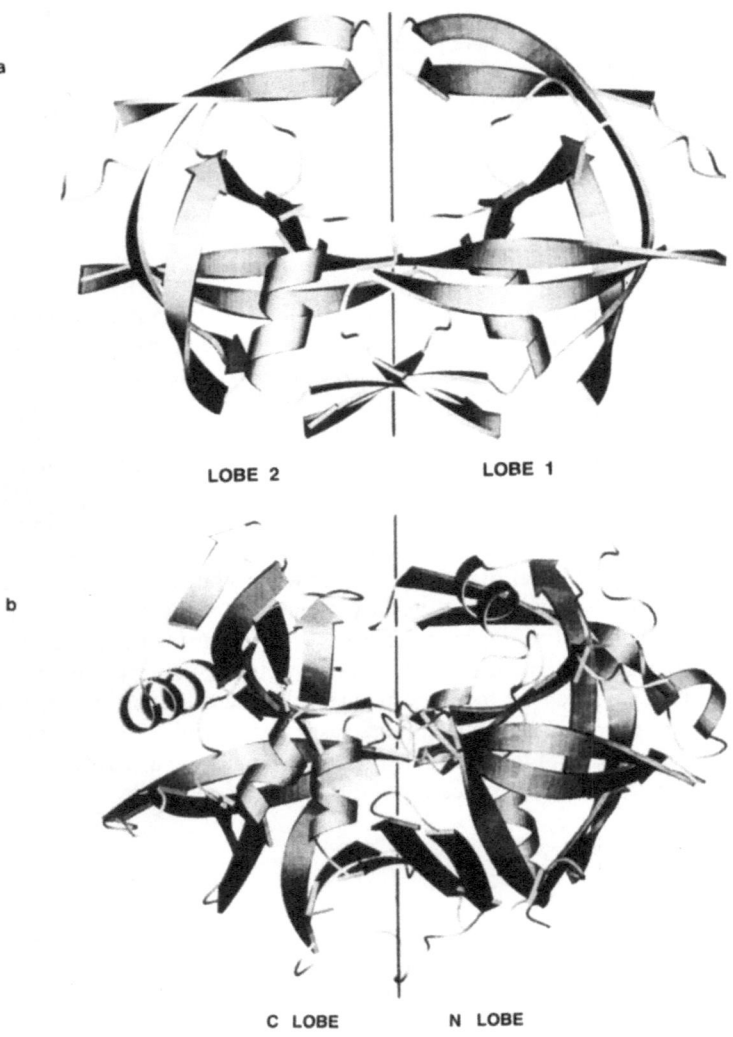

Figure 1. A schematic view perpendicular to the two-fold and along the active site cleft of a) the HIV-1 proteinase (Lapatto *et al.*, 1989) and (b) a pepsin-like aspartic proteinase (Blundell *et al.*, 1990).

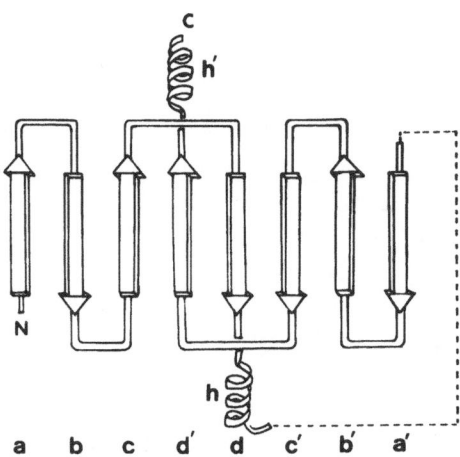

Figure 2. A schematic diagram of the strands and helices of the super-secondary structure that is common between retroviral proteinases and pepsins.

proteinases the motifs are more symmetrical and less distorted than in the pepsin-like enzymes, where both sheets are fragmented.

Much of strand a of the first motif is displaced from the main sheet and forms an anti-parallel ß-sheet with the carboxy-terminal strands of the subunit or lobe, and their equivalents in the second subunit or lobe (sheet 3). In the retroviral proteinases there is only one carboxy-terminal strand and so the inter-subunit ß-sheet contains four antiparallel strands. In the pepsin-like proteinases two carboxy-terminal strands of each lobe contribute to a six stranded antiparallel ß-sheet. Figure 1 shows that the strands of sheet 3 form the base of the enzymes below the well defined cleft. In the retroviral and pepsin-like enzymes the equivalent sheets 3 occupy the same volume but have different orientations.

SEQUENCE ALIGNMENT

Although only the sequences around the active site aspartates are obviously equivalent between the lobes of the pepsins and the subunits of the retroviral proteinases, the topological similarities between the 3-D structures allows a complete alignment of sequences. This is achieved using the computer program COMPARER (Šali & Blundell, 1990; Zhu *et al.*, 1991) which uses features of protein 3-D structures such as torsion angles, sidechain directions, accessibilities and hydrogen bonding patterns to provide an automatic alignment. This is shown in Figure 3. The single letter code for the amino acids has been modified to indicate the conformation and local environment of the sidechain (Overington *et al.*, 1990). It is evident that most amino acids that are conserved or conservatively varied are inaccessible to solvent and quite often hydrogen bonded to mainchain carbonyls or NH functions, for example Thr_{31} of HIV proteinase (Lapatto *et al.*, 1989) or Gln_{99} of pepsin (Cooper *et al.*, 1990).

One of the significant conservations in the retroviral and pepsin-like aspartic proteinases is the sequence hydrophobic-hydrophobic-glycine, for example Ile_{84}-Ile_{85}-Gly_{86} found in HIV-1 proteinase. Ile_{85}, like its equivalents in other retroviral proteinases and in pepsin-like proteinases (Figure 3), is conserved as hydrophobic because it contributes to the core. Gly_{86} is packed close to the active site residues in a way that does not allow a

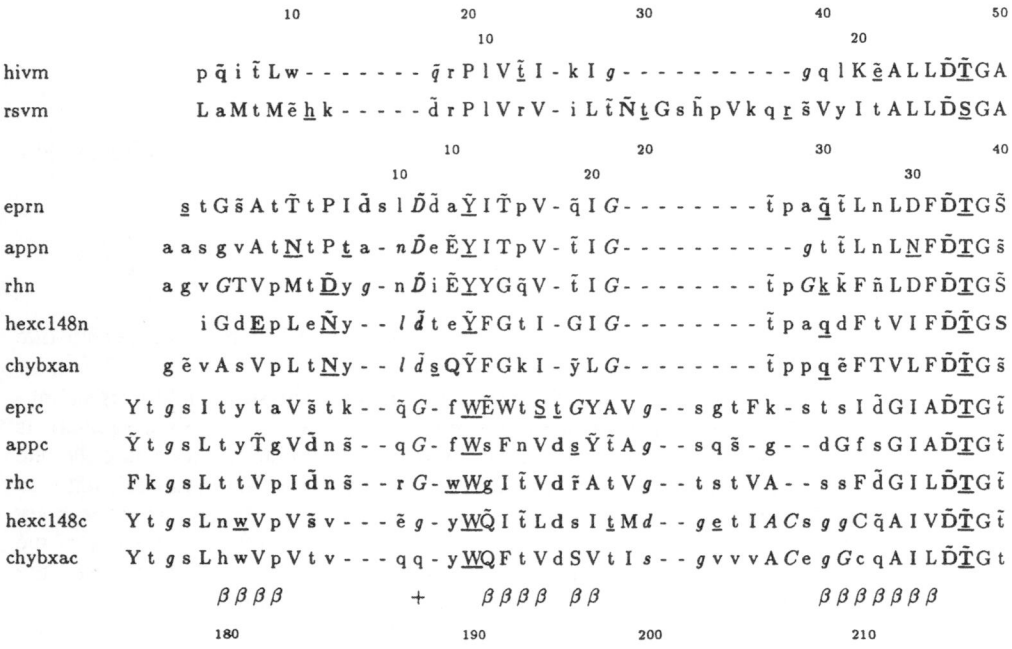

Figure 3. The alignment of sequences of aspartic proteinases achieved by comparing the three-dimensional structures using COMPARER [Šali & Blundell, 1990]. epr: endothiapepsin; app: penicillopepsin; rh: rhizopuspepsin; hexc148: hexagonal porcine pepsin; chybxa: calf chymosin; rsv: Rous sarcoma virus proteinase; hiv: human immunodeficiency virus proteinase. The last letter refers to the amino (n) or carboxy (c) terminal domains of the pepsins. The coordinates of the three-dimensional structures were obtained from the PDB databank (Bernstein *et al* , 1977). The amino acid code is the standard one-letter code formatted using the following convention (Overington *et al.*, 1990). *Italic* for positive φ; UPPER CASE for solvent inaccessible residues; lower case for solvent accessible residues; **bold type** for hydrogen bonds to mainchain amide nitrogen; <u>underline</u> for hydrogen bonds to mainchain carbonyl oxygen; tilde ~ for sidechain-sidechain hydrogen bonds.

```
              60          70          80          90          100
        30                              40          50
hivm    d̃ d T̃ V L e ẽ - - - - - - - - - - - - - - M s L p - - - - - - g r w - - k p k m i g g i - - - g
rsvm    D̃ I T̃ I I S̃ e e d̃ W P - - - - - - - - - - - t d W p - - - - - - - - - v - - m e a a n p q I h G I G

              50          60
        40          50          60          70
eprn    S̲D̃L W̃ V F S̲̃ s ẽ T̲̃ t a - - s e̲ v d g Q̲ t i Ỹ T̲̃ P s k S̃ - - t̲ t̃ A k l l s g A t W̲s̲ I s Ỹ g d̃ - - g
appn    A D̃ L W V F S̲ t̃ ẽ L p a - - s q q s̲ g H̲̃ s V Ỹ ñ P s̃ a t̃ - - - - G k e l s g Ỹ t W̲s̲ I s y g d̃ - - g
rhn     S̲D̃L W̃ I A S̲ t l C t̃ ñ̲ - - C - g s g Q̲̃ t k̲ Ỹ d P n q S̃ - - s t̃ y q a d̃ g - r̃ t W̲s̲ I s Ỹ g d̃ - - g
hexc148n S̲Ñ L W̃ V P S̲ v y C s s l A C - - s d H̲̃ ñ q F ñ P d̃ d S̃ - - s̲ t̃ F e a T̲ s - q e L s I t Ỹ g t - - -
chybxan S̲D̃ F W V P S̲ I y C k S̃ n A C - - k n H̲ q r F D̃ P r̃ k S̃ - - s̲ t̃ F q n̲ l g - k p L s I h̃ y g t - - -
eprc    t̲ L L y L p - - - - - - - - - - a t V V s a Y̲W̲ a q V s g A k s S̃ s s̃ - - - - - - - - - - - - -
appc    t̲ L L l L d̃ - - - - - - - - - - d s V V s̲ q̃ Y̲Y̲ s q V s g A q q̃ d̃ s ñ - - - - - - - - - - - - -
rhc     t̲ L L i L P - - - - - - - - - - ñ̲ n i A a s̲ V A r a Y̲ - g A s d̃ ñ̲ g - - - - - - - - - - - - -
hexc148c s̲ l L T G P - - - - - - - - - - t̃ s a I a n I Q̃ s̲ d̲ I - g A s e n s - - - - - - - - - - - - -
chybxac s̲ k L V G p - - - - - - - - - - s s d I l n I Q q a I - g A t q n q - - - - - - - - - - - - -
         β β β                              α
        220                        231        240


              110         120         130         140         150
        60                              70
hivm    g f i k V R̲̃ q̃ Ỹ - d̲ q I - - - - - - - - - - - - - l I E̲ I c - - - - - - g h k A i - - - - - - - - - - -
rsvm    g g I p M r̲ k̲ S̲ r D̃ m I - - - - - - - - - - e L G V I ñ̲ r d̃ g s̃ l Ẽ r p l l - - - - - - - - - -
        70          80                              90
        80                              90
eprn    S̃ s S̲ s G̃ d V Ỹ - - t̲ D̃ - - - - - - - - - - - - - t V s̃ V g - - - - g L t̃ V t - - - - - - - - - - - -
appn    S̃ s A s G̃ ñ V F - - t D̃ - - - - - - - - - - - - - s V t̃ V g - - - - g V t̃ A h - - - - - - - - - - -
rhn     S̃ s A s G̃ i L A - - k D̲̃ - - - - - - - - - - - - - n V n L g - - - - g L l I k - - - - - - - - - - -
hexc148n G s̲ M t G̃ i L G - - y D̃ - - - - - - - - - - - - t V q V G̃ - - - - g I s D̃ t̃ - - - - - - - - - - -
chybxan G s̃ M q G̃ i L G - - y D̃ - - - - - - - - - - - - T V t V s - - - - ñ̲ I v D̃ i - - - - - - - - - - -
eprc    - - - - - - v g - - g y V F p c s A t - - L p s F T F G V g - - - - s a r I v I p G d Y I d F g p i
appc    - - - - - - A g - - g ỹ V F d c s T̲ n - - L p d F S V s I s̃ - - - - g ỹ t A t V p G s l I n̲ Y g p S̲
rhc     - - - - - - d̃ g - - t̃ Y t̃ I s̲ c d̃ - T s̃ a F k p L v F S̃ I n̲ - - - - g a s F q V S̃ p d̃ S̃ L V F e e f
hexc148c - - - - - - d g - - e m v I s c s s i a s L p d I v F t I n - - - - g v q Y p L s P s A Y̲I̲ l q d -
chybxac - - - - - - y g - - e f d I d c d n l s y M p t V V F e I n - - - - g k m Y p L t P s A Y T s q d -
         β                    β β β β +        + β β β β
        250                        260        270
```

Figure 3. (continued).

447

```
              160           170           180           190           200
                             80
hivm    - - - - - - - g ĩ V L V G - p t - - - - - - - - - - p v N I I G - - - - - - - - - - - - - R̰ ñ L
rsvm    - - - - - - - L f P A V A - m V - - - - - - - - - r g S̃ I L G - - - - - - - - - - - - - R̰ d̰ C

              100                        110
              100           110           120           130
eprn    - - - - - - - g Q̃ A V Ẽ S̃ A k k V s - s s̃ F t e d̰ s ĩ I D G L L G L A f s̃ t l N̰ t V s p t q q k T F
appn    - - - - - - - g Q̃ A V Q̃ A A q q I s̰ - a q̃ F q q d̰ t ñ Ñ D G L L G L A F s̃ s i N̰ t V q p q̃ s q ĩ T F
rhn     - - - - - - - g Q̃ T I Ẽ L A k ĩ Ẽ a - a s F a s g - P Ñ D G L L G L G F d̃ t i T̰ ĩ V r - - g V k T̃ P
hexc148n - - - - - - - n Q̃ I F G L S̃ e t Ẽ p g s f L y y A - p F D̃ G I L G L A Y̆ p s i S̰ a s - - - g A t P V
chybxan  - - - - - - - q Q̃ T V G L S̃ t̰ q̃ Ẽ p g d v F T̰ y A - ḛ F D G I L G M A Y̆ p s̰ l A s̃ e - - - y S̃ i P V
eprc    s t g s s s C f G G I Q s S̰ a g i - - - - - - - - - g i Ñ I F G - - - - - - - - - - - - - D̲ V A
appc    g d̰̲ - g s ĩ C L G G I Q s̰ Ñ s g i - - - - - - - - - g f S I F G - - - - - - - - - - - - - D I F
rhc     - - - q g g q C i A G F G ȳ g ñ - - - - - - - - - w g F A I I G - - - - - - - - - - - - - D̲ T̲ F
hexc148c - - - d d s C t S̰̃ G F e G m d̃ v p t s - - - - - - s g ẽ L W I L G - - - - - - - - - - - - - D̲ V F
chybxac  - - - q g f C T S G F q s e n h S - - - - - - - - - q k W i L G - - - - - - - - - - - - - D V F
              β  β β β                          β β β                      α
        280           290           300
```

```
              210           220           230
              90            99
hivm    L T̲ q I - - - - - - - - - g C ĩ L ñ F
rsvm    L q g L - - - - - - - - - g L r L T N̰ l

                            120    124
              140           150           160           170    174
eprn    F d̃ ñ A k̰ a s - - L d s p V F T̃ A d̰ L g y h̰ - - - a p g t Y Ñ F G f i d̃ t ĩ a
appn    F d̃ ĩ V k̰ s s - - L a q p L F A V A L K h̃ q̰ - - - q P G v Y̲ D F G f I d̃ s̃ s̃ k
rhn     M d̃ Ñ L i s̲ g l I s ĩ p I F G V y L G K̃ a k̃ n g G g G e Y̲ I F G g y d̃ s ĩ k
hexc148n F D̃ N̰ L w̃ d̰ q̃ g l V s q D̃ L F S̰̃ V Y L S̰ n - d d̲ s g S V V L L G G i d s s y
chybxan  F D̃ Ñ M M n̲ r h l V a q d L F S̰̃ V Y̆ M D̃ ĩ d - - g q e S M L T L G a i d̃ p s̃ y
eprc    L K̃ A A - - - - - - - - - F V V F n G a t - - - - - t P t L G F A s K̰̆
appc    L K̃ S̰ Q̰ - - - - - - - - Y̆ V V F D̰ S d - - - - - - g P q̃ L G F A p Q̰ a
rhc     L K̃ Ñ N̰ - - - - - - - - Y V V F Ñ Q̃ g - - - - - - v p e V q I A p V a e
hexc148c I R̰̃ q Y̲ - - - - - - - Y̆ T̲ V F D̃ R̰ a - - - - - - ñ ñ̰ k V G L A p v a
chybxac  I R̰̃ Ẽ Y - - - - - - - Y̆ S V F D R̃ a - - - - - - n ñ l V G L A k A i
        α                     β β β β β                  β β β β
        310                         320           327
```

Figure 3. (continued).

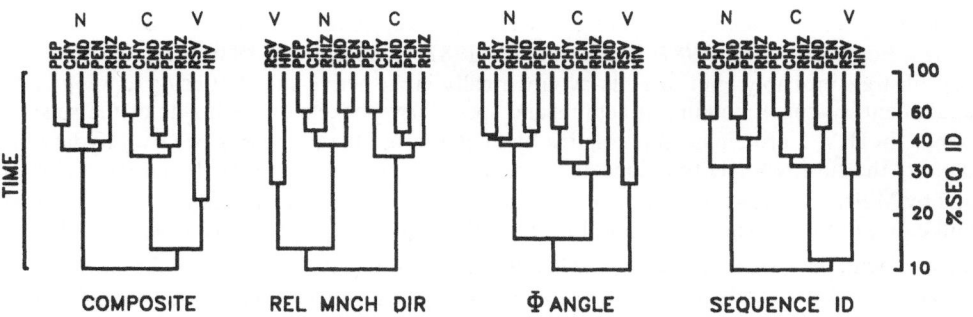

Figure 4. Phylogenetic trees derived from clustering of the lobes of aspartic proteinases on the basis of sequences and structures. These were clustered on the following criteria: COMPOSITE: all structural and sequence features; REL MNCH DIR: relative mainchain direction; PHI ANGLE: main chain torsion angle and SEQUENCE ID: percentage sequence identity.

sidechain. Ile_{84} is conserved for purposes of binding substrate (see below). This pattern is characteristic of both monomeric and dimeric aspartic proteinases. In all retroviral proteinases the following residue is an arginine. Such a basic residue is never found in the pepsins, where it is either a hydrophobic residue or an aspartic acid. In both HIV-1 proteinase and RSV-proteinase (Miller *et al.*, 1989a) this conserved arginine participates in an inter-subunit cluster of ionic and hydrogen bonding interactions, and is therefore probably important in dimer formation.

The conserved active site residues, Asp_{25}-Thr_{26}-Gly_{27}, of the HIV proteinase form a symmetrical and highly hydrogen-bonded arrangement virtually identical to that in pepsin-like aspartic proteinases (residues Asp_{32}-Thr_{33}-Gly_{34} and Asp_{215}-Thr_{216}-Gly_{217} of pepsin) as shown in Figure 3. This includes the two threonines, which are inaccessible to solvent and hydrogen-bonded so that the γ-O forms hydrogen bonds with the main chain NH and CO functions of the other subunit or lobe in a fireman's grip (Pearl & Blundell, 1984). They provide a good example of high conservation of buried, hydrogen-bonded polar residues that are important for maintenance of 3-D structure. The two aspartates lie approximately planar with their inner carboxylate oxygens hydrogen bonded to the NH functions of Gly_{27} and within hydrogen-bonding distance of each other. The conservation of the glycines appears to be a consequence of the fact that the existence of sidechains at this position would disrupt the structure of the aspartic acid sidechains.

PHYLOGENETIC TREES

The alignment of the sequences based on the topological equivalence of 3-D structures allows the construction of phylogenetic trees (Johnson *et al.*, 1990a,b; Šali & Blundell, 1990; Šali *et al.*, 1990). These depend on calculating distances based on amino acid features such as identity, conformation and sidechain acessibility for topologically equivalent residues. The trees (Figure 4) demonstrate the clear and close relationships within the equivalent classes - N-lobe, C-lobe, retroviral subunit - and indicate a slightly closer similarity between C-lobes and retroviral subunits.

INHIBITOR BINDING

The three-dimensional structures of the dimeric retroviral proteinases and the monomeric pepsins suggest a similar mode of interaction of the enzymes with substrate. Much evidence concerning this has been inferred from X-ray studies of aspartic proteinase-transition state isostere complexes (Bott *et al.*, 1982; James *et al.*, 1982; Suguna *et al.*, 1987a; Foundling *et al.*, 1987; Blundell *et al.*, 1987; Cooper *et al.*, 1989). These show that the substrate probably binds pseudo-symmetrically in the active site cleft using topologically equivalent hydrogen bonding functions and specificity pockets on each side of the scissile bond. In HIV-1 proteinase this implies by analogy that the carbonyls of Gly_{27} hydrogen-bond to the substrate NH functions of P_1 and P_2' as suggested by Blundell and Pearl (1989) and by Miller *et al.* (1989b). The analogy allowed the prediction that the conservatively varied and exposed Leu_{23}, Ala_{28}, Val_{82} and Ile_{84} form the large S_1/S_1' specificity subsites. Also by analogy with the aspartic proteinases where P_3 binds through its mainchain to the sidechain γ-O and the mainchain NH of Thr_{219} (pepsin numbering), this interaction should be mediated by Asp_{29} in HIV-proteinase. It was suggested that the mainchain NH is available for hydrogen bonding and the conservatively varied Asp sidechain might bind the NH function of P_3. However, the fact that there are two flaps in the HIV-proteinase dimer with a very different sequence when compared to the pepsins made it impossible to make useful predictions concerning their role in substrate and inhibitor binding.

This has now been resolved by X-ray analysis of several inhibitors complexed with the HIV-1 proteinase (Miller *et al.*, 1989; Fitzgerald *et al.*, 1990; Erikson *et al.*, 1990; Wilderspin *et al.*, 1990). These studies largely confirm the predicted interactions described above. However, both flaps bind the inhibitor directly through hydrogen bonds and indirectly through a water molecule. The two flaps occupy quite different positions from those occupied in the uncomplexed enzyme where they are involved in intermolecular interactions in the crystals. There is also a conformational change in the main body of the dimeric enzyme that corresponds to a hinge motion opening the active site cavity. This resembles that which is seen only in one lobe in the pepsin-like enzymes on binding an inhibitor (Šali *et al.*, 1989), in comparing pepsinogen and pepsin structures (Cooper *et al.*, 1990; Šali *et al.*, 1991) and in comparing different aspartic proteinases Šali *et al.*, 1991) as shown in Figure 5.

The hydrogen bonding at the active site residues is defined by X-ray analyses of statine-like complexes containing one hydroxy group or difluorostatone complexes that are

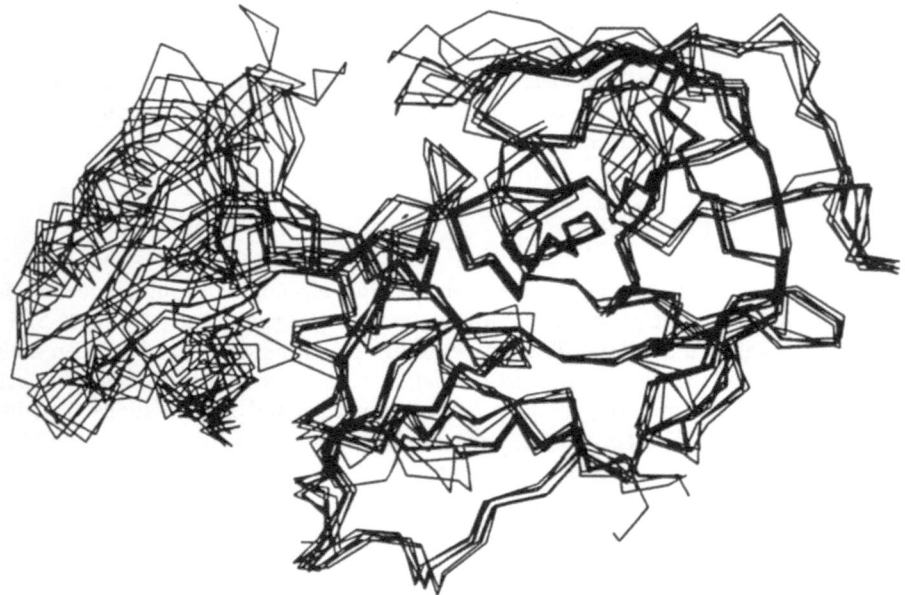

Figure 5. The superposition of aspartic proteinase structures demonstrating differing rigid group rotations of the small carboxy-terminal domain relative to the rest of the molecule.

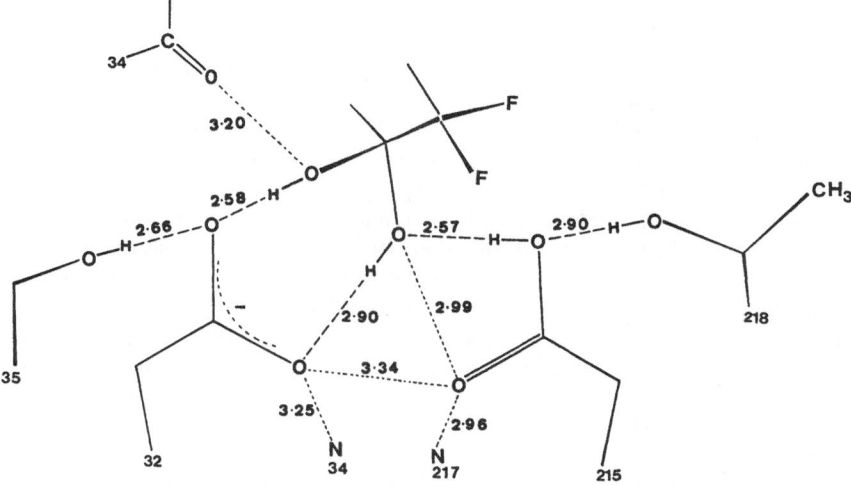

Figure 6. The stereochemistry and interactions of the tetrahedral hydrate formed from the complex of a diflourostatone complex with the enzyme endothiapepsin (Veerapandian *et al.*, 1992).

bound as a gem-diol and contain two hydroxy groups analogous to those of the putative transition state. The general arrangement is shown in Figure 6, which is derived from the X-ray analysis of a diflurostatone complex of endothiapepsin (Veerapandian *et al.*, 1992) and is discussed in another chapter of this volume (Hoover *et al.*, 1991).

Figure 6 shows that the *pro-(R)* hydroxyl oxygen occupies the same position as the corresponding hydroxyl oxygen in endothiapepsin complexes of inhibitors containing statine (e.g. Cooper *et al.*, 1989) or the hydroxyethylene dipeptide isostere (Blundell *et al.*, 1987; Veerapandian *et al.*, 1990). This oxygen atom of the inhibitor is within 3.4 Å of all four carboxyl oxygens, in an equivalent position to the oxygen atom of water in the native enzyme (Pearl & Blundell, 1984; Blundell *et al.*, 1990). The distances to the two inner carboxylate oxygens are consistent with hydrogen bonds but the shortest distance (2.57 Å) is to the outer oxygen of Asp_{215}. The second hydroxyl oxygen of the hydrate is located at an equivalent distance (2.58 Å) from the outer carboxyl oxygen of Asp_{32}. Because X-ray analyses of proteins at 2 Å resolution cannot locate hydrogens, the existence of hydrogen bonds in Figure 6 is inferred from those of the heavier atoms (Hoover *et al.*, 1991). In the preferred arrangement of hydrogen atoms a negatively charged Asp_{215} is stabilized by three hydrogen bonds, one from the statine-like hydroxyl of the hydrate, one from Thr_{218} γ-OH and one from Gly_{217} NH, whilst the inner oxygen of protonated Asp_{32} donates a hydrogen to the statine-like hydroxyl and the outer hydroxyl of the hydrate donates a further hydrogen bond to the outer oxygen of Asp_{32}.

CATALYTIC RESIDUES, TRANSITION STATE MODELS AND MECHANISM

In our preferred mechanism for proteolytic cleavage by an aspartic proteinase, the scissile-bond carbonyl is protonated by Asp_{32}, and concomitantly attacked by water polarized into a nucleophilic state by Asp_{215} (Suguna *et al.*, 1987b; Hoover *et al.*, 1991). The attack of the water on the polarized carbonyl may be additionally facilitated by enzyme induced distortion of the amide function to give a tetrahedral intermediate with a proton arrangement consistent with that of the bound hydrate as shown in Figure 6. Thus, the intermediate is

451

stabilized by hydrogen bonds from the gem-diol unit to a negatively charged Asp_{32} and from Asp_{215} to the statine-like hydroxyl oxygen.

A well organized and hydrogen-bonded water molecule can be seen in the electron density of inhibitor complexes close to this position (Veerapandian *et al.*, 1992). This could mediate the transfer of a proton from bulk solvent but only with some conformational change in the complex. Alternatively, nitrogen inversion and a rotation of about 60° about the $C(OH)_2$-N bond may allow a staggered disposition of the substrate P_1 and $P_{1'}$ alpha carbons to be maintained while the secondary amine is rendered accessible to protonation by Asp_{215}. This is attractive because a model of the intermediate state shows that the nitrogen of the amine must be within hydrogen bonding distance of the outer oxygen of Asp_{215}. This transition state gains much from the symmetry of the two aspartates in all aspartic proteinases. If amine protonation is mediated by Asp_{215}, this event would be conveniently accompanied by loss of stabilization of the tetrahedral species by removal of the hydrogen bond to the statine-like hydroxyl simultaneously as the amine is protonated. Formation and stabilization of the products would be achieved by a concomitant movement of the proton of the statine-like hydroxyl to the inner oxygen of Asp_{32}. This results in a hydrogen-bonded carboxylic acid dimer between the amino-terminal half of the product and Asp_{32}.

Thus the high resolution structure of the tetrahedral hydrate suggests a mechanism that involves the stabilization of a negatively charged aspartic acid carboxylate first at Asp_{215} and then in the intermediate at Asp_{32}. In each case the stabilization involves complete inaccessibility of the carboxylate from bulk solvent and the formation of three or four hydrogen bonds to the two carboxylate oxygens. This is similar to the environment of several other carboxylates that are conserved in pepsins (Cooper *et al.*, 1990). Site-directed mutagenesis experiments of Ser_{35} and Thr_{218} of chymosin (Mantafounis & Pitts, 1990) and pepsin (Tang, 1990) are consistent with the role of these sidechain hydroxyl hydrogen bonds in stabilizing negatively charged carboxylates that occur in the mechanism proposed. Their role is also consistent with the higher optimal pH for the catalysis by human renin with alanine at 218 and by the retroviral proteinases that have alanines at both equivalent positions.

The absence of any strong stabilization of a developing oxyanion at the peptide carbonyl argues against a possible analogy with serine proteinases, in which main chain nitrogen atoms are thought to stabilize the negative charge of an oxyanion (Blow, 1976). In the case of the aspartic proteinases the anion stabilized by the enzyme is the negatively charged carboxylate of Asp_{32}, and the intermediate is a gem diol.

ACKNOWLEDGEMENTS

We thank the Imperial Cancer Research Fund, the UK Agricultural and Food Research Council, the MRC AIDS Directed Programme and the Royal Society for support of this work. We thank our many colleagues at Birkbeck and at Pfizer for their help with this work.

REFERENCES

Bernstein, F. C., Koetzle, T. F., Williams, G. J. B., Meyer, E. F., Brice, M. D., Rodgers, J. R., Kennard, O. Shimanouchi, T. & Tasumi, M., 1977, *J. Mol. Biol.* **112**:535-542.

Blow, D. M., 1976 *Acc. Chem. Res.* **9**:145-152.

Blundell, T. & Pearl, L., 1989, *Nature* **237**:596.

Blundell, T. L., Sibanda, B. L. & Pearl, L., 1983, *Nature* **304**:273-275.

Blundell, T. L., Jenkins, J. A., Pearl, L. H. & Sewell, T. S., 1985, *in*: "Aspartic Proteinases and Their Inhibitors," Kostka, V., ed., pp. 151-161, Walter de Gruyter, Berlin.

Blundell, T. L., Cooper, J., Foundling, S. I., Jones, D. M. Atrash, B. & Szelke, M., 1987, *Biochemistry* 26:5585-5590.

Blundell, T. L., Carney, D., Gardner, S., Hayes, F., Hubbard, T., Overington, J. & Suttcliffe, M. J., 1988, *Eur J. Biochem.* 172:513-520.

Blundell, T. L., Jenkins, J. A., Sewell, B. T., Pearl, L. H., Cooper, J. B., Tickle, I. J., Wood, S. P. & Veerapandian, B., 1990, *J. Mol. Biol.* 211:919-941.

Bott, R., Subramanian, E. & Davies, D., 1982, *Biochemistry* 21:6956-6962.

Cooper, J. B., Foundling, S. I., Blundell, T. L., Boger, J., Jupp, R. A. & Kay, J., 1989, *Biochemistry* 28:8596-8603.

Cooper, J. B., Khan, G., Taylor, G., Tickle, I. J. & Blundell, T. L., 1990, *J. Mol. Biol.* 214:199-222.

Erickson, E., Neidhart, D. J., Vandrie, J., Kempf, D. J., Wang, X. C., Norbeck, D. W., Plattner, J. J., Rittenhouse, J. W., Turon, M., Widerburg, N., Kohlbrenner, W. E., Simmer, R., Helfrich, R., Paul, D. A. & Knigge, M., 1990, *Science* 249:527-533.

Fitzgerald, P. M. D., McKeever, B. M., Van Middlesworth, J. F., Springer, J. P., Heimbach, J. C., Leu, C.-T., Herber, W. K., Dixon, R. A. F. & Darke, P. L., 1990, *J. Mol. Biol.* 265:14205-14219.

Foundling, S. I., Cooper, S. I., Watson, J., Cleasby, F. E., Pearl, L. H., Sibanda, B. L., Hemmings, A., Wood, S. P., Blundell, T. L., Valler, M. J., Kay, J., Boger, J., Dunn, B. M., Leckie, B. J., Jones, D. M., Atrash, B., Hallett, A. & Szelke, M., 1987, *Nature* 327:349-352.

Hoover, D., Veerapandian, B., Cooper, J. B., Rosati, R., Dominy, B. W., Damon, D. & Blundell, T. L., 1991, this volume.

James, M. N. G., Sielecki, A. R., Salituro, F., Rich, D. H. & Hofmann, T., 1982, *Proc. Natl. Acad. Sci. U.S.A.* 79:137-6142.

James, M. N. G. & Sielecki, A., 1985, *Biochemistry* 24:3701-3713.

Johnson, M. S., Sutcliffe, M. J. & Blundell, T. L., 1990, *J. Mol. Evol.* 30:43-59.

Johnson, M. S., Šali, A. & Blundell, T. L., 1990, *Meth. Enzymology* 183:670-690.

Lapatto, R., Blundell, T., Hemmings, A., Overington, J., Wilderspin, A., Wood, S., Merson, J. R., Whittle, P. J., Danely, D. E., Geoghegan, K. F., Hawrylik, S. J., Lee, S. E., Scheld, K. G. & Hobart, P. M., 1989, *Nature* 342:299-302.

Mantafounis, D. & Pitts, J. E., 1990, *Prot. Eng.* 3:605-609.

Miller, M., Jaskólski, M., Rao, J. K. M., Leis, J. & Wlodawer, A., 1989a, *Nature* 337:576-579.

Miller, M., Schneider, J., Sathyanarayana, B. K., Toth, M. V., Marshall, G. R., Clawson, L., Selk, L., Kent, S. B. H. & Wlodawer, A., 1989b, *Science* 246:1149-1152.

Overington, J., Johnson, M., Šali, A. & Blundell, T. L., 1990, *Proc. Roy. Soc. (Lond) B* 241:132-145.

Pearl, L. H. & Blundell, T. L., 1984, *FEBS Lett.* 174:96-101.

Pearl, L. H. & Taylor, W. R., 1987, *Nature* 329:351-354.

Šali, A. & Blundell, T. L., 1990, *J. Mol. Biol.* 212:403-428.

Šali, A., Veerapandian, B., Cooper, J. B., Foundling, S. I., Hoover, D. J. & Blundell, T. L., 1989, *EMBO J.* 8:2179-2188.

Šali, A., Overington, J. P., Johnson, M. S. & Blundell, T. L., 1990, *Trends. Biochem. Sci.* 15:235-240.

Šali. A., Cooper, J. B., Hofmann, T., Veerapandian, B. & Blundell, T. L., 1991, The Proteins in press.

Suguna, K., Bott, R. R., Padlan, E. A., Subramanian, E., Sheriff, S., Cohen, G. E. & Davies, D. R., 1987a, *J. Mol. Biol.* 196:877-900.

Suguna, K., Padlan, E. A., Smith, C. W., Carlson, W. D. & Davies, D., 1987b, *Proc. Natl. Acad. Sci. U.S.A.* 84:7009-7013.

Tang, J., James, M., Sielecki, A., Jenkins, J. A. & Blundell, T. L., 1978, *Nature* 271:618-621.

Tang, J., 1990, unpublished results.

Veerapandian, B., Cooper, J. B., Šali, A., Blundell, T. L., Rosatti, R. L., Dominy, B. W., Damon, D. B., & Hoover, D., 1992, *Protein Science*, in press.

Wilderspin, A., Hemmings, H. & Whittle, P. J., 1990, unpublished results.

Wlodawer, A., Miller, M., Jaskólski, M., Sathyanarayana, B. K., Baldwin, E., Weber, I. T., Selk, L., Clawson, L., Schneider, J., Kent, S. B. H., 1989, *Science* 245:616-621.

Zhu, Z-Y., Šali, A. & Blundell, T. L., 1991, unpublished results.

THE THREE-DIMENSIONAL X-RAY CRYSTAL STRUCTURE OF HIV-1 PROTEASE

COMPLEXED WITH A HYDROXYETHYLENE INHIBITOR

Bradford J. Graves, Marcos H. Hatada, Julann K. Miller,
Mary C. Graves, Swapan Roy, Charles M. Cook, Antonin Kröhn,*
Joseph A. Martin,* and Noel A. Roberts*

Roche Research Center
Hoffmann-La Roche Inc.
Nutley, New Jersey   07110

*Roche Products Ltd.
Welwyn Garden City
Hertsfordshire AL7 3AY
England

INTRODUCTION

The aspartyl proteinase encoded within the genome of the type I human immunodeficiency virus (HIV-1 PR) is a valid and important target for the development of a therapeutic to treat HIV infections. Progress in this area has been rapid due to 1) the wealth of previous experience with other aspartyl proteinases and 2) the massive commitment by a large number of research groups worldwide. In this chapter we would like to discuss some of our efforts to develop a PR inhibitor by describing the structure of a complex between HIV-1 PR and a hydroxyethylene inhibitor. As a final note before continuing, we would like to acknowledge the significant contributions that Alex Wlodawer and his group have made to this field which have been important to the progress made not only by our group but by many others as well.

Based on work done a number of years ago, mainly on the development of renin inhibitors, several classes of compounds were identified which can be made into potent inhibitors of aspartyl proteinases. These include peptide-based inhibitors with reduced peptide bond,[1] hydroxyethylamine[2] and hydroxyethylene[3] linkages between the P_1 and P_1' residues. Since others have presented complex structures with reduced[4] and hydroxyethylamine[5] inhibitors, we thought it would be of interest to discuss the structure of a complex with a hydroxyethylene inhibitor, Ro 31-8588. The chemical structure of Ro 31-8588 is shown in Figure 1. The K_i for this inhibitor has been measured at 0.3 nM and thus Ro 31-8588 is a potent member from this class of compounds.

METHODS

Crystals of the PR:Ro 31-8588 complex are grown using the hanging drop vapor diffusion method with ammonium phosphate as the precipitant at pH 5.6. Plate-shaped crystals form which belong to the orthorhombic space group $P2_12_12$ with unit cell dimensions a = 58.92 Å, b = 87.05 Å, and c = 46.91 Å. This space group and associated unit cell constants have also been observed for crystals of the PR complex with acetyl-pepstatin reported by Fitzgerald *et al.*[6] which were grown using sodium chloride as the precipitant. The space group and cell constants are different, however, from the crystals reported by the Wlodawer group[4,5] (space group $P2_12_12_1$) and even further removed from the crystals reported by Erickson *et al.*[7] (space group $P6_1$). An interesting observation is that we have obtained both our orthorhombic form and the hexagonal form[7] under the same conditions using different inhibitors. Similarly, the hexagonal crystals with the symmetric inhibitor[7] grow under conditions used by the Wlodawer group[4] to grow their orthorhombic form. Thus, it appears the crystallization of PR:inhibitor complexes may be more dependent on the nature of the inhibitor than it is on other factors.

The orthorhombic crystals permitted the collection of diffraction data out to a resolution of 2.0 Å and the structure of the complex was solved by molecular replacement methods, as implemented in the MERLOT package,[8] using the structure of the native, uncomplexed HIV-1 PR.[9] The resulting model was refined and subsequent electron density maps showed clear density for the inhibitor. The model of the complex has been further refined using data to 2.3 Å resolution and the current R-factor is 18.9 % with no water molecules included. Subsequently, several water molecules have been located but they will be mentioned only as they contribute to the discussion.

RESULTS AND DISCUSSION

The perturbations to the protein structure caused by inhibitor binding (and presumably substrate binding) have been detailed by others.[4-7] These include the large shifts of the two "flap" regions and other correlated shifts in additional parts of the molecule. In

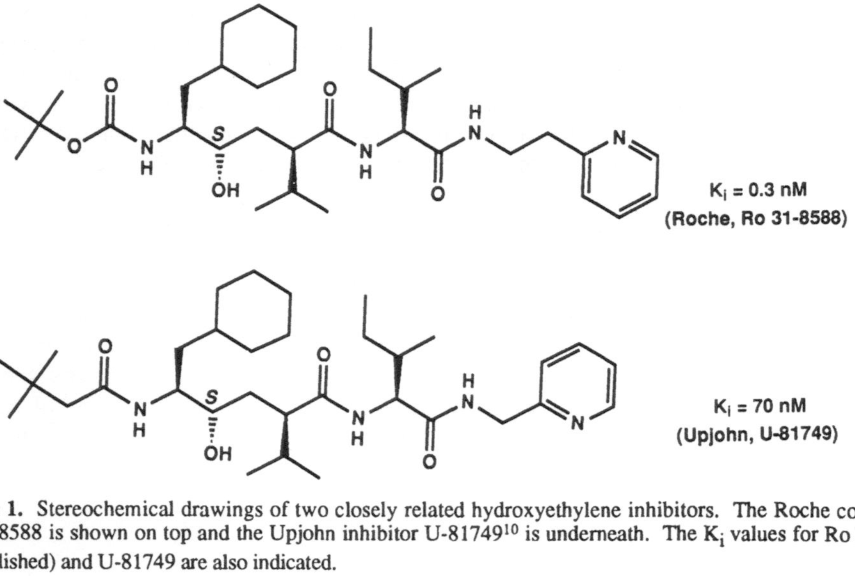

Figure 1. Stereochemical drawings of two closely related hydroxyethylene inhibitors. The Roche compound Ro 31-8588 is shown on top and the Upjohn inhibitor U-81749[10] is underneath. The K_i values for Ro 31-8588 (unpublished) and U-81749 are also indicated.

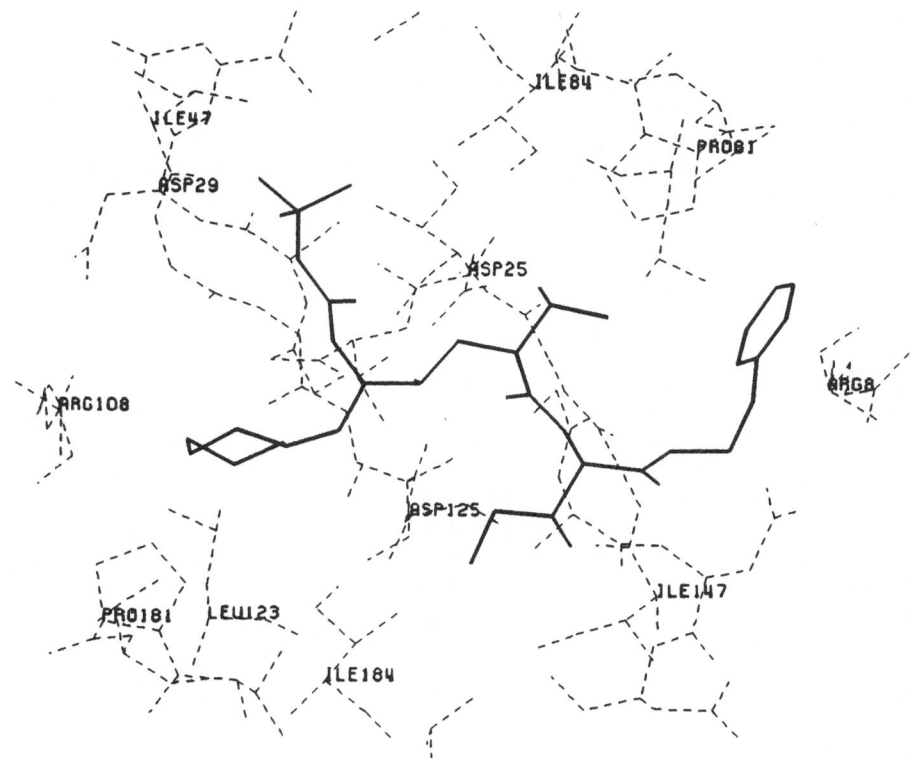

Figure 2. A plot of the HIV-1 PR:Ro 31-8588 complex. The PR protein is depicted by the dashed lines while the inhibitor is drawn with solid lines. All protein residues which have an atom that is within 5 Å of the inhibitor are included. The view is roughly looking down the pseudo two-fold axis and a few selected residues have been labelled for clarity. The numbering scheme for the residues of the protein is from 1-99 and 101-199 for the two monomers (thus Asp_{25} is two-fold related to Asp_{125}).

subsites from S_2 to S_3'. The inhibitor utilizes all of its potential hydrogen bond donors and acceptors to form an extensive hydrogen bond lattice. This includes the carbonyl oxygen atoms following the Boc and Val residues which accept hydrogen bonds from a single, well-ordered water molecule (not shown). The water molecule is positioned between the inhibitor and the leading strands of the two "flaps" and has been observed in all other HIV PR:inhibitor complexes.[4-7] The overall hydrogen bonding network for the inhibitor is shown spite of the differences in the structures of the inhibitors, it appears that the conformation of the PR in the bound state is very similar in most cases. For example, using all C_α atoms, r.m.s. deviations between the protein in the Ro 31-8588 complex and the complexes with MVT-101[4] and JG-365[5] were calculated to be on the order of 0.5 Å. Furthermore, the deviations are fairly uniform over the whole molecule so there is no single site where there is a significant difference. The same calculation comparing PR in the bound (Ro 31-8588) and unbound state[9] yields an r.m.s. deviation of 1.9 Å and even if the C_α atoms for the residues in the "flap" regions are eliminated the r.m.s. deviation is still 1.0 Å.

The prime interest in this or any other structure of a PR:inhibitor complex is the conformation of the inhibitor and the nature of its contacts with the protein. Figure 2 depicts the inhibitor and all protein atoms that are within 5 Å of the inhibitor. A few selected residues have been labelled in order to help orient the reader. The view in Figure 2 clearly shows the extended conformation that the inhibitor adopts as its sidechains occupy the protein schematically in Figure 3. In terms of group utilization on both the inhibitor and protein, this scheme is very similar to that observed in the complexes with other PR inhibitors.[4-7]

Given the similarity of the protein structures, and the protein-inhibitor contacts, one would expect that the various inhibitors would be bound in nearly the same conformation and this is exactly what is observed. Protein C_α atoms from the structures with MVT-101,[4] JG-365[5] and Ro 31-8588 were used to calculate a best molecular fit for the three complexes and the resulting relative orientation of the inhibitors is shown in Figure 4. The three inhibitors clearly adopt similar conformations and this occurs despite the fact that 1) one of the inhibitors (MVT-101) has a reduced peptide bond with no hydroxyl group to interact with the active site aspartates, 2) another (JG-365) has an additional atom in the linkage between the P_1 and $P_{1'}$ residues, and 3) Ro 31-8588 occupies one to two fewer subsites in the PR binding pocket.

Thus, what can be learned from this comparison where the three inhibitors (and probably most others that have been developed to date) bind to the PR protein in a similar manner and yet there is a difference of three orders of magnitude in their K_i values? The most obvious point is that the hydroxyl moiety which interacts with the active site aspartates is a critical component. Another point is that there is apparent flexibility in the inhibitor:PR interaction which allows for accommodation of the extra length between P_1 and P_1' in JG-365. As can be seen in Figure 4, the Pro group of JG-365 has been pushed further up into the S_1' site in order to make up for the longer linkage. Similarly, the truncated linkage in the symmetric inhibitor of Erickson *et al.*[7] appears to be tolerated (see Figure 5 of reference 7) by utilizing a more linear pathway between the C_α atoms of the P_1 and P_1' residues.

In spite of the apparent flexibility noted above, there is also some sensitivity to the PR:inhibitor interaction. This can be demonstrated by comparing Ro 31-8588 with the Upjohn inhibitor U-81749[10] which is very similar to Ro 31-8588 but whose K_i is two orders

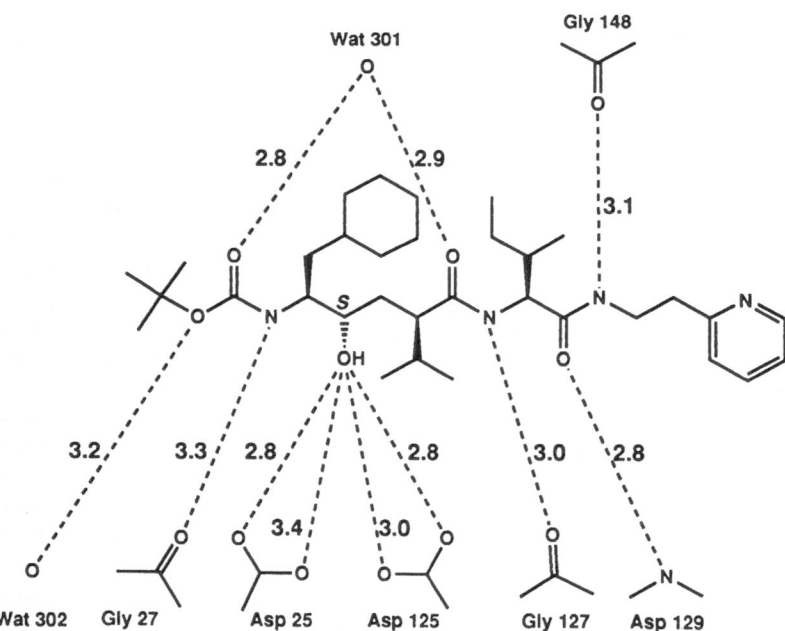

Figure 3. Schematic drawing of the hydrogen bonding pattern exhibited by Ro 31-8588. Heavy-atom contact distances are indicated as are the identities of the residues of the protein which make the hydrogen bond contacts. The numbering scheme is as in Figure 2.

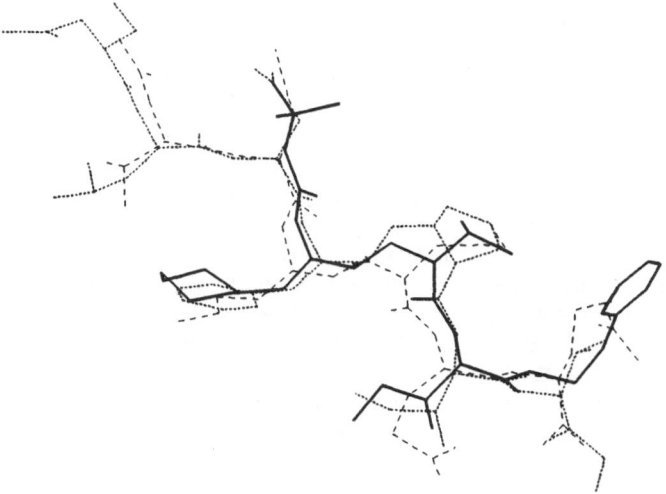

Figure 4. A plot of the overlap of the three inhibitors Ro 31-8588 (drawn in solid lines), MVT-101[4] (drawn in dotted lines), and JG-365[5] (drawn in dashed lines). The relative orientation of the inhibitors was determined by fitting the protein portions of the three complex structures as described in the text.

of magnitude greater (understandably some of this may be due to differences in assay conditions however unpublished reports verify the relative K_i values). In fact, there are only two changes in the structures of Ro 31-8588 and U-81749. As can be seen in Figure 1, the N-terminus of U-81749 has a methylene in place of the ether oxygen of the Boc group in Ro 31-8588 and at the C-terminus U-81749 has aminomethylpyridine instead of aminoethylpyridine. What difference does this make to the binding ability of the structures? Since there is no published structure with U-81749, we can not make a detailed structural comparison but what we do see in the structure with Ro 31-8588 are two interactions which U-81749 would be precluded from making. At the N-terminus, the oxygen atom of the Boc group does form a weak hydrogen bond to a water molecule (see Figure 3) which in turn forms a hydrogen bond to an arginine sidechain. At the other end, we can see from Figure 2 that the pyridine ring forms a stacking interaction with the nearby arginine residue (Arg_8). A rough estimate of the value of this second interaction was obtained by performing energy calculations at different planar separations. These calculations indicate that the stacking interaction observed here is worth approximately 2.5 kcal/mole which is equivalent to about half of what a good hydrogen bond is calculated to be. The extent to which these two interactions account for the differences in K_i for Ro 31-8588 and U-81749 is uncertain but the comparison illustrates the extraordinary sensitivity that the system can exhibit.

In conclusion, we have shown that the HIV PR protein adopts a very similar structure regardless of what type of inhibitor is bound. Likewise, the various types of inhibitors bind to the protein with similar conformations. The inhibitor Ro 31-8588 is the shortest compound described here and yet still is a very potent inhibitor. The structure of the complex with Ro 31-8588 also serves to demonstrate that very small differences in inhibitor structure can result in large differences in K_i values.

REFERENCES

1. M. Szelke, B. J. Leckie, A. Hallett, D. M. Jones, J. Sueiras, B. Atrash and A. F. Lever, Potent New Inhibitors of Human Renin, *Nature* **299**:555 (1982).

2. J. G. Dann, D. K. Stammers, C. J. Harris, R. J. Arrowsmith, D. E. Davies, G. W. Hardy and J. A. Morton, Human Renin: A New Class of Inhibitors, *Biochem. Biophys. Res. Commun.* **134**:71 (1986).

3. M. Szelke, D. M. Jones, B. Atrash, A. Hallett and B. J. Leckie, Novel Transition-State Analogue Inhibitors of Renin, *in*: "Peptides: Structure and Function, Proceedings of the 8th American Symposium," V. J. Hruby and D. H. Rich, Eds., Pierce Chemical, Rockford, IL, (1983).

4. M. Miller, J. Schneider, B. K. Sathyanarayana, M. V. Toth, G. R. Marshall, L. Clawson, L. Selk, S. B. H. Kent and A. Wlodawer, Structure of Complex of Synthetic HIV-1 Protease with a Substrate-Based Inhibitor at 2.3 Å Resolution, *Science* **246**:1149 (1989).

5. A. L. Swain, M. M. Miller, J. Green, D. H. Rich, J. Schneider, S. B. H. Kent and A. Wlodawer, X-ray Crystallographic Structure of a Complex Between a Synthetic Protease of Human Immunodeficiency Virus 1 and a Subtrate-based Hydroxyethylamine Inhibitor, *Proc. Nat. Acad. Sci., U.S.A.* **87**:8805 (1990).

6. P. M. D. Fitzgerald, B. M. McKeever, J. F. VanMiddlesworth, J. P. Springer, J. C. Heimbach, C.-T. Leu, W. K. Herber, R. A. F. Dixon and P. L. Darke, Crystallographic Analysis of a Complex Between HIV-1 Protease and Acetyl-Pepstatin at 2.0 Å Resolution, *J. Biol. Chem.* **265**:14209 (1990).

7. J. Erickson, D. J. Neidhart, J. VanDrie, D. J. Kempf, X. C. Wang, D. W. Norbeck, J. J. Plattner, J. W. Rittenhouse, M. Turon, N. Wideburg, W. E. Kohlbrenner, R. Simmer, R. Helfrich, D. A. Paul and M. Knigge, Design, Activity and 2.8 Å Crystal Structure of a C2 Symmetric Inhibitor Complexed to HIV-1 Protease, *Science* **249**: 527 (1990).

8. P. M. D. Fitzgerald, MERLOT, An Integrated Package of Computer Programs for the Determination of Crystal Structures by Molecular Replacement, *J. Appl. Crystallogr.* **21**:273 (1988).

9. A. Wlodawer, M. Miller, M. Jaskólski, B. K. Sathyanarayana, E. Baldwin, I. T. Weber, L. M. Selk, L. Clawson, J. Schneider and S. B. H. Kent, Conserved Folding in Retroviral Proteases: Crystal Structure of a Synthetic HIV-1 Protease, *Science* **245**:616 (1989).

10. T. J. McQuade, A. G. Tomasselli, L. Liu, V. Karacostas, B. Moss, T. K. Sawyer, R. L. Heinrikson and W. G. Tarpley, A Synthetic HIV-1 Protease Inhibitor with Antiviral Activity Arrests HIV-like Particle Maturation, *Science* **247**:454 (1990).

SUBSTRATE CLEAVAGE BY HIV-1 PROTEINASE

R. A. Jupp,[1] A. D. Richards,[1] L. H. Phylip,[1] J. Kay,[1] J. Konvalinka,[2]
P. Strop,[2] V. Kostka,[2] P. E. Scarborough,[3] W. G. Farmerie[3]
and B. M. Dunn[3]

[1]Department of Biochemistry
 University of Wales College ofCardiff
 P.O. Box 903
 Cardiff CF1 1ST, Wales, United Kingdom

[2]Institute of Organic Chemistry & Biochemistry
 Czechoslovak Academy of Science
 166 10 Prague 6, C.S.F.R.

[3]Department of Biochemistry and Molecular Biology
 University of Florida, Box J-245
 Gainesville, Florida   32610

INTRODUCTION

The *pol* open reading frame (ORF) of the human immunodeficiency virus (HIV-1) encodes three distinct enzymes, one of which is an aspartic proteinase (PR). Point mutations introduced at the active site of this enzyme result in an inability to form infectious virions[1] thus emphasizing the paramount importance of PR in viral maturation . Proteinase has thus become a strategic target for the development of compounds that might have therapeutic value in the treatment of AIDS.[2] By preference, such compounds should be specific for the target, HIV-PR and should not have side effects by interacting with similar enzymes present in the human body, such as have been observed for example with anti-viral agents (e.g. AZT that act against reverse transcriptase RT). The design of specific PR inhibitors is facilitated considerably by a detailed understanding of the molecular topography of the PR active site. To this end, several series of synthetic chromogenic substrates have been used to unravel the subsite preferences of this important enzyme. In some cases, the requirement for certain residues to be present in particular locations in native protein substrates in order to ensure effective hydrolysis by PR has also been examined.

Peptide 1 (Table 1) has a sequence based on the -Leu * Ala- cleavage site at one of the CA/NC junctions in the HIV polyprotein. It contains a 4-NO$_2$-phenylalanine (Nph)

Table 1 Kinetic parameters (K_m, k_{cat}) for the hydrolysis by HIV-1 PR of several chromogenic peptides containing variations in the P_1 position

	Peptide Number and Sequence	K_m, (μM)	k_{cat} (s^{-1})
1	Lys-Ala-Arg-Val-Leu * Nph-Glu-Ala-Met	22	20
2	Lys-Ala-Arg-Val-Tyr * Nph-Glu-Ala-Nle-NH$_2$	7	20
3	Lys-Ala-Arg-Val-Met * Nph-Glu-Ala-Nle-NH$_2$	15	20
4	Lys-Ala-Arg-Val-Nle * Nph-Glu-Ala-Nle-NH$_2$	15	45
5	Lys-Ala-Arg-Val- Ile * Nph-Glu-Ala-Nle-NH$_2$		~0

All reactions were carried out at 37°C in 100 mM sodium acetate buffer, pH 4.7 containing 5 mM mercaptoethanol and 4 mM EDTA plus NaCl to give a final ionic strength of 1 M. The progress of each reaction was monitored by observing the decrease in A_{300}. The precision of each individual estimation was in the range of 5-15 %. Values of k_{cat} were derived from the equation $V_{max} = k_{cat}$ [E]. The concentration of HIV-1 PR was determined by active site titration with the tight-binding inhibitor, quinoline-2-carbonyl-Asn-Phe-ψ[CHOH-CH$_2$N]-decahydro-isoquinoline-carbonyl-NH-t-butyl.[2]

reporter group in the P_1' position in place of the naturally occurring alanine residue. The use of this chromophore in substrates of mammalian and fungal aspartic proteinases has been well documented.[3,4] Thus, the hydrolysis of peptide 1 by PR could be readily followed spectrophotometrically and kinetic parameters (K_m, k_{cat}) determined under a variety of conditions.[5] The data obtained (Table 1) indicated that this peptide was a good substrate for PR with a K_m value lower than any reported previously for a synthetic substrate interacting with PR and a k_{cat} value comparable to those obtained previously for substrate hydrolysis by archetypal aspartic proteinases. Further peptides were then designed based upon this sequence but with systematic variation of the residue occupying the P_1 position (peptides 2, 3 & 4; Table 1). These were also modified slightly to replace the methionine residue in the P_4' position with a norleucine since the former is prone to oxidation.

Kinetic parameters for the hydrolysis of these substrates are also listed in Table 1. Replacement of Leu with its straight chain isomer, norleucine (Nle; peptide 4), with the equivalent Met residue (peptide 3) or with a bulkier tyrosine residue (peptide 2) in the P_1 position produced relatively small effects in either K_m or k_{cat}. By contrast, replacement with the branched isomer, isoleucine (peptide 5) resulted in a peptide that was essentially resistant to hydrolysis. These data suggest that, whereas HIV proteinase can accommodate a number of substituents in the P_1 position,[5,6] side chains branched at the ß-carbon atom are sterically unacceptable in the S_1 subsite.

The substantial decrease observed in k_{cat} (Table 1) with progressive branching of the P_1 side chain in the Nle, Leu, Ile series (peptides 4, 1, 5 ; Table 1) was further examined in a naturally occurring protein substrate for PR. The RT:IN cleavage junction in the HIV pol polyprotein is contributed by a -Leu*Phe- bond. This was readily cleaved (Figure 1) upon expression in *E. coli* of a plasmid containing a *Bgl* II-*Nde* I fragment of the *pol* ORF from the HTLV-IIIB provirus[7] which generated a PR-RT-IN polyprotein. Cleavage *in cis* by PR was monitored using (separate) polyclonal antibodies specific for IN and RT respectively. Hydrolysis of the -Leu*Phe- bond generated authentic IN with the predicted MR of 32 kDa (Figure 1, Panel b). Cleavage was also observed to have taken place at the junction between PR and RT and at the internal (-Phe*Tyr-) junction within RT to generate p51 and full length (p66) RT respectively (Figure 1, Panel a). It is thus apparent that RT is a heterodimer composed of two distinct subunits.

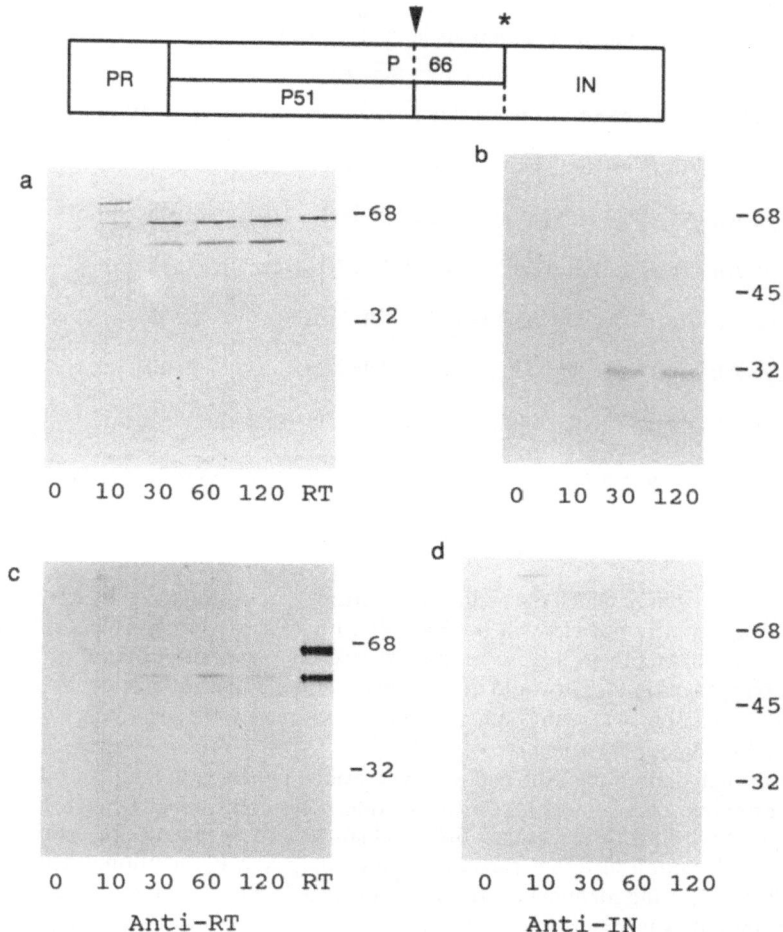

Figure 1. Processing of full length pol polyprotein with the RT:IN junction (*) as -Leu*Phe- (Upper Panels) or -Ile*Phe- (Lower Panels) bonds respectively. *E. coli* harboring the respective plasmids were induced by IPTG and aliquots (1 ml) were removed at the indicated times (between 0 and 120 min). Cell pellets were resuspended and fractionated by SDS-PAGE. After transfer to nitrocellulose membranes, immunodetection was carried out using antisera to RT (Panels a, c) or IN (Panels b, d) respectively. The ▼ symbol indicates the internal site within RT that is also cleaved by PR in one subunit to generate the p51/p66 heterodimeric RT. Markers of mol. wt. 68, 45 and 32 kDa migrated as indicated. The lanes identified as RT (in panels a and c) contain purified authentic RT either as p66/p66 or p66/p51 dimers, respectively.

Upon mutation of the leucine residue in the P_1 position of the RT:IN junction to an isoleucine, thus generating an -Ile*Phe- bond, a considerably different pattern of hydrolysis *in cis* by PR was observed (Figure 1, Lower). After immunostaining with a polyclonal antiserum to IN, no mature IN was evident (Panel d). Instead a protein with higher molecular weight of approximately 47 kDa was observed. After immunostaining with polyclonal antisera to RT, a single band was observed on Western blot co-migrating with the purified p51 marker (Figure 1, Panel c). The absence of a band corresponding to the p66 marker indicated that authentic p66 was not being produced from this mutant. The 15 kDa RNase H domain of RT would still appear to be attached to the N-terminus of IN. Thus, the

Table 2 Kinetic parameters (K_m, k_{cat}) for the hydrolysis by HIV-1 PR of several chromogenic peptides containing variations in the P_2 position

	Peptide Number and Sequence	K_m, (μM)	k_{cat}, (s^{-1})
4	Lys-Ala-Arg-Val-Nle * Nph-Glu-Ala-Nle-NH$_2$	35	32
6	Lys-Ala-Arg-Ile-Nle * Nph-Glu-Ala-Nle-NH$_2$	45	35
7	Lys-Ala-Arg-Leu-Nle * Nph-Glu-Ala-Nle-NH$_2$	25	6
8	Lys-Ala-Arg-Ala-Nle * Nph-Glu-Ala-Nle-NH$_2$	45	19
9	Lys-Ala-Arg-Asn-Nle * Nph-Glu-Ala-Nle-NH$_2$	40	0.7
10	Lys-Ala-Arg-Gly-Nle * Nph-Glu-Ala-Nle-NH$_2$		0

The assay conditions used were the same as those given in the legend to Table 1, except that the final ionic strength was 0.3 M.

mutant RT:IN junction in which the naturally-occurring Leu was replaced by a ß-branched Ile residue was apparently not cleaved by PR. Yet, under these circumstances, both of the adjacent p51/p66 bonds in RT were hydrolyzed in *both* subunits, thus generating homodimeric (p51/p51) RT. It would thus appear that the effects of side chain branching in the P_1 residue first observed with synthetic peptide substrates were accurate reflections of the processing of viral pol polyprotein.

With such definition of the nature of the residue preferred in the P_1 position in order to ensure effective cleavage of a scissile peptide bond contributed by two hydrophobic residues, the active site of PR was then further characterized by means of a systematic series of peptides each with a different substituent in the adjacent P_2 position. The data were obtained (Table 2) using identical conditions to those of Table 1, with the exception that the final ionic strength was lowered to 0.3 M in order to be more compatible with physiological values. All of the peptides examined exhibited relatively minor variations in the K_m values measured with PR. In contrast, substantial differences were detected in the k_{cat} parameter.[8] For example, ß-branched residues valine or isoleucine in the P_2 position provided substrates (peptides 4 & 6) that were hydrolyzed with k_{cat} values approximately 5-6 fold higher than that measured for the leucine containing substrate (Peptide 7; Table 2). Peptide 8 demonstrates that an alanine residue is acceptable in the P_2 position although the k_{cat} value obtained was approximately 2-fold lower than those for the isoleucine/valine containing peptides. Thus, the highest values for k_{cat} were obtained with Val, Ile or Ala in the P_2 position. The "longer" leucine side chain does not seem to be accommodated as readily when substrate interacts with the PR active site. The presence of an asparagine residue in this position resulted in a dramatic reduction of the k_{cat} value (peptide 9; Table 2) and insertion of a glycine residue (peptide 10) into the P_2 position effectively abolished hydrolysis by PR.

Once again these effects were investigated further by molecular genetic analysis of protein substrates derived from the *pol* ORF. The p51/p66 cleavage junction within RT is contributed by two hydrophobic residues (-Phe*Tyr-) analogous to those (-Nle*Nph-) in the synthetic peptides. Upon expression of a plasmid (encoding the full length PR-RT segment of the pol polyprotein) in *E. coli*, cleavage *in cis* readily generated heterodimeric p51/p66 RT (not shown). Replacement of the wild-type Thr residue that is situated in the P_2 position with valine,[9] thus creating a -Val-Phe*Tyr- junction, resulted in rapid hydrolysis to generate heterodimeric p51/p66 RT (Figure 2, Lanes 1,2 &3). Since the data in Table 2 had indicated that Val was an optimal residue for effective catalysis, this result was not unexpected. By

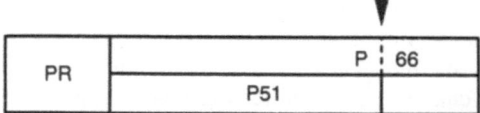

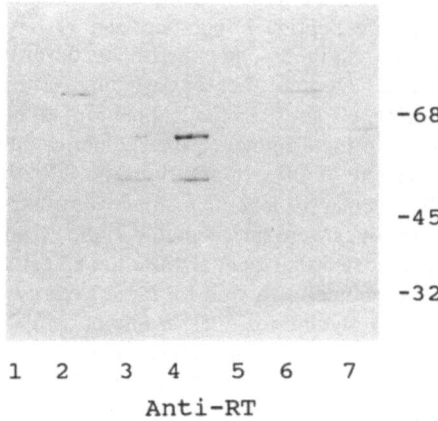

Figure 2. Processing of a PR-RT polypeptide with the internal p51/p66 junction within RT (▼) as -Val-Phe*Tyr- or -Gly-Phe*Tyr- bonds respectively. For details of the expression in *E.coli*, see Figure 1 legend. Immunostaining was with an antiserum to RT. Lanes 1, 2 and 3 and 5, 6 and 7 are samples taken 0, 10 and 30 min after IPTG induction in the -Val- and -Gly- containing mutants, respectively. Lane 4 is purified marker RT consisting of stoichiometrically equivalent amounts of p51 and p66 subunits. Molecular. weight. markers (68, 45 and 32 kDa) migrated as indicated.

contrast, insertion of a glycine residue into the P_2 position (i.e. creating the sequence -Gly-Phe*Tyr-) resulted in a slower processing of the precursor polyprotein[9] and generation of only p66 subunits under the conditions described (Figure 2, Lanes 5, 6 & 7). Thus, once again, the conclusion derived earlier from studies with synthetic peptides (Table 2) appears to reflect the situation prevailing in polyprotein processing.

The effect of various substitutions in the P_3 position was examined in turn by means of another series of (slightly different) peptides (Table 3). The data derived[6] indicate that,

Table 3. Kinetic parameters (K_m, k_{cat}) for the hydrolysis by HIV-1 PR of several chromogenic peptides containing variations in the P_3 position

	Peptide Number and Sequence	K_m, (μM)	k_{cat}, (s^{-1})
11	Ala-Thr-His-Arg-Val-Tyr * Nph-Val-Arg-Lys-Ala	15	10
12	Ala-Thr-His-Asp-Val-Tyr * Nph-Val-Arg-Lys-Ala	16	10
13	Ala-Thr-His-Glu-Val-Tyr * Nph-Val-Arg-Lys-Ala	30	15
14	Ala-Thr-His-Gln-Val-Tyr * Nph-Val-Arg-Lys-Ala	15	14
15	Ala-Thr-His-Tyr-Val-Tyr * Nph-Val-Arg-Lys-Ala	6	7
16	Ala-Thr-His-Pro-Val-Tyr * Nph-Val-Arg-Lys-Ala		0

The assay conditions were the same as those given in the legend to Table 1.

with such substrates centered upon a -Hydrophobic*Hydrophobic- scissile peptide bond, PR can accommodate a wide variety of residues in this flanking position. With the exception of a proline residue (peptide 16), the nature of the side chain of the other variants, whether (positively or negatively) charged, hydrophilic or aromatic, appeared to have little influence since the k_{cat} values measured were in the range of 7-15 sec^{-1} and K_m values were less than 30 μM.[6,8]

HIV-proteinase is somewhat unique in being able to cleave substrates containing two very different types of scissile peptide bond: between 1) -Hydrophobic*Hydrophobic- residues or 2) -Aromatic*Pro- residues. The experiments described above begin to unravel the stringent requirements that have to be met for particular residues to be present in order to achieve efficient cleavage at type 1 sites. In the P_1 position, all of the hydrophobic residues examined with the exception of the ß-branched isoleucine residue were found to be readily accommodated in the active site of PR. In addition, the RT:IN junction was found to be resistant to cleavage by PR when an isoleucine residue was present at the P_1 position of this site. Whereas ß-branched residues were unacceptable in the P_1 position, detailed analysis of the P_2 requirements of PR revealed that Ile/Val were the preferred residues in this position with Leu and Ala being accomodated less well for rapid hydrolysis of the peptide substrate. The presence of asparagine or glycine residues in this position was found to be relatively unacceptable compared to peptides containing isoleucine or valine. The nature of the residue in the P_2 position adjacent to the -Hydrophobic*Hydrophobic- scissile peptide bond thus has considerable influence on the efficiency of cleavage of that bond. The results obtained with peptides containing substitutions in the P_3 position indicate that the nature of the residue occupying this position is much less critical in substrates containing the Hydrophobic*Hydrophobic type of cleavage site. A different set of requirements may be operational for effective hydrolysis of -Aromatic*Pro- type junctions.[10]

It is clear that subtle changes in residues flanking the scissile peptide bond have considerable influence on the way in which the scissile bond is presented to the catalytic apparatus for cleavage since the dominant effect was always observed in the k_{cat} parameter with essentially trivial effects being manifested in K_m. Crystallographic analyses[11,12] have demonstrated that occupancy of the active site cylinder of PR is associated with significant movements of the two ß-hairpin loops or "flaps" which overhang the active site in the empty enzyme. This may involve precise positioning of the backbone and individual side chains of the substrate in order to achieve optimal alignment with the catalytic residues through an essential distortion of the scissile peptide bond, as proposed previously for effective hydrolysis by archetypal aspartic proteinases.[13] Considerable insight into the topology of the active site of PR has thus been forthcoming from these investigations and this can, in turn, be applied to the design of selective inhibitors of PR for anti-viral therapy.

ACKNOWLEDGEMENTS

It is a pleasure to acknowledge the substantial contributions made to the development of this work by our many colleagues in Roche/Hoffmann La Roche including Anne Broadhurst, Mary Graves, John Mills, Stuart LeGrice, Alison Ritchie and Noel Roberts.

REFERENCES

1. N. E. Kohl, E. A. Emini, W. A. Schleif, L. J. Davis, J. C. Heimbach, R. A. F. Dixon, E. M. Scolnick and I. S. Sigal, *Proc. Natl. Acad. Sci. U.S.A.* **85**:4686-4690 (1988).
2. N. A. Roberts, J. A. Martin, D. Kinchington, A. V. Broadhurst, J. C. Craig, I. B. Duncan, S. A. Galpin, B. K. Handa, J. Kay, A. Krohn, R. W. Lambert, J. H. Merrett, J. S. Mills, K. E. B. Parkes, S. Redshaw, A. J. Ritchie, D. L. Taylor, G. J. Thomas and P. J. Machin, *Science* **248**:358-361 (1990).

3. B. M. Dunn, M. Jimenez, B. F. Parten, M. J. Valler, C. E. Rolph and J. Kay, *Biochem. J.* **237**:899-906 (1986).

4. B. M. Dunn, M. J. Valler, C. E. Rolph, S. I. Foundling, M. Jimenez and J. Kay, *Biochim. Biophys. Acta* **913**:122-130 (1987).

5. A. D. Richards, L. H. Phylip, W. G. Farmerie, P. E. Scarborough, A. Alvarez, B. M. Dunn, Ph.-H. Hirel, J. Konvalinka, P. Strop, L. Pavlickova, V. Kostka and J. Kay, *J. Biol. Chem.* **265**:7733-7736 (1990).

6. J. Konvalinka, P. Strop, J. Velek, V. Cerna, V. Kostka, L. H. Phylip, A. D. Richards, B. M. Dunn and J. Kay, *FEBS Lett.* **268**:35-38 (1990).

7. L. Ratner, W. A. Haseltine, R. Patarca, K. J. Livak, B. Starcich, S. F. Josephs, E. R. Doran, J. A. Rafalski, E. A. Whitehorn, K. Baumeister, L. Ivanoff, S. R. Petteway Jr., M. L. Pearson, J. A. Lautenberger, T. S. Papas, J. Ghrayeb, N. T. Chang, R. C. Gallo and F. C. Wong-Staal, *Nature* **313**:277-284 (1985).

8. L. H. Phylip, A. D. Richards, J. Kay, J. Konvalinka, P. Strop, V. Kostka, A. J. Ritchie, A. V. Broadhurst, W. G. Farmerie, P. E. Scarborough and B. M. Dunn, *Biochem. Biophys. Res. Commun.* **171**:439-444 (1990).

9. R. A. Jupp, L. H. Phylip. J. S. Mills, S. F. J. LeGrice and J. Kay, FEBS Letts. **283**:180-184 (1991).

10. L. H. Phylip, J. T. Griffiths, J. Konvalinka, P. Strop, A. M. Gustchina, A. Wlodawer, R. J. Davenport, B. M. Dunn and J. Kay, (1991) Manuscript in submission.

11. M. Miller, B. Y. Sathyanaryana, A. Wlodawer, M. V. Toth, G. R. Marshall, L. Clawson, L. Selk, J. Schneider and S. B. H. Kent, *Science* **246**:1149-1152 (1989).

12. P. M. D. Fitzgerald, B. M. McKeever, J. F. van Middlesworth, J. P. Springer, J. L. Heimbach, C-T. Leu, W. K. Herber, R. A. F. Dixon and P. L. Darke, *J. Biol. Chem.* **256**:1420 -14219 (1990).

13. L. H. Pearl, *FEBS Lett.* **214**8-12 (1987).

THE EVALUATION OF NON-VIRAL SUBSTRATES OF THE HIV PROTEASE

AS LEADS IN THE DESIGN OF INHIBITORS FOR AIDS THERAPY

Alfredo G. Tomasselli, John O. Hui, Tomi K. Sawyer,
Suvit Thaisrivongs, Jackson B. Hester and Robert L. Heinrikson

Upjohn Laboratories
The Upjohn Company
Kalamazoo, Michigan   49001

INTRODUCTION

The aspartyl protease encoded within the *pol* gene of human immunodeficiency virus (HIV) provides a target for therapeutic intervention in the treatment of acquired immunodeficiency syndrome (AIDS). This enzyme is indispensable for processing the viral gag and gag/pol polyproteins which takes place during the final maturation step of the viral life cycle. Blocking of protease action by inhibitors[1-4] or by mutagenesis[5] results in production of immature, non-infectious viral particles. Accordingly, the past few years have witnessed a world-wide effort to discover inhibitors of the HIV protease with antiviral activity. By and large, the design of such inhibitors has been based upon the specificity of the HIV protease for its natural polyprotein substrates. This data base of information is limited, and we have sought to expand it by evaluation of non-viral proteins as substrates of the enzyme.[6-9] The present chapter describes how information thus derived can be applied to the design of protease inhibitors with potent antiviral activity.

VIRAL POLYPROTEIN CLEAVAGE SITES AS CLUES TO INHIBITOR DESIGN

HIV protease splits 7-8 bonds in liberating the structural proteins and the enzymes,[10] including itself, locked within the polyprotein format characteristic of retroviruses. Henderson *et al.*[11] have sorted these various regions into 3 classes based upon particular features of the sequences from P_4 to P_4'. Perhaps the most interesting are Class 1 substrates because they have a P_1' Pro, a residue abhorred by most proteases. In Table 1 is shown an octapeptide derived from the junction between the gag p17 and p24 proteins in which the protease cleaves the Tyr-Pro bond. This peptide, or analogs thereof, has been the basis for assay of the HIV protease in a number of laboratories.[10,12-14] In order to test the importance of the Tyr-Pro residues at P_1 and P_1', we varied residues surrounding the scissile bond according to sequences observed in other retroviral polyproteins. Table 1 shows that these

Table 1. Effect of modification of the pattern of amno acids surrounding the scissile bond P_1-P_1' of peptide substrates on the activity of HIV-1 and HIV-2 proteases. K_m is given in mM and V_{max} is given in units of µmol/min/mg.

	HIV-1 Protease			HIV-2 Protease		
Compound	K_m	V_{max}	$V_{max}/$ K_m	K_m	V_{max}	$V_{max}/$ K_m
H-Val-Ser-Gln-Asn-Tyr * Pro-Ile-Val-OH[a]	2.0	4.9	2.45	2.0	2.5	1.25
H-Thr-Phe-Gln-Ala-Tyr * Pro-Leu-Arg-Glu-Ala-OH[b]		0			0	
H-Val-Ser-Gln-Asn-Phe * Pro-Ile-Val-OH[a]	6.7	3.9	0.58	5.0	2.1	0.42
H-Tyr-Val-Ser-Gln-Asn-Phe * Pro-Ile-Val-Gln-Asn-Arg-OH[a]	1.9	3.9	2.05	1.9	4.3	2.26
H-Lys-Pro-Arg-Asn-Phe * Pro-Val-Ala-OH[c]		0			0	

[a]HIV-1 gag fragment (or analog); [b]AMV pol fragment; [c]HIV-2 gag fragment (or analog).

flanking sequences are also important in defining cleavage sites for both HIV-1 and HIV-2 proteases.

The corollary experiment is defined in Table 2; here certain changes in P_1-P_1' are tolerated and others not when flanking residues are kept constant. Thus, these unique retroviral aspartyl protease dimers[15-16] display a complex specificity for their substrates which must involve a multi-site recognition and binding process encompassing both sequence and structural determinants.

AN EXPANDED DATA BASE FOR UNDERSTANDING PROTEASE SPECIFICITY: EVALUATION OF NON-VIRAL PROTEIN SUBSTRATES

We have sought to add to the arsenal of sequences cleaved by HIV-1 and HIV-2 proteases by testing a variety of proteins as substrates for these enzymes.

Table 2. Effect of variation of the P_1-P_1' amino acids of peptide substrates on the activity of HIV-1 and HIV-2 proteases.

	HIV-1 Protease			HIV-2 Protease		
Compound[a]	K_m	V_{max}	$V_{max}/$ K_m	K_m	V_{max}	$V_{max}/$ K_m
H-Val-Ser-Gln-Asn-Tyr * Pro-Ile-Val-OH	2.0	4.9	2.45	2.0	2.5	1.25
H-Val-Ser-Gln-Asn-Phe * Pro-Ile-Val-OH	6.7	3.9	0.58	5.0	2.1	0.42
H-Val-Ser-Gln-Asn-Cha * Pro-Ile-Val-OH	20	3.9	0.20	10	0.24	0.02
H-Val-Ser-Gln-Asn-Leu * Pro-Ile-Val-OH	10	0.9	0.09	n.d.	0.01	n.d.
H-Val-Ser-Gln-Asn-Tyr * Val-Ile-Val-OH	0.6	2.7	4.50	0.6	0.08	0.13
H-Val-Ser-Gln-Asn-Ala * Ala-Ile-Val-OH		0			0	
H-Val-Ser-Gln-Asn-Leu * Val-Ile-Val-Arg-OH		0			0	

[a]All based on the HIV-1 gag region linking p17 to p24.

```
                 Succ                    CAM        CAM
                  |                       |          |
  K─── AAK-FER ──── NQM-MKS════ SKY-PNC──── IVA-CEG──── V
  1      ↑    10         32        95         ↑    112  124
```

Figure 1. Schematic representation of derivatized RNase A indicating sites of cleavage by HIV-1 and HIV-2 proteases.

Ribonuclease A

Neither ribonuclease A (RNase) nor its performic acid oxidized derivative is a substrate for HIV protease.[7] As shown in Figure 1, however, the carboxamidomethyl (CAM) derivative produced by reduction and alkylation with iodoacetamide is an excellent substrate; both proteases cleave at an Ala-CAMCys bond. Furthermore, when the CAM-RNase is succinylated, a second site of cleavage for both enzymes is generated at Succ-Lys-Phe (Figure 1). These two bonds are the first to be described in which derivatized amino acids occur in the cleavage position. Moreover, release of the carboxyl terminal peptide from CAM-RNase provides a convenient means for high volume assays of the protease;[7] this segment can be radiolabeled at His or Tyr residues and, of course, RNase is an inexpensive and readily available substrate.

Insulin B Chain

Pursuing denatured, or non-structured HIV protease substrates, we tested oxidized insulin B chain. As shown in Figure.2, both HIV-1 and HIV-2 proteases hydrolyze this peptide at sites indicated. Here we see a more restricted specificity for the HIV-2 protease, although both enzymes show the highest preference for the Glu-Ala bond (Figure 2). Charged amino acids in P_1 are unusual and a Glu in this position was unexpected. Moreover, the split at Leu-Val in the -His-Leu-Val- sequence supported the idea that renin inhibitors could find application in HIV-1 protease inhibition; renin is highly selective for the same sequence.[17]

Non-viral substrates described above provided some novel clues about protease specificity, but we wished to explore native proteins, preferably multi-domain molecules of known tertiary structure, as models of the natural viral polyprotein substrates. Such proteins could provide ideas as to conformational requisites as well as sequence preferences in defining a substrate for the HIV protease.

Pseudomonas *exotoxin derivatives*

The 3-domain exotoxin has an extended segment which connects the 2nd and 3rd domains[18] and which contains an Asn-Tyr-Pro sequence identical to that seen in natural HIV polyproteins (Table 1). We have shown, however, that the full-length exotoxin is not

```
                          HIV-1
                     ┌─────┴─────┐
                       ↓    ↓ ↓
F₁-V-N-Q-H-L-C[SO⁻₃]-G-S-H₁₀-L-V-E-A-L-Y-L-V-C[SO⁻₃]-G₂₀-E-R-G-F-F-Y-T-P-K-A₃₀
                            ↑    ↑
                          HIV-2
```

Figure 2. Schematic representation of oxidized insulin B chain indicating sites of hydrolysis by HIV proteases.

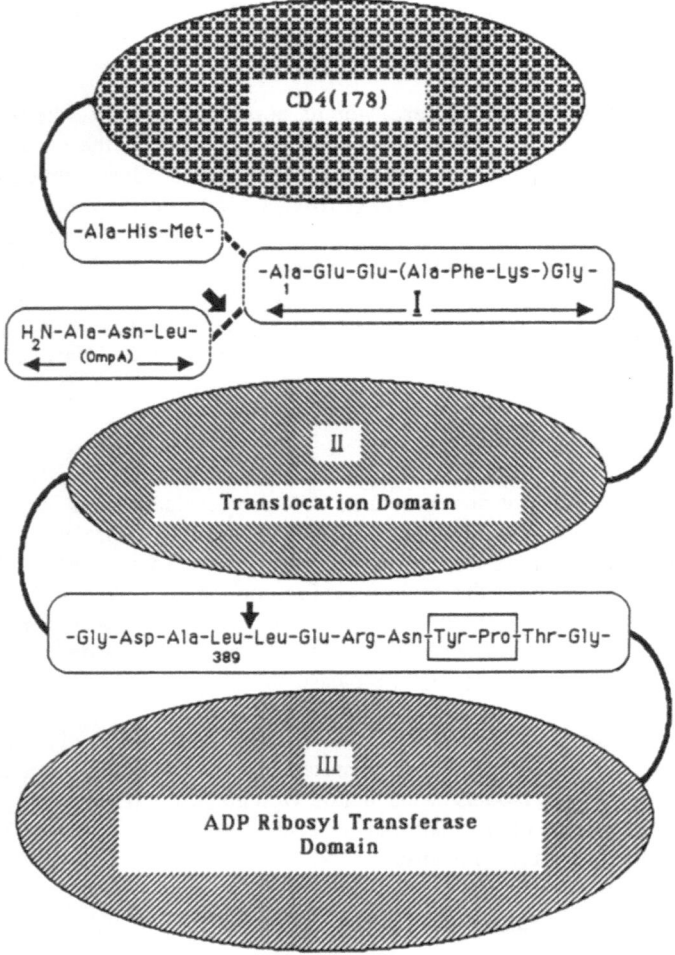

Figure 3. Schematic represention of *Pseudomononas* exotoxin derivatives PE40 and CD4PE40 with cleavage sites indicated for HIV proteases 1 and 2. Taken with permission from reference 9.

Table 3. Nonviral substrates of HIV protease: synthetic peptides modeled after cleavage sites in LysPE40

Compound	HIV-1 Protease			HIV-2 Protease		
	K_m	V_{max}	V_{max}/K_m	K_m	V_{max}	V_{max}/K_m
H-Ala-Asn-Leu * Ala-Glu-Glu-Ala-Phe-OH[a]	1.3	0.3	0.23	1.1	5.0	4.55
H-Ser-Gly-Asp-Ala-Leu * Leu-Glu-Arg-Asn-OH[a]	1.6	1.2	0.71	1.1	3.0	2.73
H-Leu-Glu-Arg-Asn-Tyr * Pro-Thr-Gly-Ala-OH[a]		0			0	
H-Val-Ser-Gln-Asn-Tyr * Pro-Ile-Val-OH[b]	2.0	4.9	2.45	2.0	2.5	1.25

[a]See Figure 3; [b]From Table 1.

hydrolyzed by either HIV-1 or 2 protease.[6,9] Removal of domain 1, or replacement of it by a truncated version of CD4 makes the interdomain segment available to both proteases; cleavage takes place not at the expected Tyr-Pro bond, however, but at the Leu-Leu bond a few residues away (Figure 3). This is the case both for the mutant protein missing domain I (PE40) as well as the chimeric CD4PE40 protein (Figure 3).

An additional site of hydrolysis was documented in PE40 at the Leu-Ala bond near the N-terminus; the Met-Ala bond in the chimeric protein was not split. It is interesting that peptides overlapping the Leu-Leu and Leu-Ala bonds were also excellent substrates of the proteases (Table 3), and that corresponding to the expected Tyr-Pro site was not. Therefore, it would appear that the Tyr-Pro bond need not be conformationally inaccessible to the proteases; its sequence, alone, is unacceptable for recognition as a substrate. As was the case in RNase (Figure 1), both bonds cleaved in the exotoxin constructs have a P_2' Glu; this has almost become a hallmark of the non-viral protein substrates.

The results obtained from analysis of the exotoxin constructs confirmed the notion that the proteases prefer cleavage sites in extended, extradomain regions, but we do not know

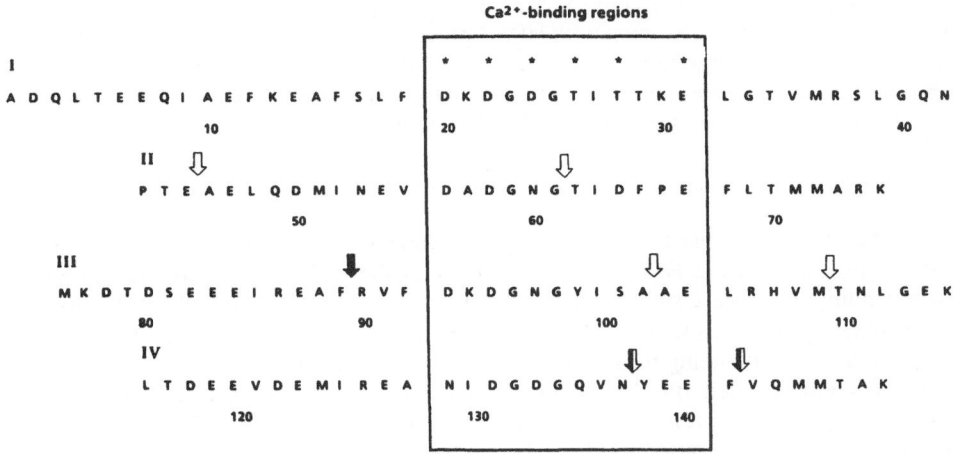

Figure 4. HIV-1 (solid arrow) and HIV-2 (unfilled arrow) protease cleavage sites in calmodulin. A half filled arrow indicates hydrolysis by both HIV-1 and HIV-2 enzymes.

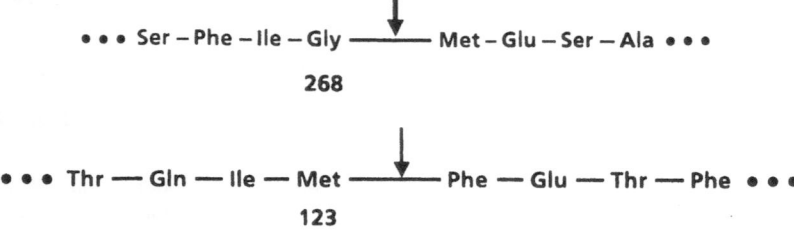

Figure 5. HIV protease cleavage sites in G-actin.

that the 3D structure of PE40 reflects that seen in the parent PE66[18] and we cannot, therefore, examine the conformation of cleavage sites.

Calmodulin

In addition to our interest in substrates with known structures, we were interested in those crucial to the well-being of the host cell. Calmodulin, a calcium shuttle to a variety of metabolically important enzymes, is one such protein, at it has the advantage of having a well defined x-ray structure.[19] That structure, however, is for the Ca^{++} bound form of the protein. In fact, calcium-calmodulin is not a substrate for either HIV-1 or HIV-2 protease. Both enzymes, however, readily cleave the protein after removal of calcium (Figure 4). This analysis adds to the repertoire of known cleavage sites and sequences, even though it tells us nothing about the conformation recognized by the enzyme since binding of calcium results in a highly constrained molecule.[19] Now, in contrast to results with insulin cleavage, we see that the HIV-2 protease is less specific than its HIV-1 counterpart. We document a cleavage by the latter enzyme at a bond with a P_1' Arg. Generally, the HIV-2 enzyme prefers bonds with small P_1-P_1' residues, in keeping with what is observed with HIV-2 polyprotein substrates.[9,11] Nevertheless, both enzymes split Asn-Tyr and Phe-Val bonds and, again, a P_2' Glu is a common element in many of the sites hydrolyzed (Figure 4).

Actin

In an attempt to discern proteins in T4 cells that might be susceptable to the protease, if ever that enzyme were to be active within the host cell, we performed 2D gel analysis of T4 extracts before and after treatment with HIV-1 protease. Actin was prominent among the few proteins with altered migration in these gels. Accordingly, we showed that pure G-actin is a substrate of both proteases (Figure 5). Interestingly, a major site of hydrolysis occurred at a site with a P_1 Gly, and both bonds split are adjacent to Glu at P_2'.

Summary

Scheme 1 summarizes some of what we have learned from this study of non-viral protein substrates of the HIV proteases. Many of these findings contradict the current understanding of protease specificity. P_1-P_1' amino acids need not be bulky or hydrophobic and residues at these positions may be even less important than those in flanking positions (e.g., Glu at P_2') in dictating the course of hydrolysis. Thus, the pattern of amino acids over the whole binding region must be considered in predicting what will or will not be a substrate of these enzymes and, although we are beginning to understand selectivity at the level of primary structure, a detailed explanation of their specificity is yet to be forthcoming. Nevertheless, studies of this kind find useful application in the design of inhibitors of HIV proteases that will, hopefully, be of value in treatment of AIDS.

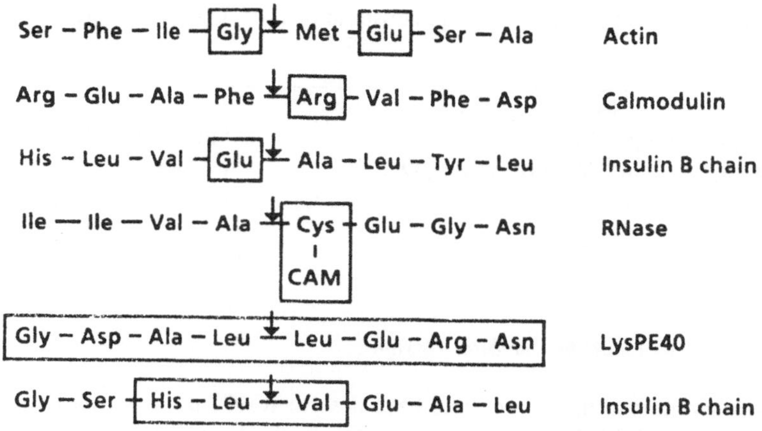

Ser — Phe — Ile —[Gly]⤒ Met —[Glu]— Ser — Ala   Actin

Arg — Glu — Ala — Phe ⤒[Arg]— Val — Phe — Asp   Calmodulin

His — Leu — Val —[Glu]⤒ Ala — Leu — Tyr — Leu   Insulin B chain

Ile — Ile — Val — Ala ⤒⌐Cys⌐ Glu — Gly — Asn   RNase
                                  |
                                CAM

[Gly — Asp — Ala — Leu ⤒ Leu — Glu — Arg — Asn]   LysPE40

Gly — Ser —⌐His — Leu ⤒ Val ⌐ Glu — Ala — Leu   Insulin B chain

Scheme 1. Some new features of HIV protease specificity revealed through study of non-viral protein substrates. The arrows indicate cleavage points.

DESIGN OF PROTEASE INHIBITORS AS ANTIVIRAL AGENTS

In order to develop a successful drug for AIDS that is based upon inhibition of the HIV protease, it is essential that the drug not only inhibit the enzyme, but that it be active against the virus. This raises questions as to the bioavailability of the drug, and its timely delivery to the appropriate site of protease activation, presumably following budding of the provirus from the host cell. Therefore, both enzyme and cellular assays are indispensable for drug development. In Scheme 2 are described two such *in vitro* cellular assays that we have employed in evaluating HIV protease inhibitors. In a non-infectious assay it is possible to measure processing of the gag-pol polyproteins to yield the p24 protein. Inhibitors shown to be effective in blocking this process can also be assayed in HIV-infected peripheral blood lymphocytes (PBL) and the Diagen company in Dusseldorf Germany has been an important resource to us for these assays.

vVK-1/CV-1 Cell Assay

Recombinant vaccinia virus expressing HIV-1 *gag-pol* genes in a monkey CV-1 cell line;

Non-infectious HIV-1-like particle formation mirrors HIV-1 replication and maturation;

Quantitative determination of HIV-1 p55 → p24 processing;
   (24 h incubation; cell lysis/SDS PAGE/immunoblotting/densitometry)

HIV-1/PBL Cell Assay (DIAGEN)

HIV-1$_{HTLV-IIIB}$ infected primary cultures of human peripheral blood lymphocytes;

Infectious HIV-1 virion formation stimulates natural retroviral life cycles;

Quantitative determination of HIV-1 p55 → p24 processing and HIV-1 RNA activity;
   (72, 96 h incubation; p24 monoclonal capture Ab-ELISA/RNA dot-blot hydridization)

Scheme 2. *In vitro* cellular assays for evaluating HIV Protease inhibitors.

Table 4. Conversion of a "viral" peptide substrate of HIV-1 protease into a highly potent inhibitor

H-Val-Ser-Gln-Asn-Tyr * Pro-Ile-Val-OH	Substrate	$K_m = 2 \times 10^{-3}$ M
H-Val-Ser-Gln-Asn-Leu * Val-Ile-Val-Arg-OH	Inhibitor	$K_i = 5 \times 10^{-4}$ M
H-Val-Ser-Gln-Asn-Leuψ[CH(OH)CH$_2$]Val-Ile-Val-OH	Inhibitor	$K_i = \; < 10^{-9}$ M

Viral Polyproteins as Leads for Inhibitor Design

Many workers in this active field have described HIV protease inhibitors that have been designed from knowledge of bonds cleaved in the natural polyproteins. One such inhibitor that has been useful to us is U-85548E, the development of which followed from our gag p17-p24 peptide substrate (Table 4). Replacement of the Tyr-Pro by a Leu-Val not only obliterated cleavage by the protease, but produced a weak inhibitor. As shown in Scheme 1, Leu-Val is a perfectly acceptable scissile bond, but not, perhaps, in the same template as one finds for Tyr-Pro. Returning now to Table 4, if we now replace the Leu-Val with a non-planar insert which mimics the presumed transition state, we generate an extremely potent HIV protease inhibitor. This peptidemimetic has been used to titrate the active site of the HIV-1 protease and thus to provide a measure of enzyme purity.[9] Moreover, in collaboration with Dr. Alexander Wlodawer we have been able to derive the structure of a complex of synthetic HIV-protease with this inhibitor. Nevertheless, this inhibitor shows no antiviral activity in the cellular assays, and holds no promise of being a drug for treatment of AIDS. Presumably, this molecule cannot gain access to the host T4 cell, or the budded proviral particle.

$$P_1 \quad P_1{'}$$

Renin Peptide Substrate	. . . . -Pro-Phe-His-Leu - Val-Ile-His- . . . .

P$_1$-P$_1{'}$ Site Modifications	Leuψ[CH(OH)CH$_2$]Val	Leuψ[CH(OH)CH(OH)]Val
	Chaψ[CH(OH)CH$_2$]Val	Chaψ[CH(OH)CH(OH)]Val

Inhibitor - U-71038   Boc-Pro-Phe-N$^\alpha$Me-His-Leuψ[CH(OH)CH$_2$]Val-Ile-Amp

K_i for renin = 70 pM K_i for HIV-1 protease = 10 nM

HIV-1 Protease Substrate Modeled $P_1 \quad P_1{'}$
    After Insulin Sequence   . . . . . -Gly-Ser-His-Leu - Val-Glu-Ala-Leu- . . . .

$K_m = 0.5$ mM, $V_{max} = 1.0$ μmol x mg^{-1} x min^{-1}

Inhibitor - U-89920E   Ac-Gly-Ser-His-Leuψ[CH(OH)CH$_2$]Val-Glu-Ala-Leu-NH$_2$
HIV-1 PR, $K_i = 10$ nM

Scheme 3. Discovery of lead inhibitors of HIV-1 Protease.

Table 5. Substrate-based inhibitors of HIV-1 protease with Leuψ[CH(OH)CH$_2$]Val P$_1$-P$_1$' site

	Structure	K$_i$, nM	% Inhibition p24 Synthesis	
			vVK-1/CV-1	HIV-1/PBL
U-85548E	Val-Ser-Gln-Asn-Leuψ[CH(OH)CH$_2$]Val-Ile-Val	< 1	0 %[a]	10 %[b]
U-89920E	Ac-Gly-Ser-His-Leuψ[CH(OH)CH$_2$]Val-Glu-Ala-Leu-NH$_2$	10	0 %[a]	n.d.
U-71038	Boc-Pro-Phe-N$^\alpha$MeHis-Leuψ[CH(OH)CH$_2$]Val-Ile-Amp	10	15 %[a]	28 %[b]
U-71017	Poa-His-Leuψ[CH(OH)CH$_2$]Val-Ile-Amp	10	61 %[a]	76 %[b]

[a]measured at 10 μM test compound; [b]measured at 1 μM test compound; Boc = *tert*-butoxycarbonyl; Poa = phenoxyacetyl; Amp = 2-aminomethylpyridine.

Non-Viral Protein Substrates as Leads For Inhibitor Design

The Upjohn Company, like many involved in the search for renin inhibitors that might find therapeutic application against hypertension, has a large number of peptidemimetic compounds with varying inhibitory activity against renin. Renin ranks among the most specific of enzymes in that it appears to have only one physiological substrate, i.e., angiotensinogen. The Leu-Val bond cleaved by renin is the tenth peptide bond in the large substrate, and the flanking sequences are depicted in Scheme 3. The Leu-Val bond has been replaced by a variety of transition state mimics including Cha(cyclohexylalanine)-Val alcohols and diols (Scheme 3). U-71038 is a lead renin inhibitor with a K$_i$ in the picomolar range. It is also a fairly potent inhibitor of the HIV-1 protease with a K$_i$ of 10 nM (Scheme 3). A Leu-Val bond preceded by His was observed to be cleaved in insulin B chain (Figure 2) and we accordingly modeled a peptidemimetic after this non-viral substrate. As shown in Scheme 3, this inhibitor, U-89920E gave a K$_i$ comparable to that of U-71038. Neither inhibitor is effective toward HIV-2 protease. As shown in Table 5, U-89920E displays no activity in either of the cellular assays described in Scheme 2. The renin inhibitor, U-71038, shows only weak activity, but truncation of the molecule to a peptide having a phenoxyacetyl blocking group yields a small inhibitor comparable in K$_i$ but quite superior in the cell assays (Table 5). With this small inhibitor as a lead, it was possible to explore further the effects of N-terminal modification and changes in the transition state insert relative both to K$_i$ and blocking of p24 synthesis. As may be seen in Table 6, the change from Leu to Cha in the P$_1$ position had a dramatic effect on enhancing activity in the cell assays, and this was even more pronounced when the hydroxyethyl insert was replaced with a diol. The end result of this line of study was U-75875, a peptidemimetic with a naphthoxyacetyl N-terminal blocking group and a Cha-Val diol insert. The properties of this inhibitor are given in Scheme 4. Like all of the other inhibitors we have studied thus far, U-75875 is more active against HIV-1 protease than the enzyme from HIV-2. U-75875 is not small, but it is highly effective in cell assays with sustained action[4] even after attempts to remove it by washing the cells (Scheme 4). Thus, it appears to be the naphthyl moiety which endows the inhibitor with properties which enable it to penetrate a lipid bilayer, perhaps one characteristic of the budded provirus, and to stay there. It should also be noted that U-75875 is effective in cell assays designed to test activity against simian immunodeficiency virus (SIV). The close similarity of SIV to HIV-2,[22] coupled with the activity of this inhibitor against the HIV-2 protease[4] (Scheme 4) would predict such activity in an SIV-based cell assay and, indeed, such is the case.

Table 6. Structure-activity analysis of U-71017 to explore the effect of N-terminal and P_1-$P_{1'}$ site modifications on inhibition of HIV-1 protease

	Structure	K_i, nM	% Inhibition p24 Synthesis	
			vVK-1/CV-1	HIV-1/PBL
U-71017	Poa-His-Leuψ[CH(OH)CH₂]Val-Ile-Amp	10	61 %[a]	76 %[b]
U-79213	Poa-His-Chaψ[CH(OH)CH₂]Val-Ile-Amp	10	95 %[a]	90 %[b]
U-88443	Poa-His-Chaψ[CH(OH)CH(OH)]Val-Ile-Amp	2	100 %[a]	n.d.
U-75875	Noa-His-Chaψ[CH(OH)CH(OH)]Val-Ile-Amp	<1	100 %[a]	100 %[b]

[a]measured at 10 µM test compound; [b]measured at 1 µM test compound; Poa = phenoxyacetyl; Amp = 2-aminomethylpyridine Cha = cyclohexylalanine; Noa = 1-naphthoxyacetyl.

U-75875

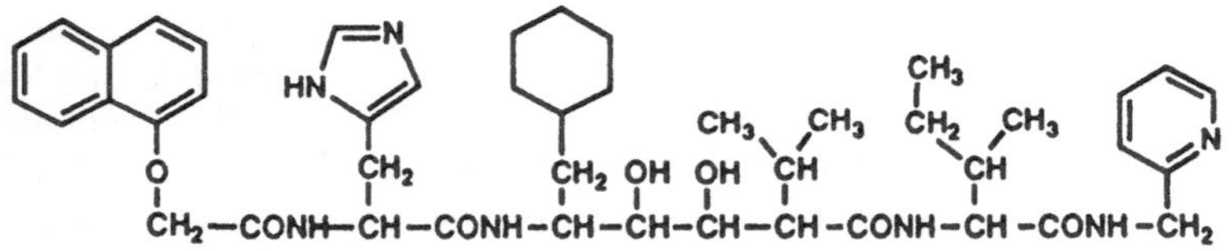

CHEMICAL PROPERTIES

Molecular formula = $C_{45}H_{61}N_7O_7$
Molecular weight = 812
cLog P = 4.7

ANTI-HIV PR PROPERTIES

HIV-1 Pr, K_i = < 1 nM (competitive)
HIV-2 PR, K_i = 30 nM

ANTI-HIV ANTIVIRAL PROPERTIES

HIV-1/PBL, IC_{50} = 3 nM (p55/RNA; 96 h); LD_{50} = >10 µM (PBL poliferation)
vVK-1/CV-1, IC_{50} = 200 nM (p55; 24 h); "irreversibility post-washout"
Sustained anti-HIV-1 activity in HIV-1$_{LAV}$/CEMX174 (1 µM U-75875; >28 days)
Sustained anti-HIV-2 activity in HIV-2$_{ST}$/CEMX174 (1 µM U-75875)
Sustained anti-SIV activity in SIV$_{MAC251}$/HUT78 (1 µM U-75875)

Scheme 4. Structure and properties of U-75875.

Table 7. Structure-activity analysis of inhibitors to probe selectivity towards HIV-1 versus HIV-2 protease

	Structure	K_i nM	
		HIV-1 PR	HIV-2 PR
U-81749E	Tba-Chaψ[CH(OH)CH$_2$]Val-Ile-Amp	70	1000
U-82159E	Chaψ[CH(OH)CH$_2$]Val-Ile-Amp	800	>1000
U-85704E	Ac-Chaψ[CH(OH)CH$_2$]Val-Ile-Amp	8	>1000
U-87401E	Ac-His-Chaψ[CH(OH)CH$_2$]Val-Ile-Amp	22	>1000
U-79213	Poa-His Chaψ[CH(OH)CH$_2$]Val-Ile-Amp	7	105
U-71017	Poa-His-Leuψ[CH(OH)CH$_2$]Val-Ile-Amp	10	107
U-76248	Poa-His-Leuψ[CH(OH)CH$_2$]Cha-Ile-Amp	40	>1000
U-71038	Boc-Pro-Phe-N$^\alpha$MeHis-Leuψ[CH(OH)CH$_2$]Val-Ile-Amp	10	>1000

Since we have pointed out the widespread occurrence of a Glu residue at P_2' in non-viral substrates, note should be made of the fact that we have made a number of inhibitors with Glu in this position. Although such compounds are effective at pH 5.5, their activity is less at neutral pH. A Gln residue in this position is generally less effective than Glu. In any case, inhibitors with Glu or Gln at P_2' show little if any antiviral activity.

CONCLUSIONS

A few conclusions from the work completed thus far in our laboratory may be summarized as follows:

1. We have greatly expanded the repertoire of known peptide bonds that are cleaved by both HIV-1 and HIV-2 proteases by evaluation of non-viral proteins as substrates.

2. We have combined information relative to HIV protease substrate specificity with knowledge accumulated by structure-activity analysis of peptidemimetic inhibitors of renin, to design templates and P_1-P_1' inserts for HIV protease lead inhibitors.

3. These lead inhibitors have been modified selectively at their N- and C-termini and in the transition-state insert to yield potent HIV protease inhibitors which are also effective HIV and SIV anti-viral agents in cell culture.

4. Because of significant differences in substrate and inhibitor specificity between HIV-1 and HIV-2 proteases, we have not yet discovered an inhibitor other than U-75875 which is effective against both. Purposeful design, however, can now lead to development of an inhibitor which will show enhanced activity toward HIV-2 protease over the HIV-1 enzyme (Table 7).

5. Our identification of a compound (U-75875) with activity against SIV[4] opens the door for testing in an animal model the concept of protease inhibition as an intervention point in AIDS therapy.

At the time of this writing, much remains to be learned about the course of AIDS infection, and the relationship of the viral life cycle to this process. A recent review by Mitsuya et al.[23] explores the many possible molecular targets available for AIDS therapy. For now, the protease remains a highly desirable target, but much remains to be understood about temporal and mechanistic aspects of its activation. The enzyme is unique in being an obligate dimer; it is inconceivable that it could function as a monomer with only half an active site. This means that conditions must arise whereby two polyproteins can come into such proximity as to allow dimerization of the protease within a polyprotein format. Other scenarios may be envisioned but, teleologically speaking, it is most reasonable that this protease activation occur only outside the host and a concentration-dependent dimerization process would provide an efficient regulatory mechanism. Premature activation would result in early processing of the viral proteins, and budding of particles with an incomplete set of molecules for viral assembly. Indeed, if there were some way to turn the protease on within the host cell, it could represent an approach to therapy. Our findings that actin and calmodulin are substrates of these proteases, and the recent demonstration by Shoeman et al.[24] that the same is true for vimentin, desmin, and glial fibrillary acidic protein would, of course, argue against such a tactic.

Conditions during budding into a small spherical particle are most propitious for bringing the polyproteins together in a highly concentrated vesicle where protease activation can take place by polyprotein dimerization. Much consideration is being given to drugs which might exert their effect by blocking the dimerization process. This tactic also suffers from the probable lack of specificity in preventing other important dimerization processes. Clearly, if one has a handle on catalytic activity and enzyme specificity in drug design, the job is made much easier and the resulting drug should carry with it an enormously enhanced margin of safety.

As is always the case with peptide drugs, bioavailability and delivery constitute substantial problems for usefulness in the clinic. Properties which make for good solubility often are undesirable from the point of view of cell penetration. Compounds that are large and highly selective are poorly absorbed, if at all, in the gut, and may be metabolically unstable as well. Rapid clearance of peptide-like drugs has always been a problem.

Overcoming these obstacles presents a formidable challenge for future research, not only in the area of the HIV protease inhibitors but with respect to peptides as drugs in general. Advances in the x-ray analysis of protease-inhibitor complexes has provided detailed structures upon which to base de novo design of non-peptide inhibitors which may bring with them improved properties of bioavailability. Realistically speaking, however, the science is not in place to do this at the present time. Increasing effort will be made in the next decade to design drugs from structural principles, and it will be interesting to see if we will be able to create effective therapeutic agents by the close of the century. In the meantime, we have become quite proficient at design of HIV protease inhibitors from a knowledge of target enzyme specificity, but much remains to be learned at the level of biology and virology about how to make them useful in the treatment of AIDS.

ACKNOWLEDGEMENTS

We would like to acknowledge Dr. W. Gary Tarpley and Tom McQuade for determining the antiviral data for all of the compounds reported in this chapter. We also thank Paula W. Lupina for her efforts in preparing this manuscript.

REFERENCES

1. T. J. McQuade, A. G. Tomasselli, L. Liu, V. Karacostas, B. Moss, T. K. Sawyer, H. L. Heinrikson and W. G. Tarpley, A synthetic HIV-1 protease inhibitor with antiviral activity arrests HIV-like particle maturation, *Science* **247**:454 (1990).

2. T. D. Meek, D. M. Lambert, G. B. Dreyer, T. J. Carr, T. A. Tomaszek, Jr, M. L. Moore, J. E. Strickler, C. Debouck, L. J. Hyland, T. J. Matthews, B. W. Metcalf and S. R. Petteway, Inhibition of HIV-1 protease in infected T-lymphocytes by synthetic peptide analogues, *Nature* **248**:358 (1990).

3. N. A. Roberts, J. A. Martin, D. Kinchington, A. V. Broadhurst, J. C. Craig, I. B. Duncan, S. A. Galpin, B. K. Handa, J. Kay, A. Krohn, R. W. Lambert, J. H. Merrett, J. S. Mills, K. E. B. Parkes, S. Redshaw, A. J. Ritchie, D. L. Taylor, G. J. Thomas and P. J. Machin, Rational design of peptide-based HIV proteinase inhibitors, *Science* **248**:90 (1990).

4. P. Ashorn, T. J. McQuade, S. Thaisrivongs, A. G. Tomasselli, W. G. Tarpley and B. Moss, An inhibitor of the protease blocks maturation of human and simian immunodeficiency viruses and spread of infection, *Proc. Nat. Acad. Sci. U.S.A.* **87**:7472 (1990).

5. N. E. Kohl, E. A. Emini, W. A. Schleif, L. J. Davis, J. C. Heimbach, R. A. F. Dixon, E. M. Scolnick and I. S. Sigal, Active human immunodeficiency virus protease is required for viral infectivity, *Proc. Nat. Acad. Sci. U.S.A.* **85**:4686 (1988).

6. A. G. Tomasselli, J. O. Hui, T. K. Sawyer, D. J. Staples, D. J. FitzGerald, V. K. Chaudhary, I. Pastan and R. L. Heinrikson, Interdomain hydrolysis of a truncated *Pseudomonas* exotoxin by the human immunodeficiency virus-1 protease, *J. Biol. Chem.* **265**:408 (1990).

7. J. O. Hui, A. G. Tomasselli, H. A. Zurcher-Neely and R. L. Heinrikson, Ribonuclease A as a substrate of the protease from human immunodeficiency virus-1, *J. Biol. Chem.* **265**:21386 (1990).

8. A. G. Tomasselli, W. J. Howe, J. O. Hui, T. K. Sawyer, I. M. Reardon, D. L. DeCamp, C. S. Craik and R. L. Heinrikson, Calcium-free calmodulin is a substrate of proteases from human immunodeficiency viruses 1 and 2, *Proteins: Structure Function and Genetics* **10**:1 (1991).

9. A. G. Tomasselli, J. O. Hui, T. K. Sawyer, D. J. Staples, C. Bannow, I. M. Reardon, W. J. Howe, D. DeCamp, C. S. Craik and R. L. Heinrikson, Specificity and inhibition of proteases from human immunodeficiency viruses-1 and 2, *J. Biol. Chem.* **265**:14675 (1990).

10. P. L. Darke, R. F. Nutt, S. F. Brady, V. M. Garsky, T. M. Ciccarone, C.-T. Leu, P. K. Lumma, R. M. Freidinger, D. F. Veber and I. S. Sigal, HIV protease specificity of peptide bond cleavage is sufficient for processing of gag and pol polyproteins, *Biochem. Biophys. Res. Commun.* **156**:297 (1988).

11. L. E. Henderson, R. E. Benveniste, R. Sowder, T. D. Copeland, A. M. Schultz and S. Oroszlan, Molecular characterization of gag proteins from simian immunodeficiency virus (SIV_{ne}), *J. Virol.* **62**:2587 (1988).

12. A. G. Tomasselli, M. K. Olsen, J. Hui, D. J. Staples, T. K. Sawyer, R. L. Heinrikson and C.-S. C. Tomich, Substrate analogue inhibition and active site titration of purified recombinant HIV-1 protease, *Biochemistry* **29**:264 (1990).

13. H.-G. Krausslich, R. H. Ingraham, M. T. Skoog, E. Wimmer, P. V. Pallai and C. A. Carter, Activity of purified biosynthetic proteinase of human immunodeficiency virus on natural substrates and synthetic peptides, *Proc. Nat. Acad. Sci. U.S.A.* **86**:807 (1989).

14. M. L. Moore, W. M. Bryan, S. A. Fakhoury, V. W. Magaard, W. F. Huffman, B. D. Dayton, T. D. Meek, L. Hyland, G. B. Dreyer, B. W. Metcalf, J. E. Strickler, J. G. Gorniak and C. Debouck, Peptide substrates and inhibitors of the HIV-1 protease, *Biochem. Biophys. Res. Commun.* **159**:420 (1989).

15. L. H. Pearl and W. R. Taylor, A structural model for the retroviral proteases, *Nature* **329**:351 (1987).

16. A. Wlodawer, M. Miller, M. Jaskólski, B. K. Sathyanarayana, E. Baldwin, I. T. Weber, L. M. Selk, L. Clawson, J. Schneider and S. B. H. Kent, Conserved folding in retroviral proteases: Crystal structure of a synthetic HIV-1 protease, *Science* **245**:616 (1989).

17. R. L. Heinrikson and R. A. Poorman, The biochemistry and molecular biology of recombinant human renin and prorenin, *in*: "Hypertension: Pathophysiology, Diagnosis, and Management" J. H. Laragh and B. M. Brenner, eds., Raven Press Ltd, New York (1990).

18. V. S. Allured, R. J. Collier, S. F. Carroll and D. B. McKay, Structure of exotoxin A of *Pseudomonas aeruginosa* at 3.0 angstrom resolution, *Proc. Nat. Acad. Sci. U.S.A.* **83**:1320 (1986).

19. Y. S. Babu, C. E. Bugg and W. J. Cook, Structure of calmodulin refined at 2.2 Å resolution, *J. Mol. Biol.* **204**:191 (1988).

20. V. Karacostas, K. Nagashima, M. A. Gonda and B. Moss, Human immunodeficiency virus-like particles produced by a vaccinia virus expression vector, *Proc. Nat. Acad. Sci. U.S.A.* **86**:8964 (1989).

21. T. D. Meek, D. M. Lambert, B. W. Metcalf, S. R. Petteway, Jr. and G. B. Dreyer, HIV-1 protease as a target for potential anti-AIDS drugs, *in*: "Design of Anti-AIDS Drugs", E. De Clercq, ed. Elsevier, New York, (1990).

22. L. Chakrabarti, M. Guyader, M. Alizon, M. D. Daniel, R. C. Desrosiers, P. Tiollais and P. Sonigo, Sequence of simian immunodeficiency virus from macaque and its relationship to other human and simian retroviruses, *Nature* **328**:543 (1987).

23. H. Mitsuya, R. Yarchoan and S. Broder, Molecular Targets for AIDS therapy, *Science* **249**:1533 (1990).

24. R. L. Shoeman, B. Honer, T. J. Stoller, C. Kesselmeier, M. C. Miedel, P. Traub and M. C. Graves, Human immunodeficiency virus type 1 protease cleaves the intermediate filament proteins vimentin, desmin, and glial fibrillary acidic protein, *Proc. Nat. Acad. Sci. U.S.A.* **87**:6336 (1990).

INTERACTION OF MUTANT FORMS OF THE HIV-1 PROTEASE

WITH SUBSTRATE AND INHIBITORS

Paul L. Darke, Nancy E. Kohl, Michelle G. Hanobik,
Chih-Tai Leu, Joseph P. Vacca, James P. Guare, Jill C. Heimbach
and Richard A. F. Dixon

Merck Sharp and Dohme Research Laboratories
West Point, Pennsylvania   19486

INTRODUCTION

The protease of HIV-1 is considered to be a prime target for the treatment of AIDS with small molecular weight inhibitors.[1,2,3] Many groups have now demonstrated that compounds designed as potent inhibitors of the enzyme *in vitro* are effective inhibitors of viral polyprotein processing and viral spread in cultured cells.[4-7] Continued effort to discover new potent inhibitors with suitable properties for use in humans is thus justified.

In order to completely understand the importance of each inhibitor--enzyme contact, we felt it desirable to examine the effects upon inhibitor binding of changes in not only inhibitor structure, but also enzyme structure. Toward that end, a series of mutant forms of the protease have been constructed with the goal of defining the energetic importance of the amino acid side chains in substrate and inhibitor binding. Amino acid residues likely to interact with either substrate or inhibitors were identified from the native crystal structure.[8] Two types of interactions of inhibitors with the enzyme have been identified: hydrophobic interaction and hydrogen bonding.[9,10] In this report we present preliminary data on the properties of the single site mutations, Arg-8-Glu, Asp-29-Ala, and Val-82-Ala of the mature 99 amino acid form of the protease. We examine the potential importance of Val_{82} in hydrophobic interactions, and Arg_8 and Asp_{29} in hydrogen bonding interactions. Although extensive mutagenesis has been performed with a precursor form of the protease[11] and other selected mutations have been evaluated qualitatively with regard to activity,[12,13,14] we have attempted to quantitatively evaluate the effect of the mutations mentioned above upon kinetic parameters.

METHODS

Synthetic Gene and Bacterial Expression

Construction of the synthetic gene for convenient mutagenesis of the 99 amino acid protease will be described in a separate communication. The gene was inserted into the protease *trp* expression vector, cultures grown and induced as previously described.[15] Expression of protein was monitored with Western analysis,[15] and clones producing HIV-1 protease were lysed and assayed for peptide hydrolase activity.

Purification

Purification attempts for each mutant protein began with application of crude supernatants to the affinity resin previously described.[16] If activity was retained on the resin, the purification scheme employed for the native form of the enzyme[16] was employed for that mutant. Otherwise, the supernatants were passed through DEAE Sephadex followed by hydrophobic interaction chromatography according to published procedures.[17] Only chromatography columns which had never been in contact with the native form of the enzyme were used, to avoid any chance of activity contamination.

Activity analysis

Kinetic parameters were determined by hydrolysis of the peptide substrate VSQN(ß-Nal)PIV at 30°C, in 50 mM sodium acetate, pH 5.5, with 0.1 % BSA present. Quenching with phosphoric acid and analysis of product formation by HPLC were performed as previously described.[17] K_m and k_{cat} determinations used 5 substrate concentrations, in duplicate, and values were estimated with the Enzfitter program. The reaction progress for each mutant was shown to be linear during the time used in these experiments (30 min) for at least one of the lower substrate concentrations. IC_{50} values for inhibitor binding were determined using a constant concentration of substrate (0.5 mg/ml, 0.425 mM) and 11 concentrations of inhibitor from 0.03 nM to 0.1 mM. Inhibitors were initially dissolved into DMSO and dilutions in DMSO were done so as to result in a final concentration of 2 % DMSO in every assay sample.

RESULTS

Activity with the substrate, VSQN(ßNal)PIV, was detected in the lysate supernatants of cultures expressing the mutants, Arg-8-Glu, Asp-29-Ala and Val-82-Ala. For Val-82-Ala, activity in the supernatant from lysed cells absorbed to the affinity resin used in purification of the native enzyme, so that our standard purification scheme[16] could be employed. Elution behavior throughout the purification was essentially like that of the native enzyme, resulting in a preparation that was approximately 80 % HIV-1 protease, as judged by Coomassie staining of 16 % SDS polyacrylamide electrophoresis gels.

Table 1. Kinetic constants of mutant forms of HIV-1 protease

Enzyme	K_m (µM)	k_{cat} (1/min)	k_{cat}/K_m	IC_{50} (nM) "679"	"676"
Native	244	1007	4.13	0.7	1096
Val-82-Ala	416	270	0.65	1.8	372
Arg-8-Glu	2397	20	0.008	8.1	*
Asp-29-Ala	2310	17	0.007	51	*

Assays were performed as described in "Methods"; *Not determined

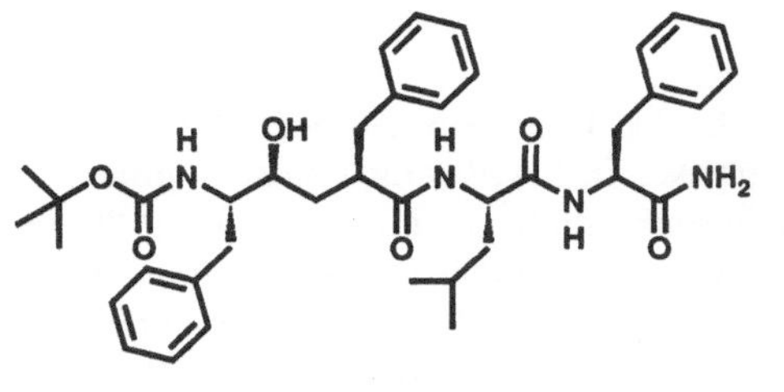

L-682,679 ("679")

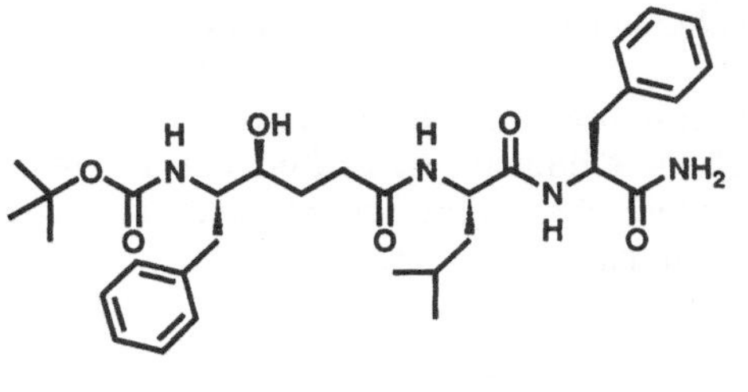

L-682,676 ("676")

Figure 1. Structures of the HIV-1 protease inhibitors.

For Arg-8-Glu and Asp-29-Ala, only 30 % and less than 10 % of the supernatant activities, respectively, absorbed to the affinity resin, so that partial purification was achieved instead by starting with DEAE Sephadex followed by hydrophobic interaction chromatography. Although a faint band could be observed in gels of these purified preparations at the appropriate molecular weight, 11,000 Da, estimation of the percent purity was not attempted. The extent of purification for Arg-8-Glu and Asp-29-Ala from the crude supernatants was approximately 30 and 20 fold, respectively. The preparations of all three mutant proteins were also found to hydrolyze peptides representing 7 of the known cleavage sites in the HIV-1 polyproteins (data not shown). In addition, supernatant of lysate from a culture expressing the inactive Asp-25-Asn mutant using the same expression system[15] showed no detectable activity with the ß-Nal substrate employed here. The appropriate peptide cleavage activities as well as inhibition by known HIV-1 protease inhibitors (described below) ensure that the activity measured is in fact due to the mutant HIV-1 protease and not a contaminating enzyme.

The kinetic constants for the purified preparations of these mutants are shown in Table 1. The Val-82-Ala substitution does not appear to be very disruptive to activity with the peptide substrate employed here. While the K_m is slightly higher than that of native enzyme, the k_{cat} value is roughly 4-fold less. The Arg-8-Glu and Asp-29-Ala mutants were much more severely affected in activity, with each having K_m about 10-fold above the native enzyme value, and k_{cat} reduced 50-fold.

The peptide mimetic, L-682,679, shown in Figure 1, has been demonstrated to be a potent competitive inhibitor of the protease capable of halting virus spread in culture.[18] The related compound, L-682,676, which lacks the P_1' benzyl moiety is 1000-fold less potent (Table 1). The reduction of hydrophobic surface in the binding site of the Val-82-Ala protease weakens the affinity of 679 while the IC_{50} of 676 is slightly decreased.

The effects of changing Arg_8 and Asp_{29} upon inhibitor binding are more pronounced, although in the case of Arg_8, not as great as might be anticipated from the activity data. For Asp-29-Ala, inhibitor binding is at least 50-fold weaker than the native enzyme, while the Arg-8-Glu mutation reduced the affinity for 679 by about 10-fold.

DISCUSSION

The mutagenesis that has been reported for HIV-1 protease has primarily been done with precursor forms expressed in *E. coli* which must be processed out of the precursor by the protease itself. In a thorough study in which each one of the 99 amino acid residues was individually changed to at least one other residue, regions of sensitivity and tolerance to changes were mapped.[11] The most sensitive areas are regions in the vicinity of the binding cleft. Interpretation of altered processing due to a given mutation is complicated by the fact that there are likely to be multiple catalytic species, incompletely processed, present. It is not clear to what extent each of the catalytic species is normally involved in the various polyprotein cleavages, either in the developing virion or in recombinant systems which allow processing to occur.

The changes that have been made in the native sequence of the mature, 99 amino acid protease have been much more limited. Aside from changes in the critical Asp_{25}, only one report has appeared in which the activity of mutants were quantified.[12] In that study, changes to Asn_{88} resulted in partial or no activity. Whether the partial activity was due to slow turnover, poor binding of substrate, or instability of the enzyme during the assay was not determined, although a weak association of subunits was suggested by size exclusion chromatography. In two other studies[13,14] Arg_{87} and Asp_{29} appeared to be critical for activity. The functional groups of Arg_{87}, Arg_8 and Asp_{29} appear close enough in the crystal structure that charge-charge interactions may stabilize the overall tertiary structure. From the work described here, it is clear that an Arg_8/Asp_{29} charge pair is not absolutely required for stability at 30°C for activity, although the Asp-29-Ala mutant activity is severely depressed.

Our interpretation of the data presented for our mutants is that Asp_{29} contributes a significant hydrogen bond to ligands, as suggested by the static picture seen with crystallography.[9,10] A charge pair of Arg_8/Asp_{29} may contribute to the overall tertiary and quaternary structure necessary for efficient catalysis.

ACKNOWLEDGEMENTS

We would like to thank Dr. Victor Garsky for the supply of ß-Nal peptide substrate.

REFERENCES

1. N. E. Kohl, E. A. Emini, W. A. Schleif, L. J. Davis, J. C. Heimbach, R. A. F. Dixon, E. M. Scolnick and I. S. Sigal, Active human immunodeficiency virus protease is required for viral infectivity, *Proc. Natl. Acad. Sci. U.S.A.* **85**:4686-4690 (1988).

2. M. I. Johnston, H. S. Allaudeen and N. Sarver, HIV proteinase as a target for drug action, *Trends Pharmacol. Sci.* **10**:305-307 (1989).

3. B. M. Dunn and J. Kay, Targets for antiviral chemotherapy: HIV-proteinase, *Antiviral Chem. and Chemotherapy* 1:3-8 (1990).

4. T. D. Meek, D. M. Lambert, G. B. Dreyer, T. J. Carr, T. A. Tomaszek, Jr., M. L. Moore, J. E. Strickler, C. Debouck, L. J. Hyland, T. J. Matthews, B. W. Metcalf and S. R. Petteway, Inhibition of HIV-1 protease in infected T-lymphocytes by synthetic peptide analogues, *Nature* **343**:90-92 (1990).

5. J. Erickson, D. J. Neidhart, J. VanDrie, D. J. Kempf, X. C. Wang, D. W. Norbeck, J. J. Plattner, J. W. Rittenhouse, M. Turon, N. Wideburg, W. E. Kohlbrenner, R. Simmer, R. Helfrich, D. A. Paul and M. Knigge, Design, activity and 2.8 angstrom crystal structure of a C2 symmetric inhibitor complexed to HIV-1 protease, *Science* **249**:527-533 (1990).

6. T. J. McQuade, A. G. Tomaselli, L. Liu, V. Karacosta, B. Moss, T. K. Sawyer, R. L. Heinrikson and W. G. Tarpley, A synthetic HIV-1 protease inhibitor with antiviral activity arrests HIV-like particle maturation, *Science* **247**:454-456 (1990).

7. P. Ashorn, T. J. McQuade, S. Thaisrivongs, A. G. Tomasselli, W. G. Tarpley and B. Moss, An inhibitor of the protease blocks maturation of human and simian immunodeficiency viruses and spread of infection, *Proc. Natl. Acad. Sci. U.S.A.* **87**:7472-7476 (1990).

8. M. A. Navia, P. M. D. Fitzgerald, B. M. McKeever, C.-T. Leu, J. C. Heimbach, W. K. Herber, I. S. Sigal, P. L. Darke and J. P. Springer, Three-Dimensional structure of the aspartyl protease from the human immunodeficiency virus, HIV-1, *Nature* **337**:615-620 (1989).

9. P. M. D. Fitzgerald, B. M. McKeever, J. P. Springer, J. C. Heimbach, C.-T. Leu, W. K. Herber, R. A. F. Dixon and P. L. Darke, Crystallographic analysis of a complex between HIV-1 protease and acetyl-pepstatin at 2.0 angstrom resolution, *J. Biol. Chem.* **264**:14209-14219 (1990).

10. M. Miller, J. Schneider, B. K. Sathyanarayana, M. V. Toth, G. R. Marshall, L. Clawson, L. Selk, S. B. H. Kent and A. Wlodawer, Structure of complex of synthetic HIV-1 protease with a substrate-based inhibitor at 2.3 angstrom resolution, *Science* **246**:1149-1152.

11. D. D. Loeb, R. Swanstrom, L. Everitt, M. Manchester, S. E. Stamper and C. A. Hutchison III, Complete mutagenesis of the HIV-1 protease, *Nature* **340**:397-400 (1989).

12. C. Guenet, R. A. Leppik, J. T. Pelton, K. Moelling, W. Lovenberg and B. A. Harris, HIV-1 protease: mutagenesis of asparagine 88 indicates a domain required for dimer formation, *Eur. J. Pharmacol.* **172**:443-451 (1989).

13. J. M. Louis, C. A. D. Smith, E. M. Wondrak, P. T. Mora and S. Oroszlan, Substitution mutations of the highly conserved arginine 87 of HIV-1 protease result in loss of proteolytic activity, *Biochem. Biophys. Res. Comm.* **164**:30-38 (1989).

14. E. Z. Baum, G. A. Bebernitz and Y. Gluzman, Isolation of mutants of human immunodeficiency virus protease based on the toxicity of the enzyme in *Escherichia coli*, *Proc. Natl. Acad. Sci.U.S.A.* **87**:5573-5577 (1990).

15. P. L. Darke, C.-T. Leu, L. J. Davis, J. C. Heimbach, R. E. Diehl, W. S. Hill, R. A. F. Dixon and I. S. Sigal, Human immunodeficiency virus protease: Bacterial expression and characterization of the purified aspartic protease, *J. Biol. Chem.* **264**:2307-2312 (1989).

16. J. C. Heimbach, V. M. Garsky, S. R. Michelson, R. A. F. Dixon, I. S. Sigal and P. L. Darke, Affinity Purification of the HIV-1 Protease, *Biochem. Biophys. Res. Commun.* **164**:955-960 (1989).

17. B. M. McKeever, M. A. Navia, P. M. D. Fitzgerald, J. P. Springer, C.-T. Leu, J. C. Heimbach, W. K. Herber, I. S. Sigal and P. L. Darke, Crystallization of the aspartyl protease from the human immunodeficiency virus, HIV-1, *J. Biol. Chem.* **264**:1919-1921 (1989).

18. S. D. Young, P. S. Anderson, P. L. Darke, L. J. Davis, S. J. DeSolms, R. A. F. Dixon, E. A. Emini, N. Gaffin, J. P. Guare, J. R. Huff, G. Robertson, W. M. Sanders, I. S. Sigal, W. Schleif, J. P. Vacca and J. M. Wiggins, The design and synthesis of HIV protease inhibitors, *Abstr. of Am. Chem. Soc.* **200**:94 (1990).

STRUCTURE-BASED INHIBITION OF HIV-1 PROTEASE ACTIVITY

AND VIRAL INFECTIVITY

Dianne L. DeCamp, Lilia M. Babé, Paul Furth, Paul Ortiz de Montellano,
Irwin D. Kuntz and Charles S. Craik

Department of Pharmaceutical Chemistry
University of California, San Francisco
San Francisco, California   94143-0446

INTRODUCTION

The protease encoded by the HIV genome is essential for viral replication, making the enzyme an important target for the therapeutic treatment of AIDS. The structure of the HIV protease[1] was used to locate compounds with molecular complementarity to the active site. The computer program DOCK[2] produced a negative image of the active site cavity that was a cylindrical shape 24 Å in length and 8 Å in diameter. The image was used to search 10,000 molecules of the Cambridge Crystallographic Database for compounds that could recognize the active site and serve as competitive inhibitors of the enzyme. One of the compounds to result from the complementarity screen was haloperidol, a known antipsychotic agent. Recombinant HIV-1 and HIV-2 proteases were expressed and purified from *E. coli*[3] and *S. cerevisiae*,[4] respectively, to assay the putative inhibitors. Haloperidol inhibits the HIV-1 and 2 proteases in a concentration-dependent fashion with a K_i of 100 μM.[5] Both haloperidol and its hydroxy derivative, hydroxy-HAL, show activity against maturation of viral polypeptides in an *E. coli* assay system.

Demonstration of the selectivity of inhibition by haloperidol as well as the ability of haloperidol to inhibit viral infection *in vitro* is essential. A useful HIV protease inhibitor should not inhibit the cellular aspartic proteases. Furthermore, it should be effective in mammalian cell culture. These characteristics of haloperidol inhibition have been investigated by examining its effect on renin and cathepsin D in addition to its ability to inhibit the infection of HeLa/T4 fused cells by virus particles.

MATERIALS AND METHODS

Enzymatic Assays

Recombinant HIV-1 protease was isolated from bacteria.[3] The enzyme was assayed as previously described[5] in 50 mM sodium acetate buffer, pH 5.5, containing 2 mM EDTA, 5 mM dithiothreitol, 1 M NaCl and 5 % DMSO. Peptide hydrolysis products were separated by HPLC and quantitated.[4] Stock solutions of haloperidol, pepstatin, and cerulenin, obtained from Sigma, or hydroxy-HAL, synthesized as described,[5] in DMSO were used in inhibition studies. Recombinant human renin was assayed with porcine angiotensinogen-(1-14) using a similar HPLC assay.[5] Bovine cathepsin D and its substrate, N-t-BOC-Phe-Ala-Ala-*p*-Nitro-Phe*Phe-Val-Leu 4-hydroxymethylpyridine ester, were purchased from Sigma. Assays were performed according to the method of Agarwal and Rich[6] in 0.5 ml microcuvettes on a Uvikon 680 spectrophotometer. Cathepsin D (0.25 μM) was added to 10 mM sodium formate buffer, pH 3.5, containing 60 μM substrate, 3 % DMSO, and varying concentrations of haloperidol. Activity was followed by measuring the increase in 310 nm absorption and could be inhibited with pepstatin.

Viral Infectivity Assay

A noncytopathic, mammalian cell assay[7] was used to test the effect of haloperidol on viral protein processing. This assay utilizes a replication defective HIV provirus (HIV-gpt) in which the envelope gene has been replaced with a drug resistance gene (gpt). Cotransfection of the gp160⁻ genome with a gp160⁺ expression vector results in packaging of the HIV-gpt genome into virions. Infection of susceptible cells with these virions bestows resistance to mycophenolic acid, which results in the growth of drug-resistant colonies. Quantitation of drug-resistant colonies allows a relatively safe measure of HIV infection.

Stock solutions (50 mM) of cerulenin or haloperidol in DMSO were added to transfected COS-7 cells and the virus-containing supernatant was collected. HeLa/T4 fused cells were incubated with supernatants from untreated and drug-treated COS-7 cells in the presence of mycophenolic acid. Viral infectivity was quantitated by counting colonies of drug-resistant HeLa/T4 cells.

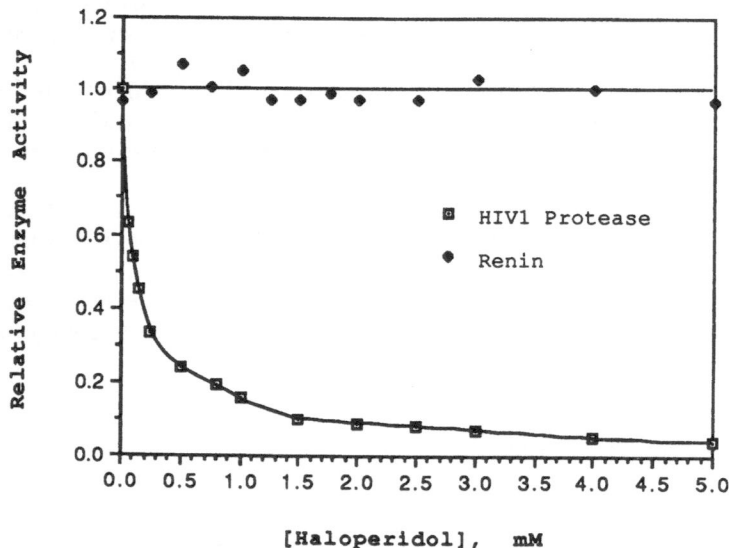

Figure 1. Inhibition of HIV-1 protease by haloperidol. Recombinant human renin, filled diamonds, and HIV-1 protease, open squares, were preincubated for five minutes with varying concentrations of haloperidol and assayed for activity as described previously[5].

Table I. HELA/T4 Infections with Virus from Transfected Cells

Treatment	% Reduction in Infectious Virions
Untreated	0
1% DMSO	0
100 μM cerulenin	80
50 μM haloperidol	0
200 μM "	30
500 μM "	40

RESULTS

Specificity of Haloperidol Inhibition

Haloperidol inhibits HIV-1 protease in a concentration-dependent manner, shown in Figure 1. The expected specificity of haloperidol for the active site of the viral protease is confirmed by its inability to inhibit human renin at concentrations as high as 5 mM (Figure 1, upper line). This is consistent with the fact that the angiotensinogen-(1-14) is not a substrate or an inhibitor of the HIV-1 protease. Renin is also not inhibited by hydroxy-HAL (data not shown). The effect of haloperidol and hydroxy-HAL on the lysosomal aspartic protease, cathepsin D, was also examined. Concentrations of up to 500 μM of either compound had no effect on the rate of substrate hydrolysis. Under the same conditions, addition of 60 μM pepstatin resulted in total loss of enzymatic activity.

In Vitro Reduction in HIV Virus Titers

The HIV-gpt virus vector was used to test the ability of haloperidol to reduce viral infectivity in cell culture. The antibiotic cerulenin was used for comparison since it is known to inhibit viral protein processing in chronically infected T cells.[8] The infectivity of virus obtained from cells incubated with various concentrations of haloperidol or cerulenin is shown in Table 1. The virus particles obtained from transfected COS-7 cells treated with 200 μM haloperidol showed a 30 % reduction in their ability to infect HeLa/T4 fused cells. Higher concentrations of haloperidol (500 μM) resulted in a dose-dependent response and 40 % reduction in infectious virions.

DISCUSSION

As an alternative to substrate-based enzyme inhibition, the 3-dimensional structure of the active site of the HIV-1 protease has been used to locate nonpeptide-based inhibitors with molecular complementarity to the active site of the enzyme. By *in vitro* assays of recombinant enzymes, we have shown that the butyrophenone derivative, haloperidol, specifically inhibits HIV-1 and HIV-2 proteases.[5] It is highly selective, having no inhibitory effect on renin at concentrations as high as 5 mM and no effect on cathepsin D at 500 μM. The results obtained in both bacterial[5] and mammalian cell assay systems suggest that haloperidol can efficiently penetrate these cells and inhibit the processing activity of the HIV-

1 protease. Due to its weak affinity for the enzyme, concentrations of haloperidol which inhibit the HIV protease *in vitro* would be toxic *in vivo*. Haloperidol itself is not a treatment for AIDS but may be a useful lead compound for the development of an effective antiviral pharmaceutical.

ACKNOWLEDGEMENTS

This work was supported by grant NIGMS-39552. D. L. D. is a recipient of N.I.H. fellowship GM 13369. L. M. B is a recipient of a University of California Task Force AIDS Fellowship F88SF122.

REFERENCES

1. A. Wlodawer, M. Miller, M. Jaskólski, B. K. Sathyanarayana, E. Baldwin, I. T. Weber, L. M. Selk, L. Clawson, J. Schneider and S. B. H. Kent, Conserved folding in retroviral proteases: Crystal structure of a synthetic HIV-1 protease, *Science* **245**:616-621 (1989).

2. R. DesJarlais, R. P. Sheridan, G. L. Seibel, J. S. Dixon and I. D. Kuntz, Using shape complementarity as an initial screen in designing ligands for a receptor binding site of known three-dimensional structure, *J. Med. Chem.* **31**:722-729 (1988).

3. L. M. Babé, S. Pichuantes, P. J. Barr, I. Bathurst, F. R. Masiarz and C. S. Craik, HIV-1 protease: bacterial expression, purification and characterization, *Proteins and Pharmaceutical Engineering, UCLA Symposium on Molecular and Cellular Biology,* **110**:71-88 (1990).

4. S. Pichuantes, L. M. Babé, P. J. Barr, D. L. DeCamp and C. S. Craik, Recombinant HIV-2 protease processes HIV-1 Pr53^{gag} and analogous junction peptides *in vitro*, *J. Biol. Chem.* **265**:13890-13898 (1990).

5. R. L. DesJarlais, G. L. Seibel, I. D. Kuntz, P. S. Furth, J. C. Alvarez, P. R. Ortiz de Montellano, D. L. DeCamp, L. M. Babé and C. S. Craik, Structure-based design of nonpeptide inhibitors specific for the human immunodeficiency virus 1 protease, *Proc. Natl. Acad. Sci. U.S.A.* **87**:6644-6648 (1990).

6. N. Agarwal and D. H. Rich, An Improved Cathepsin-D Substrate and Assay Procedure, *Anal. Biochem.* **130**:158-165 (1983).

7. K. A. Page, N. R. Landau and D. R. Littman, Construction and Use of a Human Immunodeficiency Virus Vector for Analysis of Virus Infectivity, *J. Virol.* **64**:5270-5276 (1990).

8. R. Pal, R. C. Gallo and M. G. Sarngadharan, Processing of the structural proteins of human immunodeficiency virus type 1 in the presence of monensin and cerulenin, *Proc. Natl. Acad. Sci. U.S.A.* **85**:9283-9286 (1988).

ANALYSIS OF TEMPERATURE-SENSITIVE MUTANTS OF THE HIV-1 PROTEASE

M. Manchester,[1,4] D. D. Loeb,[2,4*] L. Everitt,[2,4] M. Moody,[2,4]
C.A. Hutchison III[3,4] and R. Swanstrom[2,4]

[1]Curriculum in Genetics
[2]Department of Biochemistry
[3]Department of Microbiology and Immunology
[4]Lineberger Cancer Research Center
University of North Carolina at Chapel Hill
Chapel Hill, North Carolina   27599

*Present Address:
Department of Microbiology and Immunology
HSE401
University of California at San Francisco
San Francisco, California   94143

INTRODUCTION

Human Immunodeficiency Virus Type-1, like other retroviruses, encodes an aspartic proteinase whose activity is required for the production of infectious virions.[1-6] The protease (PR) is encoded at the 5' end of the viral *pol* gene and is responsible for cleavage of the viral gag and gag/pol precursor proteins to their mature forms.[7,8] Viruses with inactivating mutations in their protease domain yield immature, noninfectious particles containing unprocessed gag and gag/pol polyproteins;[1] for this reason the enzyme has been a prime target for structural and biochemical studies leading to the design of inhibitors of virus replication.

Recent structural and functional studies of the HIV-1 protease indicate that the enzyme has several features in common with other members of the aspartic proteinase class. The enzyme, while much smaller than related proteinases such as pepsin or renin, has emulated the bilobal structure found in larger aspartic proteinases by forming a dimer composed of identical monomers, each contributing an aspartic acid residue to the catalytic site.[9-14] Other features of HIV-1 PR are even more similar to cellular aspartic proteinases: the core of each monomer is composed of two interlocking peptide chains called the ψ-loop; and a pair of flap structures are present which fold over the catalytic site and contact the bound substrate.[11-16]

Structure and Function of the Aspartic Proteinases
Edited by B.M. Dunn, Plenum Press, New York, 1991

Saturation mutagenesis studies of HIV-1 PR have identified several mutationally sensitive domains of the protein, each of which has structural significance.[17-19] By generating random single amino acid substitutions throughout the entire protease domain, we have identified three regions of the protein in which consecutive amino acids are sensitive to non-conservative amino acid substitutions indicating that each of these regions is important for function (Figures 1, 2). The first is an eleven amino acid domain that includes the catalytic aspartic acid residue. This region interacts as part of the ψ-loop with a fifteen amino-acid domain near the C-terminus of the protein. The latter domain includes a β-turn and terminates in a short stretch of α-helix. These two mutationally sensitive regions intertwine to form the core of each subunit and the active site of the enzyme. A six amino-acid domain forming the flap structure is also mutationally sensitive, reflecting this region's role in substrate contact.[15,16,19]

Our collection of HIV-1 PR mutants has grown to over 400 single amino-acid substitutions. One approach we have taken to analyze our mutant library is to refine the assignment of mutationally sensitive domains by determining which domains are sensitive to conservative, in contrast to non-conservative, amino acid changes. In addition to highlighting the mutationally sensitive domains of the protease, our mutagenesis technique also allows us to recover mutants with such properties as altered substrate specificity and/or conditional activity. Another approach we have used is to screen a large subset of the

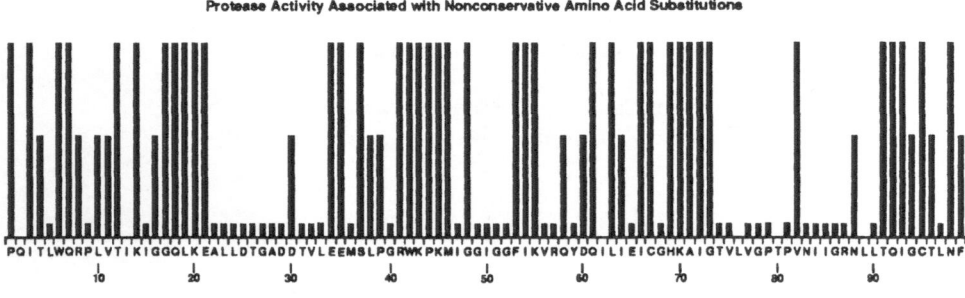

Figure 1. Protease activity associated with nonconservative and conservative amino acid substitutions. The amino acid sequence of the HIV-1 protease is shown using the single-letter amino acid designation. Vertical bars depict phenotype of either nonconservative (top) or conservative (bottom) amino acid substitutions at each position. Tallest bar, wild-type protease activity; smallest bar, negative activity; intermediate height bar, intermediate protease activity. The conservative amino acid groupings used in this analysis are A, S, T, P and G; V, L, I and M; F, Y and W; R, K and H; D, E, N and Q; and C.[21]

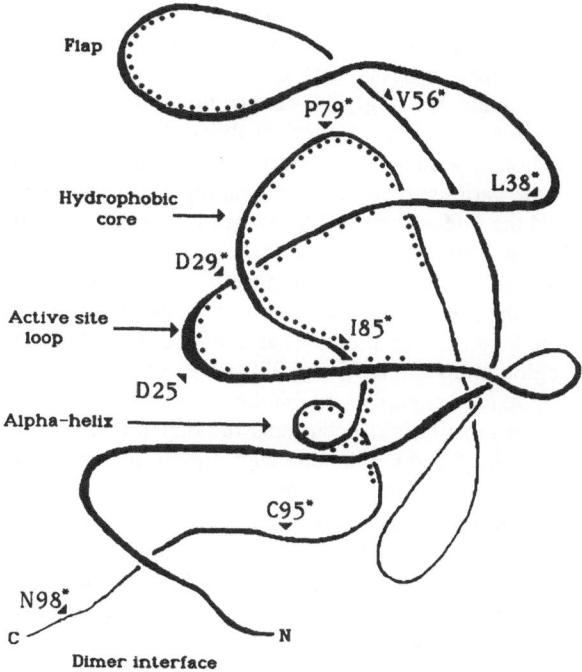

Figure 2. Line drawing of the HIV-1 protease monomer structure. N and C denote the amino and carboxy terminus, respectively. D25 denotes the catalytic aspartic acid residue. Dotted lines indicate mutationally sensitive regions. Positions of substitutions conferring a temperature-sensitive phenotype are starred. The drawing is based on the structure by Wlodawer et al.[11]

mutants for a temperature-sensitive (ts) phenotype. Temperature sensitivity of a mutant enzyme has classically been used as a measure of a particular residue's structural and functional contributions to the folded state of the protein. Studies of ts mutants of bacteriophage *T4* lysozyme have localized these types of substitutions to regions of the protein that are buried and held in a rigid conformation in the folded protein.[20] Temperature sensitive mutants are therefore helpful in pinpointing specific regions of the protein that contribute to enzyme stability.

RESULTS AND DISCUSSION

We have used a bacterial expression vector to both introduce mutations in PR and to evaluate mutant phenotypes. Mutants were generated by random site-directed mutagenesis using the expression clone pART2 which contains the *BglII-SalI pol* fragment from HIV-1 HXB2 fused to the *lac* promoter in the plasmid pIBI20 (19). Phenotype assignments were made by assaying the ability of the protease to process the pol substrate to its mature products (p11 protease; p64/51 reverse transcriptase, RT; and p34 integrase, IN) as determined by Western blot analysis. Each of the 99 amino acids of PR has been mutagenized, and we have determined the effects of both conservative and nonconservative sidechain substitutions throughout the protease domain. A compilation of these data is shown in Figure 1. As expected, fewer total residues are sensitive to conservative changes than to nonconservative changes; however, the general placement of mutationally sensitive domains remains the same. By analyzing conservative changes, we have simply narrowed the regions that are mutationally sensitive. Especially revealing is the sensitivity of the flap region. It would appear that, with the exception of two glycines at positions 49 and 52, virtually the entire flap will tolerate conservative substitutions. Further analysis of the range of substitutions allowed at each flap position should be helpful in determining which flap residues are involved in substrate contact with the various sequences representing the gag and pol processing sites.

We have tested the phenotypes of 254 protease mutants for temperature sensitivity. Each mutant was grown at 32°C and 37°C and the extent of processing evaluated by western blot analysis. Seven of the mutants, or approximately 3 % of those tested, showed increased or altered processing of the pol substrate at the lower temperature. We have localized these mutants throughout the length of the protease, as shown in Figure 2. Two of these occur in the dimer interface region at positions 95 and 98, and these substitutions may disrupt dimer formation at the restrictive temperature or inhibit the conformational change that occurs during substrate binding.[15] Two substitutions occur near the flap at residues 38 and 56, in a region that acts as a bending point for the extended flap region during substrate binding.[15] The phenotypes of these substitutions may reflect variation in flap movement at the different temperatures. Other substitutions occur in the region of the protein known as the ψ-loop: the first occurs at position 79, the top of a β-turn in the core, while the second occurs at position 85, the top of the α-helix in the core. Both of these substitutions may affect close packing of residues and disrupt the core structure of the enzyme at the restrictive temperature. The seventh mutation that confers a ts phenotype lies near the active site residue at position 29. This mutant is especially interesting in that it shows an unusual phenotype: at the restrictive temperature it shows little activity, while at the permissive temperature it will release the protease but recognizes other pol cleavage sites poorly, yielding primarily an RT/IN fusion protein.

Extensive mutagenesis of the HIV-1 PR has yielded a wealth of information regarding both the side chain requirements of individual residues, and the identification of functional domains of the enzyme. Additionally, this technique has provided a varied group of mutants with unusual phenotypes, each of which will contribute to the understanding of the activity of this important enzyme.

ACKNOWLEDGEMENT

Our work on protease mutagenesis is supported by a grant from the NIAID. M. M. is supported by a Genetics Training Grant from the NIGMS.

REFERENCES

1. N. E. Kohl, E. A. Emini, W. A. Schleif, L. J. Davis, J. C. Heimbach, R. A. Dixon, E. M. Scolnick and I. S. Sigal, Active human immunodeficiency virus protease is required for viral infectivity, *Proc. Natl. Acad. Sci. U. S. A.* **85**:4686-4690 (1988).
2. C. Peng, B. Ho, T. Chang and N. Chang, Role of human immunodeficiency virus type 1 specific protease in core maturation and viral infectivity, *J. Virol.* **63**:2550-2556 (1989).
3. H. G. Göttlinger, J. G. Sodroski and W. A. Haseltine, Role of capsid precursor processing and myristoylation in morphogenesis and infectivity of the human immunodeficiency virus type 1, *Proc. Natl. Acad. Sci. U. S. A.* **86**:5781-5785 (1989).
4. S. Crawford and S. P. Goff, A deletion mutant in the 5' part of the *pol* gene of Moloney murine leukemia virus blocks proteolytic processing of gag and pol polyproteins, *J. Virol.* **53**:899-907 (1985).

5. I. Katoh, Y. Yoshinaka, A. Rein, M. Shibuya, T. Odaka and S. Orozlan, Murine leukemia virus maturation: protease region required for conversion from "immature" to "mature" core form and for infectivity, *Virology* **145**:280-292 (1985).

6. Von der Helm, K., Cleavage of Rous sarcoma viral polypeptide precursor into internal structural proteins *in vitro* involves viral protein p15, *Proc. Natl. Acad. Sci. U.S.A.* **74**:911-915 (1977).

7. S. Oroszlan and R. B. Luftig, Retroviral proteinases, *in*: "Retroviruses: strategies for replication," R. Swanstrom and P. K. Vogt, eds., Springer, Berlin, *Curr. Top. Microbiol. Immunol.* **157**:153-185 (1990).

8. R. Swanstrom, A. Kaplan and M. Manchester, The aspartic proteinase of HIV-1, *Seminars in Virology* **1**: 175-186 (1990).

9. T. D. Meek, B. D. Dayton, B. W. Metcalf, G. B. Dreyer, J. E. Strickler, J. G. Gorniak, M. Rosenberg, M. L. Moore, V. W. Magaard and C. Debouck, Human immunodeficiency virus 1 protease expressed in *Escherichia coli* behaves as a dimeric aspartic protease, *Proc. Natl. Acad. Sci. U. S. A.* **86**:1841-1845 (1989).

10. I. Katoh, Y. Ikawa and Y. Yoshinaka, Retrovirus protease characterized as a dimeric aspartic proteinase, *J. Virol.* **63**:2226-2232 (1989).

11. A. Wlodawer, M. Miller, M. Jaskólski, B. K. Sathyanarayana, E. Baldwin, I. T. Weber, L. M. Selk, L. Clawson, J. Schneider and S. B. H. Kent, Conserved folding in retroviral proteases: Crystal structure of a synthetic HIV-1 protease, *Science* **245**:616-621 (1989).

12. M. A. Navia, P. M. Fitzgerald, B. M. McKeever, C.-T. Leu, J. C. Heimbach, W. K. Herber, I. S. Sigal, P. L. Darke and J. P. Springer, Three-dimensional structure of aspartyl protease from human immunodeficiency virus HIV-1, *Nature* **337**:615-1920 (1989).

13. M. Miller, M. Jaskolski, J. K. M. Rao, J. Leis and A. Wlodawer, Crystal structure of a retroviral protease proves relationship to aspartic protease family, *Nature* **337**:576-579 (1989).

14. R. Lapatto, T. Blundell, A. Hemmings, J. Overington, A. Wilderspin S. Wood, J. R. Merson, P. J. Whittle, D. E. Danley, K. F. Geoghegan, S. J. Hawrylik, S. E. Lee, K. G. Scheld and P. M. Hobart, X-ray analysis of HIV-1 proteinase at 2.7 Å resolution confirms structural homology among retroviral enzymes, *Nature* **342**:299-302 (1989).

15. M. Miller, B. K. Sathyanarayana, M. V. Toth, G. R. Marshall, L. Clawson, L. Selk, J. Schneider, S. B. H. Kent and A. Wlodawer, Structure of complex of synthetic HIV-1 protease with a substrate-based inhibitor at 2.3 Å resolution, *Science* **246**:1149-1152 (1989).

16. J. Erickson, D. J. Neidhart, J. Van Drie, D. J. Kampf, X. C. Wang, D. W. Norbeck, J. J. Plattner, J. W. Rittenhouse, M. Turon, N. Wideburg, W. E. Kohlbrenner, R. Simmer, R. Helfrich, D. A. Paul and M. Knigge, Design, activity and 2.8 Å crystal structure of a C2 symmetric inhibitor complexed to HIV-1 protease, *Science* **249**:527-533 (1990).

17. W. G. Farmerie, D. D. Loeb, N. C. Casavant, C. A. Hutchison III, M. H. Edgell and R. Swanstrom, Expression and processing of the AIDS virus reverse transcriptase in *Escherichia coli*, *Science* **236**:305-308 (1987).

18. D. D. Loeb, C. A. Hutchison, M. H. Edgell, W. G. Farmerie and R. Swanstrom, Mutational analysis of human immunodeficiency virus type 1 protease suggests functional homology with aspartic proteinases, *J. Virol.* **63**:111-121 (1989).

19. D. D. Loeb, R. Swanstrom, L. Everitt, M. Manchester, S. E. Stamper and C. A. Hutchison, Complete mutagenesis of the HIV-1 protease, *Nature* **340**:397-400 (1989).

20. T. Alber, D.-P. Sun, J. A. Nye, D. C. Muchmore and B. W. Matthews, Temperature-sensitive mutations of bacteriophage *T4* lysozyme occur at sites with low mobility and low solvent accessibility in the folded protein, *Biochemistry* **26**:3754-3758 (1987).

21. M. O. Dayhoff, R. M. Schwartz and B. C. Orcut, *in*: "Atlas of Protein Sequence and Structure," M. O. Dayhoff, ed., **5**:345-352 (1978).

STUDIES OF THE AUTOPROCESSING OF THE HIV-1 PROTEASE USING

CLEAVAGE SITE MUTANTS

John M. Louis,[1] Stephen Oroszlan[2] and Peter T. Mora[3]

[1]Laboratory of Cellular and Developmental Biology
NIDDK, NIH
Bethesda, Maryland   20892

[2]Laboratory of Molecular Virology and Carcinogenesis
ABL-Basic Research Program
NCI-FCRDC
Frederick, Maryland   21701

[3]Office of the Director
DCBD, NCI, NIH
Bethesda, Maryland   20892

INTRODUCTION

The distinguishing features of aspartyl proteases in retroviruses, vis à vis the semi-symmetrical aspartyl proteases in other organisms, is the paucity of the genetic information in the viruses sufficient for the synthesis of a monomeric protein, from which the homodimer is assembled with the formation of the characteristic active site structure.[1]

We have chemically synthesized and expressed the small 297 bp HIV-1 protease gene in *E. coli*, and showed the specific cleavage of the recombinant gag polyprotein and synthetic peptides used as substrates.[2,3] This approach offers a convenient method to study by mutational analysis some of the essential features of the structure necessary for activity. In addition to the Asp-Thr-Gly sequence (amino acids 25-27) in the HIV-1 protease which provides the characteristic aspartic proteinase-like active site of the dimer, there is a second highly conserved region Gly-Arg-Asp/Asn (amino acids 86-88) located near the C-terminal domain that is unique to the viral enzymes and not present in the cellular aspartyl proteases. We showed by single amino acid substitutions ($Arg_{87} \rightarrow Lys$; $Arg_{87} \rightarrow Glu$; $Asn_{88} \rightarrow Glu$) that the mutant proteases were completely devoid of activity, demonstrating the essential nature of the $Arg_{87}Asn_{88}$ residues in the HIV-1 protease.[3]

Our current mutational analysis is directed to elucidate the unique structural features in the formation of an active site structure which arises through the juxtaposition of two gag-pol polyproteins required for autoprocessing. In this communication we present initial *in vitro* analyses using purified wild-type and mutated fusion proteins as models.

Structure and Function of the Aspartic Proteinases
Edited by B.M. Dunn, Plenum Press, New York, 1991

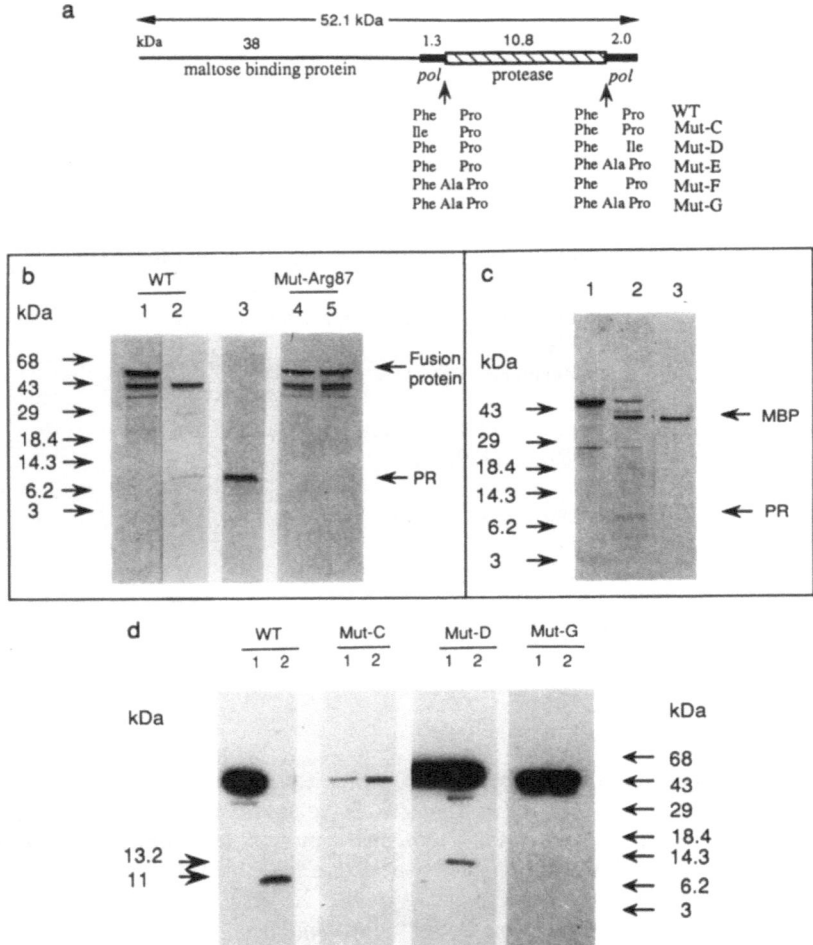

Figure 1. Autoprocessing of purified wild-type and mutated fusion proteins expressed in *E. coli*. The various wild-type and mutated fusion constructs are illustrated in 1a. Proteins were analyzed using 10-20 % gradient tricine polyacrylamide gels and stained using Coomassie blue (b and c). Molecular size standards are indicated in kDa by arrows. The relative positions of the fusion proteins, PR and the intermediate products of cleavage are also indicated. b) The wild-type and mutant fusion proteins (2 µg) before (lanes 1 & 4) and after (lanes 2 & 5) the urea denaturation/renaturation steps. Purified PR (400 ng) after cation exchange chromatography (lane 3). c) The full-length Arg87 mutant fusion protein (600 ng) before (lane 1) and after (lane 2) incubation with the purified, mature wild-type protease (12 ng). Purified, mature wild-type protease (12 ng) was undetectable by Coomassie staining (not shown). Purified MBP (0.5 µg; lane 3). d) Immunoblotting analysis of wild-type and mutant fusion proteins (2 µg) before (lanes 1) and after (lanes 2) refolding.

MATERIALS AND METHODS

The HIV-1 protease (PR) gene containing a portion of the flanking *pol* gene (12 and 19 amino acids at the N- and C-termini of the PR, respectively) was constructed in fusion with the *malE* gene of *E. coli*,[4] which codes for the maltose binding protein (MBP), as shown in Figure 1a, and was expressed in *E. coli*, JM 109. A majority (> 90 %) of the 52 kDa fusion protein that was expressed in induced cells was obtained in the soluble supernatant. The lack of self-processing allowed rapid purification of the fusion proteins by the amylose affinity column chromatography.[4]

We recovered the self-processing activity of the PR fusion protein by urea treatment followed by dialyses against decreasing concentrations of urea. Figure 1b, lane 1 shows the full-length 52 kDa fusion product of the WT clone. After the refolding procedure, the 52 kDa product was reduced in size to products migrating at the position of the maltose binding protein domain (38 kDa) and the 11 kDa PR (lane 2). Immunoblotting analysis of the same preparations before (1) and after (2) the refolding procedure confirmed the above observation (Figure 1d, WT). The cleaved MBP and HIV-1 PR were separated on a cation exchange column. By this simple method, we obtained *ca.* 1 mg/liter of ≥ 95% pure PR (Figure 1b, lane 3) from cells grown in a low density (A $_{600\,nm}$ = 0.4 to 0.5 O. D.) culture.

Processing of the Mutated PR Fusion Protein

To study the intermolecular mechanism of protease processing we used a clone which contained an inactive mutated (Arg$_{87}$ → Lys) PR gene. The mutant fusion protein was processed identically to the wild-type fusion protein, as described above. The samples analyzed before (4) and after (5) the refolding steps are shown in Figure 1b. The fusion protein containing the mutated PR did not exhibit autoprocessing; thus a conservative mutation at this position totally inactivated the enzyme. This is consistent with reports on the 3-D structure of HIV-1 PR, showing that the Arg$_{87}$ ion pairs with the conserved Asp$_{29}$ in forming the specific structure for the active site.[1]

The refolded full-length mutated fusion protein was incubated with the mature purified wild-type PR and the products of cleavage were analyzed by SDS-PAGE (Figure 1c). In the absence of the wild-type PR only the full-length product (Figure 1c, lane 1) corresponding to a size of 52 kDa and no detectable cleavage products were observed in a control incubation. In contrast, incubation of the fusion protein with the added purified wild-type PR (Figure 1c, lane 2), resulted in the cleavage of the full-length fusion protein and in the release of a product migrating to a position corresponding to the 38 kDa MBP and the 11 kDa PR (Figure 1c, lane 2). This shows that mature WT PR is capable of "trans-cleaving" a fusion protein at both the N- and C-termini of the PR. This result is consistent with suggestions based on structural studies which show that the N- and C-terminal strands of the dimer PR are organized away from the active site and unlikely to be cleaved intramolecularly.[1]

Self-Processing Activity of the Fusion Proteins Containing Mutated Cleavage Sites

Fusion proteins containing mutations at positions immediately adjacent to the N- and C-terminal amino acids of the wild-type PR were constructed and processed as mentioned before (Figure 1a & 1d). Figure 1d shows the analysis of the cleavage products before (lanes 1) and after (lanes 2) the 5 M urea denaturation / renaturation steps. After renaturation, the N-terminal mutant fusion proteins obtained from clones Mut-C (Figure 1d, lane 2) and Mut-F (not shown) revealed partial cleavage at the C-terminal site of the PR of the full length fusion protein and released a product of 50 kDa, detectable just below the 52 kDa full-length product. There was no other intermediate cleavage product detectable, demonstrating absence of cleavage at the N-terminus of the PR. The fusion protein derived from the C-terminal mutant clone (Mut-D) showed the release of a 13 kDa product through a N-terminal cleavage (Figure 1d, lane 2). A clone containing a C-terminal insertion mutation (Mut-E) showed complete processing and released a product of 11 kDa, similar to the wild-type clone (Figure 1b & D, WT, lane 2). A clone containing Ala insertion mutations at both cleavage sites (Mut-G) did not exhibit any products of cleavage (Figure 1d, lane 2). Ile insertion mutations at both cleavage sites resulted in total loss of processing activity (similar to Mut-G; Figure 1d, lane 2).

The purified fusion protein obtained from the WT clone before refolding showed very low proteolytic activity of ca. 5 pmoles/min/μg protein, which after refolding increased to a relatively high activity of ca. 1125 pmoles/min/μg protein. The PR which was ≥ 95% pure after the cation-exchange column showed a specific activity of about 8500 pmoles/min/μg. The renatured partially processed preparations derived from clones C, D, F and G showed very low activities (3-10 pmoles/min/μg protein) irrespective of the size of the flanking sequences around the PR protease. In contrast, clone Mut-E released the mature PR having a high activity similar to the wild-type.

The N-terminal mutations either Phe → Ile (Mut-C) or Phe-Ala-Pro (Mut-F) and the C-terminal mutation Pro → Ile (Mut-D) resulted in total loss of cleavage at the mutated sites and reduced cleavage at the non-mutated sites. In these mutants, after the primary cleavage at one of the non-mutated sites, apparently there must be a conformational change which prevents the cleavage at the other non-mutated site. However, Ala insertion at the C-terminus does not substantially interfere with the assembly of active enzyme and the cleavage at both termini. In contrast, Ala insertion at both ends (Mut-G) prevented cleavage at both ends. Comparisons of the processing activity between clones Mut-E and Mut-G suggested that an insertion mutation (Phe-Ala-Pro) at the C-terminal site is tolerated, provided the wild-type site is preserved at the N-terminus. These results indirectly suggest that the N-terminal cleavage precedes the cleavage at the C-terminus.

In summary, the analysis of the various mutants indicated that the cleavages occurring at the N- and the C-terminal regions of the PR are interdependent. The approach described above will be applied to mutate the protease at the interface regions to understand the mechanism of dimerization of the gag-pol precursor.

ACKNOWLEDGEMENT

This investigation was supported in part by National Cancer Institute, DHHS under contract number N01-CO-74101 with ABL.

REFERENCES

1. A. Wlodawer, M. Miller, M. Jaskólski, B. K. Sathyanarayana, E. Baldwin, I. T. Weber, L. M. Selk, L. Clawson, J. Schneider and S. B. H. Kent, Conserved folding in retroviral proteases: crystal structure of a synthetic HIV-1 protease, *Science* **245**:616 (1989).
2. J. M. Louis, E. M. Wondrak, T. D. Copeland, C. A. D. Smith, P. T. Mora and S. Oroszlan, Chemical synthesis and expression of the HIV-1 protease gene in *E. coli*, *Biochem. Biophys. Res. Comm.* **159**:87 (1989).
3. J. M. Louis, C. A. D. Smith, E. M. Wondrak, P. T. Mora and S. Oroszlan, Substitution mutations of the highly conserved Arginine 87 of HIV-1 protease results in loss of proteolytic activity, *Biochem. Biophys. Res. Comm.* **164**:30 (1989).
4. C. Guan, P. Li, P. D. Riggs and H. Inouye, Vectors that facilitate the expression and purification of foreign peptides in *Escherichia coli* by fusion to maltose-binding protein, *Gene* **67**:21 (1988).

MUTATIONAL ANALYSIS OF A NATIVE SUBSTRATE OF THE HIV-1 PROTEINASE

Kathryn Partin, Eckard Wimmer and Carol Carter

Department of Microbiology
State University of New York at Stony Brook
Stony Brook, New York   11794-8621

INTRODUCTION

Proteolytic processing of the gag/pol precursor by the human immunodeficiency virus (HIV) type 1 proteinase (PR) is essential for production of infectious viral particles. Although the sites of viral specific cleavages have been determined, the primary amino acid sequence around these scissile bonds are heterogeneous, and the determinants that direct the cleavage specificity exhibited by HIV-1 PR remain largely undefined. We performed mutational analysis of the Tyr/Pro site which produces the amino terminus of the capsid protein (CA) and the Phe/Pro site which produces the amino terminus of PR. Single-amino-acid substitutions were made, and their effects on proteolytic processing were examined by *in vitro* translation of a synthetic mRNA encoding both the mutated substrate and PR as a truncated gag/pol precursor. We found evidence for sequence determinants of cleavage at the Tyr/Pro site, and structural determinants of cleavage at the Phe/Pro site. We next investigated the role of the polyprotein structure in either impeding or facilitating cleavage at Phe/Pro. Deletion of p6*, the region in *pol* upstream of PR, resulted in improved processing of the gag cleavage sites. Our results suggest that p6* is involved in the regulation of PR activity and that release of p6* inhibition may be an activation step necessary for infectious particle maturation.

WHAT IS THE BASIS OF HIV-1 SUBSTRATE SPECIFICITY?

The HIV-1 PR recognizes at least 10 sites in gag/pol.[1] All are different in primary amino acid sequence, yet the enzyme is highly specific and will not cut apparently similar sequences derived from related viruses. We selected two cleavage sites in gag/pol for mutational analyses, Tyr/Pro at the amino terminus of capsid (CA) and Phe/Pro at the amino terminus of PR.[2] Single-amino acid substitutions were made at the cleavage site, and in the surrounding amino acids. The Tyr/Pro site was chosen because CA is the major component of the virion structure, making its production of clear biological significance, and because PR uses a single, unambiguous cleavage site at the amino terminus of CA. The Phe/Pro site was chosen because it is critical for autocatalysis of PR. Our *in vitro* analyses were facilitated by construction of a mutant (designated FS or Forced Frameshift) which contains a 4 bp insertion in the region of *gag/pol* overlap.[3] Translation of mRNA encoding this mutation in our *in vitro* system results in a shift of the reading frame from *gag* to *pol*, and circumvents the requirement for ribosomal frameshifting for expression of PR.[4] The use of polyproteins rather than small peptide substrates for our mutational studies permitted an analysis of determinants within the context of the natural substrate.

The mutants which contained a substitution of the Tyr/Pro site on the gag substrate fell into three categories. The first carried substitutions beyond positions P_5 or P_4, and exhibited a phenotype indistinguishable from wild type. The second category of mutated substrates exhibited impaired processing which resulted in accumulation of aberrant intermediates. Thus, these mutations yielded effects distal to the site at which the substitution was made, and may have made the substrates into competitive inhibitors or may have altered substrate conformation. The third category of mutated substrates was completely defective for cleavage at Tyr/Pro in the presence of 50-100 fold excess PR and exhibited a resistant p41 species which represented the MA-CA domains. Thus, radical substitutions at P_4, P_2, P_1 and $P_{2'}$, and even conservative substitution at P_2, produced an uncleavable Tyr/Pro substrate.

In contrast, the Phe/Pro site in pol demonstrated far greater tolerance to amino acid substitution. Of the 17 substitutions made at this site, none resulted in an uncleavable substrate. The exact site of processing was not determined, but our analyses indicated that cleavage occurred at a distance no greater than 5 amino acids from the Phe/Pro scissile bond.

These results indicate that the Tyr/Pro site may be restricted mainly by primary amino acid sequence, while the Phe/Pro site is primarily restricted by conformation. Single amino acid changes in the upstream region and double mutations that altered charge (and possible structure) support this hypothesis (data not shown). Such differences may not be discernible in peptide studies in which small sequences are studied. When the structure of the region was dramatically altered by deletion of the cleavage site and upstream (p6*) sequences, the efficiency of polyprotein processing was *increased* 3-fold relative to the wild type.

IS THE UPSTREAM REGION FUNCTIONING AS A REPRESSOR OF PR ACTIVITY?

The improved processing by the p6* deletion mutant was not due to increased accumulation of enzyme, since lysates of both wt and the mutant contained equivalent levels of labeled PR. Moreover, improved processing was not a result of alteration of the amino terminus of PR, since a p6* deletion mutant containing the restored natural cleavage site and surrounding amino acids also autoprocessed more efficiently than wild-type (data not shown). If, as proposed for the zymogen pepsinogen,[5] upstream sequences bind near the active site and make the binding cleft inaccessible to substrates, the active site of PR should be more accessible in the polyprotein containing the p6* deletion.

Two mutants, WT-ATG and Δp6*-ATG, which encode an Asp to Ala mutation which inactivates the active site and permits accumulation of the primary translation products[6] were constructed. Both constructs were immunoprecipitated with antibodies directed against twenty-three residues around and including the active site. Immunoprecipitation was performed under a variety of conditions, ranging from native to denaturing. As predicted by the hypothesis, the antibody directed against the active site had a higher affinity for the Δp6* polyprotein than the WT polyprotein. Quantitation by densitometry of products immunoprecipitated under native conditions indicated that this effect was 2.5-fold greater for the p6* deletion mutant compared to WT. Thus, the 3-fold improvement of polyprotein processing exhibited by the mutant correlates well with the apparent 2.5-fold increase in active site accessibility.

Figure 1 shows an alignment of pepsinogen and upstream sequences of several Lentiviruses.[7] It is intriguing that although there is no overt similarity between the "p6*" regions of Lentiviruses, they all share the presence of a short stretch of basic residues, located approximately the same distance from the Asp-Thr-Gly residues of the active site.

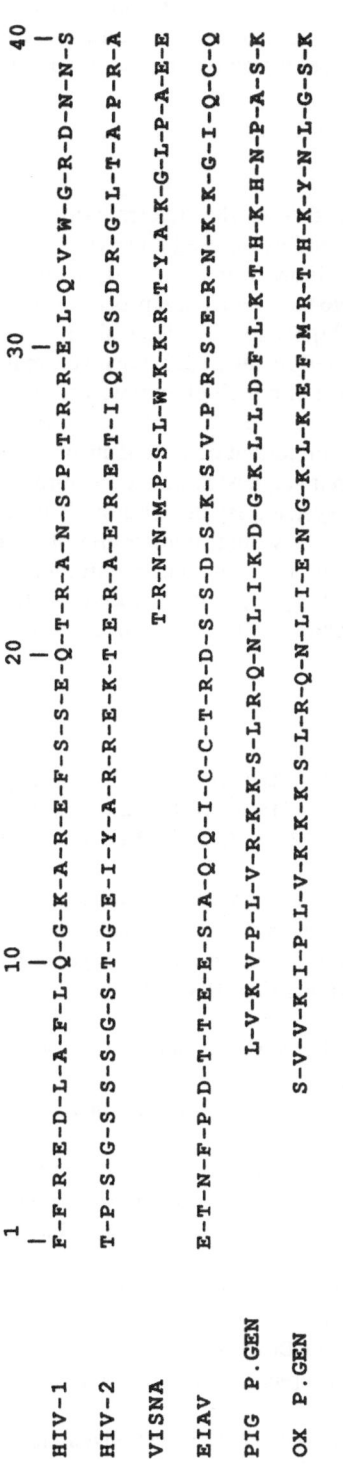
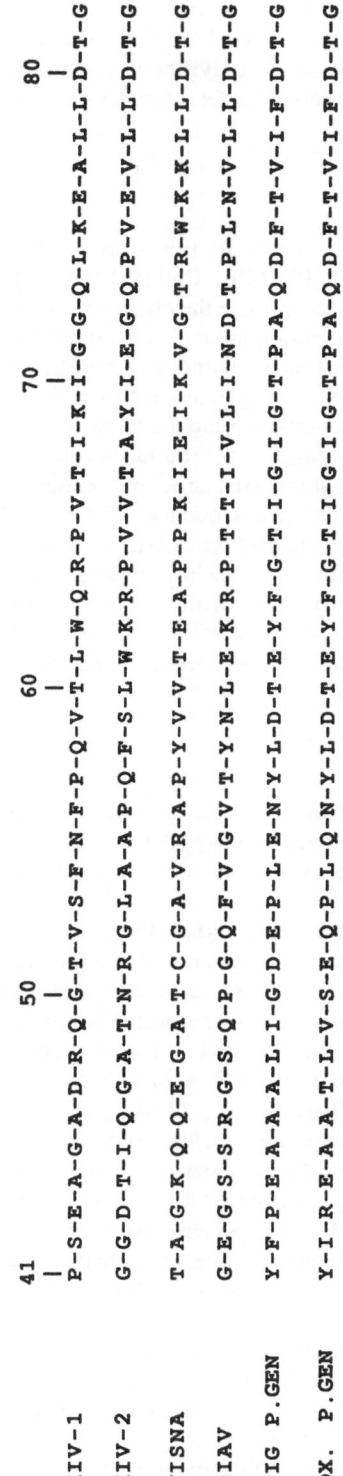

Figure 1. An alignment of pepsinogen and upstream sequences of several Lentiviruses

This similarity suggests that there may be some vestigial inhibitory activity of the p6* region by interaction with the active site. We propose a model in which p6* interaction with PR prevents efficient proteolytic processing of the gag/pol polyprotein. Definitive analysis of the contributions of the polyprotein towards efficient processing will require inclusion of the carboxyl portion of (reverse transcriptase and integrase) which are missing in our constructs.

SUMMARY

The purpose of this study was to define further the determinants of substrate specificity of HIV-1 PR. Rather than using small peptides, we used an *in vitro* system which permitted us to evaluate the effect of a mutated site within the context of its natural precursor. We made single-amino-acid substitutions around two sites which are processed by the HIV-1 PR. The Tyr/Pro site within gag appears to encode highly specific determinants which direct proteinase processing between MA and CA. The Phe/Pro site in pol, however, appears to be far more tolerant to amino acid substitutions, as none of our single-amino-acid substitutions blocked cleavage at or around this site. The increased tolerance of the Phe/Pro site may indicate that at this site, structural features are more important determinants of cleavage than primary amino acid sequence. We have shown that sequences outside of those encoding mature PR can inhibit proteolytic processing in this system. By preventing PR from cleaving itself from the polyprotein prematurely, p6* sequences would regulate morphogenesis and infectious particle formation. Late in infection, when the protein concentration of gag and gag/pol polyproteins at the cell surface becomes very high, cooperative protein-protein interactions may cause alterations of a p6*-PR interaction, relieving repression and permitting autocatalysis.

REFERENCES

1. L. E. Henderson, T. D. Copeland, R. C. Sowder, A. M. Schultz and S. Oroszlan, Analysis of proteins and peptides purified from sucrose gradient banded HTLV-III, *in:*. "Human Retroviruses, Cancer and AIDS: Approaches to Prevention and Therapy," D. Bolognesi, ed., Alan R. Liss, Inc., New York, p.135-147, (1988).
2. K. Partin, H.-G. Kräusslich, L. Ehrlich, E. Wimmer and C. Carter, Mutational analysis of a native substrate of the human immunodeficiency virus type 1 proteinase, *J. Virol.* **64**:3938-3947 (1990).
3. H.-G. Kräusslich, H. Schneider, G. Zybarth, C. A. Carter and E. Wimmer, Processing of *in vitro* synthesized gag precursor proteins of human immunodeficiency virus (HIV) type 1 by HIV proteinase generated in *Escherichia coli*, *J. Virol.* **62**:4393-4397.
4. T. Jacks, M. D. Power, F. R. Masiarz, P. A. Luciw, P. J. Barr and H. E. Varmus, Characterization of ribosomal frameshifting in HIV-1 *gag/pol* expression, *Nature (London)* **331**:280-283.
5. M. N. G. James and A. R. Sielecki, Molecular structure of an aspartic proteinase zymogen, porcine pepsinogen, at 1.8 Å resolution, *Nature (London)* **319**:33-38 (1986).
6. J. Mous, E. P. Heimer and S. F. G. LeGrice, Processing protease and reverse transcriptase from human immunodeficiency virus type 1 polyprotein in *Escherichia coli*, *J. Virol.* **62**:1433-1436 (1988).
7. B. Foltman and V. B. Pedersen, *in*: Acid Proteases, Structure, Function and Biology, J. Tang, ed., Plenum, N.Y., p. 103-127 (1977).
8. G. Myers, S. Josephs, J. A. Berzofski, A. B. Rabson, T. G. Smith and F. Wong-Staal, *in*: "Theoretical Biology and Biophysics," Los Alamos National Laboratory, USA (1989).
9. P. Sonigo, M. Alizon, K. Staskus, D. Klatzman, S. Cole, O. Danus, E. Retzel, P. Tiolais, A. Haase and S. Wain-Hobson, Equine infectious anemia virus *gag* and *pol* genes: relatedness to visna and AIDS virus, *Cell* **42**:369-382 (1985).
10. R. M. Stephens, J.W. Casey and N.R. Rice, Nucleotide sequence of the visna lentivirus: relationship to the AIDS virus, *Science* **231**:589-594 (1986).

MONOCLONAL AND POLYCLONAL ANTIBODIES:

REAGENTS FOR STUDYING HIV-1 PROTEINASE VARIANTS

Timothy J. Stoller, J. Johann Lim, Barry A. Woltizky,
and Mary C. Graves

Department of Molecular Genetics
Roche Research Center, Hoffmann-LaRoche Inc.
Nutley, New Jersey   07110

INTRODUCTION

HIV-1 and HIV-2 proteinases (PR) process the gag and gag-pol precursor polyproteins to produce the mature structural and enzymatic proteins found in infectious virus particles. These aspartic proteinases, from HIV-1 in particular, have been studied intensely since their discovery because they are attractive targets for therapeutic intervention in the treatment of AIDS. Antibodies which specifically recognize these proteinases were critical reagents in many of the experiments investigating the function of these proteinases. Here we describe the characterization of polyclonal and monoclonal antibodies against HIV-1 and HIV-2 PR.

METHODS

Antibody Production

Polyclonal antibodies 5833 and 7689 were produced by immunizing rabbits, as previously described,[1] with KLH-conjugated peptides (sequence in Figure 1). Antisera from these immunized rabbits were used directly in the experiments described below. Polyclonal antibody 6802 was raised against HIV-1 His_5 PR: a recombinant protein synthesized in bacteria composed of full-length HIV-1 PR (amino acids 1-99) extended at the amino terminus by the following sequence: Met-Arg-Gly-Ser-His-His-His-His-His. The histidine extension was added for purification purposes.[2] Antibody 6802 was used as an IgG purified fraction resuspended to 6.8 μg/μl.

Monoclonal antibodies (Mabs) against HIV-1 His_5 PR were produced by standard hybridoma techniques.[3] Culture supernatants were tested for the presence of anti-proteinase Mabs by ELISA assay and confirmed by immunoblot analysis. Ascites were raised in pristane primed Balb/c mice and Mabs purified by caprylic acid and ammonium sulfate precipitation.[4]

The HXB3 PR clone has been previously described.[1] "SF-2" is our version of the SF-2 variant: four of six changes (see Figure 1, K14R, S37N, R41K, and L63P) between HXB3 and SF-2 were introduced into the HXB3 clone by site-directed mutagenesis. The HIV-2 ROD PR plasmid expresses only the 11 kDa form of PR and was derived from plasmid p2RTL1.[5]

Immunoassays

ELISA. Peptide or pure recombinant HXB3 PR was adsorbed onto microtiter dishes. Primary antibody was as stated; secondary antibody was goat anti-mouse/rabbit conjugated to horseradish peroxidase. Color was developed with H_2O_2 and *o*-phenylenediamine and optical density determined at 490 nm.

Immunoblot. Recombinant protein was subjected to SDS-PAGE, transferred to nitrocellulose and probed with the indicated antibody. The secondary antibody was either goat anti-mouse (for Mab) or goat anti-rabbit (for 6802 or 5833) conjugated to horseradish peroxidase. Color was developed with 4-chloro-napthol and H_2O_2.

Immunoprecipitation. *E. coli* cells were incubated with media containing ^{35}S-cysteine, cells were lysed and proteins immunoprecipitated with each antibody. Immunocomplexes were concentrated with Protein G-agarose and subjected to SDS-PAGE. The gel was incubated with fluor, dried and exposed to x-ray film for 24 hours.

Sandwich ELISA. Mab was adsorbed onto microtiter dishes followed consecutively by PR, secondary antibody 6802 and goat anti-rabbit antibody conjugated to horseradish peroxidase. Color was developed with H_2O_2 and *o*-phenylenediamine and optical density determined at 490 nm.

RESULTS

A panel of Mabs and polyclonal antibodies were generated against HIV-1 and HIV-2 PR using both peptide and recombinant proteins. All antibodies function in both immunoblot and immunoprecipitation assays. Additionally, a sandwich ELISA was developed using the Mabs, in combination with antibody 6802, which is capable of detecting less than 1 ng/ml of HIV-1 PR.

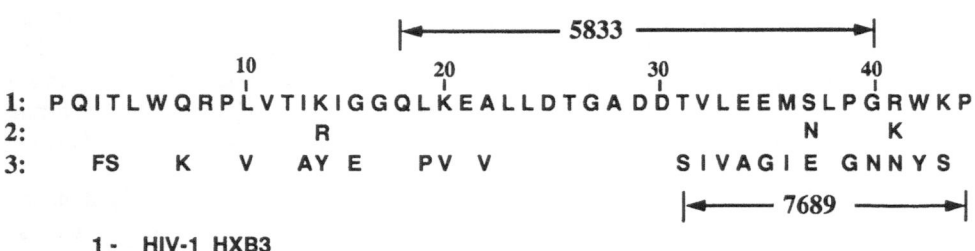

Figure 1. Comparison of the amino acid sequence of three PR variants: HIV-1 HXB3 and "SF-2" and HIV-2 ROD.

Table 1. Summary of information on the generation and characterization of the antibodies generated and tested in this study

Name	Type	Antigen		Ab recognizes			immuno-precipitate	immuno-blot
				HXB-3	"SF2"	ROD		
5833[e]	Polyclonal	HIV-1	18-40[p]	Yes	Yes	Yes	Yes	Yes
5833[l]	Polyclonal	HIV-1	18-40[p]	Yes	Yes	No	Yes	Yes
6802	Polyclonal	HIV-1	1-99[r]	Yes	Yes	Yes	Yes	Yes
7689	Polyclonal	HIV-2	31-44[p]	No	N.D.	Yes	N.D.	Yes
3B3, 3E10, 7E1, and 10B5								
	Monoclonal	HIV-1	1-99[r]	Yes	No	No	Yes	Yes

[e]early bleed; [l]late bleed; [p]peptide; [r]recombinant protein.

Immunoblot analysis showed that both general and specific antibodies were generated (summarized in Table 1). Polyclonal antibody 5833 is both a general and specific antibody. Antibody from early immunization bleeds recognized all three PR variants tested indicating a conserved section of the 18-40 region was recognized. However, antibody from hyperimmunization late bleeds failed to recognize HIV-2 ROD PR. This shift in specificity indicated the clonal emergence of high affinity antibodies to an epitope in a non-conserved section of the 18-40 region.

Rabbit polyclonal antibody 7689 was raised against peptide 31-44 of HIV-2 ROD PR. This particular peptide was chosen because it spans a region that is not conserved among the variants tested and was expected to generate an antibody specific for HIV-2 ROD PR. Immunoblot results showed that antibody 7689 recognized only HIV-2 ROD PR.

Rabbit polyclonal antibody 6802, which was raised against HIV-1 His$_5$ PR, recognized all three PR variants. The general reactivity of 6802 indicated that conserved regions between these three variants were immunogenic. Partial epitope mapping showed that 6802 recognized the regions 18-40 (peptide ELISA) and 1-50 and 67-99 (recombinant protein immunoblot) demonstrating that epitopes in both halves of HIV-1 PR are recognized. Two peptides overlapping the 18-40 peptide, 1-20 and 31-50, were only weakly recognized indicating that these two regions were not recognized when presented on a short flexible peptide.

In contrast to 6802, the Mabs generated in mice against HIV-1 His$_5$ PR recognized only the HXB3 variant of HIV-1 PR. Our results indicate that all four Mabs recognize the same epitope. Partial epitope mapping showed that the Mabs recognized an epitope in recombinant protein 1-50 but not 67-99. None of the peptides within the 1-50 region were recognized by the Mabs. Non-reactivity to HIV-2 ROD PR indicates the epitope is in a non-conserved part of the 1-50 region. The unexpected non-reactivity to the "SF-2" variant will allow further refinement in mapping the epitope as only 4 amino acids are changed from the HXB3 protein sequence (3 are within the 1-50 region).

CONCLUSIONS

We have generated antibodies which are capable of recognizing at least three different variants of HIV PR in ELISA, immunoblot and immunoprecipitation assays. We expect these antibodies, 6802 in particular, to recognize other PR variants not yet tested, thus increasing the general usefulness of these reagents. Several interesting results were observed. ELISA and immunoblot results with antibodies 5833 and 7689 suggest that the 18-40 region of HIV PR is highly antigenic. The lack of Mab reactivity to peptide 18-40 but strong reactivity to recombinant protein 1-50, suggests that the Mabs, in contrast to antibody 6802, recognize a structural conformation present in the 1-50 region. This conformation is not represented in the short linear peptides we tested.

Further evidence supporting the conformational dependence of the Mabs is the non-reactivity of the Mabs to the HIV-1 "SF-2" PR variant. The amino acids differing in "SF-2" from HXB3 in the 1-50 region (amino acids 14, 37 and 41) are located on the outside of the folded molecule and are in or near hairpin loops.[6] Amino acids changes at these positions could subtly affect the conformation and interfere with antibody recognition.

REFERENCES

1. M. C. Graves, J. J. Lim, E. P. Heimer and R. A. Kramer, An 11-kDa form of human immunodeficiency virus protease expressed in *Escherichia coli* is sufficient for enzymatic activity, *Proc. Natl. Acad. Sci. U.S.A.* **85**:2449 (1988).
2. E. Hochuli, W. Bannwarth, H. Döbeli, R. Gentz and D. Stüber, Genetic approach to facilitate purification of recombinant proteins with a novel metal chelate adsorbent, *BIO/TECH* **6**:1321 (1988).
3. G. Kohler and C. Milstein, Continuous cultures of fused cells secreting antibody of predefined specificity, *Nature (London)* **256**:495 (1975).
4. L. M. Reik, S. L. Maines, D. E. Ryan, W. Levin, S. Bandiera and P. E. Thomas, A simple, non-chromatographic purification procedure for monoclonal antibodies against cytochrome P450 isozymes, *J. Immunol. Methods* **100**:123 (1987).
5. S. F. J. LeGrice, R. Zehnle and J. Mous, A single 66-kilodalton polypeptide processed from the human immunodeficiency virus type 2 pol polyprotein in *Escherichia coli* displays reverse transcriptase activity, *J. Virol.* **62**:2525 (1988).
6. A. Wlodawer, M. Miller, M. Jaskólski, B. K. Sathyanarayana, E. Baldwin, I. T. Weber, L. M. Selk, L. Clawson, J. Schneider and S. B. H. Kent, Conserved folding in retroviral proteases: crystal structure of a synthetic HIV-1 protease, *Science* **245**:616 (1989).

CLONING, EXPRESSION AND KINETIC CHARACTERIZATION OF THE

FELINE IMMUNODEFICIENCY VIRUS PROTEINASE

W. G. Farmerie,* M. M. Goodenow+ and B. M. Dunn*

*Department of Biochemistry and Molecular Biology
+Department of Pathology
 University of Florida
 Gainesville, Florida   32610

INTRODUCTION

Biological and biochemical characterizations of non-human AIDS model systems are essential for understanding the mechanism of retrovirus-induced immune deficiency diseases. Furthermore, animal-based AIDS model systems will be important for the complete testing and evaluation of drugs and vaccines directed against diseases with retroviral etiologies. The feline immunodeficiency virus (FIV) is a member of the lentivirus family, a group of retroviruses that also includes the human immunodeficiency viruses, HIV-1 and HIV-2. As with human lentiviruses, FIV exhibits a T-lymphocyte tropism and is associated with an immunodeficiency syndrome in infected domestic cats (Pederson *et al.*, 1987, Yamamoto *et al.*, 1988, Pederson *et al.*, 1989). The merit of a feline-based system for evaluating AIDS chemotherapies will be partially dependent upon the structural and functional homology between feline and human viral proteins. A variety of HIV specific proteins have been identified as potential targets for anti-viral compounds, with much attention focused on the HIV-specific proteinase (HIV PR). At the primary structural level the human and feline viral proteinases share 29 % amino acid identity and approximately 48 % amino acid similarity, with the highest degree of homology associated with residues involved in the formation of the active site cleft. In order to evaluate the degree of functional homology between the human and feline proteinases we have cloned and expressed the FIV PR gene in *E. coli*.

CLONING AND EXPRESSION OF RECOMBINANT FIV PR

Sequences encoding FIV PR were amplified by PCR from DNA extracted from Crandell feline kidney cells infected with the Petaluma strain of FIV (Pederson *et al.*, 1987). The DNA segment encoding FIV PR was inserted into the expression vector *pT7-7* (Tabor & Richardson, 1985) in a configuration which placed the proteinase gene under transcriptional control of a *T7* RNA polymerase-specific promoter. Synthesis of recombinant FIV PR is induced by infecting cells containing the FIV PR plasmid with a lambda phage derivative

Table 1. Relative Cleavage of Peptide Substrates by FIV Proteinase

Peptide	Sequence[+]	Rate[*]
1	K-A-R-V-L*Nph-E-A-O-G-NH$_2$	0.15
2	K-A-R-V-O*Nph-E-A-O-G-NH$_2$	0.52
3	K-A-R-V-F*Nph-E-A-O-G-NH$_2$	0.86
5	K-A-R-A-O*Nph-E-A-O-NH$_2$	0.30
6	K-A-R-L-O*Nph-E-A-O-NH$_2$	0.14
7	K-A-R-I-O*Nph-E-A-O-NH$_2$	1.07
8	K-A-R-P-O*Nph-E-A-O-NH$_2$	undetected
9	K-A-R-D-O*Nph-E-A-O-NH$_2$	0.003
10	K-A-R-G-O*Nph-E-A-O-NH$_2$	undetected

[+]Nph = nitrophenylalanine; O = norleucine. [*]Rates are expressed as nmoles/sec. Reactions are performed in 0.1 M NaOAc pH 4.7, 0.05 M EDTA, 1.0 M NaCl, 140 μM peptide; 28 μg/ml refolded FIV-PR.

which expresses *T7*-RNA polymerase. The recombinant FIV PR is synthesized in the form of insoluble inclusion bodies which are isolated from the remainder of cellular proteins by centrifugation. After the inclusion bodies are solubilized in 8 M urea the proteinase is refolded into its native conformation by dialysis against three changes of 0.05 M Tris, 0.002 M ß-mercaptoethanol, pH 7.3. The resulting FIV PR preparation is suitable for spectrophotometric proteinase assays utilizing chromogenic oligopeptide substrates.

KINETIC CHARACTERIZATION OF FIV PR

We have demonstrated the advantages of water-soluble chromogenic oligopeptide substrates based on the amino acid sequence present at the HIV-1 CA/NC junction for the kinetic characterization of the HIV-1 and HIV-2 proteinases (Richards *et al.*, 1990, Phylip *et al.*, 1990). These substrates incorporate the reporter group 4-NO$_2$-phenylalanine in the P$_1$' position which allows for continuous spectrophotometric monitoring of substrate hydrolysis. The sequence at the FIV p24/p10 junction (-Lys-Met-Gln-Leu-Leu*Ala-Glu-Ala-Leu-) is reasonably homologous to the HIV-1 CA/NC junction (-Lys-Ala-Arg-Val-Leu*Ala-Glu-Ala-Met-) and suggested that oligopeptide substrates that were based on the HIV-1 CA/NC cleavage junction might be suitable substrates for the FIV PR. We therefore tested the ability of FIV PR to cleave members of a family of substrate oligopeptides that were based upon the HIV-1 CA/NC junction (Table 1). Peptides 1 through 7 were cleaved specifically by the recombinant FIV PR. Peptides 8, 9, and 10 were poor substrates for FIV PR. The relative substrate preferences shown by FIV PR are consistent with those previously determined for these same substrates with HIV-1 PR. The K$_m$ and V$_{max}$ values were compared for FIV, HIV-1 and HIV-2 proteinases (Table 2). The K$_m$ values for peptides 1 and 2 fall into the same range as the values determined for the HIV-1 and HIV-2 proteinases. The increased K$_m$ values of FIV PR for peptides 6 and 7 may be due to the truncation of these peptides at their carboxyl-termini rather than to the substitution of Leu or Ile for Val in the P$_2$ position. The comparatively lower V$_{max}$ values determined for FIV PR are probably a reflection of the lower specific activity of the FIV proteinase used in these assays.

These initial results show that FIV PR cleaves a family of substrate peptides with the same specificity and with kinetic values similar to HIV-1 and HIV-2 proteinases. The kinetic data suggest considerable functional homology in substrate preferences among this group of

Table 2. Comparison of kinetic parameters for the hydrolysis of chromogenic substrates by FIV, HIV-1 and HIV-2 proteinases

Peptide	Sequence	FIV		HIV-1		HIV-2	
		K_m	V_{max}	K_m	V_{max}	K_m	V_{max}
1	K-A-R-V-L*Nph-E-A-O-G-NH$_2$	43	0.027	6	1.00	90	0.52
2	K-A-R-V-O*Nph-E-A-O-G-NH$_2$	72	0.13	15	2.27	50	1.00
6	K-A-R-L-O*Nph-E-A-O-NH$_2$	1004	0.062	25	0.30	85	0.64
7	K-A-R-I-O*Nph-E-A-O-NH$_2$	400	0.41	45	1.73	70	0.96

Nph = nitrophenylalanine, O = norleucine, K_m values are μM, V_{max} is nmoles/sec/μg extract protein.
Kinetic parameters for FIV PR were determined in a buffer containing 0.1 M sodium acetate, 0.005 M Na$_4$EDTA, 0.005 M ß-mercaptoethanol and sufficient NaCl to give a final ionic strength of 1 M. Kinetic values for HIV-1 and HIV-2 are from work previously published (Richards *et al.,* 1990, Phylip *et al.,* 1990).

viral proteinases, an observation consistent with the conservation in the primary structures of FIV PR and HIV PR. This preliminary investigation suggests that FIV may be a useful as a model system for the testing and evaluation of drugs that are directed at interfering with the function of the viral proteinase. Clearly, additional kinetic and structural characterization of FIV PR will be necessary in order to evaluate completely the suitability of FIV as a model system for testing new anti-viral compounds.

REFERENCES

Pederson, N. C., Ho, E. W., Brown, M. L. & Yamamoto, J. K., 1987, Isolation of a T-lymphotropic virus from domestic cats with an immunodeficiency-like syndrome, *Science* **235**:790-793.

Pederson, N. C., Yamamoto, J. K., Ishido, T. & Hauser, H., 1989, Feline immunodeficiency virus infection, *Vet. Immunol. Immunopathol.* **21**:111-129.

Phylip, L. H., Richards, A. D., Kay, J., Konvalinka, J., Strop, P., Blaha, I., Velek, J., Kostka, V., Ritchie, A. J., Broadhurst, A. V., Farmerie, W. G., Scarborough, P. E. & Dunn, B. M., 1990, Hydrolysis of synthetic chromogenic substrates by HIV-1 and HIV-2 proteinases, *Biochem. Biophys. Res. Comm.* **171**:439-444.

Richards, A. D., Phylip, L. H., Farmerie, W. G., Scarborough, P. E., Alvarez, A. A., Dunn, B. M., Hirel, Ph-H., Konvalinka, J., Strop, P., Pavlickova, L., Kostka, V. & Kay, J., 1990, Sensitive, soluble chromogenic substrates for HIV-1 proteinase, *J. Biol. Chem.* **265**:7733-7736.

Tabor, S. & Ricardson, C. C., 1985, A Bacteriophage *T7* RNA polymerase/promoter system for controlled exclusive expression of specific genes, *Proc. Natl. Acad. Sci. U.S.A.* **82**:1074-1078.

Yamamoto, J. K., Sparger, E., Ho, E. W., Anderson, P. R., O'Connor, T. P., Mandell, C. P., Lowenstine, L., Munn, R. & Pederson, N. C., 1988, Pathogenesis of experimentally induced feline immunodeficiency virus infection in cats, *Am. J. Vet. Res.* **49**:1246-1258.

PROTEIN-ENGINEERED PROTEINASE OF MYELOBLASTOSIS ASSOCIATED VIRUS, AN ENZYME OF HIGH ACTIVITY AND HIV-1 PROTEINASE-LIKE SPECIFICITY

P. Štrop, M. Hořejší, J. Konvalinka, R. Škrabana, J. Velek, I. Bláha,
V. Černá, I. Pichová, L. Pavlíčková, M. Andreánsky, M. Fábry,
V. Kostka, J. Sedláček and S. Foundling*

Institute of Organic Chemistry and Biochemistry
and Institute of Molecular Genetics
Czechoslovak Academy of Sciences
Prague, Czechoslovakia

*National Cancer Institute
Frederick, Maryland

INTRODUCTION

All proteinases of avian and mammalian retroviruses belong to the family of aspartic proteinases, are of similar size and of homologous primary structure; they all act catalytically in the form of highly symmetric molecular dimers.[1] Detailed studies of retroviral proteinases were carried out on two almost identical proteinases of MAV[2,3] and RSV[4] (representing the group of avian retroviruses) and on the HIV proteinase.[5,6] The knowledge of the 3D structure,[2,4,5] catalytic activity and substrate specifity[3,6] of the MAV and the HIV proteinase has changed the notion of their general similarity since several features that distinguish each proteinase from the other were revealed. The HIV-1 proteinase has a considerably higher activity[3,6] which reflects the different conditions of the expression and action of this enzyme *in vivo*:[7] The "coding strategy" of MAV allows the expression of the proteinase from the first (*gag*) open reading frame and provides for the high (i.e. stoichiometrical) level of the relatively "weak" enzyme whereas the smaller amount of the more active HIV enzyme is a result of infrequent translational frameshift events that occur in the overlapping region of the *gag* and *pol* reading frames.[8] The substrate specificities of retroviral proteinases seem complex and the requirement for a side chain in an individual subsite of a substrate is an outcome of the combination of residues occupying other closely located subsites.[3] The two proteinases (MAV and HIV) show rather promiscuous substrate specificity, nevertheless several differences can be traced. We made an attempt to use protein engineering of the MAV proteinase to tackle directly problems of structural basis of these differences and, vice versa, to make more precise conclusions on the functional importance of the individual elements of its three dimensional structure. This article describes mutation of the MAV proteinase which resulted not only in an alteration of its substrate specificity but also in an increase of its enzymic activity - a rare case in protein engineering.

Structure and Function of the Aspartic Proteinases
Edited by B.M. Dunn, Plenum Press, New York, 1991

Table 1. Enzyme activities with peptide substrates varied in P_1, Ala-Thr-His-Gln-Val-X_1*Phe(NO_2)-Val-Arg-Lys-Ala or in P_2, Ala-Thr-His-Gln-X_2-Tyr*Phe(NO_2)-Val-Arg-Lys-Ala or in P_3, Ala-Thr-His -X_3-Val-Tyr*Phe(NO_2)-Val-Arg-Lys-Ala

Position residue		k_{cat}, sec^{-1}			K_m, μM			k_{cat}/K_m, sec^{-1}mM^{-1}		
		wt MAV	mutant MAV	HIV	wt MAV	mutant MAV	HIV	wt MAV	mutant MAV	HIV
P1	Tyr	4.9	9	14	82	7	15	59	1360	930
	Phe	3.2	7.1	13	26	13	15	120	550	870
P2	Val	4.9	9	14	82	7	15	59	1360	930
	Ile	2	3.9	11	42	7	15	48	560	730
	Ala	0.5	2.6	2	60	7	60	8	370	33
	Leu	0.2	1.5	2	120	25	30	1.7	60	67
P3	Gln	4.9	9	14	82	7	15	59	1360	930
	Arg	3.1	4.3	10	24	4	15	130	1100	670
	Asp	1.4	11	10	25	7	16	56	1570	625
	Tyr	11	6.8	6	26	1.6	7	420	4250	850
	Glu	4.7	5	15	6.2	3.3	35	760	1500	500

RESULTS AND DISCUSSION

The prediction of the side chains which could form the primary binding pocket of the dimer molecule of the enzyme (S_3 to S_3') was based on the three dimensional structures of the MAV[2] and RSV[4] proteinases, on the modeling of an MAV proteinase inhibitor (Konvalinka and Cooper, unpublished) and on the modeling of a HIV-1 proteinase - inhibitor complex.[9] Five of the amino acids of the MAV proteinase (residues 100 and 104-107) thus deduced were chosen for replacement by the alignment of its amino acids sequence with that of the HIV-1 proteinase.[9] The required site-mutation Ala$_{100} \rightarrow$Leu, Val$_{104} \rightarrow$Thr, Arg$_{104} \rightarrow$Pro, Gly$_{106} \rightarrow$Val, and Ser$_{107} \rightarrow$Asn was introduced directly into the *E. coli* expression plasmid pMG45, coding for a recombinant gag precursor of the wild-type MAV proteinase,[10] using duplexes of synthetic oligonucleotides for two-step replacement of the coding region between the *Bam*HI and *Pst*I restriction sites. The *E. coli* expression system gave similar results as with the wild-type enzyme (yields above 20 mg per liter of cultivation media)[10] and the newly acquired specificity has not abolished the capacity for the required autoprocessing of the recombinant gag precursor at the natural boundary of nucleocapsid protein and the proteinase.

A comparison of the kinetic parameters of the wild-type and mutant MAV proteinase and the HIV-1 proteinase was effected with a series of peptide substrates. Kinetic data on a superior peptide substrate of both the MAV and HIV-1 proteinase,[6] Ala-Thr-His-Gln-Val-Tyr*Phe(NO_2)-Val-Arg-Lys-Ala, and on a series of its variations in P_1, P_2, and P_3 are shown in Table 1. I t is apparent that the activity of the mutant enzyme is higher than that of the wild-type. For the prototype peptide Ala-Thr-His-Gln-Val-Tyr*Phe(NO_2)-Val-Arg-Lys-Ala , the

Table 2. Enzyme activities with the peptide substrate Lys-Ala-Arg-Val-Nle*Phe(NO$_2$)-Glu-Ala-Nle-NH$_2$

k_{cat}, sec^{-1}			K_m, μM			k_{cat}/K_m, sec^{-1}mM^{-1}		
wt MAV	mutant MAV	HIV	wt MAV	mutant MAV	HIV	wt MAV	mutant MAV	HIV
0.07	1	42	262	83	32	0.27	12	1310

parameter k_{cat}/K_m shows a 23-fold increase. The replacement of Tyr by Phe in P$_1$ results in a lesser contribution of binding (K_m) to the increase of the mutant enzyme activity. Regarding the P$_2$ position, acceptance of small hydrophopic residues, such as Val by both the wild-type MAV and HIV-1 proteinases was found in our previous studies;[3,6] Leu in this position is poorly accepted and the mutant MAV proteinase almost precisely mimics the HIV-1 proteinase (Table 1, last line of the P$_2$ series). Variations of amino acids in P$_3$ result in a substantial contribution of k_{cat} to the increased activity of the mutant except for Tyr in this position. The slight decrease in k_{cat} is more than compensated by the decrease of K_m in this case; this makes the P$_3$ Tyr substrate the best among those cleaved by the mutant enzyme. Alteration of the specificity of the mutant MAV proteinase toward that of HIV-1 can be seen from a summary of data in Table 1; values for the mutant MAV proteinase which are markedly different from the values for the wild-type enzyme are very similar to values characterizing the HIV-1 proteinase. A more explicit confirmation of the specificity shift provide kinetic measurements with Lys-Ala-Arg-Val-Nle*Phe(NO$_2$)-Glu-Ala-Nle-NH$_2$, a peptide originally developed[11] as the optimal substrate for the HIV-1 proteinase. As shown in Table 2, this substrate was cleaved much more efficiently by the mutant MAV proteinase than by the wild-type proteinase.

The site-mutation of five clustered amino acids has brought about substantial changes in the enzyme properties what indicates their high functional importance. Additional contributions to the primary binding pockets could provide amino acid residues more distant in the linear structure. A rational mutation of these residues may result in a further alteration of the kinetic and binding characteristics of the enzyme.

ACKNOWLEDGEMENT

The oligonucleotide synthesis performed by Dr. Hana Vecerkova is gratefully acknowledged.

REFERENCES

1. J. Kay and B. M. Dunn, Viral proteinases: weakness in strength, *Biochim. Biopys. Acta* **1048**:1 (1990).
2. S. I. Foundling, F. R. Salemme, B. Korant, J. J. Wendoloski, P. C. Weber, A. C. Treharne, M. C. Schadt, M. Jaskólski, M. Miller, A. Wlodawer, P. Štrop, V. Kostka, J. Sedláček and D. H. Ohlendorf, Crystal structure of a retroviral proteinase from avian myeloblastosis associated virus, *in*: "Viral proteinases as targets for chemotherapy", H. G. Krauslich, S. Oroszlan and E. Wimmer, eds., Cold Spring Harbor Laboratory, Cold Spring Harbor (1989).

3. P. Štrop, J. Konvalinka, D. Stys, L. Pavlíčková, I. Blaha, J. Velek, M. Travnicek, V. Kostka and J. Sedláček, Specificity studies on retroviral proteinase from myeloblastosis associated virus, Biochemistry, in press.

4. M. Miller, M. Jaskólski, J. K. M. Rao, J. Leis and A Wlodawer, Crystal structure of a retroviral protease proves relationship to aspartic protease family, *Nature* **337**:576 (1989).

5. M. Miller, J. Schneider, B. K. Sathyanarayma, M. V. Toth, G. R. Marshall, L. Clawson, L. Selk, S. B. H. Kent and A. Wlodawer, Structure of complex of synthetic HIV-1 protease with a substrate-based inhibitor at 2.3 Å resolution, *Science* **246**:1149 (1989).

6. J. Konvalinka, P. Štrop, J. Velek, V. Černá, V. Kostka, L. W. Phylip, A. D. Richards, B. M. Dunn and J. Kay, Sub-site preferences of the aspartic proteinase from the human immunodeficiency virus, HIV-1, *FEBS Letters* **268**:35 (1990).

7. L. H. Pearl, ed., "Retroviral proteases. Control of maturation and morphogenesis", Stockton Press, New York (1990).

8. Y. Jacks, M. D. Power, F. R. Masiarz, P. A. Luciw, P. J. Barr and H. E. Varmus, Characterization of ribosomal frameshifting in HIV-1 *gag-pol* expression, *Nature* **331**:280 (1988).

9. I. T. Weber, M. Miller, M. Jaskólski, J. Leis, A. M. Skalka and A. Wlodawer, Molecular modeling of the HIV-1 protease and its substrate binding site, *Science* **253**:928 (1989).

10. J. Sedláček, P. Štrop, F. Kapralek, V. Pecenka, V. Kostka, M. Travnicek and J. Riman, Processed enzymatically active protease (p15) of avian retrovirus obtained in an *E. coli* system expressing a recombinant precursor (Pr25), *FEBS Letters* **237**:187 (1988).

11. A. D. Richards, L. W. Phylip, W. G. Farmerie, P. E. Scarborough, A. Alvarez, B. M. Dunn, P.-H. Hirel, J. Konvalinka, P. Štrop, L. Pavlíčková, V. Kostka and J. Kay, Soluble chromogenic substrates for HIV-1 protease, *J. Biol. Chem.* **265**:7733 (1990).

p15^{gag} PROTEINASE OF MYELOBLASTOSIS ASSOCIATED VIRUS:

SPECIFICITY STUDIES WITH SUBSTRATE BASED INHIBITORS

P. Štrop, L. Pavlíčková, D. Štys, M. Souček, J. Urban, O. Hrušková,
F. Kaprálek,[*] P. Ječmen,[*] J. Sedláček[*] and V. Kostka

Institute of Organic Chemistry and Biochemistry
[*]Institute of Molecular Genetics
Czechoslovak Academy of Science
CS 166 10 Prague
Czechoslovakia

INTRODUCTION

The key role of retrovirus-encoded proteinases in limited proteolysis of polyprotein precursors, a prerequisite of the formation of mature infectious virions, makes these enzymes attractive targets of specific inhibitors. A rational basis for the design of such inhibitors requires detailed knowledge of the proteinase-inhibitor interactions at the level of three-dimensional structures. Such data are available so far only for the HIV-1 proteinase whose crystal complexes with some inhibitors were subject of X-ray studies.[1] The absence of such information on the myeloblastosis associated virus (MAV) proteinase, another retroviral aspartate proteinase, together with the need of finding a tight-binding yet hydrophilic inhibitor for both X-ray studies and titration of its active site, led us to investigate the interactions of the proteinase with various classes of custom-synthesized inhibitors.

MATERIALS AND METHODS

Proteinase Preparation

The p15^{gag} proteinase of MAV was prepared by the recombinant procedure described earlier.[2,3]

Synthesis of Inhibitors

The inhibitors were synthesized on the chloromethyl- or benzhydryl-aminopolystyrene resin by the solid phase method using the following protection groups: Boc for α-NH$_2$, Bzl for the side chains of Ser, Thr, Asp and Glu, 2,6-diClBzl for Tyr, tosyl for Arg, p-methoxyBzl for Cys, Boc for His and Z for Lys. Reduced-bond analogs were synthesized either on the resin by the method of Coy *et al*.[4] optimized by one of the authors (J. U.) or as t-Boc protected dipeptide analogs coupled to the resin by the standard

Structure and Function of the Aspartic Proteinases
Edited by B.M. Dunn, Plenum Press, New York, 1991

Table 1. Structures of inhibitors of p15^{gag} proteinase of MAV

Based on substrate

Thr Phe Gln Ala Phe   *   Pro Leu Arg Glu Ala Pro
(RSV P63/P32; pol 572/573)

	P6	P5	P4	P3	P2	P1	P1'	P2'	P3'	P4'	P5'	P6'
1	Pro	Thr	Phe	Gln	Ala	ChSta		Leu	Arg	Glu	Ala	Pro
5		Thr	Phe	Gln	Ala	Phe[CH$_2$NH]Pro		Leu	Arg	Glu	Ala	

Pro Ala Val Ser   *   Leu Ala Met Thr Met
(RSV p12/p15; gag 577/578)

	P6	P5	P4	P3	P2	P1	P1'	P2'	P3'	P4'	P5'	P6'
7		Pro	Pro	Ala	Val	CysSta		Ala	Met	Thr	Met	
8		Pro	Pro	Cys	Val	PheSta		Ala	Met	Thr	Met	
9			Pro	Cys	Val	PheSta		Ala	Met	Thr		
13			Pro	Pro	Val	PheSta		Ala	Met	Thr	Met	
14		Pro	Pro	Tyr	Val	PheSta		Ala	Met	Thr	Met	

Pro Tyr Val Gly   *   Ser Gly Leu Tyr Pro
(RSV p2b/p2c; gag 175/176)

	P6	P5	P4	P3	P2	P1	P1'	P2'	P3'	P4'	P5'	P6'
24			Boc	Tyr	Val	Sta		Ala	StaOMe			

Chimeric inhibitors

	P6	P5	P4	P3	P2	P1	P1'	P2'	P3'	P4'	P5'	P6'
28		Gly	Pro	Val	Val	PheSta		Val	Ser	Thr	Ala	
33			Pro	Cys	Val	ChSta		Val	Arg	Pro	60	
36			Pro	Cys	Val	CysSta		Val	Arg	Pro		

Palindromic inhibitors

	P6	P5	P4	P3	P2	P1	P1'	P2'	P3'	P4'	P5'	P6'
44		Thr	Phe	Gln	Ala	PheSta		Ala	Gln	Phe	Thr	
45		Thr	Thr	Phe	Gln	PheSta		Ala	Gln	Phe	Thr	

Other inhibitors

	P6	P5	P4	P3	P2	P1	P1'	P2'	P3'	P4'	P5'	P6'
48		Cys	Thr	Asn	Leu	Phe[CH$_2$NH]Pro		Ile	Ser	Pro	Ile	
50		Val	Ser	Leu	Asn	PheSta		Gln	Val	Ser	Gln	
51		Met	Ser	Leu	Asn	Sta		Val	Ala	Lys	Val	
53	His	Pro	His	Leu	Ser	Phe[CH$_2$NH]Met		Ala	Tyr			
55		His	Ser	Leu	Arg	Phe[CH$_2$NH]Phe		Arg	Leu	Pro		

Table 2. Inhibition data for inhibitors listed in Table 1

Inhibitor	IC$_{50}$, µM	K$_i$, µM
1		0.065
5	1100	15.8
7	37	0.25
8	9	0.01
9	4	0.03
13	3	0.01
14	12.5	0.01
24		0.28
28		0.24
33	60	0.04
36		100
44		0.13
45		0.026
48		4.36
50		> 40
51		> 80
53		16
55		90

procedure. For phestatine, 4-nitrophenylstatine, metstatine, cysstatine, argstatine and statine the original method[5] modified by one of the authors (L. P.) was employed.

Inhibitory Potency Assays

The residual enzyme activity was determined with the chromogenic substrate Ala-Thr-His-Gln-Val-Tyr*Phe(NO$_2$)-Val-Arg-Lys-Ala[6] in two different assays. In the HPLC assay the proteinase solution was incubated 12 min at 37°C at pH 6 with various inhibitor concentrations and the hydrolysis of the substrate was monitored by quantitation of the cleavage products resolved on a Vydac column. Alternatively, in the spectrophotometric assay the cleavage of the substrate in the presence of varying inhibitor concentrations was monitored at 305 nm. From the v_0/[I] plot the IC$_{50}$ and K$_i$ values were then calculated.

RESULTS

The inhibitors synthesized in this study can be roughly classified as falling into three groups: (a) substrate based inhibitors, (b) "chimeric" inhibitors and (c) palindromic and pseudopalindromic inhibitors. Inhibitors designed for the HIV-1 proteinase by us earlier and some inhibitors of general aspartic proteinases were also tested.

Substrate based inhibitors

These inhibitors were designed on 4 different sequences adjacent to peptide bonds cleaved in endogenous substrates (i.e. in polyprotein precursors of MAV and RSV). Synthetic 10- to 12-residue peptides which in earlier specificity assays[6] showed adequately low K_m values were chosen and the dipeptide unit containing the scissile bond was replaced by statine or its analogs. Alternatively, isosteric replacement groups for this peptide bond (CH_2-NH peptides) were used. Examples of these inhibitors are shown in Table 1 and the inhibition data is given in Table 2. The highest potency towards the proteinase was observed with inhibitors 8, 9, 11 and 13 designed on the sequence around the p12/p15 (*gag* 577/578) cut-off site in the RSV polyprotein. These inhibitors all contain at their core PheSta as the nonhydrolyzable transition state analog substituting the Ser-Leu dipeptide. This shows a distinct preference of the S_1 pocket for the aromatic or bulky hydrophobic side chain. The nature of the statine substituent is essential; the replacement of PheSta by CysSta (7) decreases the binding by one order of magnitude. The efficiency of 8, so far the best inhibitor, is six times higher than that of the inhibitor 1 based on the sequence around the p63/p32 (pol 572/573) boundary. The best values obtained with inhibitors of the unsubstituted statine series (22 and 27) are by one order of magnitude higher than those for 8. Inhibitor 5 representing reduced bond analogs is the weakest of all substrate based inhibitors.

Chimeric inhibitors

These are arbitrarily designed peptide sequences with no counterparts in the endogenous substrates of either the MAV, RSV or HIV-1 proteinase. Of the 10 inhibitors of this group tested an inhibitory potency approaching that of the best inhibitors was demonstrated for 33. The pronounced preference of the proteinase for the bulky cyclohexyl residue clearly follows from a comparison with CysSta-containing inhibitor 36 practically lacking any binding ability.

Palindromic inhibitors

Peptides whose sequence reads backward and forward the same were generally characterized by medium strong inhibitory potency, such as that of 44. The deletion of Ala from the latter upsetting the palindrome resulted in a 5-fold increase in binding ability. This may point to the possible role of residues occupying more distant positions (P_4, $P_{4'}$, P_5, $P_{5'}$).

Inhibitors of general aspartic proteinases

Some of the chymosin inhibitors (e. g. 53) are weak inhibitors only of the MAV proteinase. Other chymosin inhibitors, as well as the inhibitor of the *E. parasitica* proteinase gave values even 5 times worse.

Inhibitors of HIV-1 proteinase

Five substrate based peptides, originally designed as inhibitors of the HIV-1 and HIV-2 proteinase showed low inhibitory potency only. This clearly reflects the differences in subsite requirements of these proteinases discussed in detail in another chapter of this book.[7]

DISCUSSION

The specificity of inhibitor binding by the MAV proteinase is generally low, in analogy to substrate binding. This, together with the wide catalytic cleft providing for different binding modes, makes inhibitor design without knowledge of crystallographically defined MAV or RSV proteinase-inhibitor complexes rather difficult. Nevertheless there are some features to emerge from the kinetic data presented here. In accordance with our observations made during our earlier specificity studies,[8] the binding of inhibitors by the MAV proteinase is a result of concerted action of residues occupying positions P_3 to $P_{3'}$ rather than of an additive contribution of the individual residues. In analogy to substrates the major contribution to inhibitor binding is made by primary and secondary binding sites. A preference for bulkier hydrophobic residues in P_1 and $P_{1'}$ and for small hydrophobic side chains in P_2 and $P_{2'}$ was observed. In spite of the shortness of the cleft in the absence of inhibitor its binding is also affected by residues in P_4 and $P_{4'}$. Tight-binding inhibitors 8, 9 and 13 can be used for active site titration.

REFERENCES

1. M. Miller, J. Schneider, B. K. Sathyanarayma, M. V. Toth, G. R. Marshall, L. Clawson, L. Selk, S. B. H. Kent and A. Wlodawer, Structure of a complex of synthetic HIV-1 proteinase with a substrate based inhibitor at 2.3 Å resolution, *Science* **246**:1149 (1989).
2. J. Sedláček, P. Štrop, F. Kapralek, V. Pecenka, V. Kostka, M. Travnicek and J. Riman, Processed enzymatically active protease (p15^{gag}) of avian retrovirus obtained in an *E. coli* system expressing a recombinant precursor (Pr25$^{lac-gag}$), *FEBS Lett.* **237**:187 (1988).
3. P. Štrop, I. Pichova, V. Kostka, F. Kapralek and J. Sedláček, Isolation and characterization of p15^{gag} proteinase of myeloblastosis associated virus expressed in *E. coli*, *in:* "Proteinases of Retroviruses", V. Kostka, ed., W. de Gruyter, Berlin, New York (1989).
4. D. H. Coy, Y. Sasaki, W. A. Murphy and M. Heiman, Facile solid-phase preparation of peptides containing the CH_2NH peptide bond isostere and application to the synthesis of somatostatin (SRIF) octapeptide analogues, *in:* "Peptides 1986", D. Theodoropoulos, ed., W. de Gruyter, Berlin, New York (1987).
5. P. Jouin and B. Castro, Stereospecific synthesis of N-protected statine and its analogues *via* chiral tetramic acid, *J. Chem. Soc. Perkin Trans. I* 1177-1182 (1987).
6. S. I. Foundling, F. R. Salemme, B. Korant, J. J. Wendolski, P. C. Weber, A. C. Treharne, M. C. Schadt, M. Jaskólski, M. Miller, A. Wlodawer, P. Štrop, V. Kostka, J. Sedláček and D. H. Ohlendorf, Crystal structure of a retroviral proteinase from avian Myeloblastosis Associated Virus, *in:* "Viral Proteinases as Targets for Chemotherapy", H. G. Kräusslich, S. Oroszlan, and E. Wimmer, eds., Cold Spring Harbor Laboratory, Cold Spring Harbor (1989).
7. P. Štrop, M. Horejsi, J. Konvalinka, R. Skrabana, J. Velek, I. Blaha, V. Černá, I. Pichova, L. Pavlíčková, M. Andreansky, M. Fabry, V. Kostka and J. Sedláček, Protein-engineered proteinase of myeloblastosis associated virus: an enzyme of high activity and HIV-1 proteinase-like specificity, preceeding chapter.
8. P. Štrop, J. Konvalinka, D. Stys, L. Pavlíčková, I. Blaha, J. Velek, J. Travnicek, V. Kostka and J. Sedláček, Specificity studies on retroviral proteinase from myeloblastosis associated virus, *Biochemistry*, **30**:3437 (1991).

SCINTILLATION PROXIMITY ENZYME ASSAY

A RAPID AND NOVEL ASSAY TECHNIQUE APPLIED TO HIV PROTEINASE

Neil D. Cook,[1] Robert A. Jessop,[1] Philip S. Robinson,[1]
Anthony D. Richards[2] and John Kay[2]

[1]Assay Development Group
Amersham International plc
Cardiff CF4 7YT Wales
United Kingdom

[2]Department of Biochemistry
University of Wales College of Cardiff
P. O. Box 903
Cardiff CF1 1ST Wales
United Kingdom

INTRODUCTION

The aspartic proteinase encoded by the HIV virus is required for the processing of the structural gag and enzymatic pol precursor proteins[1] and is therefore essential for the generation of mature virus particles.[2] The proteinase itself is autocatalytically released from the gag-pol fusion precursor to yield an 11 kDa active species.[3] This crucial role in the life cycle of the HIV virus has identified this aspartic proteinase as a target for therapeutic intervention to inhibit disease progression.[4]

As with many other potential chemotherapeutic targets, the effective search for active lead compounds either by natural product screening or rational design requires a simple rapid and robust assay that is capable of testing thousands of compounds per month.

The technique of Scintillation Proximity Assay (SPA) provides a rapid simple method which requires no laborious or time-consuming separation stages. In addition, this technology is ideal for screening crude broths and natural product extracts and although based upon radioisotopic detection, requires no addition of scintillation cocktail. Samples are measured directly in an aqueous phase homogeneous assay.

METHODS

The SPA technique relies solely upon the proximity of the radioisotope to a solid scintillant. When the isotope and scintillant are close together, the radioactive disintegration is capable of stimulating the scintillant to emit light. When the isotope and scintillant are not in close proximity, the radioactive disintegration is dissipated into the aqueous medium and no light emission is observed. The effective distance required for the proximity effect is governed by the energy of radioactive emission and its path length in an aqueous environment. Fortunately, both tritium [^{3}H] and iodine [^{125}I] display favorable emission energies of 6 keV for the tritium ß particle and 35 keV for the ^{125}I Auger electron. These emissions have path lengths in aqueous media of 7 and 35 µm, respectively. These two isotopes are therefore ideal for Scintillation Proximity Assay.

To put the technology into practice we have synthesized plastic microspheres by suspension polymerization in the presence of scintillant. The scintillant is impregnated within the microspheres during the process. In order to allow the linkage of a variety of substrates, the surface of the plastic is chemically derivatized with glutaraldehyde and then reacted with streptavidin allowing biotinylated substrates to be linked to the bead. This novel technology was adopted in order to develop a rapid assay for HIV proteinase.

A hydrophilic peptide was synthesized based on information derived from substrate analog studies.[5,6,7] The sequence of this substrate was:

Ac-Tyr-Arg-Ala-Arg-Val-Phe*Phe-Val-Arg-Ala-Ala-Lys-Biotin.

The NO$_2$-Phe derivative of this peptide was found to be cleaved exclusively at the Phe*Phe bond with a K$_m$ in free solution of 20 µM.

The peptide was linked to streptavidin beads through the biotinylated lysine residue and labelled with ^{125}I in the acetylated tyrosine residue.

The substrate-containing beads were suspended in a buffer consisting of 50 mM MES, 50 mM NaH$_2$PO$_4$, 500 mM NaCl, 1 mM EDTA, 5 mM dithiothreitol and 10 % w/v glycerol, pH 5.5. The concentration of the substrate-containing beads was adjusted so that 100 µl of the suspension contained approximately 30,000 SPA counts. Reactions were started by addition of 25 µl of HIV-1 proteinase solution (to give a final concentration of ca. 5 nM) and the timecourse of each cleavage reaction monitored by repeatedly counting the assay tube in an LKB 1209 Rackbeta scintillation counter with the window settings fully open.

RESULTS AND DISCUSSION

A typical timecourse observed upon incubation of the SPA-substrate with proteinase is shown in Figure 1a. Cleavage released the ^{125}I-labelled N-terminal fragment from the bead resulting in a decrease in signal. The rate of decrease in signal was found to be directly proportional to the amount of active proteinase added (not shown). In control incubations, the stability of the SPA beads with substrate was demonstrated by prior incubation for up to 1 hr in the assay buffer, pH 5.5. Continuation of this incubation for a further 2 hrs in the absence of added proteinase revealed a satisfactory level of stability for the substrate-linked SPA beads (Figure 1a).

In order to demonstrate the simplicity of the assay and its value for inhibitor detection, separate incubations of substrate-containing SPA beads with proteinase were performed under the above conditions but with the inclusion of increasing concentrations of isovaleryl pepstatin. A dose dependent inhibition of each timecourse was observed (Figure 1a).

The assay can also be used in a stopped format by the addition of a buffer (200 µl) which adjusts the pH to 8.5. Figure 1b illustrates the inhibition by pepstatin using this stopped assay format.

The reaction was started by the addition of 10 µl of enzyme (5 nM final concentration in the assay) and incubated at room temperature for 1 hr. The reaction was terminated with bicarbonate buffer and the assay tubes were counted directly in the scintillation counter.

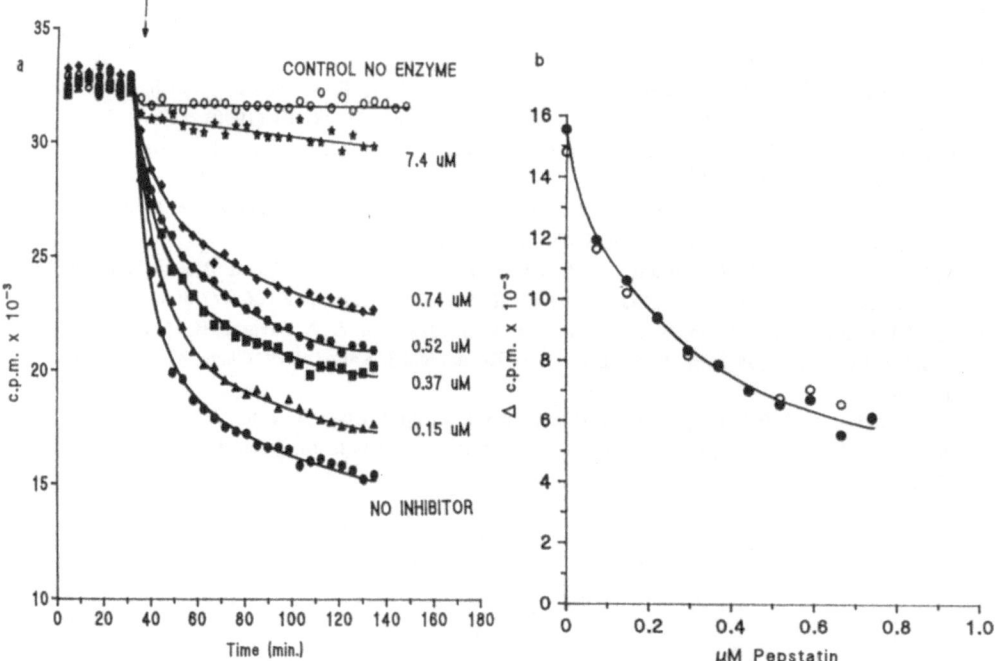

Figure 1. Scintillation proximity assay of HIV-1 proteinase. a) timecourses were performed by repeatedly counting each assay tube at room temperature in the scintillation counter. The arrow indicates the addition of enzyme (5 nM final in the assay). Pepstatin was dissolved in methanol and 10 μl samples were added to the tubes at the indicated concentrations. For the "no enzyme" control, 25 μl of buffer was added to the assay tube; b) assay tubes were incubated for 1 hr at room temperature and each assay was terminated with 200 μl of bicarbonate buffer to adjust the pH to 8.5. The closed circles depict the SPA counts determined immediately after terminating the assay. The open circles are the SPA counts measured in the same tubes 72 hrs later.

Under the assay conditions described, the IC_{50} for pepstatin was estimated to be 0.36 μM. The open circles (Figure 1b) indicate the same assay tubes counted 72 hrs later, demonstrating that no further cleavage occurs after the assay has been stopped at the higher pH.

The advantage of a Scintillation Proximity Assay for this enzyme is apparent in the simplicity of the protocol. The entire assay format involves only pipetting steps; no separation steps are required. This allows SPA assays to be automated readily so the technology is ideally suited to high throughput such as drug screening or enzyme purification procedures. This generic linkage through biotin allows the attachment of many different peptide (and other) substrates to the beads. It should therefore be possible to adapt a number of different proteinase assays to an SPA format.

ADKNOWLEDGEMENTS

A. D. R. was supported by a grant (to J. K.) from the MRC AIDS Directed Program. HIV proteinase was generously supplied by Rhône-Poulenc Santé, France.

REFERENCES

1. N. A. Roberts, J. A. Martin, D. Kinchington, A. V. Broadhurst, J. C. Craig, I. B. Duncan, S. A. Galpin, B. K. Handa, J. Kay, R. W. Lambert, J. H. Merrett, J. S. Mills, K. E. B. Parkes, J. Redshaw, A. J. Ritchie, D. L. Taylor, G. J. Thomas and P. J. Machin, *Science* **248**:338-361 (1990).
2. P. Ashorn, T. J. McQuade, S. Thaisrivongs, A. G. Tomasselli, W. G. Tarpley and B. Moss, *Proc. Natl. Acad. Sci. U.S.A.* **87**:7472-7476 (1988).
3. M. C. Graves, J. J. Lim, E. P. Heimer and R. A. Kramer, *Proc. Natl. Acad. Sci. U.S.A.* **85**:2449-2453.
4. M. I. Johnston, H. S. Allqudeen and N. Sarver, *TIPS* **10**:305-307 (1989).
5. P. L. Darke, R. F. Nutt, S. F. Brady, V. M. Garsky, T. M. Ciccarone, C.-T. Leu, P. K. Lumma, R. M. Freidinger, D. F. Veber and I. S. Sigal, *Biochem. Biophys. Res. Commun.* **156**:297-303.
6. A. D. Richards, L. H., Phylip, W. G. Farmerie, P. E. Scarborough, A. A. Alvarez, B. M. Dunn, Ph-H. Hirel, J. Konvalinka, P. Strop, L. Pavlickova, V. Kostka and J. Kay, *J. Biol. Chem.* **265**:7733-7736 (1990).
7. L. H. Phylip, A. D. Richards, J. Kay, J. Konvalinka, P. Strop, I. Blaha, J. Velek, V. Kostka, A. J. Ritchie, A. V. Broadhurst, W. G. Farmerie, P. E. Scarborough and B. M. Dunn, *Biochem. Biophys. Res. Commun.* **171**:439-444.

IMPROVED CHROMATOGRAPHIC METHOD FOR THE ASSAY

OF RETROVIRAL PROTEASES

Hitoshi Hori, Tamami Takahashi, Atsushi Kato,* Susumu Ueda*
and Hitoshi Kakidani

Biotechnology Research Laboratory
TOSOH Corporation
2743-1 Hayakawa
Ayase-shi. Kanagawa-ken 252

*Nippon Institute for Biological Science
2221-1 Shin-machi Ome
Tokyo 198
Japan

INTRODUCTION

The retroviral protease is essential for the maturation of infectious virus, and is therefore an attractive target for the design of antiviral drugs. To facilitate the screening of inhibitors of retroviral proteases, an improved chromatographic assay for the protease activity was developed.

The assay is based on the automated reversed-phase high performance liquid chromatographic analysis for the separation of the fluorogenic peptide substrate and its proteolytic reaction products. Incubation of the BLV lysate with the fluorogenic substrate, fluorescein-isothiocyanate-labeled decapaptide (FITC-PPAILPIISE), resulted in concomitant decrease in the substrate and increase in the proteolytic product, FITC-PPAIL. One cycle of HPLC analysis was completed within 15 minutes by using a non-porous resin packed column, octadecyl-NPR. This analytical procedure enables us to screen and characterize a large number of protease inhibitors, even in crude preparations.

MATERIALS AND METHODS

HPLC System

HPLC instruments used were: Pump and Controller, CCPM and PX-8010, respectively; Mixer, MX8010; UV Detector, UV-8010; Fluorescence Detector, FS-8010 and Autosampler, AS-8000. All instruments were purchased from TOSOH.

Bovine leukemia virus (BLV) was prepared from the culture fluid of fetal lamb kidney virus-producing cell line. Virus was purified by the combination of low and high speed centrifugations and was recovered on the 45 % layer of a stepwise-gradient of sucrose. Purified virus was then lysed in the solution containing 10 mM Tris-HCl (pH 7.4), 150 mM NaCl, 0.1 mM EDTA and 0.1 % NP-40. The lysate was used for the assay as an enzyme source.

Enzyme Reaction

A synthetic decapeptide substrate, Pro-Pro-Ala-Ile-Leu-Pro-Ile-Ile-Ser-Glu, was derivatized with fluorescein-isothiocyanate (FITC; DAIDO) on the peptide NH_2-terminus with conventional condensation chemistry. The fluorogenic peptide (FITCPPAILPIISE) served as a substrate. BLV protease activity was measured at 37°C in 20 mM MES (pH 6.0) containing 150 mM NaCl, 0.5% NP-40, 1 mM EDTA and 1 mM PMSF in a total volume of 50 µl. After 1 hr incubation. reactions were stopped by adding 900 µl of 0.2 % trifluoroacetic acid (TFA)/5 % acetonitrile and subjected to HPLC analysis.

HPLC Analysis

50 µl of stopped reaction samples were loaded on an octadecyl-NPR column (4.6 x 35 mm; TOSOH). Elution was carried out by using a gradient of 27.5 % - 57.5 % (vol/vol)

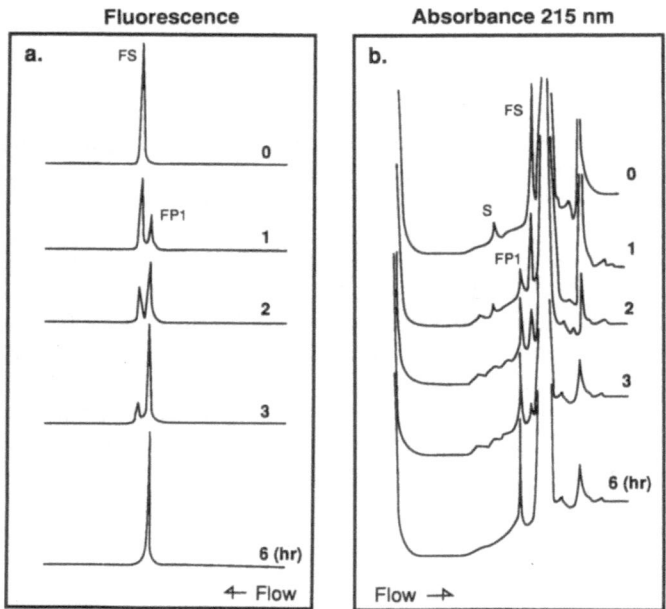

Figure 1. HPLC-UV/FL chromatograms of the time-dependent specific proteolysis of the fluorogenic peptide substrate by BLV lysate. (a) fluorescence detection (487 nm excitation >585 nm emission), (b) UV detection (215 nm). FS, fluorogenic substrate (FITC-PPAILPIISE); FP1, proteolytic product (FITC-PPAIL); S, unlabeled peptide (PPAILPIISE).

acetonitrile in 0.2 % TFA over a period of 3 min at a flow rate 1 ml/min. Absorbance at 215 nm and fluorescence intensity (487 nm excitation, 525 nm emission) were recorded.

RESULTS AND DISCUSSION

A high-performance liquid chromatographic analysis of synthetic peptide substrate-cleavage fragments has been one of the generally utilized methods for the assay of retroviral protease activity. This method seems to be suitable for studying the enzymological properties of the proteases, such as a substrate specificity. However, the method is regarded as an impractical assay for screening and characterizing large numbers of the protease inhibitors because it is time consuming.

To improve the situation, non-porous resin packed octadecyl-NPR[1,2] column was used for the HPLC analysis. By using the column, the substrate decapeptide, PPAILPIISE. and the products formed in the protease reaction were separated only in a 5 min linear gradient of 5 %-60 % acetonitrile.

The substrate decapeptide, PPAILPIISE, was then labeled with FITC to generate a fluorogenic substrate, FITC-PPAILPIISE. The fluorogenic peptide substrate was stable at least for a 6 month-storage at 4°C in the dark. The course of hydrolysis of this substrate by BLV protease was linear with incubation time (Figure 1). On the fluorescence detection chromatograms, only two peaks corresponding to the substrate (FITC-PPAILPIISE) and its proteolytic product (FITC-PPAIL) were detected, while many peaks were observed on the UV detection chromatograms. This result demonstrated that the use of the fluorogenic peptide as a substrate made it possible to assess the protease activity and its inhibition in crude reaction mixtures.

Using this system, enzymatic properties of BLV protease in BLV lysate were examined. The optimum pH and NaCl concentration of the BLV protease toward the fluorogenic substrate were 6.5 and 200 mM, respectively.

In this HPLC assay, analysis of one reaction sample was completed within 15 minutes. Coupling with an autosampler, almost 100 samples could be analyzed automatically in a day. The HPLC assay proved to be an useful tool for the screening and the characterization of a large number of protease inhibitors.

REFERENCES

1. Y. Yamasaki, T. Kitamura, S. Nakatani and Y. Kato, *J.Chromatogr.* **481**:391-396 (1989).
2. Y. Kato, S. Nakatan, T. Kitamura, Y. Yamasaki and T. Hashimoto, *J. Chromatogr.* **502**:416-422 (1990).

CLEAVAGE OF THE INTERMEDIATE FILAMENT SUBUNIT PROTEIN VIMENTIN BY HIV-1 PROTEASE: UTILIZATION OF A NOVEL CLEAVAGE SITE AND IDENTIFICATION OF HIGHER ORDER POLYMERS OF PEPSTATIN A

Robert L. Shoeman,[+] Bernd Höner,[+] Timothy J. Stoller,[*]
Elfriede Mothes,[+] Cornelia Kesselmeier,[+] Peter Traub[+]
and Mary C. Graves[*]

[+]Max-Planck-Institut für Zellbiologie
Rosenhof, D-6802
Ladenburg/ Heidelberg
Federal Republic of Germany

[*]Department of Molecular Genetics
Roche Research Center
Hoffmann-La Roche Inc.
Nutley, New Jersey   07110

INTRODUCTION

Intermediate filaments (IFs) are important constituents of the eukaryotic cell cytoskeleton. They are 10-12 nm in diameter and are composed of one or more of over 40 different subunit proteins, belonging to 5 classes of cytoplasmic and 1 class of nuclear proteins (the lamins). These proteins all possess a central rod domain, with a highly conserved amino acid sequence, which allow the IF proteins to form dimers through coiled coil interactions (much like myosin). These dimers, in turn, can polymerize into higher order structures which culminate in the final 10 nm filaments. As reviewed by Traub (1985) and Steinert and Roop (1988), much is known about the physical chemistry, molecular biology and tissue specific distribution of IF proteins; disappointingly little is known about their cellular functions. In addition to their role as cytoskeletal elements, it has been proposed that IF proteins may also participate in regulating expression of genetic information (Traub *et al.*, 1987; Chan *et al.*, 1989), although definitive proof of this role is still lacking.

As described in more detail in other contributions to this book, the HIV-1 protease (PR) is an aspartyl protease responsible for the cleavage of viral precursor polyproteins. No host cell proteins have yet been described to be cleaved during viral infection; however, recent results demonstrate that some host proteins can serve as substrate for HIV-1 PR. Here and elsewhere, we describe the cleavage *in vitro* of the IF subunit proteins vimentin, desmin and glial fibrillary acidic protein by HIV-1 PR (Shoeman *et al.*, 1990a).

Structure and Function of the Aspartic Proteinases
Edited by B.M. Dunn, Plenum Press, New York, 1991

RESULTS

Cleavage of IF subunit proteins by HIV-1 PR

Our initial experiments employed purified human and murine vimentin, murine vimentin proteolytic fragments and recombinant HIV-1 PR. As seen in Figure 1, the initial cleavage of both human and murine vimentin produced a large fragment with a M_r of 50,000. This reaction had a turnover number >50 s^{-1}, was dependent on added HIV-1 PR and could be inhibited by pepstatin A. Further incubation resulted in secondary cleavages producing three smaller peptides with M_r of 40,000-43,000 and having turnover numbers about 10-fold lower (Figure 1, longer time points). Two proteolytic fragments of murine vimentin were employed to localize these cleavages: T-vimentin, which lacks the N-terminal 70 amino acids and NT2, a smaller peptide composed of amino acid residues 1-119. As can be seen in Figure 1, T-vimentin was cleaved much more rapidly than NT2, suggesting that the primary cleavage site was located in the C-terminal tail domain. Immunoblotting with a battery of specific antibodies has confirmed that the primary cleavage removed most of the C-terminal tail domain and that the additional secondary cleavages all occurred in the N-terminal head domain (Shoeman *et al.*, 1990b). Vimentin polymerized into IFs yielded identical cleavage products; however, it was necessary to add 10 times the amount of HIV-1 PR to achieve the same extent of cleavage.

Microsequencing permitted the determination of the cleavage sites of vimentin, all of which were identical for both human and murine vimentin. The primary cleavage site was SSLNL/RETNL (cleavage between L421 and R422). This sequence resembles a hybrid between a class 1 and class 3 viral protein cleavage site (Henderson *et al.*, 1988), with the novel added feature of a charged group (R) at the P_1' position. The secondary cleavage site sequences bear little resemblance to known cleavage sites: TSRSL/YASSP56, PGGVY/ATRSS65 and ADAIN/TEFKN97 (where the sissile bond is indicated by / and the number of the last residue in the given sequence is indicated).

A decapeptide with the sequence SSLNLRETNL was found to be cleaved (between L/R) *in vitro* by HIV-1 PR with a turnover of about 1 s^{-1} (Shoeman *et al.*, 1990a). Heat-

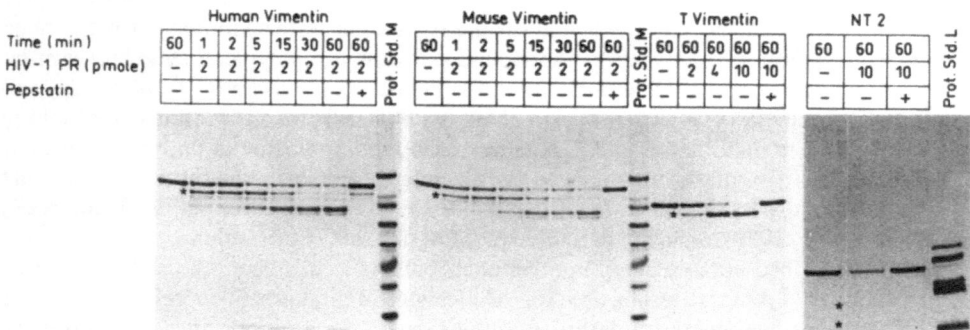

Figure 1. SDS-PAGE analysis of cleavage of vimentin and vimentin peptides by HIV-1 PR. Equivalent aliquots of reactions containing the indicated proteins (10 µg each = 185 pmol vimentin) and amounts of HIV-1 PR incubated in 100 µl of 20 mM sodium phosphate, pH 6.5, 10 mM DTT, 5 mM EGTA, 1 mM phenylmethylsulfonyl fluoride and 0.8 mM pepstatin A (where indicated). The protein standards are 94, 67, 43, 30, 20 and 14 kDa (lane Prot. Std. M) and 17, 14.4, 8.2, 6.2 and 2.5 (lane Prot. Std. L). T-vimentin contains amino acid residues 71-465 and NT2 contains residues 1-119. The large primary cleavage products are marked with an asterisk (*). (Modified from a figure in Shoeman *et al.* 1990a).

denatured vimentin was a poorer substrate for HIV-1 PR than native vimentin. This observation was somewhat surprising in light of the known ability of retroviral proteases to cleave denatured proteins (Yoshinaka & Luftig, 1981) and our results with bovine serum albumin (BSA): native BSA was not cleaved, whereas heat-denatured BSA was cleaved by HIV-1 PR (Shoeman *et al.*, 1990a). Collectively, all of these results suggest that HIV-1 PR recognizes both sequence and conformation of its substrate.

Microinjection of HIV-1 PR into human skin fibroblasts resulted in a ninefold increase in the percentage of cells with an abnormal distribution of vimentin IFs (also Shoeman *et al.*, 1990a). Later experiments employing a more highly purified and concentrated HIV-1 PR (kindly supplied by S. Roy, Hoffmann-La Roche Inc., Nutley, NJ) showed a specific effect of the HIV-1 PR on vimentin IF organization in up to 70 % of the cells. No effect was observed if these cells were microinjected with control protein solutions or cultured in the presence of pepstatin A following HIV-1 PR microinjection.

Although much is known or postulated about the role of the N-terminal head domain of IF subunit proteins, with the exception of the nuclear localization signal found in the nuclear lamins, very little is known about the function of the C-terminal tail domain. Electron microscopy of vimentin filaments demonstrated that few detectable changes were noted in filament morphology until most of the protein subunits had been cleaved within both their head and tail domains (Shoeman *et al.*, 1990b). The large primary cleavage product (M_r 50,000 Da) can form essentially normal filaments. This observation demonstrates that the tail domain of vimentin is not required for polymerization, which confirms and extends observations made with other IF subunit proteins (Kaufmann *et al.*, 1985). The smaller cleavage products (Mr 40,000-43,000 Da) additionally lack portions of the head domain and not only can not polymerize into filaments, but also are released from preformed filaments (Shoeman *et al.*, 1990b).

Polymerization of pepstatin A

During the course of our electron microscopy examination of IFs treated with HIV-1 PR in the absence and presence of pepstatin A, we noted the presence of filaments and higher order polymers of an unknown origin. Appropriate control experiments have shown that these structures were polymers of pepstatin A (Mothes *et al.*, 1990). Pepstatin A can spontaneously polymerize into a remarkable variety of higher order polymers, such as filaments, ribbons, sheets and cylinders (Figure 2). This polymerization was concentration dependent, required no additional components and occurred over the entire range of pH tested (pH 5 to pH 7.6), although at pH 5 there was a pronounced tendency for these polymers to tenaciously adhere to each other. At concentrations greater than or equal to 100 µM, pepstatin A polymerized into very long (>µm) filaments with a diameter of 9 nm and a pronounced helical twist. At pepstatin A concentrations (up to 2 mM), a larger percentage of more complex structures were seen: lateral associations of 9 nm filaments into twisted ribbons (with a width up to or more than 800 nm), which in turn formed flat sheets or, ultimately, rolled up into cylinders. Pepstatin A sequestered in these structures was no longer active as an inhibitor of HIV-1 PR (Mothes *et al.*, 1990, 1991). These observations provide an explanation for the disparate values for K_i and IC_{50} reported in the literature: at low µM concentrations, pepstatin A is soluble and behaves in a predictable fashion, while at 100 µM and above, stable polymers form in a concentration dependent fashion, making an accurate determination of active inhibitor concentration impossible. Furthermore, we have found that pepstatin A can directly associate with vimentin, probably through nonionic interactions, causing not only conformational changes in vimentin but also inducing the formation of copolymers of vimentin and pepstatin A. It is conceivable that some of the differential sensitivity of various aspartyl proteases to inhibition by pepstatin A may reflect differing ability of these molecules to engage in nonionic interactions. Thus, a potentially important

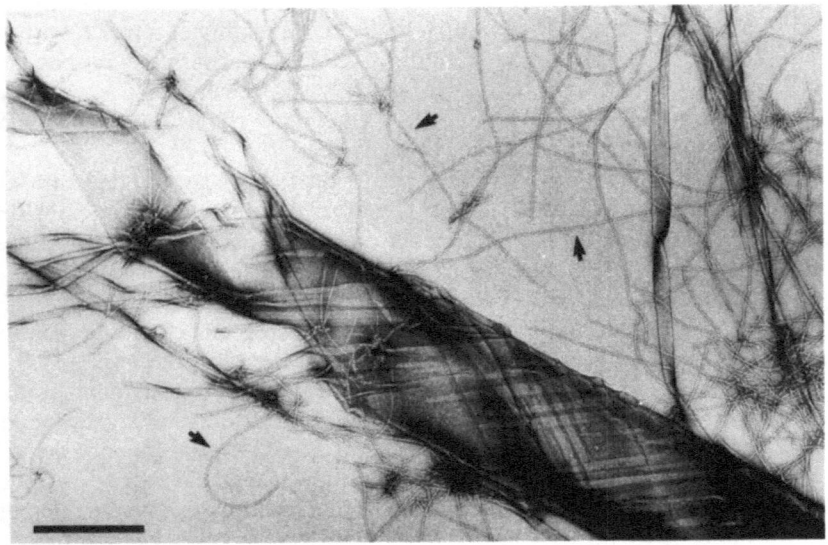

Figure 2. Higher order polymers of pepstatin A visualized by negative staining and electron microscopy. Polymers formed by pepstatin A incubated at 2 mM in 10mM Tris-acetate, pH 7.6, 6 mM ß-mercaptoethanol displayed all levels of higher order structure: 9 nm filaments (the long helical narrow strands marked by arrows) transition to small ribbons that twist and aggregate to form larger ribbons which also laterally associate and roll up into a hollow cylinder. The magnification was 44,000 X and bar represents 500 nm.

aspect of development of pepstatin A derivatives may be their engineering to enhance specific nonionic interactions with PR.

DISCUSSION

In contrast to the clear implications and consequences of our results with pepstatin A polymerization, the significance of the cleavage of IF proteins by HIV-1 PR remains an open question. It has been observed that cells chronically infected with Moloney murine sarcoma virus contain reduced amounts of vimentin and instead accumulate a highly phosphorylated and truncated form of vimentin, apparently lacking a portion of the tail domain (Singh & Arlinghaus, 1989). Since this retrovirus encodes a protease very similar to that of HIV-1 and since cellular proteases such as calpain generally cleave vimentin in the head domain, it is attractive to speculate, based on these and our *in vitro* results, that the retroviral PRs may cleave IF proteins *in vivo*. The PR of equine infectious anemia virus (another retrovirus) has been implicated to play an active role in the release of the viral genome from the capsid during infection (Roberts & Oroszlan, 1989). While the total number of PR molecules released from a single virion is probably very small (perhaps 100), there is a possibility that local degradation of IFs or other susceptible host proteins might occur in the immediate vicinity of the viral capsid during the initiation of infection. Also the cytoskeleton is, among other functions, involved in regulating exocytotic functions (reviewed by Mangeat, 1988). Can HIV-1 and other retroviruses exploit the properties of their PR to degrade IF proteins, thereby perturbing the organization of the cytoskeleton during the budding process? While this point has not been addressed experimentally, it is noteworthy that others have found that (1) cells transfected with HIV-1 PR mutants produce virions that either bud abnormally and/or possess aberrant morphology (Peng *et al.*, 1989; Göttlinger *et al.*, 1989); (2) HIV-1

PR is not dependent on virus budding for activity (Göttlinger *et al.*, 1989); and (3) sufficient amounts of HIV-1 PR are present to elicit an immune response in 70 % of infected individuals (Boucher *et al.*, 1989). These diverse results provide support for the idea that there may be additional roles of the HIV-1 and other retroviral PRs in the viral life cycle.

ACKNOWLEDGMENTS

We thank Mrs. Erika Schindler for the photograph and Mrs. Heidi Klempp for secretarial work.

REFERENCES

Boucher, C. A. B., de Jager, M. H., Debouck, C., Epstein, L. G., de Wolf, F., Wolfs, T. F. W. & Goudsmit, J., 1989, Antibody response to human immunodeficiency virus type 1 protease according to risk group and disease stage, *J. Clin. Microbiol.* **27**:1577.

Chan, D., Goate, A. & Puck T. T., 1989, Involvement of vimentin in the reverse transformation reaction, *Proc. Natl. Acad. Sci. U.S.A.* **86**:2747.

Göttlinger, H. G., Sodroski, I. G. & Haseltine, W. A., 1989, Role of capsid precursor processing and myristylation in morphogenesis and infectivity of human immunodeficiency virus type 1, *Proc. Natl. Acad. Sci. U.S.A.* **86**:5781.

Henderson, L. E., Benveniste, R. E., Sowder, R., Copeland, T. D., Schultz, A. M. & Oroszlan, S., 1988, Molecular characterization of gag proteins from simian immunodeficiency virus (SIVMne), *J. Virol.* **62**:2587.

Kaufmann, E., Weber, K. & Geisler, N., 1985, Intermediate filament forming ability of desmin derivatives lacking either the amino-terminal 67 or the carboxy-terminal 27 residues, *J. Mol. Biol.* **185**:733.

Mangeat, P. H., 1988, Interaction of biological membrane with the cytoskeletal framework of living cells, *Biol. Cell.* **64**:261.

Mothes, E., Shoeman, R. L., Schröder, R. R. & Traub, P., 1990, Polymerizing properties of pepstatin A, *J. Struct. Biol.* **105**:80.

Mothes, E., Shoeman, R. L. & Traub, P., 1991, Effect of pepstatin A on structure and polymerization of intermediate filament subunit proteins *in vitro*, *J. Struct. Biol.* **106**:64.

Peng, C., Ho, B. K., Chang, T. W. & Chang, N. T., 1989, Role of human immunodeficiency virus type 1 specific protease in core protein maturation and viral infectivity, *J. Virol.* **63**:2550.

Roberts, M. M. & Oroszlan, S., 1989, The preparation and biochemical characterization of intact capsids of equine infectious anemia virus, *Biochem. Biophys. Res. Commun.* **160**:486.

Shoeman, R. L., Höner, B., Stoller, T. J., Kesselmeier, C., Miedel, M. C., Traub, P. & Graves, M. C., 1990a, Human immunodeficiency virus type 1 protease cleaves the intermediate filament proteins vimentin, desmin, and glial fibrillary acidic protein, *Proc. Natl. Acad. Sci. U.S.A.* **87**:6336.

Shoeman, R. L., Mothes, E., Kesselmeier, C. & Traub, P., 1990b, Intermediate filament assembly and stability *in vitro*: effect and implications of the removal of head and tail domains of vimentin by the human immunodeficiency virus type 1 protease, *Cell Biol. Int. Rep.* **14**:583.

Singh, B. & Arlinghaus, R.B., 1989, Vimentin phosphorylation by p37mos protein kinase *in vitro* and generation of a 50 kDa cleavage product in Vmos-transformed cells, *Virology* **173**:144.

Steinert, P. M. & Roop, D. R., 1988, Molecular and cellular biology of intermediate filaments, *Ann. Rev. Biochem.* **57**:593.

Traub, P., 1985, "Intermediate filaments. A review," Springer Verlag, Heidelberg.

Traub, P., Plagens, U., Kühn, S. & Perides, G., 1987, Function of intermediate filaments. A novel hypothesis, *Fortschr. Zool.* **34**:275.

Yoshinaka, Y. & Luftig, R. B., 1981, A comparison of avian and murine retrovirus polyprotein cleavage activities, *Virology* **111**:239.

A NEW TYPE OF ASPARTIC PROTEINASE INHIBITORS

WITH A SYMMETRIC STRUCTURE

N. I. Tarasova, S. V. Gulnik, A. A. Prishchenko, M. V. Livantsov,
E. N. Lysogorskaya and E. S. Oksenoit

Moscow State University
Department of Chemistry
Moscow 119899
USSR

INTRODUCTION

Molecules of aspartic proteinases of mammalian and fungal origin are known to consist of one polypeptide chain forming two structurally similar domains.[1-3] The molecules in whole possess pseudosymmetric structure with two-fold axis of symmetry. In contrast to cellular aspartic proteinases, the enzymes of retroviruses can function in the form of homodimers only.[4,5] The active site is formed during dimerization so that each subunit contributes one Asp_{25} residue to the catalytic site. Thus the active molecule is perfectly symmetric. At the same time the substrates of the enzyme are not symmetric as peptide chains are always directed. So in the active site of a retroviral proteinase the productive binding of the substrate is possible both "from east to west" and vise versa, "from west to east". X-ray analysis studies have revealed that binding of inhibitors and, most likely, substrates causes significant conformational changes in aspartic proteinases.[6-8] In the case of HIV proteinase such changes were shown to disturb the symmetry of the molecule.[9,10] But this may not be necessary in the case of binding of a symmetric inhibitor, and symmetric structures may have a significant advantage both in affinity and in selectivity, as the degree of symmetry is much higher for the viral enzymes compared to cellular ones.

In designing symmetric inhibitors of aspartic proteinases two main strategies do exist. In the first peptide moieties of the inhibitors are linked "tail to tail", so that C-terminal parts of peptides are used for the linkage:-NH-CHR-CO-|Active Site|-CO-CHR-NH- . This way of design is logically more simple and more safe as a large part of asymmetric inhibitors of aspartic proteinases so far obtained were constructed by means of substitution of the nitrogen atom of the scissile peptide bond (hydroxy isostere, keto and fluoroketo isostere, statine derivatives, etc.[11]). The second approach, that seems less obvious and more risky, implies the linkage of peptides through the centered amino termini: -CO-CHR-NH-|Active Site|-NH-CHR-CO- . Such constructions could be expected to bind to the aspartic proteinases, as at least in the case of retroviral enzymes the substrate binding surface must be suited for such inverted structures because of the symmetry. We have chosen the second

Structure and Function of the Aspartic Proteinases
Edited by B.M. Dunn, Plenum Press, New York, 1991

Table 1. Substitution of amino acids in carbonyl derivatives of peptides

Peptide moiety	IC_{50}[*]			
	HIV proteinase		Human pepsin	
Pro-Ile-Val-OMe	100 µM	no inhibition	100 µM	no inhibition
Phe-Ile-Val-OMe	100 µM	no inhibition	2 µM	
nLeu-Ile-Val-OMe	100 µM	no inhibition	2.5 µM	

[*]Pepsin activity was measured with the help of fluorogenic substrate Abz-Ala-Ala-Phe-Phe-pNA (Abz = anthranilic acid)[12] at 5 µM substrate concentration in 0.1 M sodium acetate buffer, pH 3.7 in the presence of 3 % dimethylformamide (DMFA). Inhibitors were added to the assay medium from stock solution in DMFA. The presence of 3 % DMFA in the final incubation mixture had no significant effect on the enzyme activity. HIV proteinase activity was tested against fluorogenic substrate Abz-Ser-Phe-Asn-Phe-Pro-Glu-Ile-pNA (5 µM solution in 0.05 M MES-NaOH buffer pH 5.5, containing 0.7 M ammonium sulfate, 1 mM DTT, 1 mM EDTA and 2 % DMFA). nLeu = Norleucine

approach to create some new compounds that could block HIV proteinase activity. The main problem that arises in that way of intention is the size and structure of the linkage between peptide moieties. Modeling on the basis of x-ray analysis data permits only imprecise estimation of the distance between the N-terminal nitrogen atoms of the peptide parts that is necessary for the optimal location of the inhibitor in the catalytic site. And even more, it is impossible to state *a priori* that the linkage of the correct size and properties does exist. Information about the structure of numerous described inhibitors of aspartic proteinases permitted us to assume that the distance between the nitrogen atoms to be linked most likely corresponds to 1-2 carbon atoms.

EXPERIMENTAL DESIGN AND RESULTS

We have designed and synthesized compounds containing three different kinds of linkages imitating either the scissile peptide bond or a transition state.

The first type of linkage is produced by a carbonyl group positioned between the peptide wings. The second linkage is contributed by two carbonyls and can be considered a derivative of oxalic acid. The third is a phosphorus analog of the first and contains phosphate

Table 2. Substitution of amino acids in oxalyl derivatives of peptides

Peptide moiety	IC_{50}	
	HIV proteinase	Human pepsin
Pro-Ile-Val-OMe	11.7 µM	100 µM, no inhibition
Phe-Ile-Val-OMe	100 µM, no inhibition	10 µM
Phe-Ile-OMe	not determined	15 µM
Phe-OMe	not determined	100 uM, no inhibition
nLeu-Ile-Val-OMe	100 µM, no inhibition	7.5 µM

Table 3. Substitution of aminoacids in phosphorus derivatives of peptides

Peptide moiety	IC_{50}	
	HIV proteinase	Human pepsin
Pro-Ile-Val-OMe	0.37 μM	100 μM no inhibition
Pro-Ile-OMe	100 μM no inhibition	not determined
Phe-Ile-Val-OMe	100 μM no inhibition	10 μM
Phe-Ile-OMe	not determined	100 μM no inhibition
Phe-OMe	not determined	not determined
nLeu-Ile-Val-OMe	100 μM no inhibition	50 μM

moiety linked with amino groups of peptides. For all three linkages, rows of compounds were obtained with peptide moieties of different size and structure. N-terminal fragment of HIV proteinase action site between p17 and p24 of gag proteins, Pro-Ile-Val, was used as a basic sequence for varying the peptide parts of potential inhibitors. Inhibitory properties of the substances were characterized in reaction of fluorogenic peptides hydrolysis (Tables 1, 2 and 3).

The results so far obtained indicate that symmetric structures constructed from two identical peptide moieties bound to each other "head to head" can exhibit inhibitory properties against aspartic proteinases. Corresponding N-blocked peptides have shown no inhibition at concentrations up to 100 μM. So, most probably, inhibitors bind to the enzymes in a symmetric manner. Among the tested groups of compounds phosphate derivatives appeared to be the most potent inhibitors of HIV proteinase, meanwhile the activity of human pepsin was blocked most effectively by urea derivatives (Table 1). Compound (OH)PO(Pro-Ile-Val-OMe)$_2$ was found to be a specific inhibitor of human immunodefficiency virus proteinase with a moderate potency ($IC_{50} = 370$ nM). Substitution of proline residue with phenylalanine and norleucine reduced inhibitory potency dramatically, although both latter amino acids are found in P_1 position of substrates.

Further studies of the substances with such structures can benefit both the investigation of structure-function relationship in the family of aspartic proteinases and the search for effective antiviral drugs. Studies of the influence of peptide part composition on potency and specificity of inhibitors and optimization of the central part of inhibitors is in progress now.

REFERENCES

1. J. Tang, M. N. G. James, I.-N. Hsu, J. A. Jenkins and T. L. Blundell, *Nature* **271**:618-621 (1978).
2. N. S. Andreeva, A. A. Fedorov, A. E. Gustchina, R. R. Riskulov, M. G. Sufro and N. E. Shutzkever, *Mol. Biol. (Mosk)* **12**:704-717 (1978).
3. M. N. G.James and A. R. Sielecki, *J. Mol. Biol.* **163**:299-361 (1983).
4. T. D. Meek, B. D. Dayton, B. W. Metcalf, G. B. Dreyer, J. E. Strickler, J. G. Gorniak, M. Rosenberg, M. L. Moore, V. W. Magaard and C. Debouck, *Proc. Natl. Acad. Sci. U.S.A.* **86**:1841-1845 (1989).
5. M. Miller, M. Jaskólski, J. K. M. Rao, J. Leis and A. Wlodawer, *Nature* **337**:576-579 (1989).
6. K. Suguna, E. A. Padlan, C. W. Smith, W. D. Carlson and D. R. Davies, *Proc. Natl. Acad. Sci U.S.A.* **84**:7009-7013 (1987).

7. S. I. Foundling, J. Cooper, F. E. Watson, A. Cleasy, L. H. Pearl, B. L. Sibanda, A. Hemming, S. P. Wood, T. L. Blundell, M. J. Valler, C. G. Norey, J. Kay, J. Boger, B. M. Dunn, B. J. Leckie, D. M. Jones, B. Atrash, A. Hallett and M. Szelke, *Nature* **327**:349-352 (1987).

8. A. Sali, B. Veerapandian, J. B. Cooper, S. I. Foundling, D. J. Hoover and T. L. Blundell, *The EMBO Journal* **8**:2179-2188 (1989).

9. M. Miller, J. Schneider, B. K. Sathyanarayna, M. V. Toth, G. R. Marshall, L. Clawson, L. Selk, S. B. H. Kent and A. Wlodaver, *Science* **246**:1149-1152.

10. A. L. Swain, M. Miller, J. Green, D. H. Rich, S. B. H. Kent and A. Wlodawer, *Proc. Natl. Acad. Sci. U.S.A.* in press (1990).

11. M. Szelke, *in*: "Aspartic Proteinases and Their Inhibitors," V. Kostka, ed., 421-441 de Gruyter, Berlin, 1985.

12. I. Ju. Filippova, E. N. Lysogorskaya, E. S. Oksenoit, Ju. E. Komarov and V. M. Stepanov, *Bioorg. Chem. (Mosk)* **12**:1172-1180 (1986).

TIME DEPENDENT HETERODIMER FORMATION LEADS TO INHIBITION OF

HIV PROTEASE ACTIVITY

Lilia M. Babé and Charles S. Craik

Department of Pharmaceutical Chemistry
University of California, San Francisco
San Francisco, California   94143-0446

INTRODUCTION

The protease encoded by the human immunodefiency virus (HIV) is a homodimer as determined by X-ray crystallographic (Navia *et al.*, 1989, Wlodawer *et al.*, 1989) and biochemical analysis (Meek *et al.*, 1989). A four-stranded antiparallel ß-sheet generated by interdigitating N- and C-termini of the monomers dominates the dimer interface. This ß-sheet is partially stabilized by intersubunit backbone H-bonds of alternate amino acids from each of the four strands.

Each 99 amino acid monomer contributes half of the active site which includes two catalytic aspartic acids as well as threonine/serine and glycine residues, which are conserved among all aspartyl proteases for their structural role in active site geometry (Pearl & Taylor, 1987). This dimer formation generates not only the catalytic center, but also the extended substrate binding pocket.

Viral proteolytic activity is essential for the generation of infectious particles in HIV and related retroviruses (Kohl *et al.*, 1989). Therefore, agents that specifically inhibit the activity of the protease may serve as powerful antivirals (e.g. Erickson *et al.*, 1990). The requirement for the protease to dimerize provides an alternative mechanism for inhibiting enzyme activity other than active site directed inactivation. Disruption of assembled protease dimers or prevention of their formation should block viral protease activity. As a proof of this principle, we make use of defective monomers or non-identical subunits to exchange with wildtype homodimers to produce catalytically defective heterodimers.

MATERIALS AND METHODS

Cells and plasmid constructions

HIV protease was expressed in *E. coli* from plasmid pSOD/PR179 (Babé *et al.*, 1990). HIV-2 protease was expressed in *S. cerevisiae* from plasmid pHIV-2PR115 (Pichuantes *et al.*, 1990). HIV-1Δ1-5 was expressed in *S. cerevisiae* from plasmid pPR94 (Babé *et al.*, 1991).

Structure and Function of the Aspartic Proteinases
Edited by B.M. Dunn, Plenum Press, New York, 1991

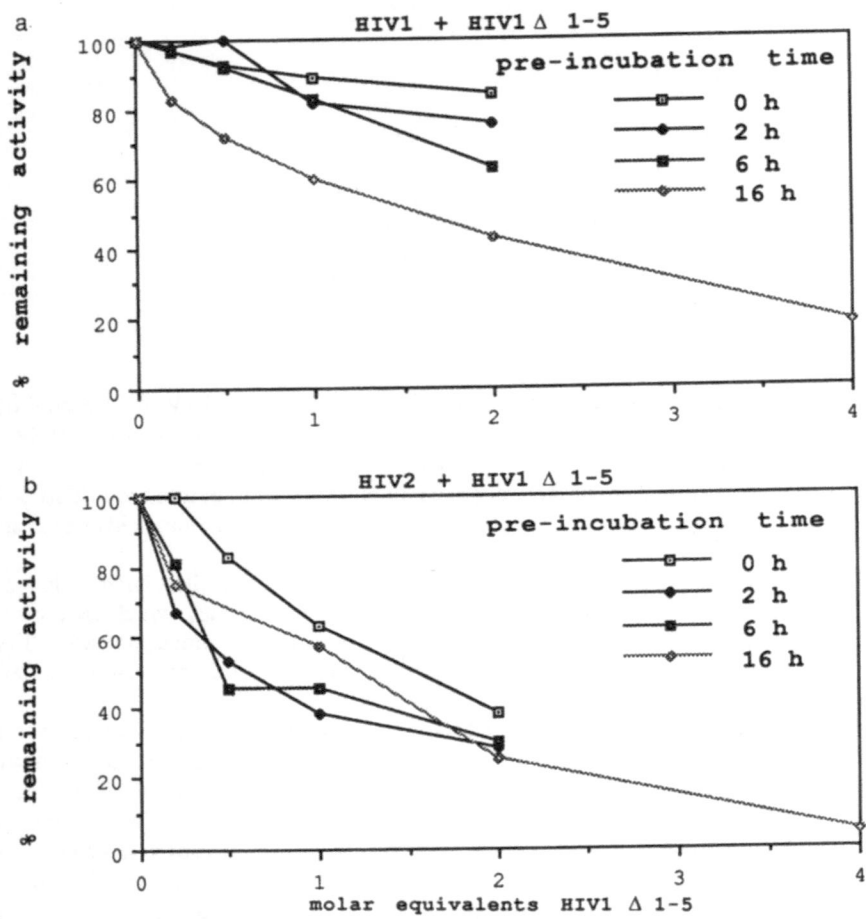

Figure 1. Time course of pre-incubation of HIV-1 (Panel a) or HIV-2 (Panel b) proteases with increasing amounts of HIV-1Δ1-5 polypeptide. Measurement of remaining proteolytic activity of mixtures using a decapeptide substrate.

Protein purification

HIV-1 protease was purified from bacteria as described previously (Babé *et al.*, 1990). The HIV-1Δ1-5 and HIV-2 proteases were purified from yeast as described elsewhere (Babé *et al.*, 1991). Following the isolation of purified proteases from reverse phase HPLC, samples from the three enzymes were lyophilyzed, denatured in 8 M urea and refolded as described previously (Tomasselli *et al.*, 1990).

Pre-incubations of proteins and in vitro *proteolytic activity assays*

Proteolysis was monitored by measuring the rate of specific hydrolysis of the decapeptide Ala-Thr-Asn-Phe*Pro-Ile-Ser-Pro-Trp which represents the junction between the protease and the reverse transcriptase in the HIV-1 gag-pol polypeptide (Lightfoote *et al.*, 1986). Cleavage at the Phe*Pro bond is monitored by detection of the two peptide products using HPLC reverse phase chromatography as described previously (Pichuantes *et al.*, 1990).

Purified HIV-1 or HIV-2 proteases, 100 ng (250 nM) were incubated for 0, 2, 6, and 16 h at 25°C with 20 ng (50 nM) to 400 ng (1 μM) of purified HIV-1Δ1-5 in 40 ul volumes. The reaction mixture consisted of: 150 mM Na acetate pH 5.5, 10 mM DTT, 1 mM EDTA, 2 % glycerol, 1 % ethylene glycol, 0.2 M urea and 1 M NaCl. A 5 μl aliquot of 2 mM decapeptide substrate was then added and the incubation proceeded for 2 h at 37°C. The samples were fractionated by HPLC as mentioned above.

RESULTS AND DISCUSSION

Incubation of the HIV-1 protease with either defective monomers (HIV-1Δ1-5) or non-identical subunits results in a gradual inactivation of the enzyme. The loss of activity is presumably due to the formation of defective heterodimers of reduced or null enzymatic activity. The deletion mutant, HIV-1Δ1-5 lacks the first 5 residues of the HIV-1 sequence. Such a species has been detected as a potential autocatalytic product of the HIV-1 enzyme and is enzymatically inactive. For these experiments, the 94 amino acid polypeptide was generated by heterologous expression. Following purification, HIV-1Δ1-5 was subjected to the refolding protocol used for HIV-1 (Tomasselli *et al.*, 1990) to induce correct folding. Incubation of the HIV-1 or HIV-2 protease with a 4-fold molar excess of HIV-1Δ1-5 leads to 80 % and 95 % inhibition of enzyme activity, respectively, when pre-incubated 16 h at 25°C.

Table I. Reduction in activity of HIV-1 protease by the addition of HIV-2 protease

Molar equivalents of HIV-2	% Reduction in activity	
	HIV-1/denat. HIV-2	HIV-1/active HIV-2
0	0	0
0.25	3	15
0.5	0	18
1.0	0	30
2.5	2	40
5.0	5	53

Presumably, the deletion mutant can form heterodimers with the wildtype polypeptide but their altered interface structures renders them enzymatically inactive. Figure 1 shows a time course of inhibition upon addition of increasing amounts of HIV-1Δ1-5 in pre-incubations of 0, 2, 6, and 16 h with HIV-1 or HIV-2 protease. The time dependence of the inhibition agrees with the expected dynamic nature of the subunit interactions and exchange.

Several differences exist in the amino and carboxyl terminal sequences of the HIV-1 and HIV-2 proteases. The results obtained from mixing samples of HIV-1 and HIV-2 protease are shown in Table I. At equimolar concentrations of enzyme, the activity is reduced by approximately 30 %, whereas at a 4 molar excess of HIV-2 protease, a reduction of 53 % is observed. This again suggests the formation of heterodimers with much reduced or no enzymatic activity. The amino acid side chain differences (Ile3Phe, Thr4Ser, Cys95Met, Thr96Ser, Phe99Leu) in the interface region of the HIV-1 and HIV-2 proteases may result in the creation of less stable heterodimers due to changes in the length and hydrophobicity of the side chains. A heat denatured HIV-2 protease sample did not inhibit the HIV-1 protease, indicating the need of a folded structure to serve as a subunit for exchange. HIV-1/2 heterodimers have been isolated by anionic exchange HPLC by virtue of their intermediate isoelectric point (Babé *et al.*, 1991).

These results indicate that inhibition of the HIV protease by heterodimer formation can be accomplished. This method of inhibition could be applied to the generation of dominant negative mutations capable of blocking proteolysis by interfering with homodimer formation. Targeting this region of the protease is attractive because the extended interface created by the N- and C-termini may be less vulnerable to mutational escape than the active site or the substrate binding regions of the protease and therefore maintained in a greater number of evolving viral strains.

ACKNOWLEDGEMENTS

This work was supported by Grant GM 39552 (C. S. C.) from the National Institutes of Health. L. M. B. is a recipient of a University of California Task Force on AIDS Fellowship F88SF122.

REFERENCES

Babé, L. M., Pichuantes, S., Barr, P. J., Bathurst, I. C., Masiarz, F. R. & Craik, C. S., 1990, HIV-1 protease: bacterial expression, purification and characterization, *in*: " Proteins and Pharmaceutical Engineering," Craik, C. S., Fletterick, R., Matthews, C. R. and Wells, J. Eds. Willey-Liss, New York.

Babé, L. M., Pichuantes, S. & Craik, C. S., 1991, Inhibition of HIV protease by heterodimer formation, *Biochem.* **30**:106.

Erickson, J., Neidhart, D. J., VanDrie, J., Kempf, D. J., Wang, X.-C., Norbeck, D. W., Plattner, J. J., Rittenhouse, J. W., Turon, M., Wideburg, N., Kohlbrenner, W. E., Simmer, R., Helfrich, R., Paul, D. A. & Knigge, M., 1990, Design, activity, and 2.8 Å crystal structure of a C_2 symmetric inhibitor complexed to HIV-1 protease, *Science* **249**:527.

Kohl, N. E., Emini, E. A., Schleif, W. A., Davis, L. J., Heimbach, J. C., Dixon, R. A. F., Scolnick, E. M. & Sigal, I. S., 1988, Active human immunodeficiency virus protease is required for viral infectivity, *Proc. Natl. Acad. Sci. U.S.A.* **85**:4686.

Lightfoote, M. M., Coligan, J. E., Folks, T. M., Fauci, A. S., Martin, M. A. & Venkatesan, S., 1986, Structural characterization of reverse transcriptase and endonuclease polypeptides of the acquired immunodeficiency syndrome virus, *J. Virol.* **60**:771.

Meek, T. D., Dayton, B. D., Metcalf, B. W., Dreyer, G. B., Strickler, J. E., Gorniak, J. G., Rosenberg, M., Moore, M. L., Magaard, V. W. & Debouck, C., 1989, Human immunodeficiency virus 1 protease expressed in *Escherichia coli* behaves as a dimeric aspartic protease, *Proc. Natl. Acad. Sci. U.S.A.* **86**:1841.

Navia, M. A., Fitzgerald, P. M., McKeever, B. M., Leu, C.-T., Heimbach, J. C., Herber, W. K., Sigal, I. S., Darke, P. L. & Springer, J. P., 1989, Three-dimensional structure of aspartyl protease from human immunodeficiency virus HIV-1, *Nature* **337**:615.

Pearl, L.H. & Taylor, W. R., 1987, A structural model for the retroviral proteases, *Nature* **329**:351.

Pichuantes, S., Babé, L. M., Barr, P. J., DeCamp, D. L. & Craik, C. S., 1990, Recombinant HIV-2 protease processes HIV-1 Pr53^{gag} and analogous junction peptides *in vitro*, *J. Biol. Chem.* **265**:13890.

Tomasselli, A. G., Olsen, M. K., Hui, J. O., Staples, T. K., Heinrikson, R. L. & Tomich, C.-S., 1990, Substrate analogue inhibition and active site titration of purified recombinant HIV-1 protease, *Biochem.* **29**:264.

Wlodawer, A., Miller, M., Jaskólski, M., Sathyanarayana. B. K., Baldwin, E., Weber, I. T., Selk, L. M., Clawson, L., Schneider, J. & Kent, S. B. H., 1989, Conserved folding in retroviral proteases: crystal structure of a synthetic HIV-1 protease, *Science* **245**:616.

MOLECULAR MODELING OF THE HIV-2 PROTEASE

Alla Gustchina,* Irene T. Weber and Alexander Wlodawer

Crystallography Laboratory
NCI-Frederick Cancer Research and Development Center
ABL-Basic Research Program
Frederick, Maryland   21702-1201

*On leave from the V.A. Engelhardt Institute of Molecular Biology
 Academy of Sciences of the USSR
 Moscow
 USSR

INTRODUCTION

The principal pathogen of acquired immunodeficiency syndrome (AIDS) in the United States is human immunodeficiency virus type 1 (HIV-1), but the disease is sometimes caused by a second variant, HIV-2. These two viruses share a similar genomic organization, indicating common evolutionary origin, but differ significantly in terms of nucleotide and amino-acid sequence, with no more than 60 % overall amino-acid identity. Elucidation of the structures of virally-encoded enzymes offers an opportunity for rational drug design and has been pursued by many groups. The structure of only one enzyme encoded by HIV-1, that of the aspartyl protease (PR), is currently known (Navia *et al.*, 1989; Wlodawer *et al.*, 1989). This enzyme is essential for viral maturation, and is the target of choice for designing new therapeutic agents.

Any drugs directed against protease need to work against many isolates, and ideally should also be effective against HIV-2. Thus, it is necessary to elucidate three-dimensional structure of the HIV-2 PR if rational drug design is to succeed. A number of different isolates of both types of the human immunodeficiency virus have been sequenced, showing considerable variations in amino acids in the PR region (Figure 1). Only 41 % of the residues are identical, but the fact that the important structural elements are highly conserved and that most substitutions occur in the surface loops was the basis for modeling the structure of HIV-2 PR from the three-dimensional structure of HIV-1 PR.

Structure and Function of the Aspartic Proteinases
Edited by B.M. Dunn, Plenum Press, New York, 1991

The HIV PR sequences corresponding to a number of known isolates of HIV-1 and HIV-2 were obtained from the Los Alamos HIV Sequence Database. The atomic coordinates of the crystal structure of the complex of HIV-1 PR with inhibitor JG-365 (Ac-Ser-Leu-Asn-Phe-ψ[CH(OH)CH$_2$N]-Pro-Ile-Val-OMe) (Swain *et al.*, 1990) were taken as a starting point for modeling of HIV-2 PR with the sequence of *rod* isolate. Amino acid side chains that were different in HIV-1 and -2 PR were built in sterically reasonable conformations and were manually optimized to prevent short contacts with neighboring residues. The conformations of the substituted residues involved in specific interactions were chosen in such a way as to allow maintaining of similar interactions. The original conformation was kept for the residues which were found on the surface and were not involved in molecular contacts. The conformation of the main chain was changed only for positions where Pro residues, or Gly residues with positive ϕ torsion angles, were replaced by another type of residue.

The atomic coordinates of the second monomer in the dimer were obtained from the first one by a superposition on the C$_\alpha$ atoms of the second monomer in the crystal structure of HIV-1 PR. For these reasons, the model of HIV-2 PR is a symmetrical dimer of two subunits related by a rotation of about 178°. For residues which are involved in either the interactions with the inhibitor or in intermolecular contacts some departure from symmetry was allowed. One additional adjustment involved the side chain of Ile$_{50}$' on the flap which was moved slightly, and the carbonyl oxygen was positioned to form a hydrogen bond interaction with NH of Gly$_{51}$. No energy minimization was performed during structure rebuilding, but energy calculation at the conclusion of the process showed values similar to those for the crystallographic model of HIV-1 PR.

The HIV-2 PR was modeled with two different inhibitors with known characteristics of binding to this enzyme, as well as to HIV-1 PR (Richards *et al.*, 1989, 1990). One of them is acetyl-pepstatin (Ac-Val-Val-Sta-Ala-Sta, where Sta is statine), a common inhibitor of aspartic proteases, and the other is a renin inhibitor, H-261 (tBoc-His-Pro-Phe-His-Leu-ψ[CHOH-CH$_2$]-Val-Ile-His). The inhibitors were positioned in a way similar to that observed in the cocrystal structures of HIV-1 PR (Miller *et al.*, 1989; Swain *et al.*, 1990).

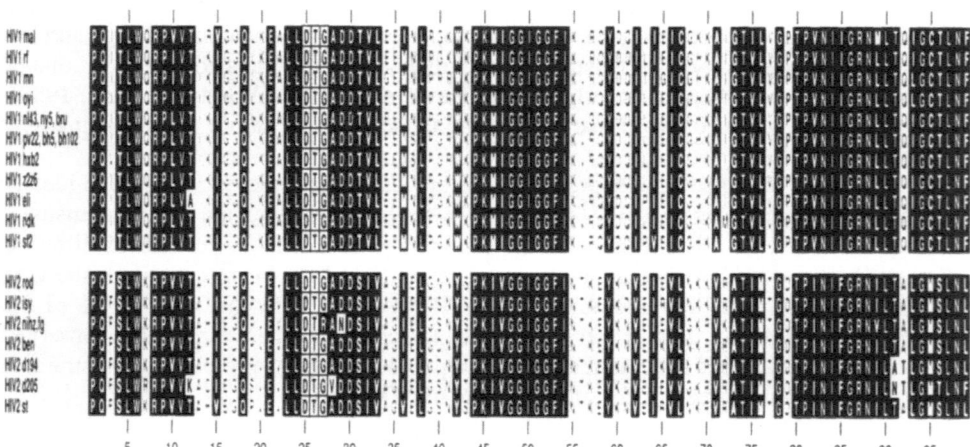

Figure 1. Sequence alignment of the proteases from HIV-1 and HIV-2. The three residues present in the active site of all aspartic proteases (DTG) are boxed. Residues conserved in proteases from both sources are shaded black, those conserved in each of the proteases are shaded grey, and residues which are not conserved are grey on white background. Those residues which appear to violate conservation are black on white (they may represent sequence errors).

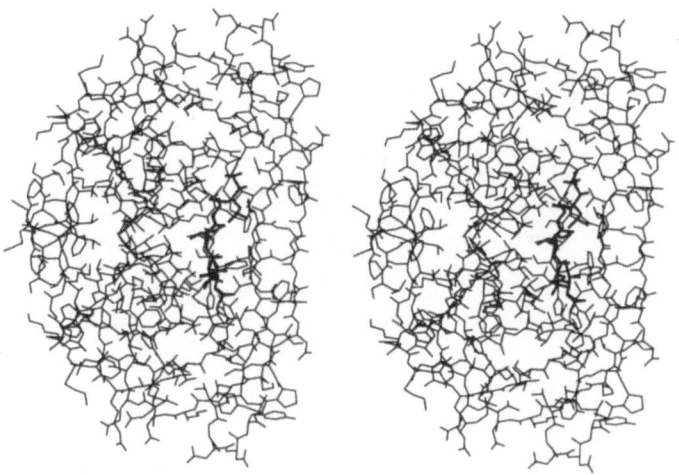

Figure 2. Stereo view of the atomic model of the molecule of HIV-2 protease. All non-hydrogen atoms belonging to the protein, as well as to both of the modeled inhibitors, are shown.

MOLECULAR MODEL OF HIV-2 PR

The structure of the model of HIV-2 PR, which is quite close to the starting coordinates of HIV-1 PR, is shown in Figure 2. The root-mean-square deviation for the 185 (out of 198) C_α atoms is 0.28 Å, and for 1-24 common atom pairs is 0.66 Å. All the important structural elements of the protease encoded by both variants of HIV have highly conserved sequences. The most conserved region is a segment containing the Asp-Thr-Gly sequence that is common to all aspartic proteases. This is part of the wide loop which forms the active site by dimerization of two subunits. The helical region, residues 86-93, is also highly conserved, due to restrictions imposed by the three-dimensional structure. The residues in these particular regions provide numerous intra- and intermolecular interactions in the active dimer and also enzyme-substrate interactions.

The other highly conserved sequence in different isolates of HIV-1 and HIV-2 PRs occurs in the flap region. The flap is a flexible structural unit in all aspartic proteases, and the residues at the tip of the flap are involved in enzyme-substrate interactions. All isolates of HIV-1 and HIV-2 PR have a conserved sequence at the tip of the flap. This sequence, -GGIGGFI-, involves four Gly residues which are very important for providing flexibility (Gustchina & Weber, 1990). Residues from the N-terminus and the C-terminus are conserved due to the formation of the four-stranded antiparallel ß-sheet that has many intersubunit interactions in the PR dimer. Conserved substitutions among the residues involved in the important structural interactions are correlated and fit very well without disturbing even compact interior formations such as the hydrophobic nucleus. On the other hand, the variable residues can form different structures on the surface of the molecule.

INHIBITOR BINDING IN HIV-1 AND HIV-2 PRs

Only four of the residues involved in enzyme-substrate interactions show differences between HIV-1 and -2 PRs. These are all conservative substitutions: Val_{32} to Ile (part of the S_2' and S_2 subsites), Ile_{47} to Val (involved in the S_2' and S_4 subsites), Leu_{76} to Met (involved in S_2 and S_2'), and Val_{82} to Ile (in the S_1' subsite). Within the different isolates of HIV-1 and -2 there are no changes in the residues forming the PR substrate binding pocket.

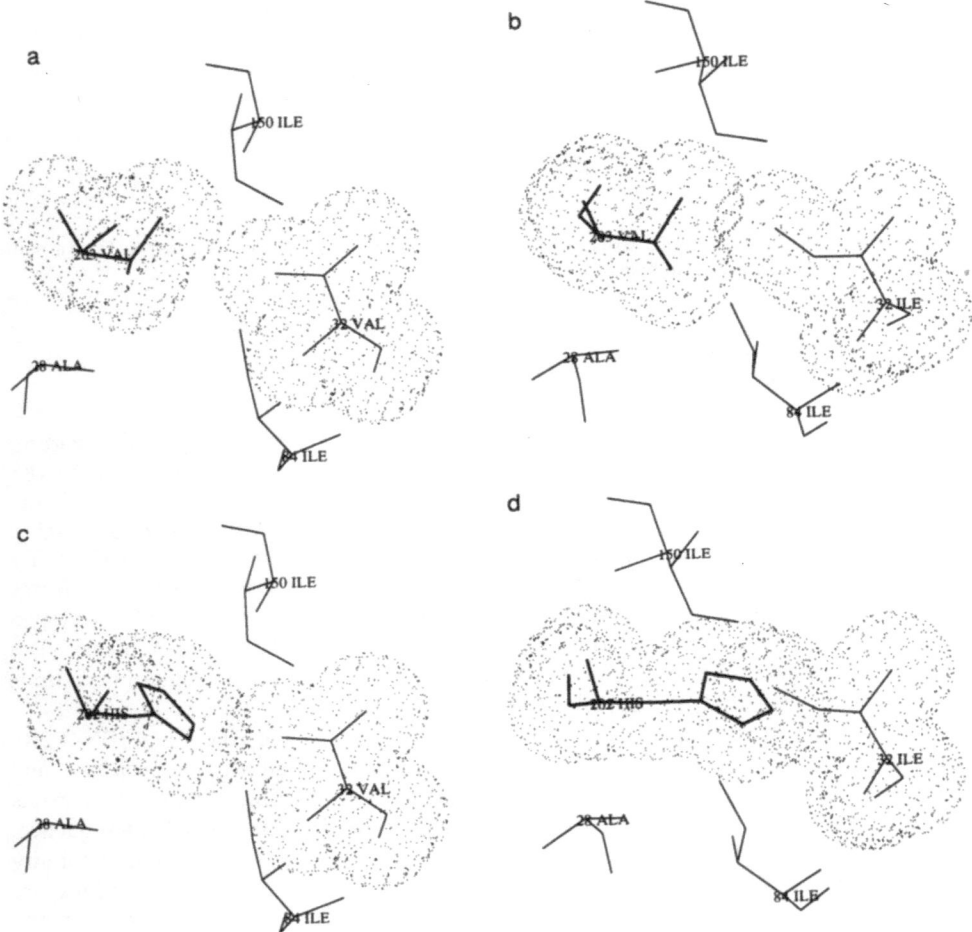

Figure 3. Contacts between residues of the protease in the S_2 binding pocket with the side chain of the inhibitor. Van der Waals surfaces are shown for residue 32 of the protein and for the P_2 side chain of the inhibitor. a) HIV-1 PR and the valine side chain of acetyl-pepstatin. b) HIV-2 PR and the valine of acetyl-pepstatin. c) HIV-1 PR and the histidine of H-261. d) HIV-2 PR and the histidine of H-261.

Modeling of the acetyl-pepstatin and renin inhibitors in the binding sites of both types of HIV PR showed that at least one of these four residues mentioned above contributes to the difference in relative inhibition constants. This is residue 32 which is Val in HIV-1 and Ile in HIV-2 PR, and forms part of the S_2 and S_2' subsites. The renin inhibitor H-261 has a larger residue in both P_2 and P_2' compared to acetyl-pepstatin: His compared to Val at P_2, and Ile compared to Ala at P_2'. Ile_{32} in HIV-2 PR appears to have unfavorably close contacts with P_2 and P_2' of H-261, while Val_{32} in HIV-1 PR does not have such close contacts. In the case of acetyl-pepstatin, on the contrary, the presence of smaller residues in P_2 and P_2' make its binding in HIV-2 PR more effective, since there are no unfavorably close contacts (Figure 3). Leu/Met_{76} also forms part of these subsites and may contribute to the hydrophobic interaction with inhibitor. However, the Met side chain is long and flexible and is not expected to interfere with inhibitor binding. The role of the substitutions of the other two residues, 47 and 82, with respect to the binding of these inhibitors is less obvious. This analysis may explain the observed differences in the K_i between these two inhibitors interacting with HIV-1 PR and HIV-2 PR (Richards *et al.*, 1989, 1990).

CONCLUSIONS

The similarity of the sequences of all isolates of HIV-1 and HIV-2 PR, and the fact that the model of HIV-2 PR preserves the important structural interactions and is characterized by low overall energy comparable to that for the crystallographic structure of HIV-1 PR, suggest that the main structural characteristics of the model should be sufficiently accurate for qualitative analysis. Examination of the PR structures with two different inhibitors identified one residue in the substrate binding site where the size of amino acid present in HIV-1 or -2 PR correlated with the relative binding constants of the inhibitors.

Comparison of the crystallographic and model structures of HIV-1 and HIV-2 PRs shows that all substitutions of residues occur in such a way that the important structural interactions can be maintained. Modeling of the substitutions in different isolates of both types of HIV PR leads to the same conclusion. These results can serve as a basis for a general approach to reveal the structural features of proteins with unknown three-dimensional structures, where for concerted substitutions in different isolates may be used to map three-dimensional structure.

Another practical result can be obtained by locating the fragments in the sequences of homologous proteins and their isolates where there are frequent substitutions. This study shows that those variable regions can form different structural surface elements and therefore can serve as epitopes for monoclonal antibodies. This serves as a possible guide to an immunological approach for separation and purification of different, but related, proteins.

REFERENCES

Gustchina, A. & Weber, I. T., 1990. *FEBS Lett.* 269:269-272.

Miller, M., Schneider, J., Sathyanarayana, B. K., Toth, M. V., Marshall, G. R., Clawson, L., Selk, L., Kent, S. B. H. & Wlodawer, A., 1989, *Science* 246:1149-1152.

Navia, M. A., Fitzgerald, P. M. D., McKeever, B. M., Leu, C., Heimbach, J. C., Herber, W. K., Sigal, I. S., Darke, P. L. & Springer, J. P., 1989, *Nature (London)* 337:615-620.

Richards, A. D., Roberts, R., Dunn, B. M., Graves, M. C. & Kay, J., 1989, *FEBS Lett.* 247:113-117.

Richards, A. D., Broadhurst, A. V., Ritchie, A. J., Dunn, B. M. & Kay, J., 1990, *FEBS Lett.* 253:214-216.

Swain, A. L., Miller, M., Green, J., Rich, D. H., Schneider, J., Kent, S. B. H. & Wlodawer, A., 1990, *Proc. Natl. Acad. Sci. U.S.A.* 87:8805-8809.

Wlodawer, A., Miller, M., Jaskólski, M., Sathyanarayana, B. K., Baldwin E., Weber, I. T., Selk, L. M., Clawson, L., Schneider, J. & Kent, S. B. H., 1989, *Science* 245:616-621.

THEORETICAL MODELS OF ASPARTIC PROTEASES: ACTIVE SITE PROPERTIES,

DIMER STABILITY AND INTERACTIONS WITH MODEL INHIBITORS

Anwar Rayan, Amit Fliess, Moshe Kotler,[1] Michael Chorev
and Amiram Goldblum

Department of Pharmaceutical Chemistry
School of Pharmacy

[1]Department of Molecular Genetics
School of Medicine
Hebrew University of Jerusalem
Jerusalem, Israel   91120

INTRODUCTION

HIV-1 PR, an aspartic proteinase (AP), is recognized now as an important target for designing effective and selective drugs which could arrest the late stages in replication of HIV-1, the causative agent of AIDS.[1] Selectivity of such enzyme inhibitors is required in addition to high affinity. The most spectacular difference between HIV-1 PR, a retroviral enzyme, and the cellular AP, is the smaller size and dimeric structure of the former. The conservation of active site residues, Asp-Thr/Ser-Gly, seems to eliminate the ability to design selective active-site inhibitors for AP.[2] This is, however, achieved by accumulated experience with the construction of novel renin inhibitors.

Theoretical approaches to the mechanisms of action and of inhibition of aspartic proteinases should be complementary to experimental work. We employ semi-empirical quantum-mechanics (QM) methods to study properties and interactions with molecular models for the active site of AP.[3,4] The more distant effect of charged side-chains is added to the QM study by "classical" electrostatics (ES). We have used X-ray results for HIV-1 PR[5,6] and of endothiapepsin [7] to construct quantum-mechanical models and their ES complements. Other AP structures [8,9] were compared for the ES effects. MNDO/H[10] was employed for all QM calculations.

RESULTS AND DISCUSSION

By studying active-site models of retroviral proteases, of human renin and of pepsins we have found[3] that the variation of the residue which follows the conserved triad, that is, Ser_{35}/Thr_{218} in pepsins to Ser/Ala in human renin and to Ala/Ala in retroviral AP, modifies the

Table 1. Local and distant effects on acidity of AP

Structure	Deprotonation enthalpy relative to 4APE, in Kcals/mol		Electrostatic stabilization in Kcals/mol	
	QM (local)	ES (side chains)	Interaction	Total charge
2APP	NC	0.2	30.1	-18
4APE	0.0	0.0	-9.2	-12
(h.renin)	2.8	NA	NA	NA
(retrov. AP)	9.3	NA	NA	NA
3APR	NC	-0.6	35.6	-8
2RSP	10.2	-3.0	-93.7	+3
3HVP	-4.6	-3.4	-67.4	+7
4HVP	8.4	-3.4	-65.8	+7
6HVP	6.2	-3.1	-66.4	+7

NC = not calculated ; NA = not applicable for models which have no coordinate sets.

local acidity so that it is lower in retroviral AP, while pepsins should have the lowest proton affinity. This is due to the ability of this fourth residue, if it is a hydrogen bond donor, to stabilize a mono-anion on the aspartic dyad. Adding electrostatic effects[11] from all the side chains in AP shows 1) the overall ES interactions of the charged side chains and terminals are stabilizing in retroviral AP compared to the pepsins; 2) all the AP are quite symmetric with respect to the ES effect on ionization of one aspartic acid *vs.* the other; and 3) the charged residues of pepsins have a stronger affinity than retroviral AP, by 2-3 kcals/mol, for the proton of the aspartic dyad. This "distant" effect in AP is opposed to the local QM effect, but is not large enough to reverse the QM trend. These results are presented in Table 1. The names of the structures are from the protein data bank. Models for human renin and for retroviral AP were constructed from 4APE. ES energies are averages for the two Asp residues.

There is a remarkable similarity in the values for ES stabilization with the different HIV-1 PR coordinates. The dimer N- and C-terminal interactions are major contributors. However, 3HVP coordinates for the active site are the only ones of retroviral PR which show a reverse QM local effect, i.e., an increase in acidity compared to 4APE. The reduced acidity of HIV-1 PR is established by the combination of local and remote effects for x-ray results and for models.

This lower acidity suggests that H-bond donors should have stronger affinity to the mono-anionic active site in HIV-1 PR than in pepsins. We modeled several small inhibitors, starting from a model of statine[4] and adding chemical functions which can contribute such H-bonding (Scheme 1).

Affinity and selectivity increase with a 1,3-diol (B), which is found to have two binding modes: (B-I), with both hydroxyls acting as donors to the "external" OD2 of Asp (-1), and (B-II), with each -OH bound to one OD2 of the dyad (Figure 1). The H-bonding of N-H to the carbonyl of Gly_{27} is lost in both binding modes. Constraining such a 1,3-diol to a relatively rigid cyclohexane structure (C) reduces the enthalpy of binding, but should lose less entropy than (B). Addition of an appropriately positioned N-H to the diol produces our model (D), which regains H-bonding to Gly_{27}. The affinities are listed in Table 2.

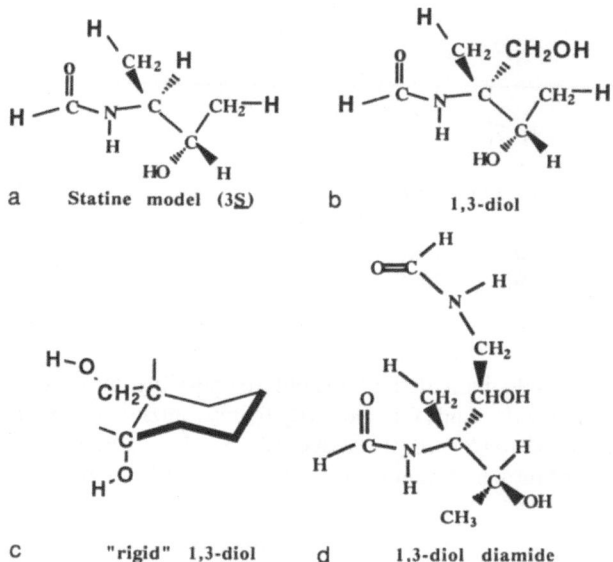

Scheme 1. Some inhibitor models in QM studies of affinity.

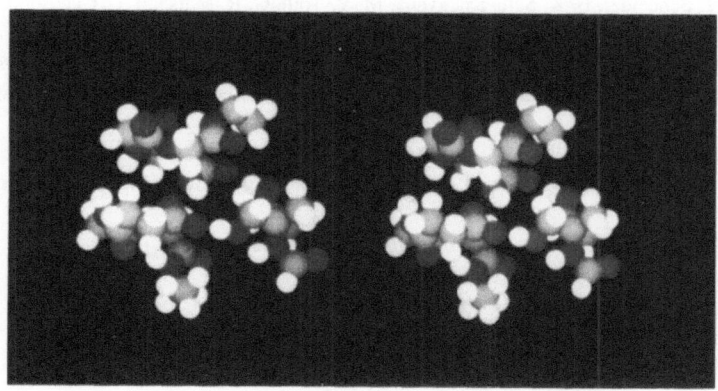

Figure 1. Binding of a 1,3-diol to an active-site model of HIV-1(B-II,3)

Table 2. Affinities of model inhibitors (Kcals/mol) to active-site models

Model of	1,3 - diol			1,3-diol	
(original coords.)	(A)	(B-I)	(B-II)	(C)	(D)
HIV-1 (4HVP)	10.7	-----	-----	14.8	22.4
HIV-1 (2RSP)	-----	24.5	23.4	-----	-----
HIV-1 (4APE)	11.1	22.8	20.4	15.9	23.5
H.REN(4APE)	10.1	19.8	19.7	-----	----
4APE	9.6	19.6	19.3	14.0	19.9

These partial results indicate that it should be possible, in principle, to design small inhibitors which fulfill the demands for selectivity by maximizing their polar interactions with active sites of AP. Larger models for the active sites of AP, including a representation of the "flaps", are required for further variations of inhibitor structures to increase their affinities.

REFERENCES

1. H. Mitsuya, R. Yarchoan and S. Broder, *Science* **249**:1533 (1990).
2. I. T. Weber, *J. Biol. Chem.* **265**:10492 (1990).
3. A. Goldblum, *Biochem. Biophys. Res. Comm.* **157**:450 (1988).
4. A. Goldblum, *FEBS Lett.* **261**:241 (1990).
5. M. Miller, J. Schneider, B. K. Sathyanarayana, M. V. Toth, G. R. Marshall, L. Clawson, L. Selk, S. B. H. Kent and A. Wlodawer, *Science* **246**:1149 (1989).
6. R. Lapatto, T. Blundell, A. Hemmings, J. Overington, A. Wilderspin, S. Wood, J. R. Merson, P. J. Whittle, D. E. Danley, K. F. Geoghegan, S. J. Hawrylik, S. E. Lee, K. G. Scheld and P. M. Hobart, *Nature* **342**:299 (1989).
7. L. Pearl and T. L. Blundell, *FEBS Lett.* **174**:96 (1984).
8. K. Suguna, E. A. Padlan, C. W. Smith, W. D. Carlson and D. R. Davies, *Proc. Nat. Acad. Sci. U.S.A.* **84**:7009 (1987).
9. M. N. G. James and A. R. Sielecki, *J. Mol. Biol.* **163**:299 (1983).
10. A. Goldblum, *J. Comp. Chem.* **8**:835 (1987).
11. E. L. Mehler, *Protein Engng.* **3**:415 (1990).

A CONSENSUS TEMPLATE FOR THE ASPARTIC PROTEINASE FOLD

Natalia S. Andreeva

V. A. Engelhardt Institute of Molecular Biology
Academy of Sciences of the USSR
117984 Moscow
USSR

INTRODUCTION

Domains of aspartic proteinase molecules have the same fold (Tang *et al.*, 1978) while their primary structures show no homology except small segments of a few amino acid residues located in the region of the active site in the three-dimensional structure. One may ask: is it possible to detect a common specific property of amino acid sequences of both domains which can explain the tendency of their chains to fold by the same manner? In other words does a consensus template for the specific aspartic proteinase fold exist? Interactions between different segments of a polypeptide chain providing formation of a central hydrophobic core of a protein molecule determine mainly the geometrical type of protein folding. If a consensus template for the specific aspartic proteinase fold really exists it must concern the formation of central hydrophobic cores in domains.

An essential part of the three-dimensional domain structure of these proteinases consists of approximately parallel layers of ß-sheets with almost orthogonal chain directions in adjacent layers. The central hydrophobic core of each domain is formed between the so-called A and B layers of these sheets (Figure 1a). The B layer in each domain is the most important region of aspartic proteinase structure. The central part of the B layer consists of the two symmetrically related ψ-shaped regions which are formed as a consequence of two wide loops interception. One of these loops in each domain contains the active aspartic acid residue. The arms of the loops have an extended ß-conformation, and altogether they comprise the four stranded ß-sheet with 5 residues in each strand. The two strands on the right are parallel to each other, as well as the two strands on the left, however the left pair of the strands is antiparallel to the right one, and the pairs are related by the intradomain two-fold axis of the pseudosymmetry. The upper surface of the B layers in both domains serves for a substrate binding, the inner surface forms an essential part of the hydrophobic core in the domains. Residue numbers for this region in pepsin are: 28-41, 100-104, 118-123 in the N-terminal domain, and 211-224, 285-289, 298-303 in the C-terminal domain.

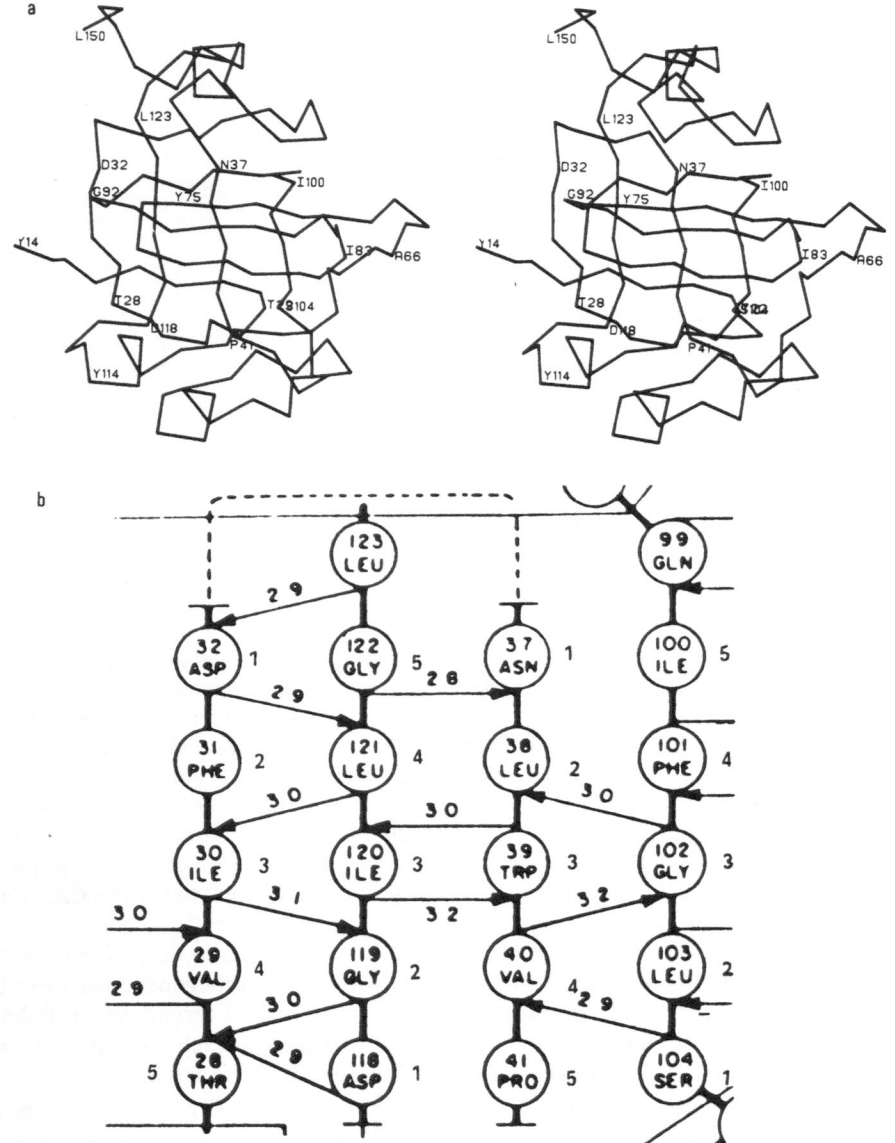

Figure 1. (a) An α-carbon stereo view of the sheets that comprise the N-terminal domain of pepsin. The strands of the B-layer are drawn vertically and the strands of the bottom A-layer are approximately orthogonal to these strands (from Sielecki *et al.*, 1990). (b) An example of identification of residue positions in the B-loop.

This region is the most conserved part of the aspartic proteinase domain structure. In actual fact, a good superposition of the N- and the C-terminal domains is observed only for these segments. Root mean square (rms) difference of C_α positions in this regions for the N- and the C-terminal domains is 1.14 Å in pepsin. On moving further away from the middle of this four stranded part of the B layer the two domains of pepsin as well as of other aspartic proteinases become more and more dissimilar (Sielecki et al., 1990). The B layer is protected from a solvent with the A-layer which consists of the two antiparallel ß-hairpins called as A-loops. These loops form an antiparallel four-stranded ß-sheet, and the inner side of this sheet oriented towards the B-layer participates in the formation of the central hydrophobic core of the domain. Residue numbers for these ß-hairpins in pepsin are: 16-28, 87-99 in the N-terminal domain, and 195-205, 257-269 in the C-terminal domain.

Is it possible to observe any special arrangement of hydrophobic residues in the A and the B-layers of aspartic proteinases? As inspection of various aspartic proteinase structures has shown, amino acid sequences of all four strands of the B-layers and the N-terminal strands of the A-loops obey a very special rule which is valid for both domains of all non-viral aspartic proteinases and for molecules of many retroviral proteases. This rule can be formulated in convenient terms if one identifies the position of a residue in the strand in relation to the loop turn, as it was proposed by Sibanda and Thornton (1985). At first, it is necessary to identify residues located at the N-terminal strand and at the C-terminal strand of a loop, in other words residues before and after the turn of this loop. The nearest to the turn residue which participates in sheet hydrogen bonding can be designated as a residue at the first position before or after the turn. The numbers of positions increase on moving further away from the turn along both arms of the loop (Figure 1b).

It is convenient to analyze at first the A-layers of aspartic proteinases. The template region of these layers comprises the N-terminal strands of the A-loops, and the strict requirement is to have hydrophobic residues at the first and at the third positions before the turns. The C-terminal strands of the A-loops are more exposed to solvent and do not obey any rule. As usual the A-loops of non-viral aspartic proteinases have tight turns and stick closely to the surface of a molecule. Table 1 contains aligned sequences of the four A-loops in the 25 non-viral aspartic proteases. All 100 sequences with one exception contain hydrophobic residues in the first and third positions before turns of these loops. One can also see some symmetrical intra- and interdomain relationships between sequences of the segments. In Figure 1a the connection between the A- and the B-layers are well visible. The C-terminal strands of the both A-loops have a rather sharp bends approximately to 90°, and their continuations after these bends participate in the B-layer. The A- and the B-layers are in a close contact at the corners near the bends, at two other corners they splay apart.

One can designate residue positions in arms of the B-loops extending the definition of residue positions proposed originally for ß-hairpins (Sibanda & Thornton, 1985). In accordance with this definition Asp_{32} takes up the first position before the turn of the active site loop, and Asn_{37} is the first residue after the turn, Ser_{104} is the first residue before the turn of the second B-loop of the N-terminal domain, while Asp_{118} is the first residue after the the turn (Figure 1b). Data on the three-dimensional structure of aspartic proteinases show that residues which occupy even positions before and after the loop turns have side chains which are oriented into the central hydrophobic core of the domains, while odd positions are occupied with residues which are oriented into the active site and substrate binding cleft. Among them residues at the third positions in all strands participate in the formation of the primary binding pockets and the first positions before the turns in the active loops are occupied with the active aspartic acid residues. These considerations are valid for both domains.

Table 1. Aligned amino acid sequences of the A-loops of aspartic proteases. In accordance with the intra- and interdomain symmetry four sequences are aligned for non-viral enzymes and two for retroviral proteases. The first line for non-viral enzymes corresponds to the loop 16-28, the second - to the loop 87-97, the third - to the loop 195-205 and the fourth - to the loop 258-268 of pepsin numbering. The first line for the eight retroviral enzymes corresponds to the loop 11-21 of HIV-1 protease, and the second - to the loop 62-72. Last ten retroviral enzymes have extended A-loops (see footnote for the Table).

positions	5	4	3	2	1	turn				1	2	3	4	
	G	T	I	G	I	G	T	P	A	Q	D	F	T	pp
	D	T	V	Q	V	G	-	-	G	I	S	D	T	
	D	S	I	T	M	D	-	-	G	E	T	I	A	
	I	V	F	T	I	D	-	-	G	V	Q	Y	P	
	G	T	I	G	I	G	T	P	A	E	D	F	T	hp
	D	T	V	Q	V	G	-	-	G	I	S	D	T	
	D	S	I	T	M	D	-	-	G	E	A	I	A	
	I	V	F	T	I	N	-	-	G	V	E	Y	P	
	G	T	I	G	I	G	T	P	A	Q	D	F	T	mpa
	D	T	V	Q	V	G	-	-	G	I	S	D	T	
	D	S	I	T	M	D	-	-	G	E	A	I	A	
	I	V	F	T	I	N	-	-	G	I	Q	Y	P	
	G	T	I	S	I	G	T	P	Q	Q	D	F	S	cp
	D	T	V	A	V	S	-	-	S	I	D	V	Q	
	D	R	V	T	V	G	-	-	N	K	Y	V	A	
	V	T	F	H	I	N	-	-	G	H	A	F	T	
	G	E	I	S	I	G	T	P	P	Q	N	F	L	rpc
	D	T	L	T	V	Q	-	-	S	I	Q	V	P	
	D	D	F	L	I	G	-	-	D	Q	A	S	G	
	L	S	F	V	L	N	-	-	G	V	Q	F	P	
	G	E	I	S	I	G	T	P	P	Q	N	F	L	hpc
	D	T	L	T	V	Q	-	-	S	I	Q	V	P	
	E	E	F	L	I	G	-	-	G	Q	A	S	G	
	L	T	F	I	I	N	-	-	G	V	E	F	P	
	G	E	I	S	I	G	T	P	P	Q	N	F	L	mpc
	D	T	L	T	V	Q	-	-	S	I	Q	V	P	
	E	E	F	L	I	G	-	-	G	Q	A	S	G	
	L	T	F	I	I	N	-	-	G	V	E	F	P	

G	K	I	Y	L	G	T	P	P	Q	E	F	T	bc
D	T	V	T	V	S	-	-	N	I	V	D	I	
D	S	V	T	I	S	-	-	G	V	V	V	A	
V	V	F	E	I	N	-	-	G	K	M	Y	P	

G	K	I	Y	I	G	T	P	P	Q	K	F	T	hc
D	T	V	T	V	S	-	-	N	I	V	D	P	
D	S	V	I	I	D	-	-	G	V	V	V	A	
A	V	F	E	I	H	-	-	G	K	K	Y	P	

G	T	I	S	I	G	T	P	P	Q	D	F	T	cep
D	T	V	T	V	A	-	-	S	L	M	D	T	
D	S	I	I	V	N	-	-	K	Q	E	I	A	
V	V	F	V	I	G	-	-	G	I	Q	Y	P	

G	E	I	G	I	G	T	P	P	Q	T	F	K	msr
D	S	V	T	V	G	-	-	G	I	T	V	T	
K	G	V	S	V	G	-	-	S	S	T	L	L	
I	S	F	N	L	G	-	-	G	R	A	Y	T	

G	E	I	G	I	G	T	P	P	Q	T	F	K	mkr
D	S	V	T	V	G	-	-	G	I	T	V	T	
K	G	V	S	V	G	-	-	S	S	T	L	L	
I	S	F	D	L	G	-	-	G	R	A	Y	T	

G	E	I	G	I	G	T	P	P	Q	T	F	K	hkr
D	I	I	T	V	G	-	-	G	I	T	V	T	
K	G	V	S	V	G	-	-	S	S	T	L	L	
I	S	F	H	L	G	-	-	K	E	Y	N	T	

G	E	I	G	I	G	T	P	P	Q	C	F	T	pcd
D	T	V	S	V	i/G	-	-	G	I	K	V	E	
N	Q	V	A	V	G	-	-	S	S	L	T	L	
V	T	V	T	L	G	-	-	G	K	K	Y	K	

G	E	I	G	I	G	T	P	P	Q	C	F	T	hcd
D	T	V	S	V	i /G	-	-	G	V	K	V	E	
D	Q	V	E	V	A	-	-	S	G	L	T	L	
I	T	L	K	L	G	-	-	G	K	G	Y	K	

Table 1. (continued)

Table 1. (continued)

G	T	I	S	I	G	S	P	P	Q	N	F	T	hce
D	Q	V	S	V	E	-	-	G	L	T	V	V	
D	N	I	Q	V	G	-	-	G	T	V	M	F	
V	T	F	T	I	N	-	-	G	V	P	Y	T	
T	P	V	T	I	G	-	-	G	T	T	L	N	pep
D	S	V	T	V	G	-	-	G	V	T	A	H	
D	S	Y	T	A	G	-	-	S	Q	S	G	D	
F	N	V	S	I	S	-	-	G	Y	T	A	T	
G	Q	V	T	I	G	T	P	G	K	K	F	N	rcp
D	N	V	N	L	G	-	-	G	L	L	I	K	
D	R	A	T	V	G	-	-	T	S	T	V	A	
L	V	F	S	I	N	-	-	G	A	S	F	Q	
G	E	V	T	V	G	T	P	G	I	K	L	K	mp
D	N	V	N	L	G	-	-	G	L	L	I	K	
K	S	I	K	I	G	-	-	T	S	T	V	A	
L	V	F	T	L	G	-	-	S	S	T	F	E	
T	P	V	Q	I	G	T	P	A	Q	T	L	N	ep
D	T	V	S	V	G	-	-	G	L	T	V	T	
T	G	Y	A	V	G	-	-	S	G	·T	F	K	
F	T	F	G	V	G	-	-	S	A	R	I	V	
I	P	V	S	I	G	T	P	G	Q	D	F	L	mmp
D	S	I	A	I	G	-	-	D	I	T	V	T	
T	G	I	T	V	D	-	-	G	S	A	A	V	
I	S	V	P	V	S	-	-	K	M	L	P	V	
I	P	V	S	I	G	T	P	G	Q	D	F	Y	mpp
T	S	I	T	V	G	-	-	G	A	T	V	K	
T	G	V	K	I	D	-	-	G	S	D	A	V	
V	S	V	P	I	S	-	-	K	M	L	L	P	
V	N	V	G	V	G	S	P	A	T	T	Y	S	ilp
D	T	V	T	L	G	-	-	S	L	T	I	.P	
Q	T	I	R	Y	G	-	-	S	S	T	S	I	
L	F	F	T	I	G	-	-	G	S	A	S	S	

Table 1. (continued)

T	D	I	T	L	G	T	P	P	Q	N	F	K	scp
D	T	L	S	I	G	-	-	D	L	T	I	P	
E	G	I	G	L	G	-	-	D	E	Y	A	E	
L	I	F	N	F	N	-	-	G	Y	N	F	T	
T	T	L	D	I	G	T	P	S	Q	S	L	T	scb
E	T	V	S	I	N	-	-	G	I	D	I	P	
Q	G	L	G	A	Q	ins		H	E	T	F	T	
Y	N	F	D	F	G	-	-	D	L	Q	I	S	
V	T	I	K	I	G	-	-	G	Q	L	K	E	h1p
I	L	I	E	I	C	-	-	G	H	K	A	I	
V	T	A	Y	I	E	-	-	G	Q	P	V	E	h2p
V	E	I	E	V	L	-	-	N	K	K	V	R	
V	T	I	K	I	G	-	-	G	Q	L	K	E	av2
I	P	V	E	I	C	-	-	G	H	K	A	I	
I	T	L	R	I	G	-	-	G	Q	P	V	T	fvp
R	R	V	Q	L	A	-	-	T	G	K	V	T	
I	T	L	K	V	G	-	-	G	Q	P	V	T	mvp
R	K	V	H	L	A	-	-	T	G	K	V	T	
I	T	L	T	V	G	-	-	G	Q	P	V	T	avp
R	K	V	H	L	A	-	-	T	G	K	V	T	
L	T	L	S	V	G	-	-	G	H	P	T	T	m7p
R	T	V	N	L	G	-	-	Q	G	M	V	T	
I	E	I	K	V	G	-	-	T	R	W	K	K	vip
V	H	L	Q	Y	K	-	-	D	K	M	I	K	
L	T	L	W	L	D	-	-	D	K	M	F	T	s1p
K	Y	L	T	W	turn			G	L	I	K		
L	H	I	Y	L	N	-	-	G	R	R	F	L	mtp
Q	P	L	R	W	turn			G	I	I	H		
V	R	V	I	L	turn			V	Y	I	T		rvp
E	L	G	V	I	turn			S	L	E_R	P		LLL
I	K	A	Q	V	turn			K	T	I	E		t1p
I	R	L	P	F	turn			I	V	L	T		
V	R	I	S	V	turn			Q	P	T	Q		t2p
I	F	L	P	F	turn			V	I	L	S		

Table 1. (continued)

V	A	V	Y	L	turn				N	Q	A	L	bvp
L	T	L	A	L	turn				I	T	I	P	
I	K	A	E	V	turn				M	I	I	E	tsp
I	R	L	P	F	turn				I	V	L	T	
C	K	A	I	I	Q	-	-	G	K	Q	F	E	hkp
E	I	L	H	C	turn				Q	E	S	T	
G	I	L	K	F	turn				L	D	L	H	cvp
L	P	I	R	L	turn				F	L	I	P	
G	R	L	Y	F	turn				I	E	L	H	vcp
I	D	L	I	I	turn				F	K	I	P	

*The A-loops of the first eight retroviral proteases in this Table have tight turns and quite definite similarity with the A-loops of non-viral enzymes. The turn of all other viral protease A-loops is more extended, therefore it is difficult to define positions of residues. The alignment of sequences with RSV protease has been partly used, however the whole alignment for the last ten proteases is only tentitative. C-terminal strand at the 4th position in all enzymes is at the corner and participates in both the A and the B layers.

**Data for sequences of non-viral enzymes were published by Foltmann (1988). References for the three additional sequences of non-viral proteases are included in the Table. For retroviral enzymes data published by Pearl & Taylor (1987) and Doolittle et al., (1989) have been used.

*** i/ - shows the position of insertions in cathepsin D loops.

pp = porcine pepsin A; hp = human pepsin A; mpa = monkey pepsin A; cp = chicken pepsin; rpc = rat pepsin C; hpc = human pepsin C; mpc = monkey pepsin C; bc = bovine chymosin; hc = human chymosin (Örd, et al.. 1990); cep = chicken embryonic pepsin; msr = mouse submandibular renin; mkr = mouse kidney renin; hkr = human kidney renin; pcd = porcine cathepsin D; hcd = human cathepsin D; hce = human cathepsin E (Azuma, et al., 1989); pep = penicillopepsin; rcp = *Rhizopus chinensis* proteinase; rnp = *Rhizopus niveous* proteinase; ep = endothiopepsin; mmp = *Mucor meihei* proteinase; mpp = *Mucor pusillius* proteinase; ilp = *Irpex lacteus* proteinase (Kobayashi, et al. 1989); scp = *Sacchromyces cerevisiae* proteinase a; scb = *Sacchromyces cerevisiae* BAR 1 proteinase; h1p = HIV-1 proteinase; h2p = HIV-2 proteinase; av2 = ARV-2 proteinase; fvp = FeLV proteinase; mvp = Mo MuLV proteinase; avp = Akv MuLV proteinase; m7p = M7 Bev proteinase; vip = visna proteinase; SRV-1 proteinase; mtv = MMTV proteinase; rvp = RSV proteinase; t1p = HTLV-1 proteinase; t2p = HTLV-2 proteinase; bvp = BLV proteinase; tsp = STLV proteinase; hkp = Herv K proteinase; cvp = CERV proteinase; and vcp = CaMV proteinase.

The requirement to have hydrophobic residues at even positions of the B-loops is strict for those positions which are located at the tight corners where layers are in a close contact. Near these corners the fourth positions before the turns of both B-loops are found. The fourth positions after the turn of both loops are located in the inner part of the double layer structure and for them the requirement to have a hydrophobic residue is also rather strict. At the same time the second positions before and after the turn of both loops are located in regions where the layers splay apart, and the nature of protecting groups from outer segments determines the type of residues at these positions. Mainly they are hydrophobic, especially at the second positions before the turns in the active loops, but the requirement cannot be strict in this case. Table 2 contains aligned amino acid sequences for the four B-loops in 25 non-viral aspartic proteinases.

Tables 1 and 2 contain also aligned amino acid sequences for 18 monomers of retroviral proteases. These enzymes were subjected to more rapid evolution than ordinary aspartic proteinases, and the A-loops in some of them have insertions, especially at the turns. Therefore one can divide retroviral proteases into two groups: enzymes of the first group follow the template pattern in the A-loops common to non-viral proteases, members of the second group have insertions in the A-loops. HIV-1 protease belongs to the first group, and RSV protease is a member of the second group. Sequences of retroviral enzymes presented in Tables have been aligned in relation to these two proteases with known three-dimensional structure (Wlodawer *et al.*, 1989, Jaskólski *et al.*, 1989). Numbers of residues for the first group correspond to the A-loops 11-21 and 62-72 of HIV-1 protease, and for the second group these loops correspond to the loops 13-33 and 81-97 of RSV enzyme. Requirements for residues of the B-loops discussed for non-viral proteinases are valid for the retroviral enzymes.

Additional requirements concern the formation of the active site, and they are the following: the first position before the turn of the active B-loops has to be occupied with the aspartic acid residue followed by a Thr-Gly sequence in adjacent turn, and the fifth positions after the turn for the second B-loops must have Gly to provide correct positioning of the active turns. Binding of hydrophobic residues of a substrate in the primary pockets of aspartic proteinases imposes additional requirements for residues which occupy the third positions in the B-loops. Altogether there are enough requirements to recognize aspartic protease sequence in unknown structure. The presence of the identical template regions in retroviral proteases of the first group and ordinary aspartic proteinases (Andreeva, 1988) made it possible to build a molecular model of HIV-1 protease (Pechik *et al.*, 1988, 1989) close to the real three-dimensional model of the enzyme (Wlodawer *et al.*, 1989).

Table 2. Aligned amino acid sequences of the B-loops of aspartic proteinases. The first line corresponds to the loop 28-41, the second - to the loop 211-224, the third - to the loop 100-122, and the fourth - to the loop 285-302 in pepsin numbering. The turns of the first two active loops contain sequences T G S S or their homologous analogs. The first B-loop of retroviral enzymes corresponds to the loop 21-34 of HIV-1 protease numbering, the second B-loop to the loop 72-86 of HIV-1 protease. For the references see footnotes to Table 1.

positions	5	4	3	2	1	turn	1	2	3	4	5	
	T	V	I	F	D	....	N	L	W	V	P	pp
	Q	A	I	V	D	....	L	L	T	G	P	
	I	F	G	L	S	......	D	G	I	L	G	
	G	F	Q	G	M..	....	L	W	I	L	G	
	T	V	V	F	D	....	N	L	W	V	P	hp
	Q	A	I	V	D	....	L	L	T	G	P	
	I	F	G	L	S	......	D	G	I	L	G	
	G	F	Q	G	M..	....	L	W	I	L	G	
	T	V	I	F	D	....	N	L	W	V	P	mpa
	Q	A	I	V	D	....	L	L	T	G	P	
	F	G	L	S	D	......	G	I	L	G		
	G	F	Q	G	M..	....	L	W	I	L	G	

Table 2. (continued)

S	V	I	F	D.......	N	L	W	V	P	cp
Q	A	I	V	D......	L	L	V	M	P	
I	F	G	L	S......	D	G	I	L	G	
G	F	E	N	M..	Q	W	I	L	G	

L	V	L	F	D......	N	L	W	V	S	rpc
Q	G	I	V	D......	L	L	V	M	P	
E	F	G	L	S......	D	G	I	M	G	
G	L	E	S	I...	L	W	I	L	G	

L	V	L	F	D......	N	L	W	V	P	hpc
Q	A	I	V	D......	L	L	T	V	P	
E	F	G	L	S......	D	G	I	M	G	
G	V	E	P	T......	L	W	I	L	G	

L	V	L	F	D......	N	L	W	V	P	mpc
Q	A	I	V	D......	L	L	T	V	P	
E	F	G	L	S......	D	G	I	M	G	
G	V	E	P	T......	Y	W	I	L	G	

T	V	L	F	D......	D	F	W	V	P	bc
Q	A	I	L	D......	K	L	V	G	P	
T	V	G	L	S......	D	G	I	L	G	
G	F	Q	S	E......	K	W	I	L	G	

T	L	V	F	D.......	D	I	W	V	P	hc
Q	A	I	L	D......	L	L	V	G	P	
T	V	G	L	S......	D	G	I	L	G	
G	F	Q	G	D......	Q	W	I	L	G	

T	V	V	F	D......	N	L	W	V	P	cep
Q	A	I	I	D......	L	V	A	G	P	
L	F	G	L	S......	D	G	I	L	G	
S	F	Q	N	S......	L	W	I	L	G	

K	V	I	F	D.......	N	L	W	V	P	msr
E	V	V	V	D......	F	I	S	A	P	
T	F	G	G	V......	D	G	V	L	G	
A	L	H	A	M..	V	W	V	L	G	

Table 2. (continued)

```
K   V   I   F   D. . . . . . N   L   W   V   P      mkr
A   V   V   V   D. . . . . . F   I   S   A   P
T   F   G   E   V. . . . . . D   G   V   L   G
A   L   H   A   M. .  . . . . V   W   V   L   G


K   V   V   F   D. . . . . . N   V   W   V   P      hkr
L   A   L   V   D. . . . . . Y   I   S   G   S
M   F   G   E   V. . . . . . D   G   V   V   G
A   I   H   A   M. .  . . . . T   W   A   L   G


T   V   V   F   D. . . . . . N   L   W   V   P      pcd
E   A   I   V   D. . . . . . L   I   V   G   Q
T   F   G   E   A. . . . . . D   G   I   L   G
G   F   M   G   M. .  . . . . L   W   I   L   G


T   V   V   F   D. . . . . . N   L   W   V   P      hcd
E   A   I   V   D. . . . . . L   M   V   G   P
V   F   G   E   A. . . . . . D   G   I   L   G
G   F   M   G   M. .  . . . . L   W   I   L   G


T   V   I   F   D. . . . . . N   L   W   V   P      hce
Q   A   I   V   D. . . . . . L   I   T   G   P
Q   F   G   E   S. . . . . . D   G   I   L   G
G   F   Q   G   L. . . . . . L   W   I   L   G


N   L   N   F   D. . . . . . D   L   W   V   F      pep
S   G   I   A   D. . . . . . L   L   L   L   B?
A   V   Q   A   A. . . . . . D   G   L   L   G
G   I   Q   S   N. . . . . . F   S   I   F   G


K   F   N   L   D. . . . . . D   L   W   I   A      rcp
D   G   I   L   D. . . . . . L   L   I   L   P
T   I   E   L   A. . . . . . D   G   L   L   G
G   F   G   Y   G. . . . . . F   A   I   I   G


K   L   K   L   D. . . . . . D   M   W   F   A      rnp
D   A   I   L   D. . . . . . L   L   L   L   P
T   I   E   L   A. . . . . . D   G   L   L   G
G   F   A   A   G. . . . . . L   A   I   L   G
```

Table 2. (continued)

T	L	N	L	D.	D	L	W	V	F	ep
D	G	I	A	D. . . .	L	L	Y	L	P	
A	V	E	S	A.	D	G	L	L	G	
G	I	Q	S	S.	I	N	I	N	F	G
L	L	L	F	D.	D	T	W	V	P	mmp
A	F	T	I	D.	F	F	I	M	P	
I	L	A	Y	V.	D	G	L	F	G	
I	I	L	P	N.	Q	Y	I	V	G	
Y	L	L	F	D.	D	T	W	V	P	mpp
A	F	T	I	D.	F	F	I	A	P	
T	L	A	Y	V.	D	G	I	F	G	
I	V	L	P	D.	Q	F	I	V	G	
S	L	L	V	D.	N	T	W	L	G	ilp
A	G	I	V	D.	L	T	L	I	A	
S	I	G	V	A.	D	G	I	L	G	
Y	L	I	V	G.	D	F	I	N	G	
K	V	I	L	D.	N	L	W	V	P	scp
G	A	A	I	D.	L	I	T	L	P	
D	F	A	E	A.	D	G	I	L	G	
A	I	T	P	M.	L	A	I	V	G	
T	V	L	F	D.	D	F	W	V	M	scb
P	V	L	L	D.	L	L	N	A	P	
Q	F	G	V	A.	S	G	V	L	G	
G	F	A	V	Q.	S	M	V	L	G	
E	A	L	L	D.	D	T	V	L	E	h1p
T	V	L	V	G.	V	N	I	I	G	
E	V	I	L	D.	D	S	I	V	A	h2p
T	I	M	T	G.	I	N	I	F	G	
E	A	L	L	D.	D	T	V	L	E	av2
T	V	L	V	G.	V	N	I	I	G	
T	F	L	V	D.	H	S	V	L	T	fvp
F	L	Y	V	P.	Y	P	L	L	G	

Table 2. (continued)

T	F	L	V	D.......	H	S	V	L	T	mvp
F	L	H	V	P.......	Y	P	L	L	G	
T	F	L	V	D.......	H	S	V	L	T	avp
F	L	H	V	P.......	Y	P	L	L	G	
T	F	L	V	D.......	H	S	V	L	T	m7p
F	L	V	V	P.......	Y	P	L	L	G	
K	L	L	V	D.......	K	T	I	V	T	vip
I	V	V	L	A.......	V	E	V	L	G	
T	G	L	I	D.......	V	T	I	I	K	s1p
P	F	V	I	P.......	V	N	L	W	G	
L	G	L	L	D.......	K	T	C	I	A	mvp
P	F	V	I	P.......	F	T	L	W	G	
T	A	L	L	D.......	I	T	I	I	S	rvp
F	P	A	V	A.......	G	S	I	L	G	
E	A	L	L	D.......	M	T	V	L	P	t1p
C	L	V	D	T.......	Q	A	I	I	G	
Q	A	L	L	D.......	L	T	V	I	P	t2p
C	L	L	D	T.......	W	T	I	I	G	
E	A	L	L	D.......	M	T	V	L	P	tsp
C	L	V	D	T.......	W	A	I	I	G	
L	M	L	V	D.......	N	T	V	L	P	bvp
I	L	V	D	T.......	W	Q	I	L	G	
E	G	L	V	D.......	V	S	I	I	A	hkp
Q	P	M	I	T.......	L	N	L	W	G	
H	C	Y	V	D.......	L	C	M	A	S	cvp
L	F	Q	Q	E.......	D	L	L	L	G	
H	C	F	V	D.......	L	C	I	A	S	vcp
V	Y	Q	Q	E.......	D	F	I	I	G	

See footnote, Table 1 for definitions.

REFERENCES

Andreeva, N., 1988, "Structure and Biosynthesis of Proteins (Puschino)," 3:97-115.

Azuma, T., Pals, G., Mohandas, T. K., Couvreur, J. M. & Taggart, R. T., 1989, *J. Biol. Chem.* 264:16748-16753.

Doolittle, R. F., Feng, D. F., Johnson, M. S. & McClure, M. A., 1989, *Quarterly Review of Biology* 64:1-30.

Foltmann, B., 1988, *in*: "Abstracts of 18th Linderstrøm-Lang Conference" pp.7-20.

Jaskólski, M., Miller, M., Rao, M. J. K., Leis, J. & Wlodawer, A., 1990, *Biochemistry* 29:5889-5898.

Kobayashi, H., Sekibata, S., Shibuya, H., Yoshida, S., Kusakabe, I. & Murakami, K., 1989, *Agric. Biol. Chem.* 53:1927-1933.

Örd, T., Kolmer, M., Villems, R. & Saarma, M., 1990, *Gene* 91:241-246.

Pechik, I. V., Gustchina, A. E., Andreeva, N. S. & Fedorov, A. A., 1988, "Structure and Biosynthesis of Proteins (Pustchino)," 3:87-96.

Pechik, I. V., Guschina, A. E., Andreeva, N. S. & Fedorov, A. A., 1989, *FEBS Lettr.* 247:118-122.

Pearl, L. & Taylor, W., 1987, *Nature (London)* 329:351-354.

Sibanda, B. L. & Thornton, J. M., 1985, *Nature (London)* 316:170-174.

Sielecki, A. R., Fedorov, A. A., Bodhoo, A., Andreeva, N. S. & James, M. N. G., 1990, *J. Mol. Biol.* 214:143-170.

Tang, J., James, M. N. G., Hsu, I.-N., Jenkins J. A. & Blundell, T. L., 1978, *Nature (London)* 271:618-621.

AUTHOR INDEX

SUBJECT INDEX

ACHPA-lactam, 326
Actin, 474
Activation, 4-7, 289-296, 297-306
    (*see also* Zymogen)
  rates, 7, 101-105
  segment, 101
Active site
  chymosin, 23-37
  HIV PR, 398
  mammalian proteinases, 143-147
  pepsin, 3-4
  Rhizopus pepsin, 230
Affinity chromatography
  barley cathepsin D, 355-356
  fungal proteinases, 259-263
  HIV PR, 501
  human cathepsin D, 291
  renin, 379-381
AIDS, *see* HIV Proteinase
Alignment, 437
    (*see also* Amino acid
    sequence)
Allozymogens, 101
Amino acid sequence, (*see also* Sequence
    alignment)
  aspartic proteinase loops, 562-571
  Ascaris pepsin inhibitor, 67, 71
  Aspergillus niger PR, 205-207
  barley proteinase, 356
  barrier proteinase, 163
  cod cathepsin D, 109
  fungal aspartic proteinases, 150-152
  HIV PR inhibitors, 433-454
  HIV PR, 508
  HIV-1 and HIV-2 PR, 550
  HIV-BLA fusion protein, 430

Amino acid sequence (continued)
  lamb and human prochymosin, 122,
    129
  Mucor pusillus PR, 234
  Pseudomonas PR, 194
  potato inhibitor, 351
  Rhizopus pepsin, 7
  Scytalidium lignicolum PR, 188
Angiotensin, 383-386
Angiotensinogen, 307-323, 387
Anodic migration, 39
Antigens, 286
Archaebacteria, 257
Ascaris, 63-73, 349
Asp$_{25}$, 398
Asp$_{32}$, 2-4, 6
Asp$_{215}$, 3
Aspartic proteinase fold, 443, 559-572
Aspergillus, 149-160
  aculeatus, 259-263
  niger var. *macrosporus*, 203-211
  ssp., 118
Assay
  chromogenic peptide, 143-147,
    276-278, 343-347, 513
  Hemoglobin digestion, 64, 107-108,
    303-305, 356
  HPLC, 529-531
  Scintillation proximity, 525-528
  viral infectivity, 490
Atlantic cod pepsin, 107-110
Autoprocessing, 396, 499-502
    (*see also* Processing)
Average temperature factors, 19-20

Bacillus, 189, 195-197